D1545023

GLYCOPROTEINS

New Comprehensive Biochemistry

Volume 29a

General Editors

A. NEUBERGER
London

L.L.M. van DEENEN
Utrecht

ELSEVIER
Amsterdam - Lausanne - New York - Oxford - Shannon - Tokyo

Glycoproteins

Editors

J. Montreuil
Université des Sciences et Technologies de Lille,
Laboratoire de Chimie Biologique,
(UMR no 111 du CNRS), 59655 Villeneuve d'Asq Cedex, France

J.F.G. Vliegenthart
Bijvoet Center for Biomolecular Research,
Department of Bio-organic Chemistry,
P.O. Box 80.075, 3508 TB Utrecht, The Netherlands

H. Schachter
Department of Biochemistry Research, Hospital for Sick Children,
555 University Avenue, Toronto, Ont. M5G 1X8, Canada

1995
ELSEVIER
Amsterdam - Lausanne - New York - Oxford - Shannon - Tokyo

Elsevier Science B.V.
P.O. Box 211
1000 AE Amsterdam
The Netherlands

ISBN 0 444 81260 1 Hardbound
ISBN 0 444 82075 2 Paperback
ISBN 0 444 80303 3 (series)

No responsibility is assumed by the publisher for any injury and/or damage to persons or property as a matter of products liability, negligence or otherwise, or from any use or operation of any methods, products, instructions or ideas contained in the material herein. Because of the rapid advances in the medical sciences, the publisher recommends that independent verification of diagnoses and drug dosages should be made.

Special regulations for readers in the USA - This publication has been registrated with the Copyright Clearance Center Inc. (CCC), 222 Rosewood Drive, Danvers, MA 01923. Information can be obtained from the CCC about conditions under which photocopies of parts of this publication may be made in the USA. All other copyright questions, including photocopying outside the USA, should be referred to the publisher.

Printed on acid-free paper

Printed in the Netherlands

Preface

The presence of sugar moieties in proteins had been the subject both of speculation and experiment for some time when Pavy (1893) carried out experiments with coagulated egg-white, and drew the conclusion that this mixture of proteins and many other proteins contained carbohydrate in a covalent combination. Several other workers confirmed the somewhat messy experiments of Pavy, but most workers in the field believed that apart from mucins and mucoids, ordinary proteins consisted only of amino acid residues and that the presence of sugars in hydrolysates was due to impurities such as mucoids. Such belief was expressed by Plimmer (1917) and by Levene (1923) in textbooks which carried considerable prestige at the time that they were published.

This was the situation that I met in 1936 when, after obtaining my Ph.D. in Biochemistry, I became interested in this problem. Almost all crystalline proteins known at that time were free of sugar with the possible exception of egg albumin. It was important therefore to find out whether egg albumin really contained covalently-bound sugar residues. As a first step I tried to separate the carbohydrate by physical means, such as prolonged dialysis, repeated ultrafiltration and crystallization. It was impossible to separate carbohydrate from the protein by any of the methods including mild chemical techniques. I will not discuss the steps taken then to hydrolyse all the peptide linkages whilst leaving the carbohydrate content intact, but I believe most of the conclusions drawn at that time were essentially correct. The first conclusion was that the carbohydrate is almost certainly combined with the peptide moiety by a covalent bond. The second conclusion was that the carbohydrate is present as *one* unit, as shown by the molecular weight estimations. A third conclusion which was implied was that the sequence of sugar residues in a glycopeptide is as constant as it is in the peptide moiety. This assumption turned out to be erroneous. At that time it would have been difficult to identify the amino acid linking the peptide to the carbohydrate component. It was a few years before paper chromatography was developed and the amino acid concerned, i.e. asparagine, gave no specific reactions. After my paper was published (1938) I spent some time trying to identify the linking amino acid. Two of the nitrogen atoms of the complex were associated with the missing linkage and I believed that either asparagine or glutamine might be involved. Indeed, I did isolate a small amount of aspartic acid from the hydrolysate of the complex but the quantity appeared to be too small to justify a definite conclusion. Within a year of the publication of my paper the Second World War had started and no long-term research was felt to be appropriate.

The problem of the linkage of egg albumin was taken up again by my colleagues and myself in 1956 and led to a paper which appeared in *Nature* in 1958. In this paper it was shown that asparagine was the linking compound. Experiments similar to our own were also described at about the same time by Cunningham and his colleagues and by Jevons. More conclusive findings were reported later by ourselves and the N-(L-β-aspartyl)-β-D-glucopyranosylamine was later synthesized with G.S. Marks. The complete identification of synthetic material with the natural glycopeptide finally clinched the structure. This type of linkage is probably the most widespread in glycoproteins. Since this work was done in the early 1960s, the number of proteins shown to have sugar residues covalently bound to the peptide structure has increased enormously. There have been other linkages demonstrated in many glycoproteins such as the linkage between galactose and serine or threonine. They are present in plants, bacteria, animals, and even in archaebacteria, and most probably it is the largest group of proteins. In recent years we have come to appreciate the fact that the glycosylation of peptides is only one, but perhaps the most widespread, example of post-translational change of protein synthesis. We can in fact divide natural products into two categories. One category represents the proteins which are controlled essentially by the sequence of bases in DNA. Any change in the sequence of bases in DNA is a mutation, and the whole process of protein synthesis tries to reduce the occurrence of mutations to a minimum. Glycosylation of peptides, like the hydroxylation of proline and glycine residues which occur in collagen, or the iodination of tyrosine residues which occurs in thyroglobulin, takes place either during the translation process of protein synthesis, or immediately afterwards. They are controlled by enzymes which themselves are products of protein synthesis. However, these non-translational processes are not subject to the same fidelity as amino acid sequence in peptides. They depend on specificity of enzymes, their location in the cell, their activity, and many other factors, which must show considerable variability. In other words, we do not expect the same constancy in composition or in structure as we would with peptides. Thus, it is likely that fatty acid residues in phospholipid might vary considerably even within one organ. Similarly, the detailed structure of the carbohydrate component in the glycopeptide might vary within certain prescribed limits. This accounts for the fact that in most, and possibly all, glycoproteins there is variability in the polysaccharide structure, while in the peptide moiety sequence is constant. This is the case even in the relatively simple case of egg albumin.

The last problems that I want to discuss concern the functions of the sugar residues in glycoproteins. The complex type of biosynthesis of the carbohydrate component of glycoproteins is truly amazing, and it is very difficult to see at present how this sequence of enzymatic reactions involving controlled addition and deletion of sugars to and from the glycopeptide is regulated. It is thus reasonable to believe that this complicated mechanism serves an important biological function. It is likely there is still much to be discovered in this field, but some general points can be made. The addition of carbohydrate to peptide structure will change the shape and size of the protein molecule. This is likely to affect the access of proteolytic enzymes, and it will almost certainly influence such factors as heat stability, solubility and many physical and chemical properties. It is also likely to affect interaction with other proteins or non-protein components of the cell, and also affect interaction with other cells. It is almost certain that glycoproteins are prominently involved with social behaviour of cells. It is thus not surprising that the lifetime of a protein in a whole organism is greatly affected by sugar residues present in the glycoprotein in question. In the case of a hormone, it is likely that whilst activity *in vitro* may not be changed by the possession of a carbohydrate group, its survival in the whole organism might be greatly modified by the sugar component of the glycoprotein.

I wish to say finally that this book contains authoritative reviews on many aspects of our present state of knowledge of glycoproteins. In this vastly expanding field, no review is likely to be exhaustive or up-to-date in a few years' time. However, I hope it will be helpful to most readers and will stimulate them to further work in this important field of biochemistry.

Professor Albert Neuberger
Charing Cross and Westminster Medical School
Department of Biochemistry
Fulham Palace Road
London W6 8RF, UK

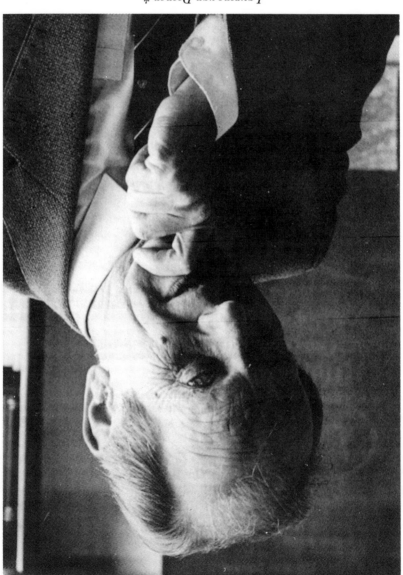

Laurens van Deenen †

In Memoriam

To our great sorrow we have to report the death of Professor Laurens van Deenen, on September 4, 1994.

Laurens van Deenen was an outstanding scientist who made great contributions to our knowledge of the structure and function of lipids of biological importance, and their interaction with other components of the cell. He also increased our understanding of the nature of membranes in biological systems and the relevant fields of enzymology. He was a pioneer full of original ideas, and carried out his work with the best available methods, always being cautious in the interpretation of his results.

Laurens van Deenen was the creator of a school which was internationally recognized and as a person was generous to his co-workers. He had wide interests outside his own discipline.

He joined the Editorial Board of *Biochimica et Biophysica Acta,* and was one of the Managing Editors from 1964–1993 and Chairman during the years 1983–1989.

From 1977 onwards, Laurens van Deenen was also involved in the editorship of the original *Comprehensive Biochemistry* series, and was one of the initiators in developing a second series, *New Comprehensive Biochemistry*, being responsible for the realization of this venture. Here his wide knowledge of biochemistry, stretching far beyond his own fields, his wisdom and his judgement were great assets.

Throughout his life, Laurens van Deenen set high standards for his own work; he was a stimulating colleague and a good friend.

London, October 1994   *Albert Neuberger*

List of contributors

Friedrich Altmann
Institut für Chemie, Universität für Bodenkultur, Gregor-Mendelstrasse 33, A-1180
Vienna, Austria

T. Bielefeldt
Institute of Organic Chemistry, University of Hamburg, Martin-Luther-King-Platz 6,
20146 Hamburg, Germany

Malgorzata Bielinska
Department of Molecular Biology and Pharmacology and Obstetrics and
Gynecology, Washington University School of Medicine, 660 S. Euclid Avenue, St.
Louis, MO 63110, USA

Irving Boime
Department of Molecular Biology and Pharmacology and Obstetrics and
Gynecology, Washington University School of Medicine, 660 S. Euclid Avenue, St.
Louis, MO 63110, USA

Inka Brockhausen
Department of Biochemistry Research, The Hospital for Sick Children, 555
University Avenue, Toronto M5G 1X8, and Biochemistry Department, University of
Toronto, Toronto M5S 1A8, Ontario, Canada

Christian Cambillau
Laboratoire de Cristallographie et de Cristallisation des Macromolécules
Biologiques, CNRS URA 1296, Faculté de Médecine Secteur-Nord, Bd. Pierre
Dramard, 13916 Marseille Cedex 20, France

Raymond T. Camphausen
Small Molecule Drug Discovery, Genetics Institute, 87 Cambridge Park Drive,
Cambridge, MA 02140, USA

Dale A. Cumming
Small Molecule Drug Discovery, Genetics Institute, 87 Cambridge Park Drive,
Cambridge, MA 02140, USA

Alan D. Elbein
Department of Biochemistry and Molecular Biology, The University of Arkansas for
Medical Sciences, Little Rock, AR, USA

Mary Catherine Glick
The Children's Hospital of Philadelphia, Department of Pediatrics, University of Pennsylvania School of Medicine, 34th Street and Civic Center Boulevard, Philadelphia, PA 19104, USA

Frank W. Hemming
Department of Biochemistry, University of Nottingham, Queen's Medical Centre, Nottingham NG7 2UH, UK

S.W. Homans
University of Dundee, Department of Biochemistry, Medical Sciences Institute, Carbohydrate Research Centre, Dundee DD1 4HN, UK

Billy G. Hudson
Department of Biochemistry and Molecular Biology, University of Kansas Medical Center, Kansas City, KS 66160-7421, USA

Frans M. Klis
Institute of Molecular Cell Biology, University of Amsterdam, BioCentrum Amsterdam, Kruislaan 318, 1098 SM Amsterdam, The Netherlands

Viktoria Kubelka
Institut für Chemie, Universität für Bodenkultur, Gregor-Mendelstrasse 33, A-1180 Vienna, Austria

L. Lehle
Lehrstuhl für Zellbiologie und Pflanzenphysiologie, Universität Regensburg, 93040 Regensburg, Germany

Leopold März
Institut für Chemie, Universität für Bodenkultur, Gregor-Mendelstrasse 33, A-1180 Vienna, Austria

Jean Montreuil
Université des Sciences et Technologies de Lille, Laboratoire de Chimie Biologique, (UMR no 111 du CNRS), 59655 Villeneuve d'Asq Cedex, France

Milton E. Noelken
Department of Biochemistry and Molecular Biology, University of Kansas Medical Center, Kansas City, KS 66160-7421, USA

Y.T. Pan
Department of Biochemistry and Molecular Biology, The University of Arkansas for Medical Sciences, Little Rock, AR, USA

H. Paulsen
Institute of Organic Chemistry, University of Hamburg, Martin-Luther-King-Platz 6, 20146 Hamburg, Germany

S. Peters
Institute of Organic Chemistry, University of Hamburg, Martin-Luther-King-Platz 6,
20146 Hamburg, Germany

Jürgen Roth
Division of Cell and Molecular Pathology, Department of Pathology, University of
Zürich, Schmelzbergstr. 12, CH-8091 Zürich, Switzerland

Harry Schachter
Department of Biochemistry Research, Hospital for Sick Children, 555 University
Avenue, Toronto, Ont. M5G 1X8, Canada

Erika Staudacher
Institut für Chemie, Universität für Bodenkultur, Gregor-Mendelstrasse 33, A-1180
Vienna, Austria

Arnd Sturm
Friedrich Miescher-Institut, Postfach 2543, CH-4002 Basel, Switzerland

Manfred Sumper
Universität Regensburg, Lehrstühl Biochemie I, Postfach 397, 93053 Regensburg,
Germany

W. Tanner
Lehrstuhl für Zellbiologie und Pflanzenphysiologie, Universität Regensburg, 93040
Regensburg, Germany

André Verbert
Laboratoire de Chimie Biologique, UMR du CNRS No. 111, Université des Sciences
et Technologies de Lille, 59655 Villeneuve d'Ascq, France

Johannes F.G. Vliegenthart
Bijvoet Center for Biomolecular Research, Department of Bio-organic Chemistry,
P.O. Box 80.075, 3508 TB Utrecht, the Netherlands

Winifred M. Watkins
Department of Haematology, Royal Postgraduate Medical School, Hammersmith
Hospital, Du Cane Road, London W12 0NN, UK

Felix T. Wieland
Institut für Biochemie I, Im Neuenheimer Feld 328, 69120 Heidelberg, Germany

Hsiang-ai Yu
Small Molecule Drug Discovery, Genetics Institute, 87 Cambridge Park Drive,
Cambridge, MA 02140, USA

Contents

Chapter 3. 3D Structure. 1. The structural features of protein–carbohydrate interactions revealed by X-ray crystallography

Chapter 3. 3D Structure. 2. Three dimensional structure of oligosaccharides explored by NMR and computer calculations

Chapter 5. Biosynthesis. 3. Biosynthesis of O-glycans of the N-acetylgalactosamine-α-Ser/Thr
linkage type
Inka Brockhausen

Chapter 5. Biosynthesis. 6. The role of polypeptide in the biosynthesis of protein-linked oligosaccharides
Raymond T. Camphausen, Hsiang-ai Yu and Dale A. Cumming

Chapter 5. Biosynthesis. 7. How can N-linked glycosylation and processing inhibitors be used to study carbohydrate synthesis and function
Y.T. Pan and Alan D. Elbein

Chapter 9. N-Glycosylation of plant proteins
Arnd Sturm . . . .

Chapter 10. Protein glycosylation in insects
Leopold März, Friedrich Altmann, Erika Staudacher and Viktoria Kubelka . . . .

Chapter 11. The glycoprotein hormone family: structure and function of the carbohydrate chains
Malgorzata Bielinska and Irving Boime 565

J. Montreuil, H. Schachter and J.F.G. Vliegenthart (Eds.), *Glycoproteins*

The history of glycoprotein research, a personal view

JEAN MONTREUIL

*Université des Sciences et Technologies de Lille, Laboratoire de Chimie Biologique
(UMR no 111 du CNRS), 59655 Villeneuve d'Asq Cedex, France*

*"It is often thought that a preoccupation with a history of one's own science is
peculiar to those approaching old age and is a special concern of investigators
who have lost the imagination to undertake original research of their own!"*
A. Neuberger

1. Evolution and revolutions of ideas in the field of glycoproteins. An overview

Glycoproteins together with glycolipids constitute the family of glycoconjugates, a term introduced in 1973 [1]. They were defined in 1908 by the Committee on Protein Nomenclature of the American Society of Biochemists as *"compounds of the protein molecules with a substance or substances containing a carbohydrate group, other than nucleic acid"*. They result from the covalent glycosidic linkage of sugar moieties called glycans with proteins (for reviews and books, see Refs. [2–24]).

Glycosylation of proteins represents one of most important post-translational events because of the ubiquity of the phenomenon. In fact, most proteins of cells and biological fluids are glycosylated, and glycoproteins are present in animals, plants, microorganisms and viruses. This explains why these components of living matter, encountered in the most varied biological media, were discovered in the early age of biochemistry. However, the advancement of knowledge of glycoproteins was very slow and, during the first decades, the acquisition of information in the field of glycoprotein chemistry was very poor. In fact, between 1865, the year of birth of glycoproteins, the birth certificate signed by Eichwald [25], and 1925, research on glycoproteins did not develop significantly so that Levene [2] could sound the alarm bell and write, "No attempt has been made to co-ordinate the older information with the new, no effort has been made to apply the newer methods of physical chemistry to the study of the molecules (glycoproteins and mucoproteins) as a whole. The subject of the chemical structure of these tissue elements seems to be a very neglected branch of science." Regarding this comment, we must confess that the situation did not evolve very much even as late as the 1970s and that the biochemistry of glycoproteins was, at that time, far behind that of nucleic acids and proteins.

The reasons for this slower development could be related, first, to the fact that carbohydrate chemistry is most complex and difficult because sugars are multifunctional molecules. We have only to realize that two different amino acids can form only two dipeptides, while two different monosaccharides can lead to more than 60 disaccharides. Consequently, carbohydrate chemists themselves were more attracted by polysaccharides like cellulose and starch which were homopolymers and substances of industrial interest. In addition, it was not necessary to solve two problems characteristic of glycoproteins: that of the complex primary structure of hetero/oligo- and polysaccharides and that of the protein moiety.

The second reason for slow progress could be found in the fact that, for a long time, carbohydrates have been regarded as reserve, energy-storage and support compounds devoid of any biological intelligence. That is why, in many countries, glycoconjugates were slow to receive their "lettres de noblesse" even if Nathan Sharon [7] asserted as early as 1975 that "We know now that the specificity of many natural polymers is written in terms of sugar residues, not of amino acids or nucleotides"; even if François Jacob wrote: "Together with nucleic acids and proteins, carbohydrates represent the third dimension of molecular biology"; even if an article published in the Herald Tribune of the 22nd of August 1990 and entitled "A spoonful of sugar makes the dividends go up" announced "People say that in five years, there will be more companies with *glyco* in their name than companies with *gene* in their name today". This prediction is now realized on the basis of the two new concepts of glycobiology and glycobiotechnology. However, we must recognize that it is often difficult to convince the public or private institutions of the importance of research on glycoconjugates. In fact, in spite of the fascinating discoveries achieved in the field of glycan primary and three-dimensional structure, in many cases the real role of biological glycoconjugate glycans is still unknown and, in many cases, we are unable to answer the question: "Why are proteins and lipids glycosylated?". On the whole, the role of many glycans remains obscure and the subject of speculation or controversy. As pointed out by Albert Neuberger in 1974 [26]: "We are now faced with the major problem of the biological function of the glycoproteins: in other words, we are asking the question how the activity of a protein, whatever it may be, is modified or fine-tuned by the presence of sugar residues".

However, we can be optimistic about the future and the progress of research in the domain of glycoconjugates in general, and of glycoproteins in particular, since the basis of the molecular biology of these compounds is now firmly established. Interest in these compounds is growing, partly due to impressive developments in the field of biotechnology. A series of crucial discoveries have pointed to the following biological roles of glycans:

(i) They intervene in the folding of proteins during their biosynthesis and favour their secretion out of the cell.

(ii) They stabilize the conformation of proteins in a biologically active form thanks to glycan–glycan or glycan–protein interactions.

(iii) They control the proteolysis of protein precursors leading to active proteins or peptides.

(iv) They protect polypeptide chains against proteolytic attack by acting as shields.

(v) By the same protective mechanism, glycans mask peptide epitopes and cause the weak antigenicity of glycoproteins. Glycans may therefore modulate recognition of viral proteins by the immune system, e.g. HIV.

(vi) They control the life-time of circulating glycoproteins and cells. Desialylation induces the capture of blood glycoproteins by hepatocytes and of erythrocytes by macrophages.

(vii) They are receptor sites for various proteins and for microorganisms, viruses, fungi and bacteria.

(viii) On the basis of the concept of membrane lectins, glycans are involved in cell–cell recognition and adhesion, cell differentiation and development, and cell-contact inhibition, thereby intervening in the social life of cells. Since the glycoconjugates of cancer cells are profoundly modified, the appearance of molecular monsters in cancer cell membranes may be associated with the asocial behaviour of cancerous cells and the metastasis phenomenon itself. In fact, the modified tumor glycans may act as antirecognition signals and explain the dispersion of cancer cells. Furthermore, they could become recognition signals for endothelial cells and play an important role in the origin of secondary tumours.

(ix) Membrane glycoprotein (as well as glycolipid) glycans are epitopes of tissue antigens, of bacteria, fungi, and parasite envelopes and of tumor associated antigens (TAA). This last discovery has opened a promising door to the immunodiagnosis and immunotherapy of cancers.

(x) Owing to a deficit in lysosomal enzymes the catabolism of glycans is deeply perturbed in the case of the genetic diseases called 'glycoproteinoses' or 'glycanoses'.

Last but not least, glycoproteins have recently entered the industrial field of genetic engineering of human glycoproteins of therapeutic interest. This gives rise to an enormous problem, because the production of recombinant human glycoproteins in non-human eukaryotic cells or in prokaryotic cells devoid of glycan biosynthesis machinery leads to the production of incorrectly glycosylated or non-glycosylated proteins. Since proteins may require proper glycosylation for therapeutic effectiveness and safety, eukaryotic cells have to be engineered to produce a recombinant glycoprotein which conforms to the native one. Unfortunately, present knowledge concerning the regulation of glycan biosynthesis is too limited to allow reliable construction of the authentic glycans of recombinant glycoproteins. This leads in many cases to sub-optimal glycosylation. The consequences may be dramatic: increase of hydrophobicity, decrease or inhibition of secretion, decrease of the stability towards heat or proteases, shortening of the *in vivo* life span of the molecules by increase of the clearance, decrease of the affinity for specific receptors and increase of antigenicity.

2. *The early history: the birth of glycoprotein biochemistry*

The biochemical era of the history of glycoproteins covers more than a century, from 1865 when Eichwald [25] provided the first evidence that various mucins "are compounds consisting of a moiety with all properties of a genuine protein and a moiety released under certain conditions as a sugar", until 1968, the date of the triumphal entry the basis of knowledge of the primary structure of the glycans. Thanks to their work, chemical, physical and enzymatic techniques for structural analyses of glycans were devised, e.g. the development in 1952 and later of several methodologies for analysis of human milk oligosaccharides (Kuhn's and Montreuil's groups), complex structures constituted of galactose, glucose, N-acetylglucosamine, fucose and N-acetylneuraminic acid.

The first determinations of glycan structures were carried out on ovine submaxillary mucin by Graham and Gottschalk in 1959 [27], and acidic mucopolysaccharides (Dorfman, Jeanloz, Jorpes, Lindahl, Meyer, Ogston and Rodén).

These early successes could be explained by the fact that the mucin glycan was a disaccharide and that mucopolysaccharides were made up of disaccharide repeating units. However, due to the complexity of glycans of other glycoproteins like orosomucoid, we had to wait until 1971 for the first structure of a glycan of the oligomannose type [28] and until 1974 for determination of the first structures of glycans of the N-acetyllactosamine type: those of human serotransferrin [29,30] and of immunoglobulins [31]. Since then, the situation has exploded rapidly and the number of known primary structures of glycans from glycoproteins of higher and lower animals, plants, microorganisms and viruses has grown immensely due to progress in the following methodologies:

(i) Refinement of isolation and purification procedures with, in particular, the introduction of affinity chromatography on immobilized lectins.

(ii) Improvement of the chemical and enzymic techniques of glycosidic bond cleavage.

(iii) Introduction of efficient methods of glycan permethylation [32] and of identification of methylated monosaccharides by GLC coupled to mass spectrometry.

(iv) Development of high-resolution mass spectrometry for direct analysis of glycans: MS-MS, FAB-MS, MALD-TOF-MS.

(v) Spectacular advances in oligosaccharide organic synthesis.

However, a decisive step in the determination of glycan primary structure was the introduction, in 1977, by Vliegenthart's and Montreuil's groups of high-field ^{1}H-NMR spectroscopy as a glycan analysis method [33–35].This event was the starting point of an impressive expansion of our knowledge in the field of glycan primary structure. Parallel to these studies and as a direct consequence of them, there was a growth of research on the metabolism of glycans. In this regard, we can remember the reflection of Harry Schachter looking at our poster presenting a series of glycan structures at the 3rd International Symposium on Glycoconjugates in Brighton held in 1975: "I know now what I shall do coming back to Toronto!" This was the starting point of the work of Harry Schachter on glycosyltransferases and substrate level control of glycan biosynthesis by the glycan itself (for review of early literature, see Ref. [36]). Earlier work had demonstrated that the N-acetylglucosamine and mannose residues of the inner-core of N-glycoprotein glycans were transferred to protein in the rough endoplasmic reticulum by a co-translational mechanism. Further N- and O-glycan biosynthesis steps occurred in the Golgi apparatus and were post-translational events. The first results obtained in this field were presented at the 2nd International Symposium on Glycoconjugates in Lille, in 1973 [1].A few years later, Stuart Kornfeld, Phil Robbins and Donald Summers established the fundamental concept of N-glycan processing (for review, see Ref. [37]).

Knowledge of glycan primary structure allowed the identification of the glycolipidoses, and the glycoproteinoses as diseases of glycoprotein catabolism, due to genetic deficits in lysosomal hydrolases. In fact, oligosaccharides accumulating in tissues and urine of patients were fragments of N-glycoprotein glycans [8,38]. This observation led to the definition of the normal lysosomal catabolic pathway of these compounds [39–41].

3. The years 1967–1969 and the birth of the molecular biology of glycoconjugates

3.1. Two discoveries change the face of the world of glycoconjugates

Owing to two fundamental discoveries which demonstrated that glycans are recognition signals and that they are modified in cancer cells, research on glycoconjugates expanded rapidly in the late 1960s.

3.1.1. Glycans are recognition signals for membrane lectins
To the first discovery are attached the names of Ashwell and Morell who demonstrated, in 1968, that glycans of blood serum glycoproteins control their life span since their desialylation induces uptake by a hepatocyte membrane lectin [42]. Thus were born (i) the concept of endogenous lectins which are universally distributed in all living organisms and in viruses; (ii) the concept of glycans as recognition and antirecognition signals which may play a role in the formidable problem of self and non-self recognition. However, we must recall that, about 10 years earlier the concept of membrane lectins had been indicated by Gottschalk et al. [43] who demonstrated that desialylation of erythrocytes abolished their agglutination by influenza virus and by Gesner and Ginsburg [44] who showed that defucosylation of lymphocytes disturbed the homing of these cells. At that time, glycoconjugate biochemists may have thought too much like pure chemists to recognize the biological importance of these findings ?

3.1.2. Glycans are modified in cancer cell membranes
The second discovery concerned the observation of Burger and Goldberg [45] and Inbar and Sachs [46] that the primary structure of glycoconjugate glycans was dramatically altered in cancer cells as revealed by the use of lectins. Many scientists believe that these modifications may be responsible for the anarchist behaviour of cancer cells.

In this regard, it is amazing to note that the same observation had already been described by Aub et al. in 1963 [47] but did not attract the attention of the scientific world probably because, with the awarding of the Nobel Prize to Jacob, Lwoff and Monod, the eyes of most researchers were turned towards nucleic acids which triggered a rush to this field similar to the Californian 'Gold Rush'. Thus an observation was buried which could have been, five years earlier, the starting point of research on the molecular biology of glycoconjugates.

Immediately after these two decisive discoveries, cell biologists and pathologists turned for help to the biochemists, eager to learn what kind of glycan structures appeared on cancer cell membranes and intervened in the cellular and molecular recognition mechanisms. Unfortunately, in 1968, research on glycoconjugates was far behind that of nucleic acids and proteins. Fortunately, in the past few years, the pharmaceutical industry has stimulated interest in glycoprotein research thanks to the above-mentioned problems encountered in the production of recombinant glycoproteins which represent a fantastic potential market: for example, in 1991, worldwide sales of recombinant erythropoietin reached $ 645 million! In my personal opinion, this situation was the clinching argument thanks to which glycoconjugates have finally been awarded their for so long awaited 'lettres de noblesse'.

To solve the central problem of conformity of recombinant glycoprotein glycans with the native ones, research must develop in two crucial fields: (i) the fractionation of glycoforms and the miniaturization of techniques for glycan primary structure determination, and (ii) the regulation of the mechanism of glycan metabolism at three levels: that of enzyme activity, that of the gene expression of glycosyltransferase and glycosidase genes and that of enzyme subcellular compartmentation. Work in this area would also help to explain the pathological metabolic deviations, in particular those which are induced by cancer at the level of membrane glycoconjugates.

Fortunately, although research on the molecular biology of glycosyltransferases and glycosidases started relatively late the pioneer work of Robert Hill, James Paulson, Harry Schachter, Nancy and Joel Schaper and others has led to rapid development in the field of glycoprotein metabolism regulation due to (i) the purification to homogeneity of glycosyltransferases and glycosidases; (ii) the use of specific inhibitors of these enzymes; (iii) the emergence of genetic engineering techniques such as site directed mutagenesis; (iv) the use of somatic cell mutants; (v) the suggestion that glycans may be markers of Evolution; and (vi) characterization and isolation of transporters present in the membranes of endoplasmic reticulum.

At this stage of the painting of our historical fresco, we have mainly related the glycan *primary structure* to the normal and pathological metabolism of glycoproteins. However, the biological roles of glycans cannot be understood only on the basis of the knowledge of glycan primary structure, but also require determination of the *three-dimensional structures* of carbohydrate–protein complexes.

3.2. Three-dimensional structure of glycans. From speculation to reality

3.2.1. The first images: speculation

For a long time the only image we had of the spatial structure of a glycan was totally speculative since it was obtained by model building (for reviews, see Refs. [8,10, 12,15]). It concerned the diantennary N-glycan of human serotransferrin the primary structure of which was determined in 1973 [1,30] and was presented by Montreuil in 1974 at the VIIth International Symposium on Carbohydrate Chemistry in Bratislava. This model was obtained by creating thermodynamically possible hydrogen bonds and led to the so-called Y-conformation [8]. However, on the basis of experimental data obtained by X-ray crystallographic studies of oligosaccharides [48,49], NMR [50,51] and EPR [52], the Y-conformation was revisited and progressively refined. This led to the T- [53], then to the 'bird'- [10] and finally to the 'broken wing'- [54] conformations. In the same way, tetra-antennary glycans could adopt umbrella- or reversed umbrella-conformations [15]. Applying the same methodologies, Vliegenthart et al. [35] gave the first images of N-glycans of the oligomannose type.

On the basis of these speculative data, the fundamental concept of the flexibility of glycan molecules that may adopt different shapes because of the considerable freedom of rotation around the glycosidic bonds, was established by Montreuil who wrote in 1980 [10]: "The double character of rigidity and flexibility that the glycans possess is completely compatible with the role of recognition signals which it is tempting to ascribe to them. They may be imagined as solidly planted on the proteins by a rigid arm. This ligand is constituted of the terminal trisaccharide which could, on account of its planar conformation, possibly penetrate into the protein to anchor the flexible antennae. The latter, because of mobility, and especially, because of their terminal sialic acid residues, could more rapidly adapt themselves to the receptor sites and fit to them". We now know that the concept of the antennae changing their conformation according to the environment fits perfectly the concept of glycans acting as recognition signals. Moreover, the double character of rigidity and flexibility of glycans plays a central role in the biological function of the carbohydrate moiety of glycoproteins, probably favouring, when necessary, an extended conformation which itself favours (i) the interaction of glycans with their own protein; (ii) glycan–glycan interactions; and (iii) the recognition of glycans by receptors like exogenous and endogenous lectins.

At the same time, Pérez, Warin and Montreuil (cited in Ref. [15]) used computer calculations to examine which structures were sterically feasible and which conformation was energetically the most favourable. The answer of the computer was: the bird-conformation. Since this first attempt, the molecular modelling of glycans has developed thanks to the improvements of computational analyses and high-resolution NMR spectroscopy. In addition, the introduction of molecular dynamics simulations has confirmed the concept of glycan flexibility [55,56].

3.2.2. The modern views: reality
Molecular modelling remains to a large extent speculative and only X-ray diffraction can furnish an exact representation of molecules in space, in particular if it is associated with NMR data. In this regard, an X-ray structure of a glycan was for a long time limited to that of the human and rabbit immunoglobulin Fc fragment determined by Deisenhofer and Huber [57,58]. This situation was the result of the difficulties encountered in obtaining crystals of glycoproteins or glycans, due to a large extent to the microheterogeneity and mobility of glycans. This explains why the immunoglobulin Fc fragment gave an interpretable X-ray diffraction pattern. In fact, the unique diantennary glycans of this molecule were 'frozen' in only one conformation due to the interaction of the α-1,6-antennae with the peptide chain and of the α-1,3 antennae of the two glycans of the Fc fragments in vis-à-vis position. However, the past few years have seen an impressive development of research in the field of X-ray diffraction studies of glycans immobilized in a fixed and immobile conformation either by interaction with lectins [59] or with their own protein moiety [60].

On the whole, results obtained from X-ray diffraction and molecular modelling have confirmed some of the earlier speculative concepts on the flexibility of the glycans and of the antennae.

4. Concluding remarks

In 1975, Montreuil concluded his general review entitled "Recent data on the structure of the carbohydrate moiety of glycoproteins. Metabolic and biological implications" [8] as follows: "The most exciting and marvellous age of the history of glycoproteins starts right now. At this time, we would like to remember the pioneers whose discoveries have established the bases of our knowledge of the physicochemical properties and structure of glycoconjugates. Not all of them have seen the fruits of the trees they planted. In this connection, I would like to conclude with Alfred Gottschalk who wrote in 1973: About the outlook of glycoprotein research, we are not at the end of all progress, but at the beginning. We have but reached the shores of a great unexplored continent".

Twenty years later, we can claim that the visionary dream of many of us is now realized. We have crossed the shores of the "great unexplored continent" announced by Alfred Gottschalk and we are discovering, filled with wonder, its fascinating secrets. We have entered the golden age of glycoconjugates.

It is hoped that the present book will represent a milestone in the progress of research on this exciting class of components of living matter.

References

1. Montreuil, J. (1974) Méthodologies concernant la structure et le métabolisme des glycoconjugués. Actes du Colloque International du C.N.R.S. no. 221, Villeneuve d'Ascq, Juin 1973, C.N.R.S., Paris.
2. Levene, P.A. (1925) Hexosamines and Mucoproteins. Longmans and Green, New York.
3. Montreuil, J. (1959) In: M. Javillier, M. Polonovski, M. Florkin, P. Boulanger, M. Lemoigne, J. Roche and R. Wurmser (Eds.), Les glycoprotéides. Traité de Biochimie Générale, vol. 1, part 2. Masson, Paris, pp. 935–1002, .
4. Balasz, E.A. and Jeanloz, R.W. (1965) The Amino Sugars, Vol. 2A: Distribution and Biological Role; R. Jeanloz and E.A. Balasz (1965) The Amino Sugars, Vol. 1B: Glycosaminoglycans, Glycoproteins and Glycosaminolipids; E.A. Balasz and R.W. Jeanloz (1966) The Amino Sugars, Vol 2B: Metabolism and Interactions; R.W. Jeanloz (1969) The Amino Sugars, Vol 1A: Chemistry of Amino Sugars. Academic Press, New York.
5. Gottschalk, A. (1966 and 1972) Glycoproteins. Their Composition, Structure and Function. Elsevier, Amsterdam.
6. Marshall, R.D. and Neuberger, A. (1970) Adv. Carbohydr. Chem., 25, 407–478.
7. Sharon, N. (1975) Complex Carbohydrates, their Chemistry, Biosynthesis and Functions, Addison-Wesley, Reading.
8. Montreuil, J. (1975) Pure Appl. Chem. 42, 431–477.
9. Horowitz, M. and Pigman W. (1977) The Glycoconjugates, Vol. 1 (1977), Vol. 2 (1978), Vol. 3 (1982), Vol. 4 (1982). Academic Press, New York.

10. Montreuil, J. (1980) Adv. Carbohydr. Chem. Biochem. 37, 157–223.
11. Lennartz, W.J. (1980) The Biochemistry of Glycoproteins and Proteoglycans, Plenum Press, New York.
12. Montreuil, J. (1982) Glycoproteins, In: A. Neuberger and L.L.M. Van Deenen L.L.M. (Eds), Comprehensive Biochemistry. Elsevier, Amsterdam, pp. 1–188.
13. Sharon, N. and Lis, H. (1982) In: H. Neurath and R.L. Hill (Eds.), The Proteins, Vol. 5. Academic Press, New York, pp. 1–144.
14. Hughes, R.C. (1983) Glycoproteins. Chapman and Hill, London.
15. Montreuil, J. (1984) Pure Appl. Chem. 56, 859–877; (1984) Biol. Cell 51, 115–132.
16. Ivatt, R.J. (1984) The Biology of Glycoproteins. Plenum Press, New York.
17. Kornfeld, R. and Kornfeld, S. (1985) Annu. Rev. Biochem. 5, 631–664.
18. Rademacher, T.W., Parekh, R.B. and Dwek, R.A. (1988) Annu. Rev. Biochem. 57, 785–838.
19. Montreuil, J. (1988) Proceedings of the IXth International Symposium on Glycoconjugates, Lille, 6–8 Juillet 1987 (1988) Biochimie, 70, pp. 1433–1706, Elsevier, Amsterdam.
20. Berger, E.G., Buddecke, E., Kamerling, J.P., Kobata, A., Paulson, J.C. and Vliegenthart, J.F.G. (1982) Experientia 38, 1129–1161.
21. Kobata, A. (1992) Eur. J. Biochem. 209, 483–501.
22. Allen, H.J. and Kisailius, E.C. (1992) Glycoconjugates: Composition, Structure and Functions. Marcel Dekker, New York.
23. Varki, A. (1993) Glycobiology 3, 97–130.
24. Lis, H. and Sharon, N. (1993) Eur. J. Biochem. 218, 1–27.
25. Eichwald, E. (1865) Ann. Chem. Pharmakol. 134, 177.
26. Neuberger, A. (1974) Biochem. Soc. Symp. 40, 1.
27. Gottschalk, A. and Graham, E.R.B. (1959) Biochim. Biophys. Acta 34, 380–391; (1960) Biochim. Biophys. Acta 38, 513–534.
28. Yamaguchi, H., Ikenaka, T. and Matsushima, Y. (1971) J. Biochem. (Jpn.) 70, 587–594.
29. Spik, G., Vandersyppe, R., Fournet, B., Bayard, B., Charet, P., Bouquelet, S., Strecker, G. and Montreuil, J. (1974) In: Ref. 1, pp. 483–500.
30. Spik, G., Bayard, B. Fournet, B., Strecker, G, Bouquelet, S. and Montreuil, J. (1975) FEBS Lett. 50, 269–299.
31. Baenziger, J., Kornfeld, S. and Kochwa, S. (1974) J. Biol. Chem. 249, 1889–1896; 1897–1903.
32. Hakomori, S.I. (1964) J. Biochem. (Jpn.) 55, 205–208.
33. Dorland, L., Haverkamp, J., Schut, B.L., Vliegenthart, J.F.G., Spik, G., Strecker, G., Fournet, B. and Montreuil, J. (1977) FEBS Lett. 77, 15–20.
34. Montreuil, J. and Vliegenthart, J.F.G. (1979) In: J.D. Gregory and R.W. Jeanloz (Eds.), Proc. IVth Int. Symp. Glycoconjugates, Woods Hole, Sept. 1977. Academic Press, New York, pp. 35–78.
35. Vliegenthart, J.F.G., Dorland, L. and van Halbeek, H. (1983) Adv. Carbohydr. Chem. Biochem. 41, 209–374.
36. Beyer, T.A., Sadler, J.E., Rearick, J.I., Paulson, J. and Hill, R.L. (1981) Adv. Enzym. 52, 23–175.
37. Kornfeld, R. and Kornfeld, S. (1985) Annu. Rev. Biochem. 54, 631–664.
38. Strecker, G. and Montreuil, J. (1979) Biochimie 61, 1199–1246.
39. Brassart, D., Baussant, T., Wieruszeski, J.M., Strecker, G., Montreuil, J. and Michalski, J.C. (1987) Eur. J. Biochem. 169, 131–136.
40. Strecker, G., Michalski, J.C. and Montreuil, J. (1988) Biochimie 70, 1505–1510.
41. Aronson, N.N. Jr. and Kuranda, M.J. (1989) FASEB J. 3, 2615–2622.
42. Morell, A.G., Irvine, R.A., Sternlieb, I., Scheinberg, I.H. and Ashwell G. (1968) J. Biol. Chem. 243, 155–159.

12

43. Gottschalk, A. (1957) Biochim. Biophys. Acta 23, 645–646; (1960) Nature 186, 949–951.
44. Gesner, B.M. and Ginsburg, V. (1964) Proc. Natl. Acad. Sci. USA 52, 750–755.
45. Burger, M.M. and Goldberg, A.R. (1967) Proc. Natl. Acad. Sci. USA 57, 359–366.
46. Inbar, M. and Sachs, L. (1969) Nature 223, 710–712; Proc. Natl. Acad. Sci. USA 63, 1418–1425.
47. Aub, J.C., Tieslau, C. and Lankester, A. (1963) Proc. Natl. Acad. Sci. USA 50, 613.
48. Warin, V., Baert, F., Fouret, R., Strecker, G., Spik, G., Fournet, B. and Montreuil, J. (1979) Carbohydr. Res. 76, 11–22.
49. Douy, A. and Gallot, B. (1980) Biopolymers 19, 493–507.
50. Bock, K., Arnarp, J. and Lönngren, J. (1982) Eur. J. Biochem. 129, 171–178.
51. Brisson, J.R. and Carver, J.P. (1983) Biochemistry 22, 1362–1368; 3680–3686.
52. Davoust, J., Michel, V., Spik, G., Montreuil, J. and Devaux, P. (1981) FEBS Lett. 125, 271–276.
53. Montreuil, J., Fournet, B., Spik, G. and Strecker, G. (1978) C.R. Acad. Sci. Paris 287D, 837–840.
54. Montreuil, J., Debray, H., Debeire, P. and Delannoy, P. (1983) In: H. Popper, W. Reutter, F. Gudat and E. Köttgen (Eds.), Structural Carbohydrates in the Liver, Falk Symposium no. 34. MTP Press, Lancaster, pp. 239–258.
55. Mazurier, J., Dauchez, M., Vergoten, G., Montreuil, J. and Spik, G. (1991) Glycoconj. J. 8, 390–399.
56. Dauchez, M., Mazurier, J., Montreuil, J., Spik, G. and Vergoten, G. (1992) Biochimie 74, 63–74.
57. Deisenhofer, J., Colman, P.M., Epp, O. and Huber, R. (1976) Z. Physiol. Chem. 357, 1421–1434.
58. Huber, R., Deisenhofer, J., Colman, P.M., Matsushima, M. and Palm, W. (1976) Nature 264, 415–420.
59. Bourne, Y., Rougé, P. and Cambillau, C. (1990) J. Biol. Chem. 265, 18161–18165; (1992) J. Biol. Chem. 267, 197–203.
60. Shaanan, B., Lis, H. and Sharon, N. (1992) Science 254, 862–866.

J. Montreuil, H. Schachter and J.F.G. Vliegenthart (Eds.), *Glycoproteins*

Primary structure of glycoprotein glycans

JOHANNES F.G. VLIEGENTHART[1] and JEAN MONTREUIL[2]

[1]*Bijvoet Center for Biomolecular Research, Department of Bio-organic Chemistry, P.O. Box 80.075, 3508 TB Utrecht, the Netherlands*
[2]*Université des Sciences et Technologies de Lille, Laboratoire de Chimie Biologique (UMR no. 111 du CNRS), 59655 Villeneuve d'Ascq Cedex, France*

1.1. Introduction

For a long time, research on the primary structure of glycans was restricted to mucins, polysaccharides of connective tissue, serum glycoproteins and egg white glycoproteins. The reasons for such a peculiar choice of compounds are found in the inherent limitations of the analytical methods. Until the 1970s, large quantities of starting material were needed, due to the low sensitivity of the applied techniques. However, as mentioned in Chapter 1, the subsequent improvements of methods for isolation and structure determination led to the characterization of thousands of primary glycan structures. In this respect, it is relevant to note that the data bank CarbBank [1] contains at the moment over 38 000 records, of which 15 000 are unique structures. On the average hundreds of novel structures have been described every year since 1991. In addition, the discovery by biologists of the fundamental importance of glycoconjugates, has contributed to the spread of the research to many other species (lower animals, parasites, yeasts, fungi and bacteria). Most of the currently known glycan structures of glycoproteins are presented in the following chapters of this book.

1.2. Intact glycoproteins

The determination of the primary structure of glycoproteins comprises the analysis of the protein sequence, the identification of the glycosylation sites, the unravelling of the glycan structures and the determination of the microheterogeneity of the glycans at each glycosylation site. For these studies it is essential that adequate starting material is available and that the glycoprotein sample to be analysed contains a collection of glycoforms similar to that occurring in the natural environment. In practice this is a difficult problem because particular glycoforms may be (partly) lost or newly created due to degradation during the isolation and fractionation procedures. For soluble glycoproteins, the standard techniques for protein isolation and purification are first followed, as reviewed in [2]. A glycoprotein that is apparently homogeneous with respect to the protein backbone, may nevertheless contain a collection of glycoforms [3]. Complete fractionation of all glycoforms is difficult to achieve, in particular when more than one glycosylation site is present and each of these sites exhibits microheterogeneity. Nevertheless, by lectin- or antibody-based affinity chromatography [2] at least partial fractionation can be obtained. High performance capillary electrophoresis [4] looks promising for realizing further separation of glycoforms. However, it can be envisaged that this technique also has inherent limitations.The isolation and purification of membrane bound glycoproteins give rise to specific problems that cannot be solved in a general way, and often detergents and/or chaotropic reagents have to be used. Glycoproteins that are membrane bound via glycosylphosphatidyl inositol (GPI) anchors are usually analysed after cleavage from the anchor. There are several ways to achieve this cleavage [5,6].

The detailed analysis of the primary structures of the glycans can hardly be carried out at the level of the intact glycoprotein. However, electrospray–mass spectrometry (ES-MS) has been shown to be a useful technique to establish post-translational modifications of proteins [7], even if the compounds have a molecular mass larger than 100 kDa. For various glycoproteins the ES-MS profile has been recorded and has yielded information about the glycoforms present, especially by coupling capillary electrophoresis as a separation technique with ES-MS as a detector [8]. For glycoproteins with a molecular mass up to 20 kDa, NMR spectroscopy can reveal many structural details [9–14]. Recently, the application of gradient-enhanced natural abundance ^{1}H-^{13}C HSQC and HSQC-TOCSY spectroscopy has been shown to be effective for the assignment of the NMR resonances of the carbohydrate chains of an intact glycoprotein [15]. By the former technique, the anomeric ^{1}H-^{13}C correlations can be derived, and the latter approach allows the deduction of correlations of the monosaccharide skeleton atoms.

1.3. Partial structures

To derive the primary structures of the entire collection of glycan chains attached to a protein, it is usually necessary to degrade the glycoprotein to partial structures like oligosaccharides, oligosaccharide–alditols or glycopeptides. For the last type of compounds it is important that the number of glycosylation sites comprised in the structure is known. There have been approaches described to obtain partial structures from N,O-glycoproteins (for a review see Ref. [16]).

However, two main problems have to be solved: (i) how to achieve a sequential, complete cleavage of N- and O-linked chains; and (ii) how to establish the positions in the protein chain of the glycosylation sites. The degradation to partial structures can be performed with chemical or enzymic methods. Hydrazinolysis is an example of a chemical method that is frequently applied to liberate N-linked chains as oligosaccharides [17,18]. Disadvantages of this methodology may be: (i) chemical modifications at the reducing end, which enhance the heterogeneity; (ii) partial cleavage of O-linked chains in N, O-glycoproteins; and (iii) removal of several essential non-carbohydrate substituents. Careful manipulation of the reaction conditions may reduce these complications and it has been claimed that it is possible to apply graded hydrozinolysis leading to sequential release of the O-linked chains followed by the N-linked chains. For each glycoprotein, the situation might be different, necessitating always a careful check of the effectiveness of the methodology. The release of O-linked chains, e.g. from O-glycoproteins in the form of oligosaccharide alditols can be realized with alkaline borohydride [19].

In the enzymic approach peptide-N^4-(N-acetyl-β-D-glucosaminyl) asparagine amidase (PNGase) is currently the most frequently used enzyme to liberate the N-chains as oligosaccharides, although endoglycosidases are also applied. Depending on the glycoprotein PNGase F or A treatment is used. After PNGase digestion of N,O-glycoproteins, the remaining O-glycoprotein can be isolated and then subjected to alkaline borohydride degradation [20]. It is essential to check the completeness of the enzymic cleavage carefully e.g. by SDS-PAGE with silver staining, or with the aid of a glycan detection kit, a Con-A test in a microtiter plate [21] or a test with another lectin in an analogous way. Often the incubation conditions like pH, the addition of chaotropic or reducing agents and for large glycoproteins a preceding, partial proteolysis have to be optimized for each glycoprotein in order to ensure a complete release of the N-linked chains.

In all cases described above the released glycans are obtained as complex mixtures of oligosaccharides or oligosaccharide alditols. This may create severe fractionation problems, that require multiple chromatographic steps. To obtain pure compounds reduction of the oligosaccharides to alditols is sometimes carried out to simplify the complexity of the mixture.

In some instances endoglycosidases are used to split off oligosaccharides from N-linked chains. The specificity of these enzymes can be rather strict, thereby limiting their applicability [22].

Various proteases are in use to degrade N-glycoproteins to glycopeptides in order to obtain pure glycopeptides for the determination of the glycosylation sites and of the site-specific (micro)heterogeneity [23]. The amino acid sequence, the glycan structure and the specificity of the protease control the size and type of the glycopeptides formed. Often incomplete and aspecific proteolytic cleavages give rise to complex mixtures of peptides and glycopeptides. This may afford redundant results, because one glycosylation site can now be represented by more than one glycopeptide. In O-, and N,O-glycoproteins often multiple O-glycosylation sites occur. These sites are frequently clustered, thereby impeding the preparation of glycopeptides containing only a single O-glycosylation site. In heavily O-glycosylated proteins the assignment of a particular glycan to a specific amino acid can be difficult.

The fractionation of partial structures has been improved enormously by the introduction of: (i) better column materials for gel permeation chromatography to achieve size fractionation [24]; (ii) lectin-affinity chromatography [25]; (iii) HPLC on anion exchange materials [26]; (iv) high pH anion exchange chromatography (HPAEC) [27], in combination with pulsed amperometric (PAD) or radiometric detection (this method is sensitive and can be applied to small amounts of material); (v) high performance capillary electrophoresis [28].

1.4. Structure determination

To define the structure of glycans completely, the following parameters have to be determined:

(i) Type and number of the constituent monosaccharides, including absolute configuration, ring size and anomeric configuration.

(ii) Sequence of the monosaccharides, including the positions of the glycosidic linkages.

(iii) Type, number and location of non-carbohydrate substituents.

A variety of methods are currently used to establish these parameters depending on the type of problems to be solved, and on the availability of material and techniques. For analysis of the monosaccharide composition, the samples are often subjected to methanolysis followed by re-*N*-acetylation, trimethylsilylation and GLC [29]. Alternatively, acid hydrolysis can be carried out followed by HPAEC-PAD [27]. Linkage analysis is carried out on partially methylated monosaccharide alditols. They are obtained by permethylation of the sample, e.g., in DMSO with a suitable base and CH$_3$I, followed by hydrolysis, e.g., in 4M trifluoro acetic acid, acetylation and analysis by GLC-MS [30,31]. Exoglycosidases are employed to gain information on the non-reducing-end monosaccharides with regard to identity and absolute and anomeric configuration. Sequential enzymic degradation with exoglycosidases can provide some insight into the structure [32]. However, these exoglycosidases are not very specific as to ring size, linkage position and branching point. Furthermore, the presence of non-carbohydrate substituents might inhibit their action. It is a prerequisite that detailed knowledge exists on the specificity of the exo-enzymes. Endoglycosidases with known substrate specificity can provide additional information [16,22]. For oligosaccharides and glycopeptides advanced mass spectrometric techniques like FAB-MS; ES-MS; MALD-MS; MALD-TOF-MS; MS-MS are suitable to obtain structural information as to branching pattern, number and length of branches and sequence. The mass spectrometric analysis of different derivatives and/or chemically modified compounds may furnish complementary information. A significant advantage of mass spectrometry is that only low amounts of material are required [7].

High resolution ^{1}H-NMR spectroscopy at 500 or at 600 MHz has become a powerful method for the identification of N- as well as O-type carbohydrate chains, due to the introduction of the structural-reporter-group concept [33–38]. In D$_2$O, the greater part of the skeleton protons of the constituting monosaccharides resonate in a narrow window between δ 3.5 and δ 3.9, thereby giving rise to a bulk signal. The signals outside this region are the structural-reporter-group signals that contain the essential information for translation into primary structures. They comprise: (i) anomeric protons; (ii) protons attached to carbon atoms close to the substitution position in a monosaccharide constituent (so-called glycosylation shifts); (iii) protons at deoxy-carbon atoms; (iv) *N*-acetyl and *N*-glycolyl methylene protons; (v) protons shifted as a result of the presence of non-carbohydrate substituents. In addition to 1D-NMR spectra, various types of 2D-NMR spectra like COSY, HOHAHA, NOESY, HMQC and HMBC are used to arrive at unambiguous assignments. The large number of ^{1}H-NMR data that are available for glycoprotein derived carbohydrate chains, has enabled the construction of a NMR database that is connected to the complex carbohydrate structure database [39]. Owing to the enormous variety of glycoprotein derived glycans that have been identified now, it has become feasible to develop oligosaccharide-finger-printing methods, facilitating reliable and fast batch control of, e.g., biotechnologically produced glycoproteins. These methods include different separation techniques like high pH anion-exchange chromatography or capillary electrophoresis [40–45]. Another possibility is offered by PAGE of oligosaccharide derivatives obtained after reductive amination with 8-aminonaphthalene 1,3,6-trisulfonic acid [46]. However, in the cases where novel structures are involved, it is essential to have pertinent structural data obtained via more than a single technique, in order to arrive at unambiguous conclusions.

The presence of non-carbohydrate substituents may give rise to specific analytical problems, starting with the isolation of the glycoprotein. It is essential that during the isolation and purification procedures the authentic non-carbohydrate substituents remain present and are not split off and do not migrate to other positions. The substituents identified so far in glycoproteins can be methyl, acetyl, lactyl, glycolyl, aminoethyl phosphonate, sulfate or phosphate groups. For the determination of the identity and location of these substituents various methods are in use, but the most detailed information is obtained from NMR spectroscopy and mass spectrometry (cf. Refs. [20,47]).

2.1. Monosaccharide constituents

For a long time, the 'classical' monosaccharides constituting the glycans were D-galactose, D-mannose, D-glucose, L-fucose, D-xylose, L-arabinofuranose, N-acetyl-D-glucosamine, N-acetyl-D-galactosamine and some sialic acids. However, refinement of the analytical methods has led to the characterization of new monosaccharides which were first considered as rare sugars, but which now appear to be more common than previously thought (for review, see Ref. [48]), e.g., 2-keto-3-deoxy-nonulosonic acid (Kdn), one of the 35 known sialic acids [49], first discovered in fish eggs [50] and recently characterized in batracian eggs [51].

The monosaccharides which are likely to be found in glycoprotein glycans, other than the above-mentioned 'classical' ones, are listed in Table I.

2.2. The glycan protein linkages

Glycans are conjugated to peptide chains by two types of primary covalent linkages: N-glycosyl and O-glycosyl linkages as listed in Table II.

Until recently, the only N-glycosyl bond characterized in glycoproteins regardless of whether they originated from animals, plants, microorganisms or viruses, was the N-acetylglucosaminyl–asparagine bond formally resulting from linking a β-N-acetylglucosamine residue to the amido group of L-asparagine, as discovered by Neuberger et al. [75] and almost simultaneously by others [76,77]. However, during recent years,

TABLE I

Monosaccharides recently identified in glycoprotein glycans [4]

Monosaccharides	Sources	Refs.
2-Acetamido-4-amino-2,4,6-trideoxyglucose	*Clostridium symbiosum*	[52]
6-Deoxylaltrose	Salmonid fish eggs	[53]
3-Deoxy-D-*glycero* -*galacto* -nonulosonic acid (Kdn)	Salmonid fish eggs	[50]
2,3-Diacetamido-2,3-dideoxy-mannuronic acid	*Bacillus stearothermophilus*	[52]
2-*O*-Methylfucose	Nematodes	[54]
Galactofuranose	Bacteria	[55,56]
	Trypanosoma	[57]
3-*O*-Methylgalactose	Yeasts	[58]
	Snail	[59]
4-*O*-Methylgalactose	Nematodes	[54]
6-*O*-Methylgalactose	Algae	[60]
Galactose-3-sulfate	Thyroglobulin	[61,62]
	Mucins of cystic fibrosis	[63]
N-Acetylgalactosamine-4-sulfate	Pituitary hormones	[64]
	Tamm Horsfall glycoprotein	[65]
	Urokinase	[66]
3-*O*-Methylglucose	*Methanothermus fervidus*	[67]
3-*O*-Methyl-*N*-acetylglucosamine	*Clostridium thermocellum*	[56]
N-Acetylglucosamine-6-sulfate	Thyroglobulin	[61,62]
Gulose	Algae	[63]
3-*O*-Methylmannose	Snail	[69]
Mannose-4-sulfate	Ovalbumin	[70]
Mannose-6-sulfate	Ovalbumin	[70]
	Slime mold	[71]
Mannose-6-phosphate Me	Slime mold	[72]
N-Acetylmannosamine	*Clostridium symbosium*	[52]
4,8-Anhydro-neuraminic acid	Edible bird's nest	[73]
8-*O*-Methyl 9-*O*-acetyl-*N*-glycolyl-neuraminic acid	Starfish	[74]
8-*O*-Methyl-7,9-*O* -diacetyl-*N*-glycolylneuraminic acid	Starfish	[74]

linkages between asparagine and other sugars such as glucose, *N*-acetylgalactosamine and L-rhamnose [48] have been characterized in bacterial glycoproteins.

Unambiguous proof for the existence in glycoproteins of O-glycosidic linkages of glycans to serine and threonine was provided in the early 1960s. The evidence was based on the alkali-lability of the sugar-β-hydroxyamino acid linkage which is split by a β-elimination reaction.

The *N*-acetylgalactosamine-serine/threonine bond was first demonstrated in the mucins [78–81] and is widely distributed in nature in the so-called mucin-type glycoproteins. The O-glycosidic carbohydrate-serine linkage was discovered by Helen Muir in 1958 [82] in chondroitin-4-sulfate and the involvement of xylose was demonstrated

TABLE II

Carbohydrate-peptide linkages in glycoproteins [a]

Type of glycoprotein	Amino acid	Monosaccharide
N-glycosylproteins		
	Asn	β-GlcNAc
	Asn	β-GalNAc
	Asn	α/β-Glc
	Asn	L-Rha
O-glycosylproteins		
Mucin type	Ser/Thr	α-GalNAc
Intracellular type	Ser/Thr	β-GlcNAc
Yeast type	Ser/Thr	*x*-Man
Rat tissues	Ser/Thr	α-L-Fuc
Proteoglycan type	Ser	β-Xyl
Worm collagen type	Ser	α-Gal
Factor IX type	Ser	β-Glc
Collagen type	OH-Lys	β-Gal
Extensin type	OH-Pro	β-L-Ara*f*
Algal type	OH-Pro	β-Gal
Glycogenin type	Tyr	α-Glc
S-layer-type[b]	Tyr	β-Glc

(a) For detailed historical review, see Refs. [48] and [105].
(b) In crystalline surface layer of *Clostridium thermohydrosulfuricum.*

in 1964 by Lindahl and Rodén in heparin [83] and in chondroitin-4-sulfate [84]. At present, the xylosyl–serine linkage is considered to be characteristic of animal proteoglycans. Other glycan–amino acid bonds are: (i) galactosyl (β1–5) hydroxylysine in guinea-pig-skin tropocollagen [85,86]; (ii) L-arabinofuranosyl (β1–4) hydroxyproline first recognized by Lamport et al. in cell walls from higher plants and in tomato extensin [87,88], by Neuberger et al. in potato lectin [89] and by others in a very wide variety of higher- and lower-plant cell tissues.

Two novel bonds have been added to the list of O-linked amino acids: the glucosyl (α1–O) tyrosine linkage in glycogenin [90] and *N*-acetylglucosaminyl (β1–O) serine in intracellular glycoproteins (nuclear pore, chromatin proteins, transcription factors and cytoplasmic inclusions); interestingly no other sugars are attached to the *N*-acetylglucosamine residue (for review, see Ref. [91]). In this regard, two additional novel bonds have been characterized. The first one originates from the non-enzymic condensation of a monosaccharide like glucose with the ε-amino group of lysine residues leading to the formation of the imino group of a Schiff's base. This reaction is called *glycation*. The second one concerns a new class of widely distributed membrane components of eukaryotic cells, parasites in particular: the glycosylphosphatidyl inositol anchor (GPI) [92,93]. This kind of attachment has been termed *glypiation*.

Recently a new type of linkage between a carbohydrate and a protein has been found, involving C-glycosylation of a specific tryptophan residue in human RNase U_s [94]. The origin of this linkage has still to be traced.

2.3. The inner-core and antenna concept

On the basis of the glycan primary structures of 14 N-glycosylproteins analyzed in 1974 and using the observation that all of them presented the common structural pattern of the mannotriose-di-*N,N'*-acetyl chitobiose (Fig. 1) linked to an asparagine residue, Montreuil [97] proposed in 1975 to designate this common and non-specific invariant fraction (*inv* fraction) *the core* and the variable oligosaccharide structure (*var* fraction) planted on the inner-core, the *antennae*. The latter are mobile and support the biological activity of the glycans.

The concept of core was further extended to various O-glycoproteins and the most classical ones are presented in Fig. 2. It is interesting to mention the hypothesis of Jett and Jamieson [96] according to which the *N*-acetylglucosamine asparagine linkage is the most primitive one and the other amino acids involved in the glycan–protein

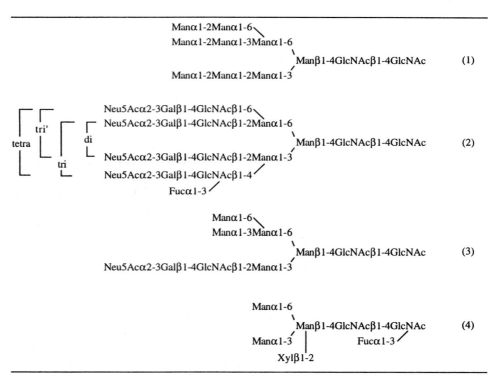

Fig. 1. Examples of the four types of N-linked carbohydrate chains of glycoproteins. 1. Oligomannose type; 2. *N*-acetyllactosamine type (complex type); 3. Hybrid type; 4. Xylose-containing type.

Type	Structure	Type	Structure
core 1	Galβ1-3GalNAc	core 4	GlcNAcβ1-6 \\ GlcNAcβ1-3GalNAc
core 2	GlcNAcβ1-6 \\ Galβ1-3GalNAc	core 5	GalNAcα1-3GalNAc
core 3	GlcNAcβ1-3GalNAc	core 6	GlcNAcβ1-6 \\ GalNAc

Fig. 2. Core structures of mucin type O-linked carbohydrate chains of glycoproteins.

linkages, except hydroxyproline could be derived from asparagine by a single-mutation of the original codons AAU and AAC into AGU and AGC for serine, ACU and ACC for threonine, AAA and AAG for lysine.

2.4. Classification and nomenclature of glycans and glycoproteins

For a long time the nomenclature and definition of glycoproteins remained chaotic because little was known before the 60s about the composition and the primary structure of glycans and about the carbohydrate–protein linkage (for review, see Refs. [97,105]). However, little by little these parameters became better defined leading to a series of definitions that are at present known as the recommendations of the Joint Commission on Biochemical Nomenclature of the International Union of Pure and Applied Chemistry (IUPAC) and the International Union of Biochemistry and Molecular Biology (IUBMB). These recommendations have been published in various periodicals. The main definitions are the following.

A glycoprotein is a compound containing carbohydrate (or glycan) covalently linked to protein. The carbohydrate may be in the form of a monosaccharide, disaccharide(s), oligosaccharide(s), polysaccharide(s), or their derivatives (e.g. sulfo- or phospho-substituted). One, a few, or many carbohydrate units may be present. Proteoglycans are a subclass of glycoproteins in which the carbohydrate units are polysaccharides that contain amino sugars. Such polysaccharides are also known as glycosaminoglycans. A glycopeptide is a compound consisting of carbohydrate linked to an oligopeptide composed of L- and/or D-amino acids. A glyco-amino-acid is a saccharide attached to a single amino acid by any kind of covalent bond. A glycosyl-amino-acid is a compound consisting of saccharide linked through a glycosyl linkage (O-, N-, or S-) to an amino acid. (The hyphens are needed to avoid implying that the carbohydrate is necessarily linked to the amino group.)

In N, O-glycoproteins the prefix N- is used for the N-glycosyl linkage to asparagine. N-linked oligosaccharides are divided into two major classes: the *N*-acetyllactosamine type containing N-acetyl-D-glucosamine, D-mannose, D-galactose, L-fucose and sialic acid; and the oligomannose type containing N-acetyl-D-glucosamine and a variable number of D-mannose residues. Structures containing both oligomannose- and N-acetyllactosamine-type oligosaccharides are designated as hybrid type. The prefix O- is used for O-glycosyl linkage to serine, threonine, hydroxylysine or bydroxyproline.

Two types of carbohydrate–peptide linkage in the same protein or peptide chain may be indicated by a combination of the prefixes. For the condensed representation of sugar chains, the non-reducing terminus of the carbohydrate chain should always be on the left-hand end. Current practice allows the use of either an extended form (a) or a condensed form (b), which allows structures to be shown in one line as well as in two or more, and in which the longest chain should always be the main chain:

(a) Extended form

$$\beta\text{-D-Gal}p\text{-}(1{\rightarrow}4)\text{-}\beta\text{-D-GlcNAc}p\text{-}(1{\rightarrow}2)\text{-}\alpha\text{-D-Man}p\text{-}(1{\rightarrow}6)\text{-}$$

$$\begin{array}{c} 3 \\ \uparrow \\ 1 \end{array}$$

$$\alpha\text{-L-Fuc}p$$

(b) Condensed form in two lines

$$\text{Gal}(\beta1{-}4)\text{GlcNAc}(\beta1{-}2)\text{Man}(\alpha1{-}6)\text{-}$$
$$|$$
$$\text{Fuc}(\alpha1{-}3)$$

or condensed form in one line

$$\text{Gal}(\beta1{-}4)[\text{Fuc}(\alpha1{-}3)]\text{GlcNAc}(\beta1{-}2)\text{-}$$

To describe large structures, the short form has been proposed:

(c) Short form

$$\text{Gal}\beta4(\text{Fuc}\alpha3)\text{GlcNAc}\beta2\text{Man}\alpha6\text{-}$$

The short form is obtained: (i) by omitting locants of anomeric carbon atoms, (ii) omitting the parentheses around the specification of linkage, and (iii) omitting hyphens.

Similarly a glycopeptide sequence, represented in the condensed form in two lines as

$$\text{Gal}(\beta1{-}3)\text{GalNAc}(\alpha1{-}O)$$
$$|$$
$$\text{-Ala-Thr-Ala-}$$

or in the condensed form in one line as

-Ala-[Gal(β1–3)GalNAc(α 1–O)]Thr-Ala-

may be written in the short form in two lines as

Galβ3GalNAcα —
　　　　　-Ala-Thr-Ala

N-linked oligosaccharides contain a common pentasaccharide core as follows:

Manα 6
　　　　　$\rangle$ Manβ4GlcNAcβ4GlcNAc
Manα 3

For the sake of uniformity, the location of substitution should be written as above in accordance with Haworth's representation of the pyranose structure of monosaccharides.

In the condensed system of symbols for sugar residues (as used in this book) the common configuration and ring size (usually pyranose) are implied in the symbol. Thus, Gal denotes D-galactopyranose; Man, D-mannopyranose; Fuc, L-fucopyranose; GlcNAc, 2-acetamido-2-deoxy-D-glucopyranose or *N*-acetyl-D-glucosamine; Neu5Ac (which may be abbreviated to NeuAc), *N*-acetylneuraminic acid. The symbol Sia stands for sialic acid, a general term that can also be used when the exact structure is unknown.

Whenever the configuration or ring size is found to differ from the common one it must be indicated by using the appropriate symbols for the extended system. The configuration of amino acids is L unless otherwise noted. Although symbols such as Gal and Man are useful in representing oligosaccharide structures they should not be used in the text to represent monosaccharides.

2.5. Microheterogeneity of glycans

In addition to genetically determined variants expressed as variations in their polypeptide chains, almost all glycoproteins reveal another form of polymorphism associated with their glycan moieties. This type of diversity is termed microheterogeneity or peripheral heterogeneity because it involves the number and positions of the most external monosaccharides in the glycan moieties. These variants, recently called *glycoforms* [3], were first characterized in α_1-acid glycoprotein from human serum by Karl Schmid [99] by electrophoresis. The microheterogeneity of α_1 acid glycoprotein was found to be due to the occurrence of di-, tri-, and tetra-antennary glycans of the *N*-acetyllactosamine type at the 5 glycosylation sites [100]. The glycans can be fucosylated [101] and sialylated at different levels. This feature is wide spread and has been observed for natural as well as for recombinant DNA glycoproteins [cf. 47]. Microheterogeneity can also be present in the O-linked glycans, of the mucin type, as shown for example for the pig zona pellucida glycoproteins [102].

The existence of microheterogeneity gives rise to a lot of interesting questions regarding the origin of this phenomenon and about its relevance for the biological functioning of the glycoforms that can be distinguished. For gaining further insight, it is at least essential to have an exact knowledge of the naturally occurring microheterogeneity at each glycosylation site. In case recombinant DNA glycoproteins are produced for therapeutic purposes it is required that the glycan chains of the produced glycoproteins are compatible with the immune system of the patient. Preferably, the glycosylation pattern should be identical or close to that of the natural glycoprotein.

2.6. Evolution

Our present knowledge of the primary structures of numerous glycans raises interesting problems from a comparative biochemical point of view and confirms some of the concepts developed by Montreuil [95,103,104]. Glycan structures from a given class or order of living organisms have common oligosaccharides and are not randomly constructed at all. On the contrary they are subject to rules based on a conservative evolution of glycosyltransferases and on the high specificity and subcellular localization of these enzymes.

3. Concluding remarks

In the area of structural glycobiology the development is going very fast. The availability of sophisticated methods for the isolation and characterization of glycoproteins and glycoprotein-derived glycans has paved the way for the study of the complete three-dimensional structure in solution of low molecular mass glycoproteins by means of a combination of NMR techniques and molecular dynamics simulations. It is in particular the insight into the spatial structures that is needed to make significant progress in the understanding of structure–function relations for carbohydrate chains. Another fascinating area that will become feasible for study is the interaction between glycoproteins and receptors. Although those systems represent usually high-molecular mass complexes, they should be studied in order to analyse in which way the carbohydrate chains are involved in the interaction processes. In situations wherein both the carbohydrate and the protein moieties participate in the interaction with receptors, it is relevant to learn if the moieties are recognized one after another or that the a 'domain type' of recognition occurs wherein both are involved at the same time. The dynamic aspects of recognition in relation to the response to the interaction form challenging goals of further work.

Abbreviations

Con A	concanavaline A
COSY	correlated spectroscopy
ES-MS	electrospray–mass spectrometry
FAB-MS	fast atom bombardment–mass spectrometry
GLC-MS	gas liquid chromatography–mass spectrometry
GPI	glycosylphosphatidyl inositol
HMBC	heteronuclear multiple bond correlation
HMQC	heteronuclear multiple quantum correlated spectroscopy
HOHAHA	homonuclear Hartmann–Hahn
HPAEC	high pH anion exchange chromatography
HPLC	high performance liquid chromatography
HSQC	^{1}H-detected heteronuclear single quantum coherence spectroscopy
MALD-MS	matrix assisted laser desorption–mass spectrometry
MALD-TDF-MS	matrix assisted laser desorption–time of flight–mass spectrometry
NMR	nuclear magnetic resonance
NOESY	nuclear Overhauser enhancement spectroscopy
PAD	pulsed amperometric detection
PNGase	peptide-N^4-(N-acetyl(β-D-glucosaminyl) asparagine amidase
S-layer	surface layer
SDS-PAGE	sodium dodecyl sulfate polyacrylamide gel electrophoresis
TOCSY	total correlated spectroscopy

References

1. Doubet, S., Bock, K., Smith, D., Darvill, A. and Albersheim, P. (1989) TIBS 14, 475–477.
2. Carlsson, S.L.R. (1993) In: M. Fukuda and A. Kobata (Eds.), Glycobiology: A Practical Approach. Oxford University Press, Oxford, UK, pp. 1–26.
3. Rademacher, T.W., Parekh, R.B. and Dwek, R.A. (1988) Annu. Rev. Biochem. 57, 785–838.
4. Rudd, P.M., Scragg, I.G., Coghill, E.C. and Dwek, R.A. (1992) Glycoconjugate J. 9, 86–91.
5. Ferguson, M.A.J. (1993) In: M. Fukuda and A. Kobata (Eds.), Glycobiology: A Practical Approach. Oxford University Press, Oxford, UK, pp. 349–383.
6. Menon, A.K. (1994) Methods in Enzymology 230, 418–442.
7. Dell, A., Reason, A.J., Khoo, K.H., Parrico, M., McDowell, R.A. and Morris, H. (1994) Methods in Enzymology 230, 108–132.
8. Kelly, J.F., Locke, S.J. and Thibault, P. (1993) Discovery 2 (2), 1–6.
9. Dill, K., Berman, E. and Pavia, A.A. (1985) Adv. Carbohydr. Chem. Biochem. 43, 1–49.
10. Brockbank, R.L. and Vogel, H.J. (1990) Biochemistry 29, 5574–5583.
11. Joao, H.C., Scragg, I.G. and Dwek, R.A. (1992) FEBS Lett. 307, 343–346.
12. Wyss, D.F., Withka, J.M., Knoppers, M.H., Sterne, K.A., Recny, M.A. and Wagner, G. (1993) Biochemistry 32, 10995–11006.
13. Withka, J.M., Wyss, D.F., Wagner, G., Arulanandam, A.R.N., Reinherz, E.L. and Recny, M.A. (1993) Structure 1, 69–81.

14. Fletcher, M.C., Harrison, R.A., Lachmann, P.J. and Neuhaus, D. (1994) Structure 2, 185–199.
15. De Beer, T., Van Zuylen, C.W.E.M., Hård, K., Boelens, R., Kaptein, R., Kamerling, J.P. and Vliegenthart, J.F.G. (1994) FEBS Lett. 348, 1–6.
16. Dwek, R.A., Edge, Chr.J., Harvey, D.J., Wormald, M.R. and Parekh, R.B. (1993) Annu. Rev. Biochem. 62, 65–100.
17. Kobata, A. and Endo, T. (1993) In: M. Fukuda and A. Kobata (Eds.), Glycobiology, A Practical Approach. Oxford University Press, Oxford, UK, pp. 79–102.
18. Patel, T.P. and Parekh, R.B. (1994) Methods in Enzymology 230, 57–66.
19. Piller, F. and Piller, V. (1993) In: M. Fukuda and A. Kobata (Eds.), Glycobiology, A Practical Approach. Oxford University Press, Oxford UK, pp. 291–328.
20. Vliegenthart, J.F.G., Hård, K., De Waard, P. and Kamerling, J.P. (1991) In: H.S. Conradt (Ed.), Protein Glycosylation: Cellular, Biotechnological and Analytical Aspects. VCH Publishers, Inc., New York, NY, USA.
21. Langeveld, J.P.M., Noelken, M.E., Hård, K., Todd, P., Vliegenthart, J.F.G., Rouse, J. and Hudson, B.G. (1991) J. Biol. Chem. 266, 2622–2631.
22. Tarentino, A.L. and Plummer, T.H. (1994) Methods in Enzymology 230, 44–57.
23. Lee, Y.C. and Rice, K.G. (1993) In: M. Fukuda and A. Kobata (Eds.), Glycobiology, A Practical Approach. Oxford University Press, Oxford, UK, pp. 127-163.
24. Kobata, A. (1994) Methods in Enzymology 230, 200–208.
25. Kobata, A. and Yamashita, K. (1993) In: M. Fukuda and A. Kobata (Eds.), Glycobiology, A Practical Approach. Oxford University Press, Oxford, UK, pp. 103–125.
26. Baenziger, J.U. (1994) Methods in Enzymology 230, 237–249.
27. Hardy, M.R. and Townsend, R.R. (1994) Methods in Enzymology 230, 208–225.
28. Linhardt, R.J. (1994) Methods in Enzymology 230, 265–280.
29. Kamerling, J.P. and Vliegenthart, J.F.G. (1989) In: A.M. Lawson (Ed.), Clinical Biochemistry; Principles, Methods, Applications. Vol. 1, Mass Spectrometry. Walter de Gruyter & Co, Berlin, Germany, pp. 175–263.
30. Cummings, R.D. (1993) In: M. Fukuda and A. Kobata (Eds.), Glycobiology, A Practical Approach. Oxford University Press, Oxford, UK, pp. 243–289.
31. Geyer, R. and Geyer, H. (1994) Methods in Enzymology 230, 86–108.
32. Jacob, G.S. and Scudder, P. (1994) Methods in Enzymology 230, 280–299.
33. Vliegenthart, J.F.G., Van Halbeek, H. and Dorland, L. (1980) In: A. Varmavuoni (Ed.), 27th Int. Congress, Pure and Appl. Chem. Pergamon Press, Elmsford, NY, USA, pp. 253–262.
34. Vliegenthart, J.F.G., Van Halbeek, H. and Dorland, L. (1981) Pure Appl. Chem. 53, 45–77.
35. Vliegenthart, J.F.G., Dorland, L. and Van Halbeek, H. (1983) Adv. Carbohydr. Chem. Biochem. 41, 209–374.
36. Kamerling, J.P. and Vliegenthart, J.F.G. (1992) Biological Magnetic Resonance 10, 1–194.
37. Hård, K. and Vliegenthart, J.F.G. (1993) In: M. Fukuda and A. Kobata (Eds.) Glycobiology, A Practical Approach. Oxford University Press, Oxford, UK, pp. 223–242.
38. Van Halbeek, H. (1994) Methods in Enzymology, 230, 132–168.
39. Van Kuik, J.A., Hård, K. and Vliegenthart, J.F.G. (1992) Carbohydr. Res. 235, 53–68.
40. Spelmann, M. (1990) Anal. Chem. 62, 1714–1722.
41. Taverna, M., Baillet, A., Biou, D., Schlütter, M., Werner, R., Fevrier, D. (1992) Electrophoresis 13, 359–366.
42. Suzuki, S., Kakehi, K. and Honda, S. (1992) Anal. Biochem. 205, 227–236.
43. Hermentin, P., Witzel, R., Doenges, R., Bauer, R., Haupt, H., Patel, T., Parekh, R.B. and Brazel, D. (1992) Anal. Biochem. 206, 419–429.
44. Hermentin, P., Doenges, R., Witzel, R., Hokke, C.H., Vliegenthart, J.F.G., Kamerling, J.P., Conradt, H.S., Nimtz, M. and Brazel, D. (1994) Anal. Biochem. 221, 29–41.

45. Rice, K.G., Takahashi, N., Namiki, Y., Tran, A.D., Lisi, P.J. and Y.C. Lee (1992) Anal. Biochem. 206, 278–287.
46. Jackson, P. (1994) Methods in Enzymology 230, 250–265.
47. Vliegenthart, J.F.G. (1994) In: K. Bock and H. Clausen (Ed.), Alfred Benzon Symposium 36, Complex Carbohydrates in Drug Research, Structural and Functional Aspects. Munksgaard, Copenhagen, Denmark, pp. 30–44.
48. Lis, H. and Sharon, N. (1993) Eur. J. Biochem. 218, 1–27.
49. Schauer, R. (1991) Glycobiology 1, 449–452.
50. Kanamori, A., Inoue, S., Iwasaki, M., Kitajima, K., Kawai, G., Yokoyama, S. and Inoue, Y. (1990) J. Biol. Chem. 265, 21811–21819.
51. Strecker, G., Wieruszeski, J.M., Michalski, J.C., Alonso, C., Leroy, Y., Boilly, B. and Montreuil, J. (1992) Eur. J. Biochem. 207, 995–1002.
52. Messner, P. and Sleytr, V.B. (1991) Glycobiology 1, 545–551.
53. Iwasaki, M., Inoue, S. and Inoue, Y. (1987) Eur. J. Biochem. 168, 185–192.
54. Khoo, K.H., Maizels, R.M., Page, A.P., Taylor, G.W., Rendell, N.B. and Dell, A. (1991) Glycobiology 1, 163–171.
55. Gerwig, G.J., Kamerling, J.P., Vliegenthart, J.F.G., Morag, E., Lamed, R. and Bayer, E.A. (1992) Eur. J. Biochem. 205, 799–808.
56. Gerwig, G.J., Kamerling, J.P., Vliegenthart, J.F.G., Morag (Morgenstern), E., Lamed, R. and Bayer, E.A. (1991) Eur. J. Biochem. 196, 115–122.
57. Moraes, C.T., Bosch, M. and Parodi, A.J. (1988) Biochemistry 27, 1543–1549.
58. Takanayagi, T., Kushida, K., Idonuma, K. and Ajisaka, K. (1992) Glycoconjugate J. 9, 229–234.
59. Van Kuik, J.A., Sybesma, R.P., Kamerling, J.P., Vliegenthart, J.F.G. and Wood, E.J. (1986) Eur. J. Biochem. 160, 621–625.
60. Schlipfenbacher, R., Wenzl, S., Lottspeich, F. and Sumper, M. (1986) FEBS Lett. 209, 57–
61. Spiro, R.G. and Bhoyroo, V.D. (1988) J. Biol. Chem. 263, 14351–14358.
62. De Waard, P., Koorevaar, A., Kamerling, J.P. and Vliegenthart, J.F.G., (1991), J. Biol. Chem. 266, 4237–4243.
63. Lamblin, G., Rahmoune, H., Wieruszeski, J.M., Lhermitte, M., Strecker, G. and Roussel, P. (1991) Biochem. J. 275, 199–206.
64. Baenziger, J.U. and Green, E.D. (1988) Biochim. Biophys. Acta 947, 287–306.
65. Hård, K., Van Zadelhoff, G., Moonen, P., Kamerling, J.P. and Vliegenthart, J.F.G. (1992) Eur. J. Biochem. 209, 895–915.
66. Bergwerff, A.A., Van Oostrum, J., Kamerling, J.P. and Vliegenthart, J.F.G. (1995) Eur. J. Biochem. in press.
67. Hartmann, E. and Koning, H. (1989) Arch. Microbiol. 151, 274–278.
68. Mengele, R. and Sumper, M. (1992) FEBS Lett. 298, 14–16.
69. Van Kuik, J.A., Sybesma, R.P., Kamerling, J.P., Vliegenthart, J.F.G. and Wood, E.J. (1987) Eur. J. Biochem. 169, 399–411.
70. Yamashita, K., Ueda, I. and Kobata, A. (1983) J. Biol. Chem. 258, 14144–14147.
71. Freeze, H.H. (1985) Arch. Biochem. Biophys. 243, 580–693.
72. Gabel, C.A., Costello, C.E. Reinhold, V.N., Kurz, L. and Kornfeld, S. (1984) J. Biol. Chem. 259, 13762–13769.
73. Pozzgay, V., Jennings, H.J. and Kasper, D.L. (1987) Eur. J. Biochem. 162, 445–450.
74. Bergwerff, A.A., Hulleman, S.H.D., Kamerling, J.P., Vliegenthart, J.F.G., Shaw, L., Reuter, G. and Schauer, R. (1992) Biochimie 74, 25–38.
75. Johansen, P.G., Marshall, R.D. and Neuberger, A. (1961) Biochem. J. 78, 518–527.
76. Nuenke, R.H. and Cunningham, L.W. (1961) J. Biol. Chem. 236, 2451–2460.

28

77. Yamashina, I. and Makino, M. (1962) J. Biochem. (Tokyo) 51, 359–364.

78. Hashimoto, Y. and Pigman, W. (1962) Ann. N.Y. Acad. Sci. 93, 541.

79. Tanaka, K., Bertolini, M. and Pigman, N. (1964) Biochem. Biophys. Res. Commun. 16, 404–409.

80. Anderson, B., Seno, H., Sampson, P., Riley, J.G., Hoffman, P. and Meyer, K. (1964) J. Biol. Chem. 239, PC 2716.

81. Bhavanandan, V.P., Buddecke, E., Carubelli, P. and Gottschalk, A. (1964) Biochem. Biophys. Res. Commun. 16, 353–357.

82. Muir, H. (1958) Biochem. J. 69, 195–204.

83. Lindahl, U and Roden, L. (1964) Biochem. Biophys. Res. Commun. 17, 254–259.

84. Roden, L. and Lindahl, U. (1965) Fed. Proc. 24, 606.

85. Butler, W.T. and Cunningham, L.W. (1955) J. Biol. Chem. 240, PC 3449.

86. Butler, W.T. and Cunningham, L.W. (1966) J. Biol. Chem. 240, 3882–3888.

87. Lamport, D.T.A. (1967) Nature 216, 1322–1324.

88. Lamport, D.T.A. and Miller, D.H. (1971) Plant Physiol. 48, 454–456.

89. Allen, A.K., Desai, N.N., Neuberger, A. and Creeth, J.M. (1978) Biochem. J. 171, 665–674.

90. Smythe, C., Caudwell, F.G., Ferguson, M. and Cohen, P. (1988) EMBO J. 7, 2681–2686.

91. Hart, G.W., Haltiwanger, R.S., Holt, G.D. and Kelly, W.G. (1989) Annu. Rev. Biochem. 58, 841–874.

92. Low., M.G., Ferguson, M.A.J., Futerman, A.H. and Silman, I. (1986) Trends Biochem. Sci. 11, 212–215.

93. Eglund, P.T. (1993) Annu. Rev. Biochem. 62, 121–138.

94. Hofsteenge, J., Müller, D.R., De Beer, T., Löffler, A., Richter, W.J. and Vliegenthart, J.F.G. (1994) Biochemistry 33, 13524–13530.

95. Montreuil, J. (1975) Pure Appl. Chem. 42, 431–477.

96. Jett, M. and Jamieson, G.A. (1971) Carbohydr. Res. 18, 466–468.

97. Kisailus, E.C. and Allen, H.J. (1992) In: H.J. Allen and E. Kisailus (Eds.), Nomenclature of Monosaccharides, Oligosaccharides, and Glycoconjugates: Composition, Structure, and Function. Marcel Dekker Inc., New York, USA, pp. 13–32.

98. Nomenclature of glycoproteins, glycopeptides and peptidoglycans:
    (a) Eur. J. Biochem. 159, 1–6 (1986), correction in 185, 485 (1989).
    (b) Glycoconjugate J. 3, 123–134 (1986).
    (c) J. Biol. Chem. 262, 13–18 (1967).
    (d) Pure Appl. Chem. 60, 1389–1384 (1988).
    (e) Spec. Period. Rep. Amino Acids Pept., Proteins, 21, 329–334 (1994).
    (f) Biochemical Nomenclature and Related Documents, A Compendium, 2nd ed. 1992. Portland Press, London, UK, pp. 84–89.

99. Schmid, K., Binette, J.P., Kamiyama, S., Pfister, V. and Takahashi (1962) Biochemistry 1, 959–956.

100. Fournet, B., Strecker, G., Montreuil, J., Vliegenthart, J.F.G., Schmid, K. and Binette, J.P. (1978) Biochemistry 17, 5206–5214.

101. Van Halbeek, H., Dorland, L., Vliegenthart, J.F.G., Montreuil, J., Fournet, B. and Schmid, K. (1981) J. Biol. Chem. 256, 5588–5590.

102. Hokke, C.H., Damm, J.B.L., Penninkhof, B., Aitken, R.J., Kamerling, J.P., and Vliegenthart, J.F.G. (1994) Eur. J. Biochem. 221, 491–512.

103. Montreuil, J., Spik, G. and Chasson, A (1962) C.R. Acad. Sci. Paris 255, 3493–3494.

104. Montreuil, J., Adam-Chasson, A. and Spik G. (1965) Bull. Soc. Chim. Biol. 47, 1867–1880.

105. Montreuil, J. (1982) In: A. Neuberger and L.L.M. van Deenen (Eds.), Glycoproteins, Comprehensive Biochemistry 19B, Part II. Elsevier, Amsterdam, pp. 1–188.

J. Montreuil, H. Schachter and J.F.G. Vliegenthart (Eds.), *Glycoproteins*
© 1995 Elsevier Science B.V. All rights reserved

3D Structure

1. The structural features of protein–carbohydrate interactions revealed by X-ray crystallography

CHRISTIAN CAMBILLAU

*Laboratoire de Cristallographie et de Cristallisation des Macromolécules Biologiques,
CNRS URA 1296, Faculté de Médecine Secteur-Nord, Bd. Pierre Dramard,
13916 Marseille Cedex 20, France*

1. Introduction

Carbohydrates have diverse roles, from the less noble, concerning their structural as well as their food storage and fuel source functions, to the most sophisticated signalling functions [1–4]. Proteins dealing with carbohydrates have to achieve degradation, transportation, synthesis, storage, but also binding, signalling, adhesion, etc. In many processes — such as fertilization, leukocytes traffic regulation, development — the precision and fidelity of the molecular recognition is the key feature for the organism to stay alive. Moreover, this fidelity is exerced toward polymers which present a vast repertoire of composition, conformations and ramifications. To quote Nathan Sharon, "the specificity of many natural polymers is written in terms of sugar residues, not of amino acids or nucleotides" [4].

X-ray crystallography has proved to be a powerful method for determining with accuracy and reliability the structural features of protein–carbohydrate interactions [5,6]. As a consequence, it allows the observation and analysis of saccharides conformation, together with NMR, providing an experimental check of the force fields involved in theoretical calculations [7–10]. The past few years have seen an impressive increase of the number, quality and diversity of structures of protein–carbohydrate complexes resolved by X-ray crystallography. In this review, we shall provide a non-exhaustive description of various molecular architectures of carbohydrate–protein complexes; the proteins participating in such complexes come from different origins, and are involved in extremely diverse functions: periplasmic receptor proteins involved in chemotaxis and transport, vegetal and animal lectins, enzymes, antibodies and glycoproteins (Table I). From these structures, a few convergent rules relevant to protein–carbohydrate interactions can be drawn. We shall successively discuss the hydrogen bonding network established by the saccharides, including the often crucial role of water molecules, and the important hydrophobic stacking interactions always present in protein–saccharide interactions. In the case of long polysaccharides and glycans, we shall examine the sugar conformation of the protein-bound saccharides.

TABLE I

List of the X-ray structures presented in this review

Protein–saccharide complex	Resolution (Å)	Reference
Arabinose binding protein/arabinose	1.7	16
Arabinose binding protein/galactose	1.9	17
Galactose binding protein/glucose	1.8	18
Maltose binding protein/maltose	2.3	19,20
Wheat germ agglutinin isolectins 1 and 2 in complex with:		
N-acetylneuramyl lactose	2.2	25
T5 sialylopentapeptide	2.0	26
Con A/α-methyl mannoside	2.8	21
Lathyrus ochrus isolectin I (LOL I) in complex with:		
α-Methyl mannoside	2.0	29
α-Methyl glucoside	2.2	29
α-Man(1–3) β-Man (1–4) GlcNAc	2.1	30
Diantennary octasaccharide	2.3	31
Muramyl-dipeptide	2.1	34
Muramic acid	2.1	34
Lathyrus ochrus isolectin II (LOL II) in complex with:		
Lactotransferrin glycofragment N2	3.3	33
Lactotransferin glycopeptide	2.8	33
Griffonia simplicifolia lectin/Lewis b	2.0	28
Erythrina corallodendron lectin/lactose	2.0	35
C-mannose lectin/oligomannoside	1.7	49
Galectin 1 (human)/lactose	2.9	55
Galectin 1 (bovine)/LacNAc	1.9	51
Galectin 1 (bovine)/diantennary octasaccharide	2.45/2.3/2.15	54
Fab/tetrasaccharide	2.05	56,57
Lysozyme/tri-NAG	1.75	58,59
Phosphorylase b/heptenitol		
Phosphorylase b/heptulose-2-P		64
Phosphorylase b/maltoheptaose	2.5	66
Sialidase/sialic acid	2.8; 1.8	67,68
α-Amylase/acarbose	2.1	72

2. Protein–carbohydrate complexes

2.1. Periplasmic receptor proteins: complexes with mono- and disaccharides

Periplasmic sugar binding proteins (PSBP), situated in the periplasmic space of Gram-negative bacteria, serve as primary receptors of the osmotic shock-sensitive uptake systems for various sugars [11]. Several structures of PSBP have been solved at high resolution and the protein–saccharide interactions have been thoroughly described by Quiocho and Vyas who established the basic rules of protein–saccharide recognition [11–15]. These receptors bear specificity for L-arabinose, galactose/glucose and maltose.

2.1.1. The arabinose binding protein (ABP)

In the arabinose binding protein (ABP) complexes with L-arabinose (1.7 Å, R = 0.137, Table I) [16] and D-galactose [17], the sugar molecules are quite buried with a solvent accessibility of only 2%. The binding site is designed in a way to reduce the differences in affinity between α- and β-anomers of arabinose and both anomers are present in the crystal structure. The Asp 90 OD2 atom is positioned in the middle of the two anomer OH positions, thus providing hydrogen bonds of comparable length and stability (Fig. 1). The position of the Asp 90 carboxyl group is held firmly by an H-bond between OD1 and NH_3 of Lys 10. Many carboxylic acids provide H-bonds directly or indirectly to the sugar: Asp 89, Asp 90, Glu 11, Glu 14 and Asp 235. Basic groups are more scarce, represented only by Lys 10 and Arg 151. Five water molecules are found in the sugar binding area, two of them being directly bound to the sugar. A hydrophobic area is formed in L-arabinose, including C-3, C-4 and C-5. This patch stacks against Trp 16 and Phe 17. The complex of α/β-D-galactose presents very similar features. Two buried water molecules (W309 and W310) are conserved for both structures in the first solvation shell (Fig. 1a). There is evidence that these molecules, forming hydrogen bonds with sugar atoms O-2 and O-5, bring the sugar and the two domains of ABP together in the cleft. In the arabinose complex, water molecule W311 replaces the galactose atom O-6 in binding Asp 89 (Fig. 1b). These water molecules have no contact with the bulk solvent.

2.1.2. The galactose/glucose binding protein (GBP)

The galactose binding protein (GBP) complex with D-glucose (1.9 Å, R = 0.146, Table I) [18], describes a receptor which presents high affinity for two different saccharides, glucose and galactose. Both anomers can also bind tightly to GBP. The active site is made of several Asp and Asn, of Arg 158 and His 152. Several aromatic residues are also in its vicinity (Fig. 2). Three residues, Asn 91, Arg 158 and Asp 236, form bidentate hydrogen bonds with glucose. All of the hydrogen-bonding groups of the glucose molecule are involved in a network of hydrogen bonds. Glucose is tightly sandwiched between Phe 16 and Trp 183. Only one water molecule (W313) is in the first binding shell. This molecule forms bidentate hydrogen bonds with glucose atoms O-3 and O-4, and is also hydrogen bonded to the amide of Phe 16.

2.1.3. The maltose binding protein (MBP)

The maltose binding protein (MBP)-maltose complex was solved to a slightly lower resolution compared with the other PSBP complexes [19] (2.3 Å, Table I). The features

32

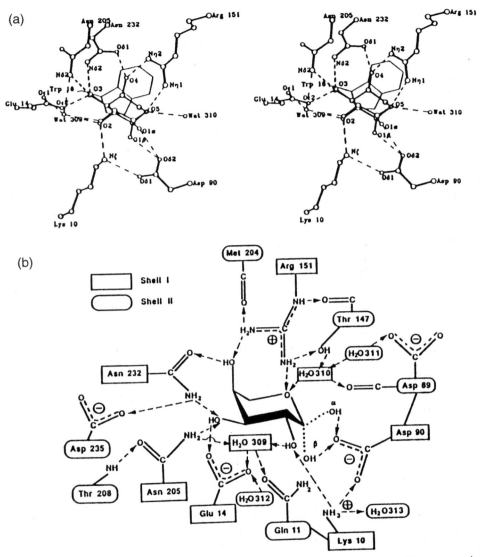

Fig. 1. Atomic interactions between L-arabinose and L-arabinose binding-protein: (a) stereoscopic view of the combining site; (b) schematic diagram of the hydrogen-bonds network (from Ref. [16]).

encountered in the above-mentioned complexes are present in this one, in spite of the fact that the saccharide is slightly more exposed (3.6%) to bulk water. Asp 14, 65 and Glu 111 and 153 as well as Arg 66 and Lys 15 build essentially the anchoring site of maltose. The aromatic stacking is much extended on both sugar residues, and involves Trp 62, 155, 230 and 340. There are five water molecules bound to maltose. In glucose 1, two water molecules are bound to O6, one to O2, and one to O3; in glucose 2, one

(a)

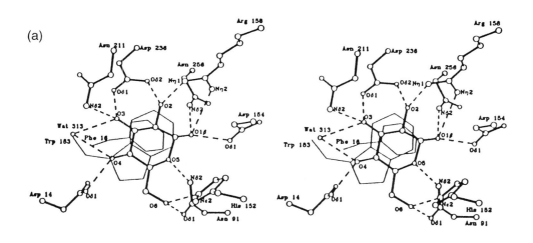

(b)

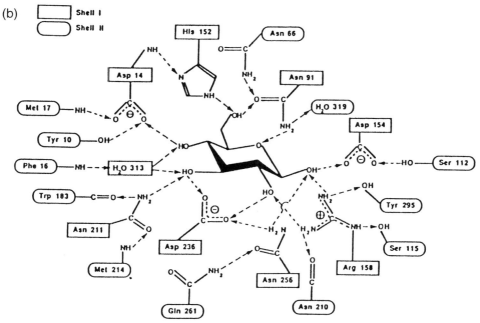

Fig. 2. Atomic interactions between D-glucose and D-galactose binding-protein: (a) stereoscopic view of the combining site; (b) schematic diagram of the hydrogen-bonds network (from Ref. [18]).

water molecule is bound to O4 (Fig. 3). All these water molecules are also hydrogen bonded to protein residues.

The carbohydrate specificity of maltodextrin binding protein MdBP was modulated

34

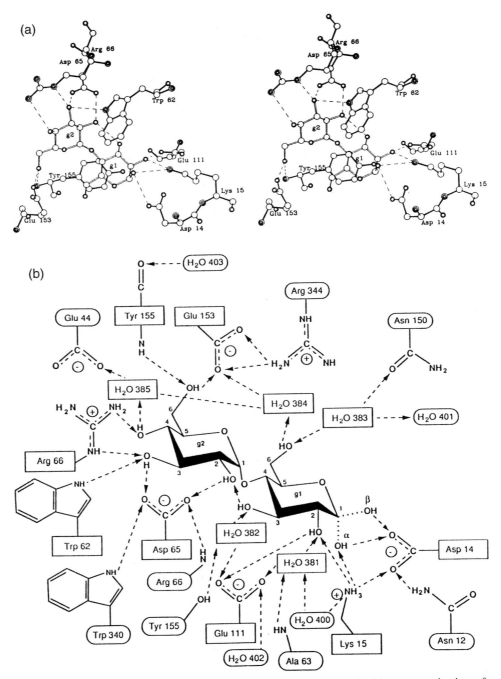

Fig. 3. Atomic interactions between maltose and maltose-binding protein: (a) stereoscopic view of the combining site; (b) schematic diagram of the hydrogen-bonds network (from Ref. [19]).

by site directed mutagenesis. The resulting stacking of tryptophane residues on the faces of the glucosyl units provides the majority of the van der Waals contacts in the complex. Two structures of mutant MdBP-maltose complexes (W230A and W232A) have been determined at 1.9 Å resolution (Table I) [20]. The MdBP mutant W230A has a 12-fold lower association constant compared with wild-type, whereas mutant W232A exhibits nearly wild-type characteristics. Both mutants reveal minor structural changes in comparison with the native MdBP. Interestingly, a single water molecule occupies the apolar cavity created by mutation W232A.

2.2. Vegetal lectins

Lectins are carbohydrate-binding proteins that have been widely used in studying the structure and function of cell surface carbohydrates. Interest in plant lectins was stimulated because they recognize certain carbohydrates at the surface of malignant cells [2]. In the plant kingdom, they are a determining factor in the symbiosis of *Rhizobia* with legumes [3]. Thus far, studies on the structure of lectin–sugar complexes have involved Con A with mannose [21], Favin with glucose [22], Pea lectin with a mannose trisaccharide [23,24], WGA isolectin with a *N*-acetylneuraminyl-lactose [25,26], *Griffonia simplicifolia* lectin with a blood determinant tetrasaccharide [27,28] and isolectin I & II (LOL I&II), isolated from the seeds of *Lathyrus ochrus* complexed with methyl mannoside and glucoside [29], with a trisaccharide [30], with a diantennary octasaccharide [31], with a glycopeptide and a glycoprotein fragment [32,33] and with muramyl peptides [34]. *Erythrina corallodendron* lectin has been solved alone or in complex with lactose [35].

2.2.1. Lathyrus ochrus isolectins

Isolectin I (LOL I), isolated from the seeds of *Lathyrus ochrus*, consists of two identical subunits each composed of a light chain (α) with 52 amino-acids, and a heavy chain (β) with 181 residues [36]. The lectin binds specifically D-mannose and D-glucose. Our studies allowed us to describe precisely the monosaccharide-binding site of LOL I [29], the binding of a trisaccharide to LOL I [30], and a complex between lectin and a diantennary octasaccharide — a great step forward in understanding lectin–glycoprotein interactions [31]. More recently, crystals of structures of isolectin II complexed with the N-glycosylated N2 fragment of human lactotransferrin (18 kDa) and with an isolated glycopeptide have been described [32], and the structures were solved recently [33] (Table I).

2.2.1.1. Monosaccharide complexes

The structures of Isolectin I from *Lathyrus ochrus* complexed to α-methyl mannoside (LOLM) and α-methyl glucoside (LOLG) have been solved (Table I). Sugar hydroxyl groups involving oxygens O3, O4, O5 and O6 are hydrogen bound to the amino acids constituting the monosaccharide binding site of the lectin: Ala 210 NH, Glu 211 NH, Asp 81 COO, Asn 125 NH2, Gly 99 NH. Asp 81 and Asn 125 interact with their second carboxylic oxygen with a Ca^{++} ion (Fig 4a,b). The saccharide is much more water-accessible than in the case of periplasmic proteins, and the network of

36

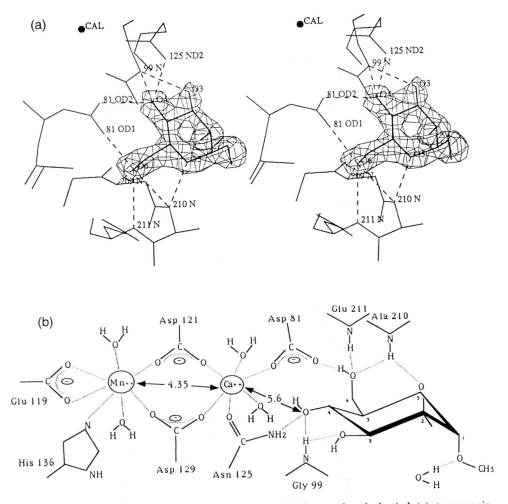

Fig. 4. Atomic interactions between α-Me mannoside and *Lathyrus ochrus* isolectin I: (a) stereoscopic view of the combining site with α-Me mannoside in its electron density map; (b) schematic diagram of the hydrogen-bonds network involving the saccharide and the Ca^{2+} and Mn^{2+} ions (from Ref. [29]).

hydrogen bonds between protein and saccharide is consequently much less extended. The saccharide also interacts with an aromatic ring, the side-chain of Phe 123 which plays an essential role in lectin saccharide binding (see further). Sugar atoms O1 (mannose) and O2 (glucose) are hydrogen-linked to water molecules which are further link to second shell water molecules. The same situation was found in the 2.8 Å resolution structure of ConA with methyl α-D-mannoside, where sugar atom O2 binds a water molecule [21]. In the LOL I native crystal structure (LOLF) [37], four well-ordered water molecules and a disordered one establish hydrogen bonds with residues involved in the binding of monosaccharides. These molecules are close to positions occupied by the monosaccharide hydroxyl groups [29], a result very often found when comparing empty and liganded combining sites.

2.2.1.2. Muramic acid and muramyl dipeptides

Besides its mannose/glucose specificity, LOL is able to accommodate two components of the bacterial cell wall, muramic acid and muramyl-dipeptide [34,38]. In both complexes, only the ring hydroxyl oxygen atoms of the bound sugar establish direct hydrogen bonds with isolectin I, as in the case of all the previously determined monosaccharide lectin complexes (Fig. 5a). In addition, the lactyl methyl group of both

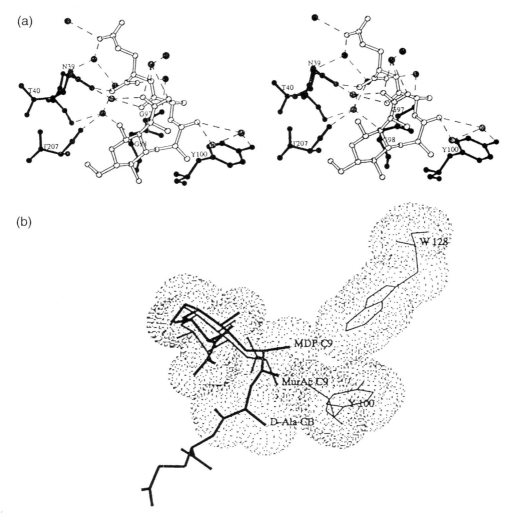

Fig. 5. (a) Stereoview of the muramyl-dipeptide molecule (MDP) bound to LOL I, including bound water molecules (spheres). (b) Hydrophobic contacts between MDP and LOL I (from Ref. [34]).

38

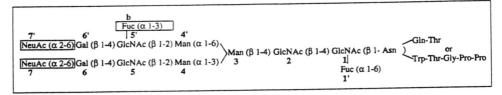

Fig. 6. Structure of the glycopeptide isolated from human lactotransferrin. Boxed parts represent the heterogeneity of the carbohydrate chain.

components strongly interacts via hydrophobic contacts with the side-chains of residues Tyr 100 and Trp 128 of isolectin I (Fig. 5b), which could explain the higher affinity of isolectin I for muramic acid as compared to glucose. These two residues, however, are not involved in the stabilization of the oligosaccharide/isolectin I complexes. The dipeptide (D-Ala-D-iGln) of the second component is in stacking interaction with the *N*-acetyl group of glucosamine and with loop Gly 97-Gly 98 of isolectin I. In addition to these van der Waals contacts, the dipeptide interacts also with the lectin via well-ordered water molecules. Superposition of the structures of the muramyl–dipeptide complex and of the muramic acid complex shows that the saccharidic ring in the dipeptide compound is tilted by about 15° in comparison with that of muramic acid. The fact that the lactyl group has the same conformation in both components reveals that the lectin is stereospecific and recognizes only diastereoisomer *S* of this group, which fits better the saccharide-binding site.

2.2.1.3. Complex with a Man(α1–3) Man (β1–4) GlcNAc trisaccharide
The structure of LOLI complexed to the Man(α1–3) Man (β1–4) GlcNAc trisaccharide (TRI) has been solved at 2.1 Å resolution (Table I, LOLT structure) [30]. The trisaccharide is a part of the complex type glycan [9,39] (Fig. 6). The mannose residue of the non-reducing end of TRI occupies exactly the same position as in the methyl α-D-mannoside complex. Moreover, the TRI conformation in the LOLT complex is very close to the one of unbound TRI [40] (Fig. 10). In contrast to the direct interaction of Man-4 of TRI with the residues of the monosaccharide-binding site, the interaction of Man-3 and GlcNAc-2 of TRI with the lectin is completely mediated by sugar–water(s)–lectin linkages (Fig. 7a, and see Fig. 6 for numbering). Most water molecules are positioned identically in the LOLT and LOLF models. In the neighbourhood of the TRI moiety, only two water molecules out of 20 are also present in the native LOLF structure, and none in the LOLM structure. These water molecules are involved in long-chain linkages between TRI and LOL I. An example of such a long chain is the chain of nine water molecules, connecting the atoms of Man-3 and GlcNAc-2 to LOL I over a 13 Å distance (Fig. 7b).

Opposite: Fig. 7. (a) Stereoview showing the superposition between the trisaccharide bound to LOL I and the unbound trisaccharide (thin). (b) Stereoview showing the trisaccharide bound to LOL I and the water molecules channel between them (water molecules are in their electron density map) [30].

(a)

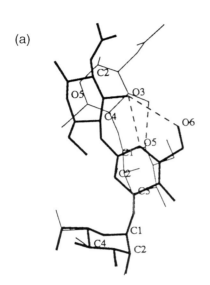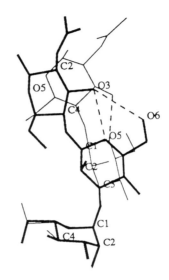

(b)

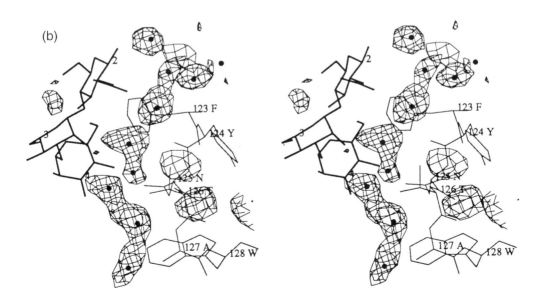

2.2.1.4. Complex with a diantennary octasaccharide

This complex [31,41] is formed with an almost complete diantennary N-acetyllac-tosamine-type glycan, a structure occurring in many glycoproteins (Fig. 6) [9,39]. The two octasaccharide molecules (OCTA) are located in clefts at each end of the long axis of the lectin (Fig. 8a). A large portion of the highly polar saccharide molecule is facing the bulk water. The tightest OCTA–lectin interaction is that of the most buried sugar residue, Man-4, in the monosaccharide binding site. Here again, the mannose position is that of the monosaccharidic mannose. The saccharide-lectin complex is stabilized by 23 hydrogen bonds, 14 directly between the protein and the saccharide and 7 by way of one water molecule. The complex is also stabilized by 68 van der Waals contacts, of which 27 are made by aromatic residues. These contacts occur mainly between aromatic rings and the sugar carbon atoms 5 and 6 (Fig. 8b). Man-4 and GlcNAc-5' interact on each side of the phenyl ring of Phe 123: they grip the ring as a clamp, whereas the two sugar residues between them, Man-3 and Man-4', are exposed to the solvent and do not interact with the lectin (Fig. 8b). Tyr 124 is stacked against Gal-6' (Fig. 8b). As a result, sugar residues 5' and 6' fit in a partly hydrophobic cleft between loop 75–78 and loop 132–134, and display numerous interactions with the lectin. Numerous water molecules are located between the octasaccharide and the lectin, improving thus the stability of the complex (Fig. 8c).

2.2.1.5. Glycan complexes

Lectins from the seeds of *Lathyrus ochrus* have a higher affinity for the diantennary complex-type glycan bearing a GlcNAc-1 linked fucose residue [42]. Crystal structures of isolectin II complexed with the N2 monoglycosyl-fragment of human lactotransfer-rin (18 kDa) and with an isolated glycopeptide have been solved (18 kDa) at 3.3 Å and 2.8 Å resolution, respectively [32,33,43]. The oligosaccharide adopts an identical extended conformation in both structures: the protein part of the glycoprotein has no influence on the binding (Figs. 9 and 10; Table III). The oligosaccharide interacts with the lectin through its common pentasaccharidic core and the α-1,3-antenna (Fig. 9). Besides the essential mannose moiety of the monosaccharide binding site, Fuc-1' of the core has a large surface of interaction with the lectin, which explains its importance in determining the glycan conformation. Therefore, the saccharidic moiety of the dian-tenna adopts a different conformation compared to the LOL-octasaccharide complex [31,43] and the fucose-1' is mandatory to achieve the proper binding conformation of the glycan (Fig. 11). This large complementarity surface area between the oligosaccha-ride and the lectin contrasts with that recently determined between the CRD domain of a C-type mammalian lectin and an oligomannoside, in which only the non-reducing terminal mannose residue interacts with the lectin [53].

Opposite: Fig. 8. (a) Stereo figure showing two octasaccharides (thick lines) complexed to the LOL I dimer (Cα plot representation, thin lines). (b) Stereoscopic view of the polar interactions between the octasaccharide moiety (thick lines) and lectin (thin lines). Asp 81, Gly 99, Asn 125 and Ala 209 make the most important polar interactions. The calcium binding site is also represented. (c) Stereoscopic view of the polar interactions between octasaccharide moiety (thick lines) and lectin (thin lines) mediated via water molecules (circles). (From Ref. [31]).

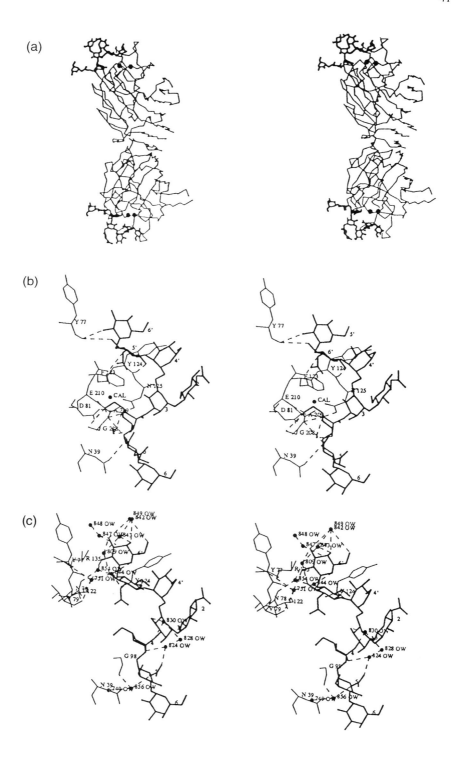

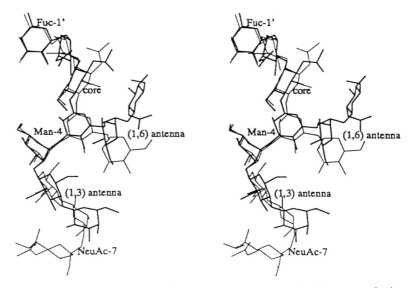

Fig. 9. Superposition of the oligosaccharide structures present in the LOL II–lactotransferrin complex (thick line) and in the LOL II–glycopeptide complex (thin line) [33].

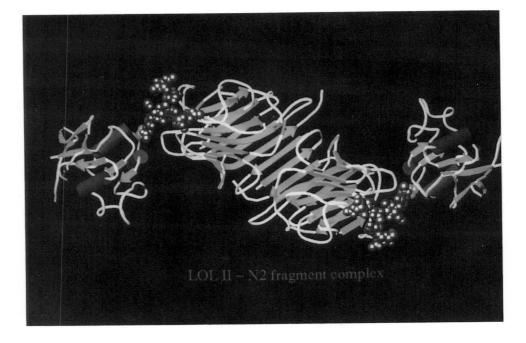

Fig. 10. Ribbon representation of the LOL II–lactotransferrin complex [33].

This complex may represent the archetype model for a better understanding of the fine specificity of other legume lectins for complex-type glycans. In particular, it allows us to understand why ConA, the most studied protein in this family, has the same affinity for fucosylated oligosaccharides compared with unfucosylated ones, and therefore opens the way for molecular biology experiments.

2.2.2. Pea lectin and Concanavalin A

Pea lectin and Jack Bean Concanavalin A are among the best characterized vegetal lectins. They are both mannose/glucose specific. Despite this fact, very little is known in structural terms of their interactions with saccharides: Con A with mannose [21], and Pea lectin with a mannose trisaccharide [23,24] in which only a mannnose residue is visible. Despite intense efforts and a high homology with LOL, Pea lectin failed to crystallize with the saccharides which gave rise to crystals in the presence of LOL (Y. Bourne and C. Cambillau, unpublished results). However, due to the similarity between LOL and Pea lectins, the structural results described in LOL are applicable to Pea. This is not completely the case with ConA as mentioned above.

2.2.3. Griffonia simplicifolia lectin

The structures of the lectin IV of *Griffonia simplicifolia* and its complex with the Lewis b O-Me human blood group determinant have been refined at 2 Å resolution (Table I) [27,28]. *Griffonia simplicifolia* lectin is Gal (α1–3) specific, with a three-dimensional structure of the monomer comparable to those of other legume lectins. However, the canonical association between both monomers is disrupted by the presence of a N-linked glycan at the dimer interface, leading to a novel quaternary structure (Fig. 12a), found also in the *Erythrina coralodendron* lectin [35]. The L-Fuc(α1–2) D-Gal(β1–3) L-Fuc(α1-4) D-GlcNAc β-O-Me tetrasaccharide is tightly bound to the lectin by an extensive network of polar and apolar interactions (Fig. 12b). Saccharide accommodation in the binding cleft displaces 8 of the 11 ordered water molecules found in the native lectin and allows formation of 12 direct lectin-saccharide hydrogen bonds. Moreover, besides the three water molecules common to both structures, 9 additional water molecules are found in the binding site. Among them, 7 water molecule establish hydrogen bonds between saccharide and lectin. A total of 31 contact distances (< 4 Å) exist between saccharide and protein. Most of them involve side-chains of six aromatic amino-acids: Tyr 105, Phe 108, His 114, Trp 133, and Tyr 223. The majority of these contacts involve Trp 133 and the β-Gal hydrophobic face as well as its C6 and O6 atoms. Saccharide accommodation produces small displacements among aromatic residues, between one and two Å. This kind of adjustment has also been reported in antibody–antigen complexes [56].

2.2.4. Erythrina corallodendron lectin

The *Erythrina corallodendron* lectin is Gal (β1–4) GlcNAc specific and presents the same type of quaternary structure as the *Griffonia simplicifolia* lectin, for the same reasons mentioned above. Its structure has been solved at 1.7 Å resolution in complex with lactose (Table I) [35]. This structure presents the peculiarity that the two sugar residues of the lactose and the seven sugar residues of the N-bound saccharide are both

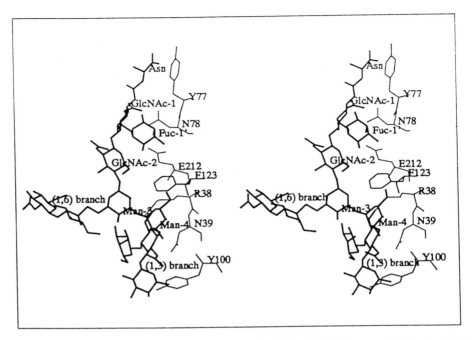

Fig. 11. Stereoview of the lectin–glycopeptide interactions in the LOL II glycopeptide structure [33].

visible in the electron density map and are well ordered (Fig. 13a). In the latter case, this is due to a network of hydrogen bonds arising from crystal packing and stabilizing the saccharide.

The N-linked saccharide Man(α1–6) Man(α1–3) [Xyl(β1–2)] Man(β1–4) GlcNAc (β1–4) [L-Fuc(α1–3)] GlcNAc(β-) adopts an extended conformation (Table II, Fig. 13a). The N-linked GlcNAc and the fucose are tightly hydrogen bound to Asp 16, Tyr 53 and Lys 55. The rest of the saccharide, trimannoside and xylose are bound by nine hydrogen-bonds to a two-fold symmetry-related lectin and to water molecules. All the dihedral angles of the saccharide have angular values similar to those found in solution, except that of the Man(α1–6)- branch.

Only the galactose of the lactose unit is directly bound to the combining site. Four hydrogen-bonds link Gal O3 and O4 to Asp 89, Asn 133, Cys 107 and Ala 218. These four residues, together with Ala 88 and Phe 131 occupy similar positions in other legume lectins (Pea, ConA, LOL) and build the primary sugar binding site (Fig. 13b). Overall, hydrophobic interactions with Ala 88, Tyr 106, Phe 131 and Ala 18, together with seven hydrogen-bonds to lectin or water molecules maintain the galactose in the binding site.

2.2.5. Wheat germ agglutinin
Wheat germ agglutinin (WGA) is a lectin of the Graminae family, which possesses

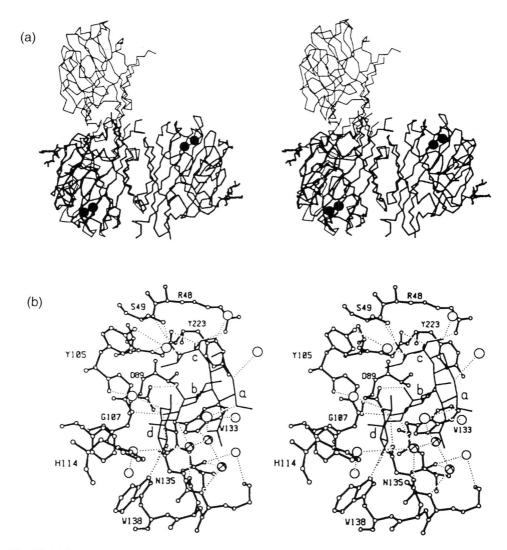

Fig. 12. (a) Stereoview of the quaternary structure of native GS lectin (thick lines) and pea lectin (thin lines). Three sugar residues per monomer have been identified on Asn 18 and are represented here. (b) Stereoview of the complex of GS lectin with Le^b-OMe in the combining site; the protein is represented in ball-and-stick form, the saccharide in stick form, water molecules in circles (from Ref. [28]).

properties distinct from other plant lectins (*Leguminosae*, etc.). For example, they are specific for two different sugars: *N*-acetyl-D-glucosamine (GlcNAc) and *N*-acetylneuraminic acid (NeuAc) [44,45]. The crystal structures of complexes of isolectins 1 and 2 of WGA with *N*-acetylneuraminyl-lactose have been described recently (Table I)

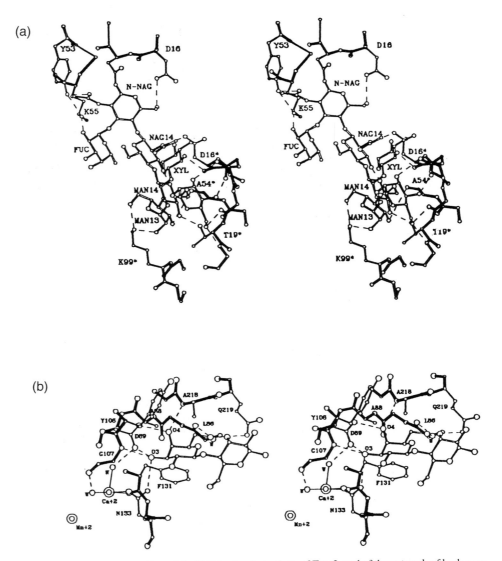

Fig. 13. (a) Stereo representation of the N-linked carbohydrate of EcorL and of the network of hydrogen-bonds stabilizing the carbohydrate; the combining site of EcorL and the network of hydrogen-bonds stabilizing the lactose (from Ref. [35]).

[25]. The NeuAc residue performs numerous hydrogen bonds with the lectin residues in site 1, whereas the two other residues, galactose and glucose, are much more exposed to the solvent. Residues interacting with sialic acid in the binding site of WGA1 are less charged than in the case of sialidase (Fig. 14). Hydrogen-bonds are established between polar residues Tyr 73, Ser 62, Ser 114 and sialic acid or galactose (Tyr 66);

TABLE II

(a) Comparison of the dihedral angles of the N-linked carbohydrate in the EcorL crystal and solution [35]. (b) Comparison of the dihedral angles of MMBP lectin bound carbohydrate in the crystal and solution [49].

Linkage	Φ/Ψ	
	X-ray	Solution
(a) EcorL:		
L-Fuc (α1–3) GlcNAc	45/19	45/30
GlcNAc(β1–4)GlcNAc	47/11	50/10
Man(β1–4)GlcNAc	41/–37	60/0
Xyl(β1–2)Man	45/19	45/30
Man(α1–6)Man	–62/76	–40/170
Man(α1–3)Man	5/–154	–15/–160
(b) MMBP:		
Man8 (α1–3) Man 5α	76/141	70/110
Man5 (α1–6) Man 4β	67/177	60/160
Man6 (α1–3) Man 4β	78/116	70/100
Man9 (α1–2) Man 6α	96/144	75/100

Glu 115 is the only charged residue hydrogen-bound to sialic acid. The same residues, namely Tyr 66, Tyr 73, and also Tyr 64 are intensively involved in stacking contacts with NeuAc, and in a much lower extent with Gal. Water molecules play also an important role in stabilizing the interaction, interacting with the water exposed galactose. One water molecule is linked to Tyr 73 in the unoccupied combining site. It repositions slightly upon sugar binding, and exhibits a lower B-factor. This water molecule (for example W180 in one of the sites) is linked tetrahedrically between NeuAc atom O4 and lectin residues Ser 114 (NH), Ser 43 (OH) and Tyr 73 (OH). It further binds to Ser 43 (NH) via water molecule 201. Another water molecule (W255) establishes a weak hydrogen bond with NeuNAc atom O8 and lectin residue Glu 115 (OE2). The last one (W277) links the saccharide internally, between glucose atom O6 and galactose atoms O1 and O2 (Fig. 21).

Structural data on WGA complexed to a glycophorin-sialoglycopeptide receptor, were also obtained recently (Table I) [26]. The receptor is made of a tetrasaccharide with two terminal sialic acids (α2–3) and (α2–6) linked. The branched tetrasaccharide has an extended, rigid conformation, and its terminal sialic acid residues occupy specific sites in domains B1 (monomer 1) and C2 (monomer 2) on opposing dimers in the crystal, respectively (Fig. 15a). The structure of the complex has demonstrated a preference of WGA for (α2–6) NeuAc in site B1 over (α2–3) NeuNAc in site C2. As suggested by these results, a positive cooperativity should exist for this binding

48

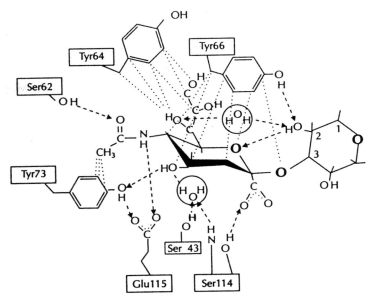

Fig. 14. Schematic representation of the interactions between the NeuAc-Gal portion of the sialyl-lactose and the residues of the combining site of WGA; hydrogen bonds are drawn as dashed lines, and potential van der Waals contacts as dotted lines (from Ref. [25]).

mechanism. Moreover, the receptor-inter-linked WGA dimers observed in the crystal lattice may represent a model of lectin aggregation (Fig. 15b).

2.3. Animal lectins

2.3.1. C-type lectins

Mammalian mannose-binding proteins (MMBP), found in serum, macrophages and liver, mediate immunoglobulin-independent defensive reactions against pathogens [46]. MMBP contains a COOH-terminal carbohydrate-recognition domain (CRD) common to the family of calcium-dependent carbohydrate-binding proteins, known as C-type animal lectins [47]. The C-type CRD of this family is defined by a sequence motif of 120 amino acids with only 30 of them conserved. This relatively low homology may explain the distinct carbohydrate-recognition specificities among the C-type lectins [47]. The crystal structures of MMBP free and complexed to an asparaginyl-octasaccharide complex has been determined at 1.7 Å resolution [48,49] (Table I). One $Man_6GlcNAc_2$-Asn glycan binds to two different protomers in the crystal lattice, resulting in a well-ordered electron density for most of the mannose residues, in an extended conformation (Fig. 16a). MMBP contains two distinct calcium ions per protomer. The key feature of the interaction is the direct ligation of calcium ion 2 by the 3- and 4-hydroxyl groups of a terminal mannose, thus forming an 8-coordinated calcium ion complex (Fig. 16b). In the native protein, the cation is 7-coordinated with

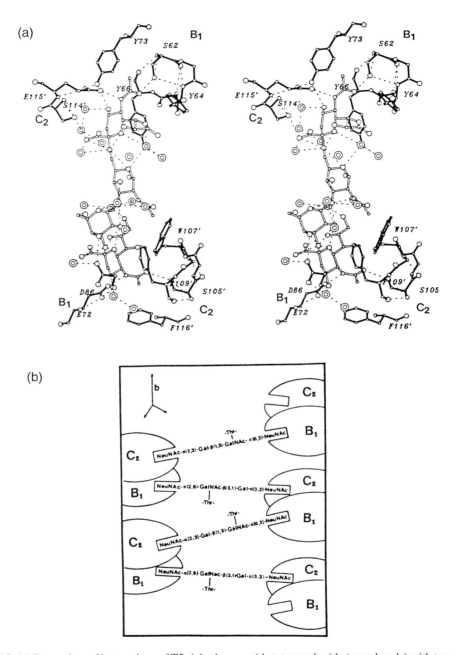

Fig. 15. (a) Stereoview of interactions of T5 sialoglycopeptide tetrasaccharide (open bonds) with two different dimers of WGA (major binding mode; H-bonds are dashed lines, water molecules open circles. (b) Schematic representation of T5 sialoglycopeptide cross-linking pattern of WGA dimers. (From Ref. [26]).

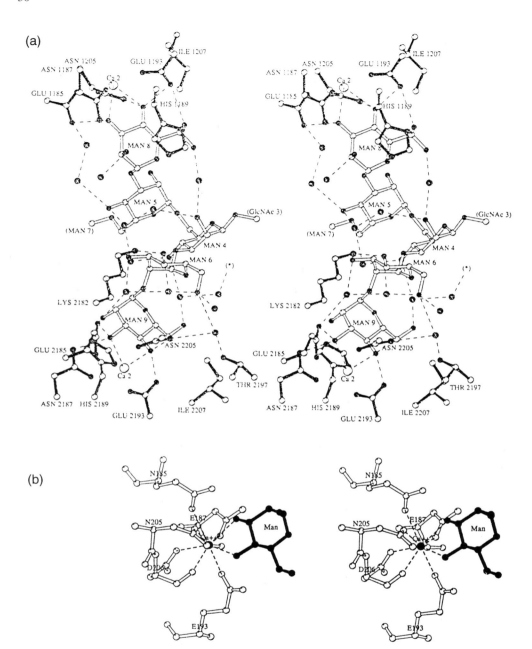

Fig. 16. (a) Stereoview of the visible part of the MBP lectin bound oligosaccharide, including bound water molecules; (open bounds are the saccharide, filled circles are water molecules; hydrogen-bonds represented with dashed lines; from Ref. [49]). (b) Stereo representation of mannose binding at the Ca^{++} site (from Ref. [6]).

a water molecule occupying the position taken up by the mannose 4-OH. Most of the side-chain atoms that participate in hydrogen bonds with the sugar are also ligated to the calcium ion, resulting in an intimately linked ternary complex of protein, Ca^{++} and sugar. Moreover, water molecules bridge hydroxyl-groups of the α-1,3-branch of the carbohydrate to the protein. A few van der Waals interactions occur between the sugar and the protein. This structure contrasts with the mode of binding of oligosaccharides to legume lectins, in which (i) the calcium ion is not directly bound to the sugar, and (ii) aromatic residues are very important in the stabilization of the complexes.

The CRD of the C-type animal lectin was modulated by protein engineering [50]. Glu 185 and Asn 187 of MMBP are at the same time Ca^{++} ligands and interacting with the bound mannose. They are conserved in the amino-acid sequences of mannose and glucose-binding proteins, but are substituted by Gln and Asp, respectively, in galactose-binding proteins, suggesting that these two positions may be a primary structural determinant of ligand-binding specificity in the C-type lectins. This double mutant binds galactose-containing ligands with an affinity 10-fold higher than that of mannose-containing ligands, though keeping the affinity for calcium intact.

2.3.2. S-type Lectins (galectins)

The 14 kDa dimeric S-lectins (or galectins) are distributed in animal tissues and belong to a family of proteins which bind to Gal(β1–4) GlcNAc or Gal(β1–3)GlcNAc terminating oligosaccharides and which have been proposed to modulate cell adhesion [52]. This protein is a member of a unique family of soluble, cation-independent carbohydrate-binding proteins referred to as 'S-type' or 'S-Lac' to distinguish it from a second lectin family, 'C-type', whose carbohydrate binding activity is dependent on the presence of calcium ions [47].

The crystal structure of a dimeric bovine spleen S-lectin LacNAc complex was determined at 1.9 Å resolution (Table I) [52]. It revealed a β-sandwich fold with the same 'jelly-roll' topology found in legume lectins [53]. The carbohydrate-binding site, located at the far ends of the dimer, has its high binding specificity for terminal galactose mediated by electrostatic interactions between conserved residues His 44, Asn 46 and Arg 48 and the sugar axial 4-hydroxyl group and by the stacking of Trp 68 against the sugar ring (Fig. 17).

Another crystal structure of a dimeric human spleen S-lectin LacNAc complex was determined at 2.8 Å resolution (Table I) [55]. The bovine and the human structures are very close and reveal the same structural features.

Three different crystal forms of the bovine heart S-lectin in complex with a diantennary octasaccharide of the complex type have been solved and refined at 2.45, 2.15 and 2.3 Å resolution, respectively. The octasaccharide and the lectin form an infinite chain of cross-linked motifs interacting only through the two N-acetyllactosamine units, located at the end of each branch of the saccharide. The octasaccharide adopts different low-energy conformations in the three structures due to variation in α-1,3- and α-1,6 glycosidic linkages (Table IV). This conformational flexibility is an opportunity for S-lectins to bind diantennary saccharides at cell surfaces with broader geometric requirements (Fig. 18) [54].

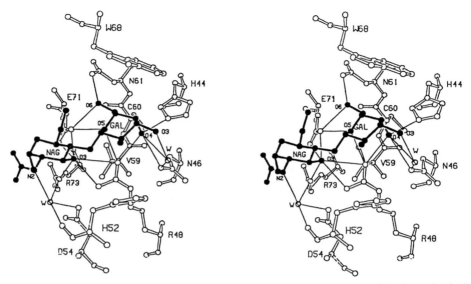

Fig. 17. Stereoview of LacNAc (filled bonds) binding in the combining site of bovine galectin 1 (from Ref. [51]).

2.4. Fab antibody fragments

The first structure of a Fab antibody fragment complexed with a polysaccharide O-antigen was reported in 1991 by Cygler et al. [56,57]. The O-antigen from pathogenic *Salmonella* is formed of repeating units of four sugars (-3)-D-Gal(α1-2) [D-Abe(α1-3)] D-Man(α1-4) L-Rha(α1-). The Fab complex involved the dodecasaccharide (3 repeating units), but only three sugar residues were clearly visible in the 2.05 Å resolution electron density map, namely abequose, mannose and galactose (Fig. 19). The Fab binding site is formed by a 8 Å deep and 7 Å wide pocket containing abequose, whereas mannose and galactose are located on the Fab surface and exposed to solvent. As a result, abequose is completely buried and accounts for the largest contact area (121 Å^2 on a total of 255 Å^2) and provides also most of the protein-saccharide contacts. All OH groups but those of galactose O6 and mannose O6 are involved in hydrogen bonds. A buried water molecule is instrumental in binding the antigen to the Fab. It displays a very strong binding to 3 hydrogen-bond acceptors (Abe O5 and Tyr 99H C=O plus Gly 96H C=O) and to 2 hydrogen-bond donors (Abe OH 4 and His 35H NH) and enhance surface complementarity. This extended net of hydrogen bonds is found in the sugar binding chemotaxis proteins, ABP and MBP. The Fab-saccharide interaction is dominated by aromatic side chains-saccharide stacking, which are also present in ABP, but to a lesser extent. Fab Trp 91L stacks against abequose C3–6 and galactose C1–2, and Trp 33H against abequose C6 and mannose C3. This stacking contributes especially to antigen selectivity.

The saccharide conformation involves low energy linkages. The D-Gal(α1–2) D-Man(α1–4) linkage is an exception, with a Φ value shifted by 40°. This may be compensated by intersaccharide hydrogen bonds.

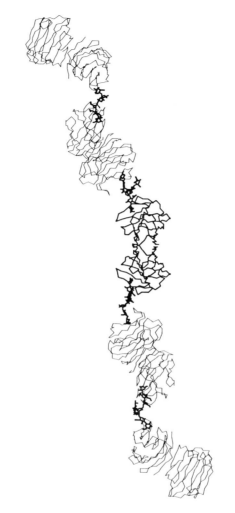

Fig. 18. Representation of two bovine galectin 1 cross-linking a biantennary octasaccharide [54].

2.5. Enzymes

Several structures of carbohydrate–enzyme complexes have been thoroughly studied, among them lysozyme, glycogen phosphorylases b, amylase and sialidase.

2.5.1. Lysozyme

Lysozyme catalyzes the hydrolysis of the β-1,4- linkage of bacterial cell wall polysac-charides. Many structures of complexes between lysozyme and saccharides (up to a tetrasaccharide) have been determined [58 and references therein], most of these at low resolution. However, even the best characterized structure, lysozyme complexed with

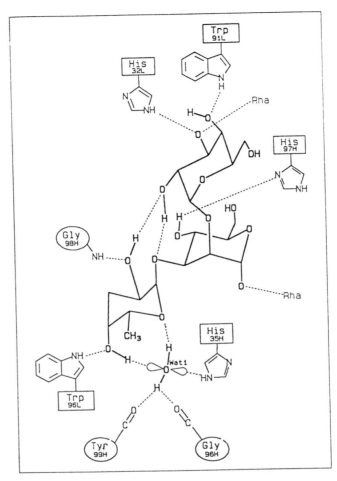

Fig. 19. Schematic representation for a trisaccharide epitope bound by the monoclonal antibody Se155.4. Circled residues make hydrogen bonds through main-chain atoms, boxed residues make hydrogen bonds through side-chain atoms; wat 1 is a buried water molecule, integral part of the complex (from Ref. [56]).

tri-NAG [58,59] at 1.75 Å resolution, could be refined to an R-factor of only 0.23 (Table I).

The contacts between tri-NAG and lysozyme sites A-C involve 6 direct hydrogen bonds, with Asn 59 NH, Asp 101 Od1, Asp 103 Od1, Ala 107 NH, and NE1 of Trp residues 62 and 63. Three water molecules identified in the electron density map interact with O1 from GlcNAc in the C site. Site C has the most important number of interactions with the saccharide (Fig. 20). The GlcNAc moiety bound to site B lies with its apolar face parallel to the indole ring of Trp 62, and contacts also Trp 63, suggesting a dominant hydrophobic interaction for this GlcNAc. The other face of the GlcNAc residues are relatively exposed to the solvent.

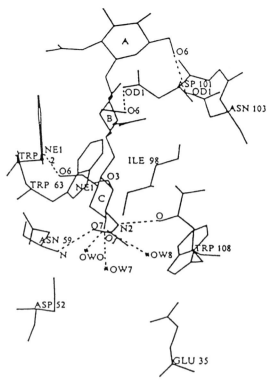

Fig. 20. (GlcNAc)₃ bound in the active-site crevice of hen egg-white lysozyme (hydrogen bonds are represented by dashed lines (from Ref. [59]).

2.5.2. Glycogen phosphorylase b

Glycogen phosphorylase [60] catalyses the reversible phosphorylation of α-1,4- linkages of glycogen, leading to the formation of glucose-1-P. This reaction occurs at the phosphorylase catalytic site, distinct from a separate glycogen binding site, called the storage site [61–63]. Regarding the phosphorylase-b -saccharides interaction, the most informative studies are the ones in which heptenitol or heptulose 2-phosphate are bound to the catalytic site [64,65], and the one in which maltoheptaose is bound to the storage site [58,66].

As in the case of ABP and GBP, monosaccharides bound to phosphorylase b active site are buried. Little or no access is given to the bulk solvent. The hydrogen bonding potential of every polar group is satisfied: no less that 14 hydrogen involving sugar OH 1–9 are established between the sugar and the protein (Fig. 21a). Only one water molecule is linked to heptenitol atom O-4, and two water molecules link the highly polar phosphorus atoms of heptulose-2-P to the protein. The number of van der Waals contacts is much reduced compared to those found in PSPs, and no stacking interaction with aromatic residues is observed.

In contrast, the glycogen storage site is situated on the surface of the phosphorylase.

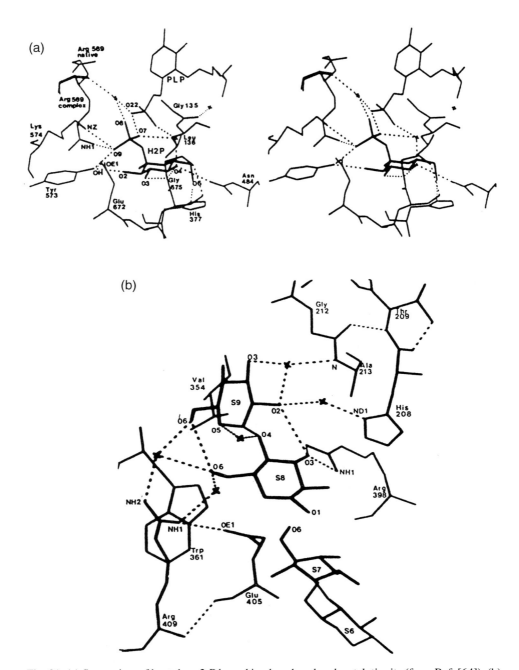

Fig. 21. (a) Stereoview of heptulose 2-P bound in phosphorylase b catalytic site (from Ref. [64]). (b) Interactions between maltoheptaose and phosphorylase b minor storage site; water molecules are shown as crosses (from Ref. [66]).

It is composed of two sub-sites, the major site and the minor site, which are almost contiguous. In the malto heptaose complex, the major site binds five sugar residues, and the minor site two (Fig. 21b). Maltoheptaose is water accessible: ten water molecules interact with the saccharide. All these water molecules hydrogen-bond to the saccharide and to protein residues. Only one stacking interaction is observed with Tyr 404. In the major site, direct interactions between saccharide and protein comprise 9 hydrogen-bonds involving sugars 3 to 7; three water molecules are hydrogen-bound, two of them establishing a linkage between the protein and the saccharide. The minor site requires special attention: among 11 saccharide–protein interactions, only one is established directly with the protein, nine being mediated via water molecules; one water molecule binds internally sugar atoms S8 O4 and S9 O5. A similar result has been found in *Lathyrus ochrus* lectin–saccharide complexes [30,31].

2.5.3. Sialidases

The single stranded RNA genome of type A and B influenza virus encodes for two surface glycoproteins, sialidase and hemagglutinin, which interact with the same sugar: sialic acid. Sialidase binding to sialic acid on the surface of the target cell is the first step of infection. Sialidase hydrolyzes the terminal glycosidic bond of sialylated cell surface glycopeptides, yielding free sialic acid. Structures of complexes between siali-dases and sialic acid have been reported at 2.8 Å [67] Ê and 1.8 Å [68], respectively (Table I). The sugar ring is distorted in boat conformation, in the same way as that found for the bound sugar in the active site of lysozyme. The sialic acid binding site is built by a majority of charged residues, mostly Arg, Glu and Asp (Fig. 22). A striking difference with most other complexes reported here, is that the saccharide ring does not

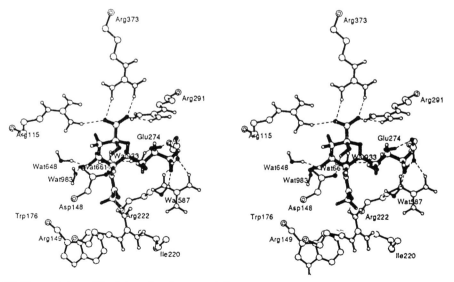

Fig. 22. Stereo picture of the interactions between sialic acid (black) and influenza B sialidase. Hydrogen bonds are represented by dotted lines (from Ref. [6]).

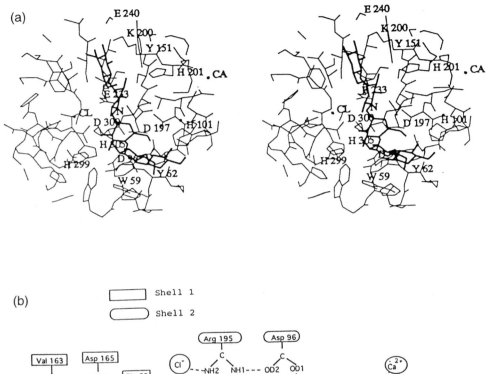

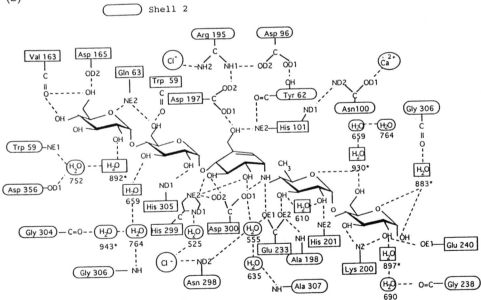

Fig. 23. (a) Stereoview of acarbose in the active site cleft of pancreatic α-amylase. (b) Schematic representation of the hydrogen bond network established between acarbose and the active site cleft residues of pancreatic α-amylase (from Ref. [72]).

establish any stacking contact with aromatic side chains of active-site residues. The only aromatic residues nearby the sialic ring are Tyr 406, hydrogen-bonded to sialic O5, and Trp 178, in loose contact (4 Å) with CH_3 of the *N*-acetyl group. This is in contrast with the sialic acid binding in WGA [25].

2.5.4. α-Amylase

Porcine pancreatic α-amylase, is an endo-type amylase that catalyses the random hydrolysis of a-D-(1,4) glucosidic bonds in amylose and amylopectin through multiple attack toward the non-reducing end [69]. The three-dimensional molecular model of pig pancreatic α-amylase has recently been described in detail [70]. On the basis of this highly refined molecular model for the native enzyme, the three-dimensional structure of a crystal of native α-amylase modified with a carbohydrate α-amylase inhibitor (acarbose) has been determined to 2.2 Å. The pseudo-tetrasaccharide, acarbose is a strong inhibitor of pig pancreatic α-amylase [71] and X-ray analysis of the crystal showed a well-defined density corresponding to five fully occupied subsites in the active site. The X-ray model clearly presents a subset of residues directly involved in binding the inhibitor [72]. Many protein–inhibitor interactions are observed in the complex structure (Fig. 23a,b), a substantial part of them being engaged with water molecules. Hydrophobic stacking interactions involve tryptophans 58 and 59, and tyrosines 62 and 151 (Fig. 23a,b).

3. Structural rules for establishing protein–carbohydrate interactions

Among these rules, hydrogen bonds are mentioned to be the main factors in conferring specificity and affinity. All the polar groups of the saccharides are involved in such bonds. Numerous van der Waals contacts are also formed, including one or two aromatic residues stacked on the sugar ring. A few water molecules are also often involved in the saccharide binding.

3.1. Hydrogen bonds and ordered water molecules

Carbohydrates are very hydrophilic molecules. They tend to have all their hydrogen binding potential satisfied. This may be achieved through an intricate network of direct hydrogen bonds between protein and saccharide. This pattern is mainly observed with mono- or disaccharides deeply buried in combining or active sites, as in the cases of PSBPs and glycogen phosphorylase. Longer saccharides tend to be more exposed to bulk solvent. As a consequence, satisfaction of their hydrogen binding potential can be fully or partially satisfied only via water molecules or via internal saccharide–saccharide binding. The number of high resolution well-refined crystallographic structures with protein, saccharides and water molecules present at the same time is still rather limited. However, it is now clear that discrete water molecules, as seen in crystallographic structures of protein–carbohydrate complexes, play a significant role in these interactions. They can establish bridges between saccharide and protein, with a length comprising one to three water molecules. Moreover, an intricate network of hydrogen

TABLE III

Comparison of glycosidic bond torsion angles of the N-linked diantennary saccharide from human lactotransferrin (Fig. 6; Ref. [33])

Glycosidic linkage	Torsion angles[a] (Φ, Ψ, ω) (°)				
	LOL II-N2	LOL II-glycopeptide Site A	LOL II-glycopeptide Site B	LOL II-glycopeptide Site C	Energy minimized Φ, Ψ, ω) (°) values[b]
GlcNAc(β1–4)GlcNAc(β1-)	−66, −120	−68, −110	−62, −102	−44, −177	−80, −110
Fuc(α1–6)GlcNAc(β1-)	−89, −153, −67	−64, −170, −68	−70, −171, −57	−72, −169, −69	−65, 155, −60
Man(β 1–4)GlcNAc(β1-)	−90, −111	−99, −96	−95, −104	−94, −106	−75, −110
Man(α1–3)Man(β–)	81, 137	94, 124	76, 128	81, 126	85, 175
Man(α1–6)Man(β1-)[c]	61, −162, −51 (−169)	39, 158, 56 (−63) −179	−179, −53, −45 (−164)	−168, −73, −33 (−154)	70, −170 (−60)
GlcNAc(β1–2)Man(α1–)[d]	63, 151; 41, 118	62, 154; 47, 127	75, 160; ND	59, 171	−90, 160/55, 140
Gal(β1–4)GlcNAc(α1–)[d]	−32, −103	−171, −114; ND	−80, −152; ND	1	−89, −150; ND
NeuAc(α2–6)Gal(β1-)	ND	ND	29, 127, 39	179, 121, −96	60, 210, 60/60, 180, −60
Fuc(α1–3)GlcNAc(β1-)	ND	−65, −151	ND	ND	−70, 145

Residue numbers are shown in Fig. 6.

[a]Torsion angles are defined by four atoms as follows: Φ = O5-C1-O1-C'X, Ψ = C1-O1-C'X,C''(X + 1). For α-1,6-linkages, Ψ = C1-O1-C'^-C'5 and ω = O1-C'6-C'5-O'5. For NeuAc residue Φ = C1-C2-O2-C'X and Ψ = C2-O2-C'X,C'(X + 1).

[b]Each linkage that is completely defined in the X-ray structure lies in the minimum of the corresponding Φ, Ψ potential surfaces calculated by semi-empirical methods [21,22].

[c]Values in parentheses are those for ωH = O1-C'C-C'5-H'5.

[d]For α-1,3- and α-1,6-branches, respectively.

ND = not determined.

bonds can be established, building a kind of 'viscous' water material around the saccharide or between saccharide and protein.

3.2. Hydrophobic stacking

The importance of stacking interactions between aromatic residues (Phe, Tyr, Trp, His) and sugars is now well established. In particular Trp residues involved in the stabilization of carbohydrate–protein interactions, have been described in nearly all the structures mentioned in this review.

3.3. Saccharide conformation

All torsion angles of the glycosidic bond of the oligosaccharides studied so far by X-ray crystallography fall in — or are close to — the energy minima derived from NMR measurements or from energy calculations on the equivalent Asn-oligosaccharides [7,8,74]. In the case of the complexes of fragment N2 and of glycopeptide with LOL II, the agreement is better in the oligosaccharide region tightly bound to LOL II compared to the more flexible parts of the oligosaccharide (Table III) [33]. This suggests that the conformation of the oligosaccharide bound to LOL is close to that of one of the several possible structures in solution: the lectin probably selects a thermo-dynamically most favourable oligosaccharide conformation. The conformation of the α-1,6- fucose is also nearly identical to that found in elastase [74] and EcorL [35].Similar agreement between free and bound oligosaccharides has been recently described for the oligosaccharide linked to EcorL (Table II) [35], for that bound to MBP (Table II) [49] and for that bound to galectin 1 (Table IV) [54].

4. Concluding remarks

X-ray structures of protein–carbohydrate complexes reported confirm well established rules of protein–carbohydrate interactions through a network of hydrogen bonds. Beyond the expected important role of hydrogen bonds in protein–saccharide interaction, all the structures reported show the importance of aromatic side-chain/saccharide stacking in stabilizing these complexes. In most complexes, the role of water molecules is also crucial in completing the hydrogen network established between saccharide and protein. Finally, all torsion angles of the glycosidic bonds of the oligosaccharide studied so far by X-ray crystallography fall in — or are close to — the energy minima derived from NMR measurements or from energy calculations on the equivalent Asn-oligosaccharides.

Abbreviations

ABP	L-arabinose binding protein
Con A	Concanavaline A

TABLE IV

Comparison of glycosidic bond torsion angles of the diantennary octa-saccharide (Fig. 6, sugars 2, 3, 4, 4′, 5, 5′, 6, 6′; Ref. [54])

Glycosidic linkage	Φ, Ψ, Ω (°) Crystal forms‡						Energy-minimised Φ, Ψ, Ω (°)†
	H	T			M		
		dimer 1	dimer 2		dimer 1	dimer 2	
Man(β1–4)GlcNAc(β1-)	–87, –115	–118, –124	–67, –69		–87, –107	ND	–75, –110
Man(α1–3)Man(β1-)	72, 144	88, –176	74, 130		88, 100	96, 135	85, 175
Man(α1–6)Man(β1-)#	62, –173, –81 (165)	72, –179, –75 (171)	147,164,–71 (173)		–164,–71 (173)	–166, –100, 56 (–61)	90, –158, 93 (–30)
GlcNAc(β1–2)Man(α1-)*	–89, 158;–86, 158	–100, 150;–98, 152	–91,137;–90,134		–91, 156;–85, 117	–74, –179;–86, 159	–90, 160
Gal(β1–4)GlcNAc(α1-)*	–79, –110;–75	–111	–75,–106;–71,–101		–67, –107;–76, –114	–81,–101;–79, –111	–76,–103;–70,–109

Torsion angles are defined by four atoms as follows: Φ = O5-C1-O1-C′X, Ψ = C1-O1-C′X-C′′X, Ψ = C1-O1-C′′X,C′′(X + 1). For α-1,6-linkages, Ψ = C1-O1-C′6-C′5 and ω = O1-C′6-C′5-O′5.

†Minimum Φ, Ψ values of corresponding potential surfaces calculated by semi-empirical methods.

#Values in parentheses are ωH = O1-C′6-C′5-H′5.

*For α-1,3- and α1,6-antennae, respectively.

‡For the hexagonal, trigonal and monoclinic crystal forms, respectively. In the case of forms T and M, two dimers are present in the asymmetric unit.

CRD	carbohydrate-recognition domain
EcorL	*Erythrina corallodendron* lectin
GBP	D-galactose/glucose binding protein
GS	*Griffonia simplicifolia*
LOL	*Lathyrus ochrus* lectin
LOL I	*Lathyrus ochrus* isolectin I
LOL II	*Lathyrus ochrus* isolectin II
LOLF	*Lathyrus ochrus* lectin, not complexed
LOLG	*Lathyrus ochrus* lectin, in complex with glucose
LOLM	*Lathyrus ochrus* lectin, in complex with mannose
LOLT	*Lathyrus ochrus* lectin, in complex with a trisaccharide
MBP	maltose binding protein
MMBP	mammalian mannose-binding proteins
MdBP	maltodextrin binding protein
OCTA	biantennary octasaccharide
PSBP	periplasmic sugar binding proteins
TRI	Man(α1–3) Man (β1–4) GlcNAc trisaccharide
WGA	wheat germ agglutinin

References

1. Montreuil, J. (1980) Adv. Carbohydr. Chem. Biochem. 37, 157–223.
2. Walker, R.A. (1988) In: T.C. Bog-Hansen and D.L.J. Freeds (Eds.), Lectins: Biology, Biochemistry, Clinical Biochemistry, Vol 6. Sigma Chemical Co., St. Louis, MO, pp. 591–600.
3. Diaz, C.L., Melchers, L.S., Hooykaas, P.J.J., Lugtenberg, B.J.J. and Kijne, J.W. (1989) Nature 338, 579–581.
4. Lis, H. and Sharon, N. (1986) Annu. Rev. Biochem. 53, 35–67.
5. Vyas, N.K. (1992) Curr. Op. Struct. Biol. 1, 732–740.
6. Bourne, Y., van Tilbeurgh, H. and Cambillau, C. (1994) Curr. Op. Struct. Biol. 3, 681–686.
7. Imberty, A., Gerber, S., Tran, V. and Pérez, S. (1990) Glycoconjugate J. 7, 27–54.
8. Imberty, A., Delage, M. M., Bourne, Y., Cambillau, C. and Pérez, S. (1991) Glycoconjugate J. 8, 456–483.
9. Mazurier, J., Dauchez, M., Vergoten, G., Montreuil, J., and Spik, G., (1991) Glycoconjugate J., 8, 390–399.
10. Montreuil, J. (1984) J. Biol. Cell. 51, 115–132.
11. Furlong, C.E. (1987) in: F.C. Neidhardt (Ed.), *Escherichia coli* and *Salmonella typhimuriam*: Cellular and Molecular Biology. American Society for Microbiology, Washington, DC, pp. 768–796.
12. Quiocho, F.A. (1986) Annu. Rev. Biochem. 55, 287–315.
13. Quiocho, F.A. (1988) Current Topics in Microbiology and Immunology 139, pp. 135–148.
14. Quiocho, F.A. (1989) Pure Appl. Chem.. 61, 1293–1306.
15. Vyas, N.K., Vyas, M.N. and Quiocho, F.A. (1991) J. Biol. Chem. 266, 5226–5237.
16. Quiocho, F.A. and Vyas, N.K. (1984) Nature, 310, 381–386.
17. Quiocho, F.A., Wilson, D.K. and Vyas, N.K. (1989) Nature 340, 404–407.
18. Vyas, N.K., Vyas, M.N. and Quiocho, F.A. (1988) Science 242, 1290–1295.
19. Spurlino, J.C., Lu, G.-Y. and Quiocho, F.A. (1991) J. Biol. Chem. 266, 5202–5219.

64

20. Spurlino, J.C., Rodseth, L.E. and Quiocho, F.A. (1992) J. Mol. Biol. 226, 15–22.

21. Derewenda, Z., Yariv, J., Helliwell, J.R., Kalb (Gilboa), A.J., Dodson, E.J., Papiz, M.Z., Wan, T. and Campbell, J. (1989) Embo J. 8, 2189–2193.

22. Reeke, G.N., Jr. and Becker, J.W. (1986) Science 234, 1108–1111.

23. Rini, J.M., Carver, J.P. and Hardman, K.D. (1986) J. Mol. Biol. 189, 259–260.

24. Rini, J.M., Hardman, K.D., Einspahr, H., Suddath, F.L. and Carver, J.P. (1993) J. Biol. Chem. 268, 10126–10132.

25. Wright, C.S. (1990) J. Mol. Biol. 215, 635–651.

26. Wright, C.S. (1992) J. Biol. Chem. 267, 14345–14352.

27. Delbaere, L.T.J., Vandonselaar, M., Prasad, L., Quail, J.W., Pearlstone, J.R., Carpenter, M.R., Smillie, L.B., Nikrad, P.V., Spohr, U. and Lemieux, R.U. (1990) Can. J. Chem. 68, 1116–1121.

28. Delbaere, L.T.J., Vandonselaar, M., Prasad, L., Quail, J.W., Wilson, K.S. and Dauter, Z. (1993) J. Mol. Biol. 230, 950–965.

29. Bourne, Y., Roussel, A., Frey, M., Rougé, P., Fontecilla-Camps, J. C. and Cambillau, C. (1990) Proteins 8, 365–376.

30. Bourne, Y., Rougé, P. and Cambillau, C. (1990) J. Biol. Chem. 265, 18161–18165.

31. Bourne, Y., Rougé, P. and Cambillau, C. (1992) J. Biol. Chem. 267, 197–203.

32. Bourne, Y., Nésa, M.P., Rougé, P., Mazurier, J., Legrand, D., Spik, G., Montreuil, J. and Cambillau, C. (1992) J. Mol. Biol. 227, 938–941.

33. Bourne, Y., Mazurier, J., Legrand, D., Spik, G., Rougé, P., Montreuil, J. and Cambillau, C. (1994) Structure 2, 209–219.

34. Bourne, Y., Ayouba, A., Rougé, P. and Cambillau, C. (1994) J. Biol. Chem. 269, 9429–9435.

35. Shaanan, B., Lis, H. and Sharon, N. (1994) Science 254, 862–866.

36. Rougé, P. and Sousa-Cavada,B. (1984) Plant Sci. 37, 21–27.

37. Bourne, Y., Abergel, C., Cambillau, C., Frey, M., Rougé, P. and Fontecilla-Camps, J. C. (1990) J. Mol. Biol. 214, 571–584.

38. Ayouba, A., Causse, H., van Damme, E.J.M., Pneumans, W.J., Bourne, Y., Cambillau, C. and Rougé, P. (1994) Biochem. System. Ecol. 22, 153–159.

39. Spik, G., Strecker, G., Fournet, B., Bouquelet, S., Montreuil, J., Dorland, L., Halkeck, H.V. and Vliegenthart, J.F.G. (1982) Eur. J. Biochem. 121, 413–419.

40. Warin, V., Baert, F., Fouret, R., Strecker, G., Fournet, B. and Montreuil, J. (1979) Carbohydr. Res. 76, 11–22.

41. Bourne, Y., Anguille, C., Fontecilla-Camps, J.-C., Rougé, P. and Cambillau, C. (1990) J. Mol. Biol. 214, 211–213.

42. Debray, H., Decout, D., Strecker, G., Spik, G. and Montreuil, J. (1981) Eur. J. Biochem. 117, 41–55.

43. Bourne, Y., Mazurier, J., Legrand, D., Spik, G., Rougé, P., Montreuil, J. and Cambillau, C. (1993) J. Cell. Biochem. 17 (suppl. A), 362.

44. Peters, B.P., Ebisu, S., Goldstein, I.J. and Flashner, M. (1979) Biochemistry 18, 5505–5511.

45. Monsigny, M., Roche, A.C., Sene, C., Magnet-Dana, R. and Delmotte, F. (1980) Eur. J. Biochem. 104, 147–153.

46. Lasky, L.A. (1992) Science 258, 964–969.

47. Drickamer K. (1988) J. Biol. Chem. 263, 9557–9560.

48. Weis, W.I., Kahn, R., Fourme, R., Drickamer, K., Hendrickson and W.A. (1991) Science 254, 1608–1615.

49. Weis, W.I., Drickamer, K. and Hendrickson, W.A. (1992) Nature 360, 127–134.

50. Drickamer, K. (1992) Nature 360, 183–186.

51. Liao, D.-I., Kapadia, G., Ahmed, H., Vasta, G.R. and Herzberg, O. (1994) Proc. Nat. Acad. Sci. 91, 1428–1432.

52. Barondes, S.H. et al. (1994) Cell 76, 597–598.

53. Richardson, J.S. (1981) Adv. Protein Chem. 34, 167–339.
54. Bourne, Y., Bolgiano, B., Liao, D.-Ing, Strecker, G., Cantau, P., Herzberg, O., Feizi, T. and Cambillau, C. (1994) Nature Struct. Biol., in press.
55. Lobsanov, Y.D., Gitt, M.A., Leffler, H., Barondes, S.H. and Rini, J.M. (1993) J. Biol. Chem. 268, 27034–27038.
56. Cygler, M., Rose, D. and Bundle, D.R. (1991) Science 235, 442–445.
57. Bundle, D.R. and Young, N.M. (1992) Curr. Opin. Struct. Biol. 2, 666–673.
58. Johnson, L.N., Cheetham, J., McLaughlin, P.J., Acharya, K.R., Barford, D. and Phillips, D.C. (1988) Curr. Topics Microbiol. Immunol. 139, 81–134.
59. Blake, C.C.F., Johnson, L.N., Mair, G.A., North, A.C.T., Phillips, D.C. and Sarma, V.R. (1967) Proc. R. Soc. Lond. (Biol.) B 167, 378–388.
60. Cohen, P. (1983) Control of Enzyme Activity, 2nd Edn. Chapman and Hall, London.
61. Metzger, B., Helmreich, E. and Glaser, L. (1967) Proc. Natl. Acad. Sci. USA 57, 994–1001.
62. Weber, I.T., Johnson, L.N., Wilson, K.S., Yeates, D.R.G., Wild, D.L. and Jenkins, J.A. (1978) Nature 274, 433–437.
63. Kasvinsky, P.J., Madsen, N.B., Fletterick, R.J. and Sygusch, J. (1978) J. Biol. Chem. 253, 1290–1296.
64. McLaughlin, P.J., Stuart, D.I., Klein, H.W., Oikonomakos, N.G. and Johnson, L.N. (1984) Biochemistry 23, 5862–5873.
65. Hadju, J., Acharya, K.R., Stuart, D.I., McLaughlin, P.J., Barford, D., Oikonomakos, N.G., Klein, H. and Johnson, L.N. (1987) EMBO J. 6, 539–546.
66. Johnson, L.N., Stura, E.A., Sansom, M.S.P. and Babu, Y.S. (1983) Biochem. Soc. Trans. II, 142–144.
67. Burmeister, W.P., Ruigrok, R.W.H. and Cusack, S. (1992) EMBO J. 11, 49–56.
68. Varghese, J.N., McKimm-Breschkin, J.L., Caldwell, J.B., Kortt, A.A. and Colman, P.M. (1992) Proteins 14, 327–332.
69. Prodanov, E., Seigner, C. and Marchis-Mouren, G. (1984) Biochem. Biophys. Chem. Commun. 122, 75–81.
70. Qian, M., Haser, R. and Payan, F. (1993) J. Mol. Biol. 231, 785–799.
71. Schmidt, D.D., Frommer, B., Junge, L., Müller, W., Wingender, W. and Truscheit, E. (1981) in: E.W. Creutzfeldt (Ed.), First International Symposium on Acarbose. Excerpta Medica, Amsterdam, pp. 5–15.
72. Qian, M., Haser, R., Buisson, G., Dúee, E. and Payan, F. (1993) Biochemistry 33, 6284–6294.
73. Brisson, J.-R. and Carver, J.P. (1983) Biochemistry 22, 3680–3686.
74. Bode, W., Meyer, E. Jr. and Powers, J.C. (1989) Biochemistry 28, 1951–1963.

J. Montreuil, H. Schachter and J.F.G. Vliegenthart (Eds.), *Glycoproteins*

3D Structure

2. Three dimensional structure of oligosaccharides explored by NMR and computer calculations

S.W. HOMANS

University of Dundee, Department of Biochemistry, Medical Sciences Institute, Carbohydrate Research Centre, Dundee DD1 4HN, UK

1. Introduction

1.1. Why study oligosaccharide conformation and dynamics?

The determination of the solution conformations of macromolecules has become an increasingly straightforward task, due to the enormous strides which have been made in recent years in both high-resolution nuclear magnetic resonance and computational chemistry. Nevertheless, such investigations can still occupy a great deal of research time and funds, to the extent that it is necessary at the outset to ask what important information one is likely to obtain from these studies. In the case of proteins and nucleic acids, the justification is often clear since a particular macromolecule can be firmly linked to a particular biological process. Until recently, oligosaccharides could not be categorised so neatly — although for many years their roles as recognition signals has been implicated, in very few cases was it possible to identify a particular oligosaccharide ligand in a recognition process. Now, however, there are an adequate number of other systems in which the oligosaccharide ligand has been defined, that we can state with confidence that a general role for oligosaccharides is in recognition. Since recognition is a 'shape' problem, it therefore follows that the three-dimensional structures of oligosaccharides will be of importance in furthering our understanding of the molecular basis of the recognition process.

1.2 Scope

In this chapter I shall attempt to summarise modern techniques and current thinking on the structure and dynamics of oligosaccharides. I shall avoid an historical perspective wherever possible, since a chapter dealing with issues such as these will appear shortly [1]. Rather, I shall focus upon work of the last few years from groups including our own which has clarified our view of the way oligosaccharides behave in solution.

Inevitably this approach will not acknowledge much important early work which laid the foundations for our current thinking, and I apologise to those whose contribution is not referenced for temporal reasons. I shall focus primarily upon the structure and dynamics of oligosaccharides in free solution, rather than studies on oligosaccharide–protein complexes — in view of the large molecular masses of most carbohydrate-binding proteins, high-resolution structures from NMR studies are conspicuous by their absence, and such investigations remain the realm of the crystallographer [2].

2. Solution conformations of oligosaccharides

2.1. Conformational parameters of oligosaccharides

At the outset it is convenient to define the various parameters which will be used throughout this article to describe the conformational properties of oligosaccharides. The absolute configuration of the monosaccharide residues which will be discussed are in the $^4C_1$ D-configuration. When two monosaccharide residues are glycosidically linked, then their relative spatial dispositions are defined by the two torsion angles φ and ψ about the glycosidic linkage, or additionally by a third torsion angle ω in the case of 1–6 glycosidic linkages. When discussing NMR-derived conformations, these angles are generally measured with respect to protons, and are defined as follows; φ_H = H1-C1-O1-CX, ψ_H = C1-O1-CX-HX (C1-O1-C6-C5 in the case of 1–6 linkages), and ω_H = O6-C6-C5-H5, where the subscripts distinguish the 'NMR definitions' of these angles from the conventional IUPAC definitions, and CX and HX refer to the aglyconic atoms.

2.2. Theoretical predictions

In computational terms, a significant advantage of oligosaccharides over other macromolecules is their small size. It is therefore possible to perform molecular mechanical energy calculations for a complete oligosaccharide, at least *in vacuo*, with only modest computational power.

2.2.1. Which forcefield?
Before any molecular mechanical studies on oligosaccharide conformations can be performed, it is necessary to possess a suitable forcefield, i.e. a consistent set of parameters which describe the energetics of the molecule in question. In a full forcefield this will include terms describing bond stretching, angular librations and torsional oscillations, together with terms describing van der Waals and Coulombic interactions, and possibly also hydrogen bonding. In the particular case of oligosaccharides, often an additional torsional term is included to take account of the angular preference of φ due to the influence of the exo-anomeric effect [3 and refs. therein].

Early conformational studies on oligosaccharides utilised a relatively simple forcefield which comprised van der Waals terms and a torsional term to account for the exo-anomeric effect, in so-called hard-sphere exo-anomeric effect (HSEA) calculations [3–9]. In order to keep the problem computationally tractable, the monosaccharide ring geometries were maintained in a fixed conformation. However, this is now recognised as an unacceptable compromise when considering oligosaccharide dynamics since torsional oscillations about glycosidic linkages are strongly underestimated [10], and with the advent of much more powerful computational platforms, it is usual nowadays to employ a full forcefield which allows for full optimisation of geometry. Several forcefields of this nature have now been described with appropriate parametrisation for oligosaccharides [11–16]. The choice of such a forcefield is often dictated by that which is most readily available, and which interfaces most conveniently with the molecular modelling package employed by the user, and fortunately differences in results between those available are not dramatic.

2.2.2. Energy minimisation

Perhaps the simplest conformational strategy which can be utilised for an oligosaccharide is energy minimisation. Energy minimisation is a nonlinear optimisation problem. Given a set of independent variables $\mathbf{x} = (x_1, x_2, x_3, ..., x_n)$ and an objective function $V = V(\mathbf{x})$, the problem is to find a set of values for the independent variables $\mathbf{x}^*$, for which the function $V(\mathbf{x}^*) = \min(V(\mathbf{x}))$. For a macromolecule with N atoms, the 3N components of $\mathbf{x}$ are the atomic coordinates and V is the potential energy. Typically, the global minimum energy conformation of an oligosaccharide is sought from an initial set of coordinates.

A fundamental difficulty with energy minimisation is the 'local minimum' problem. That is, from a set of initial coordinates, the minimiser will find the nearest minimum, i.e. where the first derivative of the function is zero, which will almost certainly not be the global minimum. A simple means by which this can be demonstrated is to minimise a series of different starting configurations for the same molecule. The result for nine different starting configurations of the trisaccharide Man(α1-3)[GlcNAc(β1-4)]Man is shown in Fig. 1. It is seen that more than one final geometry is obtained. If the molecule

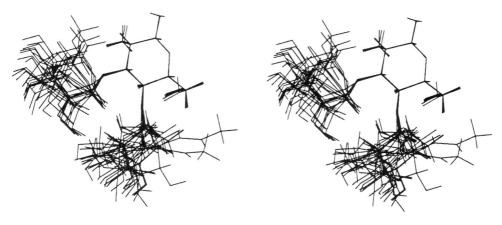

Fig. 1. Stereo view of the results from conventional minimisation of nine random starting geometries of Man(α1-3)Man.

70

under investigation is larger than a trisaccharide, the number of degrees of freedom is correspondingly greater, and thus further deviations from a single geometry can be anticipated. Thus, although straightforward energy minimisation can be undertaken literally in seconds with modern computing power, the final result is not of great value beyond relieving the initial strain in a geometry constructed manually.

2.2.3. Gridsearch calculations

In the case of simple disaccharides, an alternative strategy for finding the global minimum energy configuration is by use of gridsearch calculations. A gridsearch involves variation of two or more internal coordinates of the molecule in small, discrete steps, followed by energy minimisation at each point, holding the varied coordinate fixed. Thus, in the case of a simple disaccharide, it is possible to vary φ and ψ independently in order to determine the 'global minimum' energy configuration of the molecule with respect to these coordinates. This minimum will not however, necessarily be the true global minimum, since the remaining degrees of freedom of the system will almost certainly be fixed in a local minimum by the minimisation procedure. However, since the primary regions of conformational variation in oligosaccharides are about φ and ψ, so-called φ–ψ plots such as that shown in Fig. 2 for the disaccharide Manα1-3Man are of great value in approximating the global minimum for these simple systems.

A note of caution is in order with regard to the 'practical' aspects of φ-ψ plot generation for oligosaccharides: It is important that the starting configuration for each step is a single, good geometry for the molecule in question — if the starting geometry for each successive step is the final geometry from a previous step, there is a high

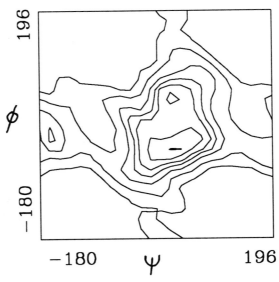

Fig. 2. Plot of φ vs. ψ for Man(α1-3)Man. The plot is contoured at 0.5, 1.0, 2.0, 5.0, 7.0 and 10.0 kcal/mol above the global minimum.

probability that one or more of the monosaccharide rings will 'flip' to an alternative configuration during minimisations in a high-energy region of the map, which will not return to the lower energy configuration in low-energy regions of the map.

Gridsearch calculations such as that shown in Fig. 2 are also known as adiabatic maps, and the name derives from the implicit assumption that the relaxation produced by energy minimisation would occur adiabatically on the timescale of the transition during actual dynamics. This may or may not be valid for all systems, and hence it is important to interpret such maps with caution. A further difficulty with gridsearch calculations is that they are computationally intensive, and rapidly become prohibitively lengthy if more than two internal coordinates are varied. Hence it is impossible to utilise this approach in order to derive the minimum energy configuration for a large oligosaccharide.

2.2.4. Simulated annealing

Given the computational expense of gridsearch methods, it is not surprising that alternative strategies have been explored for the derivation of global minimum energy configurations for oligosaccharides. One technique which has been utilised with great effect in conformational studies of proteins is dynamical simulated annealing [17]. This method is a novel application of molecular dynamics simulations (Section 3.1.1), whereby the molecule in question is equilibrated with a thermal bath at high temperature (such as ~500 K), and is then slowly cooled to a very low temperature (such as ~5 K), and finally is minimised to the 'global minimum' structure at zero Kelvin. As the molecule is slowly cooled, it experiences many local minima, into any of which it would be entrapped in conventional minimisation. However, there is a finite probability that it will escape these local minima during simulated annealing, which further increases the probability of locating the global minimum. This approach is not perfect, but is a very significant improvement on conventional minimisation, and can in principle be applied to an oligosaccharide of almost arbitrary complexity [18]. As an example, Fig. 3 shows nine initial random geometries for the disaccharide Manα1-3Man

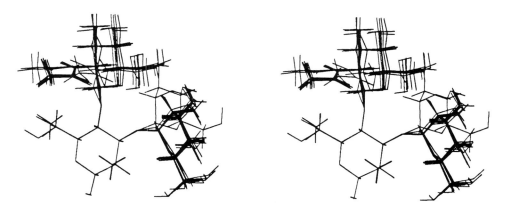

Fig. 3. Stereo view of the results from simulated annealing of nine random starting geometries of Man(α1-3)Man. These starting geometries were identical to those used in the conventional minimisation of Fig. 1.

which have each been minimised using a simulated annealing strategy. The initial nine structures are identical to those used in the generation of Fig. 1, and, in contrast to the data in Fig. 1, it is seen that all but one of these have annealed to a single family of structures with similar φ, ψ values.

2.2.5. Molecular dynamics simulations

Early computational studies on oligosaccharides using HSEA calculations led to the notion that oligosaccharides were essentially 'rigid' entities in solution, with little conformational freedom [1,3–9]. For some time it was believed that this was a true representation of the conformational behaviour of oligosaccharides in solution, since the resulting rigid geometries agreed very well with experimentally measured NOEs [19–21]. However, with the application of full molecular mechanical forcefields and consequent full optimisation of geometry, studies over the last few years have suggested that considerable torsional flexibility exists about most glycosidic linkages [22–27]. Such torsional freedom is implied from the adiabatic map of Manα1-3Man in Fig. 2. On the assumption that ~99% of molecules are bounded by the 5 kcal/mol contour at physiological temperature, then φ and ψ would be predicted to explore significant regions of conformational space under this condition. An efficient means by which the regions of conformational space available to an oligosaccharide can be determined at a given temperature is by use of molecular dynamics simulations [25,28–33]. Since the molecule is treated classically in molecular mechanics calculations, it is possible to determine the motional properties of the system given suitable initial conditions from Newton's laws of motion, and the effects of temperature can be included by coupling to a thermal bath. The advantage of the MD approach is that it can be applied to much larger structures than can be handled with gridsearch calculations, and it is also possible to include distance and angular constraints derived from NOE data (Section 2.3.1) or spin–spin coupling constants (Section 2.3.2) in the form of 'pseudo-energy' terms.

2.2.6. Calculations in vacuo vs. explicit inclusion of solvent

The discussion thus far has been based upon the tacit assumption that all calculations are performed in vacuo. Clearly this is a very poor approximation to the physiological situation where the glycan is surrounded by solvent. Calculations in vacuo are nevertheless attractive in view of the relative speed with which they may be performed.

The principal problem associated with in vacuo calculations is associated with Coulombic terms: in aqueous media there exists a distance dependent attenuation of these due to the dielectric strength of the medium. In vacuo, the dielectric constant is strictly unity, and hence the Coulombic terms (including hydrogen bonds) will appear to be much stronger than they actually are. These 'electrostatic effects' are particularly important in studies on oligosaccharides bearing a nett charge (such as sialic acids, sulfated or phosphorylated glycans), when distal parts of the molecule may artefactually become strongly associated. For this reason, the dielectric constant is usually set to 80.0, in order to simulate in part the influence of solvent water.

An alternative strategy is to include solvent water molecules explicitly into the simulation. While such simulations have been performed upon simple mono- di- and trisaccharides [13,29,30,33] which have given important insight into the nature of hydrogen bonding in oligosaccharides, even for molecules this small there is a dramatic increase in computational time since it is necessary to simulate the system with a very large number of water molecules. In addition, explicit solvent molecules tend to damp torsional oscillations in oligosaccharides, resulting in the need for prohibitively long simulations in dynamics studies [13,14] (also see Section 3.1.1). At present, therefore, simulations *in vacuo* remain the only realistic option for large oligosaccharides.

2.3. Nuclear magnetic resonance

NMR is currently the only available technique for the conformational analysis of oligosaccharides in solution at high-resolution. In essence, NMR can provide distance constraints and angular information via the Nuclear Overhauser Effect and spin–spin coupling constants, respectively.

2.3.1. Nuclear Overhauser effects

The relaxation properties of two nuclei which are NMR active are not necessarily independent of one another, under appropriate conditions, and in particular when the two nuclei are close in space (≤ 5 Å), the relaxation can be described in terms of two coupled differential equations [34]. The value of the NOE for conformational studies is that, as a result of this coupling, the NOE is dependent upon the inverse-sixth power of the distance between the two nuclei [35,36], and hence the observation of an NOE between two protons demonstrates that they approach within 5 Å of each other for a finite time. A complication in the application of the NOE for conformational studies is that its magnitude is dependent not only upon distance, but also upon the Larmor precession frequencies of the nuclei (ω), and the overall tumbling time (correlation time, τ_c) of the parent molecule. This can be overcome to a large extent by considering not the absolute magnitude of the NOE, but the relative NOE between two proton pairs, where one pair of nuclei are at a known, fixed distance with respect to each other. The latter could be the H-1 and H-2 protons within a given pyranose residue, for example. In this manner it is possible to derive a set of through-space distance constraints for a given macromolecule, and if they are sufficient in number, a solution structure can be derived from them . This approach has been utilised for many years in the conformational analysis of proteins and nucleic acids [37]. In the case of oligosaccharides, however, this approach presents two major difficulties. First, the number of distance constraints which can be obtained for an oligosaccharide is very small. Often only one or two NOEs are measurable about the primary sites of conformational variance, the glycosidic linkages, whereas it is widely recognised in conformational studies of proteins that at least ten NOE constraints are required per residue in order to generate a 'good' solution structure. The second difficulty, inextricably linked to the first, is the question of internal motion. The NOE is in principle a very precise tool when the distances between nuclei does not vary, but in the presence of internal motion (see Section 2.2.6), it is impossible to utilise the NOE in a quantitative manner by virtue of the r^{-6} dependence on distance, and attempts to do so may lead to the generation of 'virtual' conformations [23].

2.3.2. Spin–spin coupling constants

In addition to through-space coupling via the NOE, two nuclei may also couple to each other in a through-bond manner via bonding electrons. In contrast to the NOE, this interaction is directly observable in the conventional NMR spectrum in the form of multiplet splittings. Importantly, in the case of a three-bond coupling, the magnitude of the spin–spin coupling constant (J) can be empirically related to the dihedral angle formed between the outermost bonds in terms of the Karplus equation. Currently, appropriate parametrisations exist for three bond proton–proton spin–spin couplings [38] and for three bond ^{1}H–^{13}C (so-called long-range heteronuclear) spin–spin couplings [39,40].

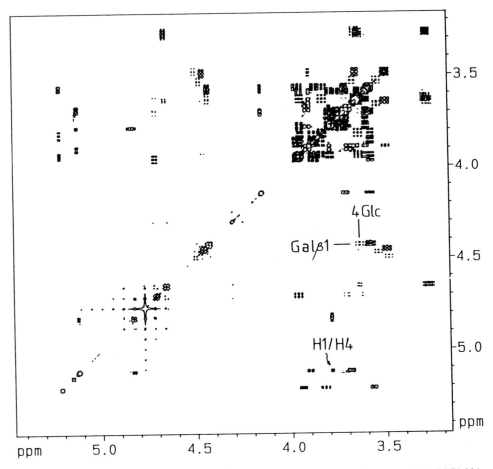

Fig. 4 Selected region from the COSY spectrum at high resolution for Gal(β1-4)[Fuc(α1-3)]GlcNAc (β1-3)Gal(β1-4)Glc. The crosspeak correlating the H-1 proton of fucose with H-4 (labelled H1–H4) can only arise from a five-bond proton–proton coupling, and a four-bond coupling across the Gal(β1-4)Glc glycosidic linkage may readily be observed as indicated. The digital resolution utilised in this spectrum was 0.28 Hz/point in each dimension.

Three-bond proton–proton coupling constants are primarily of value in the determination of the ring geometries of monosaccharides, since the dihedral angles formed by adjacent C–H bonds depend upon ring configuration. It is not widely appreciated that proton–proton couplings can extend to four bonds or more in oligosaccharides. Their magnitude is difficult to quantify, but their presence can readily be demonstrated by acquisition of a ^1H–^1H COSY spectrum at high resolution, as shown in Fig. 4 for the pentasaccharide Gal(β1-4)[Fucα1-3]GlcNAc(β1-3)Gal(β1-4)Glc. The crosspeak correlating the H-1 proton of fucose with H-4 can only arise from a five-bond proton–proton coupling, and four-bond couplings across the glycosidic linkage may readily be observed as indicated.

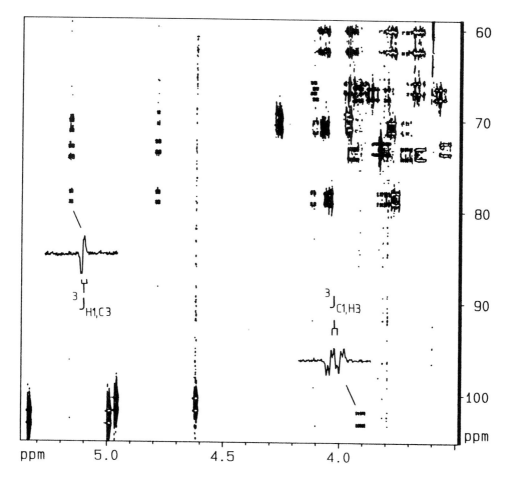

Fig. 5. HSQC spectrum for Man(α1-3)Man(α1-OMe), with (inset) expanded views of the crosspeaks correlating Manα1-H1 to -3Man-C3, and Manα1-C1 to -3Man-H3, to illustrate the method of measuring the long-range ($^3J_{CH}$) couplings between these nuclei.

Three-bond ¹H–¹³C couplings are of particular value in the determination of gly-cosidic torsion angles [41–43], notably the couplings between the anomeric proton and aglyconic carbon, and the anomeric proton and aglyconic carbon, which respectively are a direct measure of φ and ψ. The importance of such long-range couplings across the glycosidic linkage should not be underestimated, since they provide additional conformational restraints. Although, like the NOE, they cannot be used in a quantitative manner in the presence of internal motion, measured Js represent a rather kind of average than NOEs, and are hence highly complementary. Long-range couplings have not enjoyed popularity in the past, principally because techniques for their facile measurement have not existed. However, with the advent of heteronuclear long-range single quantum correlation (HSQC) methods [44], and provided that sufficient material is available (several milligrams), their measurement is relatively straightforward. As an example, we illustrate an HSQC spectrum for 5 mg of Man(α1-3)Man in Fig. 5. The long-range heteronuclear coupling constants corresponding to φ and ψ are readily detected from the antiphase splittings as shown. In this particular example, the magnitude of these long-range couplings alone gives important qualitative informa-tion on the possible extent of variation of φ and ψ, by consideration of the appropriate Karplus curve [39,40] (Fig. 6). Assuming the majority of populations exist within the 5 kcal/mol contour in the gridsearch of Fig. 2 (see Section 2.2.5), then it follows that the possible values of φ and ψ extend to ~ ±90°. However, the maximum value of the long-range heteronuclear spin–spin coupling within this range is 5.7 Hz, and since the measured values of the coupling corresponding to φ and ψ are 4.0±0.5 and 5.1±0.5 Hz respectively, then populations must be heavily weighted in favour of regions of conformational space where $^3J_{C,H}$ is large. Incidentally, these data also rule out the possibility that φ is restricted to the region ±60°, as predicted by simple HSEA calculations.

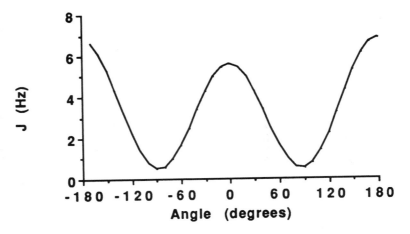

Fig. 6. Karplus curve parametrised for long-range ¹³C–¹H ($^3J_{CH}$) coupling constants across glycosidic linkages [39,40].

3. Recent results

3.1. Are oligosaccharides 'rigid' or 'flexible'?

Despite convincing early data suggesting that oligosaccharides are rather 'rigid' structures in solution, gridsearch calculations such as that shown in Fig. 1 strongly suggest that a degree of flexibility is present about the glycosidic linkage at physiological temperatures. However, the fact remains that early NOE data were interpreted on the assumption of a single, rigid conformation, apparently with complete consistency [20,21]. In order to rationalise these data it is therefore necessary to assume that an alternative model (or models) exist including internal motion which is equally consistent with the experimental data.

3.1.1. Computation of theoretical NOEs from gridsearch calculations

One means by which an alternative model can be sought is to compute theoretical NOE values from the adiabatic map with knowledge of the appropriate Boltzmann weighted populations over the relevant region of the potential surface [23,26]. Unfortunately, to date no such study has given convincing correspondence with experimental NOEs. The possible reasons for this are several fold, but the most likely explanation is that forcefields currently employed for conformational analysis of oligosaccharides are simply not accurate enough over the complete potential surface to allow the computation of theoretical NOEs. In this regard it must be remembered that any small error in distance derived from the potential surface will be magnified by the sixth power due to the distance dependence of the NOE.

3.1.2. Computation of theoretical NOEs from molecular dynamics simulations

An alternative approach is to compute theoretical NOEs from molecular dynamics simulations [18,25]. The advantage of this approach is that an estimate of the timescale of the internal motions can be obtained directly from the simulation. This knowledge is crucial to the way in which the theoretical NOEs are computed. In the case where the rate of internal motion is slow with respect to overall tumbling, the average NOE is simply an r^{-6} mean weighted by the relative populations over the potential surface [45]. In contrast, when the rate of internal motion is fast with respect to overall tumbling, as is shown to be the case from MD simulations, then the extraction of average NOE data is a much more complicated function involving angular terms (second rank spherical harmonics [45]). In our own recent studies on the Lewis-x blood group oligosaccharide [18], we used this latter approach to compute theoretical NOEs from a molecular dynamics simulation with explicit inclusion of solvent water, and theoretical NOEs were found to agree well with experimental data. Thus a rather schizophrenic situation exists whereby experimental NOE data can be interpreted both in terms of rigid geometries for oligosaccharides or equally satisfactorily by assuming a degree of torsional oscillation about glycosidic linkages.

3.1.3. Results from spin–spin coupling constant data

A possible reason for this apparent paradox is the dearth of NOE constraints in oligosaccharides, as mentioned previously. In cases where only one NOE is measurable

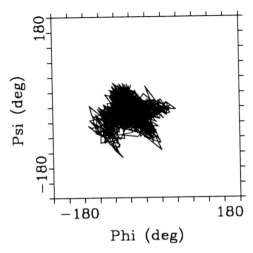

Fig. 7. *In vacuo* MD simulation for the disaccharide Man(α1-3) Man. The simulation was run for 510 ps real time, and the plot shows φ vs. ψ for the last 500 ps of the simulation. Distance restraints were used in the simulation corresponding with the three measurable NOEs across the glycosidic linkage. For further details, see Ref. [46].

across the glycosidic linkage, it is not difficult to imagine how the data can be consistent with two apparently conflicting models, since the system is underconstrained. In our own recent work, we sought to distinguish between the 'rigid' and 'flexible' models by examining a model disaccharide for which a significant number of experimental constraints could be obtained [46]. We chose the disaccharide Man(α1-3)Man, for which three NOE constraints and two heteronuclear coupling constants could be observed across the glycosidic linkage. These were incorporated in a restrained simulated annealing strategy [18] in order to determine whether these five experimental constraints were consistent with a single structure. Within experimental error, all five constraints could be incorporated into a 'rigid' model with a conformation defined by $\varphi, \psi \sim -40°$, $+20°$, which is similar to that defined previously for the same linkage [25]. We then performed a series of molecular dynamics simulations on the disaccharide, and calculated theoretical NOE and long-range $^{13}C-^{1}H$ coupling constants from them, and again within experimental error the simulated data were consistent with experiment. However, the MD simulations demonstrated significant torsional oscillations about the glycosidic linkage (Fig. 7), from which we concluded that the 'flexible' model was the correct interpretation of the data. Additional supporting evidence was obtained from ^{13}C relaxation-time measurements (Section 3.1.4).

3.1.4. Relaxation-time measurements
In an effort to demonstrate directly the presence of internal motions in oligosaccharides, several groups have utilised ^{1}H and ^{13}C relaxation time measurements [27,47–50]. The relaxation parameters T_1 and T_2 are sensitive to molecular motions if these are

on the correct timescale. The primary difficulty with these parameters is that, even if they can be measured to good accuracy, they must be interpreted in terms of a particular motional model. This limitation can be overcome, however, if the data are interpreted in the framework of the model-free approach of Lipari and Szabo [51]. In this way, two parameters are generated which characterise the internal motion, namely the internal correlation time, τ_e, which expresses the rate of internal motion, together with a generalised order parameter, S^2, which expresses the extent of motion. For a completely rigid residue $S^2 = 1$, whereas a completely mobile residue has $S^2 = 0$. Thus it is possible, in principle, to measure the extent of intramolecular motion directly by measurement of T_1 and T_2.

Results from ^1H and ^{13}C relaxation time measurements upon oligosaccharides have to date allowed the characterisation of oligosaccharide dynamics with variable success [27,47–50]. A problem with all relaxation measurements on isolated oligosaccharides is that the internal motions are often strongly coupled to the overall rotational tumbling of the molecule, under which conditions the formalism of Lipari and Szabo is not strictly applicable. Furthermore, it becomes impossible to distinguish between 'backbone' and 'sidechain' atoms when significant torsional oscillations exist about each glycosidic linkage. The net effect is that relaxation parameters tend to similar values and may become insensitive to the extent of internal motion.

One means by which these problems can be overcome is to tether the oligosaccharide in some manner to a large mass which can be described as a 'rigid' backbone on the timescale of measurement. The most convenient system in this regard is to examine the dynamics of an oligosaccharide in attachment to protein. Such studies have been performed in the past by Berman et al. on ribonuclease B [52], a glycoprotein of molecular mass ~12 kDa and with a single N-linked oligomannose glycan attached at Asn 34. These authors chose to measure ^{13}C T_1 relaxation times for the glycan moiety of this glycoprotein. The reason for this choice, rather than more sensitive measurements of proton relaxation times, is first that the glycan resonances are reasonably well resolved from those of the protein in the carbon spectrum, and most importantly, the relaxation of ^{13}C is to a good approximation dominated by the directly bonded proton, which considerably simplifies data analysis. Unfortunately, with the techniques available at that time, Berman et al. were unable to obtain quantitative relaxation data due to insufficient sensitivity. Recently, we have acquired quantitative T_1 and T_2 relaxation data for ribonuclease B using modern 'inverse-detected' ^{13}C relaxation time measurements, which has enabled us to obtain values of τ_e and S^2 for the glycan [46]. These data indicate that the GlcNAc residues of the core are relatively restricted in motion, and possess an internal correlation time which is on the order of the protein backbone ($\tau_e \sim 4.3 \times 10^{-9}$ s). In contrast, residues distal to the protein are much less restricted ($S^2 \sim 0.1–0.2$), and undergo internal reorientation an order of magnitude faster ($\tau_e \sim 0.18 \times 10^{-9}$). It is possible to rule out the notion that these motional differences are a result of steric restraints by the protein, in view of the fact that the ^1H and ^{13}C chemical shifts of the glycan in attachment to protein are identical within experimental error to those of the free glycan [52]. Thus, these data appear to provide unequivocal evidence that oligosaccharides are not 'rigid' entities, but undergo significant torsional oscillations about the glycosidic linkages. However, the relatively large S^2 values for residues close to the protein suggest that the extent of reorientation is not large, and the low S^2 values for residues distal from the protein derive from the fact that these are appended by several glycosidic linkages, each of which is undergoing reorientation. A likely scenario is that torsional oscillations about the majority of glycosidic linkages are similar to those shown in Fig. 7.

3.2. Specific examples

3.2.1. N-linked glycans

The most intensive conformational studies to date have focused upon N-linked glycans [7,9,13,14,19–25], not from the implication that these oligosaccharide types are more important than any other, but since they have historically been more readily obtainable in quantities sufficient for high-resolution NMR studies.

As mentioned above, earlier work treated the majority of glycosidic linkages in these oligosaccharides as rigid. However, even at that time it was recognised that oligosaccharides possess regions of flexibility, notably about the Manα1-6Man glycosidic linkage. It is now well accepted that this flexibility derives not only from the inherent torsion freedom of the glycosidic linkage proper, but also from the fact that several rotamers exist about the C-5–C-6 bond of the aglycon. The precise distribution of these rotamers depends upon the oligosaccharide type, but for example the predominant rotamers in a diantennary glycan such as that derived by human serotransferrin are defined by $\omega = 180°$ and $\omega = -60°$, with a marked predominance for the latter if the correct statistical analysis is performed [24]. Conversely, the $\omega = 180°$ rotamer is predominant in bisected diantennary glycans [21].

The extent of torsional variation of other glycosidic linkages is still a subject of debate. While there is little opposition to the notion that torsional variations exist, these have been predicted to be very small as in the case of blood group substances [18,53,54] (also see Section 3.2.2), to highly flexible with torsional variations on the order of $\pm70°$ and $\pm100°$ in φ and ψ respectively, for the Manα1-3Man linkage [55]. Recent work has attempted to rationalise these differences by defining a model for internal motion which is consistent with *all* experimental data, i.e. NOEs and $^3J_{CH}$ values [46]. The approach was to perform a series of molecular dynamics simulations, with NOE restraints imposed as biharmonic terms with suitable upper and lower bounds. This approach was adopted in preference to free dynamics simulations, since as mentioned above, currently available forcefields for oligosaccharides are of insufficient accuracy to predict accurately the conformational space mapped by the molecule. While it might be thought that this approach by definition constrains the resulting motional model to agree with experimental data, it should be pointed out that the distance bounds applied are very broad, in complete analogy to those utilised in structural studies of proteins (see Ref. [46] for details). Theoretical NOE and $^3J_{CH}$ values computed from these simulations were found to agree very closely with the experimentally determined values, suggesting that the motional properties of the system defined by these simulations may accurately reflect the actual dynamic properties of the glycan. As an example, the extent of torsional variation for each of the glycosidic linkages of the

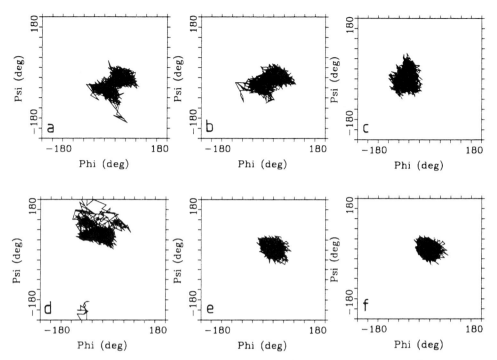

Fig. 8. Restrained MD simulation in vacuo for GlcNAc(β1-2)Man(α1-6)[GlcNAc(β1-2)Man(α1-3)]Man(β1-4)GlcNAc(β1-4)GlcNAc. The simulation was run for 510 ps real time, and each panel illustrates φ vs. ψ for each glycosidic linkage over the last 500 ps of the simulation. Panels refer to the following glycosidic linkages: (a) GlcNAc(β1-4)GlcNAc; (b) Man(β1-4)GlcNAc; (c) Man(α1-3)Man; (d) Man(α1-6)Man; (e) GlcNAc(β1-2)Man (3-antenna) (f) GlcNAc(β1-2)Man (6-antenna).

diantennary glycan GlcNAc(β1-2)Man(α1-6)[GlcNAc(β1-2)Man(α1-3)]Man(β1-4) GlcNAc(β1-4)GlcNAc computed using this procedure are shown in Fig. 8. The largest degrees of torsional oscillations are exhibited by the Manα1-6Man linkage, which appears to adopt two discrete conformational states, whereas the remaining linkages appear to exhibit torsional variations of up to ~ ±90°. It should be emphasised, however, that the major population of conformers is restricted to a region of conformational space significantly smaller than this, as is required by consideration of experimental $^3J_{CH}$ values (Section 2.3.2).

3.2.2. Blood group oligosaccharides

The conformational analysis of blood group oligosaccharides is currently of great interest in view of the roles of members of this class of glycan in cell adhesion.

A great deal of data has appeared recently on these moieties from the work of Bush and co-workers [53,54,56,57] and others [18,58]. A universal finding appears to be that blood group oligosaccharides A, H, Le[a], Le[b], and Le[x] represent a group of comparatively inflexible glycans. As an example, the torsional variations of φ and ψ for the Le[a]

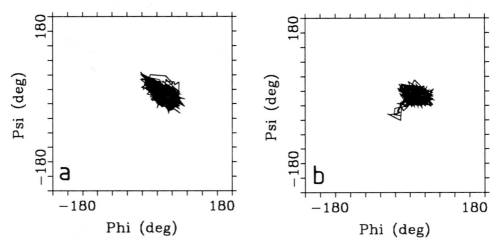

Fig. 9 Torsional variations of φ and ψ for (a) Fuc(α1-4) GlcNAc, and (b) Gal(β1-3)GlcNAc derived from the final 500 ps of a 510 ps restrained dynamics simulation *in vacuo* for the Le^a blood group.

blood group are illustrated in Fig. 9. These plots were generated using the same protocols as those employed in the generation of the data of figure 8, and by comparison it is clearly seen that Le^a is indeed a rather rigid entity in solution. An interesting finding is that the Le^a and Le^x structures can essentially be superimposed if the former is rotated by 180° about a vector normal to the C3–C4 bond of the GlcNAc ring. This reveals that the conformations of the two molecules are very similar with the fucose and galactose rings tightly stacked, thus accounting for the relative rigidity of the molecule. Interestingly, The Le^a determinant does not in general cross-react with Le^x, although instances do exist [59]. However, sialylated Le^a reacts with ELAM-1 and with monoclonal antibodies for sialylated Le^x [58]. Thus, determinants which do not cross-react must be comprised of variant regions of these glycans. In this regard, the main difference between Le^a and Le^x lies in the orientation of the GlcNAc rings such that the GlcNAc C-2 substituent of Le^a extends into the same region of space as the GlcNAc C-6 substituent of Le^x, and vice versa.

3.2.3. Glycolipids

In common with blood group substances, there has been an upsurge of interest in the solution conformations of glycolipids [60,61]. These moieties act as ligands to a variety of toxins produced by bacterial pathogens, and crystallographic data are now available for the carbohydrate-binding subunits of these toxins [63,64], in one case with carbohydrate located in the binding site [64].

Most studies on the glycan moieties of glycolipids have involved the complete molecule with lipid covalently attached, and the critical micellar concentration of these is usually very low. Hence most studies to date have involved glycolipids dissolved in aprotic solvent such as DMSO [60,61]. This approach carries the important additional advantage that the glycan can be investigated in protonated form, under which conditions the hydroxyl proton resonances can be used as additional 'long-range sensors' in conformational analysis.

In general, results of conformational studies on glycolipids provide similar conclusions to those obtained for N-linked glycans, i.e. that significant but limited torsional oscillations exist about glycosidic linkages. In a detailed study of globotriaosylceramide: Gal(α1-4)Gal(β1-4)Glc(β1-1)Cer in DMSO solution, Poppe et al. [61] reported two energy minima for the Gal(α1-4)Galβ linkage at $\varphi,\psi = -11°$, $32°$, and $-34°$, $-7°$, respectively, and similarly for the Gal(β1-4)Glcβ linkage at $\varphi,\psi = +14°$, $-30°$, and $+34°$, $+4°$. To our knowledge, detailed molecular dynamics simulations on glycolipids have not been undertaken, and hence it is difficult to determine whether the dynamic properties of these moieties will reflect those of N-linked glycans, or will adopt more rigid conformations like the blood group substances, but available evidence such as that described above suggest a dynamic behaviour more like that of N-linked glycans.

4. Conclusions

Early data suggested that oligosaccharides adopt rather rigid conformations in solution. While this was a perfectly reasonable assumption at the time, more recent data obtained with more powerful theoretical and experimental tools has forced us to conclude that oligosaccharides in fact exist in solution with significant but limited motional flexibility. An important recent finding has been that available NOE and spin-coupling constant data are, within experimental error, equally consistent with both the 'rigid' and 'flexible' models, for certain oligosaccharides at least. An element of caution is therefore required in assuming an oligosaccharide is 'flexible' or 'rigid', and it may only be possible to distinguish between the two by relaxation-time measurements.

The observation that oligosaccharides are flexible raises interesting questions regarding their roles as ligands. On the one hand, the configurational entropy gained by this flexibility will reduce the binding energy associated with the carbohydrate-protein interaction, which may explain why a large number of such interactions are of relatively low affinity. On the other hand, conformational flexibility in the ligand will lead to increased 'on' rates.

Given that a unique solution conformation cannot be defined for an oligosaccharide, the question has been posed: how can oligosaccharides act as recognition signals?. The implication is that recognition specificity cannot be maintained in a non-rigid ligand. In the opinion of the author this is a sterile argument, since small peptides, which sample regions of conformational space immeasurably greater than oligosaccharides of similar size, are able to achieve exquisite specificity. Indeed, results from x-ray crystallography of oligosaccharide–protein complexes [65–70] suggest that recognition specificity is maintained by the stereochemistry of the composite monosaccharide residues, rather than by the gross morphology of the glycan.

Acknowledgements

I thank the Lister Institute of Preventive Medicine for the award of a Centenary Research Fellowship.

Abbreviations

τ_c	correlation time
DMSO	dimethyl sulphoxide
HSEA	hard sphere exo-anomeric effect
HSQC	heteronuclear single quantum correlation
J	spin–spin coupling constant
MD	molecular dynamics
NMR	nuclear magnetic resonance
NOE	Nuclear Overhauser Effect
S^2	order parameter
T_1	spin-lattice relaxation time
T_2	spin–spin relaxation time

References

1. Homans, S.W. (1994) In: M. Fukuda (Ed.), Molecular Glycobiology, Frontiers in Molecular Biology Series, Oxford University Press, pp. 230–257.
2. Cambillau, C., (1995). In: J. Montreuil, H. Schachter and J.F.G. Vliegenthart (Eds.) Glycoproteins. New Comprehensive Biochemistry, Vol. 29a. Elsevier, Amsterdam. pp. 29–65.
3. Tvaroska, I. and Bleha, T. (1989) Adv. Carbohydr. Chem. Biochem. 45.
4. Lemieux, R.U., Delbaere, L.T.J., Koto, S. and Rao, V.S. (1980) Can. J. Chem. 58, 631.
5. Thogersen, H., Lemieux, R.U., Bock, K. and Meyer, B. (1982) Can. J. Chem. 60, 44.
6. Sabesan, S., Bock, K. and Lemieux, R.U. (1984) Can. J. Chem. 62, 1034.
7. Paulsen, H., Peters, T., Sinnwell, V., Heume, M. and Meyer, B. (1986) Carbohydr. Res. 156, 87.
8. Montreuil, J. (1984) Biol. Cell 51, 115.
9. Bock, K., Arnarp, J. and Lonngren, J. (1982) Eur. J. Biochem. 129, 171.
10. French, A. (1989) Carbohydr. Res. 188, 206
11. Scarsdale, J.N., Ram, P., Prestegard, J.H. and Yu, R.K. (1988) J. Comput. Chem. 9, 133.
12. Ha, S.N., Giammona, A., Field, M. and Brady, J.W. (1988) Carbohydr. Res. 180, 207.
13. Homans, S.W. (1990) Biochemistry 29, 9110.
14. Edge, C.J., Singh, U.C., Bazzo, R., Taylor, G.L., Dwek, R.A. and Rademacher, T.W. (1990) Biochemistry 29, 1971.
15. Jeffrey, G.A. and Taylor, R. (1980) J. Comput. Chem. 1, 99.
16. Rasmussen, K. (1982) Acta Chem. Scand. A 36, 323.
17. Nilges, M., Gronenborn, A.M. and Clore, G.M. (1988) FEBS Lett. 229, 317.
18. Homans, S.W. and Forster, M. (1992) Glycobiology 2, 143.
19. Brisson, J-R. and Carver, J.P. (1983) Biochemistry 22, 1362.
20. Brisson, J-R. and Carver, J.P. (1983) Biochemistry 22, 3671.
21. Brisson, J-R. and Carver, J.P. (1983) Biochemistry 22, 3680.

22. Homans, S.W., Dwek, R.A., Boyd, J., Mahmoudian, M., Richards, W.G. and Rademacher, T.W. (1986) Biochemistry 25, 6342.

23. Cumming, D.A. and Carver, J.P. (1987) Biochemistry 26, 6664.

24. Cumming, D.A. and Carver, J.P. (1987) Biochemistry 26, 6676.

25. Homans, S.W., Pastore, A., Dwek, R.A. and Rademacher, T.W. (1987) Biochemistry 26, 6649.

26. Imberty, A., Tran, V. and Perez, S. (1989) J. Comput. Chem. 11, 205.

27. Poppe, L. and van Halbeek, H. (1992). J. Am. Chem. Soc. 114, 1092.

28. Yan, Z-Y. and Bush, C.A. (1990) Biopolymers 29, 799.

29. Brady, J.W. (1986) J. Am. Chem. Soc. 108, 8153.

30. Brady, J.W. (1987) Carbohydr. Res. 165, 306.

31. Pollex-Kruger, A., Meyer, B., Stuike-Prill, R., Sinnwell, V., Matta, K.L. and Brockhausen, I. (1993) Glycoconj. J. 10, 365.

32. Dauchez., M., Mazurier, J., Montreuil, J., Spik, G. and Vergoten, G. (1992) Biochimie 74, 63–74.

33. Madsen, L.J., Ha, S.N., Tran, V.H. and Brady, J.W. (1990) In: A.D. French and J.W. Brady (Eds.), Computer Modeling of Carbohydrate Molecules, ACS Symposium Series.

34. Solomon, I. (1955) Phys. Rev. 99, 559.

35. Noggle, J.H. and Schirmer, R.E. (1971). The Nuclear Overhauser Effect: Chemical Applications. Academic Press, New York.

36. Neuhaus, D. and Williamson, M.P. (1991). The Nuclear Overhauser Effect.

37. Wuethrich, K. (1986) NMR of Proteins and Nucleic Acids. Wiley, New York.

38. Haasnoot, C.A.G., De Leeuw, F.A.A.M. and Altona, C. (1980) Tetrahedron 36, 2783.

39. Mulloy, B., Frenkiel, T.A. and Davis, D.B. (1988). Carbohydr. Res. 184, 39.

40. Tvaroska, I., Hricovini, M., and Petrakova, E. (1989) Carbohydr. Res. 189, 359.

41. Hamer, G.K., Balza, F., Cyr, N. and Perlin, A.S. (1978) Can. J. Chem. 56, 3109.

42. Poppe, L. and van Halbeek, H. (1991) J. Magn. Reson. 92, 636.

43. Poppe, L. and van Halbeek, H. (1991) J. Magn. Reson. 93, 214.

44. Norwood, T.J., Boyd, J., Heritage, E.J., Soffe, N. and Campbell, I.D. (1990). J. Magn. Reson. 488.

45. Tropp, J. (1980) J. Chem. Phys. 72, 6035.

46. Rutherford, T.J., Partridge, J., Weller, C.T. and Homans, S.W. (1993) Biochemistry 32, 12715

47. McCain, D.C. and Markley, J.L. (1986) J. Am. Chem. Soc. 108, 4259.

48. Kadkhodaei, M.M., Wu, H. and Brant, D. (1991). Biopolymers 31, 1581.

49. Cagas, P. and Bush, C.A. (1992). Biopolymers 32, 277.

50. Weller, C.T., McConville, M.J. and Homans, S.W. (1994) Biopolymers, in press.

51. Lipari, G. and Szabo, A. (1982) J. Am. Chem. Soc. 104, 4546.

52. Berman, E., Walters, D.E. and Allerhand, A. (1981) J. Biol. Chem. 256, 3853.

53. Yan, Z-Y. and Bush, C.A. (1990). Biopolymers 29, 799.

54. Miller, K.E., Mukhopadhyay, C., Cagas, P. and Bush, C.A. (1992). Biochemistry 31, 6703.

55. Carver, J.P., Mandel, D., Michnick, S.W., Imberty, A. and Brady, J.W. (1990). In: A.D. French and J.W. Brady (Eds.), Computer Modelling of Carbohydrate Molecules. ACS, Washington, pp. 266–280.

56. Cagas, P., Kaluarachchi, K. and Bush, C.A. (1991). Biochemistry 113, 6815.

57. Bush, C.A. and Cagas, P. (1992). Adv. Biophys. Chem. 2, 149–180.

58. Berg, E.L., Robinson, M.K., Mansson, O., Butcher, E.C. and Magnani, J.L. (1991) J. Biol. Chem. 266, 14869.

59. Feizi, T. (1991) Trends Biochem. Sci. 16, 84.

60. Poppe, L. and Dabrowski, J. (1989). J. Am. Chem. Soc. 111, 1510.

61. Poppe, L., Dabrowski, J., Lieth, C-W., Koike, K. and Ogawa, T. (1990). Eur. J. Biochem. 189, 313.

62. Scarsdale, J.N., Ram, P., Prestegard, J.H. and Yu, R.K. (1988). J. Comput. Chem. 9, 133.
63. Stein, P.E., Boodhoo, A., Tyrrell, G.J., Brunton, J.L. and Read, R.J. (1992) Nature 355, 748.
64. Sixma, T.K., Pronk, S.E., Kalk, K.H., van Zanten, B.A.M., Berghuis, A.M. and Hol, W.G. (1992) Nature 355, 561.
65. Bourne, Y., Rouge, P., and Cambillau, C. (1990) J. Biol. Chem. 265, 18161.
66. Yariv, J., Kalb (Gilboa), A.J., Papiz, M.Z., Helliwell, J.R., Andrews, S.J. and Habash, J. (1987) J. Mol. Biol. 195, 759.
67. Derewenda, Z., Yariv, J., Helliwell, J.R., Kalb (Gilboa), A.J., Dodson, E.J., Papiz, M.Z., Wan, T. and Campbell, J. (1989) EMBO J. 8, 2189.
68. Quiocho, F.A. (1986) Ann. Rev. Biochem. 55, 287.
69. Quiocho, F.A., Wilson, D.K. and Vyas, N.K. (1989) Nature 340, 404.
70. Cygler, M., Rose, D.R. and Bundle, D.R. (1991) Science 253, 442.

J. Montreuil, H. Schachter and J.F.G. Vliegenthart (Eds.), *Glycoproteins*

Chemical synthesis of glycopeptides

H. PAULSEN, S. PETERS and T. BIELFELDT

Institute of Organic Chemistry, University of Hamburg,
Martin-Luther-King-Platz 6, 20146 Hamburg, Germany

1. Introduction

Glycoproteins are composed of a polypeptide backbone which is glycosylated with one or more carbohydrate chains. The mucin type glycoproteins, which are containing an α-glycosidic linkage between 2-acetamido-2-deoxy-D-galactopyranose and the side chain hydroxy groups of L-serine or L-threonine, and the proteoglycan type constructed by an O-β-D-xylopyranosyl-(1→O)-L-seryl unit are important among the O-glycoproteins because of their biological relevance. In other glycoproteins the hydroxy groups of L-serine or L-threonine are glycosylated with α-D-mannose, β-D-galactose, β-D-glucose sugars or 2-acetamido-2-deoxy-β-D-glucose. D-Galactose residues have also been found covalently attached to other hydroxy amino acids as for instance 5-hydroxy-L-lysine or 4-hydroxy-L-proline. The core structure of the N-glycoproteins always consists of the 2-acetamido-1-*N*-(L-aspart-4-oyl)-2-deoxy-β-D-glucopyranosyl-amine linkage. The availability of glycopeptides from natural sources is limited because of the low concentration and the microheterogeneity of biological glycoconjugates. The advances in carbohydrate and peptide chemistry in the last decade has created generally applicable methodologies for the synthesis of glycopeptides as model compounds for biochemical and structural investigations. Thus glycopeptides can be made available with variations in the peptide and the carbohydrate part in superior purity and quantity.

The synthesis of glycopeptides requires a combination of synthetic methods from both carbohydrate and peptide chemistry. One of the central problems is the stereoselective connection of the carbohydrate with the amino acid or peptide. Furthermore the O-glycosidic bond is more or less labile under strong acidic or basic conditions. A treatment of O-glycopeptides with strong acids may result in a cleavage of the glycosidic bond, whereas strong bases may lead to a β-elimination of the carbohydrate. This labile character limits the variety of protecting groups that can be applied for glycopeptide synthesis. The protection schemes therefore should be chosen to enable deprotection steps under mild conditions.

Approaches to directly glycosylate a preformed polypeptide with an oligosaccharide have been described only by a few authors. These block synthetic methods were mostly unsuccessful because of the low solubility of peptides in organic solvents which were required for glycosylation reactions. The currently most effective methodology for glycopeptide synthesis is to prepare a suitably protected glycosylated amino acid and to incorporate such a building block into a peptide chain. The protection scheme must afford a selective removal of C- or N-terminal blocking groups so that the terminal carboxy or amino function of the building block can be coupled with another appropriate glycosyl-amino or amino acid. The versatility of this step by step procedure allows the synthesis of glycopeptides with variations in the peptide part and structures containing more than one carbohydrate residue as well. The introduction of the carbohydrate via an appropriate building block is applicable by both solution phase and solid phase techniques.

The synthesis of glycopeptides has been reviewed before. The publication of Garg and Jeanloz [1] considers the literature until 1985. The articles of Kunz [2] and Paulsen et al. [3] summarize the development in this field until 1988.

2. Synthesis of O-glycopeptides

2.1. Synthesis of O-glycopeptides with 2-acetamido-2-deoxy-α-D-galacto-pyranosyl linkage

2.1.1. Glycopeptide synthesis in solution

The construction of the α-glycosidic bond between 2-acetamido-2-deoxy-D-galactose and the hydroxyl group of serine or threonine has proved to be rather difficult. The synthesis of such a 1,2-*cis* glycoside can only be established with a reasonable stereo-selectivity using a sugar donor with a *non* participating neighbouring substituent at the C-2 atom [4]. Appropriate saccharide donors are derivatives of the 2-azido-2-deoxy-D-galactose. The 2-azido group does not participate during the glycosylation reaction and can later easily be converted into a N-acetyl-group [5]. The required derivatives of 2-azido-2-deoxy-D-galactose are available by azidonitration of 3,4,6-tri-O-acetyl-D-galactal 1 [6]. Treatment of the nitrate 2 with sodium acetate in acetic acid gives the tetraacetate 3 [6], which can be converted with titanic tetrabromide into the α-bromide 4. Compound 4 can subsequently be transformed into the more stable, crystalline β-chloride 5 [6–10]. Both halides are suitable glycosyl donors, however the β-chloride has been found to provide a higher yield of α-glycoside [10]. As glycosyl acceptors the serine or threonine derivatives 6 with different C- and N-terminal protecting group schemes may be utilized [9–13]. The connection of 5 with 6 in the presence of silver catalysts (silver carbonate, silver perchlorate, or silver triflate) yields the α-glycosides 7 with more or less α-stereoselectivity. The α/β ratio is dependent on the protecting scheme and reaction conditions (temperature, solvents, etc.). The corresponding β-anomer, which is usually formed in less than 15% yield can be removed by chroma-tographic methods. For the transformation of the azido-glycosylamino acid 7 into the acetamido derivative 8 different methods have been applied. The reduction of 7 with H_2S or with $NiCl_2 \cdot 6H_2O/NaBH_4$ gives the intermediate 2-amino-sugar which is sub-sequently acetylated with acetic anhydride to afford compound 8. Alternatively 7 can be treated with thioacetic acid to yield directly the acetamido derivative 8. The protec-tion schemes of the glycosylamino acids 8 (except 8a) allow a selective removal of either C- or N-terminal blocking groups. These compounds therefore are appropriate building blocks for the elongation of the peptide backbone.

AcO OAc

NaN₃ → NaN_3, [Ce(NO₃)₆]²⁻

1

AcO OAc

N_3 ONO₂

2

AcOH, NaOAc →

AcO OAc

N_3 OAc

3

TiBr₄ →

AcO OAc

N_3 Br

4

NEt₄Cl →

AcO OAc

N_3 Cl

5

+

$$HO-\overset{\displaystyle R}{\underset{}{C}H}-\overset{\displaystyle NHR^1}{\underset{CO_2R^2}{C}H}$$

6

Ag - catalyst →

AcO OAc

$$N_3 \quad O-\overset{R}{C}H-\overset{NHR^1}{\underset{CO_2R^2}{C}H}$$

7

→

AcO OAc

$$AcNH \quad O-\overset{R}{C}H-\overset{NHR^1}{\underset{CO_2R^2}{C}H}$$

8

AcO OAc

$$AcNH \quad O-\overset{R}{C}H-\overset{NHR^1}{\underset{CO_2H}{C}H}$$

9

9a R^1 = Fmoc
9b R^1 = Z
9c R^1 = Aloc

R = H or Me	R^1	R^2
6a, 7a, 8a	Z	Bzl
6b, 7b, 8b	Fmoc	Bzl
6c, 7c, 8c	Fmoc	tBu
6d, 7d, 8d	Z	All
6e, 7e, 8e	Aloc	tBu
6f, 7f, 8f	Fmoc	Phenacyl

Z = $C_6H_5 - CH_2 - O - \overset{O}{\underset{\|}{C}} -$

Aloc = $CH_2 = CH - CH_2 - O - \overset{O}{\underset{\|}{C}} -$

Fmoc = (fluorenyl) $CH_2O - CO -$

Phenacyl = $C_6H_5 - CO - CH_2 -$

Compound **8a** with the N-terminal benzyloxycarbonyl (Z) and C-terminal benzyl (Bzl) protection is suitable for the synthesis of the free glycosylamino acid [10]. The combination of the N-terminal fluoren-9-yl-methoxycarbonyl (Fmoc) group with the benzyl, tBu or phenacyl esters in **8b, 8c** or **8f** allows a selective deprotection of the N^α-amino function. The removal of the Fmoc-group could be performed with secondary amines as for instance morpholine [14] without affecting the base labile glycosidic bond. The hydrogenolysis of the C-terminal benzyl ester in **8b** in presence of the likewise labile Fmoc group is in general not possible, nevertheless compound **8b** could be transformed into the acid **9a** without affecting the Fmoc-group using a palladium catalyst and formic acid as a hydride donor [15,16]. The Fmoc/tBu **8c** or Fmoc/phenacyl **8f** combinations are particularly powerful affording completely orthogonal protection schemes. The deprotection of the tBu ester **8c** could be effected in a quantitative yield with formic acid or trifluoroacetic acid. Under these conditions the O-glycosidic bond is completely stable [11]. The phenacyl ester in **8f** is removable quantitatively with zinc/acetic acid as well [13]. The application of the Z/All combination was reported by Kunz [2]. The allyl group [17] in **8d** and the allyloxycarbonyl group (Aloc) [18,19] in **8e** have been deprotected under almost neutral condition with (PPh₃)₄Pd(0) in the presence of a nucleophile (morpholine, dimedon). The palladium catalyst is sensitive to oxidation so that the reaction must be carried out under an inert atmosphere. The combination Aloc/tBu **8e** has proved to be useful for C-terminal peptide coupling reactions with non-reactive proline derivatives which did not react in the presence of the voluminous Fmoc-group [12]. The C-terminal protecting groups in **8b**, **8c**, **8f**, **8d** and **8e** can be selectively removed to afford the unprotected compounds **9a**, **9b**, and **9c** representing appropriate building blocks for a peptide chain elongation.

10 R^1 = Ac; R^2 = tBu
11 R^1 = H; R^2 = tBu
12 R^1 = Bzl; R^2 = tBu

13

$$Su = \begin{matrix} CH_2-CO \\ | \\ CH_2-CO \end{matrix} N-$$

$$Np = NO_2-\!\!\!\!\!\bigcirc\!\!\!\!\!-$$

14 R^1 = H; R^2 = Su
15 R^1 = Me; R^2 = Np
16 R^1 = H; R^2 = Np

17 R^1 = H; R^2 = Su
18 R^1 = Me; R^2 = Np
19 R^1 = H; R^2 = Np

The protection of the carbohydrate hydroxy groups have preferably been performed with acetates which are removable under relative mild basic conditions without affecting the glycosidic bond. Alternatively, the hydroxy groups of the 2-azido-2-deoxy-D-galactose have been protected with benzyl ethers which can be deprotected in the glycopeptide under mild hydrogenolytical conditions [20]. The benzyl ether introduction has been achieved by transforming compound **4** into the tBu glycoside **10**, subsequent deacetylation to **11** and benzylation under standard conditions [20] to yield **12**. The tBu-glycoside **12** was converted into the 1-OH unprotected derivative **13** with trifluoroacetic acid. It has been demonstrated that **13** is also a suitable glycosyl donor for the connection with hydroxyamino acids. Under *in situ* activation with trifluoro-methanesulfonic acid anhydride, compound **13** reacts with the hydroxyamino acid acceptors **14**, **15** and **16** to yield the building blocks **17**, **18** and **19** [20,21]. The combination of N-terminal Fmoc-protection with the C-terminal succinimidyl or p-nitrophenyl esters is particularly interesting because the C$^\alpha$–carboxylic group of these esters is already activated for a condensation with the N$^\alpha$-amino group of another amino acid. Thus compounds **17**, **18** and **19** can directly be incorporated into a peptide chain. Reduction of the azido group in these compounds is not possible and has to be done after the construction of the glycopeptide [20,21].

Other useful glycosyl donors of 2-azido-2-deoxy-galactose derivatives are thioglycosides [13], trichloroacetimidates [22] and glycosyl fluorides [23]. It should be noted that the tert.-butyloxycarbonyl (Boc) group is not applicable as a N-terminal protecting group during glycosylation reactions because they are prone to rearrangement into glycosylurethanes [24].

In analogy to the building blocks with a monosaccharide, glycosylated amino acids with larger saccharide chains are available. The most important structure is the β-D-Gal-(1→3)-α-D-GalNAc unit. A lot of different strategies for the synthesis of such glycosylamino acids have been described [9,13,25–28]. A successful approach to synthesize a building block with this core structure has been realized as follows [27]: the azidonitrate **2** was treated with sodium in the presence of methanol to give the β-methyl glycoside **20**. A selective benzoylation of **20** yielded the 4,6-*O*-dibenzoylated sugar **21** as the main product. The mono- and tribenzoylated byproducts were separated by chromatography. Compound **21** was glycosylated with the peracetylated galactose donor **22** under promotion of trimethylsilyltrifluoromethane sulfonate (TMS-triflate) to afford the β(1→3)-linked disaccharide **23** in a high yield. Acetolysis of **23** and subsequent reaction with titanic tetrabromide gave the α-bromide **24** which is a suitable donor for the reaction with a hydroxyamino acid. In case of the α-halide **24** the stereoselectivity in the glycosylation reaction is often lower compared to the corresponding monosaccharide.

The reaction of **24** with the amino acid derivatives **6** promoted by silver carbonate or silver perchlorate gives the α-glycosylated amino acid **25** as the main product. The amount of β-anomer formation is dependent on the scheme of protecting groups. The

92

best selectivity is achieved with the combination Z/Bzl **6a** yielding the α-anomer, exclusively [26]. Unfortunately the product **25a** cannot selectively be deprotected in the amino acid part. Reduction of the azido compound **25** with H_2S and subsequent acetylation affords the glycosylamino acids **26**. The protecting schemes in **26b–26f** allow a selective removal of either the C- or N-terminus to give building blocks suitable for glycopeptide synthesis. The combinations Fmoc/tBu **26d** and Fmoc/Phenacyl **29f** have again demonstrated a high versatility [11,13].

The connection of larger oligosaccharides with hydroxyamino acids has been efficiently realized using other glycosyl donors as for instance trichloroacetimidates [29–32], thioglycosides [33] or glycosyl fluorides [23].

The above-mentioned glycosylamino acids can be used for an incorporation into a peptide chain by application of standard peptide chemistry methods. In case of a solution phase glycopeptide synthesis the construction of the peptide chain may start on either the C- or N-terminal side of the glycosylamino acid. The condensation reaction of a C-terminal deblocked glycosylamino acid with the amino group of another amino acid requires the addition of a coupling reagent for activation of the carboxylic component. 2-Ethoxy-1-ethoxycarbonyl-2,2-dihydrochinolin (EEDQ) [34] has proved to be a sufficiently activating reagent. The effectiveness was demonstrated by the synthesis of glycosylated partial structures from interleukin-2 [11]. The synthesis of the glycopeptides **40** and **41** was performed by application of the Fmoc/tBu combination. The preparation of **40** started with the connection of the two C-terminal serine residues. After Fmoc-cleavage the dipeptide **28** was reacted with the C-terminal deblocked glycosylamino acid **27**. The tripeptide **29** was again deblocked on the N-terminal side

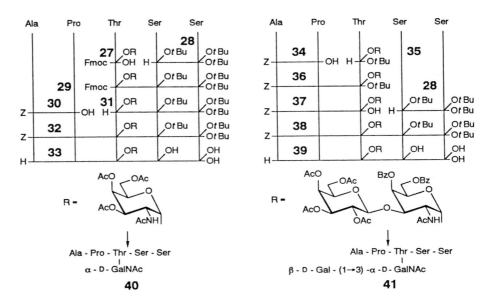

Syntheses of partial structures of interleukin 2

to yield **31** which was subsequently connected with a Z-protected alanyl-prolyl-dipeptide **30** to give the fully protected pentapeptide **32**. After removal of the protecting groups in the peptide part to yield **33** the acetyl groups in the carbohydrate moiety were split off with a catalytic amount of sodium methoxide in methanol to yield **40**. The *O*-acetyl groups are also removable with other bases as for instance hydrazine hydrate in methanol, or potassium cyanide in methanol. These mild conditions do not lead to β-elimination of the carbohydrate part.

The synthesis of the glycopeptide carrying the disaccharide unit was achieved in a similar manner. In this case the glycosylamino acid **35** was first reacted on the N-terminal side with the Z-Ala-Pro dipeptide **34**. The glycotripeptide **36** with the protecting group pattern Z/tBu could also be deblocked selectively on the C-terminus to give **37** which was condensed with the seryl–seryl–dipeptide **28** to yield **38**. Because of the more stable benzoates the removal of the carbohydrate protecting groups required slightly more basic conditions. In this case the glycopentapeptide **39** could be deblocked again without problems by using sodium methoxide in methanol to give **41**. It should be noted that other sequences may require different conditions.

The solution phase technique gives the possibility to synthesize glycopeptides of 10–12 amino acids residues which may contain even more than one carbohydrate unit. The coupling of two mono glycosylated amino acids does generally not require extended reaction times. However, the reaction of two building blocks carrying a disaccharide may need longer coupling times [15]. A series of partial structures from glycophorin containing up to four vicinal disaccharide units have been synthesized by different groups. The glycopentapeptides **42**, **43** and **44** were synthesized by application of the activated esters **17**, **18** and **19** [20,21]. In order to suppress racemization these building blocks were coupled in presence of 1-hydroxybenzotriazol (HOBt). The sequence **45** was prepared using the disaccharide glycosylamino acid **26b** with the Z/tBu combination [35,36]. The highly glycosylated sequence **46** and also **47** were synthesized employing the Fmoc/Bzl protected building block **26c** [15,16].

The sialylated sequences **48** and **49** carrying the disaccharide unit α-D-Neu5Ac-(2→6)-α-D-GalNAc are of particular interest. The syntheses of these compounds were achieved by glycosylating a Fmoc/phenacyl protected serine derivative with an 2-azido-2-deoxy-α-D-galactopyranosyl fluoride donor promoted by Cp$_2$ZrCl$_2$/AgClO$_4$ [23]. The α-anomer, which was formed as the main compound, was afterwards glycosylated with a 5-*N*-acetylneuraminic acid derivative. The disaccharide building block was transformed into a building block which was suitable for glycopeptide synthesis. The coupling reactions were carried out under activation of EEDQ as described above. After removal of the protecting groups the fully deblocked glycopeptides **48** and **49** were obtained. Ogawa et al. [31] also reported the synthesis of larger glycosylamino acid structures containing 5-*N*-acetylneuraminic acid as for instance **54** and **55**.

For immunological purposes it is often necessary to connect the glycopeptides with an artificial carrier molecule as for instance bovine serum albumin (BSA). The direct coupling of the totally deblocked glycopeptides **50** and **52** with BSA was reported by Kunz and coworkers [37]. The peptide coupling of **50** and **52** to the BSA-lysine groups was performed in aqueous solution with the water soluble coupling reagent *N*-(3-di-methylaminopropyl)-*N*′-ethylcarbodiimide hydrochloride (DAEC). The corresponding

Partial structures of glycophorin

42 Ser - Ser[*] - Thr[*] - Thr[*] - Gly   A^M - terminus

43 Ser - Ser[*] - Thr[*] - Thr[*] - Glu   A^{MC} - terminus

44 Leu - Ser[*] - Thr[*] - Thr[*] - Glu   A^N - terminus

45 Ser - Ser[**] - Thr[**] - Thr[**] - Gly   A^M - terminus

46 Thr[**] - Ser[**] - Ser[**] - Ser[**]   sequence 12 - 15

47 Val - Ser[**] - Glu - Ser[**] - Val   sequence 43 - 48

48 Ser[***] - Ser[***] - Ser[***] - Val   sequence 13 - 16

49 Ser[***] - Ser[***] - Val   sequence 14 - 16

50 Ac - Ser - Ser[*] - Thr[*] + BSA ⌐

51 Ac - Ser - Ser[*] - Thr[*] − BSA ◄⌐   A^M - terminus

52 Ac - Ser - Ser[**] - Thr[**] + BSA ⌐

53 Ac - Ser - Ser[**] - Thr[**] − BSA ◄⌐   A^M - terminus

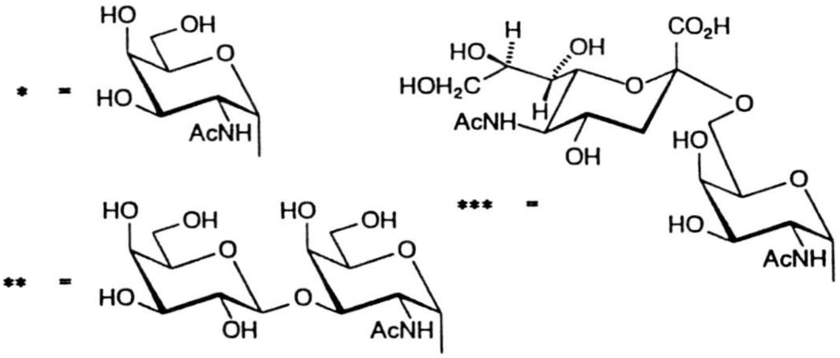

54

$$\alpha \text{ - Neu5Ac}$$
$$2$$
$$\downarrow$$
$$6$$
$$\beta \text{ - D - Gal - } (1{\rightarrow}3) \text{ - } \alpha \text{ - D - GalNAc - } (1{\rightarrow}O) \text{ - Ser}$$

55

$$\alpha \text{ - Neu5Ac}$$
$$2$$
$$\downarrow$$
$$6$$
$$\alpha \text{ - Neu5Ac - } (2{\rightarrow}3) \text{ - } \beta \text{ - D - Gal - } (1{\rightarrow}3) \text{ - } \alpha \text{ - D - GalNAc - } (1{\rightarrow}O) \text{ - Ser}$$

BSA-conjugates **51** and **53** were obtained with a high loading of either the monosac-charide or the disaccharide structures as antigenic determinants.

2.1.2. Solid phase glycopeptide synthesis

The construction of a peptide chain by solution phase techniques requires a separation and purification step after each reaction and is therefore laborious. After the development of the solid phase technique by Merrifield [38] the peptide synthesis became much more simple and faster. By attaching the growing peptide chain to an insoluble polymer the purification steps after removal of protecting groups and coupling reactions are reduced by simply washing of the polymeric support. Thus the solid phase synthesis has become the standard method in peptide chemistry. After developing the techniques for glycopeptide synthesis in solution the research has been focused to establish a method for the solid phase synthesis of glycopeptides. Although the

DMAP = 4-(dimethylamino)-pyridine, DCC = dicyclohexylcarbodiimide
HOBT = 1-hydroxybenzotriazol, Boc = t - butyloxycarbonyl, P = polymer

glycopeptides are often more labile compared to nonglycosylated peptides successful strategies were developed [39–41].

The principle of a solid phase glycopeptide synthesis is shown by the synthesis of sequence **62**. The polystyrene polymer **57**, which is derivatised with a p-alkoxybenzyl alcohol linker, is reacted with the first Fmoc-amino acid activated as symmetrical anhydride **56** under catalysis with 4-(dimethylamino)pyridine (DMAP) to afford **58**. After Fmoc-removal the next amino acid or glycosylamino acid is coupled, e.g., building block **9a** to give **59**. The carboxylic acid is activated as a 1-hydroxybenzotriazol ester which is generated in situ by reaction of **9a** with dicyclohexyl carbodiimide (DCC) and 1-hydroxybenzotriazol (HOBt). The cycle starts again with the N-terminal deprotection of the Fmoc-group followed by coupling of the Fmoc-amino acid derivatives upon completion of the desired peptide sequence **60**. The terminal amino acid may be coupled as its Boc-derivative. The glycopeptide is cleaved off the polymer by treatment with trifluoroacetic acid to yield **61**. In the final step the acetyl groups of the carbohydrate are removed by treatment with a catalytic amount of sodium methoxide in methanol yielding **62**.

A lot of different linkers and resins have been developed in peptide chemistry during the last ten years allowing the synthesis of peptide acids, amides, hydrazides and others. The cleavage of the peptides from the support is possible under acidic, basic or neutral conditions. For glycopeptide synthesis the linkers summarized in formulae **63–67** are commonly used. The p-alkoxybenzyl alcohol group **63** is suitable for synthesis of glycopeptide acids. The ester bond between linker and first amino acid is cleaved with 95% aqueous trifluoroacetic acid or 50% trifluoroacetic acid in dichloromethane [42]. The additional methoxy substituent enhances the acid lability of the linker **64**, so that

Linker	Peptide derivative	Synonym for linker or resin
HO–CH$_2$—◯—OCH$_2$–OH **63**	acid	Wang
MeO, HO–CH$_2$—◯—OCH$_2$–OH **64**	acid	Sasrin
Br–CH$_2$–CH=CH–CO$_2$H **65**	acid	Hycram
MeO, Fmoc–NH–CH$_2$—◯—O–(CH$_2$)$_4$–CO$_2$H, MeO **66**	amide	PAL
Fmoc–NH–CH—◯—OCH$_2$–OH, MeO, OMe **67**	amide	Rink

the glycopeptides anchored to that group can be cleaved off with 1–2% trifluoroacetic acid in dichloromethane [43,44]. These mild conditions may be useful if the saccharide part contains acid sensitive carbohydrate parts. The allylic anchor group **65** allows the cleavage of the synthesized glycopeptide under neutral conditions with a palladium(0)-catalyst [41]. This linker is particularly useful for the synthesis of extremely acid labile glycopeptides containing L-fucose residues. The first amino acid is connected to this linker via its caesium salt.

In many cases, especially when the synthesized glycopeptides are used for enzymatic investigations, the preparation of glycopeptide carboxamides is preferred. The charges at the C- and N-terminal end of the peptide chain are often disturbing the activity of an enzyme. Such glycopeptide amides are directly available with solid phase techniques by application of an amide linker, as for instance the PAL-group **66** [45] or the Rink-linker **67** [46]. The peptides are linked with their terminal carboxylic acid function to these linkers. The peptide amides are released by treatment with trifluoroacetic acid.

The polymeric support commonly used for solid phase glycopeptide synthesis is a polystyrene polymer crosslinked with 2% divinylbenzene. This matrix is sufficiently stable for batch synthesis during which the polymer is shaken. For continuous flow synthesis a kieselguhr supported polydimethyl acrylamide resin is recommended.

Successful solid phase glycopeptide syntheses have been described with the monosaccharide building block **9a** and the C-terminal deblocked **26d** and **26f**, respectively. The glycopentapeptides **68–70** carrying one or two GalNAc-residues were synthesized on the *p*-alkoxybenzyl alcohol linker **63** [40,47]. The glycosylated fibronectin analogues **71** and **72** were obtained from the Sasrin-linker **64** [48,49].

$$Ala - Gly - Ala - Gly - Thr - Ala$$
$$|$$
$$\alpha - D - GalNAc \qquad \textbf{68}$$

$$Ala - Thr - Val - Thr - Ala - Gly$$
$$| \qquad |$$
$$\alpha - D - GalNAc \quad \alpha - D - GalNAc \qquad \textbf{69}$$

$$Arg - Ser - Ala - Gly - Ala - Gly$$
$$|$$
$$\alpha - D - GalNAc \qquad \textbf{70}$$

$$Val - Thr - His - Pro - Gly - Tyr$$
$$|$$
$$\alpha - D - GalNAc \qquad \textbf{71}$$

$$Val - Thr - His - Pro - Gly - Tyr$$
$$|$$
$$\beta - D - Gal - (1 \rightarrow 3) - \alpha - D - GalNAc \qquad \textbf{72}$$

For the activation during solid phase glycopeptide synthesis the glycosylated amino acids were activated as 1-hydroxybenzotriazinyl esters. This procedure has the disadvantage that the hydroxybenzotriazinyl esters cannot be isolated and therefore have to be generated *in situ* each time before the coupling reaction. Other activation procedures have become more attractive in recent years especially the use of activated esters like the pentafluorophenyl (Pfp) [50] and 3,4-dehydro-4-oxo-1,2,3-benzotriazinyl (Dhbt) esters [51,52]. The Fmoc amino acids of these activated esters are crystalline substances that can be stored for a long period. The Dhbt-esters have the additional advantage that the progress of the peptide bond formation can be followed by a colour reaction [53]. The coupling of Fmoc amino acid Dhbt-esters releases the 3-hydroxy-3,4-dehydro-4-oxo-1,2,3-benzotriazine (Dhbt-OH) **74**, which protonates unreacted polymer bound amino groups **73** to give the bright yellow coloured ion pair **75**. As soon as all amino groups **73** have reacted with the carboxy group of an amino acid forming the peptide linkage the yellow colour disappears.

To assure complete peptide bond formation, the disappearance of the ion pair **75** during the reaction can comfortably be monitored with a spectrophotometer [53]. The colour reaction is also very useful in a multiple column solid phase peptide synthesis where a number of different peptides are simultaneously built up in separate columns. In this case the colour reaction in each column indicates the progress of the coupling reactions.

The concept to use activated esters, especially pentafluorophenyl esters have recently become very popular. The glycosylamino acid **9a** can be transformed with DCC and pentafluorophenol into the Pfp-ester **76** in a good yield [54,55]. The corresponding Dhbt-ester is available analogously, but due to the higher reactivity the Dhbt-ester cannot be purified by chromatography. It was demonstrated on a multiple column synthesis of 40 different mucin glycopeptides that the Pfp-ester **76** is an excellent building block [54,55]. The coupling reactions with the activated ester **76** were carried out in presence of Dhbt-OH, allowing the peptide bond formation to be monitored by the colour reaction in each column. The addition of Dhbt-OH also enhances the reactivity of **76**, probably by a transesterification into the corresponding Dhbt-ester. Applying this technology, 40 different mucin glycopeptides with variations in the peptide sequence and the glycosylation site were synthesized [54,55]. The synthesis of glycopeptides with up to three vicinal GalNAc-residues was possible without any problems.

Recently, a new strategy for the synthesis of mucin glycopeptides has been developed [56]. The synthesis of an activated ester building block was shortened by directly glycosylating the Fmoc/Pfp protected amino acid **77** with the glycosyl donor **5**. The resulting azido compound **78** can be introduced directly into a peptide chain by solid phase synthesis. Thus the number of steps is reduced. The application of the Fmoc/Pfp protection scheme also has a positive effect on the stereoselectivity of the glycosylation reaction of **5** compared to other protecting groups (for Thr $\alpha:\beta = 14:1$, for Ser $\alpha:\beta = 6:1$). The pure α-glycosides **78** are obtained by chromatographic purification [56]. The reduction of the azido group in **78** can easily be performed later while the glycopeptide is still bound to the solid support.

Formulae **81–84** exemplify the key steps of the azide strategy using the building block **78** [56]. The peptide chain was built up on a polydimethyl acrylamide resin which was derivatized with the reference amino acid norleucin and the peptide amide linker **66**. The non-glycosylated and glycosylated amino acids were coupled as suitably protected Dhbt- or Pfp-esters. After completion of the peptide synthesis, the azido group of the glycosylamino acid was reduced and N-acetylated with thioacetic acid in one step while the peptide remains on the solid support (reaction **81→82**). The progress of the reaction was followed by IR-spectroscopy by the disappearance of the azido absorption band. The purity of the final product is determined by the purity of the thioacetic acid. Commercial thioacetic acid often contains dithioacetic acid which will form a thioacetate in this reaction. It is recommended to distil the thioacetic acid at low

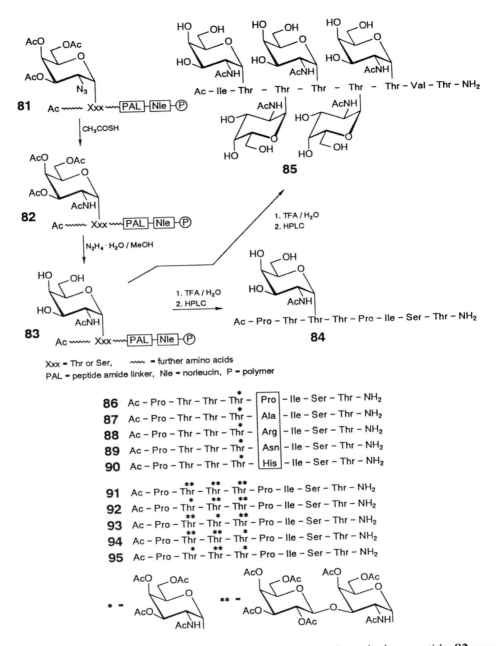

Xxx = Thr or Ser,   ⌇⌇ = further amino acids
PAL = peptide amide linker, Nle = norleucin, P = polymer

86 Ac – Pro – Thr – Thr – Thr – | Pro | – Ile – Ser – Thr – NH₂
87 Ac – Pro – Thr – Thr – Thr – | Ala | – Ile – Ser – Thr – NH₂
88 Ac – Pro – Thr – Thr – Thr – | Arg | – Ile – Ser – Thr – NH₂
89 Ac – Pro – Thr – Thr – Thr – | Asn | – Ile – Ser – Thr – NH₂
90 Ac – Pro – Thr – Thr – Thr – | His | – Ile – Ser – Thr – NH₂

91 Ac – Pro – Thr – Thr – Thr – Pro – Ile – Ser – Thr – NH₂
92 Ac – Pro – Thr – Thr – Thr – Pro – Ile – Ser – Thr – NH₂
93 Ac – Pro – Thr – Thr – Thr – Pro – Ile – Ser – Thr – NH₂
94 Ac – Pro – Thr – Thr – Thr – Pro – Ile – Ser – Thr – NH₂
95 Ac – Pro – Thr – Thr – Thr – Pro – Ile – Ser – Thr – NH₂

temperature before use. In the next step the polymer bound glycopeptide **82** was deacetylated in the carbohydrate part by treatment with hydrazine hydrate in methanol to give **83**. The final product was split off the resin with trifluoroacetic acid. After reversed phase HPLC the pure products were obtained as for instance the sequence **84** [56].

The new azide strategy was applied both in a multiple column synthesizer and in a full automatic synthesizer. The latter was chosen for sequence **85** containing five vicinal glycosylamino acids. The multiple column system was used for the simultaneous synthesis of 40 different mucin glycopeptides. A selection of these compounds are **86–90**. In these sequences the amino acids around the glycosylation site were systematically replaced by other amino acids [57], thus proving the compatibility with the standard methods from peptide chemistry.

The azide methodology is also applicable for the synthesis of mucin glycopeptides with T-antigenic structures. Coupling of the peracetylated halide **79** with the threonine derivative **77** yielded the azido building block **80** containing the desired Fmoc/Pfp combination [58]. This building block was used analogously to the above illustrated reaction scheme **81–84** to build up the triple glycosylated sequences **91–95** containing mono- and disaccharides. The largest molecule **91** was again synthesized in an automatic continuous flow peptide synthesizer, the other sequences with D-GalNAc and β-D-Gal-(1→3)-D-GalNAc residues were obtained by the multiple column technique [57]. The high efficiency and versatility of the method was demonstrated by the multiple column synthesis of more than 100 different mucin glycopeptides.

2.2. Synthesis of O-glycopeptides containing a β-D-xylopyranosyl linkage

The second important type of O-glycoprotein contains a β-glycosidic linkage between D-xylose and the hydroxy group of serine. These linkages are found mainly in the proteoglycans. In heparin, chondroitin sulfate and dermatan sulfate the protein backbone is glycosylated via the D-xylose by long carbohydrate chains. The most common core structure is the trisaccharide β-D-Gal-(1→3)-β-D-Gal-(1→4)-β-D-Xyl linked to serine. The glycosylamino acids **96** [59], **97** [60,61] and **98** [62], representing partial structures of this core unit, were synthesized previously, as well as β-D-xylopyranosyl peptides with four amino acids [63,64]. The syntheses were mostly performed by glycosylating C- and N-terminal protected serine derivatives with benzoylated glycosyl halides. The formation of these β-linked glycosides is easier to obtain compared to the mucin type structures. The formation of such a 1,2-*trans* glycoside can be performed stereoselectively with a neighbouring participating substituent at the C-2 atom of the sugar.

Recent investigations have demonstrated that larger glycopeptides of this core type can be synthesized [65]. The trichloroacetimidate method has proved to be useful for the connection of D-xylose containing oligosaccharides with serine derivatives. For instance, the trichloroacetimidate **99**, which was synthesized by reaction of the 1-OH unprotected sugar with trichloroacetonitrile, was stereoselectively coupled in a yield of 90% with the dipeptide **100** to give **101**. The orthogonal protecting group pattern of **101** allowed a selective deblocking of either the C- or the N-terminus. A Pd(0)-catalyzed cleavage of the allyl group afforded compound **102** which was coupled with a non-glycosylated Ser-Gly dipeptide. The resulting tetrapeptide was then connected with the N-terminal deprotected glycosyl unit **103**. After removal of all benzoates with

Ser
|
β - D - Xyl
96

Ser
|
β - D- Gal - (1→4) - β - D- Xyl
97

Ser
|
β - D- Gal - (1→3) - β - D- Gal - (1→4) - β - D- Xyl
98

99 + **100**

$\xrightarrow{\text{TMS - triflate}}$

101 R^1 = Z; R^2 = All
102 R^1 = Z; R^2 = H
103 R^1 = H; R^2 = All

Ser – Gly – Ser – Gly – Ser – Gly
β - D - Gal - (1→4) - β - D - Xyl   β - D- Gal - (1→4) - β - D - Xyl
104

β - D- Gal - (1→4) - β - D - Xyl
|
Ser – Gly – Ser – Gly – Ser – Gly
β - D- Gal - (1→4) - β - D - Xyl   β - D - Gal - (1→4) - β - D - Xyl
105

hydrazine in methanol the final hexapeptide **104** bearing two disaccharides was obtained [65]. The glycopeptide **105** with three glycosylation sites was synthesized analogously. It should be noted that the removal of benzoates with hydrazine in methanol did not give β-elimination nor racemisation. It is also important to know that condensation of the dipeptide fragment under activation of EEDQ gave low coupling rates. A better yield (80–90%) was obtained by using the mixed anhydride method and activating compound **102** with isobutyl chloroformate and N-methylmorpholine [65].

The trichloroacetimidate method has also been successfully applied to the synthesis of glycopeptides with longer carbohydrate chains [66]. It was shown that the tetrasaccharide trichloroacetimidate **106** is a suitable glycosyl donor to glycosylate the dipeptide **100**. The obtained building block **107** contains the same protection scheme in the amino acid part as compound **101** and was used in the same way to build up the glycohexapeptide **108** with two tetrasaccharide chains [66]. This synthesis proves to be an example of the high level of glycopeptide chemistry.

R = (CH₃)₃Si (CH₂)₂ -

β - D - GlcA - (1→3) - β - D - Gal - (1→3) - β - D - Gal - (1→4) - β - D - Xyl - (1→O) - Ser
                                                                                    |
                                                                                   Gly
                                                                                    |
                                                                                   Ser
                                                                                    |
                                                                                   Gly
                                                                                    |
β - D - GlcA - (1→3) - β - D - Gal - (1→3) - β - D - Gal - (1→4) - β - D - Xyl - (1→O) - Ser
                                                                                    |
108   Gly

2.4. Synthesis of O-glycopeptides with other saccharide units

Many examples of glycosylated serine or threonine derivatives with other carbohydrates (e.g. D-glucose, D-galactose or D-mannose) have been described. Instead of serine or threonine other hydroxyamino acids such as tyrosine or 5-hydroxyproline have been used. D-Glucose derivatives were coupled via the α-D-glucosyl bromide to serine with neighbouring participating groups to afford β-glycosides [67-69]. The 1-OH unprotected D-glucose derivatives were connected with serine and threonine compounds promoted by trifluoromethanesulfonic acid anhydride [70]. It was also demonstrated that the linkage between amino acid and carbohydrate can be built up enzymatically. The reaction of 2-nitrophenyl-β-D-glycosides with hydroxyamino acids in the presence of a β-glucosidase gave the desired β-linked glycosylamino acid building blocks [71] in moderate yield.

The trisaccharide α-D-Xyl-(1→3)-α-D-Xyl-(1→3)-D-Glc was coupled to a Z/Blz-protected serine derivative via the trichloroacetimidate in 66% yield. After removal of all protecting groups the glycosylamino acid **109** was obtained [72]. The reaction of the trisaccharide β-D-Glc-(1→4)-β-D-Glc-(1→4)-D-Glc, activated as thioglycoside under promotion of methyltriflate and a non participating substituent at C-2, with a serine acceptor yielded a mixture of anomeric glycosides [73].

α - D - Xyl - (1→3) - α - D - Xyl - (1→3) - β - D - Glc - (1→O) - L - Ser

109

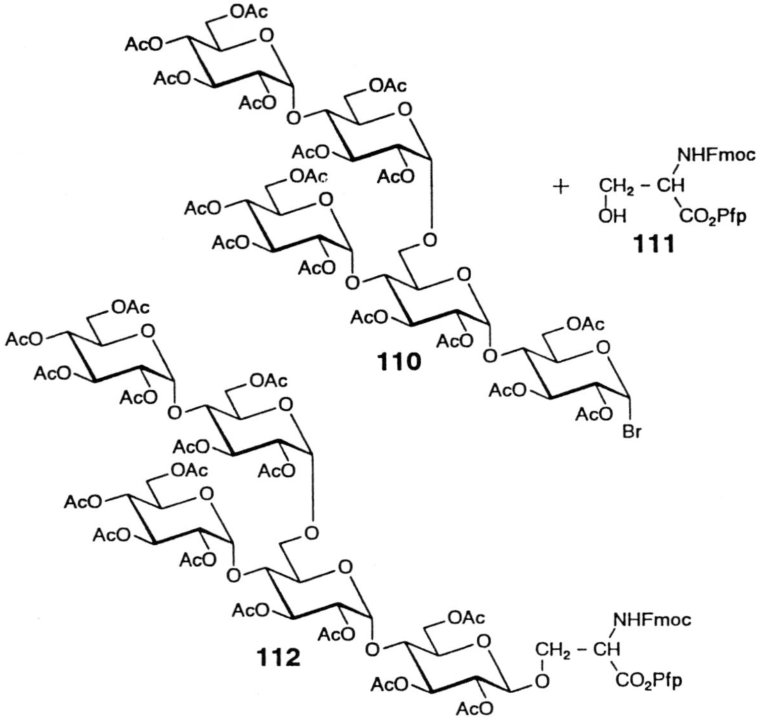

110

112

111

R^1
|
Ser - Ala - Leu - Leu - Ser - Ser - Asp - Ile - Thr - Ala - Ser - Val - Asn - Cys - Ala - Lys - Tyr
                                                                            R^2
                                                                            |

α - D - Glc - (1→4) - α - D - Glc - (1→6)
                                         α - D - Glc - (1→4) - β - D - Glc - = Glc$_5$
α - D - Glc - (1→4)

113 R^1 = Glc$_5$; R^2 = H
114 R^1 = H; R^2 = Glc$_5$

Glc$_5$
|
Ser - Pro - Glu - Leu - Phe - Glu - Ala - Leu - Gln - Lys - Leu - Phe - Lys - Hys - Ala - Tyr

115

A complex pentasaccharide, which was available by partial enzymatic degradation of starch, was acetylated and transformed into the glycosyl bromide **110**. This compound could be coupled in presence of silver triflate with the Fmoc/Pfp protected serine derivative **111**. The desired β-glycoside **112** was obtained in a yield of 42% [74]. This building block was used in an automated continuous flow solid phase glycopeptide synthesis to build up the large glycopeptides **113–115** [74]. The quantitative coupling

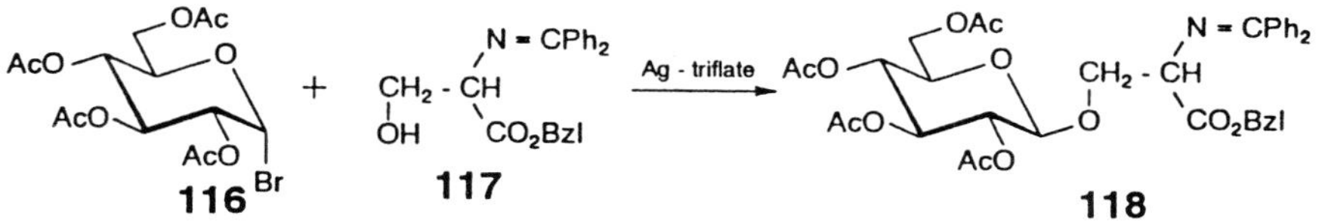

Structures 116, 117, 118, 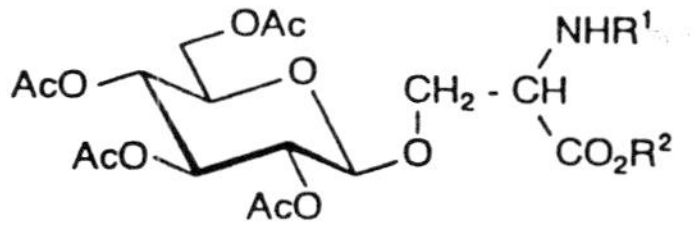 119, 120, 121, 122, 123, 124

116 Br   117   118

119 R¹ = H; R² = Bzl
120 R¹ = H; R² = H
121 R¹ = Fmoc; R² = H

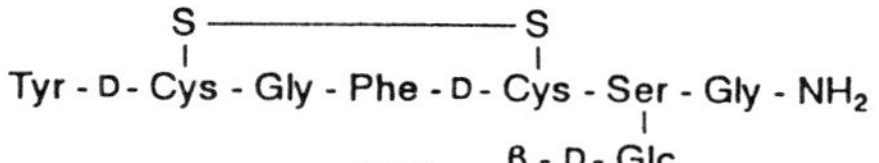

S ——————— S
Tyr - D - Cys - Gly - Phe - D - Cys - Ser - Gly - NH₂
                                    |
                              β - D - Glc

122

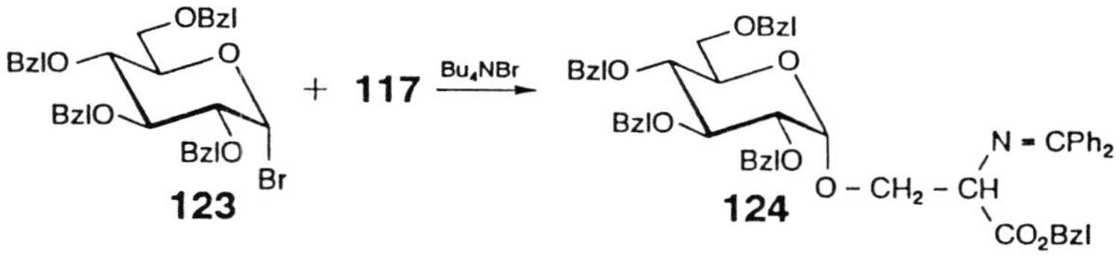

123 + 117 → 124

of all amino acids was assured by following the colour reaction of the Dhbt-esters. The Pfp-ester building block was coupled with the addition of Dhbt-OH.

The use of Schiff bases as a new type of glycosyl acceptors was reported recently [75]. The nucleophilicity of the side chain hydroxy group of **117** was suggested to be increased due to a hydrogen bond between the OH-group and the nitrogen atom of the imine group. The reaction of **116** with **117** afforded the β-glycoside **118** in 77% yield. The Schiff base **117** was also successfully glycosylated with other D-glucose containing disaccharides in yields between 70 and 80% [75,76]. For a solid phase peptide synthesis it was necessary to transform compound **118** into the Fmoc-protected derivative **121**. This was achieved by acid treatment to hydrolyse the schiff base **118** and hydro-genolysis of the benzylester. The unprotected compound **120** was then reacted with Fmoc-OSu to afford **121** [76]. The building block **121** was used for the synthesis of the ekephalin analogue **122**. The glycopeptide amide was constructed by application of the Rink-linker **67**. The Fmoc-amino acids were activated with (benzotriazole-1-yl-oxy) tris(dimethylamino)phosphonium-hexafluorophosphate (BOP) [76].

The Schiff base **117** is also a suitable glycosyl acceptor for the synthesis of 1,2-*cis* glycosides. Following the conditions of *in situ* anomerization the halide **123** was reacted with **117** in a non participating glycosylation reaction to yield the α-glycoside **124** in 56% yield [76].

The peracetylated derivative of D-glucose **125** was used to glycosylate the Z/Bzl protected hydroxyproline derivative **126**. Under promotion of borotrifluoride the β-linked glycosylamino acid **127** was obtained [77,78]. The hydrogenolytic removal of the Z- and Bzl-group afforded **128**, which was converted into the Fmoc-glycosylamino acid **129**. The building block **129** was incorporated into the tetrapeptide morphiceptin

125 + **126** → **127** R^1 = Z; R^2 = Bzl
128 R^1 = H; R^2 = H
129 R^1 = Fmoc; R^2 = H

Tyr - Pro - Phe - Hyp - NH$_2$
          |
       β - D - Glc
         130

Tyr - Pro - Phe - Hyp - NH$_2$
          |
       β - D - Gal
         131

analogue **130** by solid phase synthesis applying the PAL-linker **66** and the standard DCC/HOBt activation [78]. The glycotetrapeptide amide **131** containing a β-linked D-galactose was synthesized in the same way [78].

   Another class of O-glycopeptides can be synthesized by glycosylating the phenolic hydroxy group in tyrosine. For instance these structures occur in glycogenin, which is a primer protein in the biosynthesis of glycogen. The Fmoc/Pfp-protected tyrosine derivative **133** was glycosylated with various saccharides as for example maltose. The perbenzoylated maltosyl bromide was coupled with **133** in acetonitrile under promotion of silver triflate to afford the desired α-glycoside **134** in 33% and the corresponding β-anomer in 30% yield [79]. The building block **134** was incorporated into glycopeptide **135** by automatic continuous flow solid phase synthesis. The β-linked

132 + **133**   Ag - triflate / CH$_3$CN →

134

Glyc
  |
Phe - Ile - Tyr - Asn - Leu - Ser - Ser - Ile - Ser - Ile - Tyr - Ser - Tyr -
- Leu - Pro - Ala - Phe - Lys - Ala - Phe - Gly - Ala - Asn - Ala - Lys

135 Glyc = α - D - Glc - (1→4) - α - D - Glc -
136 Glyc = α - D - Glc - (1→4) - β - D - Glc -
137 Glyc = α - D - Glc -
138 Glyc = β - D - Glc -

glycopeptide **136** and the mono glycosylated derivatives **137** and **138** were synthesized as well [80].

The β-glycosidic linkage between D-galactose and hydroxyamino acids is also best achieved using a neighbour participating glycosylation reaction. In analogy to D-glucose, D-galactose derivatives were coupled to hydroxyamino acids by application of glycosyl halides [81,82] or the trifluoromethanesulfonic acid anhydride method [83]. The peracetylated D-galactose compound **139** has proved to be an efficient glycosyl donor. In presence of borotrifluoride **139** was reacted with the protected threonine derivatives **140** and **141** to yield the β-linked glycosylamino acids **142** [84] and **145** [85]. For an application in solid phase glycopeptide synthesis it was necessary to transform these compounds into the Fmoc-protected derivative **144**. In case of the Fmoc/All protection scheme **144** was directly available by a Pd(0)-catalyzed cleavage of the allyl ester **145**. The Z/Bzl protection requires an additional step. First the Z- and benzyl groups had to be removed to give **143** which was then converted into the building block **144**. Compound **144** was used in assembly of antifreeze glycoprotein analogues by continuous flow solid phase synthesis [84]. The introduction of the building block **144** was performed by converting the free acid into the symmetrical anhydride which was then incorporated into the growing peptide chain. The glycosylated polytripeptides **146** were synthesized with *n* ranging from 2 to 7 [84].

It was reported that the linkage between serine derivatives and D-galactose can be synthesized by enzymatic methods. For instance the galactose residue of 2-nitrophenyl-β-D-galactopyranoside [86,87] or lactose [88] has been transglycosylated to serine in presence of a β-galactosidase. The corresponding α-glycosides were made by using raffinose in combination with an α-galactosidase [88].

An α-glycosyl connection of D-mannose is fairly easily established. The silver triflate promoted reaction of the mannose disaccharide donor **147** with the threonine derivative **148** yields the α-glycoside **149** in 80%. Compound **149** was employed as building block in a solid phase synthesis with addition of Dhbt-OH. All non-glycosylated amino acids were used as their Dhbt-active esters after deprotection. The synthesis afforded the glycopeptide **150** containing of 17 amino acids and the disaccharide residue [89,90].

The glycosyl chloride [91] and oxazoline derivative [92] of 2-acetamido-2-deoxy-D-glucose was reacted with serine and threonine derivatives to the corresponding β-glycosides. Furthermore, the α-linked product is available from 2-azido-2-deoxy-D-

Gly - Phe - Tyr - Phe - Asn - Lys - Pro - Thr - Gly - Tyr - Gly - Ser - Ser - Ser - Arg - Arg - Ala
α - D - Man - (1→2) - α - D - Man   **150**

glucose in analogy to the synthesis of 2-acetamido-2-deoxy-D-galactose derivatives. The azido derivative could be used as glycosyl bromide in a non neighbouring supported reaction with serine or threonine affording the desired products after reduction of the azido group and subsequent N-acetylation [93].

O-glycosides of 2-deoxy-sugars linked to serine or threonine could be obtained by the N-iodosuccinimide reaction [94] of glycals. This is shown by conversion of the glucal **151** reacted with **152** in the presence of N-iodosuccinimide into the α-glycosyl linked product **153**. The successive reduction gives the corresponding 2-deoxy-glycoside **154** [95]. The reaction is also applicable to galactal and disaccharides like lactal, cellobial and β-D-Gal(1→3)-galactal [96,97], although the stereoselectivity of the latter is not completely guaranteed.

The N-iodosuccinimide procedure has the advantage that the use of polar, aprotic solvents like acetonitrile and acetone is also possible. Therefore peptides soluble in such solvents could be converted directly into 2-deoxy-glycopeptides. Following this strategy Fmoc-Pro-Ser-Ala-OBzl was coupled with the glucal **151** affording the corresponding α-glycoside of **154** in high yield. However, larger linear peptide fragments like Z-Phe-D-Pro-Phe-Ala-Ser-Phe-OBzl gives lower yields. In contrast the N-iodosuccinimide procedure is especially favored for the direct and stereoselective glycosylation of cyclopeptides demonstrated by the synthesis of **159**. The lactal derivative **155** was reacted with the cyclopeptide **156** in the presence of N-iodosuccinimide to give the α-glycoside **157** as the major product. Reduction of compound **157** with $NiCl_2 \cdot 6H_2O/H_3BO_3$ afforded **158** which was deblocked in the sugar part by treatment with KCN in CH_3OH/C_2H_5OH to yield the deprotected glycosylated cyclopeptide **159** [97]. Compound **159** was used for detailed conformational analysis by NMR to study the influence of glycosylation on conformation of peptides [98]. In analogy to the described procedure other glycosylated cyclopeptides like **160**, **161** and **162** were synthesized [97].

Furthermore, C-glycosides of amino acids are known [99]. Radical addition of the dehydroalanine derivative **163** to the glycosyl bromide **116** affords a diastereomeric mixture of the C-glycosides **164** (2S configuration) and **165** in a ratio of about 2.5:1 [99]. The connection between alanine and the C-1 of the saccharide residue in **164** and **165** is stable against hydrolytic and enzymatic cleavage.

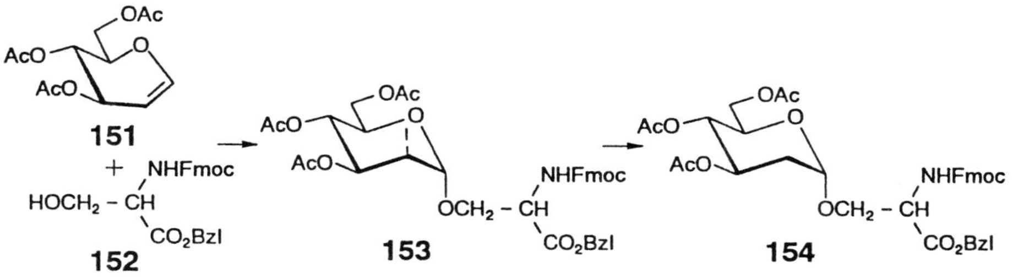

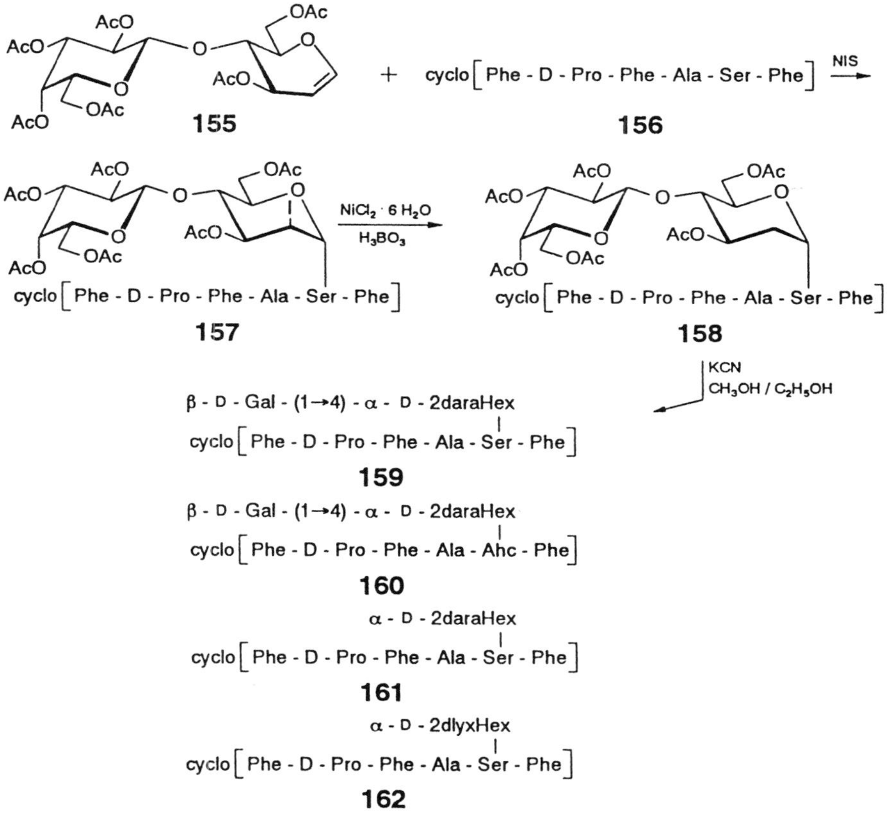

151

152

153

154

155 + cyclo[Phe - D - Pro - Phe - Ala - Ser - Phe] **156** $\xrightarrow{\text{NIS}}$

157 $\xrightarrow[\text{H}_3\text{BO}_3]{\text{NiCl}_2 \cdot 6\,\text{H}_2\text{O}}$ **158**

cyclo[Phe - D - Pro - Phe - Ala - Ser - Phe]

cyclo[Phe - D - Pro - Phe - Ala - Ser - Phe]

$\xdownarrow$ KCN / CH$_3$OH / C$_2$H$_5$OH

β - D - Gal - (1→4) - α - D - 2daraHex
cyclo[Phe - D - Pro - Phe - Ala - Ser - Phe]

159

β - D - Gal - (1→4) - α - D - 2daraHex
cyclo[Phe - D - Pro - Phe - Ala - Ahc - Phe]

160

α - D - 2daraHex
cyclo[Phe - D - Pro - Phe - Ala - Ser - Phe]

161

α - D - 2dlyxHex
cyclo[Phe - D - Pro - Phe - Ala - Ser - Phe]

162

NIS = N-iodosuccinimide

Ahc = α-amino-6-hydroxy-capronic acid

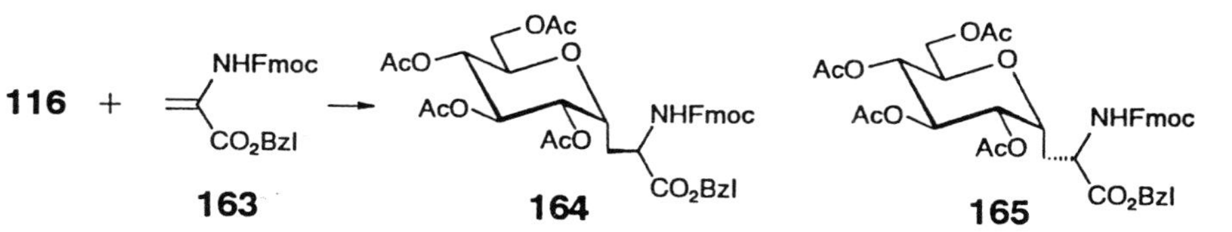

116 + **163** → **164** **165**

3. Synthesis of N-glycopeptides

3.1. Synthesis of N-glycopeptides containing 2-acetamido-2-deoxy-D-glucose β-linked to asparagine

N-glycopeptides carry an 2-acetamido-2-deoxy-D-glucosyl residue connected in a β-linkage to the amide of an asparagine residue of the protein backbone. The core structure of the N-glycopeptides consists of a pentasaccharide containing a branched (Man)$_3$ unit and a GlcNAc disaccharide attached to the asparagine. The interest in chemical synthesis of glycopeptides carrying parts of this structure has increased considerably. The review of Garg and Jeanloz [1] gives a good summary of the earlier synthetic approaches.

In contrast to the synthesis of O-glycopeptides the coupling reaction between sugar residue and amino acid requires a completely different strategy because an amide bond has to be formed instead of a glycosidic bond.

One standard method employs the β-glycosyl azide **167** obtained by treatment of the corresponding chloride **166** with silver azide [2,100] or sodium azide in formamide [2,101]. Better results are obtained by use of sodium azide in chloroform/water under phase transfer catalyzed conditions [2,102]. The following hydrogenation of the azide **167** affords the glycosylamine **168** [2]. To avoid bisglycosyl amine formation or hydrolysis during the reduction step the use of Raney-Ni [19] or Pd/C [102] is recommended instead of Pt [100] as catalyst. The labile glycosylamine **168** must be converted immediately into the N-glycoside **170** by treatment with the aspartic acid derivative **169** activated by DCC or EEDQ [1,2]. Using this strategy the Z/Bzl protected compound **169a** was converted into **170a** by DCC activation followed by hydrogenation to give the deprotected glycosylamino acid **170c** [1,103].

The glycosylamino acid derivative **170b** containing the orthogonal protecting pattern Boc/All synthesized from **169b** with EEDQ is suitable for glycopeptide synthesis. Selective carboxyl deblocking was achieved by Pd(0)-catalyzed allyl transfer to weakly basic or neutral nucleophiles or under neutral conditions by Rh(I)-catalyzed isomerization of the allyl group followed by hydrolysis of the resulting propenyl ester to give **170d**. The Boc-group was cleaved selectively from **170b** by acid treatment to give **170e**. Thereby **170b** could be elongated at the C- and N-termini resulting in glycopeptides such as **171**, **172** and **173** [104]. In combination with the Hycram-resin, **170d** was applied as building block in a solid phase synthesis. The resin bound glycopeptide **174** could be cleaved in the same way as the allyl group in **170b** due to the allyl ester bond between resin and glycopeptide to yield **175** [41].

The synthesis of the corresponding disaccharide glycoside **176** was achieved in analogy to the monosaccharide **170** using the phase transfer catalyzed method [105]. Cleavage of the allyl group followed by the reaction with a dipeptide ester and EEDQ afforded a glycopeptide carrying the disaccharide residue. After deacetylation with ammonia the glycopeptide amide **177** was obtained. Acid treatment of **177** for deblocking the N-terminus failed due to O-glycosyl bond lability. To avoid this side reaction the Boc-group has to be removed first to give **178**, demonstrating the importance of the deprotection sequence [105,106].

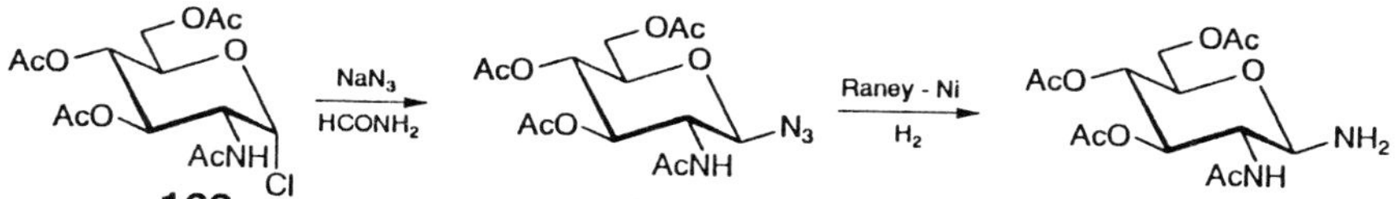

166 → **167** → **168**

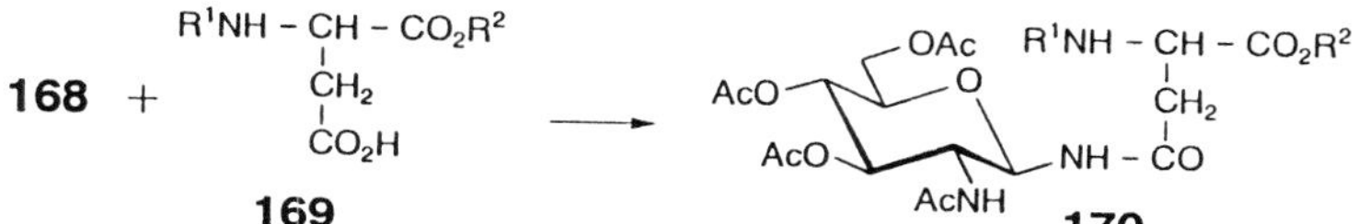

(NaN₃, HCONH₂; Raney-Ni, H₂)

168 + **169** → **170**

R¹NH – CH – CO₂R²
|
CH₂
|
CO₂H

169

	R¹	R²
169a, 170a	Z	Bzl
169b, 170b	Boc	All
170c	H	H
170d	Boc	H
170e	H	All

Asn - Thr - Ser - NH₂
|
β - N - D - GlcNAc

171

Asn - Ala - Thr - Ile - Leu - NH₂
|
β - N - D - GlcNAc

172

Bz - Gly - Thr - Asn - Ile - Ser - NH₂
|
β - N - D - GlcNAc

173

Boc - Leu - Asn - Ile - HYCRAM
|
β - N - D - GlcNAc (OAc)₃

174

→

Boc - Leu - Asn - Ile
|
β - N - D - GlcNAc (OAc)₃

175

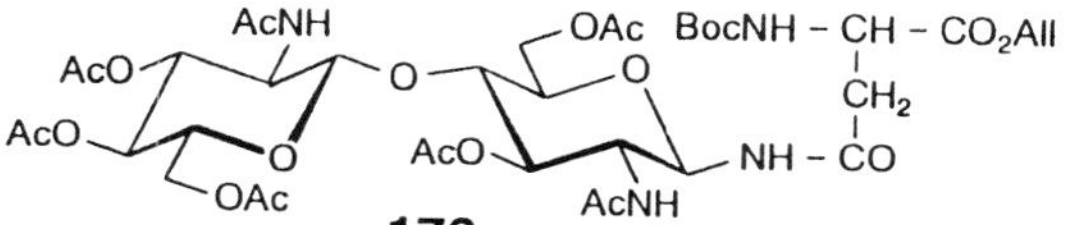

176

Boc - Asn - Phe - Thr - NH₂
|
β - D - GlcNAc - (1→4) - β - N - D - GlcNAc

177

Asn - Phe - Thr - OAll
|
β - D - GlcNAc(OAc)₃ - (1→4) - β - N - D - GlcNAc(OAc)₂

178

An alternative method employed isothiocyanate compounds [107] instead of glycosylamines to avoid the azide step [108,109]. The trisaccharide isothiocyanates **179** [108] and **180** [109] were synthesized from the corresponding oxazoline derivative or glycosyl halides by treatment with potassium thiocyanate. The following coupling under very dry conditions with the aspartic acid derivative **169** directly afforded the trisaccharide-N-glycosides [108,109].

Using natural sources the heptasaccharide **181** was isolated from the urine of swainsonine intoxicated sheep, peracetylated and converted into the glycosyl azide via the oxazoline by treatment with trimethylsilyl azide. After reduction the glycosylamine

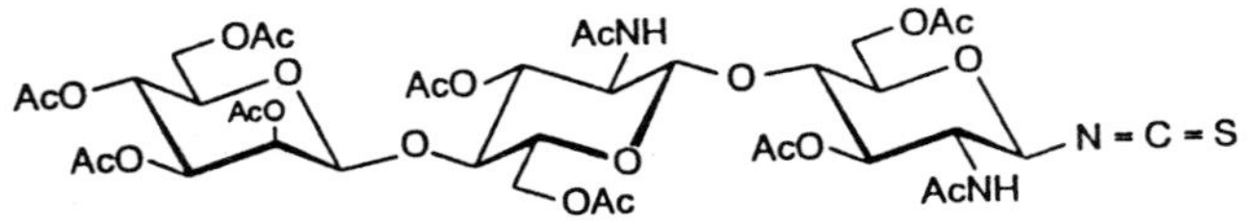
OAc
AcO
AcO
O
AcO
AcNH
O
OAc
N = C = S

179

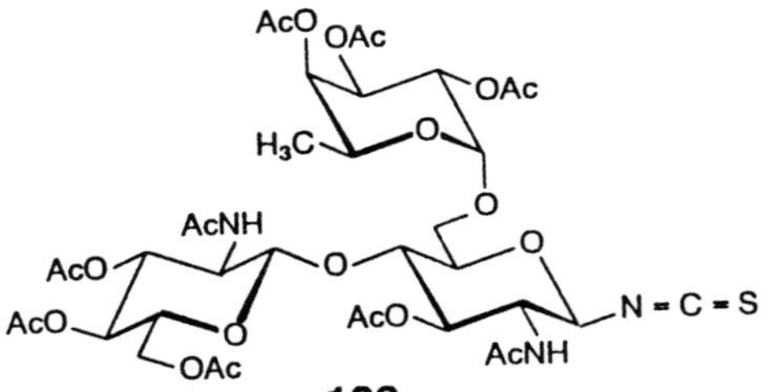
AcO
OAc
H3C
AcNH
N = C = S

180

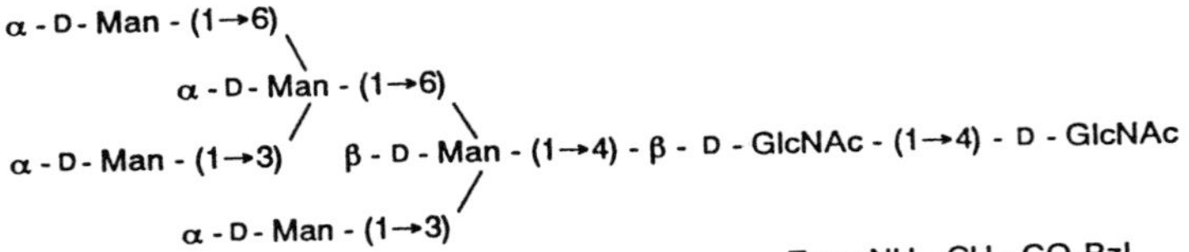
α - D - Man - (1→6)
α - D - Man - (1→6)
α - D - Man - (1→3)
β - D - Man - (1→4) - β - D - GlcNAc - (1→4) - D - GlcNAc
α - D - Man - (1→3)

181

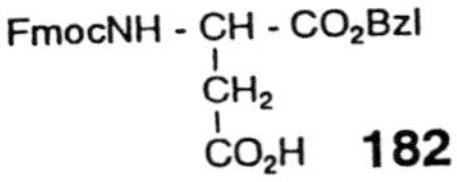
FmocNH - CH - CO2Bzl
CH2
CO2H

182

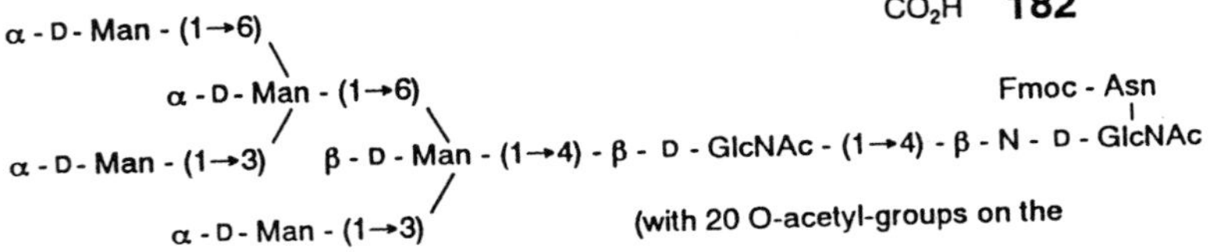
α - D - Man - (1→6)
α - D - Man - (1→6)
Fmoc - Asn
α - D - Man - (1→3)
β - D - Man - (1→4) - β - D - GlcNAc - (1→4) - β - N - D - GlcNAc
α - D - Man - (1→3)

(with 20 O-acetyl-groups on the saccharide part)

183

derivative was coupled with Fmoc-Asp-Bzl **182** in the presence of diethyl cyanophosphonate to yield the protected glycosylasparagine in 50%. Removal of the benzyl group by hydrogenolysis formed an intermediate **183** [110]. The pure fully deprotected compound could not be isolated.

The unprotected glycosylamine **184** is easily prepared by the reaction of 2-acetamido-2-deoxy-D-glucose with ammonium hydrogen carbonate. This reaction afforded a mixture of 80–90% β-glycosylamine and 10–20% starting material. After conversion into the acetylated Fmoc-derivative **185** purification by crystallization and removal of the Fmoc-group with piperidine, the glycosylamine **168** could be obtained in a pure state [111]. Treatment of the acid chloride **186** with compound **168** afforded the active ester **187** containing the protecting group pattern Fmoc/Pfp that could directly be used for solid phase glycopeptide synthesis. This building block was then applied to synthesize the N-linked glycopeptide **188**, a fragment of the enzyme glucoamylase (AMG) with 11 amino acids. The assembly was conducted on an automated peptide synthesizer employing Dhbt-esters in the continuous flow version on polyamide resin [111]. A similar building block was employed for the solid phase synthesis of the glycosylated undecapeptide **189** on 4-hydroxymethylphenoxyacetyl-norleucyl

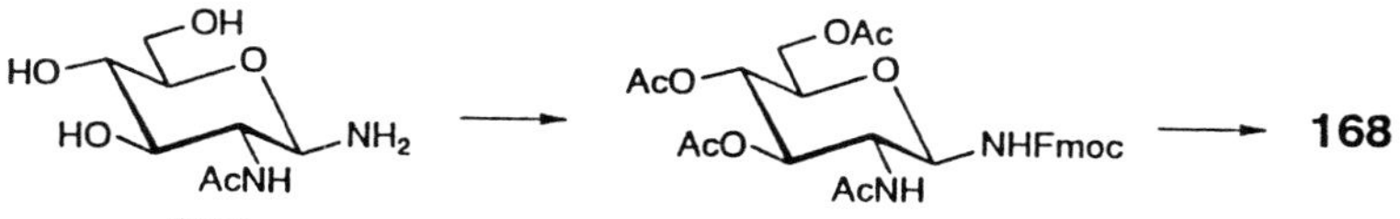

184 → **185** → **168**

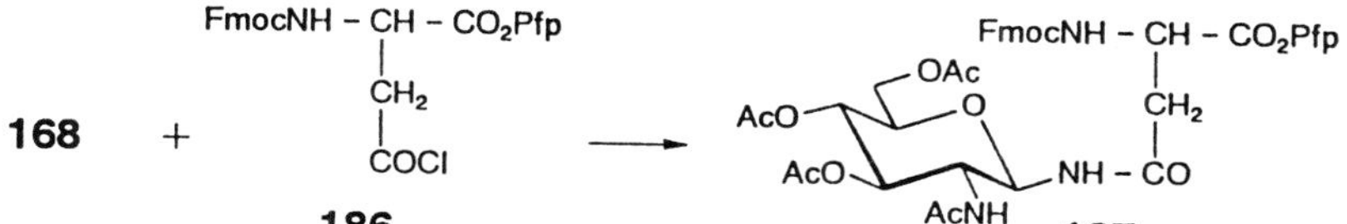

168 + **186** → **187**

Thr - His - Ala - Ala - Ser - Asn - Gly - Ser - Met - Ser - Glu
                               |
                           β - N - D - GlcNAc

188

Thr - Lys - Pro - Arg - Glu - Gln - Gln - Tyr - Asn - Ser - Thr
                               β - N - D - GlcNAc

189

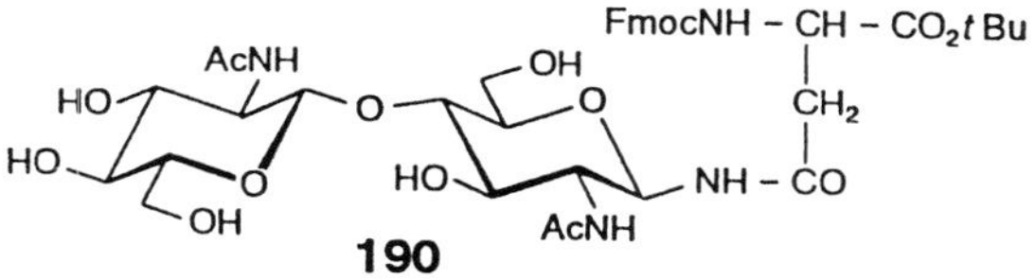

190

Gly - Lys - Ala - Tyr - Thr - Ile - Phe - Asn - Lys - Thr - Leu - Met - NH$_2$
β - D - GlcNAc - (1→4) - β - N - D - GlcNAc

191

cyclo[Phe - D - Pro - Phe - Ala - Glu - Phe]
192

cyclo[Phe - D - Pro - Phe - Ala - Gln - Phe]
              β - N - D - GlcNAc
193

derivatized kieselguhr-supported resin (Pepsyn KA) using Fmoc amino acid symmetrical anhydrides or pentafluorophenyl esters in the presence of 1-N-hydroxybenzotriazole as acylating reagents. **189** represents the 289-299 amino acid sequence of the Fc domain of IgG [112].

It was also shown that crude glycosylamines could be coupled with Fmoc-Asp-OtBu activated by PfpOH and DCC in N,N-dimethylformamide/water mixtures to give a N-glycosyl compounds containing an unprotected sugar part. Neither pentafluorophenyl esters nor the symmetrical anhydrides acylate the unprotected hydroxyl groups of the sugar moieties during peptide synthesis [113–116]. Following this procedure crude chitobiosylamine was converted into **190**. After removal of the tBu-ester with TFA, occurring without O-glycosyl bond cleavage, the unprotected disaccharide building block was used via pentafluorophenyl ester for the solid phase synthesis of the T-cell epitopic dodecaglycopeptide amide **191** [114].

Another approach to build up N-glycopeptides focused on the coupling of an oligosaccharide β-glycosylamine to an aspartic acid containing peptide. This was successfully shown by the reaction of unprotected glycosylamine **184** with the pentapeptide Ac-Tyr-Asp-Leu-Thr-Ser-NH₂ activated by BOP or HBTU. In this case activation by DCC failed completely [117].

In analogy to the synthesis of O-glycopeptides, N-glycosylation was directly achieved by coupling the glycosylamine **168** to the cyclopeptide **192** activated by DAEC/HOBt due to the conversion of the glutamic acid into the amide. After deprotection of the sugar part the glycosylated cyclohexapeptide **193** was obtained [118]. Compound **193** was used for detailed conformational studies [118].

As demonstrated by the core trisaccharide **194**, the synthesis of glycopeptides carrying L-fucose residues is difficult due to their sensitivity to hydrolysis [19]. To establish an α-glycosyl linkage between the 6-OH position of chitobiosyl azide and L-fucose the only possible glycosyl reaction is an *in situ* anomerisation without neighbour group participation. This reaction of the corresponding disaccharide with 2,3,4-tri-O-benzyl-α-L-fucopyranosyl bromide afforded **194**. After reduction and coupling with Aloc-Asp-OtBu the final deprotection of the tBu-ester with TFA simultaneously

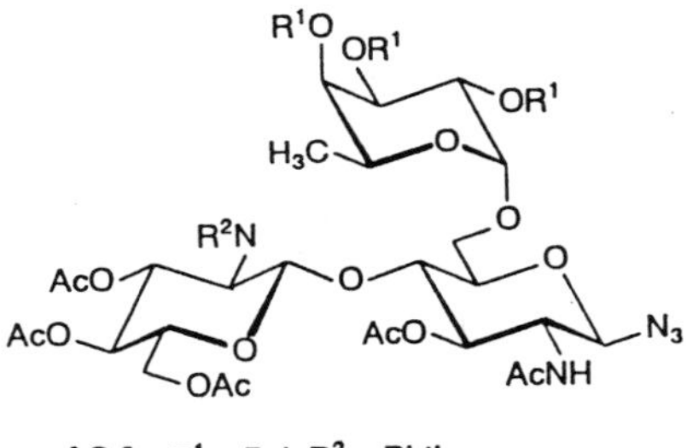

194 R¹ = Bzl; R² = Phth
195 R¹ = CH₂C₆H₄OCH₃-*p* ; R² = Phth
196 R¹ = R² = Ac

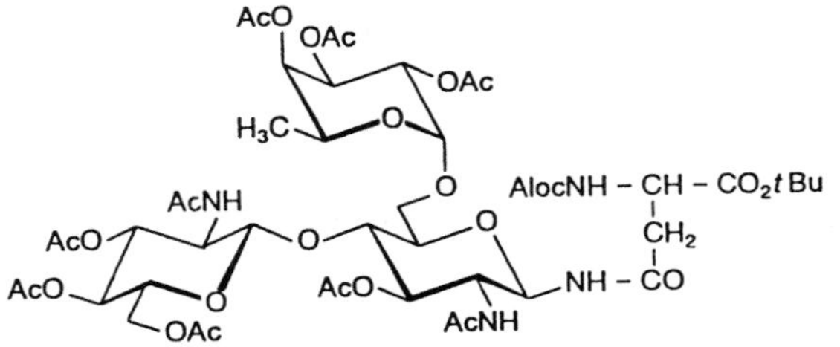

197

Aloc - Ala - Leu - Asn - Leu - Thr - Asn - *t* Bu

α - L - Fuc (OAc)₃ - (1→6)
β - D - GlcNAc (OAc)₃ - (1→4) β - N - D - GlcNAc (OAc)

198

caused hydrolysis of the L-fucose. Benzylated L-fucose is therefore not suitable for glycopeptide synthesis [19]. Substitution of the benzyl groups of L-fucose by *p*-methoxybenzyl ether allowed the selective deprotection in **195** with $(NH_4)_2Ce(NO_3)_6$ and after cleavage of the phthalimido group the conversion into the fully acetylated glycosyl azide **196**. Reduction of the azide with Raney-Ni under neutral conditions and coupling to the aspartic acid derivative afforded **197**. In **197** the glycosyl bond of L-fucose is now stable against TFA due to the rule of the influence of the protection pattern to the stability of the glycosyl bond against hydrolysis [4]. The Aloc-group in **197** was then smoothly cleaved without any glycosyl bond cleavage to elongate the N-terminus by a dipeptide. After deprotection of the C-terminus by treatment with TFA a tripeptide was finally coupled affording the glycopeptide **198** carrying the branched trisaccharide residue [19].

3.2 Synthesis of N-glycopeptides containing other sugar residues

Several other N-glycosylamino acid derivatives were synthesized using different mono- and disaccharides like D-glucose, D-galactose, D-mannose, 2-acetamido-2-deoxy-D-galactose, maltose, cellobiose, lactose or lactosamine. In general the sugar units were coupled to the aspartic acid derivative as their corresponding glycosyl-amines [1].

In the case of D-glucose the acetylated Boc-Asn(sugar)-OH **199** was introduced into the amino acid sequence 5–9 of somatostatin by the solid phase procedure to give **200** [119]. Following the same strategy as described above the crude amino sugars of D-glucose, cellobiose and maltose were employed for the synthesis of the corresponding Fmoc-Asn(sugar)-OH derivatives. The efficiency of these compounds as building blocks with unprotected hydroxyl groups was tested on the solid phase synthesis of N-glycopeptides of H-Ile-Met-Met-Asn(sugar)-Gly-NH$_2$ [115].

Furthermore, the strategy to purify glycosylamine via the acetylated Fmoc-derivative mentioned above is applicable to very different saccharides [120] as shown for instance for maltosamine. After cleavage of the Fmoc-group in **201** by treatment with morpholine the glycosylamine was coupled with the acid chloride **186** to give the building block **202**. Compound **202** was employed in a solid phase synthesis to build up an analogue of the peptide-T amide **203** [120].

An interesting synthesis of the whole structure of nephritoglucoside **205**, a glycopeptide of 21 amino acids carrying an asparagine linked core trisaccharide [121–123] composed of three glucose moieties was performed [124]. Thus the peracetylated α-glycosyl azide of the trisaccharide α-D-Glc-(1→6)-β-D-Glc-(1→6)-D-Glc was synthesized and reduced to the corresponding amine by catalytic hydrogenation using Pd/C. The following coupling with Aloc-Asp-OtBu gave an anomeric mixture of N-glycosides due to the lability of the less stable α-glycosylamine against anomerisation at C-1 carbon. After separation all acetyl groups in **204a** were replaced by Aloc-groups followed by the conversion into the N-hydroxysuccinimide active ester **204c** via **204b**. However, **204c** did not react in a coupling reaction with the Pro of the related eicosapeptide at all but cyclized itself to the succinimide derivative. In order to avoid the succinimide formation **204d** was first coupled with H-Pro-OtBu in the

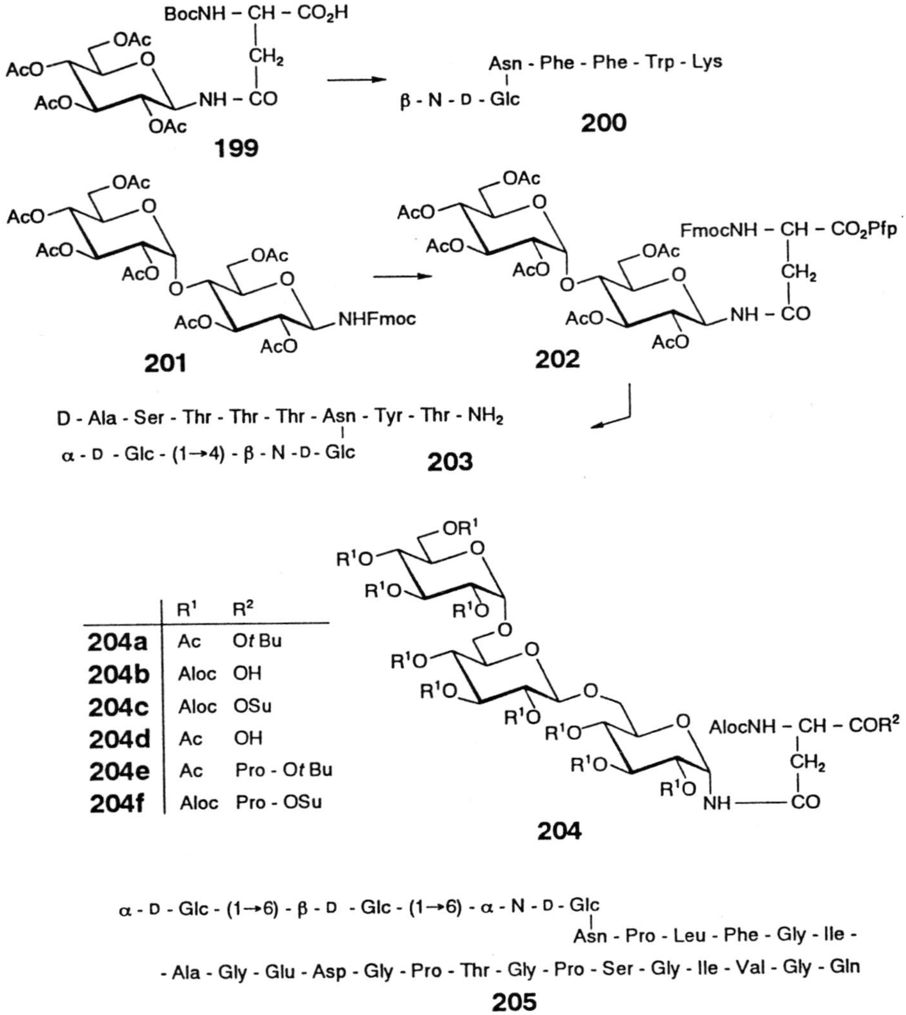
BocNH - CH - CO_2H
AcO
OAc
O
CH_2
NH - CO
199
Asn - Phe - Phe - Trp - Lys
β - N - D - Glc
200
NHFmoc
201
FmocNH - CH - CO_2Pfp
CH_2
NH - CO
202
D - Ala - Ser - Thr - Thr - Thr - Asn - Tyr - Thr - NH_2
α - D - Glc - (1→4) - β - N - D - Glc
203
OR^1
R^1O
AlocNH - CH - COR^2
CH_2
NH — CO
204
α - D - Glc - (1→6) - β - D - Glc - (1→6) - α - N - D - Glc
Asn - Pro - Leu - Phe - Gly - Ile -
- Ala - Gly - Glu - Asp - Gly - Pro - Thr - Gly - Pro - Ser - Gly - Ile - Val - Gly - Gln
205

	R^1	R^2
204a	Ac	O*t*Bu
204b	Aloc	OH
204c	Aloc	OSu
204d	Ac	OH
204e	Ac	Pro - O*t*Bu
204f	Aloc	Pro - OSu

presence DAEC/HOBT to give the glycosyldipeptide **204e**. After replacement of the acetyl groups by Aloc-groups and formation of the N-hydroxysuccinimide active ester **204f** the coupling reaction with the nonadecapeptide was successful and resulted in **205** after deprotection [124].

Abbreviations

Ac	acetyl
Ahc	α-amino-6-hydroxy-capronic acid
All	allyl

Aloc	allyloxycarbonyl
Boc	tertiary butyloxycarbonyl
BOP	(benzotriazole-1-yl-oxy)tris(dimethylamino)phosphonium hexafluorophosphate
BSA	bovine serum albumin
Bz	benzoyl
Bzl	benzyl
DAEC	N-(3'-dimethylaminopropyl)N'-ethylcarbodiimide hydrochloride
DCC	dicyclohexyl carbodiimide
Dhbt	3,4-dihydro-4-oxo-1,2,3-benzotriazinyl
DMAP	4-(dimethylamino)pyridine
EEDQ	2-ethoxy-1-ethoxycarbonyl-2,2-dihydrochinolin
Fmoc	fluoren-9-yl-methoxycarbonyl
HBTU	2-(1H-benzotriazol-1-yl)-1,1,3,3-tetramethyluronium hexafluorophosphate
HOBt	1-hydroxybenzotriazole
Me	methyl
NIS	N-iodosuccinimide
Np	p-nitrophenyl
Pfp	pentafluorophenyl
Su	succinimidyl
tBu	tertiary butyl
TFA	trifluoroacetic acid
TMS-triflate	trimethylsilyltrifluoromethane sulfonate
Triflate	trifluoromethanesulfonic acid
Z	benzyloxycarbonyl

All amino acids without the prefixes D,L are L-amino acids.

References

1. Garg, H. and Jeanloz, R.W. (1985) Adv. Carbohydr. Chem. Biochem. 43, 135–201.
2. Kunz, H. (1987) Angew. Chem. 99, 297–311; Angew. Chem. Int. Ed. Engl. 26, 294–308.
3. Paulsen, H., Adermann, K., Merz, G., Schultz, M. and Weichert, U. (1988) Starch/Stärke 40, 465–472.
4. Paulsen, H. (1982) Angew. Chem. 94, 184–201; Angew. Chem. Int. Ed. Engl. 21, 155–173.
5. Paulsen, H. and Stenzel, W. (1975) Angew. Chem. 87, 547–548; Angew. Chem. Int. Ed. Engl. 14, 558–559.
6. Lemieux, R.U. and Ratcliffe, R.M. (1979) Can. J. Chem. 57, 1244–1251.
7. Paulsen, H., Richter, A., Sinnwell, V. and Stenzel, W. (1978) Carbohydr. Res. 64, 339–364.
8. Ferrari, B. and Pavia, A.A. (1980) Carbohydr. Res. 79, C1–C7.
9. Hölck, J.-P. (1981) Dissertation Universität Hamburg.
10. Paulsen, H. and Hölck, J.-P. (1982) Carbohydr. Res. 109, 89–107.
11. Paulsen, H. and Adermann, K. (1989) Liebigs Ann. Chem. 751–769 and 771–780.
12. Paulsen, H., Merz, G. and Brockhausen, I. (1990) Liebigs Ann. Chem. 719–739.
13. Lüning, B., Norberg, T. and Tejbrant, J. (1989) Glycoconjugate J. 6, 5–19.

14. Schultheiß-Reimann, P. and Kunz, H. (1983) Angew. Chem. 95, 64; Angew. Chem. Int. Ed. Engl. 22, 62.
15. Paulsen, H. and Schultz, M. (1986) Liebigs Ann. Chem. 1435–1447.
16. Paulsen, H. and Schultz, M. (1987) Carbohydr. Res. 159, 37–52.
17. Kunz, H. and Waldmann, H. (1984) Angew. Chem. 96, 49–50; Angew. Chem. Int. Ed. Engl. 23, 71–72.
18. Kunz, H. and Unverzagt, C. (1984) Angew. Chem. 96, 426–427; Angew. Chem. Int. Ed. Engl. 23, 436–437.
19. Kunz, H. and Unverzagt, C. (1988) Angew. Chem. 100, 1763–1765; Angew. Chem. Int. Ed. Engl. 27, 1697–1699.
20. Ferrari, B. and Pavia, A.A. (1983) Int. J. Peptide Protein Res. 22, 549–559.
21. Ferrari, B. and Pavia, A.A. (1985) Tetrahedron 41, 1939–1944.
22. Grundler, G. and Schmidt, R.R. (1984) Liebigs Ann. Chem., 1826–1847.
23. Nakahara, Y., Iijima, H., Shibayama, S. and Ogawa, T. (1991) Carbohydr. Res. 216, 211–225.
24. Kolar, C., Knödler, U., Seiler, F.R., Fehlhaber, H.-W., Sinnwell, V., Schultz, M. and Paulsen, H. (1987) Liebigs Ann. Chem., 577–581.
25. Bencomo, V.V., Jaquinet, J.-C. and Sinay, P. (1982) Carbohydr. Res. 116, C9–C11.
26. Paulsen, H., Paal, M. and Schultz, M. (1983) Tetrahedron Lett. 24, 1759–1762.
27. Paulsen, H. and Paal, M. (1984) Carbohydr. Res. 135, 53–69 and 71–84.
28. Paulsen, H., Schultz, M., Klamann, J.-D., Waller, B. and Paal, M. (1985) Liebigs Ann. Chem. 2028–2048.
29. Kinzy, W. and Schmidt, R.R. (1987) Carbohydr. Res. 166, 265–276.
30. Kinzy, W. and Schmidt, R.R. (1989) Carbohydr. Res. 193, 33–47.
31. Iijima, H. and Ogawa T. (1989) Carbohydr. Res. 186, 95–106 and 107–118.
32. Toyokumi, T., Dean, B. and Hakamori, S. (1990) Tetrahedron Lett. 31, 1826–1847.
33. Paulsen, H., Rauwald, W. and Weichert, U. (1988), Liebigs Ann. Chem. 75–86.
34. Belleau, B. and Melek, G. (1968) J. Am. Chem. Soc. 90, 1651–1652.
35. Bencomo, V.V. and Sinay, P. (1983) Carbohydr. Res. 116, C9–C12.
36. Bencomo, V.V. and Sinay, P. (1984) Glycoconjugate J. 1, 5–8.
37. Kunz, H., Birnbach, S. and Wernig, P. (1990) Carbohydr. Res. 202, 207–223.
38. Merrifield, R.B. (1963) J. Am. Chem. Soc. 85, 2149–2154.
39. Lavielle, S., Ling. N.C. and Guillemin, R.C. (1981) Carbohydr. Res. 89, 221–228.
40. Paulsen, H., Merz, G. and Weichert, U. (1988) Angew. Chem. 100, 1425–1427; Angew. Chem. Int. Ed. Engl. 27, 1365–1367.
41. Kunz, H. and Dombo, B. (1988) Angew. Chem. 100, 732–734; Angew. Chem. Int. Ed. Engl. 27, 711–713.
42. Wang, S.S. (1973) J. Am. Chem. Soc. 95, 1328–1333.
43. Mergler, M., Tanner, R., Gosteli, J. and Grogg, P. (1988) Tetrahedron Lett. 29, 4005–4008.
44. Mergler, M., Nyfeler, R., Tanner, R., Gosteli, J. and Grogg, P. (1988) Tetrahedron Lett. 29, 4009–4012.
45. Albericio, F., Kneib-Cordonier, N., Biancalana, S., Gera, L., Masada, R.I., Hudson, D. and Barany, G. (1990) J. Org. Chem. 55, 3730–3743.
46. Rink, H. (1987) Tetrahedron Lett. 28, 3787–3791.
47. Paulsen, H., Merz, G., Peters, S. and Weichert, U. (1990) Liebigs Ann. Chem. 1165–1173.
48. Lüning, B., Norberg, T. and Tejbrant, J. (1989) J. Chem. Soc. Chem. Commun. 1267–1268.
49. Lüning, B., Norberg, T., Rivera-Baeza, C. and Tejbrant, J. (1991) Glycoconjugate J. 8, 450–455.
50. Kisfaludy, L. and Schön, I. (1983) Synthesis, 325–327.
51. König, W. and Geiger, R. (1970) Chem. Ber. 103, 2034–2040.

52. Atherton, E., Holder, J.L., Meldal, M., Sheppard, R.C. and Valerio, R.M. (1988) J. Chem. Soc. Perkin Trans. I, 2887–2894.

53. Cameron, L.R., Holder, J.L., Meldal, M. and Sheppard, R.C. (1988) J. Chem. Soc. Perkin Trans. I, 2895–2901.

54. Peters, S., Bielfeldt, T., Meldal, M., Bock, K. and Paulsen H. (1991) Tetrahedron Lett. 32, 5067–5070.

55. Peters, S., Bielfeldt, T., Meldal, M., Bock, K. and Paulsen, H. (1992) J. Chem. Soc. Perkin Trans. I, 1163–1171.

56. Bielfeldt, T., Peters, S., Meldal, M., Bock, K. and Paulsen, H. (1992) Angew. Chem. 104, 881–883; Angew. Chem. Int. Ed. Engl. 31, 857–859.

57. Paulsen, H., Bielfeldt, T., Peters, S., Meldal, M. and Bock, K. (1994), Liebigs Ann. Chem. 369–379 and 381–387.

58. Peters, S., Bielfeldt, T., Meldal, M., Bock, K. and Paulsen H. (1992) Tetrahedron Lett. 33, 6445–6448.

59. Lindberg, B. and Silvander, B.-G. (1965) Acta Chem. Scand. 19, 530–531.

60. Erbing, B., Lindberg, B. and Norberg, T. (1978) Acta Chem. Scand. B 32, 308–310.

61. Ekborg, G., Klinger, M., Roden, L., Jensen, J.W., Schutzbach, J.S., Huang, D.H., Krishna, N.R. and Anantharamaiah, G.M. (1987) Glycoconjugate J. 4, 255–266.

62. Ekborg, G., Curenton, T., Krishna, N.R. and Roden, L. (1990) J. Carbohydr. Chem. 9, 15–37.

63. Garg, H.G., Hasenkamp, T. and Paulsen, H. (1986) Carbohydr. Res. 151, 225–232.

64. Paulsen, H. and Brenken, M. (1988) Liebigs Ann. Chem. 649–654.

65. Rio, S., Beau, J.-M. and Jaquinet, J.-C. (1991) Carbohydr. Res. 219, 71–90.

66. Rio, S., Beau, J.-M. and Jaquinet, J.-C. (1993) Carbohydr. Res. 244, 295–313.

67. Kunz, H. and Buchholz, M. (1981) Angew. Chem. 93, 917–918; Angew. Chem. Int. Ed. Engl. 20, 894–895.

68. Buchholz, M. and Kunz, H. (1983) Liebigs Ann. Chem. 1859–1885.

69. Derevitskaya, V.A., Vafina, M.G. and Kotchetkov, N.K. (1967) Carbohydr. Res. 3, 377–388.

70. Lacombe, J.M., Pavia, A.A. and Rocheville, J.M. (1981) Can. J. Chem. 59, 473–481.

71. Turner, N.J. and Webberley, M.C. (1991) J. Chem. Soc. Chem. Commun. 1349–1350.

72. Fukase, H., Hase, S., Ikenaka, T. and Kusomoto, S. (1992) Bull. Chem. Soc. Jpn. 65, 436–445.

73. Garegg, P.J., Oscarson, S., Kvarnström, I., Niklasson, A., Niklasson, G., Svensson, S.C.T. and Edwards, J.V. (1990) Acta Chem. Scand. 44, 625–629.

74. Meldal, M., Mouritson, S. and Bock, K. (1992) in: Carbohydrate Antigens, Am. Chem. Soc. Symposium Series Vol. 519, 19–33, Washington DC.

75. Szabo, L., Li, Y. and Polt, R. (1991) Tetrahedron Lett. 32, 585–588.

76. Polt, R., Szabo, L., Treiberg, J., Li, Y. and Hruby, V.J. (1992) J. Am. Chem. Soc. 114, 10249–10258.

77. Bardaji, E., Torres, J.L., Clapes, P., Albericio, F., Barany, G. and Valencia, G. (1990) Angew. Chem. 102, 311–313; Angew. Chem. Int. Ed. Engl. 29, 291–293.

78. Bardaji, E., Torres, J.L., Clapes, P., Albericio, F., Barany, G., Rodriguez, R.E., Sacristan, M.P. and Valencia, G. (1991) J. Chem. Soc. Perkin Trans I, 1755–1759.

79. Jensen, K.J., Meldal, M. and Bock, K. (1993) J. Chem. Soc. Perkin Trans I, 2119–2129.

80. Jensen, K.J., Meldal, M. and Bock, K. (1992) In: J.A. Smith and J.E. Rivier (Eds.), Peptides 1991, Proc. 12th Ann. Pept. Symp. Piere Chemical Company, Leiden, pp. 587–588.

81. Meldal, M. and Jensen, K.J. (1990) J. Chem. Soc. Chem. Commun. 483–485.

82. Hoogerhout, P., Guis, C.P., Erkelens, C., Bloemhoff, W., Kerling, K.E.T. and van Boom, J.H. (1985) Recl. Trav. Chim. Pays-Bas 104, 54–59.

83. Lacombe, J.M. and Pavia, A.A. (1983) J. Org. Chem. 48, 2557–2563.

84. Filira, F., Biondi, L., Scolaro, B., Foffani, M.T., Mammi, S., Peggion, E. and Rocchi, R. (1990)

Int. J. Biol. Macromol. 12, 41–49.

85. De la Torre, B.G., Torres, J.L., Bardaji, E., Clapes, P., Xaus, N., Jorba, X., Clavet, S., Albericio, F. and Valencia, G. (1990) J. Chem. Soc. Chem. Commun. 965–967.

86. Sauerbrei, B. and Thiem, J. (1992) Tetrahedron Lett. 33, 201–204.

87. Holla, E.W., Schudok, M., Weber, A. and Zulauf, M. (1992) J. Carbohydr. Chem. 11, 659–663.

88. Cantacuzene, D., Attal, S. and Bay, S. (1991) Bioorganic & Medical Chem. Lett. 1, 197–200.

89. Jansson, A.M., Meldal, M. and Bock, K. (1990) Tetrahedron Lett. 31, 6991–6994.

90. Jansson, A.M., Meldal, M. and Bock, K. (1992) J. Chem. Soc. Perkin Trans I, 1699–1707.

91. Kauth, H. and Kunz, H. (1982) Liebigs Ann. Chem. 360–366.

92. Hollósi, M., Kollát, E., Lacko, J., Medzihradszky, K.F., Thurin, J. and Otvös Jr., L. (1991) Tetrahedron Lett. 32, 1531–1534.

93. Paulsen, H. and Hölck, J.-P. (1982) Liebigs Ann. Chem. 1121–1131.

94. Thiem, J., Karl, H. and Schwentner, J. (1978) Synthesis, 696–698.

95. Kessler, H., Kottenhahn, M., Kling, A. and Kolar, C. (1987) Angew. Chem. 99, 919–921; Angew. Chem. Int. Ed. Engl. 26, 888–890.

96. Kessler, H., Kling, A. and Kottenhahn, M. (1990) Angew. Chem. 102, 452–554; Angew. Chem. Int. Ed. Engl. 29, 425–427.

97. Kottenhahn, M. and Kessler, H. (1991) Liebigs Ann. Chem. 727–744.

98. Kessler, H., Matter, H. Gemmecker, G., Kottenhahn, M. and Bats, J.W. (1992) J. Am. Chem. Soc. 4805–4818.

99. Kessler, H., Wittmann, V., Köck, M. and Kottenhahn, M. (1992) Angew. Chem. 104, 877–879; Angew. Chem. Int. Ed. Engl. 31, 874–876.

100. Spinola, M. and Jeanloz, R.W. (1979) J. Biol. Chem. 245, 4158–4162.

101. Cowley, D.E., Hough, L. and Peach, C.M. (1971) Carbohydr. Res. 19, 231–241.

102. Thiem, J. and Wiemann, T. (1990) Angew. Chem. 102, 78–80; Angew. Chem. Int. Ed. Engl. 29, 80–82.

103. Marks, G.S. and Neuberger, A. (1961) J. Chem. Soc. 4872–4879.

104. Waldmann, H., März, J. and Kunz, H. (1990) Carbohydr. Res. 196, 75–93.

105. Kunz, H. and Waldmann, H. (1985), Angew. Chem. 97, 885–887; Angew. Chem. Int. Ed. Engl. 24, 883–885.

106. Kunz, H. and Waldmann, H. and März, J. (1989) Liebigs Ann. Chem. 45–49.

107. Khorlin, A.Y., Zurabyan, S.E. and Macharadze, R.G. (1980) Carbohydr. Res. 85, 201–208.

108. Kunz, H. and Günther, W. (1990) Angew. Chem. 102, 1068–1069; Angew. Chem. Int. Ed. Engl. 29, 1050–1051.

109. Lee, H.H., Baptista, J.A.B. and Krepinsky, J.J. (1990) Can. J. Chem. 68, 953–957.

110. Nakabayashi, S., Warren, C.D. and Jeanloz, R.W. (1988) Carbohydr. Res. 174, 279–289.

111. Meldal, M. and Bock, K. (1990) Tetrahedron Lett. 31, 6987–6990.

112. Biondi, L., Filira, F., Gobbo, M., Scolaro, B., Rocchi, R. and Cavaggion, F. (1991) Int. J. Peptide Protein Res. 37, 112–121.

113. Otvos Jr., L., Wroblewski, K., Kollat, E., Perczel, A., Hollosi, M., Fasman, G.D., Ertl, H.C.J. and Thurin, J. (1989) Pept. Res. 2, 362–366.

114. Otvos Jr., L., Urge, L., Hollosi, M., Wroblewski, K., Graczyk, G., Fasman, G.D. and Thurin, J. (1990) Tetrahedron Lett. 31, 5889–5892.

115. Urge, L., Kollat, E., Hollosi, L., Laczko, J., Wroblewski, K., Thurin, J. and Otvos Jr., L. (1991) Tetrahedron Lett. 32, 3445–3448.

116. Likhosterov, L.M., Novikova, O.S., Derevitskaja, V.A. and Kochetkov, N.K. (1986) Carbohydr. Res. 146, C1–C5.

117. Anisfeld, S. and Landsbury Jr., P.T. (1990) J. Org. Chem. 55, 5560–5562.

118. Kessler, H., Matter, H., Gemmecker, G., Kling, A. and Kottenhahn, M. (1991) J. Am. Chem.

Soc. 113, 7550–7563.
119. Lavielle, S., Ling, N.C. and Guillemin, R.C. (1981) Carbohydr. Res. 89, 221–228.
120. Christiansen-Brams, I., Meldal, M. and Bock, K. (1993) J. Chem. Soc. Perkin Trans I 1461–1471.
121. Sasaki, S., Tachibana, K. and Nakanishi, H. (1991) Tetrahedron Lett. 32, 6873–6876.
122. Takeda, T., Utsuno, A., Okamoto, N., Ogihara, Y. and Shibata, S. (1990) Carbohydr. Res. 207, 71–79.
123. Ogawa, T. and Nakabayashi, S. (1980) Carbohydr. Res. 86, C7–C10.
124. Teshima, T., Nakajima, K., Takahashi, M. and Shiba, T. (1992) Tetrahedron Lett. 33, 363–366.

J. Montreuil, H. Schachter and J.F.G. Vliegenthart (Eds.), *Glycoproteins*

123

Biosynthesis

1. Introduction

HARRY SCHACHTER

Department of Biochemistry Research, Hospital for Sick Children, 555 University Avenue, Toronto, Ont. M5G 1X8, Canada

Oligosaccharides, like proteins and nucleic acids, are essential biological polymers. They differ from proteins and nucleic acids in two important properties, i.e., branching and monomer linkage. Both nucleic acids and proteins are assembled as linear molecules. The monomeric units in these polymers are linked together by a single type of linkage, 3′–5′-phosphodiester bonds and amides, respectively. Linear molecules can be assembled by a template mechanism in which every new molecule is copied from a pre-existing template molecule. Oligosaccharides, on the other hand, are often highly branched molecules and their monomeric units are connected to one another by many different linkage types. Such complex structures cannot be copied from a template and must be manufactured on an assembly line in which individual components are incorporated sequentially. The assembly line for protein- and lipid-bound oligosaccharides is the endomembrane system — the endoplasmic reticulum and Golgi apparatus [1]. A series of membrane-bound glycosidases and glycosyltransferases act sequentially on the growing oligosaccharide as it moves along the lumen of the endomembrane system.

Many glycosidase and glycosyltransferase reactions are now known and these have been organized into pathways leading to the assembly of lipid-bound oligosaccharides [2,3], protein-bound N-glycans (Asn-GlcNAc N-glycosidic linkage) [1,4–14] and protein-bound O-glycans (Ser/Thr-GalNAc O-glycosidic linkage) [4,10,12,15–17]. The number of different glycan structures that have been isolated and characterized in recent years has created a dilemma for glycobiology — how and why are these structures made? Probable functions are discussed in other parts of this book. This chapter is concerned with the enzymes that synthesize N- and O-glycans and with the mechanisms that control biosynthesis. Emphasis is placed on recent contributions to the field and the reader is invited to consult the reviews listed above for earlier work.

Glycosyltransferases are enzymes which carry out the biosynthesis of oligosaccharides. The reactions catalyzed by these enzymes are:

Activated sugar or oligosaccharide donor + acceptor (R-OH or R-NH)
$\longrightarrow$
R-O(or N)-sugar (or oligosaccharide) + nucleotide, or dolichol-phosphate, or dolichol-pyrophosphate

The acceptor R can be a free saccharide, or a saccharide linked to an aglycone (e.g., a protein or lipid), or a protein or lipid. The donor can be a nucleotide-sugar, dolichol-phosphate-sugar or dolichol-pyrophosphate-oligosaccharide. The "one linkage-one glycosyltransferase" rule states that there is one distinct glycosyltransferase for every glycosidic linkage. There are, however, exceptions to this rule, e.g., the Lewis blood group-dependent α-1,3/4-fucosyltransferase. It is also possible for a particular linkage to be made by more than a single glycosyltransferase, e.g., there are at least five distinct human α-1,3-fucosyltransferases which can synthesize the Gal(β1–4)[Fuc(α1–3)] GlcNAc moiety [6,18–23]. The glycosyltransferases are usually membrane-bound and detergent treatment of the membrane preparations is required for solubilization and full expression of enzymatic activity *in vitro*. Most of the transferase assays require addition of exogenous divalent cation (usually Mn^{2+}). In some cases, Mn^{2+} can be replaced by other divalent cations. Exogenous cation is not required for the assay of sialyltransferases and several β6-GlcNAc-transferases. The metabolism and synthesis of nucleotide-sugars [24] are not considered in this book.

Glycosyltransferases usually show precise specificity for both the acceptor and nucleotide-sugar donor. Promiscuous reactions can, however, occur *in vitro*, usually at relatively slow rates. For example, the human blood group B α-1,3-Gal-transferase can transfer GalNAc instead of Gal to its acceptor substrate *in vitro* and thereby make the human blood group A epitope [25]. The consequences of such a reaction *in vivo* would clearly be disastrous. Sialyltransferases [26–28], the Lewis blood group-dependent α-1,3/4-fucosyltransferase [29] and GlcNAc-transferase I [30] have been shown to transfer derivatives of sialic acid, fucose and GlcNAc, respectively, from chemically synthesized nucleotide-sugar analogues to acceptor oligosaccharides. Sialyltransferase can transfer sialic acid in α-2,6-linkage not only to Gal-terminal compounds but also to Man, Glc, GlcNAc and GalNAc [31–35]. Commercially available bovine milk β-1,4-Gal-transferase can be used to transfer not only Gal but also GalNAc, Glc and glucosamine in β-1,4-linkage to GlcNAc-terminal oligosaccharides [36]. These reactions have proved useful as adjuncts to classical organic chemistry for the chemical-enzymatic synthesis of interesting oligosaccharides.

The large number of complex oligosaccharide structures that have been described indicates the existence of well over 100 different glycosyltransferases. The genes of only a relatively small number of mammalian glycosyltransferases have been cloned to date [37,38]. Rapid progress is being made in the molecular biology of the glycosyltransferases and glycosidases of the endomembrane assembly line. The availability of probes for the genes of these enzymes will lead inevitably to a better understanding of the functions of glycoconjugates in normal processes such as embryogenesis and development as well as in the etiology of human disease.

References

1. Kornfeld, R. and Kornfeld, S. (1985) Ann. Rev. Biochem. 54, 631–664.
2. Basu, S., Basu, M., Das, K. K., Daussin, F., Schaeper, R.J., Banerjee, P., Khan, F.A. and Suzuki, I. (1988) Biochimie 70, 1551–1563.
3. Basu, S.C. (1991) Glycobiology 1, 469–475.

4. Beyer, T.A., Sadler, J.E., Rearick, J.I., Paulson, J.C. and Hill, R.L. (1981). In: A. Meister (Ed.). Advances in Enzymology. John Wiley and Sons, New York, Vol. 52, pp. 23–175.

5. Cummings, R.D. (1992) in: H.J. Allen and E.C. Kisailus (Eds.), Glycoconjugates. Composition, Structure and Function. Marcel Dekker, Inc., New York, NY, pp. 333–360.

6. Macher, B.A., Holmes, E.H., Swiedler, S.J., Stults, C.L.M. and Srnka, C.A. (1991) Glycobiology 1, 577–584.

7. Schachter, H. and Roseman, S. (1980) in: W.J. Lennarz (Ed.), Biochemistry of Glycoproteins and Proteoglycans. Plenum Press, New York, NY, pp. 85–160.

8. Schachter, H., Narasimhan, S., Gleeson, P. and Vella, G. (1983) Can. J. Biochem. Cell Biol. 61, 1049–1066.

9. Schachter, H., Narasimhan, S., Gleeson, P. and Vella, G. (1983) Methods in Enzymology 98, 98–134.

10. Schachter, H., Narasimhan, S., Gleeson, P., Vella, G. and Brockhausen, I. (1985) in: A.N. Martonosi (Ed.), The Enzymes of Biological Membranes, Biosynthesis and Metabolism, Vol. 2. Plenum Press, New York, NY, pp. 227–277.

11. Schachter, H. (1986) Biochem. Cell Biol. 64, 163–181.

12. Schachter, H., Brockhausen, I. and Hull, E. (1989) Methods Enzymol. 179, 351–396.

13. Schachter, H. (1991) Glycobiology 1, 453–461.

14. Snider, M.D. (1984) in: V. Ginsburg and P.W. Robbins (Eds.), Biology of Carbohydrates. John Wiley and Sons, New York, NY, Vol. 2, pp. 163–198.

15. Sadler, J.E. (1984) in: V. Ginsburg and P.W. Robbins (Eds.), Biology of Carbohydrates. John Wiley and Sons, New York, NY, Vol. 2, pp. 199–288.

16. Schachter, H., Brockhausen, I. (1989) in: E. Chantler and N.A. Ratcliffe (Eds.), Mucus and Related Topics. Society for Experimental Biology, Cambridge, England, Vol. 43, pp. 1–26.

17. Schachter, H. and Brockhausen, I. (1992) in: H.J. Allen and E.C. Kisailus (Eds.), Glycoconjugates. Composition, Structure and Function. Marcel Dekker, Inc., New York, NY, pp. 263–332.

18. Mollicone, R., Gibaud, A., François, A., Ratcliffe, M. and Oriol, R. (1990) Eur. J. Biochem. 191, 169–176.

19. Tetteroo, P.A.T., de Heij, H.T., van den Eijnden, D.H., Visser, F.J., Schoenmaker, E. and van Kessel, A.H.M.G. (1987) J. Biol. Chem. 262, 15984–15989.

20. Couillin, P., Mollicone, R., Grisard, M. C., Gibaud, A., Ravisé, N., Feingold, J. and Oriol, R. (1991) Cytogenet. Cell Genet. 56, 108–111.

21. Lowe, J.B., Kukowska-Latallo, J.F., Nair, R.P., Larsen, R.D., Marks, R.M., Macher, B.A., Kelly, R.J. and Ernst, L.K. (1991) J. Biol. Chem. 266, 17467–17477.

22. Weston, B.W., Nair, R.P., Larsen, R.D. and Lowe, J.B. (1992) J. Biol. Chem. 267, 4152–4160.

23. Weston, B.W., Smith, P.L., Kelly, R.J. and Lowe, J.B. (1992) J. Biol. Chem. 267, 24575–24584.

24. Schachter, H. and Roden, L. (1973) in: W.H. Fishman (Ed.), Metabolic Conjugation and Metabolic Hydrolysis. Academic Press, New York, NY, Vol. III, pp. 1–149.

25. Greenwell, P., Yates, A.D. and Watkins, W.M. (1986) Carbohydrate Res. 149, 149–170.

26. Gross, H.J., Bunsch, A., Paulson, J.C. and Brossmer, R. (1987) Eur. J. Biochem. 168, 595–602.

27. Gross, H.J. and Brossmer, R. (1988) Eur. J. Biochem. 177 (3), 583–589.

28. Gross, H.J., Rose, U., Krause, J.M., Paulson, J.C., Schmid, K., Feeney, R.E. and Brossmer, R. (1989) Biochemistry 28, 7386–7392.

29. Gokhale, U.B., Hindsgaul, O. and Palcic, M.M. (1990) Can. J. Chem. 68, 1063–1071.

30. Srivastava, G., Alton, G. and Hindsgaul, O. (1990) Carbohydr. Res. 207, 259–276.

31. Van Pelt, J., Dorland, L., Duran, M., Hokke, C.H., Kamerling, J.P. and Vliegenthart, J.F.G. (1989) FEBS Lett. 256, 179–184.

32. Van Pelt, J., Hokke, C. H., Dorland, L., Duran, M., Kamerling, J.P. and Vliegenthart, J.F.G. (1990) Clin. Chim. Acta 187, 55–60.

126

33. Hokke, C.H., Duran, M., Dorland, L., Van Pelt, J. and Van Sprang, F.J. (1990) J. Inherited Metab. Dis. 13, 273–276.
34. Van Pelt, J., Dorland, L., Duran, M., Hokke, C. H., Kamerling, J. P. and Vliegenthart, J.F.G. (1990) J. Biol. Chem. 265, 19685—19689.
35. Hokke, C.H., van der Ven, J.G.M., Kamerling, J.P. and Vliegenthart, J.F.G. (1991) Glycoconjugate J. 8, 259.
36. Palcic, M.M. and Hindsgaul, O. (1991) Glycobiology 1, 205–209.
37. Schachter, H. (1991) Current Opinion in Structural Biology 1, 755–765.
38. Paulson, J.C. and Colley, K.J. (1989) J. Biol. Chem. 264, 17615–17618.

J. Montreuil, H. Schachter and J.F.G. Vliegenthart (Eds.), *Glycoproteins*
© 1995 Elsevier Science B.V. All rights reserved

Biosynthesis

2a. The coenzymic role of phosphodolichols

FRANK W. HEMMING

*Department of Biochemistry, University of Nottingham,
Queen's Medical Centre, Nottingham NG7 2UH, UK*

1. Dolichols and other polyprenols

Dolichols (Fig. 1) belong to one of the three groups of polyprenols that occur in living cells (Table I) [1]. No function has been ascribed to the free polyprenols of Groups 1 or 3. However, some members of group 1 give rise to the hydrophobic side chains of quinones whereby they are anchored in mitochondrial, chloroplastidic or bacterial membranes as an essential feature of their role in electron transport.

The phosphorylated polyprenols of Group 2 function in glycosyltransfer: the fully unsaturated compounds in procaryotic systems and the α-saturated compounds (dolichols) in eucaryotes. In both cases the hydrophobic polyisoprenoid chain anchors sugar phosphates in membranes to facilitate enzyme catalysed glycosylation.

2. Phosphodolichol pathway of N-glycosylation

The coenzymic role of phosphodolichols (PDol) in the N-glycosylation of proteins has been reviewed on several occasions [2,3]. This process which occurs co-translationally in the rough endoplasmic reticulum of most eucaryotic cells is summarised in Fig. 2. It can be seen that phosphodolichol acts as an acceptor for *N*-acetylglucosaminyl-1-phosphate transferase (step 1) and that further glycosylation of the product leads to an oligoglycan of 14 monosaccharide residues carried by diphosphodolichol

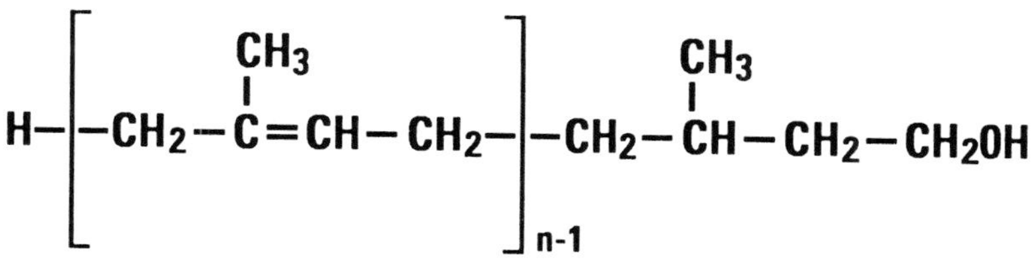

$$H-\left[-CH_2-\underset{\underset{CH_3}{|}}{C}=CH-CH_2-\right]_{n-1}-CH_2-\underset{\underset{CH_3}{|}}{CH}-CH_2-CH_2OH$$

Fig. 1. The structure of dolichol, $n = 11-12$.

TABLE I

Summary of naturally occurring polyisoprenoid alcohols

	Structure	Size range	Location
Group 1	*all trans* ΩT_{n-1}–OH	$n = 9,10$	Some leaves
Group 2	*ditrans,polycis* $\Omega T_2 C_{n-3}$OH	$n = 6$–9	Some wood plant tissue, bacteria
	$\Omega T_2 C_{n-4}$S–OH	$n = 15$–24	All eucaryotic cells (dolichols)
Group 3	*tritans,polycis* $\Omega T_3 C_{n-4}$–OH	$n = 10$–13 (–22)	Leaves
	$S_2 T_2 C_{n-5}$S–OH	$n = 19$–24	Some fungi

Note: Group 1 alcohols have *all-trans*-isoprenoid chains. Group 2 alcohols are derived from ω-*trans,trans*-triisoprenoid and group 3 are derived from ω-*trans,trans,trans*-tetraisoprenoid precursors. There a few examples where n is larger than the normal range quoted above.
Ω: terminal isoprene residue furthest from hydroxyl group (contains methyl groups *cis* and *trans* to hydrogen on the double bond).
C: isoprene residue with methyl group *cis* to hydrogen on the double bond.
T: isoprene residue with methyl group *trans* to hydrogen on the double bond.
S: dihydro isoprene residue.

(Fig. 3). The elaboration of this oligosaccharide involves phosphodolichol-mediated mannosylation (steps 8 and 4, Fig. 2) and glucosylation (steps 9 and 5, Fig. 2). Finally an oligosaccharyltransferase (step 6) transfers the completed oligosaccharide to the amido nitrogen of an asparagine residue of the triplet Asn X Ser (Thr) of an acceptor protein. The phosphodolichol pathway therefore involves at least 6 transferases capable of recognising phosphodolichol as a donor or acceptor.

3. Phosphodolichol structure and biosynthesis: influence on function

The critical features of dolichol essential for its role in the pathway continue to be explored, assisted by developments in chemical synthetic methods [4]. These methods now allow the formation of unambiguous combinations of *cis* and *trans* (Z and E) isoprene residues in dolichols of a required size which can then be phosphorylated. Chain length, *cis/trans* pattern and α-saturation are critical factors to efficient functioning. Recently [5] it has been reported that a C-3 methyl group is essential and that the phytanyl analogue (octahydro-tetraprenyl) will substitute for dolichyl derivatives efficiently [6,7].

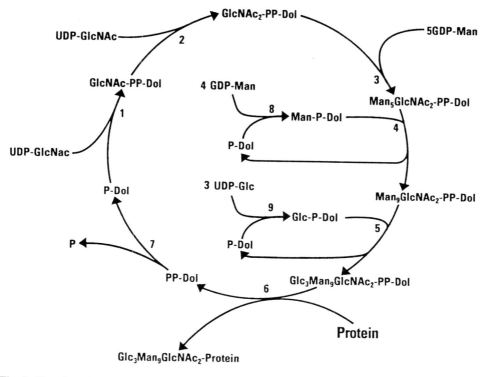

Fig. 2. The phosphodolichol pathway leading to N-glycosylation of protein. The numbers indicate specific enzymes or groups of enzymes (e.g. steps 3, 4 and 5 each summarise several enzyme-catalysed reactions).

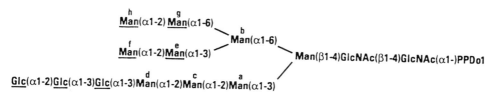

Fig. 3. The fully glycosylated oligosaccharyl diphosphodolichol. The superscript letters indicate the probable order (alphabetical) of addition of α-linked mannose residues. The underlined sugars were donated from mannosyl (or glucosyl) phosphodolichol.

The biosynthetic pathway for dolichol showing its relationship to that of ubiquinone and cholesterol is summarised briefly in Fig. 4. Details have been discussed previously [8]. It remains relevant that in most cells the metabolic flux through B is much greater than through C and D and that control of steps A and B is likely to influence the rate of

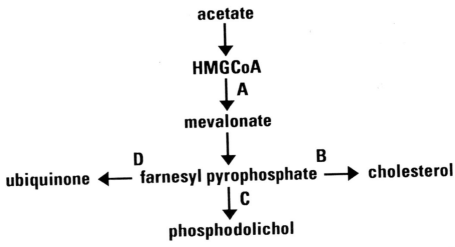

Fig. 4. Biosynthetic pathways for phosphodolichol cholesterol and ubiquinone. (The letters are added for ease of reference in the text.)

C or D (assuming that C and D are not already saturated with substrate) through changes in the concentration of farnesyl pyrophosphate.

The final stages in the biosynthesis of phosphodolichol appear to be critical to the production of a functional product. With regard to α-saturation Krag and coworkers [9] have isolated mutants of CHO cells (F2A8 and Lec9) that appear to be defective in α-saturase activity. Both mutants accumulate much increased quantities of α-saturated polyprenols but the rates of N-glycosylation of proteins are much reduced, compared with the parent cell line. An increased rate of polyprenol biosynthesis was also observed by Ekstrom et al. [10] when cell free preparations of rat liver were incubated with radioactive mevalonate in conditions in which α-saturation did not occur. These studies also revealed that continuous polyisoprene synthesis is essential for α-saturation to occur. The final product of the biosynthetic pathway may be dolichol or phosphodolichol [10]. It has been proposed that the addition of the final α-isoprene unit could in some circumstances involve isopentenol, rather than isopentenyl pyrophosphate resulting in dolichol rather than the anticipated diphosphodolichol. Recent results suggest that biosynthesis of dolichol and its derivatives occurs primarily in the peroxisomes of rat liver [11], a situation that would provide further opportunities for control of the amount of phosphodolichol available to the endoplasmic reticulum for protein N-glycosylation.

The control of the interconversion of dolichol and phosphodolichol through the activities of the CTP-dependent dolichol kinase and phosphodolichol phosphatase have been discussed on several occasions. In developing rat brain the kinase appears to be the main influence on the concentration of phosphodolichol [12]. In CHO cells there is a strong positive correlation between their growth in culture, their ability to produce glycoproteins and their content of phosphodolichol [13].

4. Phosphodolichol pathway: subcellular location

There is strong consistent evidence that in most eucaryotic cells protein N-glycosylation occurs primarily in the rough endoplasmic reticulum. Most phosphodolichol synthesis also has appeared to occur in the same organelle. This site is still accepted as the major site of both activities. However, recent reports [11] of the peroxisomes as an important organelle for dolichol biosynthesis in liver question some of the earlier evidence. Also for several years, Louisot et al. [14] have been reporting phosphodolichol pathway glycosylation activity in mitochondrial membranes after subcellular fractionation of liver homogenates. This suggestion is clearly very dependant upon the absence of contamination of mitochondrial preparations with endoplasmic reticulum membranes. The authors have undertaken extensive precautions and tests to establish this point. Recently the purification of mitochondrial mannosyl phosphodolichol synthase was reported [15,16]. The enzyme was a 30 kDa protein free of other phosphodolichol-requiring enzymes. This corresponds in size to the same enzyme described in yeast and animal microsomes [17,18]. The mitochondrial enzyme can be stabilised by phospholipids (especially sphingomyelin). Antiserum to the mitochondrial enzyme showed the enzyme to be on the cytoplasmic face of the outer mitochondrial membrane. The antiserum also cross reacted with a 30 kDa protein present in purified microsomal membranes, presumably from rough endoplasmic reticulum.

5. Phosphodolichol pathway: topography in the membrane of rough endoplasmic reticulum

The topography of the intermediates and enzymes of the phosphodolichol pathway of N-glycosylation in the membranes of the rough endoplasmic reticulum has been addressed by several workers, notably by Hirschberg [19,20]. Figure 5 summarises the work of Hirschberg et al. and of other workers which is based primarily on *in vitro* studies with sealed vesicles (cytoplasmic side out) isolated from the endoplasmic reticulum. The vesicles were probed with reagents such as proteases, lectins, glycosyltransferases and EDTA (removes essential metal ions) none of which could cross the vesicle membrane unless detergent was present. Comparative experiments, plus and minus detergent, yielded useful distributional information.

The model proposed in Fig. 5 shows the location of the transfer of N-acetylglucosamine phosphate to phosphodolichol (step 1) at the cytoplasmic face. Despite the presence of a UDPGlcNAc translocase allowing UDPGlcNAc to enter the lumenal space, the failure of 5-BrdUMP (an inhibitor of the translocase) to block step 1 supports a cytoplasmic face site. The product of step 1 has also been located on this surface. The further transfer of N-acetylglucosamine and the first five mannose residues, all directly from GDPmannose (step 2), are also clearly located at the cytoplasmic surface. It is relevant that there is no translocase for GDPmannose. Equally certain is the presence at the lumenal surface of all subsequent transfer of mannose (step 6) and glucose (step 9) to the growing oligosaccharide portion and of the final enzyme oligosaccharyl transferase (step 10). This requires the flip-flop of the $Man_5GlcNAc_2PPDol$ from the

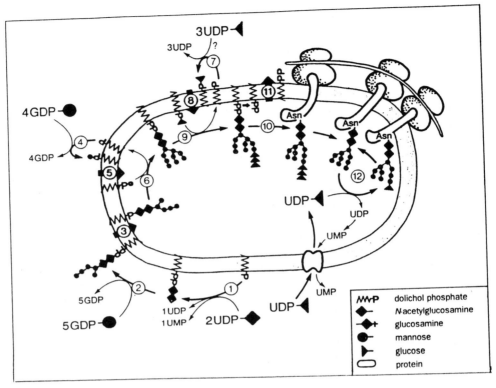

Fig. 5. Topography of N-glycosylation. The numbers are added for ease of reference in the text. From Abeijon and Hirschberg [20] with permission.

cytoplasmic to the lumenal face of the membrane (step 3). Since the evidence for synthesis of ManPDol (step 4) at the cytoplasmic face and for the transfer of mannose from mannosyl phosphodolichol to the growing oligosaccharides on disphospho-dolichol (step 6) at the lumenal face is strong the model also requires flip-flop of mannosyl phosphodolichol (step 5). Probably glucosyl phosphodolichol undergoes a similar movement across the membrane (step 8) because synthesis of this compound (step 7) is at the cytoplasmic face. It has been suggested that the UDPglucose translo-case of the rough endoplasmic reticulum provides lumenal UDPglucose for transient reglucosylation of glycoprotein (step 12) possibly in a different compartment of the endoplasmic reticulum.

The flip-flop of these hydrophilic carbohydrate groups across the hydrophobic membrane almost certainly is assisted by specific proteins and requires input of energy. In the case of mannosyl phosphodolichol the early suggestion [21] that the protein concerned could be the mannosyltransferase, based on experiments with reconstituted systems has recently been challenged by Schutzbach [22].

Some aspects of Fig. 5 have been readdressed. Using chick embryo liver mi-crosomes Kean [23] presented phosphodolichol in liposomes as acceptor for the transfer of *N*-acetylglucosamine phosphate. The recovery of newly synthesised *N*-acetylglucosaminyl diphosphodolichol in the liposomes indicated a cytoplasmic orientation of the transferase concerned. This conclusion was strengthened by the observation that trypsin (in the absence of detergent) destroyed this enzymic activity and that liposome-bound tunicamycin also inhibited the reaction.

Using the photoaffinity-labelled substrate [β-^{32}P] 5 azido UDPGlc Elbein's group have reported [24] that glucosyl phosphodolichol synthase is located at the cytoplasmic face of rat liver microsomes.

On the other hand, Deglon et al. [25] showed that when mRNA for yeast mannosyl phosphodolichol synthase was translated in the presence of dog pancreatic microsomes the enzyme was incorporated into the microsomes where it was resistant to proteolysis unless 0.6% Triton X-100 was present. They argue that a lumenal face location for the enzyme is supported by the amino acid sequence deduced from its cDNA which contains a putative yeast signal peptide at amino acid residues 1–10.

6. Mannosyl phosphodolichol: relationships to other glycosylation pathways

Mannosyl phosphodolichol also functions as a mannose donor in yeast protein O-mannosylation (Fig. 6) [26] and more recently [27] was reported to donate mannose to the PI glycan (Fig. 7) [28]. Thus in yeast three different systems call upon this donor (Fig. 8) [29] a situation which raises the question of the priority of its use. The priority between N-glycosylation and PI glycan synthesis has been explored in some animal cells. Thomas et al. [30] investigated the relationship between these processes with respect to T-cell activation. They observed that T-cell mutants defective in PI glycan synthesis could still be activated adequately. However, mutants

Yeast O-mannosylation

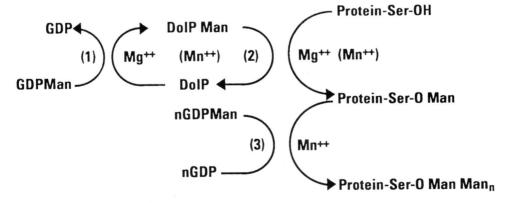

Fig. 6. The role of mannosyl phosphodolichol (DolPMan) in yeast O-mannosylation.

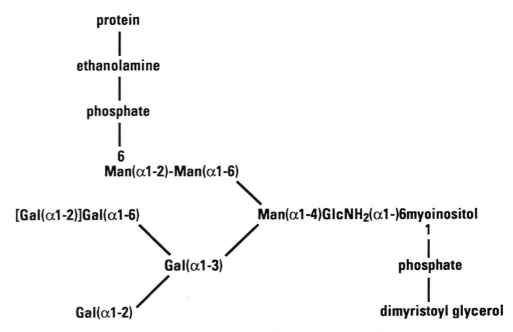

Fig. 7. Generalised structure of phosphatidylinositol glycans (PI glycans).

defective in the synthesis of mannosyl phosphodolichol resisted activation. The stable transfection of yeast mannosyl phosphodolichol synthase into the latter mutants fully restored the synthesis of mannosyl phosphodolichol and of the PI glycan in all trans-fectants. In some of these transfectants N-glycosylation was only partially corrected. A strong correlation existed between response to activation and the extent of restoration of N-glycosylation. Thus despite N-glycosylation being more important than PI glycan synthesis to T-cell activation it appears to have lower priority in terms of complete restoration of glycan synthesis. Presumably PI glycan is essential for some other important T-cell function.

Singh and Tartakoff [31] transfected a CHO mutant cell (B421), deficient in mannosyl phosphodolichol synthesis, to express the PI glycan-anchored protein, pla-cental alkaline phosphatase (AP). They found that the few transfectants expressing this enzyme also made mannosyl phosphodolichol. Upon treatment with tunicamycin the amount of mannosyl phosphodolichol synthesised increased (probably due to increased availability of phosphodolichol) and treated cells expressed PI glycan anchored AP. Fusion of PI glycan defective transfectant of B421 with a Thy-1− class E mutant thymoma deficient in mannosyl phosphodolichol synthesis resulted in hybrids, some of which both synthesised mannosyl phosphodolichol and expressed AP and Thy-1 at the surface. This again suggests an essential role for mannosyl phosphodolichol in PI glycan synthesis.

Camp et al. [32] have demonstrated that not only must this mannose donor be present

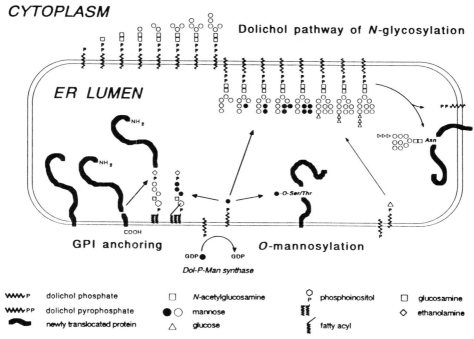

CYTOPLASM

Dolichol pathway of N-glycosylation

ER LUMEN

NH₂

NH₂

●-O-Ser/Thr

Asn

COOH

GPI anchoring   **O-mannosylation**

GDP   GDP

Dol-P-Man synthase

⌇⌇P	dolichol phosphate	□	N-acetylglucosamine	phosphoinositol	□	glucosamine
⌇⌇PP	dolichol pyrophosphate	●○	mannose	fatty acyl	◇	ethanolamine
	newly translocated protein	△	glucose			

Fig. 8. The alternative uses of mannosyl phosphodolichol in glycosylation pathways of the endoplasmic reticulum. From Orlean [29] with permission.

but it must be properly located. Their evidence suggests that in the Lec 35 mutant of CHO cells the mannosyl phosphodolichol is mislocalized for efficient PI glycan and oligo PP Dol synthesis.

7. N-acetylglucosaminyl phosphate transferase

7.1. Allosteric control by mannosyl phosphodolichol

Mannosyl phosphodolichol has been observed to act as an allosteric activator of the N-acetylglucosaminyl phosphate transferase catalysing step 1 of Figure 2. This was first reported by Kean [33] investigating direct addition of ManPDol to microsomal preparations from normal retinal cells and has subsequently been confirmed, most recently [34] using the Class E Thy-1 negative mouse lymphoma cell line which is deficient in ManPDol synthase. The inhibition of this enzyme by endogenous factors has not been reported. Tunicamycin remains the most useful complete inhibitor of protein N-glycosylation (see Chapter 5.7).

7.2. Structure–function relationships

In recent years knowledge of the structure–function relationships of several of the enzymes of the phosphodolichol pathway has benefitted from the application of molecular biology. Zhu et al. [35] (see also [36, 37]) took advantage of the tunicamycin sensitivity of step 1 and the consequent induced increase in expression of the N-acetyl-glucosaminyl phosphate transferase. They derived the full length cDNA for this enzyme from tunicamycin-resistant CHO cells. The deduced protein chain of 408 amino acids (46,197 Da) contained ten hydrophobic regions and showed 60% homology with the same enzyme cloned from yeast by Rine et al. [38].

The same enzyme has also been purified to homogeneity from bovine mammary gland [39] and cloned from mouse mammary gland [40]. The latter preparation shared 88% homology with the CHO cell cDNA. Lectin binding studies and carbohydrate analysis indicate that the enzyme is not glycosylated although four potential N-glyco-sylation sites can be identified in the sequence. When the mouse mammary gland clone was expressed in COS cells a functional enzyme was produced. The enzyme was mapped to chromosome 17.

Inspection of the amino acid sequences of the ten hydrophobic regions suggested ten passages of the endoplasmic reticulum membrane (Fig. 9) [35]. Two of these regions contained a putative dolichol recognition sequence (Albright et al. [17]) subsequently rechristened potential dolichol recognition sequence (PDRS), see e.g. [41]. The eighty-five amino acid sequence (295–380) between hydrophobic regions nine and ten was highly conserved and could well contain the catalytic site of the enzyme. An antibody to the peptide sequence 303–317 blocked enzyme activity, supporting this view.

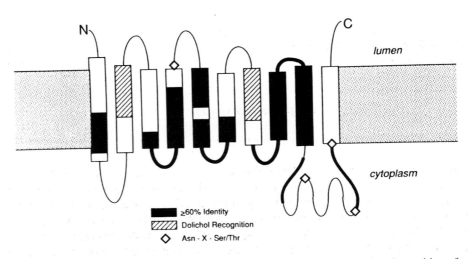

Fig. 9. A model of hamster N-acetylglucosaminyl phosphate transferase showing the position of ten hydrophobic regions in the membrane of endoplasmic reticulum. Also shown are the dolichol recognition regions. Taken from [35] with permission.

8. Potential dolichol recognition sequences (PDRSs) of proteins of the phosphodolichol pathway

The PDRSs of the N-acetylglucosaminyl phosphate transferase of CHO cells are spelled out in Table II [42] and compared with those of other dolichol-recognition proteins. These include the corresponding yeast enzyme (ALG 7), the mannosyl transferase for addition of the first mannose step 3 in Fig. 2 (ALG 1), the yeast mannosyl phosphodolichol synthase (DPM 1) a yeast non-enzymic phosphodolichol binding protein (SEC 59) and ribophorin-1. This last mentioned ribosome-binding protein was shown by Kelleher et al. [42] also to be one of the three proteins (48, 63 and 66 kDa in size) that together constitute the oligosaccharyl transferase (catalyses step 6 in Fig. 2) isolated from dog pancreas. The 63 and 66 kDa components were demonstrated immunochemically to be ribophorin I and II respectively. Only the former of these contained a putative dolichol binding site. Anti ribophorin I bound to the 63 kDa unit and blocked enzyme activity. Since Table II was first published [42] sequences showing similar homology to those in Table II have been reported for yeast α-1,3-mannosyl-transferase (Alg 2) [43] and dolichyl pyrophosphatase from Sulpholobus [44].

Also in Table II is an attempt to identify a consensus sequence homology that fits all seven proteins providing that conservative replacements of amino acids at critical

TABLE II

Sequence homology between ribophorin 1 and dolichol-binding proteins

Protein	Location	Sequence[a]														Identity to consensus no. 2
1. ALG7	79–92	**Y**	**L**	**F**	**V**	**M**	**F**	**I**	**Y**	**I**	**P**	**F**	**I**	**F**	Y	(7/7)
2. ALG1	19–32	**I**	P	**L**	**V**	**V**	**Y**	**Y**	**V**	**I**	**P**	**Y**	**L**	**F**	Y	(7/7)
3. DPM1	245–258	**I**	**L**	**F**	**I**	**T**	**F**	**W**	**S**	**I**	**L**	**F**	**F**	**Y**	V	(7/7)
4. SEC59	332–345	W	H	**F**	**I**	**I**	**F**	**L**	**L**	**I**	**I**	P	**S**	**F**	Q	(5/7)
5. GPT1	66–79	**F**	**L**	**I**	**I**	**L**	**F**	**C**	**F**	**I**	**P**	**F**	P	**F**	L	(7/7)
6. GPT2	221–234	H	**V**	**F**	**S**	**L**	**Y**	**F**	**M**	**I**	**P**	**F**	**F**	**F**	T	(6/7)
7. Ribophorin I	423–436	**F**	**Y**	**I**	**L**	**F**	**F**	**T**	**V**	**I**	**I**	**Y**	**V**	R	L	(6/7)
Consensus no. 1			L	F	V	X	F	X	X	I	P	F	X	F	Y	
Consensus no. 2		**F**	X	**F**	X	X	**F**	X	X	**I**	**P**	**F**	X	**F**		
		I		**I**			**Y**				**(I)**	**Y**		**Y**		
		Y		**L**							**(L)**					
Residue number		1		3		5		7		9		11		13		

[a] Protein sequences are in the one letter code for amino acids. In consensus no. 1, X designates amino acids normally found in membrane-spanning segments. In consensus sequence no. 2, X designates any amino acid normally found in membrane-spanning segments excluding alanine and glycine. Conservative replacements are listed below consensus sequence no. 2. Non-conservative but allowed replacements of isoleucine or leucine for proline are shown in parentheses. Residues in the sequences that match consensus sequence no. 2 are shown in bold type.
Adapted from Kelleher et al. [42] with permission.

138

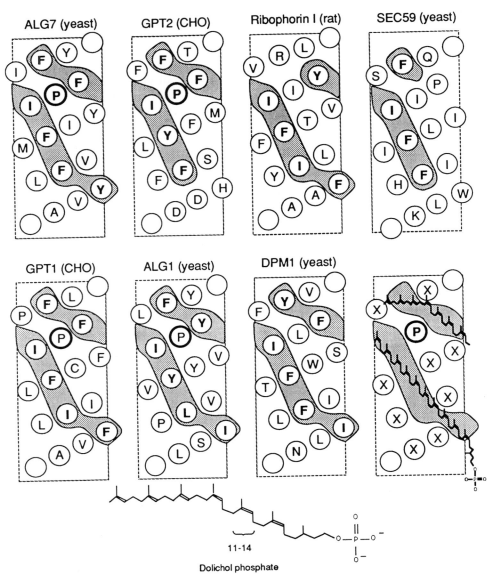

Fig. 10. A proposed helical net diagram for a dolichol binding site using the sequences of Table II. The conserved hydrophobic residues that correspond to the consensus sequences are connected by a dark band. The structure of a phosphodolichol is shown in the bottom right-hand net diagram. Taken from [42] with permission.

points are allowed and non-conservative replacement allowed at only one point. Figure 10 [42] illustrates a possible mechanism whereby the binding might occur. In this proposal fourteen of the nineteen unsaturated isoprene residues of dolichol-20 are fitted to the consensus residues of the helix. These observations gave rise to the hypothesis that the binding of phosphodolichol to these dolichol-binding sites is a characteristic feature of the function of proteins of the phosphodolichol pathway.

The presence of two dolichol-binding sites on the N-acetylglucosaminyl phosphate transferase of CHO cells stimulated the proposal that a second site may be for an effector rather than substrate of this enzyme. This would be consistent with the observation (Section 7.1 [33]) that the activity of this enzyme can be increased allosterically by increases in concentration of mannosyl phosphodolichol.

However, not all enzymes utilizing a dolichol phosphate or a glycosylated derivative as substrate contain the consensus sequence described in Table II. This appears to be absent from the dolichyl phosphate N-acetylglucosaminyl phosphate transferase of *Leishmania amazonensis* [45] and from a potential sub-unit of yeast oligosaccharyl-transferase (WBP 1) [46]. If the sequence is specific for dolichol it is also surprising to find it as part of the sialyltransferase of *Escherichi coli* [47]. These doubts have been strengthened by the application of recombinant genetic techniques to modify or remove the PDRS of some of the dolichol-using enzymes.

With regard to the DPMS of yeast, the organism remains viable after deletion of the PDRS although *in vitro* enzyme activity was much reduced [48]. Substitution of key residues of the consensus region of yeast DPMS (V for F_{250}, F for I_{253}, N for I_{253} and V for Y_{257}) had very little effect on the *in vitro* activity of this enzyme [49]. Over-expression in *E. coli* of Alg 1 (GlcNAc$_2$PPDolP-1,4-mannosyltransferase) depleted of its transmembrane domain resulted in a soluble enzyme that was still capable of using phytanyl diphospho N,N'-diacetylchitobiose as an efficient acceptor of mannose from GDPmannose [7]. The apparent Km for the phytanyl derivative remained essentially unchanged.

On the other hand when the animal N-acetylglucosaminyl phosphate transferase was subjected to these genetic techniques the activity of the enzyme after scrambling of GPT 1 (see Table II) or point mutation of key residues in GPT 2 suggested the sequence was required [41]. Expression of these modified forms of enzyme in CHO cells failed to increase microsomal enzyme activity over that of controls although they both conferred increased resistance to tunicamycin.

At present it seems impossible to generalise on the need for, or indeed presence of, a dolichol recognition sequence in enzymes utilizing dolichol-derivatives. It appears to be more important in some of these enzymes than others.

9. Oligosaccharide and oligosaccharide phosphate produced by the phosphodolichol pathway

In some circumstances oligosaccharide or oligosaccharide phosphate are products of the phosphodolichol pathway (Fig. 11). The work of Verbert et al. [50] and of Spiro [51] indicates that oligosaccharyl transferase (step 6 Fig. 2) may use water as an acceptor if appropriate protein is not available. This results in the release of free oligosaccharide — usually carrying three glucose residues (Fig. 3). Those phospho-dolichol-linked oligosaccharides that are not fully glucosylated appear to be susceptible

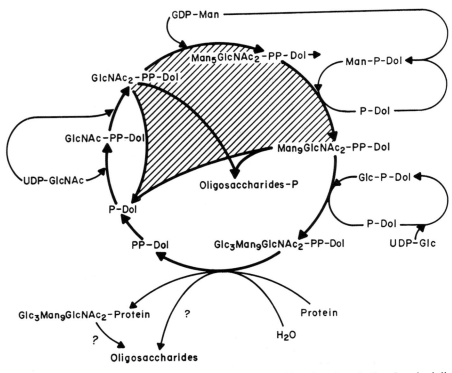

Fig. 11. The generation of oligosaccharides and oligosaccharides phosphate in the phosphodolichol pathway. After Cacan et al. [50] with permission.

to diphosphatase activity, so releasing phosphodolichol and an oligosaccharide monophosphate containing up to nine mannose residues and no glucose. This makes available phosphodolichol which may otherwise have accumulated in oligosaccharide PPDol that was not transferring oligosaccharide to protein efficiently due to the absence of glucose residues. The phosphodolichol so liberated is free to be used again in steps 8 and 9 (Fig. 2). Maintenance of step 8 may be critical for other synthesis (see Section 6).

Abbreviations

UDPGlcNAc	uridine diphosphate *N*-acetylglucosamine
5-BrdUMP	5 bromo uridine monophosphate
GDPMan	guanosine diphosphate mannose
UDPGlc	uridine diphosphate glucose
Glc	glucose
Man	mannose

GlcNAc	*N*-acetylglucosamine
PDol	phosphodolichol (dolichol phosphate)
PPDol	diphosphodolichol
ManPDol	mannosyl phosphodolichol
GlcPDol	glucosyl phosphodolichol
cDNA	deoxyribonucleic acid derived by reverse transcription of a messenger ribonucleic acid
PI glycan	phosphatidyl inositol glycan
CHO	Chinese hamster ovary
AP	alkaline phosphatase
PDRS	potential (putative) dolichol recognition sequence
ALG-7	product of yeast asparagine-linked glycosylation gene 7 (*N*-acetylglucosamine phosphate transferase)
ALG-1	product of yeast asparagine-linked glycosylation gene 1 (a mannosyl transferase)
DPM 1	dolichol phosphate mannose synthase 1
DPMS	dolichol phosphate mannose synthase
SEC 59	a phosphodolichol binding protein, the product of yeast secretory gene 59
E. coli	*Escherichia coli*
GPT 1 (& 2)	*N*-acetylglucosame phosphate transferase 1 (& 2) (see ALG-7)
PPDol	diphosphodolichol
GPI	see PI glycan
Asn	asparagine
Ser(S)	serine
Thr(T)	threonine
Y	tyrosine
M	methionine
R	arginine
L	lysine
I	isoleucine
C	cysteine
F	phenylalanine
P	proline
H	histidine
V	valine
Q	glutamine

References

1. Hemming, F.W. (1974) In: T.W. Goodwin (Ed.), Biochemistry of Lipids, Ser. 1, Vol. 4, Butterworths, London, Ch. 2, pp. 39–97.
2. Kornfeld, R. and Kornfeld, S. (1985) Annu. Rev. Biochem. 54, 631–661.
3. Hemming, F.W. (1985) In: H. Wiegandt (Ed.), Glycolipids, New Comprehensive Biochemistry, Vol. X, Elsevier, Amsterdam, Ch. 4. pp. 261–305.

4. Jaenicke, L. and Sigmund, H.-U. (1989) Chem. Phys. Lipids 51, 159–170.

5. William, I.B.H., Taylor, J.P., Webberley, M.C., Turner, N.J. and Flitsch, S.L. (1993) Biochem. J. 295, 195–201.

6. Flitsch, S.L., Pinches, H.L., Taylor, J.P. and Turner, N.J. (1992) J. Chem. Soc. Perkin Trans. 1, 2087–2093.

7. Revers, L., Wilson, I.B.J., Webberley, M.C. and Flitsch, S.L. (1994) Biochem. J., in press.

8. Hemming, F.W. (1983) In: J.W. Porter and S.L. Spurgeon (Eds.), Biosynthesis of Isoprenoid Compounds, Vol. 2, Wiley, New York, Ch. 5, pp. 305–354.

9. Rosenwald, A.G. and Krag, S.S. (1990) J. Lipid. Res. 31, 523–533.

10. Ekstrom, T.J., Chojnacki, T. and Dallner, G. (1987) J. Biol. Chem. 262, 10460–10468.

11. Ericsson, J., Appelkvist, E.-L., Thelin, A., Chojnacki, T. and Dallner, G. (1992) J. Biol. Chem. 267, 18708–18714.

12. Volpe, J.J., Sakakihara, Y. and Rust, R.S. (9187) Dev. Brain Res. 31, 193–200.

13. Kabakoff, B.D., Doyle, J.W. and Kandutsch, A.A. (1990) Arch. Biochem. Biophys. 276, 382–389.

14. Gasnier, F., Louisot, P. and Gateau, O. (1988) Biochim. Biophys. Acta 961, 242–252.

15. Gasnier, F., Morelis, R., Louisot, P. and Gateau, O. (1989) Biochim. Biophys. Acta. 925, 297–304.

16. Gasnier, F., Rousson, R., Lerne, F., Vaganay, E., Louisot, P. and Gateau-Roesch, O. (1992) Eur. J. Biochem. 206, 853–858.

17. Albright, C., Orlean, P. and Robbins, P. (1989) Proc. Natl. Acad. Sci. USA, 86, 7366–7369.

18. Jensen, J.W. and Schutzbach, J.S. (1988) Biochemistry 27, 6315–6320.

19. Hirschberg, C.B. and Snider, M.D. (1987) Annu. Rev. Biochem. 56, 63–87.

20. Abeijon, B. and Hirschberg, C.B. (1992) Trends Biochem. Sci. 17, 32–36.

21. Haselbeck, A. and Tanner, W. (1982) Proc. Nat. Acad. Sci. USA, 79, 1520–1524.

22. Schutzbach, J.S. and Zimmerman, J.W. (1992) Biochem. Cell Biol. 70, 460–465.

23. Kean, E.L. (1991) J. Biol. Chem. 266, 942–946.

24. Drake, R.R., Igari, Y., Elbein, A.D., Lester and Radominska, A. (1992) FASEB J. 6, A91.

25. Deglon, N., Krapp, A., Bron, C. and Fasel, N. (1991) Biochem. Biophys. Res. Comm. 174, 1337–1342.

26. Babczinski, P. and Tanner, W. (1973) Biochem. Biophys. Res. Commun. 54, 1119–1124.

27. Menon, A.K., Mayor, S. and Schwarz, R.T. (1990) EMBO J. 9, 4249–4258.

28. Ferguson, M.A.J. and Williams, A.F. (1988) Annu. Rev. Biochem. 57, 285–320.

29. Orlean, P. (1992) Biochem. Cell Biol. 70, 438–447.

30. Thomas, L.J., Degasperi, R., Sugiyama, E., Chang, H.M., Beck, P.J., Orlean, P., Urakaze, M., Kamitari, T., Sambrook, J.F., Warren, C.D. and Yeh, E.T.H. (1991) J. Biol. Chem. 266, 23175–23184.

31. Singh, N. and Tartakoff, A.M. (1991) Molec. Cell Biol. 11, 391–400.

32. Camp, L.A., Chauhan, P., Farrar, J.D. and Lehrman, M.A. (1993) J. Biol. Chem. 268, 6721–6728.

33. Kean, E.L. (1982) J. Biol. Chem. 257, 7952–7954.

34. Kean, E.L. (1992) Biochem. Cell Biol. 70, 413–421.

35. Zhu, X. and Lehrman, M.A. (1990) J. Biol. Chem. 265, 14250–14255.

36. Scocca, J.R. and Krag, S.S. (1990) J. Biol. Chem. 265, 20621–20626.

37. Zhu, X., Zeng, Y. and Lehrman, M.A. (1992) J. Biol. Chem. 267, 8895–8902.

38. Rine, J., Hansen, W., Hardeman, E. and Davis, R. (1983) Proc. Nat. Acad. Sci. USA, 80, 6750–6754.

39. Shailubhai, K., Dong-Yu, B., Saxena, E.S. and Vijay, I.K. (1988) J. Biol. Chem. 263, 15964–15972.

40. Rajput, B., Ma, J., Muniappa, N., Schantz, L., Naylor, S.L., Lalley, P.A. and Vijay, I.K. (1992) Biochem. J. 285, 985–992.
41. Datta, A.K. and Lehrman, M.A. (1993) J. Biol. Chem. 268, 12663–12668.
42. Kelleher, D., Kreibich, G. and Gilmore, R. (1992) Cell 69, 55–65.
43. Jackson, B.J. Kukuruzinska, M.A. and Robbins, P.W. (1993) Glycobiology 3, 357–364.
44. Meyer, W. and Schafer, G. (1992) Eur. J. Biochem. 207, 741–746.
45. Liu, X. and Chang, K.P. (1992) Mol. Cell Biol. 12, 4112–4122.
46. te Heesen, S., Knauer, R., Lehle, L. and Abei, M. (1993) EMBO J. 12, 279–284.
47. Troy, F.A. (1992) Glycobiology 2, 5–23.
48. Zimmerman, J.W. and Robbins, R.W. (1993) J. Biol. Chem. 268, 16746–16753.
49. Schutzbach, J.S., Zimmerman, J.W. and Forsee, W.T. (1993) J. Biol. Chem 268, 24190–24196.
50. Cacan, R., Villers, C., Belard, M., Kaiden, A., Krag, S., and Verbert, A. (1992) Glycobiol. 2, 127–136.
51. Spiro, M.J. and Spiro, R.G. (1991) J. Biol. Chem. 266, 5311–5317.

J. Montreuil, H. Schachter and J.F.G. Vliegenthart (Eds.), *Glycoproteins*
© 1995 Elsevier Science B.V. All rights reserved

CHAPTER 5

Biosynthesis

2b. From Glc₃Man₉GlcNAc₂-protein to Man₅GlcNAc₂-protein: transfer 'en bloc' and processing

ANDRÉ VERBERT

Laboratoire de Chimie Biologique, UMR du CNRS No. 111, Université des Sciences et Technologies de Lille, 59655 Villeneuve d'Ascq, France

1. The transfer reaction

The principal reaction of the N-glycosylation process is the transfer 'en bloc' of the pre-assembled lipid-linked oligosaccharide to an acceptor asparagine residue of the nascent protein. This reaction is mediated by an oligosaccharyltranferase.

1.1 The donor

Early *in vitro* studies comparing glucosylated and non-glucosylated oligosaccharide-PP-Dol as substrates for the oligosaccharyltransferase had shown preferential transfer of glucosylated oligosaccharides [1]. However, oligosaccharyltransferase seems to exhibit a rather broad specificity since shorter oligosaccharide-PP-Dol including GlcNAc₂-PP-Dol can be transferred to protein acceptors. This has been shown with hen oviduct [2], yeast [3], rat lymphocytes [4]. *In vivo* studies also support the idea that the presence of glucosyl residues on oligosaccharide-PP-Dol can enhance its efficiency as donor for the transfer reaction. For example a yeast mutant lacking the ability to synthesize Glc-P-Dol was shown to exhibit reduced efficiency to glycosylate the major yeast glycoprotein [5].

A striking preference for glucosylated oligosaccharide-PP-Dol has also been observed in a CHO mutant cell line unable to synthesize Man-P-Dol. In these cells, although a minor proportion of oligosaccharide-PP-Dol was glucosylated (5%) a high proportion of glucosylated glycans was recovered on the nascent protein, indicating that as soon as synthesized the glucosylated species were rapidly used for transfer to the protein acceptor sites [6].

Despite increasing evidence for preferential transfer of glucosylated oligosaccharides, there is not yet a molecular explanation for this phenomenon.

1.2. The acceptor site

Since 1967, it has been observed that the asparagine moiety must occur in an Asn-X-Thr or Asn-X-Ser sequence [7] to be a potential acceptor site. Roitsch and Lehle [8] have demonstrated that X could be any amino-acid but proline. In some cases, a cysteine in place of Ser/Thr has been found in glycosylated sequons. Although such consensus sequences have been widely recognized as an absolute requirement, they are not sufficient since many asparagine residues in such tripeptide sequences are not glycosylated. A survey of 2000 Asn-X-Ser/Thr tripeptides [9] indicated that only 312 were glycosylated. It is interesting to note that, although equal numbers of Asn-X-Thr and Asn-X-Ser were surveyed, there were three times more glycosylated Asn-X-Thr (227 sites) than Asn-X-Ser (85 sites) sequons.

A chemical basis has been brought forward to explain the role of the hydroxy amino-acid in the transfer reaction [10].The oxygen atom of the hydroxyl group serves as an electron donor to one of the hydrogen atoms of the amido-nitrogen, thus increasing its nucleophilicity to react with the C-1 of the sugar at the reducing end of the oligosaccharide donor (Fig. 1).

The fact that less than 16% of the potential glycosylation sites are glycosylated indicates that additional factors intervene. First, it is likely that the protein environment itself may interact with the oligosaccharyl-transferase giving rise to a positive or a negative control of the glycosylation by the protein itself. There is a high statistical probability for glycosylated asparagine residues to be located in peptide segments that favour the β-turn conformation [11,12]. This has been shown not to be an absolute requirement although giving favourable conditions for glycosylation.

In fact, it must be kept in mind that the transfer reaction is a cotranslational process. The transfer reaction on a given asparagine residue occurs after further addition of about 50 more amino-acids by the ribosomal machinery. Nothing is known about the folding of the protein at this stage; however Shakin-Eshleman et al. [13] demonstrated that the N-linked glycosylation efficiency can be influenced by C-terminal regions more than 68 amino acids further than the glycosylation site. This indicates that large domains of the nascent protein, even far away from the sequon, may be involved in the control of the N-glycosylation process.

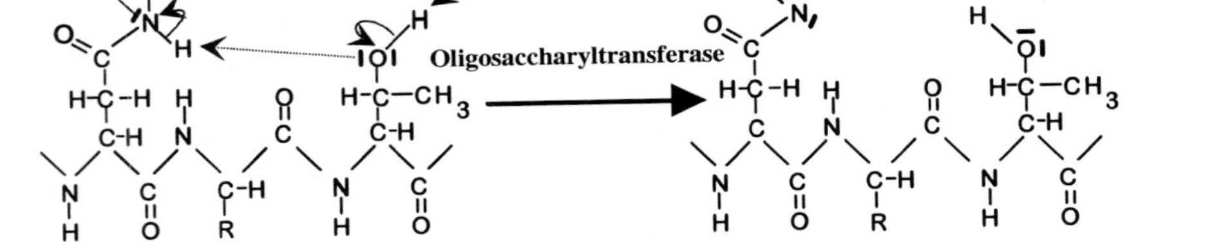

Fig. 1. Chemical mechanism of the transfer reaction according to Bause and Legler [10].

Does the presence of chaperones influence the adequate folding of the protein for appropriate glycosylation site conformation? Does the glycosylation at one site modify the protein conformation enough to influence further glycosylation at other sites? Many of these questions remain to be answered. The combination of so many factors, acting presumably in a cooperative or ordered way, renders difficult the establishment of a working biological model.

1.3. The oligosaccharyltransferase

As is the case for many membrane enzymes, the oligosaccharyltransferase (OST) is difficult to purify since rapid denaturation occurs after solubilisation with detergent and subsequent yields are poor. Because of this difficulty to purify active oligosaccharyl-transferase to homogeneity, a different approach has been undertaken using a photoaffinity acceptor peptide to label the active site of the enzyme. Despite many attempts, mainly from the group of Lennarz [9], the results have been rather discouraging. The photolabelled glycosylated tripeptide was indeed found attached to a lumenal 57 kD protein; however, this was not the OST but a glycosylation site binding protein (GSBP), which was found to be 90% identical to the protein disulfide isomerase, an abundant lumenal RER protein.

Kelleher et al. [14] have reported the copurification of the canine oligosaccharyl-transferase activity with a protein complex composed of 66, 63 and 48 kD subunits. The 66 and 63 kD subunits were found to be ribophorins I and II, abundant RER-specific proteins. The cDNA sequence of the 48 kD subunit (OST 48) has been determined [15], and found to be homologous to the cDNA sequence of WBP₁, a yeast protein necessary for oligosaccharyltransferase activity *in vivo* and *in vitro* [16–17]. Recently, Te Heesen et al. [18] reported that a 30 kD protein (SWP₁) is another essential component of the yeast transferase. The avian oligosaccharyltransferase has been purified [19] and characterized as a protein complex containing two 65 kD subunits and a 50 kD subunit the sequence of which is 92% identical to that of canine OST48 and 25% identical to yeast WBP₁ protein. The oligosaccharyltransferase from *Saccharomyces cerevisiae* has also been purified recently and is composed of six subunits: α, β, γ, δ, ε, ζ [20]. The α subunit is glycosylated, having three glycoforms of 60, 62 and 64 kD, and the β and δ subunits were shown to correspond to the 45 kD WBP₁ protein and the 30 kD SWP₁ protein, respectively. The non-glycosylated γ, ε, and ζ subunits have apparent molecular masses of 34, 16 and 9 kD respectively. These studies suggest that the oligosaccharyltransferase is a protein complex intimately associated with membrane bound ribosomes and may be a part of the machinery translocating the polypeptide chain into the RER lumen. This fits with earlier results indicating that the oligosaccharyltransferase prefers unfolded nascent polypeptide. In addition, the 66 kD ribophorin I possesses a single transmembrane domain which contains the typical consensus sequence found in all dolichol binding proteins (see Chapter 5.2a by F. Hemming).

Thus, the oligosaccharyltransferase might be considered as an element of the 'translocon', an assembly of integral and peripheral membrane proteins [21].

The transfer to the acceptor protein is considered to be the key step of the N-glyco-sylation process, but little is known about the regulation of this reaction. Hormones often influence the N-glycosylation of proteins. For example: β-adrenergic agonists stimulate glycoprotein synthesis in rat parotid acinar cells, the glycosylation of thyro-globulin in porcine thyroid cells is under the control of thyrotropin and androgens stimulate N-glycosylation in rat epididymis as estrogens do in mouse uterus. Some of these hormonal effects are mediated by cyclic AMP but the reactions that constitute the target for this type of regulation are unclear. Recently Konrad and Merz [22] have demonstrated in choriocarcinoma cells a long-term effect of cyclic AMP which medi-ates enhanced synthesis of lipid intermediates leading to increased N-glycosylation due to increased availability of the substrate of the oligosaccharyltransferase. In this model, the lipid linked oligosaccharides were transferred to the protein ten times more rapidly when the cells were pretreated with 8-bromo-cAMP. During this treatment the specific activity of the oligosaccharyltransferase remained unchanged, suggesting that the higher rate of the N-glycosylation could also be due to mass action driven by acceler-ated protein synthesis. The specific activity of the oligosaccharyltransferase does not appear to be a rate limiting factor and can therefore respond to a rapid increase of protein acceptor. Concomitantly to the N-glycosylation process, several authors [23–25] have demonstrated the release of the glycan moiety from lipid intermediates. It has been suggested that the oligosaccharyltranferase is responsible for this enzymatic hydrolysis of the oligosaccharide-lipids as a result of a transfer reaction which uses water as acceptor when protein synthesis is reduced.

2. The deglucosylation steps

Once transferred onto the protein, the glycan moiety is trimmed in a stepwise manner to the $Man_5GlcNAc_2$ structure [26,27]. The first event is the removal of the three glucose residues by the action of two different glucosidases. Glucosidase I removes the α-1,2- linked glucose residue and glucosidase II cleaves the next two α-1,3- linked glucose residues [28,29]. Glucosidase I is a transmembrane endoplasmic reticular glycoprotein with a lumenal catalytic domain [30]. Glucosidase II also appears to be lumenally oriented. This topography is in agreement with the fact that the newly translated and glycosylated proteins are sequestered within the lumen of the ER. It has been recently demonstrated [31] that glucosidase II is a two-site enzyme, each site working sequentially to cleave the two α-1,3- linked glucose residues and release the $Man_9GlcNAc_2$-protein. It is interesting to note that both enzymes can be inhibited (both in cell-free system and whole cell) by castanospermine and deoxynojirimycin and related derivatives.

Interestingly, Parodi and coworkers reported [32,33] that deglucosylated glycopro-teins can be transiently glucosylated in the endoplasmic reticulum of mammalian, plant, fungal and protozoa cells. This glucosylation of $Man_9GlcNAc_2$-protein had been mentioned earlier by Ronin and Caseti [34] and appears to be more effective with denatured glycoproteins. It is thus suggested that, in vivo, only unfolded or misfolded proteins would be reglucosylated to be retained in the endoplasmic reticulum until they

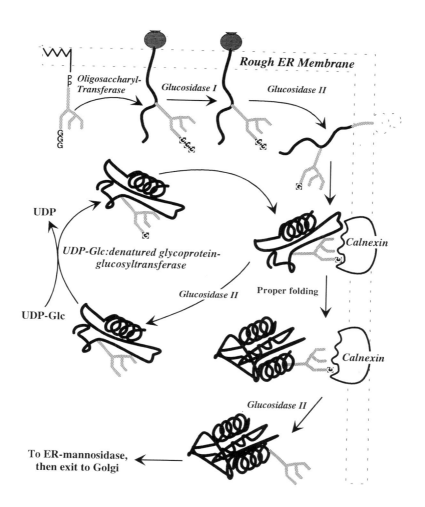

Fig. 2. The deglucosylation–reglucosylation shuttle and calnexin dependent folding. The glycoprotein is retained in the ER through a binding to calnexin, a $Glc_1Man_9GlcNAc_2$-protein specific membrane lectin, until it has been correctly folded. If a misfolded glycoprotein were inappropriately deglucosylated, it can be reglucosylated by an UDP-Glc:glucosyltransferase specific for 'denatured' glycoproteins.

adopt their correct conformation. These correctly folded glycoproteins are no longer substrates for the glucosyltransferase and can be further processed. In this connection, a glycoprotein specific chaperone, named calnexin, has recently been described [35, 36]. It is a rough ER lumenal membrane bound protein which specifically binds incompletely folded glycoproteins possessing $Glc_1Man_9GlcNAc_2$ glycan moieties (Fig. 2). Thus, the role of the deglucosylation–reglucosylation cycle may be to retain the glycoproteins in the rough ER compartment, via calnexin binding, until they have finally acquired their correct spatial conformation.

In addition, it has to be mentioned that an alternate pathway may occur when glucosidase activity is low or deficient. In this case $Glc_1Man_9GlcNAc_2$-protein may enter the Golgi where deglucosylation is achieved via the action of an endo-α-mannosidase [37] by removal of a $Glc(\alpha\text{-}1,3)$ Man disaccharide to yield a single $Man_8GlcNAc_2$ isomer.

Recently, a novel mechanism for the removal of glucose residues has been described in a post-ER compartment via deoxynojirimycin insensitive glucosidases [38].

These two alternate pathways indicate how crucial is the removal of glucose for further glycoprotein maturation.

3. Rough ER α-mannosidase

Although α-mannosidases were known to be Golgi enzymes since 1979, several lines of evidence suggested that some mannose residues could be removed from certain glycoproteins before they reach the Golgi compartment. Indeed, in 1986, Bischoff and Kornfeld [39] showed the occurrence of an ER α-mannosidase and its activity is highly specific and insensitive to the mannose analogue 1-deoxymannojirimycin, an inhibitor of certain mannosidases. This α-mannosidase could be similar to the yeast enzyme which converts the Man_9-species to the Man_8-species which can then be elongated to the typical yeast mannan structures [27].

In rat liver, this enzyme is immunologically related to the cytoplasmic α-mannosidase. It is interesting to note that the soluble form has apparent lower binding affinity for free oligosaccharides. It is known that a free oligosaccharide may adopt a variety of spatial conformations different from the one adopted by the same oligosaccharide when attached to a protein. This high specificity of the ER α-mannosidase could be related to a certain specific conformation of the glycan in its given protein environment. This could explain why some glycoprotein $Man_9GlcNAc_2$ structures (or Glc_1Man_9Glc-NAc_2 structures) might escape the action of this enzyme and reach the Golgi without being processed to Man_8-type structures.

4. Golgi mannosidase I

In the *cis*-Golgi, the glycans are further trimmed by specific mannosidases [40,41] to the $Man_5GlcNAc_2$ structure (Fig. 3). This requires the cleavage of the three remaining α-1,2-linked mannose residues. This is achieved in an ordered way, going through the formation of $Man_7GlcNAc_2$ and $Man_6GlcNAc_2$ isomers and then the final Man_5Glc-NAc_2 isomer which is the substrate for *N*-acetylglucosaminyl transferase I.

The mannosidases involved in this trimming are named mannosidases IA and/or IB. They are located in the *cis*-Golgi cisternae and, in contrast to the previous ER mannosidase, are very sensitive to deoxymannojirimycin.

In fact, several mannose removal pathways may exist if one considers the different substrates which may enter the Golgi. Beside the 'classical' $Man_8GlcNAc_2$ structure, there is the $Man_8GlcNAc_2$ isomer which comes from the previous action of endo-mannosidases (see above). Since mannosidases IA/IB can also act on glucosylated species

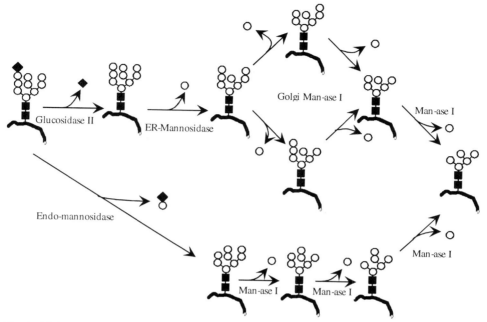

Fig. 3. Major pathway and endo-mannosidase pathway leading to the key Man$_5$GlcNAc$_2$ species (Man-ase I:Golgi mannosidase I). ◆: Glc; O: Man; ■: GlcNAc.

[42] other isomers can be found as intermediates (Glc$_1$Man$_{8-5}$GlcNAc$_2$), which are preferred substrates for the endomannosidase.

These different trimmed structures all converge towards the final Man$_5$GlcNAc$_2$ structure which is the unique substrate for GlcNAc transferase I, the key enzyme for further processing (see Chapter 5.2c by H. Schachter).

An exception has to be noted in the case of mutant cells defective in Man-P-Dol synthase in which the structure transferred onto the protein is the Glc$_3$Man$_5$GlcNAc$_2$ moiety. Trimming by RER glucosidases will lead to the 'linear' Man$_5$GlcNAc$_2$ structure which is sensitive to Golgi α-mannosidase. The final product on which GlcNAc transferase I adds the β-1,2- linked GlcNAc is the typical inner-core Man$_3$ GlcNAc$_2$ structure.

Abbreviations

CHO cell	Chinese hamster ovary cell
Glc	glucose
GlcNAc	*N*-acetylglucosamine
Man	mannose
Oligosaccharide-PP-Dol	oligosaccharide-pyrophospho-dolichol
OST	oligosaccharyltransferase
P-Dol	phospho-dolichol

References

1. Turco, S.J., Stetson, B. and Robbins, P.W. (1977) Proc. Natl. Acad. Sci. USA 74, 4411–4414.
2. Chen. W.W. and Lennarz W.J. (1977) J. Biol. Chem. 252, 3473–3474.
3. Sharma, C.B., Lehle, L. and Tanner, W. (1981) Eur. J. Biochem. 116, 101–108.
4. Hoflack. B., Debeire P., Cacan, R., Montreuil, J. and Verbert, A. (1982) Eur. J. Biochem. 124, 527–531.
5. Ballou, L., Gopal, P., Krummel, B., Tammi, M. and Ballou C. (1986) Proc. Natl. Acad. Sci. USA, 83, 3081–3085.
6. Stoll, J., Cacan, R., Verbert, A. and Krag, S.S. (1992) Arch. Biochem. Biophys. 299, 225–231.
7. Marshall, R.D. (1967) Abstract or the 7th Int. Congr. Biochem. 3, 573–574.
8. Roitsch, T. and Lehle, L. (1989) Eur. J. Biochem. 181, 525–529.
9. Kaplan, H.A., Welply, J.K. and Lennarz W.J. (1987) Biochim. Biophys. Acta, 906 161–173.
10. Bause, E. and Legler, G. (1981) Biochem. J. 195, 639–644.
11. Aubert, J.P., Biserte G. and Loucheux-Lefebvre M.H. (1976) Arch. Biochem. Biophys. 175, 410–418.
12. Beeley J.G. (1977) Biochem. Biophs. Res. Commun. 76, 1051–1055.
13. Shakin-Eshleman, S.H., Wunner, W.H. and Spitalnik, S.L. (1993) Biochemistry 32, 9465–9472
14. Kelleher, D.J., Kreibich, G. and Gilmore, R. (1992) Cell 69, 55–65.
15. Silberstein,S., Keller, D.J. and Gilmore R. (1992) J. Biol. Chem., 267, 23658 – 23663.
16. Te Heesen, S., Rauhat, R., Aebersold, A., Abelson, A., Aebi, M. and Clark, M.W. (1991) Eur. J. Cell. Biol. 58, 8–18.
17. Te Heesen, S., Janielzky, B., Lehle, L. and Aebi, M. (1992) EMBO J. 11, 2071–2075.
18. Te Heesen, S., Knauer, S., Lehle, L. and Aebi, M. (1993) EMBO J. 12, 279–284.
19. Kumar, V., Heinemann, F.S. and Ozols, J. (1994). J. Biol. Chem. 269, 13451–13457.
20. Kelleher, D.J. and Gilmore, R., (1994) J. Biol. Chem. 269, 12908–12917.
21. Gilmore, R. (1993) Cell 75, 589–592.
22. Konrad, M. and Merz W.E. (1994) J. Biol. Chem. 269, 8659–8666.
23. Cacan, R., Hoflack, B. and Verbert, A. (1980) Eur. J. Biochem. 105, 473–479.
24. Anumala, K.R and Spiro, R. (1983) J. Biol. Chem., 258, 15274– 15282.
25. Cacan, R., Villers, C., Belard, M., Kaiden A., S.S. and Verbert, A. (1992) Glycobiology 2, 177–136.
26. Kornfeld, R.C and Kornfeld, S. (1985) Annu. Rev. Biochem. 54, 631–664.
27. Moremen, K.W., Trimble, R.B. and Hercovics, A. (1994) Glycobiology, 4, 113–125.
28. Shailubhai, K., Pratta M.A. and Vijay, I.K. (1987) Biochem J. 247, 555–562.
29. Saxena, S.K., Shailubhai K., Dong-Yu, B. and Vijay I.K. (1987) Biochem. J. 247, 563–570.
30. Shailubhai, K., Pukazhenthi B.S., Saxena, E.S., Varma, G.M. and Vijay I.K. (1991) J. Biol. Chem. 266, 16587–16593.
31. Alonso, J.M., Santa-Cecilia, A. and Calvo, P. (1991) Biochem. J. 278, 721–727.
32. Ganan, S., Cazzulo, J.J. and Parodi A. (1991) Biochem. 30, 3098–3104.
33. Sousa, M.C., Ferrero-Garcia, M.A. and Parodi A. (1992) Biochem. 31, 97–105.
34. Ronin. C. and Caseti, C. (1981) Biochim. Biophys. Acta 674, 58–64.
35. Ou, W-J., Cameron, P.H., Thomas, D.Y. and Bergeron J.J.M. (1993) Nature 364, 771–776.
36. Hammond, C. and Helenius, A. (1993) Current Biology 3, 884–886.
37. Lubas, W.A. and Spiro, R.G. (1987) J. Biol. Chem. 262, 3775–3781.
38. Suh, K., Gabel, C.A. and Bergmann, J.E. (1992) J. Biol. Chem. 267, 21671–21677.
39. Bischoff, J. and Kornfeld, R. (1986) J. Biol. Chem. 261, 4758–4765.
40. Kornfeld, S., Li, E. and Tabas, I. (1978) J. Biol. Chem. 253, 7771–7778.
41. Tulsiani, D.R.P. and Touster, O. (1988) J. Biol. Chem. 263, 5408–5417.
42. Lubas W.A. and Spiro, R.G. (1988) J. Biol. Chem. 263, 3990–3998.

J. Montreuil, H. Schachter and J.F.G. Vliegenthart (Eds.), *Glycoproteins*

153

Biosynthesis

2c. Glycosyltransferases involved in the synthesis of N-glycan antennae

HARRY SCHACHTER

Department of Biochemistry Research, Hospital for Sick Children, 555 University Avenue, Toronto, Ont. M5G 1X8, Canada

1. Introduction

Oligosaccharides attached to protein by way of an asparagine-*N*-acetylglucosamine (Asn-GlcNAc) linkage are known as N-glycans and are classified into three main types, high mannose, hybrid and complex [1]. Complex N-glycans are usually terminated at their non-reducing ends with characteristic trisaccharides (sialic acid–Gal–GlcNAc–) which Jean Montreuil [2] has named "antennae" to emphasize that the functional role for these terminal structures may be the transmission of biological recognition signals. The antennary hypothesis has inspired a great deal of interest in the structure, function and biosynthesis of the antennae of branched complex N-glycans [3,4].

Previous sections of this chapter [5,6] have discussed the assembly of dolichol–pyro-phosphate–oligosaccharide and the transfer of oligosaccharide $Glc_3Man_9GlcNAc_2$ from this donor to an Asn residue in the polypeptide to initiate the synthesis of N-glycans [1,7,8]. This is followed by oligosaccharide processing within the lumen of the endoplasmic reticulum and Golgi apparatus, i.e., removal of glucose and mannose residues to form high-mannose N-glycans. $Man_5GlcNAc_2$-Asn-X is converted to hybrid and complex N-glycans when antennae are initiated by addition of GlcNAc residues to the N-glycan core (Fig. 1). These GlcNAc residues are usually elongated and termi-nated by addition of Gal, GlcNAc, GalNAc, Fuc and sialyl residues; addition of sulfate, phosphate or other non-carbohydrate groups can also occur. The discussion in this chapter is restricted to the glycosyltransferases involved in the assembly of sialyl–Gal–GlcNAc attennae. The addition of other sugars is discussed elsewhere in this volume [9–11]. This chapter will emphasize recent work on glycosyltransferase molecular biology. Many reviews are available which deal with the kinetic properties, substrate specificity, purification, cell biology, molecular biology and other aspects of the glycosyltransferases [1,3,12–24]. An especially useful review has recently been published by Kleene and Berger [25] who have compiled comprehensive and up-to-date tables on the purification, immunochemistry, cloning strategies, domain structures, mRNA transcripts, genomic organization and Golgi targeting signals of the glycosyltransferases.

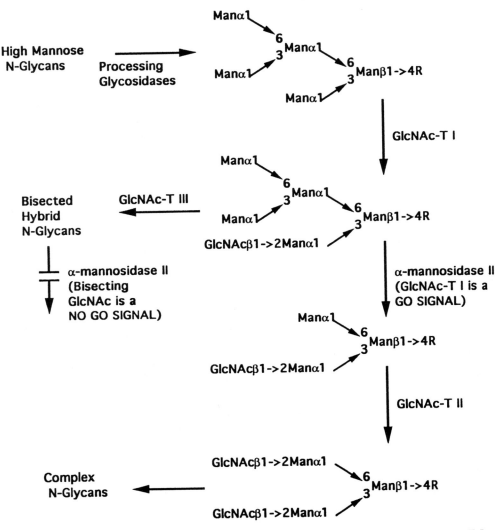

Fig. 1. Conversion of high mannose N-glycans to hybrid and complex N-glycans. GlcNAc-T I, UDP-GlcNAc:Man(α1–3)R [GlcNAc to Man(α1–3)] β-1,2-GlcNAc-transferase I. GlcNAc-T II, UDP-GlcNAc: Man(α1–6)R [GlcNAc to Man(α1–6)] β–1,2-GlcNAc-transferase II.

2. Biosynthesis of N-glycan antennae

The key enzyme for the conversion of high-mannose to complex and hybrid N-glycans is UDP-GlcNAc:Man(α1–3)R [GlcNAc to Man(α1-3)] β-1,2-GlcNAc-transferase I (EC 2.4.1.101, Fig. 1). The presence of a β-1,2-linked GlcNAc residue at the non-reducing terminus of the Man(α1–3)Man(β1–4)GlcNAc arm is essential for subsequent action of several enzymes in the processing pathway (Fig. 1), i.e., α-3/6-mannosidase II, GlcNAc-transferases II, III and IV, and the α-1,6-fucosyltransferase which adds fucose in (α1–6) linkage to the Asn-linked GlcNAc [17,26).

N-glycan antennae can be elongated and terminated in a variety of ways [1,27]. The most common antenna is sialyl (α2–6)Gal(β1–4)GlcNAc- formed by the action of UDP-Gal:GlcNAc-R β-1,4-Gal-transferase (E.C. 2.4.1.38/90) and CMP-sialic acid: Gal(β1–4)GlcNAc-R α-2,6-sialyltransferase (ST6N) (E.C. 2.4.99.1). Sialic acid can also be incorporated in α-2,3 linkage to Gal by CMP-sialic acid:Gal(β1–3/4)GlcNAc-R α-2,3-sialyltransferase (ST3N) (E.C. 2.4.99.6). The Gal(β1–4) residue may be replaced by a Gal(β1–3) residue or by a GalNAc(β1–4) residue. The GalNAc residue may be sulfated. In some species, the terminal sialic acid residue may be replaced by a Gal(α1–3) residue due to the action of UDP-Gal:Gal(β1–4)GlcNAc-R α-1,3-Gal-transferase (E.C. 2.4.1.124/151). N-glycans may be elongated by the addition of linear or branched poly-*N*-acetyllactosamine chains; the synthesis of these chains is discussed in another section [9]. The genes for GlcNAc-transferases I, II, III and V, α3/6-mannosidase II, β-1,4-Gal-transferase, the N-glycan α-2,6- and α-2,3-sialyltransferases and α-1,3-Gal-transferase have been cloned (Table I).

3. Galactosyltransferases

3.1. UDP-Gal:GlcNAc-R β-1,4-Gal-transferase (E.C. 2.4.1.38/90)

UDP-Gal:GlcNAc-R β-1,4-Gal-transferase has been studied in detail by many different groups and the reader is referred to previous comprehensive reviews [19,20,28,29] for the purification, kinetics, substrate specificity and mechanism of action of this enzyme. The enzyme is ubiquitous and is responsible for the synthesis of the Gal(β1–4)GlcNAc moiety in N-glycans, O-glycans and glycolipids. The enzyme is the only glycosyltransferase known to alter its substrate specificity in the presence of another protein, i.e., it complexes with α-lactalbumin to form lactose synthetase, the enzyme responsible for the synthesis of the milk sugar lactose in the mammary gland. β-1,4-Gal-transferase has been purified from several sources [28] and the bovine enzyme was the first mammalian glycosyltransferase to be cloned [30,31].

3.1.1. Cloning of β-1,4-Gal-transferase cDNA and genomic DNA

Partial cDNA clones for bovine β-1,4-Gal-transferase were first isolated in 1986 [30, 31]. Full-length bovine cDNA (Table I) was subsequently isolated by several groups [32–34]. The full-length 4.5 kb mRNA has a 5′-untranslated region of about 600 bp, a coding region of 1206 bp and a 3′-untranslated region of 2.7 kb. Comparison of the bovine sequences reported by different laboratories [30–32] show several base substitutions and an in-frame three nucleotide deletion. These sequence differences do not alter the reading frame and cause conservative amino acid changes; they have been attributed to genetic polymorphism.

Human milk β-1,4-Gal-transferase was purified and partial amino acid sequence data was used to design a 60-mer 'optimal' probe [35,36]. Screening of a human liver

156

TABLE I

Domain structures of cloned mammalian and avian glycosyltransferases

Glycosyltransferase	Species	Domains (number of amino acids)			Total number of amino acids	3'-untranslated sequence (kb)	Chromosome localization	Refs.
		N-terminal cytoplasmic	Signal/ anchor domain	C-terminal lumenal/ catalytic				
UDP-Gal:GlcNAc-R β-1,4-Gal-transferase (EC 2.4.1.38; EC 2.4.1.90)	Human	23	20	354	397	2.9	9p13-p21	35,37
	Murine	11/24	20	355	386/399	2.6	4	38,39
	Bovine	11/24	20	358	389/402	2.6–2.8	U18	30–34
Blood group B UDP-Gal:Fuc(α1–2)Gal-R α-1,3-Gal-transferase (EC 2.4.1.37)	Human	16	21	316	353		9q34	24,83
UDP-Gal:Gal(β1–4)GlcNAc-R α-1,3-Gal-transferase (EC 2.4.1.151)	Bovine	6	16	346	368	2.4		78
	Murine	41	16	337	394	2.1	2	79
	Murine	6	16	337	359			82
	Murine	6	16	315	337			
	Murine	6	16	327	349			
	Murine	6	16	349	371			
CMP-sialic acid: Gal(β1–4)GlcNAc-R α-2,6-sialyltransferase (EC 2.4.99.1)	Rat	9	17	377	403	2.8		90
	Human	9	17	380	406		3q27-q28	93–96
	Chick	9	17	387	413	0.6		124
CMP-sialic acid:	Rat	8	20	346	374	~1.0		122

Gal(β1–3/4)GlcNAc-R α-2,3-sialyltransferase I (EC 2.4.99.6)	Human	8	20	347	375			125
CMP-sialic acid: Gal(β1–3/4)GlcNAc-R α-2,3-sialyltransferase II	Human	8	18	303	329	0.6		127
CMP-sialic acid: Gal(β1–4)GlcNAc-R α-2,3-sialyltransferase STZ	Human	7	18	307	332			130
Brain sialyltransferase STX	Rat	6	17	352	375			128,129
CMP-sialic acid: Gal(β1–3)GalNAc-R α-2,3-sialyltransferase (EC 2.4.99.4)	Porcine	11	16	316	343	~1.0		208
	Mouse	7	16	314	337	0.4		91
CMP-sialic acid:GalNAc-R α-2,6-sialyltransferase I (EC 2.4.99.3)	Chick	16	21	529	566			209
GDP-Fuc:β-galactoside α-1,2-Fuc-transferase (blood group H) (EC 2.4.1.69)	Human	8	17	340	365	2.2	19	210
Lewis GDP-Fuc: Gal(β1–3/4)GlcNAc-R α-1,4/3-Fuc-transferase III (EC 2.4.1.65)	Human	15	19	327	361	0.9	19	211,212

(continued)

TABLE I (*continuation*)

Glycosyltransferase	Species	Domains (number of amino acids)			Total number of amino acids	3′-untranslated sequence (kb)	Chromosome localization	Refs.
		N-terminal cytoplasmic	Signal/anchor domain	C-terminal lumenal/catalytic				
GDP-Fuc: Gal(β1–4)GlcNAc-R α-1,3-Fuc-transferase V (plasma type)	Human	15	19	340	374		19	213
GDP-Fuc: Gal(β1–4)GlcNAc-R α-1,3-Fuc-transferase VI	Human	14	20	324	358		19	214
GDP-Fuc: Gal(β1–4)GlcNAc-R α-1,3-Fuc-transferase IV (myeloid type)	Human	22	34	349	405		11q	215
Blood Group A UDP-GalNAc:Fuc(α1–2)Gal-R α-1,3-GalNAc-transferase (EC 2.4.1.40)	Human	16	21	316	353		9q34	216
UDP-GalNAc:G_{M3}/G_{D3} β-1,4-GalNAc-transferase (EC 2.4.1.92)	Human	7	18	536	561	>0.8		217

Enzyme	Species						Chromosome	Reference
UDP-GalNAc:polypeptide (GalNAc to Thr) α-GalNAc-transferase (EC 2.4.1.41)	Bovine	8	20	531	559	>0.5		218,219
UDP-GlcNAc:Man(α1–3)R β-1,2-GlcNAc-transferase I (EC 2.4.1.101)	Human / Rabbit / Murine	4 / 4 / 4	25 / 25 / 25	416 / 418 / 418	445 / 447 / 447	1.09 / 1.06 / 0.99	5q35 / / 11	141–143 / 138,139 / 144,145
UDP-GlcNAc:Man(α1–6)R β-1,2-GlcNAc-transferase II (EC 2.4.1.143)	Human / Rat	9 / 9	20 / 20	418 / 413	447 / 442		14q21	146 / 160
UDP-GlcNAc:R_1-Man(α1–6)-[GlcNAc(β1–2)Man(α1–3)]-ManβR_2(GlcNAc to Manβ) β-1,4-GlcNAc-transferase III (EC 2.4.1.144)	Rat / Human	5 / 5	16 / 16	515 / 510	536 / 531	>0.5	22q13.1	135 / 172
UDP-GlcNAc:R_1-Man(α1–6)R_2 [GlcNAc to Man(α1–6)] β-1,6-GlcNAc-transferase V (EC 2.4.1.155)	Rat / Human	13 / 13	17 / 17	710 / 711	740 / 741	2.3	2q21	196 / 197
O-glycan core 2 UDP-GlcNAc: Gal(β1–3)GalNAc-R (GlcNAc to GalNAc) β-1,6-GlcNAc-transferase (EC 2.4.1.102)	Human	9	23	396	428	0.6	9q21	206
Blood Group I UDP-GlcNAc: GlcNAc(β1–3)Gal(β1–4)-GlcNAc-R (GlcNAc to Gal) β-1,6-GlcNAc-transferase	Human	6	19	375	400		9q21	205

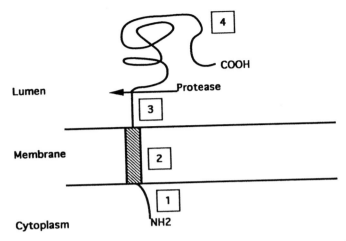

Fig. 2. Domain structure of the glycosyltransferases. All the mammalian glycosyltransferases which have been cloned to date are type II transmembrane glycoproteins (N_{in}/C_{out}) (Table I). (1) Short N-terminal cytoplasmic segment. (2) Trans-membrane non-cleavable signal/anchor sequence. (3) Stem region which can be cleaved by proteases to release a catalytically active soluble enzyme. (4) The globular C-terminal catalytic domain.

cDNA library yielded a 982 bp fragment encoding the C-terminal 261 amino acids of the enzyme. This 982 bp fragment was used to probe a human placenta cDNA library [37]; two over-lapping cDNA clones were isolated which together yielded a 1.4 kb sequence containing the full coding region (Table I).

Shaper et al. [38] used a bovine β-1,4-Gal-transferase cDNA probe to screen murine cDNA libraries. A series of overlapping clones were obtained which yielded a full-length cDNA (Table I) spanning 4038 bp and containing a 1197 bp coding region, a 2582 bp 3′-untranslated region and an 88 bp poly(A) tract. Nakazawa et al. [39] obtained a full-length murine 3.9 kb cDNA using a similar strategy.

A murine genomic DNA library was screened with a murine β-1,4-Gal-transferase cDNA probe and several clones were characterized [40]. The gene contains 6 exons and is at least 50 kb in size. Exon 1 encodes the 5′-untranslated region, the short N-terminal cytoplasmic domain, the transmembrane segment and the first 94 amino acids of the catalytic domain (Fig. 2). Exon 6 encodes the C-terminal segment of the protein and the entire 3′-untranslated region.

A human genomic library was screened with a murine cDNA probe and several clones were analyzed [41]. The human gene has a structure similar to the murine gene containing 6 exons spanning over 50 kb of genomic DNA. Southern analysis indicated the presence of only a single gene.

The chromosomal localization of the β-1,4-Gal-transferase gene has been determined in four species (Table I) (42–45).

Comparison of the amino acid sequences of bovine, human and murine β-1,4-Gal-transferases shows conservation of a four Pro sequence in the stem region (Fig. 2) and of all seven Cys residues. There is a single conserved potential N-glycosylation site.

The greatest variability occurs in the stem region; the sequence similarity is over 90% in regions where functional domains have been located, i.e., the transmembrane anchor and C-terminal catalytic domains. No significant sequence similarities have as yet been found between β-1,4-Gal-transferase and other proteins; a short hexapeptide region has been suggested as a potential UDP-Gal binding site on the basis of similarity to UDP-Gal:Gal(β1–4)GlcNAc-R (Gal to Gal) α-1,3-Gal-transferase (see below).

3.1.2. Domain structure of glycosyltransferases

Full-length β-1,4-Gal-transferase is a type II integral membrane protein (N_{in}/C_{out} orientation) with a short amino-terminal cytoplasmic domain, a non-cleavable signal/anchor transmembrane domain and a long carboxy-terminal catalytic domain within the endomembrane lumen (Table I). This domain structure has been found for all the glycosyltransferase genes cloned to date (Table I, Fig. 2) [15,18].

Size heterogeneity has been observed for several glycosyltransferases purified from various tissues [29,32,46–51]. For example, soluble and enzymatically active β-1,4-Gal-transferase purified from bovine milk [32] was shown to have undergone proteolysis at the stem region between the transmembrane and catalytic domains (Fig. 2). Data such as this as well as expression of truncated recombinant enzymes (see below) indicate that glycosyltransferases can be cleaved at the stem region to yield active soluble enzyme. Although the stem region shows relatively low amino acid sequence conservation between species, there are some common features which indicate that this domain serves as a flexible tether to hold the catalytic domain within the Golgi lumen, i.e., the presence of Pro and Gly residues and of consensus sites for N-glycosylation.

It appears, therefore, that the amino-terminal and transmembrane domains are not required for catalytic activity. The transmembrane domain is, however, essential for targeting to the Golgi apparatus and anchoring to the Golgi membrane. Truncated glycosyltransferases that have been cleaved in the stem region are more amenable to purification than the full-length proteins.

3.1.3. Alternate transcription initiation sites

Northern blots on mRNA from bovine, human, mouse and rat tissues usually show 2–3 bands, the largest being about 4.2–4.5 kb. Murine β-1,4-Gal-transferase cDNA has two in-frame ATG codons at its 5′-end and a termination codon 18 bp upstream of the first ATG. Analysis of mRNA from a murine mammary cell line showed the presence of two sets of transcripts differing in length by about 200 bp [38,40,52]. The first transcription initiation site was upstream of the first ATG codon and the second site was between the two ATG codons indicating that at least two promoters control the gene; Northern analysis shows two transcripts at 3.9 and 4.1 kb. Translation from the two in-frame ATG codons predicts proteins of 399 and 386 amino acids respectively. The long 4.1 kb transcript should produce only the 399 amino acid long form of the protein but could also produce the short form by a process called 'leaky scanning' in which an internal Met codon is used as the translation initiation point. The short 3.9 kb transcript can only result in the short protein species. The two Met codons are conserved in the bovine, human and murine sequences. Analysis of mRNA from Madin–Darby bovine kidney cells showed two sets of transcripts very similar to those described for murine mammary cells encoding proteins of 389 and 402 amino acids (Table I) [34].

In vitro translation experiments were carried out with bovine β-1,4-Gal-transferase transcripts in the absence of microsome membranes [34]. The protein products were analyzed by sodium dodecyl sulfate-polyacrylamide gel electrophoresis. Only a single protein product of the predicted size was obtained from the short and long transcripts respectively proving that the long transcript initiated translation only at the up-stream Met codon and that 'leaky scanning' did not occur. When *in vitro* translation was carried out in the presence of dog pancreas microsomes, both the long and short proteins increased in size by about 3 kDa. Endoglycosidase H treatment removed most of this extra material indicating that both proteins had been glycosylated. It was concluded that both proteins contain a non-cleavable transmembrane segment and are oriented as type II integral membrane proteins (Fig. 2) with the large C-terminal catalytic domain within the lumen of the microsomes. Protease protection experiments confirmed this orientation [34].

There has been some recent divergence of opinion on the functional significance of the long and short transcripts of β-1,4-Gal-transferase. A series of papers from the laboratory of Barry Shur [20,53–56] have suggested that the long form may be preferentially targeted to the cell surface. However, other workers have concluded that both the long and short forms are targeted to the Golgi apparatus [19,57–62].

Although most of the intra-cellular β-1,4-Gal-transferase is located in the Golgi apparatus, there is general agreement that a small proportion of the total intracellular enzyme is on the cell surface of some cells [20]; this cell surface β-1,4-Gal-transferase may play a role in cell–cell recognition [20,56,63,64]. Studies on RNA from a variety of tissues indicated that the relative abundance of the short and long forms of β-1,4-Gal-transferase mRNA correlates with β-1,4-Gal-transferase specific activities in the Golgi and plasma membrane respectively. Both stable and transient transfections have been carried out with DNA constructs encoding either the short or long forms of β-1,4-Gal-transferase; some reports indicated no preferential targeting [57,58,60,61] whereas Shur's group reported that the long form is targeted to the plasma membrane fraction at a level 2–4 fold higher than the short form [20,55]. These discrepancies may be due to the different systems used by the various groups and to the difficulty of establishing by immunohistochemical localization that a small fraction of the total intra-cellular enzyme is or is not present on the cell surface. However, Masibay et al. [60] used subcellular fractionation methods to separate Golgi and plasma membrane fractions and found that the the long and short isoforms of β-1,4-Gal-transferase gave identical distribution patterns for these fractions.

Harduin-Lepers et al. [62] have recently shown that the long and short forms of β-1,4-Gal-transferase are expressed in a tissue-specific manner in the mouse and provide a mechanism for regulation of enzyme levels. Tissues which express relatively low levels of β-1,4-Gal-transferase (such as brain) are under the control of the upstream promoter and contain only the longer 4.1 kb transcript. Most somatic mouse tissues express intermediate levels of enzyme and are under the control of both promoters; the ratio of the 4.1 kb to 3.9 kb transcripts is about 5:1. The short 3.9 kb transcript is the predominant mRNA species in tissues which express β-1,4-Gal-transferase at very high levels such as parietal yolk sac cells and lactating mammary gland [54–56,62]; lactating mammary gland has a ratio of the 4.1 kb to 3.9 kb transcripts of 1:10 indicating that the enzyme in this tissue is primarily under the control of the downstream promoter [62].

Both promoters are weak and are unable to effect expression of chloramphenicol acetyltransferase (CAT) in transient expression experiments unless an enhancer element is incorporated into the construct [62]. The upstream promoter appears to be a typical constitutive 'house-keeping' promoter. It lacks classical CAAT and TATA boxes and has six upstream GC boxes (cis-acting positive regulatory elements) under the control of the Sp1 transcription factor. Between the two promoters are three more GC boxes, at least two mammary gland-specific positive regulatory elements and a negative cis-acting regulatory element. It is suggested that in the mammary gland, mammary gland specific transcription factors bind to their regulatory elements and deactivate the negative control thereby causing a marked increase in transcription of the 3.9 kb mRNA.

Murine male germ cells (spermatogonia, pachytene spermatocytes and round spermatids) do not express the 3.9 kb transcript [65]. Spermatogonia show a strong 4.1 kb signal, pachytene spermatocytes show very faint bands at 2.9, 3.1 and 4.1 kb, and round spermatids show strong bands at 2.9 and 3.1 kb. The two truncated transcripts and the 4.1 kb transcript all produce the same 399-amino acid long form of β-1,4-Gal-transferase protein. Truncation is due to the use of alternate polyadenylation signals. The 2.9 kb truncated transcript (and probably also the 3.1 kb transcript) has an additional 5′-terminal untranslated 559 bp sequence not present in the 4.1 kb transcript [66]. This extra sequence is immediately upstream and contiguous to the 4.1 kb transcriptional start site suggesting the presence of a germ cell-specific promoter which regulates expression of β-1,4-Gal-transferase in haploid round spermatids.

3.1.4. Expression of recombinant β-1,4-Gal-transferase

Several groups have reported the expression of recombinant β-1,4-Gal-transferase. Transient transfection of full-length cDNA into monkey kidney COS-7 cells under the control of a simian virus 40 promoter [33] or into COS-1 cells under the control of a human cytomegalovirus promoter [67] resulted in the expression of active intra-cellular membrane-bound enzyme. Although human β-1,4-Gal-transferase is a glycoprotein [28], Aoki et al. [68] were able to express in E. coli the unglycosylated but enzymatically active soluble catalytic region of human β-1,4-Gal-transferase lacking the N-terminal and transmembrane domains (Fig. 2). They used the pIN-III-ompA2 vector which has the ompA signal peptide driven by combined lpp and lac promoters. The vector also has the lac operator so that fusion protein synthesis is induced by isopropyl-β-D-thiogalactoside. The ompA signal peptide directs the fusion protein to the outer membrane where the signal is cleaved by bacterial signal peptidase with resulting secretion of the expressed protein into the periplasmic space and culture medium. The enzyme from the culture medium was purified to homogeneity in a single step using a GlcNAc–Sepharose affinity column. These findings indicate that the E. coli periplasm can provide an environment which allows proper folding of the protein even in the absence of glycosylation and that the catalytic activity requires neither the N-terminal nor transmembrane domains (Fig. 2).

A putative UDP-binding site was identified (68) by reacting purified human milk β-1,4-Gal-transferase with [¹⁴C]4-azido-2-nitrophenyluridylpyrophosphate in the presence of ultra-violet light. A labeled 53-amino acid peptide was isolated after cyanogen bromide cleavage of the protein. Site-directed mutagenesis in this region of the protein suggested three Tyr and two Trp residues as potential sites for UDP-Gal and/or acceptor binding.

Bovine β-1,4-Gal-transferase was expressed as a glutathione-S-transferase fusion protein in *E. coli* using the pGEX-2T vector [69]. Recombinant enzyme accumulated within insoluble inclusion bodies and required guanidine HCl and a mixture of oxidized and reduced glutathione for regeneration of enzyme activity. The amino terminal 129 amino acid residues could be deleted without loss of enzyme activity.

A protease-defective strain of *Saccharomyces cerevisiae* (BT 150) was used to express full-length cDNA of HeLa cell β-1,4-Gal-transferase [70]. The enzyme was not secreted. Enzyme was purified from cell lysates with a 24% recovery; a typical preparation yielded about 0.001 mg enzyme with a specific activity of 4.7 units/mg.

3.1.5. Targeting to the Golgi apparatus

All Golgi-resident proteins identified to date belong to the type II integral membrane protein category (Table I, Fig. 2) [15,18,71]. β-1,4-Gal-transferase has been localized by immuno-electron microscopy to the trans-cisternae of the Golgi apparatus [28]. Several recent reports have investigated the glycosyltransferase domains responsible for Golgi targeting. The general approach has been to construct chimeric cDNAs encoding hybrid proteins in which various domains of β-1,4-Gal-transferase are connected to reporter proteins that would normally not be retained in the Golgi apparatus [57,58]. Following transfection and transient expression of these hybrid constructs in mammalian cells, the intracellular destination of the reporter protein is determined by immunofluorescence and immuno-electron microscopy.

Nilsson et al. [57] found that the human β-1,4-Gal-transferase cytoplasmic domain did not cause retention in the Golgi. The transmembrane domain without the cytoplasmic domain did result in Golgi retention of the reporter protein but there was also some reporter protein detected at the cell surface. The cytoplasmic domain from either β-1,4-Gal-transferase or from the reporter protein, in combination with the β-1,4-Gal-transferase transmembrane domain, resulted in complete Golgi retention. The transmembrane domain clearly contains the signal for Golgi retention while the cytoplasmic domain plays a non-specific accessory role. As few as 10 amino acids on the lumenal half of the transmembrane domain were sufficient to localize the reporter protein to the Golgi. Immuno-electron microscopy using gold particles localized the reporter protein to the *trans*-Golgi.

Teasdale et al. [58] carried out similar experiments with bovine β-1,4-Gal-transferase. Removal of most of the cytoplasmic tail did not alter the Golgi localization of the enzyme in either transiently transfected COS-1 cells or stably transfected murine L cells. Replacement of the cytoplasmic tail and transmembrane domain of β-1,4-Gal-transferase with the cleavable signal peptide from influenza hemagglutinin led to rapid secretion of the enzyme from transfected COS-1 cells. Replacement of the transmembrane domain of β-1,4-Gal-transferase with the signal/anchor domain of the plasma membrane-localized human transferrin receptor resulted in transport of β-1,4-Gal-transferase to the cell surface of transfected COS-1 cells. Finally, replacement of the internal signal sequence of ovalbumin, which is normally secreted, with the transmembrane domain of β-1,4-Gal-transferase resulted in retention of ovalbumin in the Golgi of transfected COS-1 cells.

Aoki et al. [59] used the C-terminal portion of human chorionic gonadotropin as a reporter molecule for localization of Gal-transferase; this technique allowed them to distinguish Gal-transferase produced by transient expression of their constructs in COS cells from endogenous Gal-transferase. Immunofluorescence localized the recombinant protein primarily to the Golgi apparatus although over-expression led to some fluorescence throughout the cytoplasm. Deletions of large portions of the cytoplasmic tail or stem region of Gal-transferase did not alter its Golgi localization. Site-directed mutagenesis of the transmembrane segment identified a 5-amino acid region required for Golgi retention; in particular residues Cys^{29} and His^{32} in this region of the signal-anchor domain were shown to be essential. It was suggested that these two amino acids may be involved in a metal-binding site involved in Golgi retention.

Russo et al. [61] found that both the long and short forms of β-1,4-Gal-transferase targeted to the Golgi apparatus after stable transfection into Chinese hamster ovary (CHO) cells. Immunolocalization at the electron microscope level showed both forms of the enzyme to be located in the trans-region of the Golgi. No immunoreactivity was detected at the cell surface. They showed that the cytoplasmic tail, the transmembrane domain and 8 amino acids of the stem region of Gal-transferase were sufficient for Golgi retention of a reporter protein. Both the short and long fusion proteins targeted the reporter protein to the Golgi apparatus with no evidence of immunoreactivity at the cell surface. Although a truncated form of β-1,4-Gal-transferase lacking the transmembrane segment can be expressed in *E. coli* in an enzymatically active non-glycosylated form [68], transient expression of a similar construct in COS-7 cells [72] resulted in mRNA synthesis but no enzyme protein. It is possible that the lack of a transmembrane segment prevents targeting of the enzyme to the Golgi apparatus with consequent degradation of the protein in the cytoplasm. This effect was used [60] to show that the cytoplasmic domain was not required for Golgi targeting but that removal of even the first five amino acids at the amino-terminal side of the transmembrane domain prevented Golgi localization and led to enzyme inactivation. The cytoplasmic and transmembrane domains of bovine UDP-Gal:Gal(β1–4)GlcNAc-R α-1,3-Gal-transferase could replace the equivalent regions of the bovine β-1,4-Gal-transferase without any apparent effect on Golgi targeting. However, when the equivalent regions of CMP-sialic acid:Gal(β1–4)GlcNAc-R α-2,6-sialyltransferase were used, no enzyme activity was obtained unless flanking sequences encoding part of the neck region were also included in the construct. The role of the neck region in sialyltransferase targeting has also been reported by Munro [73] as is discussed below.

Masibay et al. [60] point out that the transmembrane domains of plasma membrane-targeted proteins are broader and more hydrophobic than those of Golgi-targeted proteins. Replacement of amino acids in the transmembrane domain of β-1,4-Gal-transferase with isoleucine residues or insertion of extra isoleucine residues caused a significant shift of targeting from the Golgi to the cell surface.

The data indicate that the transmembrane domain of β-1,4-Gal-transferase contains the Golgi retention signal but the cytoplasmic domain also plays a role in targeting.

3.2. UDP-Gal:Gal(β1–4)GlcNAc-R (Gal to Gal) α-1,3-Gal-transferase (E.C. 2.4.1.124/151)

Gal(α1–3) occupies the terminal non-reducing position on N-glycan antennae and is an uncharged alternative to sialic acid. Galili [74–77] has carried out extensive studies on the evolution of the Gal(α1–3)Gal(β1–4)GlcNAc epitope. The epitope is absent from fish, amphibian, reptile and bird fibroblasts but has a wide distribution in non-primate mammals, lemurs and New World monkeys. The epitope is absent from Old World monkeys, apes and humans. Anti-α-1,3-Gal antibodies are natural antibodies present in human serum and constitute about 1% of circulating IgG. The appearance of aberrant α-1,3-Gal-transferase activity in human cells is a possible trigger for some forms of auto-immune disease [75,77].

3.2.1. Cloning of α-1,3-Gal-transferase cDNA and genomic DNA

Joziasse et al. [78] isolated a 1.8 kb cDNA encoding bovine α-1,3-Gal-transferase. The domain structure of the enzyme is similar to that described above for β-1,4-Gal-transferase (Table I, Fig. 2). The only amino acid sequence similarity noted between the α-1,3- and β-1,4-Gal-transferases is a Cys(X)$_{5-6}$Lys(or Arg)AspLysLysAsnAsp(or Glu) sequence downstream from the sequence identified by Aoki et al. [68] as a possible UDP-binding site. Analysis of calf thymus RNA showed at least two sets of transcripts (3.6–3.9 kb) indicating the presence of at least two promoters but only a single protein is encoded.

Larsen et al. [79] used an approach to cloning murine α-1,3-Gal-transferase cDNA that required neither the purification of the enzyme nor the availability of antibodies to the enzyme. They used a transient expression screening approach involving gene transfer of a murine cDNA library ligated into the mammalian expression vector pCDM7 [80,81] which replicates at high efficiency in mammalian cells expressing the SV40 large T antigen. The vector is capable of high level transient expression of DNA. The screening procedure was based on identifying cells with the product of α-1,3-Gal-transferase action on their cell surfaces. COS-1 cells were selected because they express the SV40 large T antigen, they do not express α-1,3-Gal-transferase, and their cell surfaces carry the oligosaccharide substrate for α-1,3-Gal-transferase. A plasmid (pCDM7-αGT) was obtained capable of directing expression of *Griffonia simplicifolia* isolectin-I-B$_4$ binding activity in COS-1 cells. The cDNA insert of pCDM7-αGT was 1.5 kb in size and encoded a 394-amino acid protein with a domain structure typical of glycosyltransferases (Table I, Fig. 2). Northern analysis showed a message of 3.6 kb in murine F9 teratocarcinoma cells.

Joziasse et al. [82] screened a murine cDNA library with a bovine probe for α-1,3-Gal-transferase and isolated a cDNA encoding a 337-amino acid protein (Table I). The deviation from the previously described murine sequence [79] is due to (i) assignment of a downstream ATG as the translation initiation codon thereby giving a cytoplasmic tail of 6 amino acids rather than 41, and (ii) a 22-amino acid deletion in the stem region. Analysis of mouse mRNA by a polymerase chain reaction (PCR) approach identified four distinct messages (all about 3.7 kb in size) encoding four different protein products which differ from one another in the stem region (Table I). The 359-amino acid protein (Table I) corresponds to the cDNA previously isolated by Larsen et al. [79]. Although there may be more than one bovine α-1,3-Gal-transferase transcript (see above), PCR analysis of bovine mRNA [82] showed only a single 368-amino acid protein product (Table I). The 368-amino acid bovine protein shows a strong similarity (about 80%) to the 371-amino acid murine protein (Table I); the main difference is a 3-amino acid deletion in the stem region of the bovine protein. Although α-1,3-Gal-transferase has a domain structure typical of the glycosyltransferases (Table I, Fig. 2), the only sequence similarity is an approximately 55% identity of the human α-1,3-Gal-transferase pseudogene to the human blood group A and B transferases [83]. The Cys(X)$_{5-6}$Lys(or Arg)AspLysLysAsnAsp(or Glu) consensus sequence suggested as a UDP-Gal-binding site (see above) appears as Cys(X)$_5$GlnAspLysLysHisAsp in murine α-1,3-Gal-transferase.

The α-1,3-Gal-transferase is not normally expressed in human tissues. No α-1,3-Gal-transferase mRNA can be detected in human and Old World monkey tissues. However, Southern analysis of human genomic DNA has revealed sequences homologous to α-1,3-Gal-transferase using bovine [78] and murine [84] probes.

Two homologous genes were obtained from a human genomic DNA library probed with bovine α-1,3-Gal-transferase cDNA [85]. Both genes are non-functional pseudogenes; clone HGT-2 is an intron-less processed pseudogene located on chromosome 12 and clone HGT-10 is an unprocessed pseudogene on chromosome 9. Murine α-1,3-Gal-transferase maps to a region of chromosome 2 with linkage homology to human chromosome 9q22-ter [86]. Fluorescent *in situ* hybridization has localized HGT-2 to human chromosome 12q14-q15, and HGT-10 to chromosome 9q33-q34 [87].

HGT-2 has an uninterrupted sequence 81% similar to the complete bovine coding sequence but contains multiple frame-shifts and nonsense codons in all three reading frames. HGT-10 contains a sequence similar to the N-terminal and transmembrane domains of the bovine enzyme and is flanked by introns at both ends. The C-terminal segment of this gene was not cloned by Joziasse et al. [85] but may correspond to a clone isolated by Larsen et al. [84] which contains a 703 bp region 82% similar to the murine α-1,3-Gal-transferase C-terminal coding region; this gene contains two frame-shift mutations and a nonsense codon that disrupt the reading frame.

The C-terminal region of HGT-2 shows about 39% sequence similarity to the analogous regions of human blood group A α-1,3-GalNAc-transferase and human blood group B α-1,3-Gal-transferase. Clone HGT-10 is located on human chromosome 9q33-q34 and the gene locus encoding the allelic blood group A and B transferase genes is on human chromosome 9q34. These findings indicate that all four genes are derived from the same ancestral gene by gene duplication and subsequent divergence.

The non-functional human pseudogenes probably replaced a functional α-1,3-Gal-transferase gene in Old World monkeys some time after the divergence of New World and Old World monkeys [76,85].

Joziasse et al. [82] isolated five overlapping murine genomic clones for α-1,3-Gal-transferase and determined that the gene consists of 9 exons spanning at least 35 kb. The genomic structure resembles that of the α-2,6-sialyltransferase and β-1,4-Gal-transferase genes in that all are relatively large and are divided into 6 or more exons. In each case, the cytoplasmic domain, the transmembrane domain and part of the stem region (Fig. 2) are on a single exon while another exon contains the translation termination codon and all of the relatively long 3'-untranslated region.

Analysis of the mouse gene showed that the four transcripts for α-1,3-Gal-transferase (Table I) are due to alternative splicing of a message transcribed from a single gene. Splicing produces mRNAs which either (i) contain both exons 5 and 6, (ii) contain only exon 5, (iii) contain only exon 6, or (iv) contain neither exons 5 or 6. The four transcripts were found in all mouse tissues tested except for male germinal cells which had no detectable mRNA.

3.2.2. Expression of recombinant α-1,3-Gal-transferase

The cDNA encoding the putative catalytic domain of murine α-1,3-Gal-transferase (lacking the N-terminal and transmembrane segments) was fused in-frame downstream to the cDNA encoding a secretable form of the IgG binding domain of *Staphylococcus aureus* protein A in a mammalian expression vector [79]. Transfection of COS-1 cells with this fusion construct resulted in the secretion of α-1,3-Gal-transferase activity into the culture medium proving that the signal for retention of transferase in the Golgi apparatus resides in the N-terminal and transmembrane segments. The secreted and soluble fusion protein bound specifically to an IgG-Sepharose column. The IgG-Sepharose-bound material was enzymatically active.

High-level expression of enzymatically active α-1,3-Gal-transferase was achieved using an insect cell expression system [88]. Recombinant baculovirus containing the entire coding sequence of bovine α-1,3-Gal-transferase under the control of the poly-hedrin promoter was used to express the enzyme in insect cells. The cells were lysed and enzyme was purified by a single step using UDP-hexanolamine-Sepharose affinity chromatography. The expressed enzyme accounted for about 2% of the total cellular protein and had a specific activity 1000-fold higher (> 0.1 μmoles/min/mg) than that found in calf thymus gland. Both the recombinant and calf thymus enzymes incorporate Gal in (α1–3) linkage to the terminal Gal of Gal(β1–4)Glc(NAc) on oligosaccharides, glycolipid and glycoproteins and do not act on Fuc(α1–2)Gal- terminated compounds thereby distinguishing them from blood group B α-1,3-Gal-transferase.

Stable transfection of the murine α-1,3-Gal-transferase gene into Chinese hamster ovary (CHO) cells [89] was used to show that α-2,3-sialyltransferase and α-1,3-Gal-transferase compete for Gal(β1–4)GlcNAc termini in these cells and therefore probably reside in the same sub-cellular compartment.

4. Sialyltransferases

The 'one linkage–one glycosyltransferase' rule suggests that there are at least 12 different sialyltransferases required for the synthesis of all the known sialylated structures [1,46,90,91]. N-glycan synthesis involves primarily two of these enzymes, CMP-sialic acid:Gal(β1–4)GlcNAc-R α-2,6-sialyltransferase (ST6N) (E.C. 2.4.99.1) and CMP-sialic acid:Gal(β1–3/4)GlcNAc-R α-2,3-sialyltransferase (ST3N) (E.C.

2.4.99.6); these enzymes have been purified and substrate specificities have been determined [49,92]. The more recent work on the molecular biology of these enzymes is described below.

4.1. CMP-sialic acid:Gal(β1–4)GlcNAc-R α-2,6-sialyltransferase (ST6N) (E.C. 2.4.99.1)

4.1.1. Cloning of α-2,6-sialyltransferase cDNA and genomic DNA

Screening of a rat liver cDNA λgt11 expression library with antibody to α-2,6-sialyltransferase yielded a single positive clone [90]. The clone was used to obtain overlapping cDNA clones encoding a 403 amino acid protein with a type II integral membrane protein domain structure typical of the glycosyltransferases (Table I, Fig. 2).

The coding region of rat liver α-2,6-sialyltransferase was used to isolate human sialyltransferase cDNA from cDNA libraries [93,94]. Human α-2,6-sialyltransferase cDNA has also been isolated by expression cloning; screening was carried out either for the appearance of antigen CDw75 [95] or of the antigen HB-6 on the host cell surface [96]. The sialyltransferase is essential for the synthesis of the CDw75 and HB-6 antigens. The human enzyme is 406 amino acids long and has a domain structure similar to the rat enzyme (Table I).

The rat and human α-2,6-sialyltransferase amino acid sequences show 87.6% similarity [93]. Comparison of the sequences for human cDNAs isolated from four different tissues show minor differences in both the coding and non-coding regions [96]. Although there is no obvious sequence similarity between the α-2,6-sialyltransferases and β-1,4-Gal-transferases, the enzymes have the same domain structure (Table I, Fig. 2) and both have seven Cys residues. One Cys is in the transmembrane domain of both enzymes and four other Cys residues are in equivalent positions in the catalytic domains of both enzymes.

Northern analysis of rat tissues show a tissue-specific distribution of mRNAs [97]. An abundant mRNA of 4.3 kb was detected only in liver, a 4.7 kb mRNA was seen in many tissues and a 3.6 kb mRNA was seen only in kidney. The 4.7 kb mRNA differs from the 4.3 kb transcript only in containing a longer 5'-untranslated region; both transcripts encode the same α-2,6-sialyltransferase protein. Northern analyses of human tissues show mRNAs between 4 and 5 kb containing a long 3'-untranslated region [94,96].

The organization of the rat α-2,6-sialyltransferase gene resembles that of the UDP-Gal:GlcNAc-R β-1,4-Gal-transferase gene [98–100] and the UDP-Gal:Gal(β1–4)GlcNAc-R (Gal to Gal) α-1,3-Gal-transferase gene [82] although there is no sequence similarity between the three enzymes. The rat α-2,6-sialyltransferase gene spans at least 80 kb of genomic DNA and contains at least 11 exons. There are at least three promoters responsible for the production of five or more messages from a single gene sequence [97,99–101]. Promoter P_L produces a liver-specific 4.3 kb mRNA (exons 1 to 6). Promoter P_C is constitutive in several tissues and produces a 4.7 kb mRNA (exons –1 to 6). Promoter P_K is restricted to the kidney and produces several 3.6 kb mRNAs missing the 5'-half of the coding sequence (kidney-specific exons K1, K2 and K3 and exons 4 to 6). Although the 3.6 kb messages produce a protein in kidney

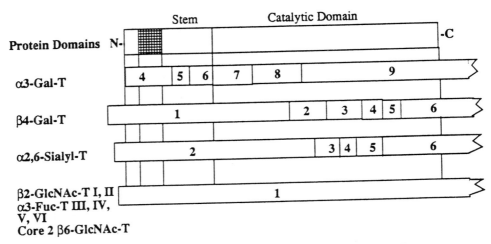

Fig. 3. Comparison of the protein domains of several glycosyltransferases with their exon structures (adapted from Ref. [13]). The hatched region is the transmembrane domain of the protein. The numbered boxes represent the exons. Some genes (e.g., UDP-Gal:Gal(β1–4)GlcNAc-R α-1,3-Gal-transferase, CMP-sialic acid:Gal(β1–4)GlcNAc-R α-2,6-sialyltransferase and UDP-GlcNAc: Man(α1–3R [GlcNAc to Man(α1–3)] β-1,2-GlcNAc-transferase I) have at least one up-stream exon containing only non-coding sequences.

that is immunologically cross-reactive with α-2,6-sialyltransferase, this protein does not appear to encode an active α-2,6-sialyltransferase [102]. Mouse kidney expresses only one 4.7 kb mRNA suggesting that the 3.6 kb transcripts in rat kidney are not essential for kidney function [99].

Human α-2,6-sialyltransferase exons Y, Z and I to VI [103] are homologous to rat exons −1, 0 and 1 to 6 respectively. The position of human exon X on genomic DNA is not known and no equivalent exon has been found in the rat genome. There is good evidence that the human gene also has at least three promoters.

Comparison of the α-2,6-sialyltransferase gene with the genes for UDP-Gal: GlcNAc-R β-1,4-Gal-transferase and UDP-Gal:Gal(β1–4)GlcNAc-R (Gal to Gal) α-1,3-Gal-transferase indicates that all are relatively large and are divided into 6 or more exons. In each case, a single exon contains the cytoplasmic domain, the transmembrane domain and part of the stem region (Figs. 2 and 3). The 3′-terminal exon is very large and contains the translation termination codon and all of the relatively long 3′-untranslated region. The catalytic domain is divided between several small exons (Fig. 3). Multiple transcripts are produced from all three genes, at least in some species, due either to more than one site of transcription initiation, or alternative exon splicing, or alternative polyadenylation signals, or a combination of these processes.

Tissue-specific expression of rat α-2,6-sialyltransferase appears to be regulated at the transcriptional level [97]. The 4.3 kb mRNA is expressed only in liver under the control of the liver-specific promoter P_L; the 4.7 kb mRNA is expressed in lesser amounts (10–100-fold less) in liver and several other tissues. This correlates with a

10–100-fold higher activity of α-2,6-sialyltransferase activity in liver relative to other rat tissues. The levels of the 4.7 kb mRNA vary 50–100 fold in various rat tissues and these levels also correlate with enzyme activity.

The use of alternative promoters also appears to control the tissue-specific expression of human α-2,6-sialyltransferase [103]. Two different human α-2,6-sialyltransferase cDNAs have been isolated (see above), clone STP from a placenta library [93] and clone STB from a B-cell library [95]. STB and STP are identical over the entire coding region and over part of the 5′-untranslated region (UTR) immediately upstream of the coding region. The clones diverge in the remainder of the UTR, probably due to an exon–intron junction [104]. STB is found only in mature B cells while STP is found in all human cell types tested [104] suggesting that alternate splicing controls tissue-specific expression.

Cell type-specific transcription of genes is controlled by DNA-binding proteins called transcription factors which themselves have a restricted pattern of expression. Promoter P_L was demonstrated to be about 50-fold more active in a hepatoma cell line (HepG2) known to express α-2,6-sialyltransferase than in a cell line (Chinese hamster ovary) which does not express this enzyme [98]. The P_L promoter sequence contains consensus binding sites for liver-restricted transcription factors [105]. Transcription factors HNF-1α (hepatocyte nuclear factor 1α), DBP (D-binding protein) and LAP (liver-enriched transcriptional activator protein) were found to be strong activators of promoter P_L and are most likely responsible for the high level expression of α-2,6-sialyltransferase in rat liver.

The hepatic acute phase response is accompanied by increased levels of α-2,6-sialyltransferase in liver and in the circulation [106–110]. Dexamethasone is believed to mediate this response. Dexamethasone was shown to cause 3–4-fold enrichment in α-2,6-sialyltransferase mRNA in rat hepatoma cells and a corresponding increase in enzyme activity [111]; the induction was mediated by a transcriptional enhancement mechanism [112]. A similar enhancement of α-2,6-sialyltransferase, but not α-2,3-sialyltransferase, was observed in Fisher rat fibroblasts cultured in the presence of dexamethasone due to an increased transcription rate of the 4.7 kb mRNA encoding this enzyme [113]. Thyrotropin decreases the level of α-2,6-sialyltransferase mRNA in rat thyroid cells [114].

4.1.2. Expression of recombinant α-2,6-sialyltransferase

Stable transfection [115] of α-2,6-sialyltransferase cDNA into Chinese hamster ovary (CHO) cells, which do not express α-2,6-sialyltransferase activity, permitted synthesis of the NeuAc(α2–6)Gal(β1–4)GlcNAc structure indicating that recombinant α-2,6-sialyltransferase competed effectively with endogenous α-2,3-sialyltransferase and other glycosyltransferases for available Gal(β1–4)GlcNAc termini.

Transformation of COS cells with an α-2,6-sialyltransferase expression plasmid generated three distinct cell surface sialic-acid dependent antigens (HB-6, CDw75 and CD76) [96]. The sialyltransferase was shown to be present in the Golgi apparatus and no enzyme was detected at the cell surface [96,116]; the suggestion that the CD75 antigen is cell surface sialyltransferase [95] has been withdrawn [117].

4.1.3. Targeting to the Golgi apparatus

Rat α-2,6-sialyltransferase has been localized to the *trans*-Golgi cisternae and the *trans*-Golgi network of hepatocytes, hepatoma cells and intestinal goblet cells but may be more diffusely distributed in the Golgi apparatus of other cell types [118]. N-terminal analysis of purified catalytically active soluble enzyme showed proteolytic cleavage in the stem region 38 amino acids downstream of the C-terminal end of the transmembrane domain [90,119]. Inflammation has been shown to cause the release of soluble and catalytically active α-2,6-sialyltransferase into the serum due to cleavage of the enzyme at the stem region by a cathepsin D-like protein [106–108]. These studies indicate that the amino-terminal cytoplasmic tail, the transmembrane segment and the stem region are not needed for catalysis but may be required for Golgi targeting and anchoring.

Replacement of the cytoplasmic tail, transmembrane domain and 31 amino acids from the stem region of α-2,6-sialyltransferase with a cleavable signal peptide sequence [118] resulted in secretion of active enzyme from the CHO host cell. The secreted protein was mainly resistant to endo-β-*N*-acetylglucosaminidase H digestion indicating the formation of complex N-glycans; intra-cellular enzyme was digested by the endoglycosidase. Control cells transfected with cDNA encoding full-length α-2,6-sialyltransferase did not secrete enzyme.

Fusion proteins between partial α-2,6-sialyltransferase sequences and a reporter protein [73] were used to prove that the transmembrane segment of α-2,6-sialyltransferase is necessary and sufficient to prevent movement of the reporter protein to the cell surface. However, constructs with either the cytoplasmic or stem region from α-2,6-sialyltransferase in addition to the α-2,6-sialyltransferase transmembrane domain showed less leakage to the plasma membrane than constructs lacking these regions. Charged residues flanking the transmembrane domain are important for effective retention in the Golgi. It was shown that a poly(Leu)$_{17}$ but not a poly(Leu)$_{23}$ transmembrane domain can target to the Golgi provided a part of the α-2,6-sialyltransferase stem region is also present. The findings that the stem region contains Golgi targeting information and that the spacing between the stem and cytoplasmic domains is important for targeting have been confirmed by another group [120]. However, other workers [121] found that the 17-residue transmembrane domain of sialyltransferase was sufficient for complete Golgi retention and found no evidence to support the presence of Golgi targeting sequences in either the cytoplasmic tail or the stem region of sialyltransferase.

4.1.4. The sialylmotif

Comparison of CMP-sialic acid:Gal(β1–4)GlcNAc-R α-2,6-sialyltransferase (ST6N), CMP-sialic acid:Gal(β1–3)GalNAc-R α-2,3-sialyltransferase (ST3O) and CMP-sialic acid:Gal(β1–3/4)GlcNAc-R α-2,3-sialyltransferase I (ST3N I, see below) reveals a region of 55 amino acids with extensive homology (40% identity; 58% conservation) in the middle of the catalytic domain [122]. There is a second 23 residue region of similarity near the COOH-terminus [123]. Each similarity region contains a conserved Cys residue suggesting that these two Cys residues may form a disulphide bridge in the native proteins. None of the elements of the conserved motif (sialylmotif) are found in other glycosyltransferases.

4.1.5. Chicken CMP-sialic acid:Gal(β1–4)GlcNAc-R α-2,6-sialyltransferase

The polymerase chain reaction (PCR) was performed with chicken cDNA as template and degenerate primers based on the sialylmotif (see above) [124]. The PCR product was used to screen a chick embryo cDNA library and the entire coding region of chicken CMP–sialic acid:Gal(β1–4)GlcNAc-R α-2,6-sialyltransferase was isolated (Table I). The catalytic domain of the enzyme was expressed by transient transfection of COS-7 cells and showed the same substrate specificity, kinetics and branch specificity as the mammalian enzyme. The chicken enzyme showed 56-61% amino acid identity in comparison to the human, rat and mouse enzymes.

4.2. CMP-sialic acid:Gal(β1–3/4)GlcNAc-R α-2,3-sialyltransferases I and II (ST3N I, E.C. 2.4.99.6, and ST3N II)

4.2.1. Cloning of α-2,3-sialyltransferase I cDNA

Rat liver CMP-sialic acid:Gal(β1–3/4)GlcNAc-R α-2,3-sialyltransferase I was purified and tryptic peptides were sequenced by liquid secondary ion mass spectrometry (LSIMS) followed by tandem mass spectrometry on a four-sector machine using high energy collision-induced dissociation analysis (CID) [122]. Degenerate oligonucleotide sense and anti-sense primers based on the amino acid sequence were used to carry out PCR on rat liver cDNA. Screening a rat liver cDNA library with one of the PCR products yielded two clones with inserts of 2.1 and 1.5 kb respectively.

The 2.1 kb clone contained a complete 1122 bp open reading frame encoding a 374 amino acid protein with all the peptide sequences revealed by mass spectrometry. The protein shows the domain structure typical of the glycosyltransferases (Fig. 2, Table I). The amino terminus of the purified enzyme was in the stem region 49 residues downstream from the initiation Met residue showing that this enzyme preparation was due to proteolysis and that the transmembrane domain is not required for enzyme activity.

Based on the sequence information of the sialylmotif (see above), two degenerate oligonucleotides were synthesized and the polymerase chain reaction (PCR) was carried out with human placental cDNA as template [125]. The resulting PCR product was used to screen a human placental cDNA library and a clone was obtained encoding the entire open reading frame of human CMP-sialic acid:Gal(β1–3/4)GlcNAc-R α-2,3-sialyltransferase I (Table I). There is 91% homology between the rat and human enzymes at the nucleotide level and 97% at the amino acid level. Northern analysis showed single messages at 2.7 and 2.4 kb in human and rat tissues respectively. Rat colon showed an additional 2.0 kb message. These messages were differentially expressed with especially high levels in skeletal muscle.

4.2.2. Expression of recombinant α-2,3-sialyltransferase I

A fusion protein with the catalytic domain of rat α-2,3-sialyltransferase I (lacking the cytoplasmic tail and transmembrane segment) and the cleavable insulin signal sequence was transiently expressed in COS cells [122]. Active enzyme was secreted into the medium proving that the cytoplasmic tail and transmembrane segment are not required for enzyme activity but are necessary for targeting to the Golgi apparatus.

Transient expression of the similarly truncated human placental enzyme in COS-1 cells resulted in secretion of an active enzyme which, like the rat enzyme, showed preferential activity towards type 1 chains, Gal(β1-3)GlcNAc [125]. The data indicate that this human placental α-2,3-sialyltransferase is different from the previously reported human placental enzyme which shows preferential activity towards type 2 chains, Gal(β1-4)GlcNAc [126].

4.2.3. CMP-sialic acid:Gal(β1–3/4)GlcNAc-R α-2,3-sialyltransferase II (ST3N II)

A novel stable expression cloning approach and selection with a cytotoxic lectin were used to clone a human CMP-sialic acid:Gal(β1–3/4)GlcNAc-R α-2,3-sialyltransferase which differs from the previously cloned α-2,3-sialyltransferases in showing a 3:1 preferential activity towards type 2 chains, Gal(β1-4)GlcNAc, relative to type 1 chains, Gal(β1-3)GlcNAc [127]. This enzyme is referred to as α-2,3-sialyltransferase II in this chapter. A human melanoma cDNA plasmid expression library was stably transfected into a human Burkitt lymphoma cell line and clones were selected which were resistant to the cytotoxic action of *Ricinus communis* agglutinin 120 (RCA$_{120}$). Plasmids encoding α-2,3-sialyltransferase II were rescued from these lectin-resistant cells (Table I). Human α-2,3-sialyltransferases I and II have only 34% homology over-all and are distinct entities; however, α-2,3-sialyltransferase II has the sialylmotif typical of all sialyltransferases cloned to date (see above). The relationship of α-2,3-sialyltransferase II to a similar human placental enzyme [126] and its role in the *in vivo* synthesis of the sialyl-Lewis[x] determinant remain to be established.

4.3. Brain specific sialyltransferase STX

PCR with degenerate primers based on the ends of the highly conserved N-terminal segment of the sialylmotif was used to clone additional rat cDNA sequences [128,129]. Newborn rat brain cDNA was used as a template for PCR in one of these experiments and yielded a novel 150 bp sialylmotif sequence. This PCR fragment was used to screen a newborn rat brain cDNA library and a novel sialyltransferase (STX) was discovered. STX has the type II transmembrane protein domain structure typical of all previously cloned glycosyltransferases (Table I, Fig. 2). A fusion construct was made in which the amino-terminal and transmembrane domains of STX were replaced with the cleavable insulin signal sequence followed by the protein A IgG binding domain. This construct was expressed in COS-1 cells and a 85 kDa fusion protein was secreted, adsorbed to IgG-Sepharose and the beads were assayed for sialyltransferase activity using a variety of acceptors (oligosaccharides, glycoproteins, gangliosides). No enzyme activity was detected. Northern analysis of rat tissues revealed a 5.5 kb message that was found only in newborn rat brain; adult rat brain and newborn kidney did not show the message indicating that STX is both tissue-specific and developmentally regulated, unlike the other cloned sialyltransferases which are differentially expressed in several tissues. It is suggested that STX makes a developmentally regulated brain-specific structure that may as yet not have been described. Alternatively, the enzyme may have been denatured during expression or may not encode a sialyltransferase.

4.4. CMP-sialic acid:Gal(β1–4)GlcNAc-R α-2,3-sialyltransferase (STZ/SAT-3)

PCR with human placental cDNA as template and degenerate primers based on the sialylmotif was used to clone a novel α-2,3-sialyltransferase STZ (Table I) which could transfer sialic acid in α-2,3-linkage to the terminal Gal of Gal(β1–4)GlcNAc or Gal(β1–3)GalNAc of oligosaccharide, glycoprotein and glycolipid acceptors [130]. The enzyme could not act on Gal(β1–3)GlcNAc termini and was therefore different from the CMP-sialic acid:Gal(β1–3/4)GlcNAc-R α-2,3-sialyltransferases I and II described above. It is suggested that the enzyme may be the glycolipid α-2,3-sialyltransferase SAT-3 [131,132]. The enzyme appears to exist in three forms differing at the amino-terminal end.

5. N-Acetylglucosaminyltransferases

The characteristics of six *N*-acetylglucosaminyltransferases (GlcNAc-T I to VI) involved in the synthesis of N-glycan antennae have been reviewed [17,26]. Four of these enzymes have been purified, GlcNAc-T I [50,133], GlcNAc-T II [51,134], GlcNAc-T III [135] and GlcNAc-T V [136,137]. Recent work on the kinetic properties of these enzymes and on the cloning of the genes is described below.

5.1. UDP-GlcNAc:Man(α1–3)R [GlcNAc to Man(α1–3)] β-1,2-GlcNAc-transferase I (EC 2.4.1.101)

5.1.1. Cloning of β-1,2-GlcNAc-transferase I cDNA and genomic DNA
Polymerase chain reaction (PCR) amplification of rabbit liver cDNA using degenerate oligonucleotide primers based on the amino acid sequence of rabbit liver UDP-GlcNAc:Man(α1–3)R [GlcNAc to Man(α1–3)] β-1,2-GlcNAc-transferase I was used to clone a 2.5 kb cDNA encoding a protein with the type II integral membrane protein domain structure typical of all glycosyltransferases cloned to date (Fig. 2; Table I) [138–140].

A human genomic DNA library was screened with rabbit liver cDNA and a 4.5 kb hybridizing DNA fragment was obtained [140–142]. The gene encoding human β-1,2-GlcNAc-transferase I was also isolated by an expression screening strategy involving sequential stable transfection of human genomic DNA into Lec1 Chinese hamster ovary (CHO) mutant cells which lack β-1,2-GlcNAc-transferase I activity [143]. The β-1,2-GlcNAc-transferase I was rescued by screening for human repetitive Alu sequences that are not present in CHO DNA. A human cDNA library was screened with the genomic DNA and a 2.6 kb cDNA was obtained.

A 4.1 kb mouse genomic DNA fragment containing the entire coding region of β-1,2-GlcNAc-transferase I was isolated by screening a library with human genomic DNA [144]. A mouse F9 teratocarcinoma cell cDNA library was screened with a human β-1,2-GlcNAc-transferase I DNA probe and a 2.7 kb cDNA was isolated [145] containing a 363 bp 5′-untranslated sequence and the complete coding region and 3′-untranslated sequences. The human and mouse genes have been designated MGAT1 and *Mgat-1* respectively.

Comparison of the human genomic DNA sequence [140–142] with the human cDNA sequence [143], and of the mouse genomic DNA sequence [144] with the mouse cDNA sequence [145] shows that 126 bp and 121 bp, for human and mouse respectively, of the 5′-untranslated region and all of the coding and 3′-untranslated regions are on a single 2.46 kb exon. The remaining 5′-untranslated sequence of the human gene is on a different exon (or exons) at least 5 kb upstream [142]; the presence of at least 2 upstream exons has been shown for the mouse gene [145].

The human gene appears to have at least two promoters. The genes for α-2,6-sialyltransferase, β-1,4-Gal-transferase and α-1,3-Gal-transferase (Fig. 3) also have multiple promoters and exons several kb upstream of the coding region but the coding regions cover several exons and the genes span 35 kb or more of genomic DNA.

The β-1,2-GlcNAc-transferase I gene has been localized to human chromosome 5 by Southern analysis of human-hamster somatic cell hybrids [142]. Fluorescent *in situ* hybridization [145,146] has localized the gene to human chromosome 5q35. Southern analysis [144] of the DNA from the progeny of various mouse genetic crosses indicated

```
                                                                         *
human:  MLKKQSAGLVLWGAILFVAWNALLLLFFWTRPAPGRPPSVSALDGDPASLTREVIRLAQD
mouse:  -----—T---------I--G----------------L--D---GD----------H--E-
rabbit: -----------------------------V-S-L--DN---D----------------

                                                                         *
human:  AEVELERQRGLLQQIGD..ALSSQRGRVPTAAPPAQPRVPVTPAPAVIPILVIACDRSTV
mouse:  --A-----------KEhy--WR--W----V----W-------S-VQ------------
rabbit: --------------REhh--W---WK----------H-----P--------------

        *                    *
human:  RRCLDKLLHYRPSAELFPIIVSQDCGHEETAQAIASYGSAVTHIRQPDLSSIAVPPDHRK
mouse:  ------------------R------------V-----T----------N---Q-----
rabbit: ----------------------------------V--------------N---Q-----

                                                                         *
human:  FQGYYKIARHYRWALGQVFRQFRFPAAVVVEDDLEVAPDFFEYFRATYPLLKADPSLWCV
mouse:  ----------------I-NK-K---------------------Q------RT-------
rabbit: ----------------I-HN-NY-------------------Q---------------

human:  SAWNDNGKEQMVDASRPELLYRTDFFPGLGWLLLAELWAELEPKWPKAFWDDWMRRPEQR
mouse:  -----------S-K----------------D--------------------------
rabbit: -----------S-K-------------------------------------------

              *
human:  QGRACIRPEISRTMTFGRKGVSHGQFFDQHLKFIKLNQQFVHFTQLDLSYLQREAYDRDF
mouse:  K------------------------------------------P----------Q-------
rabbit: K----V-------------------------------------P----------Q-------

human:  LARVYGAPQLQVEKVRTNDRKELGEVRVQYTGRDSFKAFAKALGVMDDLKSGVPRAGYRG
mouse:  --Q----------------Q-----------S----------------------------
rabbit: ----------------------------------------------------------

human:  IVTFQFRGRRVHLAPPPTWEGYDPSWN
mouse:  ----------------Q--T-------
rabbit: ----L----------Q--D------T
```

Fig. 4. Comparison of the amino acid sequences of human, mouse and rabbit UDP-GlcNAc:Man(α1–3) R [GlcNAc to Man(α1–3)] β-1,2-GlcNAc-transferase I, using the l-letter code. The transmembrane segment is under- and over-lined. The 5 Cys residues are indicated by an asterisk.

linkage of the mouse β-1,2-GlcNAc-transferase I gene (*Mgat-1*) with markers on mouse chromosome 11. No recombinants were detected between *Mgat-1* and the gene *Il-3* encoding interleukin-3 indicating that these two genes are closely linked. Southern analysis indicated that there is only a single copy of the gene in the haploid human [141] and mouse [145] genomic DNA.

Northern analysis of rabbit liver [139], human liver [142], human HL-60 cells [143], other human cell lines (unpublished data), both wild type and Lec1 Chinese hamster ovary cells and several mouse tissues and embryonic stem cells [144,145] all revealed a major transcript at about 3 kb. Mouse tissues showed significant variation in expression levels with the highest being in kidney. Some mouse [144,145] and human tissues (unpublished) show two transcripts of about 3.0 and 3.4 kb whereas mouse brain shows a single transcript at 3.4 kb. The functional significance of this tissue-specific variation in transcription remains to be determined.

The protein is not N-glycosylated because there are no Asn-X-Ser(Thr) sequons but is probably O-glycosylated [147]. There is no sequence homology to any other known glycosyltransferase. The amino acid sequences of the human, rabbit and mouse enzymes are 92% identical (Table I; Fig. 4). Most of the differences between the species occur in the stem region. The transmembrane and catalytic domains are highly conserved. A similar observation was made for β-1,4-Gal-transferase. It is interesting that the catalytic domains of both β-1,2-GlcNAc-transferase I and β-1,4-Gal-transferase have 5 Cys residues which are conserved between species.

5.1.2. Expression of recombinant β-1,2-GlcNAc-transferase I

In vitro transcription-translation of the rabbit liver cDNA resulted in the synthesis of a 52 kDa protein with β-1,2-GlcNAc-transferase I activity [139]. Transient transfection of Lec1 Chinese hamster ovary cells, which lack β-1,2-GlcNAc-transferase I, with a human genomic DNA fragment encoding β-1,2-GlcNAc-transferase I gave expression of enzyme activity [142].

The bacterial expression vector pGEX was used to generate an in-frame construct encoding a fusion protein comprising the lumenal domain of rabbit β-1,2-GlcNAc-transferase I fused to the carboxyl terminus of glutathione-S-transferase [148,149]. Triton X-100 extraction of *E. coli* transformed with this construct solubilized only 5% of the fusion protein. Enzymatically active soluble β-1,2-GlcNAc-transferase I was obtained by purification on a glutathione-agarose affinity column. The bacterial expression vector pET.3b was used to express a fusion protein containing the lumenal domain of mouse β-1,2-GlcNAc-transferase I [145]; NP-40 extraction of *E. coli* transformed with this construct yielded soluble active enzyme although the bulk of the fusion protein appeared to be present as an inactive intra-cellular aggregate.

High-level expression of enzymatically active β-1,2-GlcNAc-transferase I has been achieved using an insect cell expression system [150,151]. A baculovirus shuttle vector was constructed in which the short N-terminal cytoplasmic tail and transmembrane segment of the rabbit β-1,2-GlcNAc-transferase I cDNA were replaced with an in-frame cleavable signal sequence. The modified β-1,2-GlcNAc-transferase I gene was inserted into the genome of *Autographa californica* nuclear polyhedrosis baculovirus (AcNPV) and *Spodoptera frugiperda* (Sf9) insect cells were infected with the recombinant baculovirus. High level expression of enzymatically active and soluble enzyme was achieved under the control of the polyhedrin promoter. About 90% of the enzyme was secreted into the growth medium at a concentration of about 1–5 mg/liter, indicating that the cytoplasmic tail and transmembrane domain are required for Golgi retention. Recombinant rabbit enzyme has been purified to near homogeneity by affinity chromatography. The specific activity of recombinant mouse GlcNAc-transferase I from crude *E. coli* lysates [145] or of affinity-purified recombinant rabbit enzyme from *E. coli* [148] was about 0.001–0.002 µmoles/min/mg compared to 2 µmoles/min/mg for the purified insect cell enzyme suggesting that solubilization of the *E. coli* enzyme may have led to denaturation. The recombinant rabbit enzyme from insect cells was shown to incorporate GlcNAc in (β1–2) linkage to the Man(α1–3)Manβ1- arm of the N-glycan core [152,153].

5.1.3. Targeting to the Golgi apparatus

A polyclonal antibody was generated by injecting mice with purified glutathione-S-transferase/rabbit β-1,2-GlcNAc-transferase I fusion protein [148]. Murine L cells stably transfected with rabbit cDNA for β-1,2-GlcNAc-transferase I showed localization of the transferase in Golgi cisternae by immunofluorescence; no cell surface expression was detected. Immunoelectron microscopy localized the transferase in medial-Golgi cisternae. A hybrid construct encoding a protein with the 31 amino-terminal amino acids of β-1,2-GlcNAc-transferase I fused to ovalbumin was expressed transiently in COS-7 cells and stably in murine L cells. Ovalbumin is a secretory protein but the fusion protein remained in the medial-Golgi cisternae. These experiments show that the cytoplasmic tail and transmembrane segment of β-1,2-GlcNAc-transferase I cause retention in medial-Golgi cisternae. Expression of a hybrid protein in which the transmembrane domain of a type II membrane surface protein was replaced with the transmembrane domain of human β-1,2-GlcNAc-transferase I showed that the transferase transmembrane domain was by itself sufficient to retain a polypeptide in the Golgi [154].

5.1.4. The effect of homozygous 'knock-out' of the GlcNAc-transferase I gene in transgenic mouse embryos

β-1,2-GlcNAc-transferase I plays a key role in controlling the conversion of high mannose N-glycans to hybrid and complex structures. Lec1 Chinese hamster ovary mutant cells which lack a functional GlcNAc-transferase I and cannot make complex and hybrid N-glycans are nevertheless viable and grow with a normal doubling time indicating that complex N-glycans are not essential for the survival and proliferation of mammalian cells in culture [155]. In an attempt to study the biological roles of hybrid and complex N-glycans in the intact animal, e.g., in processes such as development and differentiation, the mouse *Mgat-1* gene locus has been mutated via homologous recombination in embryonic stem (ES) cells and the mutated allele has been introduced into the mouse germ line [156]. Live-born heterozygotes were obtained and mated but no living homozygous mutant mice were obtained. Homozygous mutant embryos die prenatally at about 10 days of gestation.

Extracts of homozygous wild type, heterozygous transgenic and homozygous transgenic 9.5-day embryos were assayed for GlcNAc-transferase I activity. The hetero-zygotes showed enzyme levels that were 50% of the wild type levels indicating co-dominant expression of the *Mgat-1* gene; the homozygous transgenic embryos showed no enzyme activity. Analysis of glycopeptides from these embryos with several lectins (concanavalin A, ricin and L-phytohemagglutinin) showed a significant in-crease in the ratio of high-mannose to complex N-glycans in the homozygous trans-genic embryos.

Development of mouse embryos to E9.5 constitutes approximately 40% of gestation time with the beginning stages of organogenesis underway. The homozygous trans-genic embryos were generally smaller than controls and harbored fewer somites. The acquisition of left-right body plan asymmetry, apparent in heart loop and tail orienta-tion, was randomized in the absence of complex N-glycans. This result may reflect the phylogenetic basis for the emergence of the *Mgat-1* gene and complex N-glycans in higher eukaryotes that acquire left-right body plan asymmetry in embryogenesis.

Embryonic lethality may have occurred by the loss of vascularization potential as Mgat-1⁻/Mgat-1⁻ embryos displayed abnormal vasculature with accumulations of red blood cells in unexpected compartments. Moreover multiple hemorrhages were appar-ent by E10.5. Neural tube formation was also impaired although neural epithelial cells appeared to differentiate normally. Complex N-glycans appear to be required for specific morphogenic processes that are initiated by cell–cell interactions among differentiated cell types. These findings indicate the importance of GlcNAc-transferase I to normal mammalian developmental and morphogenic processes.

5.1.5. Substrate requirements for β-1,2-GlcNAc-transferase I

The conversion of high-mannose to complex and hybrid N-glycans is controlled by GlcNAc-transferase I which, *in vivo*, converts [Man(α1–6){Man(α1–3)}Man(α16)] [Man(α1–3)]Man(β1–4)GlcNAc(β1–4)GlcNAc-Asn-X to [Man(α1–6){Man(α1–3)} Man(α1–6)][**GlcNAc(β1–2)**Man(α1–3)]Man(β1–4)GlcNAc (β1– 4) GlcNAc-Asn-X. GlcNAc-transferase I acts on acceptors with the general formula R_1(α1–6)[Man(α1–3)]Man(β1–4)GlcNAc-R_2, e.g., Man(α1–3)Man(β1–4) GlcNAc ($K_{m(app)}$ is 4.5 mM) and oligosaccharides in which R_1 is Man- ($K_{m(app)}$ is 0.4–0.6 mM), or Man(α1–6) [Man(α1–3)]Man- ($K_{m(app)}$ is 0.25 mM). The R_2 group can be (β1–4)[+/–Fuc(α1–6)]GlcNAc-Asn-X, or (β1–4)GlcNAc, or H. GlcNAc-transferase I can also act on the simple trisaccharide Man(α1–6)[Man(α1–3)]Manβ1-R where R can be a hydrophobic group like octyl [133,157]. Analogues of such trisaccharides are amenable to organic synthesis and have been used [50,133,157–159] to delineate the substrate requirements of GlcNAc-transferase I (Fig. 5).

Removal of the 2′-OH group of Man′(α1–3) (2′-deoxy analogue) or substitution of the 2′-OH with a methyl group prevents both catalysis and binding suggesting that the 2′-OH forms an essential hydrogen bond with the enzyme as it transfers a GlcNAc residue to this position. The 4′-OH group of Man′(α1–3) is also essential. Substitution of the 6′-OH [Man′(α1–3)] with a bulky group prevents catalysis but not binding. Removal of the 4-OH of the Man(β1–4) residue (4-deoxy analogue), conversion of the 4-OH$_{eq}$ to a 4-OH$_{ax}$ group or substitution with a bisecting GlcNAc residue prevent both binding and catalysis indicating that this OH forms an essential hydrogen bond with

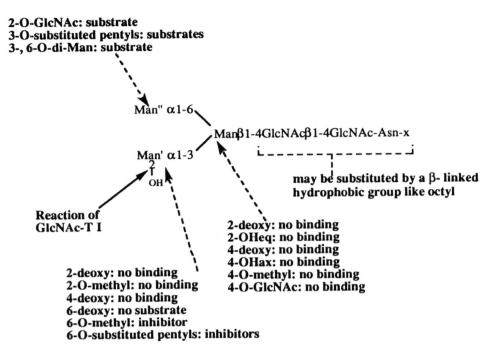

2-O-GlcNAc: substrate
3-O-substituted pentyls: substrates
3-, 6-O-di-Man: substrate

Man" α1-6

Manβ1-4GlcNAcβ1-4GlcNAc-Asn-x

Man' α1-3

OH

may be substituted by a β- linked hydrophobic group like octyl

Reaction of GlcNAc-T I

2-deoxy: no binding
2-OHeq: no binding
4-deoxy: no binding
4-OHax: no binding
4-O-methyl: no binding
4-O-GlcNAc: no binding

2-deoxy: no binding
2-O-methyl: no binding
4-deoxy: no binding
6-deoxy: no substrate
6-O-methyl: inhibitor
6-O-substituted pentyls: inhibitors

Fig. 5. Substrate requirements of GlcNAc-transferase I. The structure shown is Man(α1–6)[Man(α1–3)]Man(β1–4)GlcNAc(β1–4)GlcNAc-Asn, the substrate for GlcNAc-transferase I. Indicated on the figure are the effects of deletions or substitutions at various OH groups in the structure. "No binding" indicates the structure is neither a substrate nor an inhibitor.

the catalytic site of the enzyme. The same is true for the 2-OH [Man(β1–4)]. In contrast to the Man'(α1–3) and Man(β1–4) residues, the 2″-OH, 3″-OH and 6″-OH groups of Man″(α1–6) can be substituted with various groups without affecting catalysis. Groups capable of reacting covalently with the protein (i.e., epoxy, diazirino and iodo-acetamido groups) have been placed at the 6′-OH and 3″-OH positions ([133,159] and in preparation). The compounds with substituents at the 3″-OH are all excellent substrates. However, the 6′-O-(epoxy)pentyl- and 6′-O-(diazirino)pentyl- derivatives are reversible inhibitors and the 6′-O-(iodoacetamido)pentyl- derivative is an irreversible inhibitor of GlcNAc-transferase I (in preparation).

Kinetic analysis of purified rabbit GlcNAc-transferase I [50] indicates a largely ordered sequential mechanism with UDP-GlcNAc binding to the enzyme first and UDP leaving last. The V_{max} of the highly purified enzyme is 19.8 units/mg and the $K_{m(app)}$ is 0.25 mM for [Man(α1–6){Man(α1–3)}Man(α1–6)][Man(α1–3)]Man(β1–4) GlcNAc (β1–4)GlcNAc-Asn-X and 0.08 mM for UDP-GlcNAc. The optimum pH for GlcNAc-transferase I is about 5.6. The activation effect of Triton X-100 on the pure enzyme is not dramatic and shows a broad concentration dependency. The addition of 2-mercaptoethanol (1–10 mM) to the enzyme incubation has no effect on enzyme activity. The enzyme shows an absolute requirement for divalent cation and the optimum concentration of Mn^{2+} is between 20 and 100 mM. The effectiveness of other divalent cations at 20 mM on activating GlcNAc-transferase I is as follows: $Co^{2+} > Mn^{2+} > Mg^{2+} > Ni^{2+} = Cd^{2+} >> Ca^{2+} = Ba^{2+} > Sr^{2+}$.

5.2. UDP-GlcNAc:Man(α1–6)R [GlcNAc to Man(α1–6)] β-1,2-GlcNAc-transferase II (E.C. 2.4.1.143)

5.2.1. Cloning of β-1,2-GlcNAc-transferase II cDNA and genomic DNA

UDP-GlcNAc:Man(α1–6)R [GlcNAc to Man(α1–6)] β-1,2-GlcNAc-transferase II (E.C. 2.4.1.143) was purified from rat liver [51] and partial amino acid sequences were used to design synthetic oligonucleotide probes. Screening of rat liver cDNA libraries yielded a cDNA clone encoding a 442 amino acid protein with the type II transmembrane protein domain structure typical of glycosyltransferases (Fig. 2) [160]. The C-terminal 389 amino acids were linked in frame to the cleavable signal sequence of the Il-2 receptor and expressed under control of the Rous sarcoma virus promoter in COS-7 cells. A similar construct carrying GlcNAc-transferase II cDNA out of frame was used as a negative control. A 77-fold enhancement of enzyme activity in the culture medium over the negative control was detected at 72 hr after transfection in transient expression experiments. These data verify the identity of the cloned cDNA sequences for rat GlcNAc-transferase II and demonstrate that the COOH-terminal region of the polypeptide chain includes the catalytic site.

A rat GlcNAc-transferase II cDNA probe was used to screen a rat genomic DNA library and several clones were isolated [161]. The nucleotide sequence of a genomic DNA region spanning 3.3 kb was determined revealing a 1326-nucleotide open reading frame that codes for the entire GlcNAc-transferase II polypeptide chain and is not interrupted by introns. The single-exon GlcNAc-transferase II gene is flanked by a GC-rich 5'-untranslated region. The 3'-untranslated region contains multiple polyadenylation signals and ATTTA motifs.

A 1.2 kb probe from rat liver cDNA encoding GlcNAc-transferase II was used to screen a human genomic DNA library [146]. Two overlapping sub-clones containing 5.5 kb of genomic DNA were isolated and analyzed. One of the clones contains a 1341 bp open reading frame encoding a 447 amino acid protein with the type II transmembrane protein domain structure typical of all previously cloned glycosyltransferases (Fig. 2, Table I). Northern analyses of several human lymphocyte lines showed a major band at about 3 kb and fainter bands at 2 and 4.5 kb. There is no sequence homology to any previously cloned glycosyltransferase including human β-1,2-GlcNAc-transferase I. There is a 90% identity between the amino acid sequences of the catalytic domains of human and rat GlcNAc-transferase II. The entire coding regions of the human and rat enzymes are on single exons. The human GlcNAc-transferase II gene is on chromosome 14q21 [146].

5.2.2. Substrate requirements for β-1,2-GlcNAc-transferase II
The only effective substrate for GlcNAc-transferase II is the structure Man(α1–

6)[GlcNAc(β1–2)Man(α1–3)]Manβ-R where R can be 1–4GlcNAc(β1–4)[+/– Fuc(α1–6)]GlcNAc-Asn-X, or 1-4GlcNAc(β1–4)GlcNAc, or 1–4GlcNAc, or hydrophobic groups like $(CH_2)_8COOCH_3$ or octyl [134,157,162–165]. The enzyme incorporates a GlcNAc residue into the 2-position of the Man(α1–6) residue. The GlcNAc(β1–2)Man(α1–3)Manβ moiety is essential for activity indicating that GlcNAc-transferase I must act before GlcNAc-transferase II. The presence of a bisecting GlcNAc residue or galactosylation of the GlcNAc(β1–2)Man(α1–3)Manβ moiety prevent GlcNAc-transferase II action [134].

The substrate requirements of GlcNAc-transferase II have been determined by using a series of synthetic substrate analogues [134,152,153,157,162–166]. The available data are summarized in Fig. 6. The enzyme attaches GlcNAc in (β1–2) linkage to the 2$'''$-OH of the Man$'''$(α1–6) residue (Fig. 6). The 2$'''$-deoxy analogue is a competitive inhibitor (K_i = 0.13 mM) indicating that the 2$'''$-OH is not essential for binding to the enzyme. The 2$'''$-O-methyl compound does not bind to the enzyme presumably due to steric hindrance. The 3$'''$-, 4$'''$- and 6$'''$-OH groups are not essential either for binding or catalysis since the 3$'''$-, 4$'''$- and 6$'''$-deoxy and the 3$'''$-O-methyl derivatives are all good substrates. Increasing the size of the substituent at the 3$'''$-OH position to pentyl

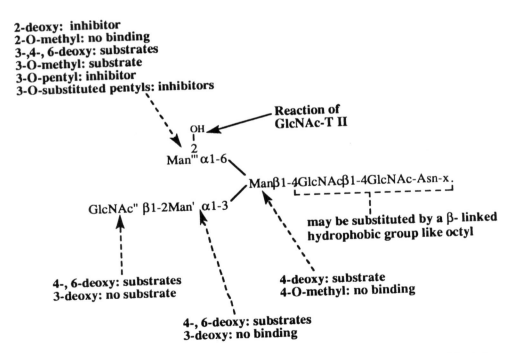

Fig. 6. Substrate requirements of GlcNAc-transferase II. The structure shown is Man(α1–6)[GlcNAc(β1–2)Man(α1–3)]Man(β1–4)GlcNAc(β1–4)GlcNAc-Asn, the substrate for GlcNAc-transferase II. Indicated on the figure are the effects of deletions or substitutions at various OH groups in the structure. "No binding" indicates the structure is neither a substrate nor an inhibitor.

and substituted pentyl groups causes competitive inhibition (K_i = 1.0–2.5 mM). Compounds with a photolabile 3‴-O-(4,4-azo)pentyl group and a 3‴-O-(5-iodo-acetamido)pentyl group are irreversible inhibitors of GlcNAc-transferase II [152,153, 166]. These inhibitors may prove useful as active site reagents. The data indicate that none of the hydroxyls of the Man‴(α1–6) residue are essential for binding although the 2‴- and 3‴-OH face the catalytic site of the enzyme since bulky groups at these positions respectively prevent binding and catalysis.

The 4-OH group of the Manβ residue is not essential for binding or catalysis since the 4-deoxy derivative is a good substrate; the 4-O-methyl derivative does not bind. This contrasts with GlcNAc-T I which cannot bind to the 4-deoxy-Manβ- substrate analogue. The data are compatible with previous observations that a bisecting GlcNAc at the 4-OH position prevents both GlcNAc-T I and GlcNAc-T II catalysis. However, in the case of GlcNAc-T II, the bisecting GlcNAc prevents binding due to steric hindrance rather than to removal of an essential OH group.

The 3′-OH of the Man′(α1–3) and the 3″-OH of the GlcNAc″(β1–2) are essential groups for GlcNAc-T II since the respective deoxy derivatives do not bind to the enzyme. The 4′-, 6′-, 4″- and 6″-deoxy analogues are all good substrates. The trisaccharide GlcNAc(β1–2)Man(α1–3)Manβ-O-octyl is a good inhibitor (K_i = 0.9 mM). These data and previous observations that the GlcNAc(β1–2)Man(α1–3)Manβ moiety is an essential substrate requirement for GlcNAc-transferases II, III and IV, α-mannosidase II and core α-1,6-fucosyltransferase suggest that the trisaccharide may be a binding site for all of these enzymes.

In summary, the Man(α1–6) residue, which is the site of catalytic action, is not a strong binding site for the enzyme. Rather, the essential binding site is on the other arm of the branched substrate and requires the 3′-OH of the Man′(α1–3) residue and the 3″-OH of the GlcNAc″(β1–2) residue.

Kinetic analysis with the purified rat enzyme [134] indicates a largely ordered sequential mechanism with UDP-GlcNAc binding to the enzyme first and UDP leaving last. The V_{max} of the highly purified rat liver enzyme was 27.5 µmoles/min/mg and the $K_{m(app)}$ was 0.19 mM for Man(α1–6)[GlcNAc(β1–2)Man(α1–3)]Manβ-R and 0.96 mM for UDP-GlcNAc. The optimum pH for GlcNAc-transferase II is about 6.0 to 6.5. The addition of 2-mercaptoethanol (5 mM) to the enzyme incubation has no effect on enzyme activity. The enzyme shows an absolute requirement for divalent cation and the optimum concentration of Mn^{2+} is between 10 and 15 mM.

5.3. UDP-GlcNAc:R_1-Man(α1–6)[GlcNAc(β1–2)Man(α1–3)]Man(β1–4)R_2
[GlcNAc to Man(β1–4)] β-1,4-GlcNAc-transferase III (E.C. 2.4.1.144)

GlcNAc-transferases I and II are widely distributed in tissues and cells and form the basic di-antennary N-glycan structure. There are four other GlcNAc-transferases (III–VI) which can then act to produce more highly branched N-glycans (Fig. 7). These enzymes have a more variable tissue distribution. UDP-GlcNAc:R_1-Man(α1–6) [GlcNAc(β1–2)Man(α1–3)]Man(β1–4)R_2 [GlcNAc to Man(β1–4)] β-1,4-GlcNAc-transferase III incorporates a bisecting GlcNAc residue into the N-glycan core [167]. It is of interest because it causes a 'stop' signal, i.e., several enzymes involved in

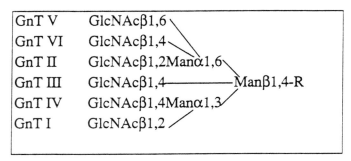

Fig. 7. GlcNAc-transferases I to VI incorporate GlcNAc residues into the Man(α1–6)[Man(α1–3)] Manβ-R N-glycan core.

N-glycan synthesis cannot act on bisected substrates (see above). The enzyme has also been shown to be elevated in various types of rat hepatoma [168–171].

5.3.1. Cloning of β-1,4-GlcNAc-transferase III cDNA

The enzyme was purified [135] from rat kidney, a relatively rich source of the enzyme [170], by a 9-step procedure giving a 1.5% yield. The most effective purification steps were affinity chromatography on agarose columns with immobilized UDP or oligosac-charide substrate. The final purification was 153,000-fold with a specific activity of 5 μmoles/min/mg. Tryptic peptides were purified by HPLC and sequenced. Polymerase chain reaction amplification of rat kidney cDNA using degenerate oligonucleotide primers based on the amino acid sequences yielded three fragments of 0.16, 0.49 and 0.63 kb [135]. These fragments were further amplified in a plasmid vector and one was used to probe a rat kidney cDNA library. A 2.7 kb cDNA sequence was obtained (from three overlapping clones) with a 1608 bp open reading frame encoding a 536 amino acid protein with the type II integral membrane protein domain structure typical of all the glycosyltransferases cloned to date (Fig. 2; Table I). Transient expression of this cDNA in both COS-1 and HeLa cells yielded GlcNAc-transferase III activity. There is no apparent sequence similarity to any known glycosyltransferase. The stem region is enriched in Pro residues as is the case for several other glycosyltransferases. The enzyme has three putative N-glycosylation sites and adheres to Con A-Sepharose suggesting that it is a glycoprotein. Northern analysis showed a strong 4.7 kb message in rat kidney and a weak message in rat liver.

A human fetal liver cDNA library was screened with a rat GlcNAc-transferase III probe to yield cDNA encoding the full-length human enzyme (Table I) which is 91% homologous to the rat enzyme at the amino acid level [172]. Transient expression in COS-1 cells yielded active enzyme (12–29 nmoles/mg/h). Analysis of the human GlcNAc-transferase III gene shows that the entire coding region is on a single exon on chromosome 22q.13.1.

5.3.2. Substrate requirements for β-1,4-GlcNAc-transferase III

GlcNAc-transferase III transfers a bisecting GlcNAc residue in β-linkage to the 4-OH of the Manβ residue of R_1Man(α1–6)[GlcNAc(β1–2)Man(α1–3)]Manβ-R_2 where R_1 can be GlcNAc(β1–2) or Man(α1–6)[Man(α1–3)] and R_2 can be 1–4GlcNAc(β1–4) [+/–Fuc(α1–6)]GlcNAc-Asn-X,, or 1–4GlcNAc(β1–4)GlcNAc, or hydrophobic groups like $(CH_2)_8COOCH_3$ [165,167,173,174]. The GlcNAc(β1–2)Man(α1–3)Manβ moiety is essential for activity indicating that GlcNAc-transferase I must act before GlcNAc-transferase III. The enzyme has a relatively low activity towards Man(α1–6)[GlcNAc (β1–2)Man(α1–3)]Man(β1–4)GlcNAc-R. Galactosylation of the GlcNAc (β1–2)Man (α1–3)Manβ moiety prevents GlcNAc-transferase III action. A bisecting GlcNAc residue can be added to [Man(α1–6){Man(α1–3)}Man(α1–6)][GlcNAc(β1–2)Man (α1–3)]Man(β1–4)GlcNAc(β1–4)GlcNAc-Asn-X. This prevents GlcNAc-transferase I-dependent α-3/6-mannosidase II action and routes the pathway towards the synthesis of bisected hybrid structures. [GlcNAc(β1–2)(6-deoxy)Man(α1–6)] [GlcNAc(β1–2) (4-deoxy)Man(α1–3)]Manβ-octyl is a specific substrate for GlcNAc-transferase III since the OH groups required for the actions of GlcNAc-transferases I, II, IV and V are all unavailable [165,174].

The $K_{m(app)}$ for the enzyme from hen oviduct [167,173] is 0.23 mM for GlcNAc(β1–2) Man(α1–6)[GlcNAc(β1–2)Man(α1–3)]Manβ-R and 1.1 mM for UDP-GlcNAc. The optimum pH for GlcNAc-transferase III is between 6.0 and 7.0. The enzyme shows an absolute requirement for divalent cation and the concentration of Mn^{2+} shows a broad optimum at about 12 mM. Addition of Triton X-100 at concentrations ranging from 0.13 to 1.3% (v/v) all stimulate enzyme activity about 8-fold.

5.4. UDP-GlcNAc:R_1Man(α1–3)R_2 [GlcNAc to Man(α1–3)] β-1,4-GlcNAc-transferase IV (E.C. 2.4.1.145)

GlcNAc-transferase IV [175] adds a GlcNAc in (β1–4) linkage to the Man(α1–3) Manβ- arm of GlcNAc(β1–2)Man(α1–6)[GlcNAc(β1–2)Man(α1–3)]Man(β1–4) GlcNAcβ-R where R can be 1–4[+/–Fuc(α1–6)]GlcNAc-Asn-X, or 1-4GlcNAc. The GlcNAc(β1–2)Man(α1–3) moiety is essential for activity indicating that GlcNAc-transferase I must act before GlcNAc-transferase IV. The enzyme has a relatively low activity towards Man(α1–6)[GlcNAc(β1–2)Man(α1–3)]Man(β1–4)GlcNAcβ-R. GlcNAc-transferase IV cannot act on bisected substrate acceptors. Galactosylation of the GlcNAc(β1–2)Man(α1–6)Manβ arm reduces activity by about 66% whereas galacto-sylation of the GlcNAc(β1–2)Man(α1–3)Manβ arm abolishes GlcNAc-transferase IV action completely. The enzyme has not been purified.

The optimum pH for GlcNAc-transferase IV under standard assay conditions is 7.0. The optimum Triton X-100 concentration is at 0.125% (v/v) but it is difficult to maintain proportionality of activity to enzyme concentration under these conditions; the standard assay is therefore carried out at 0.5% (v/v) Triton X-100. The optimum Mn^{2+} concentration is 12 mM.

Simple oligosaccharides of the form GlcNAc(β1–2)Manα-R can be converted to GlcNAc(β1–4)[GlcNAc(β1–2)]Manα-R by hen oviduct extracts, possibly due to the action of GlcNAc-transferase IV [176]. This activity is named GlcNAc-transferase VI′ (see below).

5.5. UDP-GlcNAc:R₁Man(α1–6)R₂ [GlcNAc to Man(α1–6)] β6-GlcNAc-transferase V (E.C.2.4.1.155)

One of the most common alterations in transformed or malignant cells is the presence of larger N-glycans due primarily to a combination of increased GlcNAc branching, sialylation and poly-*N*-acetyllactosamine content [177–180]. This topic will be discussed in detail elsewhere in this book but it is important to emphasize at this point that GlcNAc-transferase V plays a major role in these effects. Polyoma virus, Rous sarcoma virus- and T24 H-ras-transformed cells were shown to have significantly increased GlcNAc-transferase V activity [181–185]. Increased N-glycan β6-branching is now established as a consistent finding in many animal models of tumor progression and acquisition of metastatic potential, and in human melanomas and carcinomas of breast, colon and ovaries [180]. Poly-*N*-acetyllactosamine chains have been shown to carry cancer-associated antigens and the initiation of these chains is favoured on GlcNAc (β1–6)Man(α1–6)Manβ- branches of N-glycans [180,186,187].

5.5.1. Properties of GlcNAc-transferase V

GlcNAc-transferase V [173,188] adds a GlcNAc in β1–6 linkage to the Man(α1–6)Manβ arm of GlcNAc(β1–2)Man(α1–6)[GlcNAc(β1–2)Man(α1–3)]Man(β1–4)GlcNAcβ-R to form [**GlcNAcβ1–6** {GlcNAc(β1–2)}Man(α1–6)] [GlcNAc(β1–2) Man(α1–3)]Man(β1–4)GlcNAcβ-R. Simple trisaccharides of the form GlcNAc(β1–2) Man (α1–6)Man(or Glc)β-R can be converted to GlcNAcβ1-6[GlcNAc(β1–2)] Man(α1–6) Man(or Glc)β-R by the enzyme [189,190]. This enzyme requires the prior action of GlcNAc-transferase II which in turn requires the prior action of GlcNAc-transferase I. The compound GlcNAc(β1–2)(6-deoxy)Man(α1–6)Glcβ-octyl was shown to be a competitive inhibitor of GlcNAc-transferase V with a K_i of 0.07 mM [182,191]; this compound lacks the OH to which the enzyme attaches GlcNAc.

GlcNAc-transferase V cannot act on substrates which contain a bisecting GlcNAc, or in which either the GlcNAc(β1–2)Man(α1–3)- or GlcNAc(β1–2)Man(α1–6)- arms are substituted with a Gal(β1–4) residue [136,173,188]. The importance of the 4-OH on the terminal GlcNAc of the substrate GlcNAc(β1–2)Man(α1–6)Glcβ-octyl is indicated by the fact the removal of this OH (deoxy analogue) or substitution of the OH by various groups leads to an inactive substrate suggesting that the OH may be required for hydrogen bond formation with the enzyme [192]. Although a bisecting GlcNAc attached to the 4-OH of the β-linked Man residue on a di-antennary substrate prevents enzyme action, 4-O-methyl or 4-deoxy analogues are excellent substrates, e.g., GlcNAc(β1–2)Man(α1–6)(4-deoxy)Glcβ-octyl and GlcNAc(β1–2)Man(α1–6)(4-*O*-methyl)Glcβ-octyl [190] and GlcNAc(β1–2)Man(α1–6)[GlcNAc(β1–2)Man(α1–3)] (4-deoxy)Manβ-octyl and GlcNAc(β1–2)Man(α1–6)[GlcNAc(β1–2)Man(α1–3)] (4-*O*-methyl)Manβ-octyl [193].

The presence of a core Fuc(α1–6) residue inhibits enzyme activity minimally and the GlcNAc-terminal tri-antennary compound is a better substrate than the di-antennary compound [136]. Neither the 3 nor the 4-OH of the Man(α1–3) residue of bi-antennary substrate are essential for activity [193]. The 4-OH of the Man(α1–6) residue of GlcNAc(β1–2)Man(α1–6)Glcβ-octyl is not important for binding or catalysis since the (4-deoxy)Man compound is a good substrate; however, the (4-O-methyl)Man derivative can bind but is not a substrate probably because the methyl group interferes with the ability of the enzyme to catalyze the transfer reaction, a process the authors have called "steric exclusion" [194].

Analogues of the substrate GlcNAc(β1–2)Man(α1–6)Glcβ-octyl have been synthesized in which the rotation about the C5–C6 bond of the Glc residue has been prevented by linking O4 and C6 with an ethylene bridge [195]. The resulting fused-ring trisaccharides represent conformationally restricted analogues of the "gt" and "gg" conformations of the substrate. The gg analogue is a good substrate but the gt compound is not thereby indicating that GlcNAc-transferase V acts on a specific conformation of the acceptor. The data support the hypothesis that carbohydrate-binding proteins such as glycosyltransferases and lectins interact with highly specific three-dimensional forms of their carbohydrate ligands.

The optimum pH for GlcNAc-transferase V from baby hamster kidney (BHK) cells is about 6.0 [184]. The $K_{m(app)}$ was 0.18 mM for GlcNAc(β1–2)Man(α1–6)Manβ-O-$(CH_2)_8COOCH_3$ and 0.25 mM for UDP-GlcNAc. GlcNAc-transferase V is active both with and without the addition of exogenous Mn^{2+} [173]; Mn^{2+} is usually left out of the incubation mixture to reduce interference from other GlcNAc-transferases.

5.5.2. Purification of GlcNAc-transferase V

GlcNAc-transferase V has been purified from a Triton X-100 extract of rat kidney acetone powder [137] and from the culture medium of a human lung cancer cell line [136]. The rat kidney enzyme was purified 450,000-fold, with a 26% yield, to a specific activity of 19 µmol/min/mg using only two affinity chromatography steps, UDP-hexanolamine-Sepharose and an inhibitor column (synthetic oligosaccharide inhibitor-spacer-bovine serum albumin-Sepharose). The pure enzyme showed two major bands (69 and 75 kDa) on SDS-PAGE. The human lung cancer enzyme was purified 20,000-fold from culture medium without the use of detergents, with a 37% yield, to a specific activity of about 0.14 µmol/min/mg. Several chromatography steps, including UDP-hexanolamine-Sepharose, were used but the most effective step was affinity chromatography on a column of GlcNAc(β1–2)Man(α1–6) [GlcNAc(β1–2)Man(α1–3)] Man(β1–4) GlcNAc (β1–4)GlcNAc-Asn-Sepharose which gave a 147-fold purification. Two bands at 73 and 60 kDa were seen on SDS-PAGE; tryptic peptide maps from these two bands were almost identical indicating that the 60 kDa protein was probably a proteolytic product.

5.5.3. Cloning of GlcNAc-transferase V cDNA

Degenerate oligonucleotide primers were synthesized on the basis of rat kidney GlcNAc-transferase V amino acid sequence data and used to obtain cDNA encoding full-length rat GlcNAc-transferase V (Table I) by a reverse transcriptase-polymerase chain reaction (RT-PCR) approach [196]. GlcNAc-transferase V is the largest glycosyltransferase cloned to date (Table I). Transient expression of the cDNA in COS-7 cells yielded active enzyme (0.6 nmoles/mg/h). A similar approach was used to clone cDNA and genomic DNA for human GlcNAc-transferase V [197]. There is 97% homology between the amino acid sequences of the human and rat enzymes. The human gene is on chromosome 2q21.

5.6. UDP-GlcNAc:R_1(R_2)Man(α1–6)R_3 [GlcNAc to Man(α1–6)] β-1,4-GlcNAc-transferase VI

GlcNAc-transferase VI [198] adds a GlcNAc residue in (β1–4) linkage to the Man(α1–6) arm of [GlcNAcβ1–6{GlcNAc(β1–2)}Man(α1–6)][GlcNAc(β1–2)Man(α1–3)] Manβ-R to form [GlcNAcβ1–6{**GlcNAc(β1–4)**}{GlcNAc(β1–2)}Man(α1–6)] [GlcNAc (β1–2)Man(α1–3)]Manβ-R. The synthetic oligosaccharides GlcNAcβ1–6 [GlcNAc (β1–2)]Man(α1–6)Manβ-(CH$_2$)$_8$COOCH$_3$, GlcNAcβ1–6[GlcNAc(β1–2)] Man(α1–6) Manβ-methyl and GlcNAcβ1–6[GlcNAc(β1–2)]Manα-methyl are also excellent substrates for the enzyme. Compounds lacking the β-1,2-linked terminal GlcNAc, such as GlcNAcβ1–6Man(α1–6)Manβ-(CH$_2$)$_8$COOCH$_3$ and GlcNAcβ1–6 Manα-methyl, are not acceptors.

The minimum structural requirement for GlcNAc-transferase VI is the trisaccharide GlcNAcβ1–6[GlcNAc(β1–2)]Manα-, i.e., the enzyme requires the prior actions of GlcNAc-transferases I, II and V. Unlike N-glycan GlcNAc-transferases I to V, GlcNAc-transferase VI can act on both bisected and non-bisected substrates. The enzyme has been demonstrated in hen oviduct, liver and colon, and in other avian tissues (duck and turkey), but not in mammalian tissues. The enzyme has not been purified.

The pH optimum for hen oviduct GlcNAc-transferase VI is between 7.0 and 8.4. Triton X-100 stimulates the enzyme at concentrations between 0.125 and 1.25%. The enzyme shows an absolute requirement for divalent cation; Mn^{2+} gives the greatest stimulation of activity, Co^{2+}, Mg^{2+} and Ca^{2+} substitute partially for Mn^{2+}, while Ni^{2+}, Zn^{2+} and Cd^{2+} show no significant stimulation (concentration = 12.5 mM).

The optimum Mn^{2+} concentrations with GlcNAcβ1–6[GlcNAc(β1–2)]Man(α1–6) Manβ-(CH$_2$)$_8$COOCH$_3$ and GlcNAcβ1–6[GlcNAc(β1–2)]Manα-methyl as substrates are at the relatively high concentrations of 100 and 75 mM MnCl$_2$ respectively. GlcNAc-transferases III and IV are inhibited at Mn^{2+} concentrations above 50 mM suggesting that these enzymes are different from GlcNAc-transferase VI.

The $K_{m(app)}$ values for hen oviduct GlcNAc-transferase VI activity at 100 mM Mn^{2+} are 0.09 mM for GlcNAcβ1–6[GlcNAc(β1–2)]Man(α1–6)Manβ-(CH$_2$)$_8$COOCH$_3$ and 0.6 mM for UDP-GlcNAc.

5.7. UDP-GlcNAc:GlcNAc(β1–2)Manα-R [GlcNAc to Manα-] β-1,4-GlcNAc-transferase VI'

Hen oviduct membranes contain at least three β-1,4-GlcNAc-transferase activities that attach a GlcNAc residue in (β1–4) linkage to a Man residue of the N-linked oligosaccharide core, i.e., (i) β-1,4-GlcNAc-transferase III which adds a bisecting GlcNAc to form the **GlcNAc(β1–4)**Manβ- moiety, (ii) β-1,4-GlcNAc-transferase IV which adds GlcNAc to the Man(α1–3)- arm to form the **GlcNAc(β1–4)**[GlcNAc(β1–2)]Man(α1–3)Manβ- moiety and (iii) β-1,4-GlcNAc-transferase VI which adds GlcNAc to the Man(α1–6)- arm to form the GlcNAcβ1–6[**GlcNAc(β1–4)**][GlcNAc(β1–2)]Man(α1–6) Manβ- moiety. Hen oviduct extracts also transfer GlcNAc to GlcNAc(β1–2) Man(α1–6) Manβ-R to form **GlcNAc(β1–4)**[GlcNAc(β1–2)]Man(α1–6)Manβ-R where R is

$(CH_2)_8COOCH_3$ or methyl. This β-1,4-GlcNAc-transferase activity, termed β-1,4-GlcNAc-transferase VI' [176], is strongly inhibited at 100 mM $MnCl_2$ and is therefore different from GlcNAc-transferase VI. Finally, hen oviduct extracts transfer GlcNAc to GlcNAc(β1–2)Manα-methyl to form **GlcNAc(β1–4)**[GlcNAc(β1–2)] Manα-methyl (176). The latter activity and GlcNAc-transferases IV and VI' share similar kinetic characteristics suggesting that, at least in hen oviduct, GlcNAc-transferases IV and VI' may be a single enzyme.

When GlcNAc(β1–2)Man(α1–6)Man(or Glc)β-R is used as a substrate for GlcNAc-transferase VI', the nature of the R group has an important influence on enzyme activity. Compounds in which R is allyl, methyl, p-nitrophenyl or 8-methoxycarbonyloctyl are excellent substrates but the activity is drastically reduced if R is GlcNAc or Glcβ-allyl. This may explain why hen oviduct membranes convert the di-antennary substrate GlcNAc(β1–2)Man(α1–6)[GlcNAc(β1–2)Man(α1–3)]Manβ-R to [GlcNAc(β1–2) Man (α1–6)][{**GlcNAc(β1–4)**}{GlcNAc(β1–2)}Man(α1–3]Manβ-R in a branch-specific manner [i.e., Man(α1–3)- arm but not Man(α1–6)- arm] but lose branch specificity when confronted with a mono-antennary substrate. Conformational factors may cause mono-antennary compounds with an additional Glc or GlcNAc unit to retain features of a di-antennary substrate and this may inhibit GlcNAc-transferase VI' action, i.e., incorporation of GlcNAc into the Man(α1–6)-arm is not expressed when assayed *in vitro* with the di-antennary substrates that are present *in vivo*.

6. Discussion

6.1. The domain structure of glycosyltransferases

Although there is relatively little sequence similarity between most of the glycosyl-transferases cloned to date, these enzymes (Table I) are of approximately the same length and share the same type II integral membrane glycoprotein domain structure (Fig. 2) [15,18]. The hydrophobic trans-membrane sequence near the amino-terminus serves to target and anchor the enzyme to the Golgi membrane. The short amino-terminal cytoplasmic domain carries positive charges which are probably necessary for correct orientation in the membrane. The large globular catalytic domain sits within the Golgi lumen to act on glycoproteins traversing the endomembrane assembly line. The catalytic domain requires neither the cytoplasmic nor the transmembrane domains for activity. The stem region is often rich in Pro residues and may serve as a flexible tether for the catalytic domain. Transferases purified from natural sources (β-1,4-Gal-trans-ferase, several sialyltransferases, GlcNAc-transferases I, II, III and V) often exhibit multiple catalytically active molecular weight forms; it is probable that these forms represent shorter versions of the native enzymes which have been cut in the stem region since the full-length membrane-bound enzyme forms are difficult to purify. It is not clear whether proteolysis occurs *in vivo* or during *in vitro* purification.

Comparison of protein domain and exon structures (Fig. 3) [13] indicates that there is no common pattern of intron-exon structure and no correlation between exons and protein domains. This suggests that exon shuffling did not play a role in the development of these genes. However, the functional domains of the large catalytic unit remain unknown although some interesting work has been done, e.g., the distinction between the α-lactalbumin and acceptor binding sites of β-1,4-Gal-transferase from the UDP-Gal binding site [199,200]. An accurate assessment of correlations between exons and functional protein domains must await the three-dimensional structures of the transferases and an understanding of how amino acid residues that are widely separated from one another come together to create the substrate-binding and catalytic sites of the enzymes.

6.2. Localization of glycosyltransferases to the Golgi apparatus

Evidence has been presented above that the transmembrane domain is essential for targeting glycosyltransferases to the Golgi apparatus but that other domains may play an accessory role. Sequence comparisons of the transmembrane domains of different glycosyltransferases show no similarities other than the presence of charged residues flanking the transmembrane domain [19]. For example, the transmembrane domain controls retention of both α-2,6-sialyltransferase and β-1,4-Gal-transferase in the trans-Golgi although there is no sequence similarity between these enzymes. The differences in the Golgi targeting signals of these two enzymes may explain the immunohistochemical evidence that α-2,6-sialyltransferase is localized more distally in the Golgi (trans-Golgi network) than β-1,4-Gal-transferase (trans-Golgi) [16].

Double immunolabeling of Golgi enzymes has been used recently to confirm that different glycosyltransferases target to different cisternae of the Golgi apparatus. Evidence for the separation of α-2,6-sialyltransferase and β-1,4-Gal-transferase in the Golgi apparatus was obtained by double immunofluorescent staining of the two enzymes in various human cell lines treated with Golgi-disturbing agents [201]. In a similar study, HeLa cells were stably transfected with a DNA expression vector encoding the complete human GlcNAc-transferase I protein tagged at its COOH-terminal end with an 11 amino acid myc epitope [202]. This cell line expressed GlcNAc-transferase I that could be detected with monoclonal antibody to the myc epitope. Double labeling of this cell line was achieved by using the anti-myc antibody and an antibody to the endogenous human UDP-Gal:GlcNAc β-1,4-Gal-transferase. Immunogold electron microscopy and confocal microscopy showed that Gal-transferase was in the trans-cisternae and trans-Golgi network while GlcNAc-transferase I was located both in the medial- and trans-cisternae. There was appreciable overlapping of the two enzymes in the trans-cisterna of the Golgi apparatus.

A possible mechanism for the specific retention of proteins within the Golgi apparatus is the presence of a unique retention signal or three-dimensional signal patch on every protein and a unique Golgi membrane receptor for every such signal. A simpler mechanism has been suggested. The Golgi-targeting domains may cause homo-oligomerization of the glycosyltransferase or hetero-oligomer formation with other glycosyltransferases. The large aggregate may be unable to enter budding vesicles either because of its size or due to interaction with the Golgi membrane lipid bilayer [203].

Evidence for such hetero-oligomers or 'kin oligomers' has recently been obtained [147,203,204]. Expression plasmids were constructed encoding hybrid proteins in which an endoplasmic retention signal was attached to each of GlcNAc-transferase I, α-mannosidase II and β-1,4-Gal-transferase [147]. Expression of each of these constructs in turn in HeLa cells showed that retention of either GlcNAc-transferase I or α-mannosidase II in the endoplasmic reticulum resulted in retention of endogenous α-mannosidase II or GlcNAc-transferase I respectively. Expression of β-1,4-Gal-transferase, however, had no effect on either of these enzymes. The evidence indicates [204] that GlcNAc-transferase I and α-mannosidase II do not bind to each other but rather that both enzymes bind directly or indirectly to a protein-containing intercisternal matrix that is specific for medial-Golgi enzymes. β-1,4-Gal-transferase does not bind to this matrix. The matrix has been isolated as an insoluble pellet by extraction of rat liver Golgi preparations with a low-salt 2% Triton X-100 buffer solution [204]. This pellet contains most (77–86%) of the GlcNAc-transferase I and α-mannosidase II activities; enzymes can be solubilized by extraction with 0.15 M NaCl.

6.3. Genomic organization

Joziasse [13] has classified mammalian glycosyltransferase genes into two categories (Fig. 3): (i) single-exonic and (ii) multi-exonic. The entire coding regions for human α-1,3-Fuc-transferases III to VI, human and mouse β-1,2-GlcNAc-transferase I, human and rat β-1,2-GlcNAc-transferase II and human core 2 β6-GlcNAc-transferase are within a single exon whereas the coding regions for human and rat α-2,6-sialyltransferase, human and murine β-1,4-Gal-transferase and murine α-1,3-Gal-transferase are distributed over 5 or more exons. Many of these genes have a single long (2–3 kb) 3'-terminal exon which carries the entire relatively long 3'-untranslated region, the translational stop site and part of the 3'-terminal coding region. Some of these genes have at least one 5'-terminal exon several kilobases upstream of the translational start site.

There are at least four glycosyltransferase gene families [13,23,25]: (i) the α-1,3-Fuc-transferases, (ii) the α-1,3-Gal-transferases, (iii) the β6-GlcNAc-transferases and (iv) the sialyltransferases. The human α-1,3-Fuc-transferase family presently includes five members, α-1,3-Fuc-transferases III to VII, which are similar in primary sequence over most of their length. Three of the genes (α-1,3-Fuc-transferases III, V and VI) are syntenic on human chromosome 19 suggesting that the family arose by gene duplication and subsequent divergence.

The α-1,3-Gal-transferase family includes the enzymes which transfer Gal in (α1–3) linkage to the Gal residue of Gal(β1–4)GlcNAc-R (murine and bovine) and Fuc(α1–2) Gal-R (human blood group B) and the human blood group A α-1,3-GalNAc-transferase. The family also includes three human non-functional pseudogenes (HGT-2, HGT-10 and blood group O). The genes for the human blood group A and B transferases and for the human HGT-10 pseudogene are all located on human chromosome 9q33–34 suggesting evolution via gene duplication and subsequent divergence. However, Joziasse [13] points out that the exon structure of UDP-Gal:Gal(β(1–4) GlcNAc-

R α-1,3-Gal-transferase is consistent with the involvement of exon shuffling in the evolution of these genes.

The β6-GlcNAc-transferase family has at least two members [205,206] localized to human chromosome 9q21 suggesting gene duplication and divergence. The sialyltransferase gene family has 12 or more members, of which 7 enzymes have been cloned (ST6N, ST3N I, ST3N II, STZ, STX, ST3O and ST6O I, Table I). These enzymes share a sialylmotif which has been discussed in previous sections of this chapter and which has enabled the cloning of several of the enzymes.

6.4. Temporal and spatial control of glycosyltransferase activity

Glycoconjugate structures vary greatly during development and differentiation, in disease processes and between different normal tissues. Substrate-level factors such as the organization of the transferases on the Golgi membrane and the substrate specificities of the transferases are important in regulating the synthesis of complex carbohydrates within a specific tissue at a specific time [1,17,26,207]. However, temporal and spatial variations in glycoconjugate synthesis are controlled by transcriptional and translational factors.

Transcription of rat and human α-2,6-sialyltransferase and murine β-1,4-Gal-transferase genes appears to be under the control of multiple promoters [10]. This reflects the dual functions of these enzymes, i.e., to be constitutive ('house-keeping') enzymes in most tissues but under the control of tissue-specific transcription factors in others. The genes for UDP-Gal:Gal(β1–4)GlcNAc-R α-1,3-Gal-transferase and β-1,2-GlcNAc-transferase I also have multiple promoters but the tissue-specificity and functional aspects of these promoters remain to be determined. Multiple transcriptional start sites, alternate transcript splicing and alternate polyadenylation sites have been implicated in the production of different transcripts from the same gene. Different transcripts may produce either the same protein or several isoforms [13].

The absence of the β-1,2-GlcNAc-transferase I gene has little effect on the ability of cells to survive in culture but stops mouse embryogenesis at 10 days (see above). This suggests that perturbation of the glycosylation machinery to determine function must be carried out on the intact organism, e.g., by using the transgenic mouse model or administration of glycosylation inhibitors. Such studies will elucidate not only the mechanisms of glycoconjugate biosynthesis but also the elusive problem of function.

Acknowledgments

This work was supported by grants from the Medical Research Council of Canada and the Protein Engineering Network of Centres of Excellence (PENCE) of Canada.

References

1. Kornfeld, R. and Kornfeld, S. (1985) Annu. Rev. Biochem. 54, 631–664.
2. Montreuil, J. (1974) Pure Appl. Chem. 42, 431–477.
3. Rademacher, T.W., Parekh, R.B. and Dwek, R.A. (1988) Annu. Rev. Biochem. 57, 785–838.
4. Varki, A. (1993) Glycobiology 3, 97–130.
5. Hemming, F. (1995) in: J. Montreuil, H. Schachter and J.F.G. Vliegenthart (Eds.), Glycoproteins. New Comprehensive Biochemistry, Vol. 29a. Elsevier, Amsterdam, pp. 127–143.
6. Verbert, A. (1995) in: J. Montreuil, H. Schachter and J.F.G. Vliegenthart (Eds.), Glycoproteins. New Comprehensive Biochemistry, Vol. 29a. Elsevier, Amsterdam, pp. 145–152.
7. Struck, D.K. and Lennarz, W.J. (1980) in: W.J. Lennarz (Ed.), The Biochemistry of Glycoproteins and Proteoglycans. Plenum Press, New York, NY, pp. 35–83.
8. Abeijon, C. and Hirschberg, C.B. (1992) Trends Biochem. Sci. 17, 32–36.
9. Brockhausen, I. (1995) in: J. Montreuil, H. Schachter and J.F.G. Vliegenthart (Eds.), Glycoproteins. New Comprehensive Biochemistry, Vol. 29a. Elsevier, Amsterdam, pp. 201–259.
10. Glick, M.C. (1995) in: J. Montreuil, H. Schachter and J.F.G. Vliegenthart (Eds.), Glycoproteins. New Comprehensive Biochemistry, Vol. 29a. Elsevier, Amsterdam, pp. 261–280.
11. Watkins, W. (1995) in: J. Montreuil, H. Schachter and J.F.G. Vliegenthart (Eds.), Glycoproteins. New Comprehensive Biochemistry, Vol. 29a. Elsevier, Amsterdam, pp. 313–390.
12. Hart, G.W. (1992) Curr. Opin. Cell Biol. 4, 1017–1023.
13. Joziasse, D.H. (1992) Glycobiology 2, 271–277.
14. Macher, B.A., Holmes, E.H., Swiedler, S.J., Stults, C.L.M. and Srnka, C.A. (1991) Glycobiology 1, 577–584.
15. Paulson, J.C. and Colley, K.J. (1989) J. Biol. Chem. 264, 17615–17618.
16. Roth, J. (1987) Biochim. Biophys. Acta 906, 405–436.
17. Schachter, H. (1986) Biochem. Cell Biol. 64, 163–181.
18. Schachter, H. (1991) Curr. Opin. Struct. Biol. 1, 755–765.
19. Shaper, J.H. and Shaper, N.L. (1992) Curr. Opin. Struct. Biol. 2, 701–709.
20. Shur, B.D. (1991) Glycobiology 1, 563–575.
21. Snider, M.D. (1984) in: V. Ginsburg and P.W. Robbins (Eds.), Biology of Carbohydrates. John Wiley and Sons, New York, NY, Vol. 2, pp. 163–198.
22. Stanley, P. (1992) Glycobiology 2, 99–107.
23. Van den Eijnden, D.H. and Joziasse, D.H. (1993) Curr. Opin. Struct Biol 3, 711–721.
24. Yamamoto, F., Clausen, H., White, T., Marken, J. and Hakomori, S. (1990) Nature 345, 229–233.
25. Kleene, R. and Berger, E.G. (1993) Biochim. Biophys. Acta 1154, 283–325.
26. Schachter, H. (1991) Glycobiology 1, 453–461.
27. Cummings, R.D. (1992) in: H.J. Allen and E.C. Kisailus (Eds.), Glycoconjugates. Composition, Structure and Function. Marcel Dekker, Inc., New York, NY, pp. 333–360.
28. Strous, G.J. (1986) Crit. Rev. Biochem. 21, 119–151.
29. Schachter, H. and Roseman, S. (1980) in: W.J. Lennarz (Ed.), Biochemistry of Glycoproteins and Proteoglycans. Plenum Press, New York, NY, pp. 85–160.
30. Narimatsu, H., Sinha, S., Brew, K., Okayama, H. and Qasba, P.K. (1986) Proc. Nat. Acad. Sci. USA 83, 4720–4724.
31. Shaper, N.L., Shaper, J.H., Meuth, J.L., Fox, J.L., Chang, H., Kirsch, I.R. and Hollis, G.F. (1986) Proc. Nat. Acad. Sci. USA 83, 1573–1577.
32. D'Agostaro, G., Bendiak, B. and Tropak, M. (1989) Eur. J. Biochem. 183, 211–217.
33. Masibay, A.S. and Qasba, P.K. (1989) Proc. Natl. Acad. Sci. USA. 86(15), 5733–5737.
34. Russo, R.N., Shaper, N.L. and Shaper, J.H. (1990) J. Biol. Chem. 265, 3324–3331.

35. Appert, H.E., Rutherford, T.J., Tarr, G.E., Wiest, J.S., Thomford, N.R. and McCorquodale, D.J. (1986) Biochem. Biophys. Res. Communs. 139, 163–168.

36. Appert, H.E., Rutherford, T.J., Tarr, G.E., Thomford, N.R. and McCorquodale, D.J. (1986) Biochem. Biophys. Res. Communs. 138, 224–229.

37. Masri, K.A., Appert, H.E. and Fukuda, M.N. (1988) Biochem. Biophys. Res. Commun. 157, 657–663.

38. Shaper, N.L., Hollis, G.F., Douglas, J.G., Kirsch, I.R. and Shaper, J.H. (1988) J. Biol. Chem. 263, 10420–10428.

39. Nakazawa, K., Ando, T., Kimura, T. and Narimatsu, H. (1988) J. Biochem. (Tokyo) 104, 165–168.

40. Hollis, G.F., Douglas, J.G., Shaper, N.L., Shaper, J.H., Stafford-Hollis, J.M., Evans, R.J. and Kirsch, I.R. (1989) Biochem Biophys. Res. Commun. 162, 1069–1075.

41. Mengle-Gaw, L., McCoy-Haman, M.F. and Tiemeier, D.C. (1991) Biochem. Biophys. Res. Commun. 176, 1269–1276.

42. Shaper, N.L., Shaper, J.H., Hollis, G.F., Chang, H., Kirsch, I.R. and Kozak, C.A. (1987) Cytogenet. Cell. Genet. 44, 18–21.

43. Shaper, N.L., Shaper, J.H., Peyser, M. and Kozak, C.A. (1990) Cytogenet. Cell Genet. 54, 172–174.

44. Shaper, N.L., Shaper, J.H., Bertness, V., Chang, H., Kirsch, I.R., Hollis, G.F. (1986) Somat. Cell. Molec. Genet. 12, 633–636.

45. Duncan, A.M.V., McCorquodale, M.M., Morgan, C., Rutherford, T.J., Appert, H.E. and McCorquodale, D.J. (1986) Biochem. Biophys. Res. Communs. 141, 1185–1188.

46. Beyer, T.A., Sadler, J.E., Rearick, J.I., Paulson, J.C. and Hill, R.L. (1981). In: A. Meister (Ed.) Advances in Enzymology. John Wiley and Sons, New York, NY, Vol. 52, pp. 23–175.

47. Elhammer, A. and Kornfeld, S. (1986) J. Biol. Chem. 261, 5249–5255.

48. Sugiura, M., Kawasaki, T. and Yamashina, I. (1982) J. Biol. Chem. 257, 9501–9507.

49. Weinstein, J., DeSouza-e-Silva, U. and Paulson, J.C. (1982) J. Biol. Chem. 257, 13845–13853.

50. Nishikawa, Y., Pegg, W., Paulsen, H. and Schachter, H. (1988) J. Biol. Chem. 263, 8270–8281.

51. Bendiak, B. and Schachter, H. (1987) J. Biol. Chem. 262, 5775–5783.

52. Shaper, J.H., Hollis, G.F. and Shaper, N.L. (1988) Biochimie. 70, 1683–1688.

53. Lopez, L.C. and Shur, B.D. (1988) Biochem. Biophys. Res. Communs. 156, 1223–1229.

54. Lopez, L.C., Maillet, C.M., Oleszkowicz, K. and Shur, B.D. (1989) Mol. Cell. Biol. 9(6), 2370–2377.

55. Lopez, L.C., Youakim, A., Evans, S.C. and Shur, B.D. (1991) J. Biol. Chem. 266, 15984–15991.

56. Evans, S.C., Lopez, L.C. and Shur, B.D. (1993) J. Cell Biol. 120, 1045–1057.

57. Nilsson, T., Lucocq, J.M., Mackay, D. and Warren, G. (1991) EMBO J 10, 3567–3575.

58. Teasdale, R.D., D'Agostaro, G. and Gleeson, P.A. (1992) J. Biol. Chem. 267, 4084–4096.

59. Aoki, D., Lee, N., Yamaguchi, N., Dubois, C. and Fukuda, M.N. (1992) Proc. Nat. Acad. Sci. USA 89, 4319–4323.

60. Masibay, A.S., Balaji, P.V., Boeggeman, E.E. and Qasba, P.K. (1993) J. Biol. Chem. 268, 9908–9916.

61. Russo, R.N., Shaper, N.L., Taatjes, D.J. and Shaper, J.H. (1992) J. Biol. Chem. 267, 9241–9247.

62. Harduin-Lepers, A., Shaper, J.H. and Shaper, N.L. (1993) J. Biol. Chem. 268, 14348–14359.

63. Miller, D.J., Macek, M.B. and Shur, B.D. (1992) Nature 357, 589–593.

64. Hathaway, H.J. and Shur, B.D. (1992) J. Cell Biol. 117, 369–382.

65. Shaper, N.L., Wright, W.W. and Shaper, J.H. (1990) Proc. Natl. Acad. Sci. USA 87, 791–795.

66. Harduin-Lepers, A., Shaper, N.L., Mahoney, J.A. and Shaper, J.H. (1992) Glycobiology 2, 361–368.

67. Nakazawa, K., Furukawa, K., Kobata, A. and Narimatsu, H. (1991) Eur. J. Biochem. 196, 363–368.
68. Aoki, D., Appert, H.E., Johnson, D., Wong, S.S. and Fukuda, M.N. (1990) EMBO J 9, 3171–3178.
69. Boeggeman, E.E., Balaji, P.V., Sethi, N., Masibay, A.S. and Qasba, P.K. (1993) Protein Eng. 6, 779–785.
70. Krezdorn, C.H., Watzele, G., Kleene, R.B., Ivanov, S.X. and Berger, E.G. (1993) Eur. J. Biochem. 212, 113–120.
71. Moremen, K.W. and Robbins, P.W. (1991) J. Cell Biol. 115, 1521–1534.
72. Masibay, A.S., Boeggeman, E. and Qasba, P.K. (1992) Mol. Biol. Rep. 16, 99–104.
73. Munro, S. (1991) EMBO J. 10, 3577–3588.
74. Galili, U., Shohet, S.B., Kobrin, E., Stults, C.L.M. and Macher, B.A. (1988) J. Biol. Chem. 263, 17755–17762.
75. Galili, U. (1989) Lancet 2, 358–361.
76. Galili, U. and Swanson, K. (1991) Proc. Natl. Acad. Sci. USA 88, 7401–7404.
77. Galili, U. (1992) in: C.A. Bona, A.K. Kaushik (Eds.), Molecular Immunobiology of Self-Reactivity. Marcel Dekker, Inc., New York, NY, pp. 355–373.
78. Joziasse, D.H., Shaper, J.H., Van den Eijnden, D.H., Van Tunen, A.J. and Shaper, N.L. (1989) J. Biol. Chem. 264, 14290–14297.
79. Larsen, R.D., Rajan, V.P., Ruff, M.M., Kukowska-Latallo, J., Cummings, R.D. and Lowe, J.B. (1989) Proc. Natl. Acad. Sci. USA 86, 8227–8231.
80. Aruffo, A. and Seed, B. (1987) Proc. Nat. Acad. Sci. USA 84, 8573–8577.
81. Seed, B. and Aruffo, A. (1987) Proc. Nat. Acad. Sci. USA 84, 3365–3369.
82. Joziasse, D.H., Shaper, N.L., Kim, D., Van den Eijnden, D.H. and Shaper, J.H. (1992) J. Biol. Chem. 267, 5534–5541.
83. Yamamoto, F. and Hakomori, S. (1990) J. Biol. Chem. 265, 19257–19262.
84. Larsen, R.D., Rivera-Marrero, C.A., Ernst, L.K., Cummings, R.D. and Lowe, J.B. (1990) J. Biol. Chem. 265, 7055–7061.
85. Joziasse, D.H., Shaper, J.H., Jabs, E.W. and Shaper, N.L. (1991) J. Biol. Chem. 266, 6991–6998.
86. Joziasse, D.H., Shaper, N.L., Shaper, J.H. and Kozak, C.A. (1991) Somatic Cell Mol Genet. 17, 201–205.
87. Shaper, N.L., Lin, S., Joziasse, D.H., Kim, D. and Yang-Feng, T.L. (1992) Genomics 12, 613–615.
88. Joziasse, D.H., Shaper, N.L., Salyer, L.S., Van den Eijnden, D.H., Van der Spoel, A.C. and Shaper, J.H. (1990) Eur. J. Biochem. 191, 75–83.
89. Smith, D.F., Larsen, R.D., Mattox, S., Lowe, J.B. and Cummings, R.D. (1990) J. Biol. Chem. 265, 6225–6234.
90. Weinstein, J., Lee, E.U., McEntee, K., Lai, P.-H., Paulson, J.C. (1987) J. Biol. Chem. 262, 17735–17743.
91. Lee, Y.C., Kurosawa, N., Hamamoto, T., Nakaoka, T., Tsuji, S. (1993) Eur. J. Biochem. 216, 377–385.
92. Weinstein, J., DeSouza-e-Silva, U. and Paulson, J.C. (1982) J. Biol. Chem. 257, 13835–13844.
93. Grundmann, U., Nerlich, C., Rein, T. and Zettlmeissl, G. (1990) Nucleic Acids Res. 18, 667.
94. Lance, P., Lau, K.M. and Lau, J. (1989) Biochem. Biophys. Res. Commun. 164, 225–232.
95. Stamenkovic, I., Asheim, H.C., Deggerdal, A., Blomhoff, H.K., Smeland, E.B. and Funderud, S. (1990) J. Exp. Med. 172, 641–643.
96. Bast, B.J.E.G., Zhou, L.J., Freeman, G.J., Colley, K.J., Ernst, T.J., Munro, J.M. and Tedder, T.F. (1992) J. Cell Biol. 116, 423–435.

97. Paulson, J.C., Weinstein, J. and Schauer, A. (1989) J. Biol. Chem. 264(19), 10931–10934.

98. Svensson, E.C., Soreghan, B. and Paulson, J.C. (1990) J. Biol. Chem. 265, 20863–20868.

99. Wen, D.X., Svensson, E.C. and Paulson, J.C. (1992) J. Biol. Chem. 267, 2512–2518.

100. Wang, X.-C., O'Hanlon, T.P., Young, R.F. and Lau, J.T.Y. (1990) Glycobiology 1, 25–31.

101. O'Hanlon, T.P., Lau, K.M., Wang, X.C. and Lau, J.T.Y. (1989) J. Biol. Chem. 264, 17389–17394.

102. O'Hanlon, T.P. and Lau, J.T.Y. (1992) Glycobiology 2, 257–266.

103. Wang, X.-C., Vertino, A., Eddy, R. L., Byers, M. G., Jani-Sait, S.N., Shows, T.B. and Lau, J.T.Y. (1993) J. Biol. Chem. 268, 4355–4361.

104. Aasheim, H.C., Aas-Eng, D.A., Deggerdal, A., Blomhoff, H.K., Funderud, S. and Smeland, E.B. (1993) Eur. J. Biochem. 213, 467–475.

105. Svensson, E.C., Conley, P.B. and Paulson, J.C. (1992) J. Biol. Chem. 267, 3466–3472.

106. Lammers, G. and Jamieson, J.C. (1988) Biochem. J. 256, 623–631.

107. Lammers, G. and Jamieson, J.C. (1989) Biochem. J. 261(2), 389–393.

108. Lammers, G. and Jamieson, J.C. (1990) Comp. Biochem. Physiol. 95B, 327–334.

109. Kaplan, H.A., Woloski, B.M.R.N.J., Hellman, M. and Jamieson, J.C. (1983) J. Biol. Chem. 258, 11505–11509.

110. Lombart, C., Sturgess, J. and Schachter, H. (1980) Biochim. Biophys. Acta 629, 1–12.

111. Wang, X.C., O'Hanlon, T.P. and Lau, J.T.Y. (1989) J. Biol. Chem. 264, 1854–1859.

112. Wang, X.C., Smith, T.J. and Lau, J. (1990) J. Biol. Chem. 265, 17849–17853.

113. Vandamme, V., Pierce, A., Verbert, A. and Delannoy, P. (1993) Eur. J. Biochem. 211, 135–140.

114. Grollman, E.F., Saji, M., Shimura, Y., Lau, J. and Ashwell, G. (1993) J. Biol. Chem. 268, 3604–3609.

115. Lee, E.U., Roth, J. and Paulson, J.C. (1989) J. Biol. Chem. 264 (23), 13848–13855.

116. Munro, S., Bast, B.J.E.G., Colley, K.J. and Tedder, T.F. (1992) Cell 68, 1003–1003.

117. Stamenkovic, I., Sgroi, D. and Aruffo, A. (1992) Cell 68, 1003–1004.

118. Colley, K.J., Lee, E.U., Adler, B., Browne, J.K. and Paulson, J.C. (1989) J. Biol. Chem. 264, 17619–17622.

119. Paulson, J.C., Weinstein, J., Ujita, E.L., Riggs, K.J. and Lai, P.-H. (1987) Biochem. Soc. Trans. 15, 618–620.

120. Dahdal, R.Y. and Colley, K.J. (1993) J. Biol. Chem. 268, 26310–26319.

121. Wong, S.H., Low, S.H. and Hong, W. (1992) J. Cell Biol. 117, 245–258.

122. Wen, D.X., Livingston, B.D., Medzihradszky, K.F., Kelm, S., Burlingame, A.L. and Paulson, J.C. (1992) J. Biol. Chem. 267, 21011–21019.

123. Drickamer, K. (1993) Glycobiology 3, 2–3.

124. Kurosawa, N., Kawasaki, M., Hamamoto, T., Nakaoka, T., Lee, Y. C., Arita, M. and Tsuji, S. (1994) Eur. J. Biochem. 219, 375–381.

125. Kitagawa, H. and Paulson, J.C. (1993) Biochem. Biophys. Res. Commun. 194, 375–382.

126. Nemansky, M. and Van den Eijnden, D.H. (1993) Glycoconjugate J. 10, 99–108.

127. Sasaki, K., Watanabe, E., Kawashima, K., Sekine, S., Dohi, T., Oshima, M., Hanai, N., Nishi, T. and Hasegawa, M. (1993) J. Biol. Chem. 268, 22782–22787.

128. Livingston, B., Kitagawa, H., Wen, D., Medzihradsky, K., Burlingame, A. and Paulson, J.C. (1992) Glycobiology 2, 489.

129. Livingston, B.D. and Paulson, J.C. (1993) J. Biol. Chem. 268, 11504–11507.

130. Kitagawa, H. and Paulson, J.C. (1994) J. Biol. Chem. 269, 1394–1401.

131. Basu, S.C. (1991) Glycobiology 1, 469–475.

132. Basu, M., Basu, S., Stoffyn, A. and Stoffyn, P. (1982) J. Biol. Chem. 257, 12765–12769.

133. Möller, G., Reck, F., Paulsen, H., Kaur, K.J., Sarkar, M., Schachter, H. and Brockhausen, I. (1992) Glycoconjugate J. 9, 180–190.

134. Bendiak, B. and Schachter, H. (1987) J. Biol. Chem. 262, 5784–5790.
135. Nishikawa, A., Ihara, Y., Hatakeyama, M., Kangawa, K. and Taniguchi, N. (1992) J. Biol. Chem. 267, 18199–18204.
136. Gu, J., Nishikawa, A., Tsuruoka, N., Ohno, M., Yamaguchi, N., Kanagawa, K. and Taniguchi, N. (1993) J. Biochem. 113, 614–619.
137. Shoreibah, M.G., Hindsgaul, O. and Pierce, M. (1992) J. Biol. Chem. 267, 2920–2927.
138. Sarkar, M., Hull, E., Simpson, R.J., Moritz, R.L., Dunn, R. and Schachter, H. (1990) Glycoconjugate J. 7, 380.
139. Sarkar, M., Hull, E., Nishikawa, Y., Simpson, R.J., Moritz, R. L., Dunn, R. and Schachter, H. (1991) Proc. Natl. Acad. Sci. USA 88, 234–238.
140. Schachter, H., Hull, E., Sarkar, M., Simpson, R.J., Moritz, R.L., Höppener, J.W.M. and Dunn, R. (1991) Biochem. Soc. Trans. 19, 645–648.
141. Hull, E., Schachter, H., Sarkar, M., Spruijt, M.P.N., Höppener, J.W.M., Roovers, D. and Dunn, R. (1990) Glycoconjugate J. 7, 468.
142. Hull, E., Sarkar, M., Spruijt, M.P.N., Höppener, J.W.M., Dunn, R. and Schachter, H. (1991) Biochem. Biophys. Res. Commun. 176, 608–615.
143. Kumar, R., Yang, J., Larsen, R.D. and Stanley, P. (1990) Proc. Natl. Acad. Sci. USA 87, 9948–9952.
144. Pownall, S., Kozak, C.A., Schappert, K., Sarkar, M., Hull, E., Schachter, H. and Marth, J.D. (1992) Genomics 12, 699–704.
145. Kumar, R., Yang, J., Eddy, R.L., Byers, M.G., Shows, T.B. and Stanley, P. (1992) Glycobiology 2, 383–393.
146. Tan, J., D'Agostaro, G.A.F., Bendiak, B.K., Squire, J. and Schachter, H. (1993) Glycoconjugate J. 10, 232–233.
147. Nilsson, T., Hoe, M.H., Slusarewicz, P., Rabouille, C., Watson, R., Hunte, F., Watzele, G., Berger, E.G. and Warren, G. (1994) EMBO J. 13, 562–574.
148. Burke, J., Pettitt, J.M., Schachter, H., Sarkar, M. and Gleeson, P.A. (1992) J. Biol. Chem. 267, 24433–24440.
149. Uhlén, M. and Moks, T. (1990) Methods in Enzymology 185, 129–143.
150. Sarkar, M. and Schachter, H. (1992) Glycobiology 2, 483.
151. Sarkar, M. (1994) Glycoconjugate J. 11, 204–209.
152. Reck, F., Paulsen, H., Brockhausen, I., Sarkar, M. and Schachter, H. (1992) Glycobiology 2, 483.
153. Reck, F., Springer, M., Paulsen, H., Brockhausen, I., Sarkar, M. and Schachter, H. (1994) Carbohydrate Res. 259, 93–101.
154. Tang, B.L., Wong, S.H., Low, S.H. and Hong, W. (1992) J. Biol. Chem. 267, 10122–10126.
155. Stanley, P. (1989) Mol. Cell. Biol. 9, 377–383.
156. Metzler, M., Gertz, A., Sarkar, M., Schachter, H., Schrader, J.W. and Marth, J.D. (1994) EMBO J. 13, 2056–2065.
157. Kaur, K.J., Alton, G. and Hindsgaul, O. (1991) Carbohydrate Res. 210, 145–153.
158. Vella, G.J., Paulsen, H. and Schachter, H. (1984) Can. J. Biochem. Cell Biol. 62, 409–417.
159. Paulsen, H., Reck, F. and Brockhausen, I. (1992) Carbohydrate Res. 236, 39–71.
160. D'Agostaro, G.A.F., Zingoni, A., Simpson, R.J., Moritz, R.L., Schachter, H. and Bendiak, B.K. (1993) Glycoconjugate J. 10, 234.
161. Petrarca, C., Bendiak, B.K. and D'Agostaro, G.A.F. (1993) Glycoconjugate J. 10, 235.
162. Paulsen, H., Van Dorst, J.A.L.M., Reck, F. and Meinjohanns, E. (1992) Liebigs Ann. Chem. 513–521.
163. Paulsen, H. and Meinjohanns, E. (1992) Tetrahedron Lett. 33, 7327–7330.
164. Srivastava, G., Alton, G. and Hindsgaul, O. (1990) Carbohydrate Res. 207, 259–276.

165. Kaur, K.J. and Hindsgaul, O. (1992) Carbohydrate Res. 226, 219–231.
166. Reck, F., Meinjohanns, E., Springer, M., Wilkens, R., Van Dorst, J.A.L.M., Paulsen, H., Möller, G., Brockhausen, I. and Schachter, H. (1994) Glycoconjugate J. 11, 210–216.
167. Narasimhan, S. (1982) J. Biol. Chem. 257, 10235–10242.
168. Narasimhan, S., Schachter, H. and Rajalakshmi, S. (1988) J. Biol. Chem. 263, 1273–1281.
169. Ishibashi, K., Nishikawa, A., Hayashi, N., Kasahara, A., Sato, N., Fujii, S., Kamada, T. and Taniguchi, N. (1989) Clin. Chim. Acta 185, 325–332.
170. Nishikawa, A., Gu, J., Fujii, S. and Taniguchi, N. (1990) Biochim. Biophys. Acta Gen. Subj. 1035, 313–318.
171. Miyoshi, E., Nishikawa, A., Ihara, Y., Gu, J., Sugiyama, T., Hayashi, N., Fusamoto, H., Kamada, T. and Taniguchi, N. (1993) Cancer Res. 53, 3899–3902.
172. Ihara, Y., Nishikawa, A., Tohma, T., Soejima, H., Niikawa, N. and Taniguchi, N. (1993) J. Biochem. 113, 692–698.
173. Brockhausen, I., Carver, J. and Schachter, H. (1988) Biochem. Cell Biol. 66, 1134–1151.
174. Alton, G., Kanie, Y. and Hindsgaul, O. (1993) Carbohydrate Res. 238, 339–344.
175. Gleeson, P.A. and Schachter, H. (1983) J. Biol. Chem. 258, 6162–6173.
176. Brockhausen, I., Möller, G., Yang, J.M., Khan, S.H., Matta, K.L., Paulsen, H., Grey, A.A., Shah, R.N. and Schachter, H. (1992) Carbohydrate Res. 236, 281–299.
177. Smets, L.A. and Van Beek, W.P. (1984) Biochim. Biophys. Acta 738, 237–249.
178. Dennis, J.W., Laferte, S., Waghorne, C., Breitman, M.L. and Kerbel, R.S. (1987) Science 236, 582–585.
179. Dennis, J.W. and Laferte, S. (1988) in: C.L. Reading, S.-I. Hakomori and D.M. Marcus (Eds.), Altered Glycosylation in Tumor Cells. Alan R. Liss Inc., New York, NY, Vol. 79, pp. 257–267.
180. Dennis, J.W. (1992) in: M. Fukuda (Ed.), Cell Surface Carbohydrates and Cell Development. CRC Press, Boca Raton, FL, pp. 161–194.
181. Yamashita, K., Tachibana, Y., Ohkura, T. and Kobata, A. (1985) J. Biol. Chem. 260, 3963–3969.
182. Palcic, M.M., Ripka, J., Kaur, K.J., Shoreibah, M., Hindsgaul, O. and Pierce, M. (1990) J. Biol. Chem. 265, 6759–6769.
183. Dennis, J.W., Kosh, K., Bryce, D.-M. and Breitman, M.L. (1989) Oncogene 4, 853–860.
184. Arango, J. and Pierce, M. (1988) J. Cell. Biochem. 37, 225–231.
185. Crawley, S.C., Hindsgaul, O., Alton, G., Pierce, M. and Palcic, M.M. (1990) Anal. Biochem. 185, 112–117.
186. Cummings, R.D. and Kornfeld, S. (1984) J. Biol. Chem. 259, 6253–6260.
187. van den Eijnden, D.H., Koenderman, A.H.L. and Schiphorst, W.E. C.M. (1988) J. Biol. Chem. 263, 12461–12471.
188. Cummings, R.D., Trowbridge, I.S. and Kornfeld, S. (1982) J. Biol. Chem. 257, 13421–13427.
189. Hindsgaul, O., Tahir, S.H., Srivastava, O.P. and Pierce, M. (1988) Carbohydrate Res. 173, 263–272.
190. Srivastava, O.P., Hindsgaul, O., Shoreibah, M. and Pierce, M. (1988) Carbohydrate Res. 179, 137–161.
191. Hindsgaul, O., Kaur, K.J., Srivastava, G., Blaszczyk-Thurin, M., Crawley, S.C., Heerze, L.D. and Palcic, M.M. (1991) J. Biol. Chem. 266, 17858–17862.
192. Kanie, O., Crawley, S.C., Palcic, M.M. and Hindsgaul, O. (1993) Carbohydrate Res. 243, 139–164.
193. Paulsen, H., Meinjohanns, E., Reck, F. and Brockhausen, I. (1993) Liebigs Ann. Chem. 737–750.
194. Khan, S.H., Crawley, S.C., Kanie, O. and Hindsgaul, O. (1993) J. Biol. Chem. 268, 2468–2473.
195. Lindh, I. and Hindsgaul, O. (1991) J. Am. Chem. Soc. 113, 216–223.

196. Shoreibah, M., Perng, G.S., Adler, B., Weinstein, J., Basu, R., Cupples, R., Wen, D., Browne, J.K., Buckhaults, P., Fregien, N. and Pierce, M. (1993) J. Biol. Chem. 268, 15381–15385.
197. Saito, H., Nishikawa, A., Gu, J.G., Ihara, Y., Soejima, H., Wada, Y., Sekiya, C., Niikawa, N. and Taniguchi, N. (1994) Biochem. Biophys. Res. Commun. 198, 318–327.
198. Brockhausen, I., Hull, E., Hindsgaul, O., Schachter, H., Shah, R.N., Michnick, S.W. and Carver, J.P. (1989) J. Biol. Chem. 264(19), 11211–11221.
199. Yadav, S. and Brew, K. (1990) J. Biol. Chem. 265, 14163–14169.
200. Yadav, S.P. and Brew, K. (1991) J. Biol. Chem. 266, 698–703.
201. Berger, E.G., Grimm, K., Bächi, T., Bosshart, H., Kleene, R. and Watzele, M. (1993) J. Cell. Biochem. 52, 275–288.
202. Nilsson, T., Pypaert, M., Hoe, M.H., Slusarewicz, P., Berger, E.G. and Warren, G. (1993) J. Cell Biol. 120, 5–13.
203. Nilsson, T., Slusarewicz, P., Hoe, M.H. and Warren, G. (1993) FEBS Lett. 330, 1–4.
204. Slusarewicz, P., Nilsson, T., Hui, N., Watson, R. and Warren, G. (1994) J. Cell Biol. 124, 405–413.
205. Bierhuizen, M., Mattei, M.G. and Fukuda, M. (1993) Genes. Dev 7, 468–478.
206. Bierhuizen, M.F.A. and Fukuda, M. (1992) Proc. Natl. Acad. Sci. USA 89, 9326–9330.
207. Schachter, H. (1995) in: J. Montreuil, H. Schachter and J.F.G. Vliegenthart (Eds.), Glycoproteins. New Comprehensive Biochemistry, Vol. 29a. Elsevier, Amsterdam, pp. 281–286.
208. Gillespie, W., Kelm, S. and Paulson, J.C. (1992) J. Biol. Chem. 267, 21004–21010.
209. Kurosawa, N., Hamamoto, T., Lee, Y.C., Nakaoka, T., Kojima, N. and Tsuji, S. (1994) J. Biol. Chem. 269, 1402–1409.
210. Larsen, R.D., Ernst, L.K., Nair, R.P. and Lowe, J.B. (1990) Proc. Natl. Acad. Sci. USA 87, 6674–6678.
211. Kukowska-Latallo, J.F., Larsen, R.D., Rajan, V.P., Lowe, J.B. (1990) FASEB J. 4, A1930.
212. Kukowska-Latallo, J.F., Larsen, R.D., Nair, R.P., Lowe, J.B. (1990) Genes Dev. 4, 1288–1303.
213. Weston, B.W., Nair, R.P., Larsen, R.D. and Lowe, J.B. (1992) J. Biol. Chem. 267, 4152–4160.
214. Weston, B.W., Smith, P.L., Kelly, R.J. and Lowe, J.B. (1992) J. Biol. Chem. 267, 24575–24584.
215. Lowe, J.B., Kukowska-Latallo, J.F., Nair, R.P., Larsen, R.D., Marks, R.M., Macher, B A., Kelly, R.J. and Ernst, L.K. (1991) J. Biol. Chem. 266, 17467–17477.
216. Yamamoto, F., Marken, J., Tsuji, T., White, T., Clausen, H., Hakomori, S. (1990) J. Biol. Chem. 265, 1146–1151.
217. Nagata, Y., Yamashiro, S., Yodoi, J., Lloyd, K.O., Shiku, H. and Furukawa, K. (1992) J. Biol. Chem. 267, 12082–12089.
218. Hagen, F.K., Vanwuyckhuyse, B. and Tabak, L.A. (1993) J. Biol. Chem. 268, 18960–18965.
219. Homa, F.L., Hollander, T., Lehman, D.J., Thomsen, D.R. and Elhammer, Å. P. (1993) J. Biol. Chem. 268, 12609–12616.

J. Montreuil, H. Schachter and J.F.G. Vliegenthart (Eds.), *Glycoproteins*
© 1995 Elsevier Science B.V. All rights reserved

Biosynthesis

3. Biosynthesis of O-glycans of the N-acetylgalactosamine-α-Ser/Thr linkage type

INKA BROCKHAUSEN

Research Institute, The Hospital for Sick Children, 555 University Avenue, Toronto M5G 1X8, and Biochemistry Department, University of Toronto, Toronto M5S 1A8, Ontario, Canada

1. Introduction

This chapter deals with the biosynthesis of oligosaccharides (O-glycans) linked to polypeptide O-glycosidically via GalNAcα-Ser/Thr (mucin-type) linkages. Glycoproteins in which GlcNAc (cytoplasmic or nuclear glycoproteins), mannose (yeast glycoproteins) or other sugars are O-glycosidically linked will not be discussed here. The biosynthesis of N-glycans and O-glycans share common features. Both types of chains may occur on the same molecule, and many glycosyltransferases, for example, those adding blood group antigens, are common to the syntheses of both types of chains. In contrast to the biosynthesis of N-glycans, O-linked oligosaccharides are not preassembled on a dolichol derivative in the endoplasmic reticulum [1] but sugars are added individually from nucleotide sugar donors. O-glycans do not appear to be processed by glycosidases, and the addition of the first sugar to the peptide, GalNAc, occurs mainly in the *cis*-Golgi without a requirement for a specific amino acid sequon. The terminal structures of O-glycans often resemble those of N-glycans and certain glycolipids but may be functionally distinct, probably due to characteristic differences in the presentation to their biological environments.

The synthesis of the frame-work of oligosaccharide chains regulates the expression of functional terminal carbohydrate structures. Thus a control at early steps of the biosynthetic pathways may have a great impact on the structures, properties and biological functions of O-glycans.

O-Glycosylation commonly changes during development, differentiation, growth and in disease. In addition, various tissues and species express characteristic O-glycans associated with various biological functions. Genetic defects in the biosynthetic pathways of O-glycans are rare, possibly because the development of a multicellular organism is dependent on interactions with cellular carbohydrate.

2. Structures and distributions of O-Glycans
(GalNAcα-Ser/Thr-linked-oligosaccharides)

The sugars commonly found in O-glycans are GalNAc, Gal, GlcNAc, sialic acid and fucose. Gal, GlcNAc and GalNAc may be sulfated and contribute with sialic acid to the acidity of O-glycans.

O-glycans are found on many mammalian and non-mammalian soluble and membrane-bound glycoproteins and on proteoglycans. Mucins, the major class of glycoproteins carrying O-glycans, are large molecules found in mucous secretions (Section 3.1) and carry many of the blood group antigens and recognition signals required for intercellular and molecular interactions as well as cancer-associated and differentiation antigens [2,3,4]. More than 80% of the molecular weight of mucins may consist of O-glycans, thus most of the Ser and Thr residues are glycosylated. The oligosaccharides range in size from a single monosaccharide to about 20 residues and mucin O-glycans usually exhibit an extreme structural heterogeneity with several hundred different oligosaccharide chains per molecule. Oligosaccharides on serum glycoproteins or those synthesized in non-mucin secreting cells are generally shorter and restricted to the core 1 and 2 structures (Section 2.1) and not as abundant; in addition they are often found on specific Ser or Thr residues.

O-glycans may be released from glycoproteins by hydrazine [5], or in the reduced form containing terminal GalNAcol by β-elimination in alkaline borohydride solution. The common O-glycan Galβ1-3GalNAc- may be exposed by sialidase treatment and can be released by a specific endo-*N*-acetylgalactosaminidase (O-glycanase). In addition, an endo-*N*-acetylgalactosaminidase from Streptomyces has recently been shown to release more complex O-glycans [6]. After release from the protein, O-glycans may be purified by ion exchange chromatography to separate acidic and neutral fractions, by electrophoresis, paper chromatography, gel filtration, thin layer chromatography, high pressure liquid chromatography, affinity chromatography on lectin or antibody columns, or by other techniques. Structural analyses have been carried out using mass spectrometric methods, methylation analysis and nuclear magnetic resonance techniques. Electrospray mass spectrometry has proven to be a superior and sensitive method for the analysis of glycopeptides without prior release of carbohydrate [7].

The simplest oligosaccharide, and the only sugar common to all O-glycans, is GalNAc. GalNAc-Ser/Thr is often referred to as the Tn antigen (Table I). The sialylated Tn antigen has the structure SA(α2–6)GalNAcα-Ser/Thr and is common in submaxillary mucins and other glycoproteins, but appears to be increased in some cancers.

The core structures may be substituted by monosaccharides such as SA or may be elongated by repeating linear or branched sequences of GlcNAc and Gal residues, or may be fucosylated, sialylated, sulfated and contain blood group or tissue antigens. There may also be unusual epitopes at the nonreducing termini such as GlcNAc(α1–3 or 4).

2.1. O-glycan core structures

There are seven core classes of O-glycans containing GlcNAc, Gal or GalNAc substitutions of GalNAc-:

TABLE I

Carbohydrate blood group and tissue antigens occurring on O-glycans

Antigen	Structure
Type 1 chain	Galβ1-3GlcNAc-
Type 2 chain	Galβ1-4GlcNAc-
A	GalNAcα1-3Galβ- 　　　　｜α1-2 　　　　Fuc
B	Galα1-3Galβ- 　　　｜α1-2 　　　Fuc
Linear B	Galα1-3Gal-
H	Galβ- ｜α1-2 Fuc
T	Galβ1-3GalNAcα-Ser/Thr
Tn	GalNAcα-Ser/Thr
Sialyl-Tn	SAα2-6GalNAcα-Ser/Thr
S^d (Cad)	GalNAcβ1-4 Galβ- 　　　　｜α2-3 　　　　SA
i	Gal1-4GlcNAcβ1-3Galβ-
I	Galβ1-4GlcNAc 　　　｜β1-6 Galβ1-4GlcNAcβ1-3Galβ-
Le^a	Fuc ｜α1-4 Galβ1-3GlcNAcβ1-3Gal-
Le^b	Fuc ｜α1-4 Galβ1-3GlcNAcβ1-3Gal- ｜α1-2 Fuc

(continued)

TABLE I (*continuation*)

Antigen	Structure
Le^x	Galβ1-4GlcNAcβ1-3Gal- | α1-3 Fuc
Le^y	Galβ1-4GlcNAcβ1-3Gal- | α1-2 | α1-3 Fuc Fuc
Sialyl-Le^x	SAα2-3Galβ1-4GlcNAcβ1-3Gal- | α1-3 Fuc

Core 1:	Gal(β1–3)GalNAcα-Ser/Thr-R
Core 2:	GlcNAc(β1–6)[Gal(β1–3)]GalNAcα-Ser/Thr-R
Core 3:	GlcNAc(β1–3)GalNAcα-Ser/Thr-R
Core 4:	GlcNAc(β1–6)[GlcNAc(β1–3)]GalNAcα-Ser/Thr-R
Core 5:	GalNAc(α1–3)GalNAcα-Ser/Thr-R
Core 6:	GlcNAc(β1–6)GalNAcα-Ser/Thr-R
Core 7:	GalNAc(α1–6)GalNAcα-Ser/Thr-R

Core 1 and core 2 are the most common core structures in mucins as well as in other secreted and cell surface glycoproteins. Core 1, or T-antigen, is not usually exposed in glycoproteins but is monosialylated (SA(α2–3)Gal(β1–3)GalNAcα- and SA(α2–6)[Gal(β1–3)]GalNAcα-) or disialylated (SA(α2–6)[SA(α2–3)Gal(β1–3)]GalNAcα-). O-glycans with core 3 and 4 have only been found on mucins. Core 5 occurs in glycoproteins from several species, and was found in human adenocarcinoma [8] and meconium [9,10].

To date, oligosaccharides with core 6 have only been reported on human glycoproteins, including meconium, and ovarian cyst mucins [11]. It cannot be excluded that core 6 structures arise due to degradation by human β-galactosidase acting on Gal(β1–3) and thus converting core 2 to 6. A different structure has previously been designated as core 6 (Gal(β1–6)[Gal(β1–3)] GalNAcα-) [12] and [13]. However, the latter unusual structure has only been reported in human gastric mucin and was not based on modern methods of structural analysis [14]. Core 7 has recently been described in bovine submaxillary mucin [15].

With the exception of core 7 which to date has only been found as the disaccharide, all core structures may occur unsubstituted, or elongated, and may carry terminal antigens.

2.2. Elongated O-glycan structures

Elongation of O-glycans involves the addition of repeating units of GlcNAc and Gal residues in β-linkages to form Gal(β1–4)GlcNAc(β1–3)Gal- structures (i antigen, type 2 chain, poly-*N*-acetyllactosamines), or type 1 chains (Table I) or branched structures Gal(β1–4)GlcNAc(β1–6) [Gal(β1–4)GlcNAc(β1–3)] Gal-β- (I antigen, branched poly-*N*-acetyllactosamines, Table I). Linear, unbranched GlcNAc(β1–6)Galβ- structures have also been found.

Elongated structures form the backbone of O-glycans and may be substituted with various terminal carbohydrates.

2.3. Terminal structures

Terminal structures typically found on O-glycans of mammalian glycoproteins include SA(α2–3 and –6), Fuc(α1–2, –3 and –4), GalNAc(α1–3 and –6), GalNAc(β1–4), GlcNAc(α1–4), Gal(α1–3) and sulfated residues. A number of unusual sugars and linkages are found in species such as fish, reptiles, birds and insects. Sialic acids, sulfate and Fuc(α1–3 or –4) residues may be attached to terminal as well as to internal sugar residues. Many of these structures are of the Lewis and ABO blood group system (Table I) and are recognized by specific anti-blood group antibodies. As new antigenic determinants are added to the growing oligosaccharide chain, internal antigens may be masked, for example, the blood group A determinant will mask the H determinant, and the I antigen will mask the i antigen.

3. Functions of O-glycans

Considering the vast number of O-glycan chains, relatively few functions of specific structures are known. Carbohydrate heterogeneity may be of advantage to an organism as it allows the binding of a variety of molecules and microorganisms rather than selecting for a few. The role of O-glycans may be studied after removal of carbohydrate chains, inhibition of biosynthesis, use of carbohydrate-binding proteins such as lectins or antibodies, use of mutant cell lines lacking enzymes involved in O-glycan processing or nucleotide–sugar metabolism, introduction into cells of genes coding for glycosyl-T, or expression of mammalian glycoproteins in systems differing in glycosylation potential. A general inhibitor for O-glycan biosynthesis corresponding to tunicamycin which blocks N-glycan biosynthesis, is not yet known. UDP-GalNAc analogues may inhibit polypeptide α-GalNAc-T (Section 6) and other GalNAc-T, and substrate analogues have been used that may inhibit specific glycosyltransferases acting on O-glycans. Kuan et al. [16] have employed the substrate for core 1 β3-Gal-T (Section 7), Gal-NAcα-benzyl, -phenyl or -nitrophenyl to inhibit mucin oligosaccharide biosynthesis in the mucin producing colon cancer cell line LS174. GalNAc-benzyl was found to enter cells and to be metabolized to larger O-glycan structures while synthesis of oligosaccharide chains of mucins was deficient. The mechanism of inhibition is probably the competition with the natural substrates. This method is expected to inhibit the formation of core 1 to 4 which are the cores most commonly found in mucins. However,

GalNAcα-benzyl is not a substrate for all elongation reactions. For example, Gal-NAcα-benzyl is not a substrate for α6-SA-T I (Section 12.1), thus the formation of the sialyl-Tn epitope is not expected to be inhibited, and it is not known if GalNAc-benzyl would inhibit core 5, 6 and 7 synthesis.

The role of O-glycans has recently been reviewed by Varki [17]. O-glycans contribute to the physical, chemical, antigenic and conformational properties of a glycoprotein. For example, the high glycosylation of mucins results in a drastic change of protein conformation and chain extension [18]. The carbohydrate may affect stability of proteins and reduce the susceptibility to proteolytic degradation. The assembly of protein subunits may be supported by the presence of O-glycans, as shown for GM-CSF [19]. The intracellular transport and cell surface expression of certain receptors may be effected by O-glycans (see Section 3.5). The carbohydrate of antifreeze glycoprotein is responsible for its ability to lower the freezing point of arctic fish serum [20]. The processing of prehormones may be regulated by O-glycans [21].

In addition, O-glycans have also been shown to play a critical role in fertilization, cell adhesion, functions in the immune system and receptor functions as discussed in the following sections. Bacteria, viruses and other microbes may bind to O-glycans.

3.1. Biological roles of mucins

Internal epithelial surfaces are usually covered with a layer of mucus containing polymers of glycoproteins of high molecular weight called mucins. Due to the high number of O-glycan chains mucins in solution readily form a highly viscous gel that functions to protect the mucus epithelium from chemical, physical and microbial impact.

Mucin gene expression is organ- and species-specific and is regulated in disease [22]. Mucin peptides contain O-glycosylated repetitive regions (tandem repeats) which are similar in sequence but differ among mucin molecules, and are rich in Ser, Thr and Pro, extensively O-glycosylated and resistant to proteolysis. There are also less glycosylated regions which are Cys-rich and involved in the folding of molecules by disulfide bonding [23]. These regions may also have N-glycosylation sites. All mucins appear to show similar physical properties independent of the type of O-glycan chains present. However, the finding that mucin O-glycan structures are organ- and species-specific suggests characteristic biological roles. These may reside in the property of O-glycans to bind microbes and lectin-like molecules, antibodies or cells expressing lectins.

Adhesion to carbohydrate epitopes attached to mucins has been demonstrated for a number of microorganisms. It is possible that binding of the two pathogenic bacterial strains *Pseudonomas aeruginosa* [24] and *Pseudonomas cepacia* [25] to mucin plays an important role as an early step in the lung pathology of cystic fibrosis [26].

Mucin oligosaccharides may contain tumour-associated or developmentally regulated antigens and may be biologically active in soluble body fluids (milk, blood) as well as on cell surfaces. In many diseases, there may be quantitative changes or the appearance of new, inappropriately expressed or oncofoetal antigens. In colon cancer, for example, a general loss of mucin O-glycans and the exposure of the T-antigen (O-glycan core 1, Section 2.1) has been observed. There may also be exposure of mucin peptide regions. In addition, there appear to be fewer and shorter O-glycans [27], and a loss of sulfation with an increase in sialylation [28].

Metastasis may be related to the presence and type of O-glycans in mucins. Mucin-rich variants of the human colonic cancer cell line HT29 showed increased metastatic potential in nude mice, and organ selectivity of metastasis depended on O-glycan structures [29]. In cancer, mucin carbohydrates may mask normal antigens or contain antigens that are recognized by selectins. Interactions with cells via selectins may be important in the metastatic process, either by promoting selectin-mediated invasion, or by blocking cell–cell interactions through circulating mucins. New approaches for cancer therapy are the neutralization of mucins or mucin fragments by antibodies directed against peptide or carbohydrate. In addition, immunization with mucins containing cancer-specific epitopes may trigger an enhanced immune response against cancer cells [30,31].

3.2. Cell adhesion and mammalian lectins

Cell adhesion molecules recognizing carbohydrate ligands may play an important role in the inflammatory process, trafficking of lymphocytes and other cell–cell interactions. E-, L- and P-selectins recognize SA-Le^x (Table I) and related antigens [32].

Although Le^x and SA-Le^x structures are also found on glycolipids and N-glycans, inhibition experiments revealed that the ligand for E-selectin binding is not located on N-glycans but on O-glycans since binding could be inhibited by GalNAcα-benzyl which inhibits O-glycan extension, but not by N-glycosylation inhibitors [33]. Human epithelial cancer cell lines express ligands for E-selectins (ELAM-1) and showed ELAM-1 dependent adhesion to vascular endothelial cells [33,34], inhibitable by anti-ELAM-1 antibodies. However, several leukaemia cell lines expressing the same ligands did not show binding, suggesting that the presence of the ligand alone is not sufficient for binding, but that underlying structures carrying the ligands or other cell surface antigens may be involved. Human epithelial cancer cells and human leukemic cell lines have been studied for their ELAM-1 mediated adherence to vascular epithelium. All of the cancer cells but only 3 of 12 leukemia lines bound to epithelium concomitant with the expression of Le antigens on their cell surfaces [34]. This binding may be important in the invasion of cancer cells in the process of metastasis.

Only a small proportion of membrane O-glycans appear to contain the ligand for P-selectin since P-selectin binding of HL60 cells is destroyed by alkaline borohydride treatment as well as O-sialoglycoprotease from *P. hemolytica* without significantly affecting total cell surface sialyl-Le^x expression [35]. This suggests that O-glycoproteins may provide effective ligands with sialyl-Le^x structure that may be found on a 'clustered saccharide patch'. The main candidates for the presentation of these ligands are therefore cell surface glycoproteins such as leukosialin [36], glycophorin, CD34 [37], MUC1 [38] and GLYCAM 1 [4].

The ligand on high endothelial venules is a mucin-type glycoprotein which is preferentially expressed in lymph nodes and recognized by L-selectins [39,40].

Sulfated ligands have been reported to bind to rat L-selectin with higher affinity than sialylated structures [41] suggesting a functional role for sulfated mucin-type oligosaccharides.

3.3. Haematopoietic system

Various cell types of the haematopoietic and immune systems carry characteristic carbohydrate structures implying specific functions of carbohydrates. For example, there are differences in the quantitative distribution of O-glycans between erythroid, myeloid and T-lymphoid cell lines, between lineages and upon differentiation [42]. The cell surfaces of erythrocytes and leukocytes are abundant in O-glycans covalently attached to cell surface glycoproteins such as glycophorin and leukosialin. At least the first sugar of O-glycans, GalNAc, is necessary for the expression of recombinant human red cell glycophorin on the cell surfaces of the ldlD mutant of CHO cells (see Section 3.6) [43].

It has been suggested that the O-glycans of leukosialin are involved in leukocyte functions by binding to antigen presenting B-cells. In addition, antibodies against leukosialin induce changes in the proliferation of monocyte-dependent T-cells and natural killer cell activity [36]. T-lymphocyte activation is accompanied by a conversion of sialylated core 1 to sialylated core 2 O-glycans of leukosialin (see Section 9.1) [44].

Mucins carrying $SA(\alpha2–6)GalNAc-$ residues such as ovine submaxillary mucin have been shown to have an immunomodulatory role since they partially inhibit natural killer cell toxicity against erythroleukaemia K562 cells in the presence of ammonium [45].

Desialylated ovine submaxillary mucin as an immunogen (containing GalNAc-) provided protection against highly invasive tumour cells in mice [30]. Injection into cancer patients of partially desialylated ovine submaxillary mucin induces the formation of antibodies directed against sialyl-Tn [31]. This may boost the immune response against tumour cells containing the epitope.

Monoclonal antibodies directed against Tn-antigen blocked infectivity of the human immunodeficiency virus and fusion of infected and uninfected cells. Binding of monoclonal antibodies to virus was inhibitable by Tn antigen, suggesting that the virus contains this epitope [46]. The sialyl-Tn antigen (Table I) of the HIV GP 120 appears to be a neutralization target for antibodies and could represent an appropriate antigen for anti-HIV immunization [47]. Since the epitope is found on many cancer cells, immunization with Tn or sialyl-Tn linked to human serum albumin may also protect against cancer [48].

3.4. Fertilization

The binding of sperm to the cell surface zona pellucida glycoproteins of the mouse egg appears to be mediated by O-glycans. Alkali treatment releases O-glycans and destroys receptor activity. B. Shur and his group [49,50] demonstrated that a cell surface Gal-T is implicated in the binding of mouse sperm to terminal GlcNAc residues present on the egg surface during fertilization. Upon addition of UDP-Gal, the Gal-T can galactosylate zona pellucida glycoprotein ZP3 and thereby destroy its own receptor. Anti-Gal-T antibodies inhibited sperm-egg binding. The amounts and structure of this β4-Gal-T are regulated during spermatogenesis [51].

In contrast, Wassarman and his group showed that Galα- residues on O-glycans of ZP3 were essential for sperm binding in the mouse since α-galactosidase and Gal-oxidase as well as release of O-glycans by alkaline borohydride treatment destroyed the sperm receptor [52,53]. The pig analogue of mouse ZP3 was analyzed and found to contain O-glycans clustered in specific domains of the heavily glycosylated glycoprotein [54].

3.5. Receptor functions

Many cell surface glycoproteins contain O-glycans and may depend on them for their stability, folding, cell surface expression, intracellular transport, secretion or functions. For example, the stable expressions of human interleukin-2 receptor and the major antigen envelope protein of Epstein–Barr virus, the human granulocyte colony-stimulating factor, and the cell surface expression of human decay accelerating factor depend on O-glycosylation [55].

The mutant CHO cell line (ldlD) which has a reversible defect in 4-epimerase and is therefore deficient in UDP-Gal and UDP-GalNAc synthesis proved to be very useful to study the role of glycans in the functions and cell surface expressions of glycoproteins. By the exogenous addition of Gal or GalNAc, the deficient UDP-Gal and UDP-GalNAc pools may be respectively replenished and O-glycans may be fully (Gal and GalNAc addition) or partially (GalNAc addition) restored [56]. Using this cell line, the stability, cell surface expression and functions of the low density lipoprotein receptor [56, 57] were shown to depend on the presence of O-glycans although it is not clear which O-glycans were responsible. The O-glycans of apolipoprotein ApoA-II modify its association with lipid and high density lipoproteins [58], and thus the distribution of the apolipoprotein within HDL subfractions. O-glycans therefore appear to play important roles in lipoprotein metabolism. Decay-accelerating factor, a glycosylphospatidyl-inositol anchored cell surface glycoprotein involved in the inactivation of complement components, was also studied in ldlD cells. The O-glycans on the factor were apparently not required for activity, but for protection of this glycoprotein from proteolysis and its cell surface expression [59]. Coyne et al. [60] demonstrated by site-directed mutagenesis of the decay-accelerating factor that O-glycans may be involved in the spatial presentation of functional domains.

O-glycan deficiency on recombinant human interleukin-2 receptor resulted in missorting of the receptor in the Golgi or during transport from the Golgi to the cell surface [61].

4. Intracellular localization and transport of glycosyltransferases and their substrates

The intracellular organization of glycosyltransferases, their cofactors and substrates is important for proper O-glycan processing and is further discussed elsewhere. Enzymes appear to be arranged in the Golgi sequentially as expected for an assembly line in which substrate is vectorially transported from cis to trans Golgi compartments. In an alternative model, glycosyl-T may exist in close proximity in a complex where they act sequentially on a substrate with high efficiency.

210

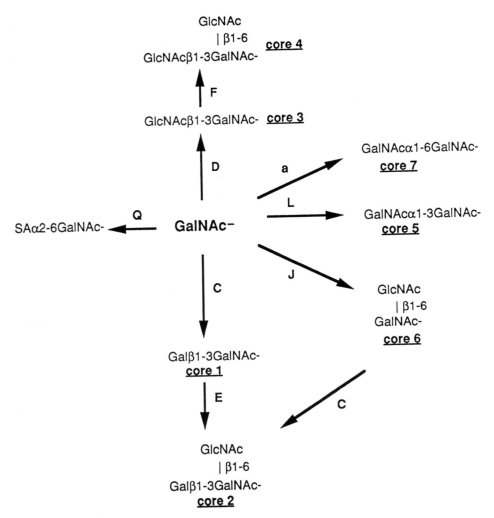

Fig. 1. Composite pathways in O-glycan biosynthesis. (A) Synthesis of O-glycan core structures. a, core 7 α6-GalNAc-T; reaction C, core 1 β3-Gal-T; reaction D, core 3 β3-GlcNAc-T; reaction E, core 2 β6-GlcNAc-T; reaction F, core 4 β6-GlcNAc-T; reaction L, core 5 α3-GalNAc-T; reaction J, core 6 β6-GlcNAc-T; reaction Q, α6-SA-T I. Paths C,D, E, F and Q are well established. Path J may be specific for human tissues. Paths a, J, and L require further investigation.

DNA sequences reveal that all glycosyltransferases have a type II membrane protein structure with a short amino-terminal end directed towards the cytoplasm, an uncleaved membrane signal anchor which is embedded in the Golgi membrane, a stem region with little defined secondary structure but rich in Ser/Thr and Pro residues suggesting that these are possible sites for O-glycosylation, and a carboxy-terminal globular peptide region containing the catalytic site. Thus glycosyltransferase reactions take place inside

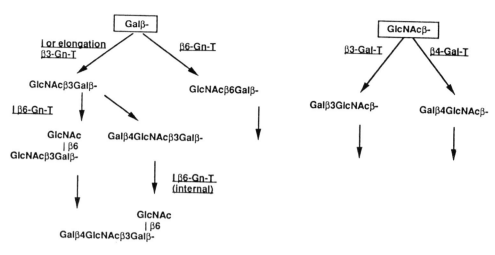

Fig. 1. (B) Elongation reactions. Gal(β1–3) or Gal(β1–4) residues of O-glycans are elongated or branched by various GlcNAc-transferases; GlcNAc(β1–3) or GlcNAc(β1–6) residues are elongated by Gal-transferases. Their reactions form the backbone of O-glycans.

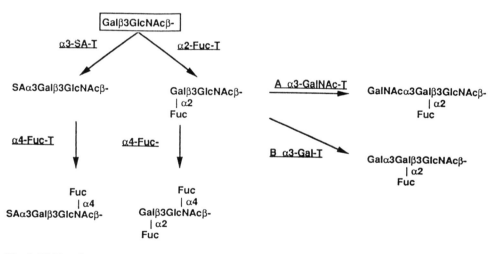

Fig. 1. (C) Termination reactions of type 1 chains. Type 1 chains may be sialylated by α3-SA-T (possibly the one also acting on N-glycans) or fucosylated by α2-Fuc-T or α4-Fuc-T. This may be followed by the blood group A or B-dependent transferases.

the lumen of the Golgi. Proteolytic cleavage at the stem region destroys Golgi targetting signals and produces soluble glycosyl-T.

The targetting signals retaining glycosyltransferases in these Golgi compartments still need to be defined but are probably due to the sequence, hydrophobicity, three-di-

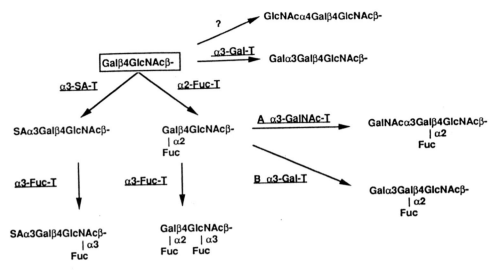

Fig. 1. (D) Termination reactions of type 2 chains. Type 2 chains may be sialylated by α3-SA-T (possibly the one also acting on N-glycans) or fucosylated by α2-Fuc-T or α3-Fuc-T. This may be followed by the blood group A or B-dependent transferases. Type 2 chains may also obtain other α-linked terminal sugars.

mensional arrangement of the peptide and spacing between regions adjacent to the membrane anchor.

From an inspection of the composite schemes (Fig. 1) it appears that many glyco-syltransferases compete with each other for a common substrate. The diversity of O-glycan chains indicates that this competition actually takes place. However, competition is only possible if the intracellular localization of glycosyltransferases allows access to the substrate. Only a few enzymes of the O-glycan pathways have been localized to specific compartments. Most of the evidence suggests that O-glycan intitiation occurs in the *cis*-Golgi and enzymes involved in subsequent processing are located in later Golgi compartments. However, it appears that there is no fixed localization for any of the glycosyltransferases and there may be variations and diffuse distributions throughout the endomembrane system, depending on cell type and growth and differentiation status. A proposed scheme for the intracellular organization of O-glycan synthesis is shown in Fig. 2.

4.1. Initiation of O-glycosylation

Studies on isolated peptidyl-tRNA containing GalNAc suggested that GalNAc incorporation can be a relatively early event although most of the GalNAc transfer occurs post-translationally [62,63].

The ionophore monensin interferes with movement of glycoproteins within the Golgi apparatus. Monensin completely blocked GalNAc incorporation into murine hepatitis virus A59 glycoproteins and herpes simplex virus type 1 glycoprotein [64].

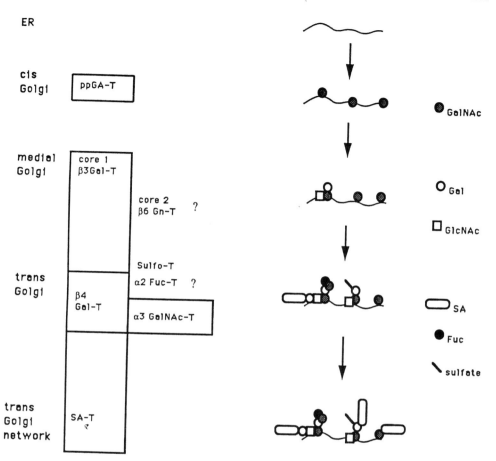

Fig. 2. Proposed intracellular organization of O-glycan synthesis within the Golgi apparatus. O-glycan synthesis is initiated by GalNAc transfer to peptide, mainly in the *cis* Golgi. Chains grow by the addition of GlcNAc or Gal residues to form cores 1 to 4 in the medial Golgi; some sialylation may also occur. In the *trans* Golgi further sugars are added to continue elongation, and to form sulfated structures, blood group determinants and terminal sugars. Sialylation continues throughout the trans Golgi network. It is not known if sulfate is added before or after sialic acid. There is an overlap of transferases, depending upon the cell type and growth conditions. Although there appears to be a gradient with early acting enzymes in the *cis* Golgi to late acting enzymes in the *trans* Golgi and *trans*-Golgi network, it is possible that synthesis is achieved by glycosyltransferase complexes which differ in composition, depending on their position in the Golgi.

The conclusion was that GalNAc incorporation into polypeptide is a post ER event. Other studies, however, showed that herpes simplex virus type 1 glycoprotein C appeared to be O-glycosylated before its entry into the Golgi, and the subsequent addition of sugars occurred much later [65]. Thus the effect of monensin may depend on the cell type and the given conditions.

Additional evidence for a pre-Golgi GalNAc transfer comes from studies on the E1 glycoprotein of mouse hepatitis virus-A59 which buds from transitional compartments between the ER and the *cis* Golgi. It was shown that GalNAc is added to E1 even at low temperature, when the assembled virus cannot enter the Golgi [66].

Carraway's group [67] has studied the biosynthesis of the major sialomucin from rat mammary ascites tumor cells by pulse-chase labeling experiments. Sialomucin containing O-glycan core 1 was bound and precipitated with *Peanut* agglutinin in less than 5 minutes after labeling whereas the appearance of glucosamine-containing glycoprotein at the cell surface had a half time of >4 hours. The kinetics of amino acid and sugar labeling suggested that O-linked carbohydrate was added either co-translationally or several minutes after translation and further O-glycan initiation occured from the rough endoplasmic reticulum throughout the endomembrane system to the appearance of the glycoprotein at the plasma membrane. Thus only a fraction of the O-glycosylation sites are glycosylated in early compartments, but the glycosylation is completed after transfer to a second and later compartment, and possibly throughout the entire pathway [68].

Jokinen et al. [69] suggested that in the synthesis of core 1 structures of glycophorin A in K562 cells, GalNAc was incorporated very early while Gal was added much later. Using the same cell line, Piller et al. [70] showed that initiation of O-glycosylation followed N-glycosylation in early Golgi compartments but galactosylation and sialylation were later events.

Electronmicroscopy studies by Roth [71], using Gold-labeled *Helix Pomatia* lectin, localized terminal GalNAcα- residues of intestinal goblet cells of rat and chicken in the *cis*-Golgi compartment. The *trans*-Golgi compartment was also labeled, probably due to the presence of GalNAcα- in the blood group A determinant.

Similarly, Deschuyteneer et al. [72] determined the subcellular localization of porcine submaxillary gland apomucin by electron microscopy using anti-mucin antibody, and GalNAcα-residues using *Helix pomatia* lectin. Apomucin was detected early in the ER but GalNAc primarily in the *cis*-Golgi apparatus.

By subcellular fractionation of rat liver, Abeijon and Hirschberg [73] found that the specific activity of UDP-GalNAc:polypeptide α-GalNAc-transferase was enriched in the Golgi fraction, but was not detected in membranes from the rough and smooth endoplasmic reticulum. UDP-GalNAc was transported at a 4 to 6-fold higher rate into the lumen of Golgi vesicles than into vesicles from smooth and rough endoplasmic reticulum. Thus endoplasmic reticulum of rat liver may contain a small amount of UDP-GalNAc, but not the enzyme necessary for initial glycosylation.

Hanover et al. [74] and others showed that O-glycosylation is initiated later in the biosynthetic pathway. O-glycosylation of human chorionic gonadotrophin (hcG) in a human choriocarcinoma cell line was shown to occur shortly before secretion, presumably in the Golgi.

The human T-cell leukaemia line Jurkat expresses leukosialin with a high number of GalNAc-Ser/Thr residues reactive with *Salvia sclarea* lectin [75]. Monensin treatment of Jurkat cells did not inhibit GalNAc transfer to leukosialin apoprotein whereas lowering the temperature to 15°C inhibited all O-glycan initiation. This suggests that initiation of O-glycan synthesis starts in the cis Golgi and not in the ER or in transitional vesicles between ER and Golgi.

The distribution of *Helix pomatia* binding sites has been found to vary between chondrocytes of different ages from the *cis*-Golgi location in primary cultures to ER location upon further culturing [76]. This indicates that the localization of initial glycosylation may be cell- and differentiation-dependent.

4.2. Elongation of O-glycans

Core 1 β3-Gal-T synthesizing core 1 (Fig. 1A, Section 7) may be localized mainly to the medial and *trans*-Golgi compartments. Elhammer and Kornfeld [77] subfractionated mouse lymphoma BW 5147 cell homogenates and found that core 1 β3-Gal-T migrated in the same fractions as β4-Gal-T. Andersson and Ericsson [78] found that the endoplasmic reticulum also contained a significant amount of activity which differed slightly in its properties. Lectin staining with Gal(β1–3)GalNAc binding *Peanut* agglutinin localized core 1 oligosaccharides in the medial and *trans*-Golgi [76]. These results suggest that core 1 β3-Gal-T is localized in later compartments than polypeptide α-GalNAc-T but acts before or in the same compartments as transferases adding terminal sugars.

GlcNAc-T involved in O-glycan synthesis may be assumed to be in intermediate medial Golgi compartments in analogy to the N-glycan pathway [79]. Spielman et al. [80] studied the formation of core 2 oligosaccharides attached to rat mammary tumour cell surface sialo-mucin. Monensin reduced the size of O-glycan chains attached to the mucin but did not inhibit initial GalNAc transfer and did not prevent core 2 formation and sialylation but galactosylation. This indicates that some of the core 2 β6-GlcNAc-T and sialylation reactions occur in early compartments not affected by monensin.

The β4-Gal-T has not only been found in the *trans*-Golgi but also in soluble body fluids and on cell surfaces. Immunolabeling studies showed that β4-Gal-T is located in the *trans*-Golgi [71,81].

It is possible that O-glycan SA-Ts are localized in the *trans*-Golgi and the *trans*-Golgi network which is distributed throughout the cytoplasm. α6-SA-T of the N-glycan pathway as well as SA residues have been localized in the *trans*-Golgi and *trans*-Golgi network by electronmicroscopy using antibodies against α6-SA-T and lectins [71,82]. However, two α3-SA-T involved in glycolipid synthesis were separated by density centrifugation into two compartments, probably *cis* and *trans*-Golgi, where the later acting SA-T coincided with β4-Gal-T activity, presumably in the *trans*-Golgi compartment [83].

As expected for a terminal glycosylation event, blood group A α3-GalNAc-T (Section 12.5) and its product, blood group A determinant, are found in the *trans*-Golgi compartment. However, intestinal goblet and absorptive cells differed in that absorptive cells showed a much more diffuse distribution of the blood group A determinant throughout the central and *trans*-Golgi stack [82,84].

Another terminal event is O-glycan sulfation. Radiolabeling studies showed that sulfation of rat mammary adenocarcinoma cell surface mucin probably occurs in the same compartment as β4-galactosylation, but before sialylation [85]. This mucin was shown to contain sulfate attached to the GlcNAc residue of neutral and sialylated core 2 structures. Brefeldin A and monensin, which interfere with Golgi organization, inhibit the appearance of mature sulfated mucin chains which have been processed beyond the GalNAc-Ser/Thr residues [86], indicating that sulfation occurs in compartments later than *cis* Golgi.

In summary, there is evidence for the parallel occurrence of different glycosyltransferases as well as a gradient towards the *trans*-Golgi network of early to late acting enzymes. Thus a continuous model of O-glycan processing may be proposed as an assembly line of glycosyltransferase complexes that differ slightly in composition to complete the synthesis characteristic of a specific subcompartment.

Mucins often have a high proportion of incomplete chains lacking some of the elongating and terminal sugars. This may be due to incomplete biosynthesis, for example when the system is overloaded with a high amount of mucin peptide synthesis. It is also possible that this is a normal mechanism and has functional significance.

Glycoproteins may be recycled after they reach the cell surface. About one fifth of the Golgi proteins were shown to originate from the cell surface through recycling in glycosylation deficient ldlD mutant CHO cells [87]. During recycling a small number of carbohydrate chains may be added as has been shown for SA in human erythroleukaemia K562 cells and mouse T-cell lymphoma EL-4 cells [88] and rat hepatocytes [89]. This recycling may be viewed as a regeneration system which continues to finish incomplete chains or restores carbohydrate masking or recognition signals that have been modified by external glycosidases.

Several glycosyl-T have been localized to the cell surface [50]. These enzymes are not expected to function as transferases *in vivo* due to the lack of nucleotide sugars. It has been suggested that these enzymes are lectins involved in interactions with the cellular environment.

Factors affecting the transport and metabolism of sugar nucleotides may also affect glycosylation reactions. The regulation of nucleotide-sugar transport into the Golgi by specific antiport-transport systems plays a major role in controlling sugar-donor substrate availability and therefore glycosyltransferase activities. Sugar nucleotides are synthesized in the cytosol (UDP-GlcNAc, UDP-GalNAc, UDP-Gal, GDP-Fuc) or in the nucleus (CMP-SA) and then are transported into the Golgi by specific antiport systems which remove the free nucleotides [90]. Free nucleotides are products of glycosyltransferase reactions and may act as inhibitors for glycosylation reactions.

5. *Factors controlling O-glycan biosynthesis*

O-glycans are assembled by the controlled sequential addition of individual sugars by glycosyltransferases (Table II). The species-, tissue- and growth-specific synthesis of hundreds of different O-glycans is regulated by the complex interplay of many factors most of which are not well understood.

It appears that there is a number of possible pathways to construct complex O-glycan chains guided by the substrate specificity and intracellular arrangement of glycosyltransferases.

Proposed biosynthetic pathways for common O-glycans based on published O-glycan structures and reported glycosyltransferase activities are shown in Fig. 1.

TABLE II

Enzymes involved in O-glycan biosynthesis

Enzyme name		Reaction catalyzed	EC number
polypeptide α-GalNAc-T	UDP-GalNAc: polypeptide α-GalNAc-transferase	A,B	EC 2.4.1.41
core 5 α3-GalNAc-T	UDP-GalNAc: GalNAc-R α1,3-GalNAc-transferase	L	
Cad or S^d β4-GalNAc-T	UDP-GalNAc: SAα2-3Galβ-R (GalNAc to Gal) β1,4-GalNAc-transferase	Z	
A α3-GalNAc-T	UDP-GalNAc: Fucα1-2Galβ-R (GalNAc to Gal) α1,3-GalNAc-transferase	X	EC 2.4.1.40
B α3-Gal-T	UDP-Gal: Fucα1-2Galβ-R (Gal to Gal) α1,3-Gal-transferase	Y	EC 2.4.1.37
β4-Gal-T	UDP-Gal: GlcNAc-R β1,4-Gal-transferase	O	EC 2.4.1.38
β3-Gal-T	UDP-Gal: GlcNAc-R β1,3-Gal-transferase	P	
core 1 β3-Gal-T	UDP-Gal: GalNAc-R β1,3-Gal-transferase	C	EC 2.4.1.1.
i β3-GlcNAc-T	UDP-GlcNAc: Galβ1-4-R β1,3-GlcNAc-transferase	N	EC 2.4.1.149
I β6-GlcNAc-T	UDP-GlcNAc: GlcNAcβ1-3Galβ1-4-R (GlcNAc to Gal) β1,6-GlcNAc-transferase	G,H,I	EC 2.4.1.150
core 2 β6-GlcNAc-T	UDP-GlcNAc: Galβ1-3GalNAc-R (GlcNAc to GalNAc) β1,6-GlcNAc-transferase	E	EC 2.4.1.102
core 3 β3-GlcNAc-T	UDP-GlcNAc: GalNAc-R β1,3-GlcNAc-transferase	D	EC 2.4.1.147
core 4 β6-GlcNAc-T	UDP-GlcNAc: GlcNAcβ1-3GalNAc-R (GlcNAc to GalNAc) β1,6-GlcNAc-transferase	F	EC 2.4.1.148
elongation β3-GlcNAc-T	UDP-GlcNAc: Galβ1-3(R$_1$)GalNAc-R$_2$ (GlcNAc to Gal) β1,3-GlcNAc-transferase	M	EC 2.4.1.146
O-glycan α3-SA-T	CMP-SA: Galβ1-3GalNAc-R α2,3-SA-transferase	S	EC 2.4.99.4
α6-SA-T I	CMP-SA: R$_1$-GalNAc-R$_2$ (SA to GalNAc) α2,6-SA-transferase	Q	EC 2.4.99.3
α6-SA-T II	CMP-SA: SAα2-3Galβ1-3GalNAc-R (SA to GalNAc) α2,6-SA-transferase	R	EC 2.4.99.7
α2-Fuc-T	GDP-Fuc: Galβ-R α1,2-Fuc-transferase	T	EC 2.4.1.69
α3-Fuc-T	GDP-Fuc: Galβ1-4GlcNAcβ-R (Fuc to GlcNAc) α1,3-Fuc-transferase	U,V	EC 2.4.1.152
α3/4-Fuc-T	GDP-Fuc: Galβ1-4/3GlcNAcβ-R (Fuc to GlcNAc) α1,3/4-Fuc-transferase	U,V,W	EC 2.4.1.65

The substrate specificities of glycosyl-T control the direction of pathways and limit the possible number of O-glycan structures. The microenvironment within the membrane and the lumen of the Golgi may control enzyme activities and specificities in a tissue-specific fashion.

The amounts of final oligosaccharide products are determined by the relative activities of competing glycosyltransferases and their intracellular distribution, as well as the transport rate of substrates through the Golgi. For example, if core 1 β3-Gal-T and core 3 β3-GlcNAc-T were in the same compartment competing for GalNAc-substrates, the relative activities towards various GalNAc- sites would determine if oligosaccharides are based on core 1 or core 3. This could have an influence on the three-dimensional presentation and possibly the biological activities of sugars attached to the core.

The addition of α-linked sugars often terminates chains. Sialylation of GalNAc-R by α6-SA-T I (Section 12.1) to form SA(α2-6)GalNAc terminates chains since none of the enzymes that have been studied can act on this structure. At least five enzymes compete for GalNAc-R (Fig. 1A) *in vitro*. SA-Ts are assumed to be present in the *trans*-Golgi compartment while the core synthesizing enzymes are in earlier compartments, thus the intracellular localization would favour O-glycan core synthesis independently of high α6-SA-T activity.

5.1. Genetic control of glycosyltransferases

There are many examples for the structural and enzymatic differences between species, tissues, cell types and growth conditions, suggesting transcriptional regulation [91]. Even the expression of transferase activities that appear to be essential since they occur in almost every mammalian cell (for example core 1 β3-Gal-T and core 2 β6-GlcNAc-T) is regulated during cellular differentiation and growth. The expression of acceptor substrates is similarly regulated. There are distinct differences in mucin gene expression between species and tissues, and in disease [12].

Very little is known about the genetic regulation of enzymes involved in O-glycan processing since cDNA for only four glycosyltransferases specific for O-glycan pathways have been cloned to date, i.e. polypeptide α-GalNAc-T, core 2 β6-GlcNAc-T and α3- and α6-SA-T..

The control of the gene expression of glycosyltransferases acting on N- and O-glycans is discussed in detail in other chapters of this book.

5.2. Regulation of glycosyltransferase activities

Glycosyltransferases catalyze the transfer of monosaccharides from nucleotide sugar donor substrates to Ser or Thr of the peptide or to another sugar acceptor substrate.

Enzyme activities may be regulated by substrate concentrations, metal ion concentrations, membrane composition, posttranslational modifications or cofactors. Most of the UDP-dependent transferases require Mn^{2+} or another divalent cation with the exception of β6-GlcNAc-T that can act in the presence of EDTA, although it is possible that these enzymes contain tightly bound metal ions. SA-T and Fuc-T do not require but may be stimulated by divalent cations. Certain metal ions may have a strong inhibitory effect. Since glycosyltransferases are membrane-bound, they are usually stimulated by detergents or membrane lipid components *in vitro*. Association with other transferases may also stimulate activities [92].

Processing enzymes are often glycoproteins themselves and enzyme glycosylation may influence stability, transport and activity, and could be a possible factor in feedback regulation or compensatory mechanisms. Other types of regulation by post-translational mechanisms include phosphorylation, for example for β4-Gal-T (Section 11.3). There are tissue-specific factors affecting glycosyltransferase specificities as exemplified by the interaction of α-lactalbumin and β4-Gal-T which changes the specificity resulting in the synthesis of lactose instead of poly-*N*-acetyllactosamine chains.

5.3. Glycosyltransferase substrate specificity

The specificities of glycosyltransferases direct the biosynthesis of O-glycans in predetermined pathways. Specificities of glycosyl-T often vary between tissue or cell types suggesting that there may be tissue-specific factors affecting enzyme recognition of their substrates; alternatively, different enzyme species occur in various tissues.

Glycosyl-T usually can make only one type of linkage between sugars. An exception is the α3/4 Fuc-T which can add Fuc in either α3 or α4-linkage to GlcNAc, and the core 2 β6-GlcNAc-T M which synthesizes core 2, core 4 and I antigen. A particular linkage, however, may be made by several different enzymes, for example, the GlcNAc(β1–3)Gal linkage can be made by at least 2 enzymes. In addition, isoenzymes may exist. *In vitro*, many substrates can be acted upon by a particular enzyme with various efficiencies as long as the minimum substrate structure essential for binding and catalysis is present.

The specificities of glycosyltransferases, i.e. the specific recognition of structural features of the substrate, result in STOP and GO signals for subsequent reactions. For example, SA α-2–6-linked to GalNAc of O-glycans is a STOP signal for all other reactions. Conversely, the insertion of a critical residue may be required to create an essential GO signal for other enzymes. For example, the addition of SA(α2–3) to Gal (Section 12.3) of core 1 structures is a GO signal for the addition of SA(α2–6) by α6-SA-T II. At the same time, the branching reaction by core 2 β6-GlcNAc-T to form the sialylated core 2 structure, is inhibited. The sialylated core 2 structure can only be formed by branching prior to sialylation. STOP and GO signals may be viewed as a control mechanism to ensure the synthesis of a great variety of chains with defined structures. Generally, in the O-glycan paths a three-linked residue has to be introduced before a 6-linked branch can be added. Thus core 3 is synthesized before core 4 and this results in very few core 6 chains (Fig. 1A).

The peptide portion of substrates controls peptide glycosylation as well as O-glycan processing. Glycosylation may be due to selective accessibility due to protein folding, or due to direct recognition of protein structure. Site-directed processing to create characteristic structures at each glycosylation site may be due to differential accessibility for various processing enzymes; alternatively, it may be due to a direct recognition of the peptide in combination with carbohydrate binding. In addition, carbohydrate chains may adopt site-specific carbohydrate conformations due to interactions with the peptide, and processing enzymes may select preferred conformations of their substrates [93].

6. O-glycosylation: polypeptide α-GalNAc-transferase

The action of UDP-GalNAc:polypeptide α-N-acetylgalactosaminyltransferase (polypeptide α-GalNAc-T, EC 2.4.1.41) is common to the synthesis of all O-glycan chains. The enzyme is probably expressed in all mammalian cells. Polypeptide α-GalNAc-T catalyzes the transfer of GalNAc from UDP-GalNAc in α-linkage to Ser or Thr of the peptide backbone in the presence of Mn^{2+}:

Reaction A, polypeptide α-GalNAc-T:

Ser-R $\longrightarrow$ **GalNAcα-Ser-R**

Reaction B, polypeptide α-GalNAc-T:

Thr-R $\longrightarrow$ **GalNAcα-Thr-R**

It is not yet known if reactions A and B are catalyzed by the same enzyme. The enzyme purified from porcine submaxillary glands [94] and bovine colostrum appears to act essentially only on Thr [95] (Reaction B). An efficient transfer to Ser, reaction A, has not yet been demonstrated *in vitro*, suggesting the existence of at least two different polypeptide α-GalNAc-T. Alternatively, a factor affecting specificity may exist or the transfer to Ser may require protein acceptors that are different in structure, size or conformation than those studied *in vitro*.

Evidence for two different polypeptide α-GalNAc-T also comes from studies by Matsuura et al. [96] who investigated the synthesis of an oncofoetal epitope in human fibronectin recognized by a monoclonal antibody. GalNAc attached to the Thr residue of the V-T-H-P-G-Y sequence creates the epitope probably by inducing the required three-dimensional conformation since carbohydrate or peptide alone did not form the epitope. Polypeptide α-GalNAc-T from human hepatoma HUH-7 cells, human foetal fibroblasts WI-38 and human epidermoid carcinoma A431 cells, placenta and hepatoma tissue transferred GalNAc to peptides and fibronectin containing the above peptide sequence, but not to oncofoetal fibronectin (containing GalNAc linked to Thr in the V-T-H-P-G-Y sequence). The enzyme in normal tissue from liver, lung, spleen and kidney did not show activity this sequence, but incorporated GalNAc into de-O-glycosylated normal fibronectin with newly exposed Ser or Thr residues, indicating that the enzyme from normal tissue has a different substrate specificity.

Based on amino terminal sequence of purified bovine colostrum polypeptide α-Gal-NAc-T lacking the transmembrane region, cDNA coding for the enzyme was isolated from a bovine intestinal [97] and bovine placental [98] cDNA library. Homa et al. [97] found an open reading frame encoding a protein of 559 amino acids (predicted molecular weight of 64,000 Da). The protein has the typical type II membrane protein domains but no significant homology to other glycosyltransferases. Two of three Asn-X-Ser/Thr sequences present are potential N-glycosylation sites (N-R-S-P is not expected to be N-glycosylated). Active enzyme immunoprecipitated from expressing Sf9 insect cells had a molecular weight of 67,000 Da, indicating posttranslational processing. Additional evidence for the existence of more than one enzyme comes from the presence of at least two mRNA species [97] of ~4.1 and 3.2 kb in bovine tissues and of 4.8 and 3.9 kb in human tissues, as well as of ~1.5 kb in skeletal muscle. Hagen et al. [98] reported an open reading frame encoding 519 amino acids for the placental enzyme. The recombinant enzymes transfer GalNAc with high efficiency to Thr but act poorly on Ser.

The bovine colostrum enzyme is a glycoprotein containing two N-glycans, mostly of the complex type [99] and is different from the porcine submaxillary enzyme since the latter could not be purified on bovine submaxillary apomucin linked to Sepharose, but bound to 5Hg-UDP-GalNAc-thiopropyl-Sepharose [94].

Polypeptide α-GalNAc-T has been purified from ascites hepatoma AH 66 cells [100] with a molecular weight of 56,000 Da by affinity chromatography on bovine submaxillary apomucin coupled to Sepharose. A soluble form of molecular weight 70,000 Da has been purified from bovine colostrum and a membrane bound form of about 71,000 Da from mouse lymphoma BW5147 cells [99], from lactating bovine mammary gland (200,000 Da molecular weight) [101] and from bovine submaxillary gland (1320 fold) [94]. Divalent cation is essential for activity with Mn^{2+} being the most effective cation. The reason for these different sizes of enzyme protein may be proteolytic cleavage and posttranslational processing.

The peptide structure of the substrate plays an important role in the activity of the enzyme. The ability of various peptides and apoglycoproteins to accept GalNAc has been studied by many laboratories with both crude and purified enzyme preparations [102, 103,104]. Cruz and Moscarello [105] demonstrated that two sites (Thr[95] and Thr[98]) in the human myelin basic protein sequence (T[95]-P-R-T[98]-P-P-P-S-) that are not glycosylated in vivo were glycosylated by pig submaxillary gland α-GalNAc-T. The acetylated partial sequence Ac-T-P-P-P has the minimum size for polypeptide α-Gal-NAc-activity and is an excellent substrate for the enzyme, whereas many other short peptides including (T-C-P-P-P)$_2$ from the hinge region of IgA are not [95,104,106]. Baby hamster kidney cells appear to have only the activity catalyzing reaction B since substitution of Thr with Ser in T-P-P-P resulted in inactive substrate [104].

It is clear that polypeptide α-GalNAc-T requires a specific recognition signal in the acceptor peptide. However, attempts to define this specific sequence have been largely unsuccessful [107]. Since Pro is usually found near O-glycosylation sites, Pro may serve to expose Ser/Thr residues or confer a three-dimensional structure to the peptide that provides this recognition signal. By examination of peptide sequences around glycosylation sites, O'Connell et al. [108] found that Pro, Ala, Ser or Thr were often present at 3 amino acids towards the N-terminal (-3 position) and 1, 3 and 6 amino acids towards the C-terminal end (+1, +3, +6 positions), whereas charged amino acids at these positions were usually not found. Statistically, Ser and Thr are equally glycosylated in glycoproteins and almost all Ser and Thr in mucins are glycosylated. Using sequences of 12 amino acids of a Thr-*O*-glycosylation site of human von Willebrand factor, O'Connell et al. [95] determined that polypeptide α-GalNAc-T activity from bovine colostrum was dependent on the composition of the peptide as well as the positions of amino acids. Pro appeared to be important at the +3 but not at other positions. An erythropoietin-derived sequence containing Ser showed no activity although analogous peptides containing Thr were active, indicating that the bovine colostrum enzyme has essentially only Thr α-GalNAc-T activity.

Our studies on the bovine colostrum enzyme (unpublished) showed that the presence of GalNAc and Gal(β1–3) GalNAc residues on glycopeptide substrates was inhibitory to the activity suggesting that subsequent multiple glycosylation is more difficult than initial glycosylation. Clearly, the three-dimensional structure of a substrate is an important factor for polypeptide α-GalNAc-T activity.

Elhammer et al. [109] systematically investigated O-glycosylation sites according to a cumulative enzyme specificity model. The results infer that polypeptide α-Gal-NAc-T binds to 8 amino acids in the substrate, 3 residues preceding and 4 following the Ser or Thr to be glycosylated. Using synthetic peptide substrates the bovine colostrum enzyme was 35 times more effective with Thr than with Ser-containing acceptors. These statistical studies have been carried out with non-mucin glycoproteins. The results may, therefore, not be applicable to the polypeptide α-GalNAc-T in goblet cells which is involved in mucin synthesis and may have a different specificity.

Other O-glycosylation reactions also appear to require the presence of Pro residues in the substrate such as O-glycosylation of Ser residues of yeast glycoproteins [110] and the addition of O-GlcNAc to proteins [111] by the cytosolic peptide GlcNAc-T. O-glycosylation by polypeptide GalNAc-transferase from porcine or bovine submaxillary glands and hen oviduct is not mediated by a dolichol or lipid intermediate [1]. This is in contrast to the O-glycosylation of yeast mannoproteins where mannose is transferred from dolichol-phospho-mannose, possibly by two different mannosyl-T that synthesize the mannose-O-linkage [112].

7. Synthesis of O-glycan core 1: core 1 β3-Gal-transferase

UDP-Gal:GalNAc-R β-1,3-Gal-transferase (EC 2.4.1.122), core 1 β3-Gal-T, synthesizes O-glycan core 1, the cancer-associated T-antigen [113], by adding a β-1–3-linked Gal to GalNAc-R (Fig. 1A) where R may be one of many membrane-bound or secreted glycoproteins or mucins, or synthetic hydrophobic groups:

Reaction C, core 1 β3-Gal-T:

GalNAcα-R $\longrightarrow$ **Gal(β1–3)GalNAcα-R**

Core 1 β3-Gal-T is abundant in mammalian and nonmammalian cells and tissues but varies in activities among cell types [114]. The enzyme requires divalent cation such as Mn^{2+} or Cd^{2+}, is inhibited by Zn^{2+} and is not affected by α-lactalbumin [115]. The enzyme is clearly different from the β3-Gal-T in the albumen gland of the snail [116] which adds Gal to GalNAc(β1–4)GlcNAc-R but cannot act on GalNAcα-R, and from other β3-Gal-T (Section 11.4) acting on GlcNAc termini.

Core 1 β3-Gal-T has been partially purified from detergent extracts of swine trachea [117]. The enzyme bound to an affinity column of asialo Cowper's gland mucin (containing GalNAc- and Gal(β1–3)GalNAc-) bound to Sepharose. Furukawa and

Roth [118] reported the partial purification from chick-embryo liver of β4-Gal-T (Section 11.3) and core 1 β3-Gal-T. The two enzymes behaved similarly on several affinity columns but could be separated by affinity chromatography using GlcNAc-Sepharose and asialo-agalacto-α1-acid glyoprotein-Sepharose as ligand for the β4-Gal-T and asialo-sheep submaxillary mucin-Sepharose as ligand for the core 1 β3-Gal-T. We have partially purified the enzyme from Triton extracts of rat liver homogenates about 17,000 fold using 5-Hg-UDP-GlcNAc-thiopropyl-Sepharose as an affinity column (unpublished). The enzyme is extremely stable, even in crude fractions, but has not been purified to homogeneity and its gene has not yet been cloned.

The specificity of rat liver core 1 β3-Gal-T has been investigated extensively [119]. The enzyme recognizes the 2-N-acetyl group and the 3- and 4- but not the 6-hydroxyl of GalNAc. Since the 6-hydroxyl of GalNAc is not required for activity, it is not surprising that core 1 β3-Gal-T from rat liver [119] and from human uterus and normal and cancerous ovarian tissue [120] can act on core 6, GlcNAc(β1–6)GalNAcα-R, to synthesize core 2. This may be a physiological path in human tissues containing core 6 oligosaccharides, although due to lack of core 6 β6-GlcNAc-T (Section 9.2) in most tissues core 1 must be the more common intermediate in the synthesis of core 2, i.e. the 3-linked residue is added first.

Although 6-deoxy-GalNAc-R and core 6 are effective substrates, core 1 β3-Gal-T does not act on SA(α2–6)GalNAcα-R, the product of SA-T I, indicating that SA is blocking the transfer of Gal, probably by introducing an unfavourable charge. The addition of SA by SA-T I, however, is possible after formation of core 1.

Processing of O-glycans to form core 1 is directed by the peptide structures of substrates. The rat liver core 1 β3-Gal-T acts readily on various synthetic and glyco-protein derived GalNAc-peptides containing either four or more amino acids or protective groups to eliminate charges near the glycosylation site. The activity is strongly influenced by the composition, length and sequence of the peptide backbone of the substrate and the attachment position and number of sugar residues present. A systematic analysis of glycopeptide substrates which are variants of the human intestinal MUC2 mucin repeat sequences [121] showed that Pro in the +1 position relative to the GalNAc-Thr substrate site is unfavourable. Negatively charged amino acid residues at the −1 position result in excellent substrates with low K_M values, suggesting that the enzyme binding is favoured by negative charges in that position (unpublished). Probably six amino acids as well as the GalNAc residue can be accommodated in the binding site [122]. Since the conformation of GalNAcα-Thr unit appears to be very similar in various glycopeptides [123] it is the direct recognition of the peptide moiety that influences enzyme activity.

Core 1 β3-Gal-T, but not polypeptide α-GalNAc-T, is deficient in erythrocytes (Tn-erythrocytes) from patients with permanent mixed-field polyagglutinability. Gly-cophorin A from these erythrocytes contains O-glycans with reduced amounts of SA and Gal [124,125] whereas only 4% of patient lymphocytes appeared to be affected [126]. The increased binding of anti-Tn antibodies to these Tn-erythrocytes demonstrates that the O-glycan pathways are terminated mainly at the GalNAc stage. Although the defect was found to be stable for >1 year in clones of affected lymphocytes [126], the condition is reversible with inducers of gene expression such as Na-butyrate or 5-azacytidine which increase enzyme activity as well as the cell surface appearance of core 1, the T-antigen. This demonstrates that the lack of core 1 β3-Gal-T expression in Tn-erythrocytes is not due to a disrupted gene but due to gene repression specific for Tn-erythrocytes [127].

Another cell type that shows premature termination of O-glycan synthesis is the human T-lymphoblastoid cell line Jurkat. Jurkat cells were found to lack core 1 structures and core 1 β3-Gal-T [75] and therefore represent a model for functional studies of core 1 oligosaccharides. The mechanisms causing core 1 β-Gal-T deficiency in Jurkat cells appears to be different from the mechanism described above for Tn-erythrocytes since sodium-butyrate and 5-azacytidine could not cause the appearance of core 1 on the surface of Jurkat cells [127].

Our studies on glycosyltransferase activities in colon cancer and normal colonic tissue [128] showed that the increase of unsubstituted core 1, the T-antigen, in colon cancer is not due to the increase in core 1 β3-Gal-T. Core 3 synthesis (Section 8) competes with core 1 synthesis *in vitro*, the decreased core 3 β3-GlcNAc-T activity in colon cancer could thus be responsible for the increased T-antigen. O-glycan α3-SA-T, an enzyme that masks the T-antigen, is increased while α6-SA-T I (Section 12.1) is decreased in colon cancer. The mechanism for T and Tn antigen appearance may differ in various types of cancer. Zhuang et al. [129] found an inverse correlation between the ratio of *Helix pomatia* lectin to *Peanut* lectin binding and core 1 β3-Gal-T activity in human breast cancer, and suggested that the increased appearance of the Tn antigen is caused by low activities of the core 1 β3-Gal-T. An alternative mechanism for the appearance of the T-antigen in cancer may be the redistribution of polypeptide α-Gal-NAc-T throughout later compartments of the O-glycan pathways [68].

Although core 1 β3-Gal-T is ubiquitous, it appears to be regulated during differentiation. Human colonic adenocarcinoma Caco-2 cells show decreased activity of the enzyme after spontaneous differentiation to enterocytes [130].

8. Synthesis of O-glycan core 3: core 3 β3-GlcNAc-transferase

Core 3 is synthesized by UDP-GlcNAc: GalNAcα-R β-1,3-GlcNAc-T (EC 2.4.1.147), core 3 β3-GlcNAc-T, which catalyzes the transfer of GlcNAc in β-1,3 linkage from UDP-GlcNAc to GalNAcα-R in the presence of Mn^{2+}:

Reaction D, core 3 β3-GlcNAc-T:

GalNAc-α-R   ⟶   **GlcNAc(β1–3)GalNAc-α-R**

R may be phenyl, benzyl or peptide [122,131]. In most tissues, synthesis of core 3 precedes core 4 synthesis (Fig. 1A), although a low activity of core 3 β3-GlcNAc-T to form the core 4 structure has been reported using core 6, GlcNAc(β1-6)GalNAcα-benzyl [131, 132] as the substrate. Core 3 β3-GlcNAc-T and high levels of core 4 β6-GlcNAc-T activity have only been found in mucin secreting tissues. In rat colonic mucosa the β6-GlcNAc-transferase synthesizing core 4 (Section 9.1.2) proceeds at a much faster rate than the β3-GlcNAc-transferase making core 3. Thus the relatively low activity and limited tissue distribution of core 3 β3-GlcNAc-T are the limiting factors in the synthesis of core 4. Core 3 β3-GlcNAc-T activity is present in colonic tissues and, based on the occurrence of core 3 structures, must also be present in foetal and adult human colon [9,10,133], human bronchial tissue [134] and human and bovine salivary glands [15,135] and sheep stomach [136]. Unusual core 3 structures are found for example in the jelly coat of the mexican axelotl [137].

Human colonic mucins have been reported [133] to contain oligosaccharides exclusively with core 3 structures. This finding is not in agreement with the abundance of glycosyltransferases synthesizing cores 1, 2 and 4 in human colon [128,131]. If, however, core 1 β3-Gal-T and core 2/4 β6-GlcNAc-T were present in later compartments than core 3 β3-GlcNAc-T, a high proportion of core 3 structures could be expected.

Specificity studies on rat colonic core 3 β3-GlcNAc-T (unpublished) indicate that the enzyme has an absolute requirement for all of the hydroxyls and the 2-N-acetyl group of the GalNAcα-R substrate. It is therefore not surprising that the enzyme cannot act well on compounds in which the 6-hydroxyl of GalNAc is substituted, i.e. it acts poorly on core 6 oligosaccharides and cannot act after the addition of α-2,6-linked SA to GalNAc [131]. The synthesis of the sialylated core 3 structure, SA(α2–6)[GlcNAc(β1–3)] GalNAc-R, therefore must occur via the action of α6-SA-T on GlcNAc(β1–3)GalNAc-R, although this remains to be demonstrated *in vitro*. It is also unknown whether galactosylation of core 3 by β4-Gal-T (Section 11.3) occurs before or after sialylation by α6-SA-T (Section 12.1).

Core 3 β3-GlcNAc-T activity is reduced in colon cancer tissue and it is possible that only selected cell types are affected. We noted that the activity is not detectable in cancer cell lines and appears to be selectively turned off since other β3-GlcNAc-T are still active [128]. However, certain cancer cells may retain a small amount of the activity since colon cancer CL.16E cells synthesize core 4 [138].

The tissue distribution and specificity of the core 3 β3-GlcNAc-transferase are different from those of the elongation β3-GlcNAc-T (Section 11.1) which elongates cores 1 and 2, and from those of the β3-GlcNAc-T synthesizing the i antigenic determinant (Section 11.2), indicating that these are different enzymes.

9. Synthesis of GlcNAc(β1–6)Galβ- and GlcNAc(β1–6)GalNAc linkages by β-1,6-GlcNAc-transferases

Studies on the substrate specificity of β-1,6-GlcNAc-T in various tissues and recent gene cloning [139,140] suggest that there is a family of β-1,6-GlcNAc-T that arose by gene duplication and subsequent divergence. At least one of these enzymes is regulated during cellular differentiation.

To date, four different types of β-1,6-GlcNAc-T activities involved in the synthesis of GlcNAc(β1–6)Gal(NAc) linkages have been identified that are distinguished by their acceptor substrate specificities:

1. Core 2 β6-GlcNAc-T L synthesizing only core 2 (reaction E).
2. Core 2 β6-GlcNAc-T M synthesizing core 2, core 4 and the I antigen (reactions E,F,G,H).

3. I β6-GlcNAc-T (internal) synthesizing the internal I antigen (reaction I).

4. β6-GlcNAc-T synthesizing linear GlcNAc(β1-6)Gal(NAc) structures (reactions J, K).

Various cell types may contain different ratios of these enzyme activities. Core 2 β6-GlcNAc-T L apparently occurs in many non-mucin secreting tissues, e.g. leukocytes [141]. Core 2 β6-GlcNAc-T M is associated with most mucin secreting tissues [131,141], whereas the I β6-GlcNAc-T (internal) activity occurs in selected tissues such as rat intestines, and in human serum. Activities synthesizing linear GlcNAc(β1–6)Gal and GlcNAc(β1–6)GalNAc structures have been described in Novikoff ascites tumour cells, ovarian tissue and may be a characteristic of human cells [11].

All of the known β-1,6-GlcNAc-T, including GlcNAc-T V acting on N-glycans, can act in the presence of EDTA and do not require the addition of Mn^{2+}. This may be due to metal ions tightly bound to the enzyme. Only core 2 β6-GlcNAc-T M has been purified [142], and the cDNA for core 2 β6-GlcNAc-T L [139] and one of the I synthesizing activities [140] has been cloned.

9.1. Synthesis of O-glycan core 2, 4 and the I antigen

9.1.1. Core 2 β6-GlcNAc-transferases L and M

UDP-GlcNAc:Gal(β1–3)GalNAc-R (GlcNAc to GalNAc) β-1,6-N-acetylglucosaminyltransferase (EC 2.4.1.102), core 2 β6-GlcNAc-T, synthesizes core 2 (Fig. 1A), GlcNAc (β1–6) [Gal(β1–3)] GalNAc-R from Gal(β1–3)GalNAc-R in many tissues:

Reaction E, core 2 β6-GlcNAc-T:

$$\text{GlcNAc(β1–6)}$$
$$|$$

Gal(β1–3)GalNAcα-R $\longrightarrow$ Gal(β1–3)GalNAcα-R

The enzyme acts on substrates where R is hydrogen (free reducing oligosaccharide), synthetic hydrophobic groups or peptide [141,143,144].

The enzyme has been characterized from dog submaxillary gland [143,145], rabbit small intestine [146] and bovine trachea [147]. The conditions optimal for activity vary between tissues. The enzyme does not require divalent cation but, depending on the tissue is stimulated or inhibited by various divalent cations [141,145–147]. The enzyme was partially purified from pig trachea by affinity chromatography on Gal(β1–3)Gal-NAc-containing pig Cowper's gland mucin and was stabilized by the presence of Cowper's gland mucin [148]. Ropp et al. [142] purified the activity to apparent homogeneity from Golgi fractions of bovine tracheal epithelium. A 134,000 fold purification was achieved by affinity chromatography on UDP-hexanolamine immobilized on monoaldehyde activated agarose at high (20 μmol/ml) ligand density. It has not been established if other UDP-binding proteins were bound to this column under the given conditions. However, the activity is assumed to be pure based on the presence of one band on electrophoresis of 69,000 Da and the high specific activity of 70 μmol/min/mg. The activity was stabilized in the presence of Galβ1–3GalNAcα-benzyl substrate or by freezing at $-70°C$. The K_M value for this substrate of 0.36 mM is similar to the value obtained for microsomes or tissue homogenates [141]. Kinetic analysis showed, similar to other glycosyltransferases, that the nucleotide sugar is bound first, and the free nucleotide is released last, in an ordered sequential mechanism. The enzyme has been shown to synthesize core 2, core 4 and the I antigen (reactions E,F,G) and is therefore the core 2 β6-GlcNAc-T M which forms an exception to the one enzyme–one linkage rule.

Although core 2 β6-GlcNAc-T activity is present in most tissues there are distinct specificity differences between the enzymes from various species and cell types [141]. In mucin secreting tissues such as rat or human colon or pig stomach the activity appears to catalyze branching reactions E, F, G and H [131,141,144]. The enzyme has an absolute requirement for the 4- and 6-hydroxyl of GalNAc and the 6-hydroxyl of the Gal residue of Gal(β1–3) GalNAcα-benzyl substrates, but accepts substrates lacking other hydroxyls or the N-acetyl group of the sugar rings. This explains that the M-enzyme may accommodate three different substrates, Gal(β1–3) GalNAcα-, GlcNAc(β1–3) GalNAcα-, and GlcNAc(β1–3) Galβ-, to form core 2, core 4 and I antigen, respectively.

The enzyme from leukocytes which has the characteristics of core 2 β6-GlcNAc-T L catalyzes only the conversion of core 1 to core 2 and, as expected, has a more restricted substrate specificity. The enzyme from acute myeloid leukaemia (AML) cells has an absolute requirement for the 2-N-acetyl group and the 4- and 6-hydroxyls of GalNAc as well as the 4- and 6-hydroxyls of the Gal residues of Gal(β1–3) GalNAcα-benzyl substrates. The substrate recognition of the M and L enzyme show slight variations between cell types and species, suggesting that there are either tissue-specific enzyme proteins or tissue- and species-specific factors exist affecting enzyme specificity.

In most tissues, there is no activity towards GalNAc-Ser/Thr-mucin indicating that the core 1 β3-Gal-T must act before core 2 β6-GlcNAc-T to form core 2. The presence of a SA residue in SA(α2–3)Gal(β1–3)GalNAc- or SA(α2–6) [Gal(β1–3)] GalNAc-inhibits the action of the core 2 β6-GlcNAc-T. A GlcNAc(β1–3) residue linked to Gal of Gal(β1–3)GalNAc- substrates makes the conversion to core 2 improbable but favours the synthesis of the I antigen so that chains become committed to core 1 structures (Fig. 3). Fuc added in α-1,2-linkage to Gal of O-glycan core 1 represents a GO signal for the formation of blood group A and B but blocks the conversion to O-glycan core 2 [141,143,144].

Cloning of cDNA coding for core 2 β6-GlcNAc-T L has been achieved by Bierhuizen and Fukuda [139]. A cDNA library from HL60 promyelocytic cells was transfected into CHO cells constructed to stably express leukosialin and polyoma virus large T antigen. This made a sensitive detection of enzyme product possible by specific antibody directed against a sialylated core 2 hexasaccharide and a leukosialin epitope. Recombinant enzyme was secreted from a transient expression system as a soluble fusion protein with a signal peptide sequence and the IgG binding domain of Staphylococcus aureus protein A. The enzyme therefore binds to immobilized IgG, and may be purified by this method. Northern analysis of mRNA from HL60 cells showed one major band at ~2.1 Kb and minor bands at ~3.3 and 5.4 Kb. No bands were detected in K 562 erythroleukaemia cells and the recipient CHO cells.

228

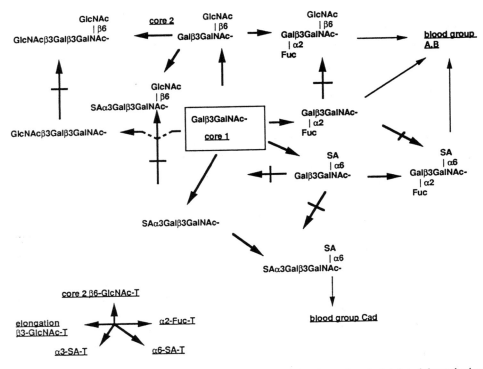

Fig. 3. Processing pathways of O-glycan core 1. Core 1 may be fucosylated, sialylated, branched or elongated. Sialylation of GalNAc prevents core 2 formation and sialylation or elongation of Gal but does not prevent fucosylation. Fucosylation prevents core 2 formation and sialylation but is a GO signal for blood group A and B attachment. Sialylation of Gal is a GO signal for sialylation of GalNAc by α6-SA-T II but a STOP signal for core 2 β6-GlcNAc-T. The Cad determinant may be added to the disialo-core 1 structure. The elongated core 2 structure can only be formed by elongation of core 2 but not by core 2 β6-GlcNAc-T acting on the elongated core 1.

The gene for the enzyme was localized to chromosome 9 q21 [140], near the locus for the blood group A and B transferases. The secreted enzyme as well as the enzyme from CHO cells transfected with cDNA coding for and stably expressing non-secretable protein were incapable of synthesizing core 4 and I branches indicating that it is the L-type enzyme. The cDNA encodes a protein of 428 amino acids (predicted molecular weight ~50,000) with the typical type II membrane protein domain structure but no apparent homology to other transferases and with two potential N-linked glycosylation sites (Asn-X-Ser/Thr-X, with X not being Pro).

Core 2 β6-GlcNAc-T L, and possibly other β6-GlcNAc-T, appear to be regulated during growth and differentiation and may be used to assess the growth and differentiation status of a cell. Activation of lymphocytes with anti CD3 antibody and interleukin-2 [44] or differentiation of mouse F9 teratocarcinoma cells by retinoic acid and other agents [149] cause an increase in core 2 β6-GlcNAc-T activity. Caco-2 human adenocarcinoma cells that spontaneously differentiate in culture show an increase of the core 2 β6-GlcNAc-T L upon differentiation [130]. Treatment of CHO cells by butyrate but not other differentiating agents caused the core 2 β6-GlcNAc-T activity to increase 16-fold in a timed progression with cAMP levels rising before activity increases [150]. In butyrate and cholera toxin treated cells, a 9-fold increase in K_M for UDP-GlcNAc but 80-fold increase in V_{max} was observed. Cholera toxin alone caused no stimulation of the activity indicating that elevated cAMP levels alone were not sufficient. The induction was prevented by inhibitors of protein synthesis and protein kinase inhibitors. This may suggest that the enzyme is subject to complex regulation involving transcriptional and translational controls.

The enzyme activity is elevated in lymphocytes and platelets from children with the Wiscott–Aldrich immunodeficiency syndrome [151, 152, 153] and appears to be responsible for the high expression of core 2 oligosaccharides in leukosialin on lymphocytes from these patients [153]. High activities are found in various leukaemias such as chronic myelogenous leukaemia (CML) and AML [154]. AML and CML leukocytes are in early stages of differentiation, whereas normal granulocytes are more mature cells. The core 2 β6-GlcNAc-transferase was found to be four-fold increased over normal values in granulocytes from CML patients and eighteen-fold in leukocytes from AML patients [154].

Compared with mature normal lymphocytes, T-lymphocytic leukaemia cells from patients with acute lymphocytic leukaemia (ALL) and chronic lymphocytic leukaemia (CLL) express increased amounts of core 2 structures on leukosialin, correlating with increased core 2 β6-GlcNAc-T activity. Normal T-lymphocytes and K562 cells are very low in the core 2 β6-GlcNAc-T and have negligible amounts of core 2 structures [155].

These studies suggest that cellular maturation is associated with the loss of core 2 expression, and the increase of core 2 β6-GlcNAc-T in leukaemia may be due to the relative immaturity of cells.

Heffernan et al. [149] suggested that β6-GlcNAc-T controls the level of poly-N-acetyllactosamines found in glycoproteins since the core 2 GlcNAc may provide preferred attachment sites for polylactosamine chains. However, the levels of this branching activity do not always correlate with the poly-N-acetyllactosamine content of a cell which is therefore controlled by other factors as well. Metastatic sublines from nonmetastatic mammary adenocarcinoma SP1 cells showed an increase in core 2 β6-GlcNAc-T correlating with an increase in O-linked poly-N-acetyllactosamine content. A similar correlation was not found in rat 2 fibroblasts upon T24H-ras oncogene transfection which caused an increase in core 2 β6-GlcNAc-T but a decrease in O-linked poly-N-acetyllactosamine content and decreased synthesis of poly-N-acetyllactosamine chains by i β3-GlcNAc-T [156].

It has been shown that human urinary chorionic gonadotrophin (hCG) from normal pregnant women and from women with hydatiform moles have almost entirely sialylated core 1 O-glycans whereas hCG from choriocarcinoma patient urine was very rich in sialylated core 2 O-glycans [157,158]. Patients with invasive moles had intermediate levels of these core 2 O-glycans. The suggested explanation is an increase in the ratio of core 2 β6-GlcNAc-T to O-glycan α3-SA-T (Section 12.3) but there are many other possible explanations. Breast cancer, in contrast, exhibits less core 2 than normal [159]. Mucins isolated from normal human milk contain neutral and monosialylated core 2 structures. Mucin derived from human breast cancer cell line BT-20 contained neutral and sialylated core 1 as the major oligosaccharides [159] which is consistent with previous reports of increased T-antigen in breast cancer [113]. Although the enzyme has not been measured, this suggests that core 2 β6-GlcNAc-T may be reduced in breast cancer. To rule out an effect of cell culturing, the results should be confirmed with cancer tissue.

Our studies on core 2 β6-GlcNAc-T in colon cancer showed that the activity is slightly reduced in cancer tissue. Interestingly, about half of the cultured colon cancer cells tested show a change from the M-type to the L-type core 2 β6-GlcNAc-T compared to normal colon which has the M-type. It is possible that cancer tissue switches from M enzyme to L enzyme expression. Alternatively, colonic tissue may contain both the L and M enzyme, and cancer cells may selectively lose the M enzyme. The tissue-specific expression of the two enzymes in various ratios is corroborated by the tissue-dependent varying ratios of core 2 to core 4 activities [131]. A third possibility is that the L and M activity are due to the same enzyme protein which is modified by a factor affecting specificity, and this factor is changed in cancer.

A β6-GlcNAc-T from mouse kidney has been found to act on glycolipid containing Gal(β1-3)GalNAcβ-structure to form GlcNAc(β1–6) [Gal(β1–3)] GalNAc(β1–3) Gal(α1–4) lactosyl-ceramide [160]. The expression of this globoside correlated with enzyme activity in inbred strains of mice. Interestingly, the mouse kidney β6-GlcNAc-T was much more active in male than in female mice. An inbred mouse strain with a defect in a single autosomal gene lacked the expression of this activity which apparently controls the expression of the GL-Y (Gal(β1–4) [Fuc(α1–3)] GlcNAc-Gb5-ceramide) antigen. Core 2 β6-GlcNAc-T from canine submaxillary glands also acts on Gal(β1–3)GalNAcβ- [143]; therefore the mouse kidney activity may be the same or similar than core 2 β6-GlcNAc-T L or M and mice strains deficient in this enzyme may be a good model to study the biological role of core 2.

Another β6-GlcNAc-T activity which, based on specificity, could be the M type core 2 β6-GlcNAc-T and is active in the presence of EDTA was described in marsupial mammary tissue [161] and appears to be involved in the synthesis of oligosaccharides found in milk. The activity, like the pig gastric enzyme [144], acts on Gal(β1–3)Galβ1-R to form the Gal(β1–3) [GlcNAc(β1–6)] Galβ1-R branch but cannot act on Gal(β1–3)GlcNAc(β1–3)Gal(β1–4)Glc. During lactation periods in the tammar wallaby, the activity reaches a peak around 150 days postpartum and is down-regulated after this time.

9.1.2. Core 4 β6-GlcNAc-transferase activity

The activity synthesizing core 4 from core 3 (Fig. 1A), UDP-GlcNAc: GlcNAc(β1–3) GalNAc- (GlcNAc to GalNAc) β-1,6-GlcNAc-T (EC 2.4.1.148), core 4 β6-GlcNAc-T, occurs in parallel with core 2 β6-GlcNAc-T activity in ratios characteristic of the tissue type:

Reaction F, core 4 β6-GlcNAc-T:

$$\text{GlcNAc(β1–6)}$$
$$|$$
$$\text{GlcNAc(β1–3)GalNAcα-R} \longrightarrow \text{GlcNAc(β1–3)GalNAcα-R}$$

The core 4 structure is more restricted in its occurrence than core 2, due to the limiting and tissue-specific activity of core 3 β3-GlcNAc-T. Core 4 occurs on bovine and human submaxillary mucins, sheep gastric mucin [136], human respiratory tract mucin [15,134,162,163] and, based on the presence of the transferases assembling cores 3 and 4, should be present in colonic mucins from many species. Core 4 β6-GlcNAc-T always accompanies core 2 β6-GlcNAc-T activity, e.g. in the purified enzyme from bovine trachea [142] and the two activities are therefore believed to be due to the same enzyme [131,142,43].

9.1.3. Blood group I β6-GlcNAc-transferase

The I antigen is a developmentally regulated antigen that appears on adult human erythrocytes where it replaces the foetal i antigen [2]. The structure also occurs on many mucins, e.g. human respiratory mucin [164]. The branching reaction to form the branch of the I antigen (Fig. 1B) is catalyzed by UDP-GlcNAc:GlcNAc(β1–3)Galβ-R (GlcNAc to Gal) β6-GlcNAc-T, I β6-GlcNAc-T:

Reaction G, I β6-GlcNAc-T:

$$\textbf{GlcNAc(}\boldsymbol{\beta}\textbf{1–6)}$$
$$|$$

GlcNAc(β1–3)Galβ-R $\longrightarrow$ GlcNAc(β1–3) Galβ-R

Reaction G has been demonstrated in pig gastric mucosa by Piller et al. [165] using GlcNAc(β1–3)Gal(β1–4)Glcβ-methyl as acceptor for β6-GlcNAc-transferase activity and in Novikoff ascites tumour cells [166]. The enzyme is also active with GlcNAc(β1–3)Galβ-methyl as the substrate and apparently can act on N-glycans, O-glycans and glycolipids and may be the same as core 2 β6-GlcNAc-T M.

Pig gastric mucosal extracts have also been shown to synthesize a GlcNAc(β1–6) branch to form GlcNAc(β1–6) [GlcNAc(β1–3)] Gal(β1–3) [+/–GlcNAc(β1–6)] GalNAcα-benzyl [144]:

Reaction H, I β6-GlcNAc-T:

+/–GlcNAc(β1–6)   **GlcNAc(β1–6)**   +/–GlcNAc(β1–6)
| | |
GlcNAc(β1–3)Gal(β1–3)GalNAcα-R $\longrightarrow$ GlcNAc(β1–3) Gal(β1–3) GalNAcα-R

No GlcNAc(β1–3)Galβ1–3 [GlcNAc(β1–6)] GalNAcα-benzyl product was detected in these experiments suggesting that elongation of core 1 (see below) by the β3-GlcNAc-T prevents the formation of core 2 structures.

cDNA cloning and expression of a I β6-GlcNAc-T was achieved by Bierhuizen et al. [140]. The cDNA expression library from PA-1 human teratocarcinoma cells was transfected into CHO cells expressing the Polyoma large T antigen and leukosialin but not the I antigen. The expression of I antigen on the surfaces of transfected cells was detected with anti I antibodies. ^{14}C-Gal labeled cell surface carbohydrate was shown to contain GlcNAc(β1–6)Gal branches and less linear poly-N-acetyllactosamines. The cDNA sequence of the enzyme revealed homology in the putative catalytic domain to core 2 β6-GlcNAc-T but not to other glycosyltransferases. Both I β6-GlcNAc-T and core 2 β6-GlcNAc-T genes were localized to chromosome 9, q21 [140]. There are four N-glycosylation sites (Asn-X-Ser/Thr-X, X not Pro) in the molecule. Because of the close relationship of the two genes for these β6-GlcNAc-T, one may conclude that they arose by gene duplication, similar to the family of α3-Fuc-T. Since no activity of the recombinant enzyme could be detected, its specificity appears to be different from the known I β6-GlcNAc-T specificities.

Gu et al. [167] measured I β6-GlcNAc-T in rat liver tissues using pyridylamine derivatives of lactotetraose oligosaccharides as substrates. Whereas several tissues including the liver of LEA control rats showed I β6-GlcNAc-T activity, liver of hepatitis and hepatoma-predisposed LEC rats had very low levels of activity. This lack of activity correlated with I antigen expression in the liver of LEC rats. Thus the difference between the two strains of rats was apparent in the inability of the LEC rat livers to synthesize the I antigen. The enzyme was probably lost during embryogenesis since significant activity was found only in fetal liver of LEC rats. The disappearance of I β6-GlcNAc-T activity in the liver of LEC rats is contrary to the observation that the I antigen replaces the i antigen during the development of foetal erythrocytes to the adult type [2]. The I β6-GlcNAc-T synthesizing reaction G can only act when GlcNAc is not substituted by a Gal(β1–4) residue.

Another enzyme activity from human serum [168] adds preferably an internal branch to substrates containing a Gal(β1–4) substituted GlcNAc residue (Fig. 1B):

Reaction I, I β6-GlcNAc-T (internal):

$$\text{Gal}(\beta1{-}4)\text{GlcNAc}(\beta1{-}3)\text{Gal}\beta\text{-R}_2 \longrightarrow \begin{array}{c} \textbf{GlcNAc(}\boldsymbol{\beta}\textbf{1--6)} \\ | \\ \text{Gal}(\beta1{-}4)\,\text{GlcNAc}(\beta1{-}3)\text{Gal}\beta\text{-R}_2 \end{array}$$

This activity provides an alternate pathway to the synthesis of highly branched poly-*N*-acetyllactosamine chains. Rat intestine [169] contains two β6-GlcNAc-T synthesizing the I branch, one with the specificity for nonreducing GlcNAc(β1–3)Galβ-R termini (catalyzing reaction G), and for Gal(β1–3)GalNAc-R and may therefore be the core 2 β6-GlcNAc-T M. The other activity synthesizes an internal branch; it requires Galβ1–4 substitution of the GlcNAc(β1–3) residue and differs from the enzyme described by Leppänen et al. [168] in that it is inhibited with SA(α2–3)Gal(β1–4), Gal(β1–3), or GlcNAc(β1–3)Gal(β1–4) substituted substrates. Changes at the reducing end of the molecule were found not to be important. The two activities eluted in different fractions from a QAE-Sepharose column and showed different kinetic properties towards various substrates.

9.2. β-1,6-GlcNAc-transferases synthesizing linear GlcNAc(β1–6)Gal and GlcNAc(β1–6)GalNAc structures

Core 6 has only been detected in human glycoproteins such as meconium [10] and human ovarian cyst [170] suggesting the existence of a human-specific β-1,6-GlcNAc-T acting on GalNAc without prior substitution of GalNAc by a Gal(β1–3) or

GlcNAc(β1–3) residue. An activity has been described in human ovarian tissue [132] that catalyzes the synthesis of core 6 (Fig. 1A):

Reaction J, core 6 β6-GlcNAc-T:

GalNAcα-R $\longrightarrow$ **GlcNAc(β1–6)GalNAcβ-R**

We could not demonstrate the presence of this enzyme in human colon. It is possible, that when using GalNAc-R as a substrate *in vitro*, core 6 may arise due to β-hexosaminidase degradation of core 4. Similarly, O-glycans on human mucins with core 6 may be formed by degradation of the common core 2. Human cells have been shown to contain an active β-galactosidase acting on core 2 structures [154]. The reason for the low abundance of core 6 in O-glycans may be high activity of the core 1 β3-Gal-transferase that converts core 6 to core 2. Elongation reactions for core 6 have not yet been studied.

Novikoff ascites tumour cells [166] have been shown to catalyze the synthesis of a linear GlcNAc(β1-6)Gal- structure (reaction K):

Reaction K, I β6-GlcNAc-T:

Galβ-R $\longrightarrow$ **GlcNAc(β1–6) Galβ-R**

This linear I β6-GlcNAc-T differs from the I β6-GlcNAc-T by its response to Triton-X-100 although this may be a cell dependent phenomenon. A similar enzyme was described in mouse T-lymphoma cells that synthesizes the linear I structure in glycolipids [171].

It is not known if the enzyme that synthesizes core 6 (reaction J) also catalyzes reaction K. The various species of β6-GlcNAc-T that exist probably have homologous regions because they bind UDP-GlcNAc and similar acceptor structures, and synthesize the GlcNAc(β1–6) linkage, and because homology has been found between the two cloned β6-GlcNAc-T. cDNA cloning and expression are required to resolve the question of how many species exist.

10. Synthesis of O-glycan cores 5 and 7

Core 5, GalNAc(α1–3)GalNAcα-, has been found in mucins from several mammalian [135] and non-mammalian species [172], and in human meconium [10] and glycoproteins from colonic adenocarcinoma [8]. UDP-GalNAc: GalNAc-mucin α3-GalNAc-T, core 5 α3-GalNAc-T, catalyzes the following reaction:

Reaction L, core 5 α3-GalNAc-T:

GalNAc-R $\longrightarrow$ **GalNAc(α1–3)GalNAc-R**

The only report on the biosynthesis of core 5 is by Kurosaka et al. [173] who have shown that detergent extracts of one of six human intestinal cancerous tissues contain an enzyme capable of synthesizing a disaccharide on asialo bovine submaxillary mucin *in vitro*. Enzyme product was released by alkaline borohydride and comigrated with

GalNAc(α1–3) GalNAc-ol, and could be converted to GalNAc-ol by digestion with α-N-acetylgalactosaminidase.

The enzyme acts in the presence of Mn^{2+}, but the exact linkage in the GalNAcα-GalNAc- enzyme product has yet to be demonstrated. It is not known, how the core 5 α3-GalNAc-T differs from the blood group A α3-GalNAc-T. Since blood group A is much more common than core 5, these two structures are probably not made by the same enzyme. However, since the A transferase can synthesize the B antigen under certain conditions (Section 12.5) it is conceivable that the A transferase may experience a change in specificity in the intestine in early foetal life as well as in cancer, possibly by association with certain cofactors, and synthesize the oncofoetal core 5 structure. Nothing is known about the elongation and sialylation reactions for core 5. It is possible that the sialylated core 5 structure, SAα2–6 [GalNAc(α1–3)] GalNAc-R, is synthesized by SA-T I from core 5.

There is no report to date on the formation of core 7 [15] by a core 7 α6-GalNAc-T. Based on the tissue distribution of cores 5 and 7 one may speculate that two separate enzymes are involved.

11. Elongation of O-glycan cores

The elongation of O-glycan cores involves addition of GlcNAc in β-1,3 and β-1,6 linkages and Gal residues in β-1,3 and β-1,4 linkages to form linear and branched poly-N-acetyllactosamine structures (Fig. 1B,C,D) which may then further be acted upon by glycosyltransferases adding terminal sugar residues.

11.1. Elongation β3-GlcNAc-transferase

We have found in pig gastric mucosa an elongation enzyme [174,175], UDP-GlcNAc:+/–GlcNAc(β1–6) [Gal(β1–3)] GalNAc-R (GlcNAc to Gal) β-1,3-GlcNAc-T (EC 2.4.1.146), elongation β3-GlcNAc-T, which catalyzes the elongation of core 1 and 2 structures with an unsubstituted Gal(β1–3) residue:

Reaction M, elongation β3-GlcNAc-T:

+/–GlcNAc(β1–6)   +/–GlcNAc(β1–6)
       | |

Gal(β1–3) GalNAcα-R $\longrightarrow$ **GlcNAc(β1–3)Gal(β1–3) GalNAcα-R**

R may be hydrophobic synthetic groups or peptide. The enzyme is relatively restricted in its distribution [11] and is present in tissues synthesizing elongated O-glycans such as human intestinal tissue and colon cancer cell lines [128]. The activity is relatively low in non-mucin secreting cells, such as leukocytes [154] and in colon cancer cells [128]. The enzyme shows a strict requirement for divalent cation (Mn^{2+} or Co^{2+}). The metal ion requirements, tissue distribution, and substrate specificity of this enzyme differ from those of the i β3-GlcNAc-T (Section 11.2) which is ubiquitous, and from those of the core 3 β3-GlcNAc-T (Section 8). These enzymes are therefore probably different although they share requirements for divalent cations, and bind to UDP-GlcNAc and terminal Gal- or GalNAc- in the acceptor substrate. The elongation β3-GlcNAc-T from pig trachea [148] was partially separated from core 2 β6-GlcNAc-T by affinity chromatography on Galβ1–3GalNAc-containing pig Cowper's gland mucin. Further purification and cloning is necessary to distinguish and characterize the β3-GlcNAc-T.

11.2. Blood group i β3-GlcNAc-transferase

The i blood group antigen or poly-*N*-acetyllactosamine structure is a developmentally regulated determinant on erythrocytes that is made in early foetal life [2]. Poly-*N*-acetyllactosamines linked to O-glycans occur on many secreted and membrane-bound glycoproteins and often carry terminal carbohydrate recognition sequences. β4-Gal-T (Section 11.3) and UDP-GlcNAc:Gal(β1–4)GlcNAc-R (GlcNAc to Gal) β-1,3-GlcNAc-T (EC 2.4.1.149), i β3-GlcNAc-T, are the two enzymes that act sequentially to build up these poly-*N*-acetyllactosaminoglycan chains (Fig. 1B). The i β3-GlcNAc-T catalyzes the following reaction in the presence of Mn^{2+}:

Reaction N, i β3-GlcNAc-T:

Gal(β1–4)GlcNAc-R $\longrightarrow$ **GlcNAc(β1–3)**Gal(β1–4)GlcNAc-R

The enzyme is ubiquitous and has been described in many different cell types, including colon cancer cells, lymphoid and myeloid cells [11,130,154,171,176–182]. The activity is relatively high in human serum, suggesting that i β3-GlcNAc-T is one of the glycosyltransferases that are released from Golgi membranes by proteolytic cleavage.

The same enzyme is thought to act on N-glycans, O-glycans and glycolipids. Like the other β3-GlcNAc-T, i β3-GlcNAc-T exhibits surprisingly low *in vitro* activity, although structural data indicate that the enzyme must be very active *in vivo*. In the synthesis of poly-*N*-acetyllactosamine chains, the i β3-GlcNAc-T must be limiting since the β4-Gal-T is always present in high activity.

The enzyme has been partially purified from Novikoff cell ascites fluid [183]. The enzyme bound to ConA Sepharose, suggesting that it contains N-glycans. The activity requires divalent cation with Mn^{2+} being optimal. Although the enzyme acts with variable efficiency on different glycoproteins and oligosaccharides, it can elongate Gal residues of a great variety of acceptors containing Galβ1–4GlcNAc termini but not the Gal residues of core 1 or core 2.

Yousefi et al. [156] found that T24 ras-oncogene transfection of rat 2 fibroblasts resulted in decreased polylactosamine content and i β3-GlcNAc-T activity. Leukocytes from patients with AML show high activity [154]. This indicates that i β3-GlcNAc-T is subject to regulation. The gene has not yet been cloned.

The i β3-GlcNAc-T differs from the elongation β3-GlcNAc-T (Section 11.1, EC 2.4.1.146) acting on Gal(β1–3)GalNAc- and from the core 3 β3-GlcNAc-T by its tissue distribution and specificity.

Yet another i β3-GlcNAc-T activity synthesizing the GlcNAc(β1–3)Gal linkage in GlcNAc(β1–3)Gal(β1–4)Glcβ-ceramide (lactotriaosylcermide) can be distinguished from the i β3-GlcNAc-T synthesizing poly-N-acetyllactosamine chains by its metal ion activation, pH optimum, kinetics and is found in the myeloid but not lymphoid cell lineages [179]. This suggests that there is also a family of β3-GlcNAc-T.

11.3. β-1,4-Galactosyl-transferase

UDP-Gal:GlcNAc-R β-1,4-Gal-T (EC 2.4.1.38; EC 2.4.1.90), β4-Gal-T, is ubiquitous and has been purified and cloned from several tissues and has been dealt with in detail elsewhere. The enzyme catalyzes the formation of N-acetyllactosamine (Fig. 1B) in the presence of Mn^{2+}:

Reaction O, β4-Gal-T:

GlcNAcβ-R $\longrightarrow$ **Gal(β1–4)GlcNAcβ-R**

β4-Gal-T is apparently involved in the synthesis of the N-acetyllactosamine structure in most glycoconjugates. Studies of enzyme kinetics suggest that the enzyme reaction occurs in an ordered mechanism by binding first Mn^{2+}, then UDP-Gal, and then GlcNAc [184].

The specificity of the enzyme is regulated in mammary glands by binding to α-lactalbumin which changes the kinetics of β4-Gal-T to favour the synthesis of lactose in milk. The enzyme is ubiquitous in relatively high activity and has been purified and the genes have been cloned from several species. The enzyme shows great homology between species but there is no homology to other glycosyltransferases. The putative UDP-Gal binding site can be labeled with uv-reactive[14C]4-azido -2-nitrophenylu-ridylylpyrophosphate [185]. Site-directed mutagenesis identified three Tyr and two Trp as important residues of the UDP-Gal binding site. The acceptor binding site may be labeled with a uv reactive GlcNAc derivative [186].

The β4-Gal-T contains several N-glycosylation sites. Expression of active enzyme in *E. coli* suggests that the carbohydrate is probably not involved in catalysis [185].

It is possible that a β4-Gal-T associated protein kinase is involved in posttransla-tional regulation since β4-Gal-T may be activated by protein phosphorylation [187].

11.4. β-1,3-Galactosyl-transferases

Another elongating enzyme, UDP-Gal:GlcNAcβ-R β-1,3-Gal-T, β3-Gal-T, has been characterized and purified from pig trachea [188,189]. The enzyme catalyzes the fol-lowing reaction:

Reaction P, β3-Gal-T:

GlcNAcβ-R $\longrightarrow$ **Gal(β1–3)GlcNAcβ-R**

This enzyme is clearly different from the β4-Gal-T and core 1 β3-Gal-T. The β3-Gal-T bound to UDP-hexanolamine-Sepharose and could be separated from other Gal-T activities by Sephacryl S-200. The activity is not influenced by α-lactalbumin and requires Mn^{2+}, Co^{2+} or Ni^{2+}. The physiological function of this enzyme appears to be the synthesis of type 1 chains on glycoproteins and glycolipids, because it adds Gal(β1–3) to GlcNAc residues of O-glycan core 3 and other structures (Fig. 1B) with terminal β-GlcNAc residues.

Human adenocarcinoma cells HCT 15 do not express type 1 chains (Table I). Upon stable or transient transfection of HCT cells with cDNA from human colonic adenocarcinoma cells (COLO 205), Sherwood et al. [190] achieved the expression of β3-Gal-T synthesizing type 1 chains. It remains to be shown if this enzyme is the same as the one described by Sheares et al. [188].

Messer and Nicholas [191] also described a β3-Gal-T in the mammary gland of tammar wallaby which acts on the Gal residue of lactose. The enzyme is not influenced by bovine or tammar α-lactalbumin in contrast to the β4-Gal-T in the same tissue which is stimulated by α-lactalbumin to synthesize lactose. Both the latter β3-Gal-T and β4-Gal-T appear to be subject to developmental regulation in the mammary tissue of the tammar wallaby. A β3-Gal-T with similar specificity has been reported in human kidney [192] and in the albumen gland of the snail [193]. Several other Gal-T exist, e.g. snails also contain a β3-Gal-T which adds Gal to GalNAc in β1–3-linkage [116] to form oligosaccharides of hemocyanin. The relationship between all of these β3-Gal-T remains to be established.

12. Terminal glycosylation

O-glycans may carry special terminal epitopes at the nonreducing end. Many of these are recognized as blood group or tissue antigens, for example the Lewis, ABO, Cad and S^d blood groups. Terminal sugars include SA, Fuc, Gal, GalNAc and GlcNAc, usually in α-linkage. A number of unusual terminal structures with yet unknown significance occur in various species. These include GlcNAc(α1–4)Gal which is proposed to be a duodenal-specific structure [194]. The addition of these α-linked sugars to terminal or internal residues of O-glycan chains may mask underlying antigens, or produce new antigens, and may terminate chain growth although many of the possible internal branching reactions have not been investigated. Certain α-linked sugars are also found in internal positions, such as GalNAc(α1–3) residues as part of the core 5 structures and SA as part of the polysialic acid structures.

SA residues are present on glycoconjugates as O- or N- acyl and other derivatives [195,196] that appear in a developmentally regulated and tissue-specific fashion. SA is part of several cancer-associated antigens and plays an important role in the cell surface properties of cells. SA may cover underlying antigens and prevent cells or glycoproteins from binding to mammalian lectins, although some lectins recognize SA as part of their ligands.

SA modifications are made by enzymes that act either on the glycoconjugate or on CMP-SA and maybe abnormally expressed in malignancies. N-glycolyl-neuraminic acid is a common structure in mucins from many species but does not occur normally in humans; it may appear in cancer cells as Hanganutziu–Deicher antigen, for example on mucins in breast cancer [197]. N-glycolyl-neuraminic acid is formed by CMP-SA hydroxylase in the soluble cellular portion, an NAD[P]H-dependent monooxygenase that acts on CMP-SA but not on the glycoconjugate to form CMP-N-glycolyl-neuraminic acid [198]. A sialic acid derivative commonly found in nonmammalian species is 2-keto-3-de-oxy-D-glycero-D-galacto-nonulosonic acid (Kdn). For example, Kdn replaces N-acetyl-neuraminic acid in the disialylated core 1 structures in rainbow trout eggs [199].

Chains of repeating SA residues, poly-sialic acids containing internal SA residues, are found on fish egg mucins [200] and neural cell adhesion molecules and appear to modulate cell–cell interactions.

SA residues may be present on O-glycans in characteristic linkages, for example, SA(α2–6) GalNAc (sialyl-Tn antigen), SA(α2–3) Gal(β1–3) GalNAc- and SA(α2–6) [SA(α2–3) Gal(β1–3)] GalNAc- (sialylated core 1). These are synthesized by O-gly-can-specific enzymes. Other SA linkages resemble those found on N-glycans and glycolipids, i.e. SA(α2–3) Gal(β1–4) GlcNAc- and SA(α2–8) SA-. The α6-SA-T (EC 2.4.99.1) and α3-SA-T (EC 2.4.99.6) acting on Galβ- residues of N-glycans have been purified and their genes have been cloned and expressed [201–203]. This α6-SA-T is subject to tissue-specific expression [91,204]. It remains to be shown if these two enzymes (covered in detail elsewhere) also act on Gal(β1–3/4) GlcNAcβ- termini of O-glycans.

Blood group antigens on O-glycans are probably synthesized by the same enzymes that act on N-glycans and glycolipids (Figs. 1C,D and 3).

12.1. α6-Sialyl-transferase I

SA(α2–6) GalNAc- chains are often found in submaxillary mucins [11], and in cancer as the sialyl-Tn antigen [205]. CMP-SA: GalNAc-R α6-SA-T I (EC 2.4.99.3), α6-SA-T I, catalyzes the synthesis of the SA(α2-6)GalNAc linkage:

Reaction Q, α6-SA-T I:

$$\text{SA}(\alpha2\text{–}6)$$
$$|$$
$$R_1\text{-GalNAc}\alpha\text{-}R_2 \longrightarrow R_1\text{- GalNAc-}R_2$$

where R_1 can be H, Gal(β1–3), or SA(α2–3)Gal(β1–3), and where R_2 must be peptide. The enzyme does not act on synthetic nitrophenyl, benzyl or phenyl derivatives. The inhibitor for O-glycosylation, GalNAcα-aryl [16] is therefore not expected to affect sialylation of GalNAc, since it is not a substrate for α6-SA-T I.

SA(α2–6)GalNAc- is a chain terminating structure since it is not a substrate for any of the known glycosyltransferases. α6-SA-T I can act after core 1 β3-Gal-T (Section 7) and after O-glycan α3-SA-T (Section 12.3), but the reverse is not possible (Fig. 3). However, sialylation may not always be a chain terminating reaction. α2-Fucosylation or the presence of the blood group A or B determinant on the Gal residue of core 1 inhibits α6-SA-T I, but the α2-Fuc-T can act on SA(α2–6) [Gal(β1–3)] GalNAc-R. Thus in the pathways to sialylated blood group A or B structures (Fig. 3) α6-SA-T I must act first, followed by α2-Fuc-T.

α6-Sialylation of core 1 prevents formation of core 2 by core 2 β6-GlcNAc-T and inhibits elongation of core 1 by elongation β3-GlcNAc-T [174] (Fig. 3).

It is unknown which α6-SA-T is responsible for sialylation of cores 3 and 5.

α6-SA-T I has been described in submaxillary glands of several species. The enzyme has been purified 117,000 fold from porcine submaxillary glands using affinity chromatography on CDP-hexanolamine-Sepharose columns [206,207]. Several molecular weight species of the enzyme were found between about 60,000 and 160,000 Da. The enzyme has no requirement for cation and acts on mucin (with GalNAc-chains) or antifreeze glycoprotein (with Gal(β1–3)GalNAc- chains). Kinetic studies using these two substrates suggested either a random equilibrium mechanism or an ordered mechanism where the acceptor binds first [207]. It is possible that the glycoprotein substrate structure may have an influence on the kinetic mechanism.

The cDNA for several SA-T species, probably including α6-SA-T I has recently been isolated [208].

Cell surface hypersialylation is a common observation in transformed and metastatic cells [209,210] as well as in leukaemia cells [42,211,212]. Sialyl-Tn antigen is rarely found normally but is found in cancer tissues and is associated with a poor prognosis [205,213,214]. Although the mechanism of increase in sialyl-Tn is not known, it is possible that this structure arises due to high α6-SA-T activity or relatively low activity of core synthesizing enzymes, or by intracellular rearrangements, allowing α6-SA-T I to act first.

12.2. α6-Sialyl-transferase II

Two different enzymes, α6-SA-T I and II, may sialylate GalNAc; the main difference between these two enzymes is their tissue distribution and the fact that α6-SA-T I requires peptide while α6-SA-T II requires the prior insertion of a SA(α2–3) residue. It is unknown how α6-SA-T I and II are related.

CMP-SA:SA(α2–3) Gal(β1–3) GalNAc-R (SA to GalNAc) α6-SA-T II (EC 2.4.99.7), α6-SA-T II, catalyzes the following reaction:

Reaction R, α6-SA-T II:

$$
\begin{array}{c}
\textbf{SA(α2–6)} \\
\mid \\
\text{SA(α2–3)Gal(β1–3)GalNAcα-R} \longrightarrow \quad \text{SA(α2–3)Gal(β1–3)GalNAcα-R}
\end{array}
$$

Prior sialylation of the core 1 substrate is required for α6-SA-T II action, thus the synthesis of the disialylated core 1 tetrasaccharide that is present in many glycoproteins proceeds by the 'three-before-six' rule, i.e., O-glycan α3-SA-T (Section 12.3) acts before α6-SA-T I or II. α6-SA-T II, present in fetal calf liver microsomes and brain tissues, cannot act on GalNAc-R but acts on SA(α2–3) Gal(β1-3) GalNAc-R, where R may be peptide or a synthetic phenyl derivative [215,216].

Inhibition studies with N-ethylmaleimide showed that the α6-SA-T II acting on asialofetuin but not the O-glycan α3-SA-T contains sulfhydryl groups that are important for activity. Since the inhibition could be prevented by CMP-SA, these sulfhydryl groups may be near the CMP-SA binding site [217].

The embryonic form of neural cell adhesion molecules, N-CAM, in the brain contain long SAα2-8SA linked polysialic acid chains which are involved in the regulation of cellular interactions. After birth, the α8-SA-T activity decreases leading to disappearance of polysialic acid. Several SA-T activities acting on asialofetuin, and especially the activity of α6-SA-T (probably α6-SA-T II), measured by the incorporation of SA into native fetuin (containing SA(α2-3)Gal(β1-3)GalNAc-), were also found to depend on ontogenic development and are high in embryonic rat brain but low in the adult brain [218]. The latter activity, however, is not the main determinant for total SA in brain glycoproteins, since glycoproteins from adult brain contain a higher amount of total SA, possible due to the greater abundance of glycoproteins.

12.3. O-glycan α3-Sialyl-transferase

CMP-SA: Gal(β1-3)GalNAc-R α3-SA-T (EC 2.4.99.4), O-glycan α3-SA-T, catalyzes the following reaction (Fig. 3):

Reaction S, O-glycan α3-SA-T:

Gal(β1-3)GalNAc-R $\longrightarrow$ **SA(α2-3)** Gal(β1-3) GalNAc-R

The enzyme from pig submaxillary gland or liver can act on a variety of Gal-terminal acceptors where R is peptide, and in contrast to the α6-SA-T I R can be a synthetic group such as phenyl-derivatives and even the free reducing sugar [141,219,220]. The enzyme acts poorly on Gal(β1-3)GlcNAc, or Gal(β1-4)Glc, but can readily act on Gal(β1-3) GalNAc- termini of gangliosides. However, the O-glycan α3-SA-T does not act on lactosylceramide to form SA(α2-3) Gal(β1-4) Glcβ-ceramide; the enzyme responsible for this reaction (EC 2.4.99.9) is therefore different.

O-glycan α3-SA-T is present in many tissues including liver, submaxillary gland, leukocytes and brain tissue from various species. Using CDP-Sepharose as an affinity matrix, the enzyme has been purified 90,000 fold from pig submaxillary gland [206] as two molecular weight species of about 50,000 Da. One of the two species with the higher Stokes radius bound detergent and may contain the membrane-spanning region of the protein, although there appeared to be no kinetic difference between the two forms.

The kinetic analysis by Rearick et al. [219] suggests a random equilibrium mechanism. Substrate specificity studies using Gal(β1-3) GalNAcα-benzyl derivatives indicate that the enzyme from placenta and from AML cells [141] has an absolute requirement only for the 3-hydroxyl of the Gal residue, and has a partial requirement for the 4-hydroxyl of Gal and the 2-N-acetyl group of GalNAc. Thus it can act on substrates with 6-substituted Gal or GalNAc, e.g. core 2 structure, although it does not tolerate the 6-substitution of GalNAc by SA. The 3-deoxy-Gal-derivative of the substrate is a competitive inhibitor for this enzyme [141] but the corresponding 3-deoxy-Gal substrate is not an inhibitor for the α3-SA-T acting on N-glycans [221].

Reboul et al. [222] postulated that the O-glycan α3-SA-T may be regulated by a phosphorylation/dephosphorylation mechanism. O-glycan α3-SA-T (as measured by the incorporation of SA into desialylated fetuin) from C6 rat glioma cells was deactivated by dephosphorylation with alkaline phosphatase. Inhibition of protein phosphatase by okadaic acid and Calyculin A as well as Ca/calmodulin antagonists also led to enzyme inhibition, possibly by a complex concerted mechanism.

The enzyme from C6 glioma cells appears to depend on N-glycosylation for full activity. Activity was inhibited by treating cells with tunicamycin; enzyme extract from C6 cells bound to ConA Sepharose, and the partially purified enzyme was inactivated by peptide-N-glycosidase [223]. However, it is possible that the N-glycans of an associated glycoprotein are required for activity.

The cDNA for O-glycan α3-SA-T has been cloned from a pig submaxillary gland cDNA library, using amino acid sequence information from the purified pig liver enzyme [224]. Two forms with molecular weights 45 and 48 KDa were purified and found to be two proteolytic products of SA-T, cleaved at the N-terminal end. This suggests, as has been shown for other transferases, that the N-terminal end including the stem region is not necessary for catalytic activity. The enzyme has the typical type II membrane protein structure and 4 potential N-linked glycosylation sites. A 45 amino acid stretch is highly homologous to other SA-T and corresponds to a sequence that overlaps two exons of the N-glycan α6-SA-T gene. Northern analysis shows tissue specific expression of the enzyme and only one size of mRNA (5.7 Kb) in various rat tissues [224]. Subsequently, Chang and Lau (unpublished) cloned the cDNA for the human α3-SA-T and consistently observed larger mRNA of 6.8 Kb size in human cells and tissues.

Chromosomal, immunological and enzymatic analysis of mouse myeloid cell–human CML leukocyte hybrids showed that the expression of O-glycan α3-SA-T and cell surface SA-Le^x is associated with the presence of human chromosome 11 [225] although the enzyme is not likely to be involved in the synthesis of this determinant. It is therefore possible that the gene for O-glycan α3-SA-T or a regulatory protein is on chromosome 11. Although the α3-SA-T acting on N-glycans is believed to be involved in the synthesis of the SA-Le^x determinant, this enzyme did not correlate with SA-Le^x expression. A similar correlation with human chromosome 11 was found for the α3-Fuc-T involved in SA-Le^x expression [226].

The SA(α2–3)Gal(β1–3)GalNAc- structure synthesized by O-glycan α3-SA-T may be a receptor for virus binding and be involved in the infectivity [11]. Sendai virus-mediated erythrocyte hemagglutination and infection of bovine kidney cells can be prevented by removing SA, and is restored with O-glycan α3-SA-T whereas other SA-T are ineffective [227]. Since viruses often have different affinities for N- or O-acyl SA derivatives, not only the linkages between sugars but also the type of SA may play a role in virus infections.

Sialoadhesins are present on macrophages and are known to function in removing erythrocytes and lymphocytes from the circulation. The product of O-glycan α3-SA-T,

SA(α2–3)Gal(β1–3)GalNAc-, has been identified as a ligand for a mouse sialoadhesin which is involved in interactions between different cells of the haematopoietic system [228].

The expression of O-glycan α3-SA-T appears to be regulated during growth and differentiation. O-glycans containing the SA(α2–3)Gal- moiety vary between cells at different stages of myeloid differentiation [42]. The O-glycan α3-SA-T also appears to be very active in leukaemia-derived cell lines during differentiation [229].

Granulocytes from patients with CML compared with normal granulocytes showed more sialylation and less *Peanut* lectin binding, reversible by sialidase treatment or chemotherapy, indicating more sialylation of core 1. The O-glycans isolated from AML and CML cells have been found to be shorter, more sialylated and contain less elongation and branching compared to normal granulocytes [230]. The basis for the hypersialylation and changed cell surface properties of these leukaemia cells could be the increase in core 2 β6-GlcNAc-T providing more substrate for sialylation (see above) [154] as well as an increase in O-glycan α3-SA-T [211,212]. The elongation β3-GlcNAc-T (Section 11.1) that is present in normal granulocytes is undetectable in leukaemic cells [154] but the enzymes synthesizing poly-*N*-acetyllactosamine chains are highly active in AML. These findings can only partially explain the structural aberrations found in leukaemic cells, thus there must be complex control factors regulating cell surface sialylation.

Saitoh et al. [231] investigated O-glycan biosynthesis in HL60 cell lines resistant to retinoic acid and 6-thioguanidine induced differentiation. The SA(α2–3)Gal(β1–3)GalNAc- structure on leukosialin as well as the O-glycan α3-SA-T activity were much more prevalent in the wild type cell line than in the altered cells, suggesting a role for O-glycan α3-SA-T in cellular differentiation.

In situ hybridization by Gillespie et al. [232] using digoxigenin-labeled RNA probes, showed that the enzyme is expressed in human T-lymphoblastoid cell lines and thymocytes, and is regulated during maturation of thymocytes. The expression of the O-glycan α3-SA-T was inversely proportional to the *Peanut* lectin binding sites since it masks the Gal(β1–3)GalNAc- ligands for *Peanut* lectin. Cortical thymocytes which stain with *Peanut* lectin are more immature and upon maturation change to *Peanut* lectin negative mature medullary thymocytes while simultaneously expressing increased O-glycan α3-SA-T. Thus thymocytes differ from granulocytes where less mature cells from CML patients exhibit an increased O-glycan α3-SA-T activity and poor *Peanut* lectin binding [212].

Ha-ras oncogene transfection into FR3T3 rat fibroblasts caused a decreased expression of the O-glycan α3-SA-T and an increased expression of the N-linked type α6-SA-T [233]. No specific inhibitor was found but a reduction of V_{max} occurred. The conclusion was that there was less O-glycan α3-SA-T protein produced upon Ha-ras transfection. Since there was no concomitant increase in *Peanut* lectin binding to cell surfaces, the expression of sialylated O-glycans may not be affected or cell surface O-glycans of core 1 may have been processed differently. For example, core 2 β6-GlcNAc-T which would also decrease *Peanut* lectin binding, may be elevated upon oncogene transfection [156].

12.4. Fucosylation of O-glycans

Normal and tumour cells may express various types of Fuc-T and many different O-glycan chains carrying Fuc residues. These fucosylated structures may designate part of blood group antigens or ligands for selectins and change during growth and in some diseases. Detailed discussions of Fuc-T are found elsewhere.

12.4.1. α-1,2-Fuc-transferase
There are at least two enzymes that synthesize the blood group H (O) determinant. The H-dependent and secretory gene-dependent GDP-Fuc: Galβ-R α-1,2-Fuc-T (EC 2.4.1.69), α2-Fuc-T, catalyze the following reaction:

Reaction T, α2-Fuc-T:

$$\text{Gal}\beta\text{-R} \longrightarrow \textbf{Fuc}(\alpha\textbf{1--2}) \text{ Gal}\beta\text{-R}$$

Phenyl β-D-galactoside is a useful specific substrate for the enzyme since it is inactive with α-1,3/4 or 6- Fuc-T. The blood group H enzyme attaches Fuc in α1-2 linkage to the terminal Gal residue of any β-D-galactoside, irrespective of the subterminal sugars. Structures like Galβ1–3(Fucα1–4)GlcNAc-R (the Le^a determinant) or Gal(β1–4) [Fuc(α1–3)] GlcNAc-R are not acceptors. Thus the Lewis blood group determinants (Table I) are added after α2-fucosylation (Fig. 1C,D). The H- and Se-gene-dependent transferases have different properties, affinities for their substrates and tissue distribution. The secretor gene-dependent α2-Fuc-T is expressed in secretory tissues in many species and is responsible for the addition of the H determinants on mucins, whereas the H-dependent transferase is expressed in certain haematopoietic cells and is absent from secretory tissues and fluids. Both enzymes are present in human serum. In the human intestines, α2-Fuc-T is expressed at high levels in the stomach and in the upper intestines and at lower levels in the colon. α2-Fuc-T is active in lymphocytes but not in granulocytes [234].

Although α2-Fuc-T does not require divalent cations, it is stimulated by Mn^{2+} and other cations. The human tracheal enzyme is strongly inhibited by Co^{2+}, Zn^{2+}, and Cd^{2+} and N-ethylmaleimide as well as GMP, GDP, GTP [235].

α2-Fuc-T, presumably the Se-dependent enzyme, has been highly purified to homogeneity from pig submaxillary glands using affinity chromatography on GDP-hexanolamine-agarose [236]. The enzyme was present as two molecular weight species of 55,000 and 60,000 Da and acts preferably on Galβ1–3GalNAc- substrates. Stimulation of activity by divalent metal ions shows that the pig submaxillary enzyme may differ from the human tracheal enzyme [237]. Kinetic studies suggest that catalysis proceeds by either a random equilibrium mechanism or by an ordered mechanism with GDP-Fuc binding first [237]. Employing S-Sepharose, GDP-hexanolamine-Sepharose, high pressure liquid chromatography and other steps, the secretor gene-dependent α2-Fuc-T was highly purified from human serum by Sarnesto et al. [238]. The enzyme appears as a multimeric complex of 150,000 Da with subunits of 50,000 Da each.

The H-dependent α2-Fuc-T can be distinguished by its electrophoretic mobility, binding properties and kinetics from the secretor type enzyme and has been purified 10,000,000-fold by the same group [239]. The most effective purification step was GDP-hexanolamine-Sepharose. The enzyme also appears to assemble in three or four subunits of about 50,000 Da each. The enzyme bound to wheat germ agglutinin, indicating that it is a glycoprotein. The H-dependent α2-Fuc-T purified from human blood group O plasma by Kyprianou et al. [240] showed a molecular weight of 158,000 Da. The 200,000 fold purification involved Phenyl Sepharose and GDP-column chromatography.

The gene encoding human H-dependent enzyme has been cloned and expressed in several mammalian systems, and has been localized to chromosome 19 [241–243].

12.4.2. α-1,3-Fuc-transferases

A family of related GDP-L-Fuc:Gal(β1–4)GlcNAc (Fuc to GlcNAc) α3-Fuc-T (EC 2.4.1.152), α3-Fuc-T I to VII, exist that synthesize the neutral and sialylated Le^x determinants on type 2 chains where α3-Fuc-T I and II have been studied in CHO cells, and α3-Fuc-T III to VI in human cells:

Reaction U, α3-Fuc-T I, II,III, IV, V,VI:

Gal(β1–4) GlcNAcβ-R $\longrightarrow$ Gal(β1–4) GlcNAcβ-R
$\qquad\qquad\qquad\qquad\qquad\qquad\qquad\qquad\quad$ |
$\qquad\qquad\qquad\qquad\qquad\qquad\qquad\quad$ **Fuc (α1–3)**

Reaction V, α3-Fuc-T I,III, V, VI and VII:

SA(α2–3) Gal(β1–4) GlcNAcβ-R $\longrightarrow$ SA(α2–3) Gal(β1–4) GlcNAcβ-R
$\qquad\qquad\qquad\qquad\qquad\qquad\qquad\qquad\qquad\qquad\qquad\qquad\qquad\quad$ |
$\qquad\qquad\qquad\qquad\qquad\qquad\qquad\qquad\qquad\qquad\qquad\quad$ **Fuc(α1–3)**

The sialylated or neutral Lewis antigens commonly occur in O-glycans (Table I) [11]. α3-Fuc-T activities synthesizing these structures are ubiquitous but individual enzymes of this family of Lewis-dependent and Lewis-independent α3-Fuc-T exhibit tissue- and disease-specific expression (Fig. 1C,D). The enzyme can be assayed using substrate derivatives containing 2-O-methyl-substituted terminal Gal to avoid incorporation into Gal by α2-Fuc-T [244].

α3-Fuc-Ts can be distinguished by their substrate specificities, properties and tissue distribution. Mollicone et al. [245] divided α3-Fuc-T activities into five types according to their specificity and metal ion activation patterns: 1. myeloid cell (acting on neutral substrates); 2. brain; 3. plasma and liver (acting on neutral and sialylated substrates); 4. intestine, gall bladder, kidney and milk (α3/4-Fuc-T); and 5. stomach. This division may become difficult when tissues contain more than one type, normally or during differentiation or disease. In addition, properties of the same enzyme may vary according to the tissue source.

The CHO cell mutant LEC 11 contains α3-Fuc-T I which can act on neutral and sialylated substrates (reactions U, V). The CHO cell mutant LEC 12 contains an activity (α3-Fuc-T II) which is restricted to neutral chains (reaction U). α3-Fuc-T I does not bind well to GDP-hexanolamine-Sepharose, and in contrast to α3-Fuc-T II in LEC 12, is strongly inhibited by N-ethylmaleimide, indicating that a sulfhydryl group is essential for activity [246]. α3-Fuc-T I and II also differ in specificity since α3-Fuc-T II prefers the more terminal GlcNAc residues in glycolipid substrates with several potential acceptor sites [247]. α3-Fuc-T I resembles human α3-Fuc-T V in its properties (see below).

GDP-Fuc: Gal(β1–4/3) GlcNAc- (Fuc to GlcNAc) α-1,3/4-Fuc-T (EC 2.4.1.65), α3-Fuc-T III, is an α3/4-Fuc-T which in addition to reaction U and V catalyzes the following reaction:

Reaction W, α3/4-Fuc-T (α3-Fuc-T III):

Fuc(α1–4)
|
Gal(β1–3) GlcNAcβ-R $\longrightarrow$ Gal(β1–3) GlcNAcβ-R

This enzyme is capable of attaching Fuc in α-1,4 or 3 linkage to the subterminal GlcNAc of unsubstituted, α2-fucosylated or sialylated Gal(β1–3/4)GlcNAc-R structures but does not act on Gal(β1–3)GalNAc-R, thus it catalyzes reactions U, V and W. This activity represents another exception to the 'one linkage-one glycosyltransferase' rule. The enzyme does not require, but is stimulated by divalent cations and can be assayed using substrate derivatives of type 1 chains containing 2-Fuc- or 2-O-methyl-substituted terminal Gal to avoid incorporation into Gal by α2-Fuc-T. Submaxillary glands, gastric mucosa, cervical epithelium and milk of individuals with the Lewis (*Le*) gene contain the activity but not human serum, irrespective of the Lewis genotype, or lymphocytes and granulocytes [234]. The enzyme has been purified over 500,000-fold to apparent homogeneity from human milk by affinity chromatography on GDP-hex-anolamine-agarose [248]. Two molecular species with 53,000 and 51,000 Da molecular weight were identified. A similar purification procedure by Eppenberger-Castori et al. [249], however, yielded an enzyme of 98,000 Da molecular weight, probably a dimer. The activity bound to lentil lectin and ConA Sepharose and showed diffuse bands on SDS polyacrylamide gels, and several bands on isoelectric focusing, indicating that it is a glycoprotein.

Kinetic analysis indicates that a single active site is involved in both α-1,3 and 4-Fuc incorporation. The human α3/4-Fuc-T cDNA has been cloned by gene transfer into mammalian cells and screening enzyme product in expressing cells with anti SSEA-1 antibodies [250]. The gene codes for a protein of 361 amino acids and was found to be localized to chromosome 19, similar to that of the H-dependent α2-Fuc-T.

α3-Fuc-T IV (reaction U) is present in myeloid cells, granulocytes, lymphoblasts, monocytes and brain and cannot act on SA(α2–3)Gal(β1–4)GlcNAc, similar to the enzyme in the LEC12 mutant Chinese hamster ovary cell line (α3-Fuc-T II) [246], but can act on α2-fucosylated (H) substrates. The enzyme seems to be expressed in an early stage of development. Thus in the synthesis of SA-Le^x (Table I), sialylation probably occurs before fucosylation (Fig. 1D). Neither the α3-SA-T acting on N-glycans nor α2-Fuc-T can act on Gal(β1–4) [Fuc(α1–3)] GlcNAc- and therefore both of these enzymes act before the α3-Fuc-T.

The human gene for α3-Fuc-T IV has been cloned by homology screening [251] and codes for a protein of 405 amino acids. The catalytic domain of these Fuc-T contains regions that are conserved among species. α3-Fuc-T IV has been purified from human neuroblastoma cells [252] employing immobilized acceptor Gal(β1–4) GlcNAc. The enzyme was present as two species of molecular weight 45,000 and 40,000 Da.

F9 embryonal carcinoma cells contain an α3-Fuc-T (probably IV) that was purified by Muramatsu et al. [253]. The pure enzyme showed a molecular weight of 65,000 and differs from the milk enzyme by several criteria. The milk enzyme has a molecular weight of 44 000 Da and showed different binding characteristics to SP-Sephadex which could be explained by protein differences due to proteolytic degradation.

α3-Fuc-T V is present in serum, milk and liver [254] and can act on neutral and sialylated substrates (reactions U, V), and with less efficiency on α2-fucosylated substrates [247, 255]. The human gene for α3-Fuc-T V was cloned by Weston et al. [255] and codes for a protein of 374 amino acids. The gene is highly homologous and closely linked to that of α3/4-Fuc-T III at chromosome 19. α3-Fuc-T V acts on neutral and sialylated substrates. The human serum enzyme was purified employing GDP-Sepharose and has a molecular weight of 45,000 Da and is a glycoprotein since it bound to wheat germ agglutinin [256].

The binding site of α3-Fuc-T V has been investigated by Holmes [257]. Labeling of the purified enzyme with the photo-reactive inhibitor ^{125}I-GDP-hexanolamine-4-azidosalicylic acid could be prevented by GDP-fucose. Pyridoxal phosphate competitively inhibited the enzyme; after labeling with ^{3}H-pyridoxal phosphate, radioactivity was isolated as pyridoxal-lysine. The author concluded that Lys must have been an amino acid which was essential for catalytic activity.

α3-Fuc-T is increased in the sera of patients with various malignancies [55]. Chandrasekaran et al., [258] found that the enzyme (probably α3-Fuc-T V) from human ovarian tissues could be assayed with neutral, sialylated and sulfated acceptors. Extracts from normal ovarian tissue could utilize 3-sulfate-Gal(β1–4)GlcNAcβ-benzyl similarly but were not active with 3-sulfate-Gal(β1–3)GlcNAcβ-benzyl. This could mean that cancer is expressing an α4-Fuc-T which is not present in normal tissue.

Yet another enzyme of this family, α3-Fuc-T VI, exists with a specificity similar to α3-Fuc-T V in that it can act on neutral and sialylated substrates but not on Fuc(α1–2) Gal(β1–4)GlcNAc (reactions U and V). The human gene for α3-Fuc-T VI has been cloned and was localized close to the chromosome 19 locus of other α3-Fuc-T [259–261].

12.5. Blood group A-dependent α3-GalNAc-transferase and blood group B-dependent α3-Gal-transferase

Blood group A and B determinants (Table I) that occur on many mucins are synthesized from the H-determinant by A-dependent α3-GalNAc-transferase and B-dependent α3-Gal-T, respectively, which are discussed in detail elsewhere.

Blood group A determinant is synthesized by UDP-GalNAc:Fuc(α1–2)Gal-R (Gal-NAc to Gal) α-1,3-GalNAc-T (EC 2.4.1.40), A-dependent α3-GalNAc-T:

Reaction X, A-dependent α3-GalNAc-T:

$$\text{GalNAc}(\alpha 1\text{--}3)$$
$$|$$

Fuc(α1–2) Galβ-R $\longrightarrow$ Fuc(α1–2) Galβ-R

The enzyme can act on the Fuc(α1–2)Gal- structure attached to N- or O-glycan chains or glycolipids of type 1 or type 2 chains. The blood group ABO determinant is also found on the Gal residue of sialylated core 1 structures, indicating that the enzyme acts on a great variety of substrates (Fig. 1C,D, Fig. 3).

The A-dependent GalNAc-transferase has been purified from human serum, from porcine submaxillary glands and from human lung and gastric and duodenal scrapings [262–265]. The enzyme from porcine submaxillary gland was purified to homogeneity by chromatography on UDP-hexanolamine-agarose. The human gut enzyme was purified 125,000-fold on UDP-hexanolamine-agarose and octyl-Sepharose and showed a molecular weight of 40,000 Da.

The blood group B determinant is synthesized by UDP-Gal:Fuc(α1–2)Gal-R (Gal to Gal) α3-Gal-T (EC 2.4.1.37):

Reaction Y, B-dependent α3-Gal-T:

$$\text{Gal}(\alpha 1\text{--}3)$$
$$|$$

Fuc(α1–2)Galβ-R $\longrightarrow$ Fuc(α1–2)Galβ-R

Both the blood group B determinant and the 'linear B determinant' lacking Fuc, are commonly found in non-human mucins. The linear B structure (Table I) may be synthesized by a similar, homologous α3-Gal-T (EC 2.4.1.124/151) [266] (discussed elsewhere. The B-dependent α3-Gal-transferase has been purified 400,000-fold from human serum [267].

Both A and B transferases have very similar substrate specificity, i.e., they require the H determinant at the non-reducing terminus of the substrate. They were shown to have overlapping specificities *in vitro* and can synthesize the other blood groups under certain conditions, e.g. the B enzyme can use UDP-GalNAc as the nucleotide donor although this does not seem to be a problem *in vivo* but may account for the anomalous expression of blood group antigenic activity by certain human tumours [268,269].

The genes for both A and B transferases have been cloned [270,271]. The similarity of the two enzyme proteins is explained by the finding that they differ only by four amino acids and the genes by a few base pairs. Blood group O is due to the absence of A- or B-dependent transferases [270,271].

12.6. Blood group Cad or S^d-dependent β-1,4-GalNAc-transferases

Several β-1,4-GalNAc-T may exist that differ in substrate specificity and synthesize the Cad and Sd^a antigens found on mucins and glycoproteins by acting on sialylated

O- and N-glycans and glycolipids with the sequences SA(α2–3)Gal(β1–4)GlcNAc or SA(α2–3)Gal(β1–3)GalNAc. Blood Group Cad- or Sd^a-dependent UDP-GalNAc: SA(α2–3)Galβ-R (GalNAc to Gal) β4-GalNAc-transferases, β4-GalNAc-T, catalyze the following reaction:

Reaction Z, Cad-dependent β4-GalNAc-T:

$$\textbf{GalNAc(}\boldsymbol{\beta}\textbf{1–4)}$$
$$|$$
$$\text{SA}(\alpha2\text{–}3)\ \text{Gal}\beta\text{-R} \quad \longrightarrow \quad \text{SA}(\alpha2\text{–}3)\ \text{Gal}\beta\text{-R}$$

R may be the backbone structure of Tamm–Horsfall glycoprotein or glycophorin or oligosaccharide [272–275]. The presence of the sialyl residue is essential for β4-Gal-NAc-T activity (Fig. 3). SA(α2–3)Gal(β1–4)Glc and SA(α2–3)Gal(β1–4)GlcNAc-glycoprotein but not ganglioside G$_{M3}$ (SA(α2–3)Gal(β1–4)Glc-ceramide) are acceptors for the enzyme from human and guinea-pig kidney and human plasma [276]. It has not been established how many enzymes are involved in the synthesis of these determinants.

Conzelmann and Kornfeld [277] reported the presence of O-glycans with Gal-NAc(β1–4)Gal- termini in a mouse cytotoxic T lymphocyte line. The Cad determinant and the β4-GalNAc-T is lacking in a mutant mouse cytotoxic cell line resistant to the GalNAc binding *Vicia villosa* lectin [278].

The expression of β4-GalNAc-T is regulated during differentiation and development and differs among functionally distinct T-cell clones in the mouse. The enzyme is present in human colonic adenocarcinoma Caco-2 cells but not in a number of other colon cancer cell lines [279,280]. The activity increased with enterocytic differentiation of Caco-2 cells. In the human intestine there is a gradient of increasing activity towards proximal segments.

The expression of the enzyme was found to be drastically reduced in colon cancer [280].

In the rat colon, the activity is under developmental control [218] and is very low at birth but increases after two weeks, in contrast to the α6-SA-T acting on N-glycans which was only high in the newborn rat.

A partial purification of the Cad or Sd^a-dependent β4-GalNAc-T from human urine of Sd^a individuals has been achieved by Serafini-Cessi et al. [281]. The urinary enzyme does not require detergents for activity, and is probably a proteolytic degradation product of the kidney enzyme. The β4-GalNAc-T present in human kidney [282] is dependent on Mn^{2+} in the assay and acts preferably on N- and also on O-linked chains and sialyl-lactose but not on glycophorin A or free reduced SA(α2–3)Gal(β1–3)Gal-NAc-ol. The human plasma enzyme also acts poorly on glycophorin A but the guinea-pig kidney and the human urinary enzyme can incorporate GalNAc into native glycophorin A, while the human kidney enzyme cannot use glycophorin at all [283]. A similar activity requiring SA(α2–3)Gal termini and acting on glycolipids was reported in human stomach [284]. Glycoprotein acceptors were not studied, it is possible that the stomach enzyme is the same as the kidney enzyme.

It is evident that a family of related β4-GalNAc-T must exist that synthesize Cad or Sd^a determinants on O-glyans and on other glycoconjugates. Purification and cloning will hopefully help to resolve the different species and elucidate the role of Cad or Sd^a determinants.

13. Sulfation of O-glycans

Sulfate may be linked to various positions at terminal or internal sugar residues of O-linked oligosaccharides, for example to the 3-position of Gal(β1–4) in human meconium glycoproteins [9] or respiratory mucin [3] or to the 6-position of Gal(β1–3 or 4) or the GlcNAc(β1–3 or 6) residues of human tracheobronchial mucins [285–287] and ovomucin [288]. Interestingly, the same sugar residue may be substituted with sulfate and sialic acid, for example in mucin from the human colon cancer line CL.16E [138].

A portion of probably most mucins is sulfated; structural analysis, however, has been difficult in the past due to the lability of the sulfate–sugar linkages. Mucin oligosaccharides from various species and organs have been found to be sulfated and may in addition carry SA and thus be highly acidic [286]. The functions of these sulfated chains are largely unknown, it is possible that they serve a protective role. The number of sulfated chains in mucins vary in cancer and inflammation. Increased secretion of sulfated mucins is observed in cystic fibrosis [289].

The sulfate donor substrate, 3'-phospho-adenosine-5'-phosphosulfate (PAPS) is transported into the Golgi via a transporter protein analogous to the transport of nucleotide sugars [290]. Chlorate inhibits PAPS synthesis and thus glycoprotein sulfation [291] and may be used to study sulfate functions.

Several sulfo-T have been identified that act on termini of glycolipids and proteoglycans. It is not known if any of these sulfo-T are also involved in mucin sulfation. Assuming a linkage specificity of sulfo-T, there may be at least three sulfo-T acting on O-glycans. Carter et al. [292] described a sulfo-T activity in the Golgi fraction of rat stomach acting on rat gastric mucin. A crude preparation of the enzyme transferred sulfate from PAPS to the GlcNAc residue of the elongated core 1 structure to form 6-sulfate-GlcNAc(β1–3)Gal(β1–3)GalNAc- although other linkages may have been formed as well [292]. Using low molecular weight substrates we showed that sulfo-T in various mucin secreting tissues (unpublished) act on core 1, 2 and 6 substrates. Further studies are required to establish and characterize the sulfo-T acting on O-glycans.

14. Concluding remarks

A number of enzymes exclusively involved in O-glycan biosynthesis have been purified with refined methods and the cDNA for four of these enzymes have been cloned. It appears that a number of glycosyltransferase families exists, e.g. SA-T, β3-Gal-T, β3-GlcNAc-T, β6-GlcNAc-T, β4-GalNAc-T and α3-Fuc-T families. The differences in enzyme activities between tissues contribute to the extraordinary and tissue-specific variety of complex O-glycan chains found in nature. Many more enzymes involved in the biosynthesis of O-glycans remain to be detected and characterized, and their role and control mechanisms to be elucidated. The refinement of structural analyses and further gene cloning should advance our knowledge of O-glycan structures and their biosynthesis and provide the long awaited tools to study their functions.

Abbreviations

ER	endoplasmic reticulum
Fuc	L-fucose
Gal	D-galactose
GalNAc	N-acetyl-D-galactosamine
GalNAcol	N-acetylgalactosaminitol
GlcNAc	N-acetyl-D-glucosamine
SA	sialic acid
T	transferase

References

1. Babczinski, P. (1980) FEBS Lett. 117, 207–211.
2. Feizi, T. (1985) Nature 314, 53–57.
3. Lamblin, G., Lhermitte, M., Klein, A., Houdret, N., Scharfman, A., Ramphal, R. and Roussel, P. (1991) Am. Rev. Respirat. Dis. 144, S19–S24.
4. Dowbenko, D., Andalibi, A., Young, P.E., Lusis, A.J. and Lasky, L. A. (1993) J. Biol. Chem. 268, 4525–4529.
5. Patel, T., Bruce, J., Merry, A., Brigge, C., Womald, M., Jaques, A. and Parek, R. (1993) Biochemistry 32, 679–693.
6. Ishii-Karakasa, I., Iwase, H., Hotta, K., Tanaka, Y. and Omura, S. (1992) Biochem. J., 288, 475–482.
7. Reinhold, V.N., Reinhold, B.B. and Chan, S. (1994) In: B. Karger (Ed.), Methods in Enzymology. Academic Press, NY. In press.
8. Kurosaka, A., Nakajima, H., Funakoshi, J., Matsuyama, M., Nagayo, T. and Yamashina, I. (1983) J. Biol. Chem. 258, 11594–11598.
9. Capon, C., Leroy, Y., Wieruszeski, J.-M., Ricart, G., Strecker, G., Montreuil, J. and Fournet, B. (1989) Eur. J. Biochem. 182, 139–152.
10. Hounsell, E.F., Lawson, A.M., Feeney, J., Gooi, H.C., Pickering, N.J., Stoll, M.S., Lui, S.C. and Feizi, T. (1985) Eur. J. Biochem. 148, 367–377.
11. Schachter, H. and Brockhausen, I. (1992) In: H. Allen and E. Kisailus (Eds.), Glycoconjugates, Composition, Structure and Function. Marcel Dekker, New York, pp. 263–332.
12. Roussel, P., Lamblin, G., Lhermitte, M., Houdret, N., Lafitte, J.-J., Perini, J.-M., Klein, A. and Scharfman, A. (1988) Biochimie 70, 1471–1482.
13. Strous, G.J. and Dekker, J. (1992) Crit. Rev. Biochem. Mol. Biol. 27, 57–92.
14. Slomiany, B.L., Zdebska, E. and Slomiany, A. (1984) J. Biol. Chem. 259, 2863–2869.
15. Chai, W., Hounsell, E., Cashmore, G.C., Rosankiewicz, J.R., Bauer, C.J., Feeney, J., Feizi, T. and Lawson, A.M. (1992) Eur. J. Biochem. 203, 257–268.
16. Kuan, S.-F., Byrd, J.C., Basbaum, C. and Kim, Y.S. (1989) J. Biol. Chem. 264, 19271–19277.
17. Varki, A. (1993) Glycobiology 3, 97–130.

18. Shogren, R., Gerken, T.A. and Jentoft, N. (1989) Biochemistry 28, 5525–5536.
19. Oh'Eda, M., Hasegawa, M., Hattori, K., Kuboniwa, H., Kojima, T., Orita, T., Yomonou, K., Yamazaki, T. and Ochi, N. (1990) J. Biol. Chem. 265, 11432–11435.
20. Davies, P.L. and Hew, C.L. (1990) Biochemistry of antifreeze proteins. FASEB J. 4, 2460–2469.
21. Birch, N.P., Estivariz, F.E., Bennett, H.P.J. and Loh, Y.P. (1991) FEBS Lett. 290, 191–194.
22. Ho, S.B., Niehans, G.A., Lyftogt, C., Yan, P.S., Cherwitz, D.L., Gum, E.T., Dahiya, R. and Kim, Y.S. (1993) Cancer Res. 53, 641–651.
23. Forstner, J.F. and Forstner, G.G. (1993) In: E.D. Jacobson, L.R. Johnson et al. (Eds.), Physiology of the Gastrointestinal Tract. 3rd Edn., Raven Press, NY. In press.
24. Ramphal, R., Carnoy, C., Fiebre, S., Michalski, J.-C., Houdret, N., Lamblin, G., Strecker, G. and Roussel, P. (1991) Infection Immun. 59, 700–704.
25. Sajjan, U.S., Corey, M., Karmali, M.A. and Forstner, J.F. (1992) J. Clin. Invest. 89, 648–656.
26. Lamblin, G., Rahmoune, H., Wieruszeski, J.-M., Lhermitte, M., Strecker, G. and Roussel, P. (1991) Biochem. J. 275, 199–206.
27. Boland, C.R. (1982) Proc. Nat. Acad. Sci. USA 79, 2051–2055.
28. Corfield, A.P., Do Amaral Corfield, C., Wagner, S.A., Warren, B.F., Mountford, R.A., Bartolo, D.C.C. and Clamp, J.R. (1992) Biochem. Soc. Trans. 20, 95S.
29. Bresalier, R.S., Niv, Y. and Kim, Y.S. (1991) J. Clin. Invest. 87, 1037–1045.
30. Singhal, A., Fohn, M. and Hakomori, S.-I. (1991) Cancer Res. 51, 1406–1411.
31. O'Boyle, K.P., Zamore, R., Adluri, S., Cohen, A., Kemeny, N., Welt, S., Lloyd, K.O., Oettgen, H.F., Old, L.J. and Livingston, P.O. (1992) Cancer Res., 52, 5663–5667.
32. Varki, A. (1992) Current Biol. 4, 257–266.
33. Kojima, N., Handa, K., Newman W. and Hakomori, S.-I. (1992) Biochim. Biophys. Res. Comm. 182, 1288–95.
34. Takada, A., Ohmori, K., Yoneda, T., Tsuyoka, K., Hasegawa, A., Kiso, M. and Kannagi, R. (1993) Cancer Res., 53, 354–361.
35. Norgard, K.E., Moore, K.L., Diaz, S., Stults, N.L., Ushiyama, S., McEver, R.P., Cummings, R.D. and Varki, A. (1993) J. Biol. Chem. 268, 12764–12774.
36. Fukuda, M. (1991) Glycobiology 1, 347–356.
37. Sutherland, D.R., Watt, S.M., Dowden, G., Karhi, K., Baker, M.A., Greaves, M.F. and Smart, J.E. (1988) Leukemia 2, 793–803.
38. Lancaster, C.A., Peat, N., Duhig, T., Wilson, D., Taylor-Papadimitriou, J. and Gendler, S. (1990) Biochem. Biophys. Res. Comm. 173, 1019–1029.
39. Lasky, L.A., Singer, M.S., Dowbenko, D., Imai, Y., Henzel, W.J., Grimley, C., Fennie, C., Gillett, N., Watson, S.R. and Rosen, S.D. (1992) Cell 69, 927–938.
40. Seed, B. (1992) Current Biol. 2, 457–459.
41. Suzuki, Y., Toda, Y., Tamatani, T., Watanabe, T., Suzuki, T., Nakao, T., Murase, K., Kiso, M., Hasegawa, A., Tadano-Aritomi, K., Ishizuka, I. and Miyasaka, M. (1993) Biochem. Biophys. Res. Comm. 190, 426–434.
42. Carlsson, S.R., Sasaki, H. and Fukuda, M. (1986) J. Biol. Chem. 261, 12787–12795.
43. Remaley, A.T., Ugorski, M., Wu, N., Litzky, L., Burger, S.R., Moore, J.S., Fukuda, M. and Spitalnik, S.L. (1991) J. Biol. Chem., 266, 24176–24183.
44. Piller, F., Piller, V., Fox, R.I. and Fukuda, M.A. (1988) J. Biol. Chem. 263, 15146–15150.
45. Ogata, S., Maimonis, P.J. and Itzkowitz, S.H. (1992) Cancer Res., 52, 4741–4746.
46. Hansen, J.E.S., Nielsen, C., Arendrup, M., Olofsson, S., Mathiesen, L., Nielsen, J.O. and Clausen, E. (1991) J. Virol. 65, 6461–6467.
47. Hansen, J.-E.S., Clausen, H., Hu, S.L., Nielsen, J.O. and Olofsson, S. (1992) Arch. Virol. 126, 11–20.

252

48. MacLean, G.D., Reddish, M., Koganty, R.R., Wong, T., Gandhi, S., Smolenski, M., Samuel, J., Nabholtz, J.M. and Longenecker, B.M. (1993) Cancer Immunol. Immunother. 36, 215–222.

49. Lopez, L., Bayna, E.M., Litoff, D., Shaper, N.L., Shaper, J.H. and Shur, B.D. (1985) J. Cell Biol. 101, 1501–1510.

50. Shur, B.D. (1989) Biochim. Biophys. Acta 988, 389–409.

51. Shaper, N.L., Shaper, J.H., Peyser, M. and Kozak, C.A. (1990) Cytogenet. Cell Genet. 54, 172–174.

52. Florman, H.M. and Wassarman, P.M. (1985) Cell 41, 313–324.

53. Bleil, J.D. and Wassarman, P.M. (1988) Proc. Nat. Acad. Sci. USA 85, 6778–6782.

54. Yurewicz, E.C., Pack, B.A. and Sacco, A.G. (1992) Mol. Reprod. Develop. 33, 182–188.

55. Brockhausen, I. (1993) Crit. Rev. Clin. Lab. Sci. 30, 65–151.

56. Kingsley, D.M., Kozarsky, K.F., Hobbie, L. and Krieger, M. (1986) Cell 44, 749–759.

57. Kozarsky, K., Kingsley, D. and Krieger, M. (1988) Proc. Nat. Acad. Sci. USA 85, 4335–4339.

58. Remaley, A.T., Wong, A.W., Schumacher, Meng, M., Brewer, H. B. and Hoeg, J. (1993) J. Biol. Chem. 288, 6785–6790.

59. Reddy, P., Caras, I. and Krieger, M. (1989) J. Biol. Chem. 264, 17329–17336.

60. Coyne, K.E., Hall, S.E., Thompson, E.S., Arce, M.A., Kinoshita, T., Fujita, T., Anstee, D.J., Rosse, W. and Lublin, D.M. (1992) J. Immunol. 149, 2906–2913.

61. Kozarsky, K.F., Call, S.M., Dower, S.K. and Krieger, M. (1988) Mol. Cell. Biol. 8, 3357–3363.

62. Strous, G.J.M. (1979) Proc. Nat. Acad. Sci. USA 76, 2694–2698.

63. Johnson, W.V. and Heath, E.C. (1986) Biochemistry 25, 5518–5525.

64. Niemann, H., Boschek, B., Evans, D., Rosing, M., Tamura, T. and Klenk, H. (1982) EMBO J. 1, 1499–1504.

65. Serafini-Cessi, F., Dall'olio, F., Malagolini, N. and Campadelli-Fiume, G. (1989) Biochem. J. 262, 479–484.

66. Tooze, S.A., Tooze, J. and Warren, G. (1988) J. Cell Biol. 106, 1475–1487.

67. Spielman, J., Rockley, N.L. and Carraway, K.L. (1987) J. Biol. Chem. 262, 269–275.

68. Carraway, K.L. and Hull, S.R. (1989) BioEssays 10, 117–121.

69. Jokinen, M., Abdersson, L.C. and Gahmberg, C.G. (1985) J. Biol. Chem. 260, 11314–11321.

70. Piller, V., Piller, F., Klier, F.G. and Fukuda, M. (1989) Eur. J. Biochem. 183, 123–135.

71. Roth, J. (1987) Biochim. Biophys. Acta 906, 405–436.

72. Deschuyteneer, M., Eckhardt, A., Roth, J. and Hill, R.L. (1988) J. Biol. Chem. 263, 2452–2459.

73. Abeijon, C. and Hirschberg, C.B. (1987) J. Biol. Chem. 262, 4153–4159.

74. Hanover, J.A., Elting, J., Mintz, G.R. and Lennarz, W.J. (1982) J. Biol. Chem. 257, 10172–10177.

75. Piller, V., Piller, F. and Fukuda, M. (1990) J. Biol. Chem. 265, 9264–9271.

76. Perez-Vilar, J., Hidalgo, J. and Velasco, A. (1991) J. Biol. Chem. 266, 23967–23076.

77. Elhammer, A. and Kornfeld, S. (1984) J. Cell Biol. 98, 327–331.

78. Andersson, G.N. and Eriksson, L.C. (1981) J. Biol. Chem. 256, 9633–9639

79. Dunphy, W.G., Brands, R. and Rothman, J.E. (1985) Cell 40, 463–472.

80. Spielman, J., Hull, S.R., Sheng, Z., Kanterman, R., Bright, A. and Carraway, K.L. (1988) J. Biol. Chem. 263, 9621–9629.

81. Suganuma, T., Muramatsu, H., Muramatsu, T., Ihida, K., Kawano, J.-I., Murata, F.J. (1991) Histochem. Cytochem. 39, 299–309.

82. Roth, J., Taatjes, D.J., Weinstein, J., Paulson, J.C., Greenwell, P. and Watkins, W.M. (1986) J. Biol. Chem. 261, 14307–14312.

83. Trinchera, M., Pirovano, B. and Ghidoni, R. (1990) J. Biol. Chem. 265, 18242–18247.

84. Roth, J., Greenwell, P. and Watkins, W.M. (1988) Eur. J. Cell Biol. 46, 105–112.

85. Hull, S.R. and Carraway, K.L. (1989) J. Cell. Biochem. 40, 67–81.

86. Roth, J. (1984) J. Cell Biol. 98, 399–406.
87. Huang, K.M. and Snider, M.D. (1993) J. Biol. Chem., 268, 9302–9310.
88. Reichner, J.S., Whiteheart, S.W. and Hart, G.W. (1988) J. Biol. Chem. 263, 16316–16326.
89. Kreisel, W., Hildebrandt, H., Mössner, W., Tauber, R. and Reutter, W. (1993) Biol. Chem./ Hoppe-Seyler 374, 255–263.
90. Abejon, C. and Hirschberg, C.B. (1992) Trends Biochem. Sci. 17, 32–36.
91. Paulson, J.C., Weinstein, J. and Schauer, A. (1989) J. Biol. Chem. 264, 10931–10934.
92. Moscarello, M., Mitranic, M.M. and Vella, G. (1985) Biochim. Biophys. Acta 831, 192–200.
93. Carver, J.P. and Cumming, D.A. (1987) Pure Appl. Chem. 59, 1465–1476.
94. Wang, Y., Abernethy, J.L., Eckhardt, A.E. and Hill, R.L. (1992) J. Biol. Chem. 267, 12709–12726.
95. O'Connell, B.C., Hagen, F.K. and Tabak, L.A. (1992) J. Biol. Chem. 267, 25010–25018.
96. Matsuura, H., Greene, T. and Hakomori, S.-I. (1989) J. Biol. Chem. 264, 10472–10478.
97. Homa, F.L., Hollander, T., Lehman, D.J., Thomsen, D.R. and Elhammer, Å.P. (1993) J. Biol. Chem. 268, 12609–12616.
98. Hagen, F.K., Van Wuyekhuysen, B. and Tabak, L.A. (1993) J. Biol. Chem. 268.
99. Elhammer, A. and Kornfeld, S. (1986) J. Biol. Chem. 261(12), 5249–5255.
100. Sugiura, M., Kawasaki, T. and Yamashina, I. (1982) J. Biol. Chem. 257, 9501–9507.
101. Takeuchi, M., Yoshikawa, M., Sasaki, R. and Chiba, H. (1985) Agric. Biol. Chem. 49, 1059–1069.
102. Briand, J.P., Andrews, S.P., Cahill, E., Conway, N.A. and Young, J.D. (1981) J. Biol. Chem. 256, 12205–12207.
103. Cottrell, J.M., Hall, R.L., Sturton, R.G. and Kent, P.W. (1992) Biochem. J. 283, 299–305.
104. Hughes, R.C., Bradbury, A.F. and Smyth, D.G. (1988) Carbohydr. Res. 178, 259–269.
105. Cruz, T.F. and Moscarello, M.A. (1983) Biochim. Biophys. Acta 760, 403–410.
106. Young, J.D., Tsuchiya, D., Sandlin, D.E. and Holroyde, M.J. (1979) Biochemistry 18, 4444–4448.
107. Wilson, I.B.H., Gravel, Y. and von Heijne, G. (1991) Biochem. J. 275, 529–534.
108. O'Connell, B., Tabak, L.A. and Ramasubbu, N. (1991) Biochem. Biophys. Res. Comm. 180, 1024–1030.
109. Elhammer, Å.P., Poorman, R.A., Brown, E., Maggiora, L.L., Hoogerheide, J.G. and Kézdy, F.J.J. (1993) Biol. Chem. 268, 10029–10038.
110. Lehle, L. and Bause, E. (1984) Biochim. Biophys. Acta 799, 246–251.
111. Haltiwanger, R.S., Holt, G.D. and Hart, G.W. (1990) J. Biol. Chem. 265, 2563–2568.
112. Herscovics, A. and Orlean, P. (1993) FASEB J. 7, 540–550.
113. Springer, G.F., Desai, P.R., Wise, W., Carlstedt, S.C., Tegtmeyer, H., Stein, R. and Scanlon, E.F. (1990) In: R.B. Herberman and D.W. Mercer (Eds.), Immunodiagnosis of Cancer. Marcel Dekker.
114. Wilson, J.R., Deinhart, J.A. and Weiser, M.M. (1987) Biochim. Biophys. Acta 924, 332–340.
115. Cheng, P. and Bona, S.J. (1982) J. Biol. Chem. 257, 6251–6258.
116. Mulder, H., Schachter, H., de Jong-Brink, M., van der Ven, J.G. M., Kamerling, J.P. and Vliegenthart, J.F.G. (1991) Eur. J. Biochem. 201, 459–465.
117. Mendicino, J., Sivakami, S., Davila, M. and Chandrasekaran, E.V. (1982) J. Biol. Chem. 257, 3987–3994.
118. Furukawa, K. and Roth, S. (1985) Biochem. J. 227, 573–582.
119. Brockhausen, I., Möller, G., Pollex-Krüger, A., Rutz, V., Paulsen, H. and Matta, K.L. (1992) Biochem. Cell Biol. 70, 99–108.
120. Chandrasekaran, E.V., Jain, R.K. and Matta, K.L. (1992) J. Biol. Chem. 267, 19929–19937.
121. Gum, J.R., Byrd, J.C., Hicks, J.W., Toribara, N.W., Lamport, D.T.A. and Kim, Y.S. (1989) J.

Biol. Chem. 264, 6480–6487.

122. Brockhausen, I., Möller, G., Merz, G., Adermann, K. and Paulsen, H. (1990) Biochemistry 29, 10206–10212.

123. Paulsen, H., Pollex-Krüger and A. Sinnwell, V. (1991) Carbohydr. Res. 214, 199–226.

124. Berger, E.G. and Kozdrowski, I. (1978) FEBS Lett. 93, 105–108.

125. Cartron, J., Andrev, J., Cartron, J., Bird, G.W.G., Salmon, C. and Gerbal, A. (1978) Eur. J. Biochem. 92, 111–119.

126. Thurnher, M., Clausen, H., Fierz, W., Lanzavecchia, A. and Berger, E. (1992) Eur. J. Immunol. 22, 1835–1842.

127. Thurnher, M., Rusconi, S. and Berger, E. (1993) J. Clin. Invest. 91, 2103–2110.

128. Vavasseur, F., Yang, J., Dole, K., Corfield, A. and Brockhausen, I. (1992) Glycobiology 2, 6.10.

129. Zhuang, D., Yousefi, S. and Dennis, J.W. (1991) Cancer Biochem. Biophys. 12, 185–198.

130. Brockhausen, I., Romero, P. and Herscovics, A. (1991) Cancer Res. 51, 3136–3142.

131. Brockhausen, I., Matta, K.L., Orr J. and Schachter, H. (1985) Biochemistry 24, 1866–1874.

132. Yazawa, S., Abbas, S.A., Madiyalakan, R., Barlow, J.J. and Matta, K.L. (1986) Carbohydrate Res. 149, 241–252.

133. Podolsky, D.K. (1985) J. Biol. Chem. 260, 8262–8271.

134. Breg, J., van Halbeek, H., Vliegenthart, J.F.G., Klein, A., Lamblin, G. and Roussel, P. (1988) Eur. J. Biochem. 171, 643–654.

135. Savage, A.V., Donoghue, C.M., D'Arcy, S.M., Koeleman, C.A.M. and van den Eijnden, D.H. (1990) Eur. J. Biochem. 192, 427–432.

136. Hounsell, E.F., Wood, E., Feizi, T., Fukuda, M., Powell, M.E. and Hakomori, S.-I. (1981) Carbohydr. Res. 90, 283–307.

137. Strecker, G., Wieruszeski, J.-M., Michalski, J.-C., Alonso, C., Leroy, Y., Boilly, B. and Montreuil, J. (1993) In press.

138. Capon, C., Laboisse, C. L., Wieruszeski, J.-M., Maoret, J.-J., Augeron, C. and Fournet, B. (1992) J. Biol. Chem. 267, 19248–19257.

139. Bierhuizen, M.F. and Fukuda, M. (1992) Proc. Nat. Acad. Sci. USA 89, 9326–9330.

140. Bierhuizen, M.F.A., Mattei, M.-G. and Fukuda, M. (1993) Genes Develop. 7, 468–476.

141. Kuhns, W., Rutz, V., Paulsen, H., Matta, K.L., Baker, M.A., Barner, M., Granovsky, M. and Brockhausen, I. (1993) Glycoconjugate J. 10, 381–394.

142. Ropp, P.A., Little, M.R. and Cheng, P.-W. (1991) J. Biol. Chem. 266, 23863–23871.

143. Williams, D., Longmore, G.D., Matta, K.L. and Schachter, H. (1980) J. Biol. Chem. 255, 11253–11261.

144. Brockhausen, I., Matta, K.L., Orr, J., Schachter, H., Koenderman, A.H.L. and van den Eijnden, D.H. (1986) Eur. J. Biochem. 157, 463–474.

145. Williams, D. and Schachter, H. (1980) J. Biol. Chem. 255, 11247–11252.

146. Wingert, W.E. and Cheng, P. (1984) Biochemistry 23, 690–697.

147. Cheng, P., Wingert, W.E., Little, M.R. and Wei, R. (1985) Biochem. J. 227, 405–412.

148. Sangadala, S., Sivakami, S. and Mendicino, J. (1991) Mol. Cell. Biochem. 101, 125–143.

149. Heffernan, M., Lotan, R., Amos, B., Palcic, M., Takano, R. and Dennis, J.W. (1993) J. Biol. Chem. 268, 1242–1251.

150. Datti, A. and Dennis, J.W. (1993) J. Biol. Chem. 268, 5409–5416.

151. Greer, W.L., Higgins, E., Sutherland, D.R., Novogrodsky, A., Brockhausen, I., Peacocke, M., Rubin, L.A., Dennis, J.W. and Siminovitch, K.A. (1989) Biochem. Cell Biol. 67, 503–509.

152. Higgins, E.A., Zhuang, D.L., Brockhausen, I., Siminovitch, K.A. and Dennis, J.W. (1991) J. Biol. Chem. 266, 6280–6290.

153. Piller, F., Le Deist, F., Weinberg, K.I., Parkman, R. and Fukuda, M. (1991) J. Exp. Med. 173, 1501–1510.

154. Brockhausen, I., Kuhns, W., Schachter, H., Matta, K.L., Sutherland, R. and Baker, M.A. (1991) Cancer Res. 51, 1257–1263.

155. Saitoh, O., Piller, F., Fox, R.I. and Fukuda, M. (1991) Blood 77, 1491–1499.

156. Yousefi, S., Higgins, E., Daoling, Z., Pollex-Krüger, A., Hindsgaul, O. and Dennis, J.W. (1991) J. Biol. Chem. 266, 1772–1782.

157. Amano, J., Nishimura, R., Mochizuki, M. and Kobata, A. (1988) J. Biol. Chem. 263, 1157–1165.

158. Cole, L.A. (1987) J. Clin. Endocrinol. Metab. 65, 811–813.

159. Hull, S.R., Bright, A., Carraway, K.L., Abe, M., Hayes, D.F. and Kufe, D.W. (1989) Cancer Comm. 1, 261–267.

160. Sekine, M., Hashimoto, Y., Inagaki, F., Yamakawa, T. and Suzuki, A. (1990) J. Biochem. 108, 103–108.

161. Urashima, T., Messer, M. and Bubb, W.A. (1992) Biochim. Biophys. Acta, 1117, 223–231.

162. Klein, A., Carnoy, C., Wieruszeski, J.-M., Strecker, G., Strang, A.-M., van Halbeek, H., Roussel, P. and Lamblin, G. (1992) Biochemistry 31, 6152–6165.

163. Klein, A., Carnoy, C., Lamblin, G., Roussel, P., van Kuik, J.A. and Vliegenthart, J.F.G. (1993) Eur. J. Biochem. 211, 491–500.

164. van Kuik, A., de Waard, P., Vliegenthart, J.F.G., Klein, A., Carnoy, C., Lamblin, G. and Roussel, P. (1991) Eur. J. Biochem. 198, 169–182.

165. Piller, F., Cartron, J.P., Maranduba, A., Veyrieres, A., Leroy, Y. and Fournet, B. (1984) J. Biol. Chem. 259, 13385–13390.

166. Koenderman, A.H.L., Koppen, P.L. and van den Eijnden, D.H. (1987) Eur. J. Biochem. 166, 199–208.

167. Gu, J., Nishikawa, A., Matsuura, N., Sugiyama, T. and Taniguchi, N. (1992) Jpn. J. Cancer Res. 83, 878–884.

168. Leppänen, A., Penttilä, L., Niemelä, R., Helin, J., Seppo, A., Lusa, S. and Renkonen, O. (1991) Biochemistry 30, 9287–9296.

169. Gu, J., Nishikawa, A., Fuji, S.,Gasa, S. and Taniguchi, N. (1992) J. Biol. Chem. 267, 2994–2999.

170. Wu, A.M., Kabat, E.A., Nilsson, B., Zopf, D.A., Gruezo, F.G. and J. Liao (1984) J. Biol. Chem. 259, 7178–7186.

171. Basu, M. and Basu, S. (1984) J. Biol. Chem. 259, 12557–12562.

172. Wieruszeski, J.M., Michalski, J.C., Montreuil, J., Strecker, G., Peter-Katalinic, J., Egge, H., van Halbeek, H., Mutsaers, J.H.G.M. and Vliegenthart, J.F.G. (1987) J. Biol. Chem. 262, 6650–6657.

173. Kurosaka, A., Funakoshi, I., Matsuyama, M., Nagayo, T. and Yamashina, I. (1985) FEBS Lett. 190, 259–262.

174. Brockhausen, I., Williams, D., Matta, K.L., Orr J. and Schachter, H. (1983) Can. J. Biochem. Cell Biol. 61, 1322–1333.

175. Brockhausen, I., Orr, J. and Schachter, H. (1984) Can. J. Biochem. Cell Biol. 62, 1081–1090.

176. Yates, A.D. and Watkins, W.M. (1983) Carbohydrate Res. 120, 251–268.

177. Piller, F. and Cartron, J. (1983) J. Biol. Chem. 258, 12293–12299.

178. Zielenski, J. and Koscielak, J. (1983) FEBS Lett. 158, 164–168.

179. Stults, C.L.M. and Macher, B.A. (1993) Arch. Biochem. Biophys. 303, 125–133.

180. Holmes, E.H., Hakomori, S. and Ostrander, G.K. (1987) J. Biol. Chem. 262, 15649–15658.

181. Hosomi, O., Takeya, A. and Kogure, T. (1989) Jpn. J. Med. Sci. Biol. 42, 77–82.

182. van den Eijnden, D.H., Winterwerp, H., Smeeman, P. and Schiphorst, W.E.C.M. (1983) J. Biol. Chem. 258, 3435–3437.

183. van den Eijnden, D.H., Koenderman, A.H.L. and Schiphorst, W.E.C.M. (1988) J. Biol. Chem.

263, 12461–12471.

184. Morrison, J.F. and Ebner, K.E. (1971) J. Biol. Chem. 246, 3977–3984.

185. Aoki, D., Appert, H.E., Johnson, D., Wong, S.S. and Fukuda, M.N. (1990) EMBO J. 9, 3171–3178.

186. Ats, S., Lehmann, J. and Petry, S. (1992) Carbohydr. Res. 233, 125–139.

187. Bunnell, B.A., Adams, D.E. and Kidd, V.J. (1990) Biochem. Biophys. Res. Comm. 171, 196–203.

188. Sheares, B.T., Lau, J.Y.T. and Carlson, D.M. (1982) J. Biol. Chem. 257, 599–602.

189. Sheares, B.T. and Carlson, D.M. (1983) J. Biol. Chem. 258, 9893–9898.

190. Sherwood, A.L., Greene, T.G. and Holmes, E.H. (1992) J. Cell. Biochem. 50, 165–177.

191. Messer, M. and Nicholas, K.R. (1991) Biochim. Biophys. Acta 107, 79–85.

192. Bailly, P., Piller, F. and Cartron, J.-P. (1986) Biochim. Biophys. Res. Comm. 141, 84–91.

193. Joziasse, D.H., Damen, H.C.M., de Jong-Brink, M., Edzes, H.T. and van den Eijnden, D.H. (1987) FEBS Lett. 221, 139–144.

194. van Halbeek, H., Gerwig, G.J., Vliegenthart, J.F.G., Smits, H.L., van Kerkhof, P.J.M. and Kramer, M.F. (1983) Biochim. Biophys. Acta 747, 107–116.

195. Varki, A. (1992) Glycobiology 2, 25–40.

196. Schauer, R. (1982) Adv. Carbohydr. Chem. Biochem. 40, 131–234.

197. Devine, P.L., Clark, B.A., Birell, G.W., Layton, G.T., Ward, B.G., Alewood, P.F. and McKenzie, I.F.C. (1991) Cancer Res., 51, 5826–5836.

198. Shaw, L., Schneckenburger, P., Carlsen, J., Christiansen, K. and Schauer, R. (1992) Eur. J. Biochem. 206, 269–277.

199. Kanamori, A., Inoue, S., Iwasaki, M., Kitajima, K., Kawai, G., Yokoyama, S. and Inoue, Y. (1990) J. Biol. Chem. 265, 21811–21819.

200. Iwasaki, M., Inoue, S. and Troy, F.A. (1990) J. Biol. Chem. 265, 2596–2602.

201. Weinstein, J., DeSouza-e-Silva U. and Paulson, J.C. (1982) J. Biol. Chem. 257, 13835–13844.

202. Weinstein, J., DeSouza-e-Silva, U. and Paulson, J.C. (1982) J. Biol. Chem. 257, 13845–13853.

203. Weinstein, J., Lee, E.U., McEntee, K., Lai, P.-H. and Paulson, J.C. (1987) J. Biol. Chem. 262, 17735–17743.

204. Wang, X.-C., O'Hanlon, T.P., Young, R.F. and Lau, J.T.Y. (1990) Glycobiology, 1, 25.

205. Yonezawa, S., Tachikawa, T., Shin, S. and Sato, E. (1992) Am. J. Clin. Pathol. 98, 167–174.

206. Sadler, J.E., Rearick, J.I., Paulson, J.C. and Hill, R.L. (1979) J. Biol. Chem. 254, 4434–4443.

207. Sadler, J.E., Rearick, J.I. and Hill, R.L. (1979) J. Biol. Chem. 254, 5934– 5941.

208. Kurosawa, N., Hamamoto, T., Lee, Y.-C., Nakaoka, T. and Tsuji, S. (1993) Glycoconjugate J. 10, S 213.

209. Warren, L. (1974) Am. J. Pathol. 77, 69–76.

210. Saitoh, O., Wang, W.-C., Lotan, R. and Fukuda, M. (1992) J. Biol. Chem. 267, 5700–5711.

211. Baker, M.A., Taub, R.N., Kanani, A., Brockhausen, I. and Hindenburg, A. (1985) Blood 66, 1068–1071.

212. Baker, M.A., Kanani, A., Brockhausen, I., Schachter, H., Hindenburg, A. and Taub, R.N. (1987) Cancer Res. 47, 2763–2766.

213. Nakasaki, H., Mitomi, T., Noto, T., Ogoshi, K., Hanaue, H., Tanaka, Y., Makuuchi, H., Clausen, H. and Hakomori, S.-I. (1989) Cancer Res. 49, 3662–3669.

214. Itzkowitz, S.H., Dahiya, R., Byrd, J.C. and Kim, Y.S. (1990) Gastroenterology 99, 431–442.

215. Bergh, M.L.E. and van den Eijnden, D.H. (1983) Eur. J. Biochem. 136, 113–118.

216. Bergh, L.E.M., Hooghwinkel, G.J.M. and van den Eijnden, D.H. (1983) J. Biol. Chem. 258, 7430–7436.

217. Baubichon-Cortray, H., Broquet, P., George, P. and Louisot, P. (1989) Glycoconjugate J. 6, 115–127.

218. Dall'olio, F., Malagolini, N., di Stefano, G., Ciambella, M. and Serafini-Cessi, F. (1990). Biochem. J. 270, 519–524.
219. Rearick, J.I., Sadler, J.E., Paulson, J.C. and Hill, R.L. (1979) J. Biol. Chem. 254, 4444–4451.
220. Bergh, M.L.E., Hooghwinkel, G.J.M. and van den Eijnden, D.H. (1981) Biochim. Biophys. Acta 660, 161–169.
221. Hindsgaul, O., Kaur, K.J., Srivastava, G., Blaszczyk-Thurin, M., Crawley, S.C., Heerze, L.D. and Palcic, M.M. (1991) J. Biol. Chem. 226, 17858–17862.
222. Reboul, P., George, P., Geoffrey, J., Louisot, P. and Broquet, P. (1992) Biochim. Biophys. Res. Comm. 186, 1575–1581.
223. Broquet, P., George, P., Geoffrey, J., Reboul. and Louisot, P. (1991) Biochem. Biophys. Res. Comm. 178, 1437–1443.
224. Gillespie, W., Kelm, S. and Paulson, J.C. (1992) J. Biol. Chem. 267, 21004–21010.
225. de Heij, H.T., Tetteroo, P.A.T., van Kessel, A.H.M.G., Schoenmaker, E., Visser, F.J. and van den Eijnden, D.H. (1988) Cancer Res. 48, 1489–1493.
226. Tetteroo, P.A.T., de Heij, H.T., van den Eijnden, D.H., Visser, F.J., Schoenmaker, E. and Geurts van Kessel, A.H.M. (1987) J. Biol. Chem. 262, 15984–15999.
227. Markwell, M.A.K. and Paulson, J.C. (1980) Proc. Nat. Acad. Sci. USA 77, 5693–5697.
228. Crocker, P.R., Kelm, S., Dubois, C., Martin B., McWilliams, A.S., Shotton, D.M., Paulson, J.C. and Gordon, S. (1991) EMBO J. 10, 1661–1669.
229. Kanani, A., Sutherland, D.R., Fibach, E., Matta, K.L., Hindenburg, A., Brockhausen, I., Kuhns, W., Taub, R.N., van den Eijnden D. and Baker, M.A. (1990) Cancer Res. 50, 5003–5007.
230. Fukuda, M., Carlsson, S.R., Klock, J.C. and Dell, A. (1986) J. Biol. Chem. 261, 12796–12806.
231. Saitoh, O., Gallagher, R.E. and Fukuda, M. (1991) Cancer Res. 51, 2854–2862.
232. Gillespie, W., Paulson, J., Kelm, S., Pang, M. and Baum, L. (1993) J. Biol. Chem. 268, 3801–3804.
233. Delannoy, P., Pelczar, H., Vandamme, V. and Verbert, A. (1993) Glycoconjugate J. 10, 91–98.
234. Greenwell, P., Ball, M G. and Watkins, W.M. (1983) FEBS Lett. 164, 314–317.
235. Cheng, P.-W. and de Vries, A. (1986) Carbohydr. Res. 149, 253–261.
236. Beyer, T.A., Sadler, J.E. and Hill, R.L. (1980) J. Biol. Chem. 255, 5364–5372.
237. Beyer, T.A. and Hill, R.L. (1980) J. Biol. Chem. 255, 5373–5379.
238. Sarnesto, A., Köhlin, T., Hindsgaul, O., Thurin, J. and Blaszcyk-Thurin, M. (1992) J. Biol. Chem. 267, 2737–2744.
239. Sarnesto, A., Köhlin, T., Thurin, J. and Blaszczyk-Thurin, M. (1990) J. Biol. Chem. 265, 15067–15075.
240. Kyprianou, P., Betteridge, A., Donald, A.S.R. and Watkins, W.M. (1990) Glycoconjugate J. 7, 573–588.
241. Larsen, R.D., Ernst, L.K., Nair, R.P. and Lowe, J.B. (1990) Proc. Nat. Acad. Sci. USA 87, 6674–6678.
242. Ernst, L.K., Rajan, V.P., Larsen, R.D., Ruff, M.M. and Lowe, J.B. (1989) J. Biol. Chem. 264, 3436–3447.
243. Rajan, V.P., Larsen, R.D., Ajmera, S., Ernst, L.K. and Lowe, J.B. (1989) J. Biol. Chem. 264, 11158–11167.
244. Madiyalakan, R., Yazawa, S., Abbas, S.A., Barlow, J.J. and Matta, K.L. (1986) Anal. Biochem. 152, 22–28.
245. Mollicone, R., Candelier, J.-J., Mennesson, B., Couillin, P., Venot, A. and Oriol, R. (1992) Carbohydr. Res. 228, 265–276.
246. Campbell, C. and Stanley, P. (1984) J. Biol. Chem. 259, 11208–11214.
247. Howard, D.R., Fukuda, M., Fukuda, M.N. and Stanley, P. (1987) J. Biol. Chem. 262, 16830–16837.

248. Prieels, J., Monnom, D., Dolmans, M., Beyer, T.A. and Hill, R.L. (1981) J. Biol. Chem. 256, 10456–10463.
249. Eppenberger-Castori, S., Lötscher, H. and Finne, J. (1989) Glycoconjugate J. 6, 101–114.
250. Kukowska-Latallo, J. F., Larsen, R.D., Nair, R.P. and Lowe, J.B. (1990) Genes Devel. 4, 1288–1303.
251. Lowe, J.B., Kukowska-Latallo, J.F., Nair, R.P., Larsen, R.D., Marks, R.M., Macher, B.A., Kelly, R.J. and Ernst, L.K. (1991) J Biol. Chem. 266, 17467.
252. Foster, C.S. and Glick, M.C. (1991) J. Biol. Chem. 266, 3526–3531.
253. Muramatsu, H., Kamada, Y. and Muramatsu, T. (1986) Eur. J. Biochem. 157, 71– 75.
254. Mollicone, R., Gibaud, A., Francois, A., Ratcliffe, M. and Oriol, R. (1990) Eur. J. Biochem. 191, 169–176.
255. Weston, B.W., Nair, R.P., Larsen, R.D. and Lowe, J.B. (1992). J. Biol. Chem. 267, 4152.
256. Sarnesto, A., Köhlin, T., Hindsgaul, O., Vogele, K., Blaszcyk-Thurin, M. and Thurin, J. (1992) J. Biol. Chem. 267, 2745–2752.
257. Holmes, E.H. (1992) Arch. Biochem. Biophys. 296, 562–568.
258. Chandrasekaran, E.V., Jain, R.K. and Matta, K.L. (1992) J. Biol. Chem. 267, 23806–23814.
259. Koszdin, K.L. and Bowen, B.R. (1992) Biochem. Biophys. Res. Comm. 187, 152–157.
260. Weston, B.W., Smith, P., Kelly, R.J. and Lowe, J.B. (1992) J. Biol. Chem. 267, 24575–24584.
261. Nishihara, S., Nakazato, M., Kudo, T., Kimura, H., Ando, T. and Narimatsu, H. (1993) Biochem. Biophys. Res. Comm. 190, 42–46.
262. Nagai, M., Dave, V., Kaplan, B.E. and Yoshida, A. (1978) J. Biol. Chem. 253, 377–379
263. Navaratnam, N., Findlay, J.B.C., Keen, J.N. and Watkins, W.M. (1990) Biochem. J. 271, 93–98.
264. Schwyzer, M. and Hill, R.L. (1977) J. Biol. Chem. 252, 2338–2345.
265. Schwyzer, M. and Hill R.L. (1977) J. Biol. Chem. 252, 2346–2355.
266. Joziasse, D.H., Shaper, J.H., Jabs, E.W. and Shaper, N.L. (1991) J. Biol. Chem. 266, 6991–6998.
267. Nagai, M., Dave, V., Meunsch, H. and Yoshida, A. (1978) J. Biol. Chem. 253, 380–381.
268. Greenwell, P., Yates, A.D. and Watkins, W.M. (1986) Carbohydrate Res. 149, 149–170.
269. Yates, A.D., Feeney, J., Donald, A.S.R. and Watkins, W.M. (1984) Carbohydrate Res. 130, 251–260.
270. Yamamoto, F.-I. and Hakomori, S.-I. (1990) J. Biol. Chem. 265, 19257–19262.
271. Yamamoto, F.-I., Clausen, H., White, T., Marken, J. and Hakomori, S.-I. (1990) Nature 345, 229–233.
272. Blanchard, D., Capon, D.C., Leroy, Y., Cartron, J.P. and Fournet, B. (1985) Biochem. J. 232, 813–818.
273. Blanchard, D., Cartron, J.P., Fournet, B., Montreuil, J., van Halbeek, H. and Vliegenthart, J.F.G. (1983) J. Biol. Chem. 258, 7691–7695.
274. Donald, A.S.R., Soh, C.P.C., Yates, A.D., Feeney, J., Morgan, W.T.J. and Watkins, W.M. (1987) Biochem. Soc. Trans. 15. 606–608.
275. Herkt, F., Parente, J.P., Leroy, Y., Fourne, B., Blanchard, D., Cartron, J.P., van Halbeek, H. and Vliegenthart, J.F.G. (1985) Eur. J. Biochem. 146, 125–129.
276. Takeya, A., Hosomi, O. and Kogure, T. (1987) J. Biochem. (Tokyo) 101, 251–259.
277. Conzelmann, A. and Kornfeld, S. (1984) J. Biol. Chem. 259, 12536–12542.
278. Conzelmann, A. and Bron, C. (1987) Biochem. J. 242, 817–824.
279. Malagolini, N., Dall'Olio, F. and Serafini-Cessi, F. (1991) Biochem. Biophys. Res. Comm. 180, 681–686.
280. Malagolini, N., Dall'Olio, F., Di Stefano, G., Minni, F., Marrano, D. and Serafini-Cessi, F. (1989) Cancer Res. 49, 6466–6470.
281. Serafini-Cessi, F., Malagolini, N. and Dall'Olio, F. (1988) Arch. Biochem. Biophys. 266(2),

573–582.

282. Piller, F., Blanchard, D., Huet, M. and Cartron, J. (1986) Carbohydrate Res. 149, 171–184.

283. Serafini-Cessi, F., Dall'Olio, F. and Malagolini, N. (1986) Carbohydrate Res. 151, 65–76.

284. Dohi, T., Nishikawa, A., Ishizuka, I., Totani, M., Yamaguchi, K., Nakagawa, K., Saitoh, O., Ohshiba, S. and Oshima, M. (1992) Biochem. J. 288, 161–165.

285. Mawhinney, T.P., Adelstein, E., Morris, D.A., Mawhinney, A.M. and Barbero, G.J. (1987) J. Biol. Chem. 262, 2994–3001.

286. Mawhinney, T.P., Landrum, D.C., Gayer, D.A. and Barnero, G.J. (1992) Carbohydr. Res. 235, 179–197.

287. Mawhinney, T.P., Adelstein, E., Gayer, D.A., Landrum, D.C. and Barbero, G.J. (1992) Carbohydr. Res. 223, 187–207.

288. Strecker, G., Wieruszeski, J.-M., Martel, C. and Montreuil, J. (1987) Glycoconjugate 4, 329–337.

289. Chace, K.V., Leahy, D.S., Martin, R., Carubelli, R., Flux, M. and Sachdev, G.P. (1983) Clin. Chim. Acta 132, 143–155.

290. Perez, M. and Hirschberg, C.B. (1986) Biochim. Biophys. Acta 864, 213–222.

291. Baeuerle, P.A. and Huffner, W.B. (1986) Biochem. Biophys. Res. Comm. 141, 870–877.

292. Carter, S.R., Slomiany, A., Gwozdzinski, K., Liau, Y.H. and Slomiany, B.L. (1988) J. Biol. Chem. 263, 11977–11984.

J. Montreuil, H. Schachter and J.F.G. Vliegenthart (Eds.), *Glycoproteins*

Biosynthesis

4a. Gene regulation of terminal glycosylation

MARY CATHERINE GLICK

The Children's Hospital of Philadelphia, Department of Pediatrics,
University of Pennsylvania School of Medicine,
34th Street and Civic Center Boulevard, Philadelphia, PA 19104, USA

1. Introduction

The genetic control of glycoprotein biosynthesis is not yet fully appreciated. The exquisite specificity of the glycosyltransferases in coordinating glycoprotein synthesis is discussed elsewhere in this volume [1]. The available substrate for each enzyme is one of the major factors in this synthesis. Does this then imply that all (an estimated 100 or more) of the trimming enzymes and glycosyltransferases are expressed and active in all tissues at all times? Based on the limited amount of information available, this is not the case. Obviously the glycosyltransferase must be expressed for synthesis to proceed. However, other factors must provide controlling parameters to the cell to negotiate terminal glycosylation which may be required for biological function.

A number of biological events, normal and pathological, appear to be coordinated with terminal glycosylation. Therefore, disregarding the complexities of total oligosaccharide synthesis, the events leading to terminal glycosylation in the biosynthesis of N-linked glycans will be discussed here. Sialyl and fucosyltransferases will be used as examples, since in both of these cases regulation at the gene level has been examined. Genetic regulation of blood group substances will be discussed elsewhere in this volume [2]. Pathology leading to altered terminal glycosylation will be discussed briefly, however, see Kobata [3] for a comprehensive review. For additional information including comparative genomic structural data with the other glycosyltransferases the reader is referred to Schachter [1]. In addition, the reviews of Joziasse [4]; Schachter [5,6]; Shaper and Shaper [7]; de Vries and van den Eijnden [8]; and Watkins et al. [9] deal with aspects of this subject.

2. Normal regulation

The following sections point out several biological processes which are accompanied by exquisite regulation of terminal glycosylation.

2.1 Development

In 1978, Solter and Knowles [10] described an antigen which they followed during development of the mouse embryo. This antigen was stage specific, arising at the compaction stage and disappearing in the adult. The antigen, SSEA-1, was subsequently found to be Gal(β1,4)[Fuc(α1,3)]GlcNAc, present on glycolipids and glycoproteins (reviewed in Foster and Glick [11]). Soon thereafter fucosyl residues α-1,3-linked to GlcNAc were shown by biochemical techniques to be present on human neuroblastoma [12] and subsequently on a variety of other tumors [9,13]. The issue of whether the tumor cells reexpress an embryonic function or whether another enzyme is overexpressed for the synthesis of glycoproteins containing fucosyl residues in α-1,3-linkage has not been resolved to date. It has been shown however that substitution by terminal fucosyl residues affects the conformation of the oligosaccharide antennae [14].

The first attempt to define the developmental α-1,3-fucosyltransferase came with the cloning of α-1,3/4-fucosyltransferase by Lowe and his colleagues [15]. The techniques used by Lowe to clone by transfection procedures bypassed the usual enzyme purification, followed by protein sequencing and oligonucleotide screening of libraries. Indeed Lowe and his colleagues have been foremost in defining the variety of human α-1,3-fucosyltransferases which could potentially direct the synthesis of SSEA-1. However, the story which unfolded was more complicated than previously suspected. Of the four human α-1,3-fucosyltransferases already cloned (Table I) it is not certain which is expressed in development or whether there is yet another candidate.

Related to developmental glycosylation changes is the sialylation of the neural cell adhesion molecule (N-CAM). N-CAM contains a homopolymer of α-2,8-linked sialic acid, polysialic acid. This highly charged molecule extends from the surface of cells as they migrate from the neural crest. N-CAM has been extensively studied [16] and it is reported that the length of the sialyl polymers contributes to migration, the longer polymers being present in the embryo whereas the shorter polymers are present in the adult. As with the presence of Fuc(α1–3)GlcNAc, the expression of polysialic acid has been shown in several human tumors (reviewed in [17]). Again the question is asked whether a specific polysialyltransferase is developmentally regulated and subsequently reexpressed in certain tumors or do several polysialyltransferases exist. Since short oligomers exist on N-CAM from adults, it is possible that an elongating enzyme is present in the early stages of development. Although the bacterial polysialyltransferase system has been extensively studied [17], definition of the human enzyme(s) is only beginning. A sialyltranferase modulated with development was recently cloned but the expressed protein had no activity with the substrates examined [18].

The development of thymocytes is characterized by conversion of PNA+ cells to PNA− cells. Peanut agglutinin (PNA) reacts with the less mature cells which subsequently lose this binding capacity as they develop [19]. The PNA− cells have been characterized as containing NeuNAc(α2–3)Gal(β1–3)GalNAc residues [20]. Another example is the maturation of B lymphocytes. In human mature B lymphocytes, a sialylated cell surface epitope CDw75 is expressed when α-2,6-sialyltransferase is induced by B cell activation [21].

TABLE I

Substrate requirements of human α-1,3-fucosyltransferases

Substrate	Relative activity (percentage)[e]											
	FucT-III[a]	FucT-IV[a]		FucT-V[a]		FucT-VI[a]		Neuro-blastoma[b]	Plasma[c]	Neutrophils[c] Normal	Neutrophils[c] CML	Liver[d]
Gal(β1–4)GlcNAc	100	100	100	100	100	100	100[f]	100	100	100	100	100
Gal(β1–4)Glc	145	3	2	11	65	<1	20	10	–[g]	–	–	–
NeuNAc(α2–3)Gal(β1–4)GlcNAc	56	<1	0	115	45	110	45	0	147	46	2	133
Fuc(α1–2)Gal(β1–4)Glc	254	6	18	42	105	<1	60	27	–	–	–	–
Fuc(α1–2)Gal(β1–4)GlcNAc	–	–	–	–	–	–	–	140	162	102	163	120
Lacto-N-biose	420	<1	0	10	10	<1	4	0	0	0	0	0
Reference	15	86	68	25	68	64	68	40	9	9	9	9

[a] cDNA expressed in COS cells.
[b] Purified to electrophoretic homogeneity from neuroblastoma cells.
[c] Purified from plasma and neutrophils, CML from a patient with chronic myelogneous leukemia. Relative activities of a purified serum enzyme were not reported although the K_m values were determined [89].
[d] Purified from liver.
[e] Expressed as percentage of activity using Gal(β1–4)GlcNAc as substrate.
[f] Terminology of Ref. [68].
[g] Not reported.

2.2. Tissue specificity

Terminal glycosyltransferases do not fit completely into the category of either house-keeping or tissue specific genes. In the case of α-2,6-sialyltransferase, most tissues express a constitutive enzyme but the level of expression and isotype varies due to multiple and distinct promoter regions [22]. On the other hand, the tissue specific expression of α-1,3-fucosyltransferase is thought to be regulated by multiple fucosyl-transferase genes which encode catalytically similar enzymes. These genes are differ-entially expressed in different tissues and have different substrate specificities [4].

In an extensive study of crude tissue preparations, Oriol and his colleagues [23] concluded that α-1,3-fucosyltransferases with different biochemical characteristics were present in different tissues, terming the enzymes plasma, myeloid and Lewis. Another category of lung has been added [24]. This led to the unfortunate terminology for α-1,3-fucosyltransferases using these tissue locations. It remains to be shown by molecular biology techniques whether or not this classification is valid. Due to the presence of several α-1,3-fucosyltransferases in the same tissue, it may not be possible to classify the enzymes from crude homogenates. Moreover, even purification and assignment of substrate specificity (Table I) or function to the particular purified enzyme do not preclude other α-1,3-fucosyltransferases in the same tissue, *in vivo*. For these reasons, the terminology proposed by Weston et al. [25] of assigning numbers to the cloned human α-1,3-fucosyltransferases (FucT-III to -VI) will be used here (Table I). FucT-I and -II were derived from CHO mutants [26,27] and are not discussed, and FucT-III, the Lewis type enzyme, is discussed by Watkins in this volume [2].

Paulson and his colleagues [22] showed an unusual tissue pattern for Gal(β1–4)GlcNAc α-2,6-sialyltransferase cDNA and the molecular mechanisms which could lead to the expression of the enzyme are described in Section 4 of this chapter. A tissue specific expression of Gal(β1–3)GalNAc α-2,3-sialyltransferase has been shown which differs from that of Gal(β1–4)GlcNAc α-2,6-sialyltransferase, with the tissues synthesizing O-linked glycans showing the highest expression [28]. In contrast, the highest mRNA level for Gal(β1–4)GlcNAc α-2,3-sialyltransferase was detected in brain with only low levels found in submaxillary gland, a reverse of the expression of α-2,6-sialyltransferase [29]. Therefore the hypothesis is presented that tissue-specific transcription elements regulate the expression of terminal glycosyltransferases [30].

2.3. Chromosomal location

The chromosomal locations of several glycosyltransferases have been assigned by use of somatic cell hybrids or *in situ* hybridization. The results show disparate locations for the different glycosyltransferases (Table II). This may be surprising considering the exquisite specificity and orderly action of these enzymes in glycoprotein synthesis [6]. To further stress the chromosomal independence of the glycosyltransferases the loca-tion of GlcNAcT-I which resides on chromosome 5 [31] is also given in Table II. Thus far with the exception of the ABH gene transferases [2] there appears to be little relationship to chromosomal location and the order of glycoprotein synthesis. Interest-ingly, the ABH gene transferases are present on chromosome 9 along with the non

TABLE II

Chromosome location of human glycosyltransferases

Glycosyltransferase	Chromosome	Ref.
α-2,6-NeuNAcT (Gal(β1–4)GlcNAc)	3q27–q28	59
α-2,3-NeuNAcT (Gal(β1–3)GalNAc)[a]	11	90
α-1,3-FucT-III, -V, -VI	19	25,64
α-1,3-FucT-IV	11	25
α-1,2-FucT	19	70
β-1,4-GalT	9p13	91
GlcNAcT-I	5q35	31
α-1,3-GalT Pseudogene		32
non-processed	9q33–q34	
processed	12q14–q15	

[a] Or a regulatory factor.

processed pseudogene of α-1,3-galactosyltransferase [32] and this group of transferases shows sequence similarity. It is suggested therefore that the homologous sequence provides evidence for exon recycling [4].

The FucT-VI gene is located only 13 kb 3′ to the FucT-III gene on chromosome 19 [33]. Several families of α-1,3-fucosyltransferases exist with different chromosomal locations (Table II) and the location may be directly related to function. These fucosyl residues are not universally found on glycoproteins. The homology or lack thereof of the two categories of α-1,3-fucosyltransferase genes which are located on chromosomes 19 and 11 (Table II) provide evidence for the postulate that a distant gene duplication event led to two α-1,3-fucosyltransferase gene families on two different chromosomes with diverse interior sequences [4].

Evolution may also play a regulating role for terminal glycosylation. For example, α-1,3-galactosyltransferase is inactive in catarrhine cells [34] which at least in humans are left with pseudogenes (Table II).

3. Pathological regulation

A few examples of pathological conditions which appear to be closely coordinated with changes in terminal glycosylation are described. It should be noted however that the genetic regulation of a particular glycosyltransferase bringing about the glycosylation change has not yet been fully delineated. Attempts to control aberrant glycosylation by way of glycosyltransferase inhibitors have been reviewed [35].

3.1. Oncogenesis

Glycoproteins and glycolipids have been shown to be altered in transformed and tumor cells [36]. Although a number of specific glycoproteins and glycolipids have been reported, the overall change is to more highly branched oligosaccharides on glycoproteins and less complicated oligosaccharides on glycolipids. The main issue remains whether or not these alterations which appear to correlate with oncogenic transformation are due to overexpression of certain glycosyltransferases to bring about the complexity of oligosaccharide antennae observed on the glycoproteins. The focus here will be on α-1,3-fucosyltransferases and sialyltransferases which show aberrant patterns of expression as represented by their final products, the membrane glycoproteins.

After transfection of mouse NIH-3T3 with H-*ras* oncogene, terminal glycosylation was shown to change from Gal(α1–3)Gal(β1–4) ···· terminated oligosaccharides to NeuNAc(α2–3)Gal(β1–4) ···· [37]. Prior to transformation the glycoproteins of NIH-3T3 cells contained a high percentage of tetra-antennary oligosaccharides which were capped with α-1,3-Gal residues. NIH-3T3 cells are immortalized and thus may express a first step of transformation, the presence of tetra-antennary oligosaccharides. The second step characterized phenotypically by growth in soft agar and tumor formation and in this case brought about by oncogenic transformation would be a negative expression of α-1,3-galactosyltransferase with a positive expression of α-2,3-sialyltransferase. The regulatory factors involved in the expression of the enzymes in this particular system (other 3T3 cells may not be exactly similar [38,39]) could provide insights into the glycosylation changes brought about by oncogenic transformation in addition to control of terminal glycosylation.

α-1,3-Fucosyltransferases have been demonstrated in a number of human tumors with a variety of substrate specificities [13,40]. The differences are to the extent that although α-1,3-fucosyltransferase has been suggested as diagnostic for certain tumors it would probably depend on the substrate utilized. Therefore at this point there does not appear to be one particular α-1,3-fucosyltransferase universally expressed in human tumors. For example, Foster et al. [40] purified an α-1,3-fucosyltransferase to electrophoretic homogeneity from human neuroblastoma cells and the purified enzyme showed substrate specificities quite divergent from the enzyme described in other tumors [24]. That is, it utilized Fuc(α1–2)Gal(β1–4)GlcNAc as the preferred substrate examined and did not utilize NeuNAc(α2–3)Gal(β1–4)GlcNAc (Table I). The neuroblastoma enzyme corresponds in substrate specificity most closely to the cloned FucT-IV expressed in COS cells (Table I). Other enzymes from human tumors utilize other substrates with preference some transferring to Gal(β1–4)GlcNAc and many transferring to NeuNAc(α2–3)Gal(β1–4)GlcNAc (reviewed in Ref. [8,9]). Many of the studies have been performed on crude or partially purified extracts of the tumor cells so it is hard to make a comparison with the purified enzymes from neuroblastoma or CML cells (Table I).

Several α-1,3-fucosyltransferases have been shown in ovarian cancer by the use of synthetic and natural substrates [41]. The striking substrate differences between the enzymes derived from soluble or particulate fractions of the tumor in addition to those found elevated in the serum of tumor-bearing patients accentuates the complexity of the types of α-1,3 and α-1,2-fucosyltransferases expressed during oncogenesis.

The expression of α-1,3-fucosyltransferases in leukemia cells compared to the enzymes expressed in normal leukocytes has been extensively studied by Watkins and her colleagues and has been reviewed [9]. In several leukemias, α-1,3-fucosyltransferases have been purified and their biochemical properties related to normal leuko-cytes. The leukemia cells have substrate specificities different from the normal leuko-cytes (Table I). This led to the suggestion that either activation of a silent gene or other alterations at the genetic level may regulate the expression of two different fucosyl-transferases albeit perhaps from the same gene family [9,42].

3.2. Metastasis

The metastatic spread of tumor cells is associated not only with glycosylation changes [3] but also with many other factors [43]. Only terminal glycosylation changes are discussed below.

Gasic and Gasic [44] observed that injection of mice with neuraminidase reduced the number of tumors formed by TA3 cells when compared with non-treated mice. Decades later many additional observations have been made linking sialylation to the metastatic properties of cells [36] but no one has yet reported any controls at the genetic level. Most recently the presence of polymers of α-2,8-linked sialic acid has been reported on human tumors [17,45]. If an analogy can be made to the developmental properties of polysialylated N-CAM then perhaps polysialic acid on the surface of tumors contrib-utes to the metastatic properties [46]. Indeed, in the case of human neuroblastoma, α-2,8-sialic acid has been found in polymers of 55 or greater sialyl residues [45].

As stated above, Fuc(α1–3)GlcNAc has also been observed on glycoproteins in human tumors. A relationship of α-1,3-fucosyltransferase with the metastatic proper-ties of ovarian cancer has been suggested [41]. Interestingly α-1,2-fucosyltransferase is also highly expressed in the ovarian tumors. Clinically a number of glycoproteins are used as markers to detect various carcinomas with antibodies to Fuc(α1–3/4)GlcNAc (reviewed in [11]).

3.3. Inflammatory response

α-1,3-Fucosyltransferase is partially responsible for conferring biological specificity on molecules which react with the selectin family of cell adhesion molecules [47]. Although four members of the human α-1,3-fucosyltransferase gene family have been cloned and their substrate specificities partially described (Table I) it is not at present known which of these is responsible for the particular biologically functional surface molecules of the inflammatory process [48].

3.4. Genetic diseases

In a number of genetic diseases terminal glycosylation is altered due to an enzyme deficiency associated with the biosynthesis of core oligosaccharides. These include carbohydrate deficient glycoprotein (CDG) syndrome [49], I cell disease, and aspartyl-glycosaminuria [50]. There are a number of genetic diseases in which altered glycosy-lation may be an indirect effect of the genetic determinants of the disease, however the altered glycosylation may lead in part to the pathology found in the disease. These include cystic fibrosis [51] and galactosemia [52, 53]. In cystic fibrosis, a reciprocal relationship has been described for fucose and sialic acid (reviewed in [54]).

4. Molecular mechanisms

The molecular mechanisms for the regulation of glycosyltransferases at the level of the genes are only beginning to unfold. The regulation of α-2,6-sialyltransferase has been studied extensively and it appears that tissue specific regulatory elements provide controls for the enzyme [30,55]. On the other hand, gene duplication rather than alternate splicing may provide some of the multiple forms of α-1,3-fucosyltransferase [4]. The hypothesis proposed by Harduin-Lepers et al. [56] for β-1,4-galactosyltransferase provides a mechanism depicting the extensive interplay of tissue specific regulatory factors. Although one function of β-1,4-galactosyltransferase is to provide lactose, and therefore may require unique regulation, it may be that this hypothesis [56] provides a universal form of regulation for glycosyltransferases.

4.1. Exons and regulatory elements

Schematic representation of the organization of the rat α-2,6-sialyltransferase and the human α-1,3-fucosyltransferase genes are shown in Fig. 1. The α-2,6-sialyltransferase gene has multiple promoters and eight exons whereas the α-1,3-fucosyltransferase has only one exon. In the case of the α-2,6-sialyltransferase gene a constitutive promoter

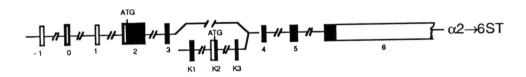

Fig. 1. Organization of α-2,6-sialyl and α-1,3-fucosyltransferase genes. The intron–exon structure of the rat α-2,6-sialyltransferase (ST) and the human α-1,3-fucosyltransferase (FT) genes is represented. Black boxes represent protein coding sequence exons, open boxes 5'- and 3'-untranslated exonic sequence. The translational initiation site is indicated (ATG). α-2,6-ST: alternative promoter usage in rat kidney produces three different transcripts. These transcripts contain one or two of the exons (K1, K2, K3) that localize between exon 3 and exon 4 on the α-2,6-ST genomic map, and are absent from the constitutively expressed transcript. The transcripts that incorporate K2 contain an in-frame translational start site [30,57,58]. Exons X and Y occur in the human sequence downstream from exon −1 [59]. α-1,3-FT: the schematic containing one exon is representative of the cloned gene sequences [64]. Modified with permission from Ref. [4].

is thought to be expressed in all tissues [22,30]. A liver specific promoter has been shown to be under the influence of liver specific regulatory elements [55]. Each initiates transcription of mRNA of different sizes. In addition, a kidney specific promoter has been described which produces three different transcripts [57,58] and human B cells contain three exons not seen in rat cells or human hepatoma cells [59].

The liver enriched transcription factors HNF-1, DBP and LAP were shown to be capable of *trans*-activating transcription of the α-2,6-sialyltransferase promoter [55]. Since these factors are also active in the synthesis of albumin by rat liver a more detailed regulation may be necessary for specific synthesis of the glycosyltransferase which provides terminal glycosylation.

In the rat kidney two exons, K_1 and K_2, which are localized between exons 3 and 4, have been found to yield two distinct size classes of mRNA (Fig. 1, Table III) which are absent from constitutively expressed transcripts of α-2,6-sialyltransferase. O'Hanlon and Lau [60] found that the larger transcript (4.7 kb) when transfected into CHO cells yields glycoproteins containing sialic acid in α-2,6-linkage as determined by reactivity with SNA, a lectin specific for sialic acid in α-2,6-linkage [61]. In contrast, they found that two other mRNAs which were smaller transcripts (3.6 kb) did not express surface glycoconjugates detectable with SNA even though the CHO transfectants contained detectable transcripts. In fact these smaller transcripts are responsible for the predominant α-2,6-sialyltransferase protein in rat kidney, and are not observed in other rat tissues or in mouse kidney. Although there is speculation on the function of the divergent transcripts no activity has yet been detected when these sialyltransferase homologues are expressed [60].

The rat and human α-2,6-sialyltransferase genomic sequences show conservation of intron/exon boundaries and extensive inter-species sequence similarity. It appears however that human cells of the B cell lineage contain three upstream exons not seen in mRNA isotypes expressed in human hepatoma cells (Table III). These exons result in B-lymphoblastoid cells expressing an isotype containing the one exon sequence whereas the mature B cell lines express a high level of slightly smaller mRNA as a result of the other exon [59]. Thus a complicated series of controls must be necessary to express α-2,6-sialyltransferase by the multiple promoters on a single gene. These controls in turn must be linked not only to the tissue specific expression of the sialyltransferase but also to the environment invoked expression of the enzyme when terminal sialylation is needed.

The history of β-1,4-galactosyltransferase should be noted at this point. Two size forms of β-1,4-galactosyltransferase have been studied and since biological observations yielded a functional enzyme at the surface of cells [62] and in the *trans*-Golgi [63] it was thought that the longer form (mRNA 4.1.kb) was directed to the membrane and the shorter form (mRNA 3.9 kb) to the cytoplasm. In a series of experiments (summarized in [56]) the regulatory factors controlling the expression of both forms of the enzyme were described. The cellular requirements for the enzyme correlated with the transcriptional start site used. When increased transcripts were observed in lactating mammary gland the 3.9 kb transcription start site was used, whereas the 4.1 kb start site or a combination of both were used in other cell types. The *cis*-acting elements were studied and a model (Fig. 2) was proposed with the suggestion that the distal region

TABLE III
Some characteristics of cloned glycosyltransferases

Cloned transferase	Substrate specificity[a]	Sequence homology[b]	Species and tissue	Prominent mRNA transcripts (kb)	Ref.
Sialyltransferases	(CMP-NeuNAc)				
α-2,3-NeuNAcT	Gal(β1-3/4)GlcNAc	15%[c]	rat brain	2.5	29
α-2,3-NeuNAcT	Gal(β1-3)GalNAc	15%[c]	sal. gland	5.7	28
α-2,3-NeuNAcT	Gal(β1-3)GlcNAc[d]	–[e]	human placenta	2.0	67
	Gal(β1-4)GlcNAc[d]		fetal heart	2.0	
α-2,6-NeuNAcT	Gal(β1-4)GlcNAc/Glc	15%[c]	rat all tissues	4.7[f]	22
			liver	4.3	60
			kidney	3.6	59
			human B cells	Novel[g]	66
α-2 6-NeuNAcT	GalNAc	–[e]	chicken embryo	3.0	
Fucosyltransferases	(GDP-Fuc)				
α-1,3-FucT-III	Gal(β1-4)GlcNAc	high	human	multiple genes	15
α-1,3-FucT-IV	(and substituents	less	human		83
α-1,3-FucT-V	as in Table I	high	human		25
α-1,3-FucT-VI		high	human		64
α-1,2-FucT	Gal(β1-4)Glc	none	human		70
Galactosyltransferases	(UDP-Gal)				
β-1,4-GalT	Glc/GlcNAc...		rat somatic cells	4.1[h]	56
			mammary	3.9	
α-1,3-GalT	Gal(β1-4)GlcNAc	none	murine all tissues	4 different	92

[a]Nucleotide sugar in parenthesis. [b]Within gene family. [c]Same stretch of amino acids for all. [d]Glycoprotein or glycolipid. [e]cDNA cloned using the sialyl motif information [18]. [f]Not always prominent but constitutive. [g]Mature B cells and lymphoblastoid cell lines utilize exons not present in other cells to produce novel mRNA isotypes. [h]Constitutive but lower in brain, male germ cells and mammary gland.

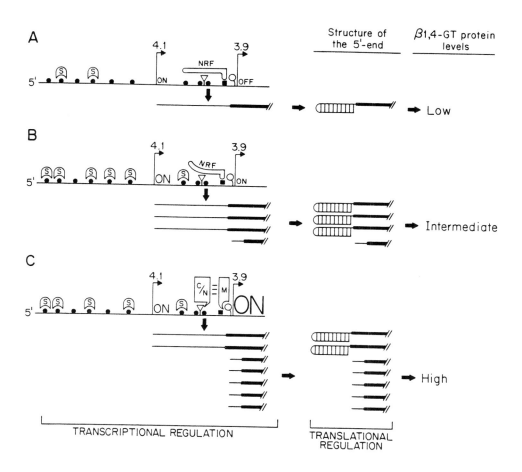

Fig. 2. A model depicting mechanisms of transcriptional and translational regulation of β-1,4-galac-tosyltransferase (β1,4-GT) gene [56]. In "Transcriptional Regulation", A–C: the murine β1,4-GT genomic fragment containing the Spl sites (●), the negative element (■), the half-palindromic CTF/NFI site (open inverted triangle) and the MAF sequence (open circle). The Spl binding protein (S) and a putative negative regulatory factor (NRF) are shown interacting with DNA. C: The *trans*-acting factors CTF/NFI (C/N) and MAF (M) are shown interacting with their respective binding sites; the parallel lines between them indicate their cooperative interaction. The position of the 4.1 and 3.9 kb transcriptional start sites are indicated by the upward bent arrows. The 5'-end of either the 4.1 or 3.9 kb transcript is shown directly below the DNA. The ratio of the 4.1 to the 3.9 kb transcript is actually 5:1 in B and 1:10 in C. The designations 'on' and 'off' and their relative heights indicate the status of transcript initiation at either the 4.1 or 3.9 kb site. In "Translational Regulation", the secondary structure at the 5'-end of either the 4.1 or 3.9 kb transcript is shown. The predicted β1,4-GT protein levels resulting from translation of the different transcript pools are indicated. A: Cells expressing low levels of β1,4-GT mRNA (e.g. brain); B: cells expressing intermediate level s (e.g. other somatic cell types); and C: cells expressing high levels (e.g. lactating mammary gland). From Harduin-Lepers et al. [56] with permission.

functions as a housekeeping promoter while the proximal region functions as a tissue-specific promoter, in this case mammary tissue. The differential initiation of these promoters regulates the enzyme level. It was also concluded that both forms are resident *trans*-Golgi enzymes. This appears contrary to the experiments of Evans et al. [62] showing that overexpression of the long form of β-1,4-galactosyltransferase without the catalytic domain resulted in loss of cell adhesion and spreading. Alternatively both reports may be correct and the movement of β-1,4-galactosyltransferase to the cell surface has as yet an unknown regulation controlled in some manner by the events of cell adhesion. Although the membrane associated β-1,4-galactosyltransferase form has yet to be resolved, the study of the enzyme has provided a hypothesis to explain the turn on of the enzyme in lactation when large amounts of lactose under hormonal control are required for nutrition (Fig. 2).

Since α-1,3-fucosyltransferases contain only mono-exonic coding sequences (Fig. 1) the diverse substrate specificities (Table I) which are thought to be correlated with tissue specificity must be achieved by transcriptional regulation of multiple genes rather than by differential splicing of the primary transcripts from a single gene [64]. The gene specific probes are now available and delineation of the regulatory processes in different tissues is anticipated in the near future.

Terminal glycosylation is achieved in several ways at the DNA level. α-2,6-Sialyltransferase expresses tissue differences probably as a result of multiple and distinct promoters on a single gene. On the other hand, α-1,3-fucosyltransferase has several genes which encode catalytically similar enzymes to give tissue specificity. The proposed coordinated utilization of regulatory elements for the expression of β-1,4-galactosyltransferase when needed may provide a unifying mechanism for the genetic regulation of the glycosyltransferases (Fig. 2), particularly when specific terminal glycosylation is necessary.

4.2. Domain structure and DNA sequence homology

The glycosyltransferases which promote terminal glycosylation and the other glycosyltransferases thus far described have structural features in common: a short cytoplasmic amino terminal segment, an anchor hydrophobic domain, and a catalytic domain. There is a stem region between the transmembrane and catalytic domains [4,5,65]. Although these gross domains are similar, there is very little homology among the different glycosyltransferases at the DNA level. There is some homology among the α-1,3-fucosyltransferases and limited homology between α-2,6 and α-2,3-sialyltransferases. This is surprising since the enzymes have binding sites for similar or even the same nucleotide sugars and acceptor substrates (Table III).

The efforts to find the Golgi retention signal for the glycosyltransferases have not been as yet successful. The transmembrane domain is recognized for each glycosyltransferase sequence as a hydrophobic sequence, however, there appears to be a complete lack of a common motif that could function as a retention signal for the glycosyltransferases. The possibility that the signal depends on a three dimensional structure rather than a linear amino acid sequence has not yet been investigated [7].

As summarized in Table III, with a possible exception of the sialyltransferases, two points can be made regarding the known general genomic sequences of the glycosyl-transferases: (1) The binding site for the nucleotide sugar is not common amongst the different glycosyltransferases, moreover it does not appear to be common even among glycosyltransferases utilizing the same nucleotide sugar. (2) The site for substrate binding is not similar even among transferases utilizing the same site such as GlcNAc. This monosaccharide is utilized by β-1,4-galactosyltransferase or by α-1,3-fucosyl-transferases, but in the latter case can only be utilized after β-1,4-galactosyltransferase has acted (Table III). Nothing in the DNA sequences of these two enzymes suggests homology. Indeed when one examines the various α-1,3-fucosyltransferases which differ largely in catalytic sites [64] it is impossible to discern common sequences to allow for the enzyme to transfer fucose to a GlcNAc which has specific substitutions such as NeuNAc(α2–3)Gal(β1–4), Fuc(α1–2)Gal(β1–4) or repeating units of Gal(β1–4)GlcNAc (Tables I and III).

A comparison of the cDNA sequences of Gal(β1–3)GalNAc α2,3-sialyltransferase with Gal(β1–4)GlcNAc α-2,6-sialyltransferase shows an unusual homology not seen among other glycosyltransferases (Table III). The sequences are nonhomologous except for a region of 45 amino acids which is 65% homologous. It is suggested that these two enzymes are derived from a common gene. It remains to be determined whether or not this region is the nucleotide sugar binding site [28]. The sequence of Gal(β1–3/4)GlcNAc α-2,3-sialyltransferase supports this hypothesis since a comparison with the other two sialyltransferases reveals 58% conservation between all three enzymes, although representing only 15% of the total sequence. Since the homologous region is near the center of the catalytic domain it has been proposed that it represents at least a structure necessary for enzyme activity [29]. Thus, three cloned sialyltrans-ferases have a protein motif not seen in the primary structures of other glycosyltrans-ferases [29]. Recently several additional sialyltransferases have been cloned by PCR using primers based on the sialylmotif [18,66,67].

As suggested, the homologies within enzyme families are not predictable. There appears to be lack of homology between α-1,3- and β-1,4-galactosyltransferases, however, as discussed in [2], the blood group B-dependent α-1,3-galactosyltransferase and the blood group A-dependent α-1,3-GalNActransferase differ by only a four point mutation even though the enzymes use different nucleotide sugars. On the other hand, a degree of homology exists among FucT-III, -V and -VI, but less so with FucT-IV [64]. The most marked sequence differences between FucT-IV and the others are found at the NH2-terminal ends of the catalytic domains. The catalytic domain of FucT-III is 96 and 95% identical with FucT-V and -VI, but less identical with FucT-IV. The FucT-IV catalytic domain is 73, 72, and 70% identical to the corresponding position in FucT-III, -V, and -VI genes, respectively. In addition, the FucT-IV probe detects other DNA sequences leading to the suggestion that the family has other as yet unknown members [64].

In spite of some degree of homology among the α-1,3-fucosyltransferases, the substrate specificity is extremely selective. Overexpression of cDNA for α-1,3-fuco-syltransferases allowed the determination of substrate specificity (Table I). When these are compared with the results of Koszdin and Bowen [68] (Table I), also expressing cDNA in COS cells, the substrate specificities are not exactly the same. The reasons for this disparity are not obvious but caution against comparing results not utilizing the same substrates and experimental conditions even in overexpressing cells. On the other hand, it is surprising that there is no homology with α-1,2-fucosyltransferase [69,70] since not only the sugar nucleotide, but also the oligosaccharide substrates in some cases are the same (Table III). There are several α-1,2-fucosyltransferases in human serum [9,71,72] and at least one has been expressed in ovarian cancer cells [41], however, demonstration of the homology among these enzymes will have to await further molecular cloning [70].

In examining the substrate specificity of human purified α-1,6-fucosyltransferase which transfers Fuc to the Asn-linked GlcNAc, it was found that Asn was not required since the trimannosyl core alone served as substrate [73]. This was surprising since the enzyme acts on a residue immediately adjacent to the protein. It will be interesting to determine the DNA sequence of this enzyme and compare the catalytic site to those of the α-1,3-fucosyltransferases which also require more than one monosaccharide for activity. Other glycosyltransferases may lack some of the substrate specificity of the fucosyltransferases in that they utilize monosaccharides; β-1,4-galactosyltransferase is an example. It is clear that, in addition to the elucidation of genomic sequences, the structures of the binding sites of the crystalline enzymes are necessary to resolve these issues.

It is concluded that currently there appears to be little evidence that homologous structural relationships amongst and within the enzyme families may be a factor in control of terminal glycosylation. A recent comprehensive review discusses the similarities and differences of all the cloned glycosyltransferases [74].

4.3. Hormones

A few cases have been reported where hormonal regulation of a glycosyltransferase at a molecular level has been shown to correlate to terminal glycosylation. Thyrotropin has been reported to down-regulate α-2,6-sialyltransferase, but not α-2,3-sialyltransferase [75] in thyroid cells, FRTL-5. In the liver, glucocorticoids modulated α-2,6-sialyltransferase [76]. Therefore depending on the tissue α-2,6-sialyltransferase may be modulated by tissue specific hormones. In the rat thyroid cell line, when thyrotropin modulated mRNA for α-2,6-sialyltransferase, a marked decrease in membrane bound sialic acid in α-2,6-linkage was found.

The control of β-1,4-galactosyltransferase in the mammary gland is by α-lactalbumin which is in turn controlled by lactogenic hormones [77]. As described (Fig. 2) the differential initiation of promoters appears to regulate the level of two isoforms of the enzyme.

4.4. Expression of cDNA

cDNAs from α-1,3-fucosyltransferases have been expressed in COS and CHO cells (Table I). The product of the expressed enzyme was detected with antibodies to the surface glycoproteins of the expressing cells [64]. Similarly, when α-2,6-sialyltransferase [78] and α-1,3-galactosyltransferase [79] genes were expressed in CHO cells their products were detected by lectin binding of the surface glycoconjugates. Interestingly in the latter case, α-1,3-galactosyl residues were reciprocal with α-2,3-sialyl residues [79] similar to the finding in mouse cells after oncogenic transformation [37].

5. Direct relationship of molecular mechanisms to structure/function

The tissue specific expression of β-1,4-galactosyltransferase has been related to lactose synthesis in mammary tissue. A hypothesis, as described in Fig. 2, has been presented for the genetic regulation of two isoforms of the enzyme [56]. Although presented as a hypothesis, β-1,4-galactosyltransferase is the best characterized and only detailed example of genetic regulatory elements controlling a glycosyltransferase in a manner directly related to a biological function. The final product does not however have to integrate into a glycoprotein on the cell surface so the mechanism may be less demanding than terminal glycosylation.

The maturation of thymocytes that is followed by the conversion of PNA$^+$ cells to PNA$^-$ cells [19] has been studied using Gal(β1–3)GalNAc α-2,3-sialyltransferase probes. It was shown that the regulated expression of this gene accounted for the developmental change [20]. α-2,6-Sialyltransferase which could have affected PNA negativity was not expressed. The developmental significance of this highly specific sialylation remains to be shown.

Similarly the maturation of human B cells is accompanied by an elevation of Gal(β1–4)GlcNAc α-2,6-sialyltransferase mRNA [21] and the appearance of CD$_w$75, an antigen of human mature B-cells. Tissue specificity exists within the human B-cell lineage in that two distinct α-2,6-sialyltransferases are present as a result of three exons [59]. A basal level of mRNA is maintained by exons (Y+Z) whereas on maturation of the B-cell population, the exon (X) sequence is synthesized at a high level. Since this latter mRNA synthesis coincides with the appearance of CD$_w$75, a sialylated marker of human mature B-cells, the mRNA expression appears to correlate with terminal glycosylation. This is an example of alternative splicing controlling expression.

The fact that α-1,3-fucosyltransferase was cloned when searching for the ligand for the selectin, ELAM-1 demonstrates a role in this selectin interaction [80]. The ligand for some selectins contains sialic acid linked α-2,3 to galactose, therefore an α-2,3-sialyltransferase must also be involved in the biological specificity [47,48]. Although overexpression of the cDNA of these enzymes relates to function, the actual genetic regulatory mechanisms have not as yet been elucidated.

6. Additional controls

Does the transcription of mRNA and the translation of large amounts of a glycosyltransferase control terminal glycosylation? No, not in all cases. An example is α-1,3-fucosyltransferase in a human erythroleukemic (HEL) cell line. Kannagi, et al. [81] found that HEL cells did not react with an antibody specific to Fuc(α1–3)GlcNAc, therefore it was assumed that no or an insignificant amount of α-1,3-fucosyltransferase was present in these cells. However, it was found that high activity of the enzyme was present even though the surface and cell glycoproteins did not contain Fuc(α1–3)GlcNAc [82]. In the case of HEL cells and F9 cells [83] the required substrate was present, although others found that lack of a substrate was a controlling factor [84]. A number of other factors have been reported as contributing to the mechanism of terminal glycosylation including the reciprocal relationship of the glycosyltransferases [9] and the microenvironment of the membrane [85].

A different example of failure to express terminal glycosylation at the cell surface in the presence of a glycosyltransferase comes from the expression system of α-1,3-fucosyltransferase. When FucT-IV cDNA obtained from a human leukocyte library was expressed in COS-1 cells, the overexpressing cells did not serve as a ligand for ELAM-1 [86]. However, when α-1,3-fucosyltransferase cDNA obtained from an HL-60 library was expressed in CHO cells, the overexpressing cells served as a ligand for ELAM-1 [87]. Even more complicated is the fact that when FucT-III (Table I) is overexpressed in COS cells the overexpressing cells will provide an ELAM-1 ligand suggesting little correlation between *in vivo* and *in vitro* expression [88]. The case can be made for additional specificity not maintained in the catalytic domain but elsewhere on the gene. Indeed, at this point it has not yet been established which α-1,3-fucosyltransferase is utilized in the inflammatory process, development or tumorigenesis. Nevertheless it appears that factors in addition to transcription/translation of the glycosyltransferase may contribute to the control of the surface expression of terminal carbohydrate residues on the glycoprotein.

7. Conclusions

Terminal glycosylation of surface glycoproteins provides an extraordinarily flexible means of contributing to a variety of biological functions. An attractive control mechanism may be in the genetic regulation of specific glycosyltransferases. However, if promoters and regulatory elements which are tissue specific are utilized to regulate the expression of glycosyltransferases in general then an argument must be made for another level of control when specific and unusual modifications of the basic glycoproteins are required. The hypothesis for modulating the expression of β-1,4-galactosyltransferase (Fig. 2) as required for specific function may provide a partial solution. However, the controlling mechanisms required in development, for example, may be quite different from those controlling expression in tumor cells and other pathological conditions.

On one hand the expression of a particular glycosyltransferase in controlled amounts is regulated by genetic factors; on the other hand, the presence of the enzyme even in high amounts does not necessarily delineate a particular glycosylation pattern. As pointed out, few genetic controls have been related to structure or function of glycoproteins. Orchestration of function in an orderly manner is the ultimate goal of determining regulatory factors. If progress continues at the rapid pace of the past five years, this goal may be attainable at least for some of an increasing number of glycoproteins whose terminal glycosylation specifies a biological or pathological function.

Acknowledgements

Supported by NIH grants.

Abbreviations

GDP-L-Fuc: *N*-acetyl-β-D-glucosaminide α-1,3-fucosyltransferase, termed α-1,3-fucosyltransferase in general or when the enzyme has not been cloned; and FucT-III to -VI when discussing specific cloned enzymes.

References

1. Schachter, H. (1995) in: J. Montreuil, H. Schachter and J.F.G. Vliegenthart (Eds.), Glycoproteins. New Comprehensive Biochemistry, 29a. Elsevier, Amsterdam, pp. 281–286.
2. Watkins, W. (1995) in: J. Montreuil, H. Schachter and J.F.G. Vliegenthart (Eds.), Glycoproteins. New Comprehensive Biochemistry, 29a. Elsevier, Amsterdam, pp. 313–390.
3. Kobata, A. (1995) in: J. Montreuil, H. Schachter and J.F.G. Vliegenthart (Eds.), Glycoproteins and Disease. New Comprehensive Biochemistry, Amsterdam, in prep.
4. Joziasse, D.H. (1992) Glycobiology 2, 271–277.
5. Schachter, H. (1991) Curr. Opin. Struct. Biol. 1, 755–765.
6. Schachter, H. (1992) Trends Glycosci. Glycotech. 4, 241–250.
7. Shaper, J.H. and Shaper, N.L. (1992) Curr. Opin. Struct. Biol. 2, 701–709.
8. De Vries, T. and van den Eijnden, D.H. (1992) Histochem. J. 24, 761–770.
9. Watkins, W.M., Skacel, P.O. and Johnson, P.H. (1993) In: P.J. Garegg and A.A. Lindberg (Eds.), Carbohydrate Antigens. ACS Symposium Series 519. American Chemical Society, Washington, DC, pp. 34–63.
10. Solter, D. and Knowles, B.B. (1978) Proc. Natl. Acad. Sci. U.S.A. 75, 5565–5569.
11. Foster, C.S. and Glick, M.C. (1988) In: A.E. Evans, G.J. D'Angio, A.G. Knudson and R.C. Seeger (Eds.) Advances in Neuroblastoma Research 2. Alan R. Liss, Inc., New York, pp. 421–432.
12. Santer, U.V. and Glick, M.C. (1980) Biochem. Biophys. Res. Commun. 96, 219–226.
13. Hakomori, S. (1989) Adv. Cancer Res. 52, 257–331.
14. Montreuil, J. (1983) Biochem. Soc. Trans. 11, 134–136.
15. Kukowska-Latallo, J.F., Larsen, R.D., Nair, R.P. and Lowe, J.B. (1990) Genes Dev. 4, 1288–1303.
16. Edelman, G.M. (1985) Ann. Rev. Biochem. 54, 135–169.
17. Troy, F.A. (1992) Glycobiology 2, 5–23.
18. Livingstone, B.D. and Paulson, J.C. (1993) J. Biol. Chem. 268, 11504–11507.
19. Reisner, Y., Linker-Israeli, M. and Sharon, N. (1976) Cell. Immunol. 25, 129–134.
20. Gillespie, W., Paulson J.C., Kelm, S., Pang, M. and Baum, L.G. (1993) J. Biol. Chem. 268, 3801–3804.
21. Stamenkovic, I., Sgroi, D., Aruffo, A., Sy, M.S. and Anderson, T. (1991) Cell 66, 1133–1144.
22. Paulson, J.C., Weinstein, J. and Schauer, A. (1989) J. Biol. Chem. 264, 10931–10934.
23. Mollicone, R., Gibaud, A., François, A., Ratcliffe, M., Oriol, R. (1990) Eur. J. Biochem. 191, 169–176.

278

24. Macher, B.A., Holmes, E.H., Swiedler, S.J., Stults, C.L. and Srnka, C.A. (1991) Glycobiology 1, 577–584.
25. Weston, B.W., Nair, R.P., Larsen, R.D. and Lowe, J.B. (1992) J. Biol. Chem. 267, 4152–4160.
26. Howard, D.R., Fukuda, M., Fukuda, M.N. and Stanley, P. (1987) J. Biol. Chem. 262, 16830–16837.
27. Potvin, B., Kumar, R., Howard, D.R. and Stanley, P. (1990) J. Biol. Chem. 265, 1615–1622.
28. Gillespie, W., Kelm, S. and Paulson, J.C. (1992) J. Biol. Chem. 267, 21004–21010.
29. Wen, D.X., Livingston, B.D., Medzihradszky, K.F., Kelm, S., Burlingame, A.L. and Paulson, J.C. (1992) J. Biol. Chem. 267, 21011–21019.
30. Svensson, E.C., Soreghan, B. and Paulson, J.C. (1990) J. Biol. Chem. 265, 20863–20868.
31. Hull, E., Sarkar, M., Spruijt, M.P., Höppener, J.W., Dunn, R. and Schachter, H. (1991) Biochem. Biophys. Res. Commun. 176, 608–615.
32. Shaper, N.L., Lin, S.P., Joziasse, D.H., Kim, D.Y. and Yang-Feng, T.L. (1992) Genomics 12, 613–615.
33. Nishihara, S., Nakazato, M., Kudo, T., Kimura, H., Ando, T. and Narimatsu, H. (1993) Biochem. Biophys. Res. Commun. 190, 42–46.
34. Galili, U. and Swanson, K. (1991) Proc. Natl. Acad. Sci. USA 88, 7401–7404.
35. Khan, S.H. and Matta, K.L. (1992) In: H.J. Allen and E.C. Kisailus (Eds.), Glycoconjugates: Composition, Structure and Function. Marcel Dekker, Inc., New York, pp. 361–378.
36. Alhadeff, J.A. (1989) Crit. Rev. Oncol. Hematol. 9, 37–107.
37. Santer, U.V., DeSantis, R., Hård, K.J., van Kuik, J.A., Vliegenthart, J.F.G., Won, B. and Glick, M.C. (1989) Eur. J. Biochem. 181, 249–260.
38. Easton, E.W., Bolscher, J.G.M. and van den Eijnden, D.H. (1991) J. Biol. Chem. 266, 21674–21680.
39. Vandamme, V., Cazlaris, H., Le Marer, N., Laudet, V., Lagrou, C., Verbert, A. and Delannoy, P. (1992) Biochimie 74, 89–99.
40. Foster, C.S., Gillies, D.R.B. and Glick, M.C. (1991) J. Biol. Chem. 266, 3526–3531.
41. Chandrasekaran, E.V., Jain, R.K. and Matta, K.L. (1992) J. Biol. Chem. 267, 23806–23814.
42. Gillies, D.R.B., Foster, C.S. and Glick, M.C. (1991) In: A.E. Evans, A.G. Knudson, R.C. Seeger and G.J. D'Angio (Eds.) Advances in Neuroblastoma Research 3. Wiley-Liss, Inc., New York, pp. 301–307.
43. Fidler, I.J. (1985) Cancer Res. 45, 4714–4726.
44. Gasic, G. and Gasic, T. (1962) Proc. Natl. Acad. Sci. USA 48, 1172–1177.
45. Livingston, B.D., Jacobs, J.L., Glick, M.C. and Troy, F.A. (1988) J. Biol. Chem. 263, 9443–9448.
46. Glick, M.C., Livingston, B.D., Shaw, G.W., Jacobs, J.L. and Troy, F.A. (1991) In: A.E. Evans, A.G. Knudson, R.C. Seeger and G.J. D'Angio (Eds.) Advances in Neuroblastoma Research 3. Wiley-Liss, Inc., New York, pp. 267–274.
47. Cummings, R.D. and Smith, D.F. (1992) BioEssays 14, 849–856.
48. Paulson, J.C. (1992) In: J. Harlan and D. Liu (Eds.), Adhesion: Its Role in Inflammatory Disease. Stockton Press, New York, pp. 19–42.
49. Yamashita, K., Ideo H., Ohkura T., Fukushima K., Yuasa, I, Ohno, K. and Takeshita, K. (1993) J. Biol. Chem. 268, 5783–5789.
50. Beaudet, A.L. and Thomas, G.H. (1989) In: C.R. Scriver, A.L. Beaudet, W.S. Sly and D. Valle (Eds.), The Metabolic Basis of Inherited Disease. MacGraw-Hill Inc., New York, pp. 1603–1621.
51. Scanlin, T. F. (1988) In: A.P. Fishman (Ed.). Pulmonary Diseases and Disorders. McGraw-Hill, New York, pp. 1273–1294.
52. Jaeken, J., Kint, J. and Spaapen, L. (1992) Lancet 340, 1472–1473.

53. Segal, S. (1989) In: C.R. Scriver, A.L. Beaudet, W.S. Sly and D. Valle (Eds.), The Metabolic Basis of Inherited Disease. McGraw-Hill, New York, pp. 453–480.
54. Wang, Y.M., Hare, T.R., Won, B., Stowell, C.P., Scanlin, T.F., Glick, M.C., Hard, K., van Kuik, J.A. and Vliegenthart, J.F.G. (1990) Clin. Chim. Acta. 188, 193–210.
55. Svensson, E.C., Conley, P. and Paulson, J.C. (1992) J. Biol. Chem. 267, 3466–3472.

56. Harduin-Lepers, A., Shaper, J.H. and Shaper, N.L. (1993) J. Biol. Chem. 268, 14348–14359.
57. Wen, D.X., Svensson, E.C. and Paulson, J.C. (1992) J. Biol. Chem. 267, 2512–2518.
58. Wang, X., O'Hanlon, T.P., Young, R.F. and Lau, J.T.Y. (1990) Glycobiology 1, 25–31.
59. Wang, X., Vertino, A., Eddy, R.L., Byers, M.G., Jani-Sait, S.N., Shows, T.B. and Lau, J.T.Y. (1993) J. Biol. Chem. 268, 4355–4361.
60. O'Hanlon, T.P. and Lau, J.T.Y. (1992) Glycobiology 2, 257–266.
61. Shibuya, N., Goldstein, I.J., Broekaert, W.F., Nsimba-Lubaki, M., Peeters, B. and Peumans, W.J. (1987) Arch. Biochem. Biophys. 254, 1–8.
62. Evans, S.C., Lopez, L.C. and Shur, B.D. (1993) J. Cell Biol. 120, 1045–1057.
63. Huang, K.M. and Snider, M.D. (1993) J. Biol. Chem. 268, 9302–9310.
64. Weston, B.W., Smith, P.L., Kelly, R.J. and Lowe, J.B. (1992) J. Biol. Chem. 267, 24575–24584.
65. Paulson, J.C. and Colley, K.J. (1989) J. Biol. Chem. 264, 17615–17618.
66. Kurosawa, N., Hamamoto, T., Lee, Y-C., Nakaoka, T., Kojima, N. and Tsuji, S. (1994) J. Biol. Chem. 169, 1402–1409.
67. Kitagawa, H. and Paulson, J.C. (1994) J. Biol. Chem. 269, 1394–1401.
68. Koszdin, K.L. and Bowen, B.R. (1992) Biochem. Biophys. Res. Commun. 187, 152–157.
69. Ernst, L.K., Rajan, V.P., Larsen, R.D., Ruff, M.M. and Lowe, J.B. (1989) J. Biol. Chem. 264, 3436–3447.
70. Larsen, R.D., Ernst, L.K., Nair, R.P. and Lowe, J.B. (1990) Proc. Natl. Acad. Sci. USA 87, 6674–6678.
71. Sarnesto, A., Köhlin, T., Thurin, J. and Blaszczyk-Thurin, M. (1990) J. Biol. Chem. 265, 15067–15075.
72. Sarnesto, A., Köhlin, T., Hindsgaul, O., Thurin, J. and Blaszczyk-Thurin, M. (1992) J. Biol. Chem. 267, 2737–2744.
73. Voynow, J.A., Kaiser, R.S., Scanlin, T.F. and Glick, M.C. (1991) J. Biol. Chem. 266, 21572–21577.
74. Kleene, R. and Berger, E.G. (1993) Biochem. Biophys. Acta 1154, 283–325.
75. Grollman, E.F., Saji, M., Shimura, Y., Lau, J.T.Y. and Ashwell, G. (1993) J. Biol. Chem. 268, 3604–3609.
76. Wang, X.C., Smith, T.J. and Lau, J.T.Y. (1990) J. Biol. Chem 265, 17849–17853.
77. Brew, K. (1970) Essays-Biochem. 6, 93–118.
78. Lee, E.U., Roth, J. and Paulson, J.C. (1989) J. Biol. Chem. 264, 13848–13855.
79. Smith, D.F., Larsen, R.D., Mattox, S., Lowe, J.B. and Cummings, R.D. (1990) J. Biol. Chem. 265, 6225–6234.
80. Bevilacqua, M.P. and Nelson, R.M. (1993) J. Clin. Invest. 91, 379–387.
81. Kannagi, R. Papayannopoulou, T., Nakamoto, B., Cochran, N.A., Yokochi, T., Stamatoyannopoulos, G. and Hakomori, S. (1983) Blood 62, 1230–1241.
82. Giuntoli II, R.L., Stoykova, L.I., Gillies, D.R.B. and Glick, M.C. (1994). Eur. J. Biochem. 225, 159–166.
83. Romero, P.A., Way, T. and Herscovics, A. (1993) Biochem J. 296, 253–257.
84. Holmes, E.H., Hakomori, S. and Ostrander, G.K. (1987) J. Biol. Chem. 262, 11331–11338.
85. Holmes, E.H. and Macher, B.A. (1993) Arch. Biochem. Biophys. 301, 190–199.
86. Lowe, J.B., Kukowska-Latallo, J.F., Nair, R.P., Larsen, R.D., Marks, R.M., Macher, B.A.,

Kelly, R. and Ernst, L.K. (1991) J. Biol. Chem. 266, 17467–17477.

87. Goelz, S.E., Hession, C., Goff, D., Griffiths, B., Tizard, R., Newman, B., Chi-Rosso, G. and Lobb, R. (1990) Cell 63, 1349–1356.

88. Lowe, J.B., Stoolman, L.M., Nair, R.P., Larsen, R.D., Berhend, T.L. and Marks, R.M. (1990) Cell 63, 475–484.

89. Sarnesto, A., Köhlin, T., Hindsgaul, O., Vogele, K., Blaszczyk-Thurin, M. and Thurin, J. (1992) J. Biol. Chem. 267, 2745–2752.

90. de Heij, H.T., Tetteroo, P.A.T., van Kessel, A.H.M., Schoenmaker, E., Visser, F.J. and van den Eijnden, D.H. (1988) Cancer Res. 48, 1489–1493.

91. Shaper, N.L., Shaper, J.H., Bertness, V., Chang, H., Kirsch, I.R. and Hollis, G.F. (1986) Somat. Cell Mol. Genet. 12, 633–636.

92. Joziasse, D.H., Shaper, N.L., Kim, D., van den Eijnden, D.H. and Shaper, J.H. (1992) J. Biol. Chem. 267, 5534–5541.

J. Montreuil, H. Schachter and J.F.G. Vliegenthart (Eds.), *Glycoproteins*
© 1995 Elsevier Science B.V. All rights reserved

Biosynthesis

4b. Substrate level controls for N-glycan assembly

HARRY SCHACHTER

Department of Biochemistry Research, Hospital for Sick Children, 555 University Avenue,
Toronto, Ont. M5G 1X8, Canada

1. The endomembrane assembly line and substrate level controls

Processing and biosynthesis of protein-bound oligosaccharides differ in several important respects from the biosynthesis of nucleic acids and proteins. Both of the latter polymers are synthesized as linear molecules in which the bonds between the individual monomer subunits are always the same and assembly can therefore be carried out by a template-directed mechanism. In contrast, protein-bound glycans are often highly branched and the linkages between individual monomers can vary greatly. The assembly of such a molecule cannot be template-directed and requires an elaborate assembly line [1] provided by the endomembrane system (rough endoplasmic reticulum and Golgi apparatus). Processing enzymes (glycosidases and glycosyltransferases) are anchored to the membranes of the system and nascent glycoproteins are processed as they pass through the lumen.

The processing of Asn-GlcNAc N-glycans and Ser/Thr-GalNAc O-glycans occurs within the lumen of the endomembrane system although recent work has shown that O-linked GlcNAc can be added to protein on the cytoplasmic side [2–4]. The message encoding the peptide backbone of nascent glycoprotein destined for the lumen contains a signal sequence for attachment of the polyribosome to membrane. N-glycans are added as the nascent peptide enters the lumen by the action of an oligosaccharyl-transferase [5,6] and processing begins. A variety of 'substrate level' factors operate throughout processing to control N-glycan assembly, i.e., the location on the membrane, the substrate specificity and the relative efficiency of every one of the many processing enzymes, and the availability within the lumen of all the substrates and co-factors needed for processing. It is beyond the scope of this chapter to discuss all these factors, e.g., we will not discuss the cytoplasmic synthesis of nucleotide-sugar precursors [7] nor the transport proteins in the endoplasmic reticulum and Golgi apparatus which control movement of nucleotide-sugars into the lumen [8,9]. The discussion will be limited to the mechanisms whereby enzyme substrate specificities channel metabolic flow; the GlcNAc-transferases which initiate the synthesis of N-glycan antennae will be used as illustrative examples [10].

2. The roads to highly branched complex N-glycans

The enzymatic steps leading from dolichol pyrophosphate $Glc_3Man_9GlcNAc_2$ to complex N-glycan antennae have been discussed elsewhere [1,5,6,10]. There are five positions at which complex N-glycan antennae may be initiated by the actions of GlcNAc-transferases I, II, IV, V and VI [10]; in addition, complex N-glycans may be 'bisected' by a GlcNAc β-1,4-linked to the β-linked Man of the core due to the action of GlcNAc-transferase III.

GlcNAc-transferase I catalyzes the first committed step towards the synthesis of hybrid and complex N-glycans [10]. If GlcNAc-transferase III acts on the product of GlcNAc-transferase I before α-mannosidase II to form the bisected hybrid structure, the pathway is committed to hybrid structures because α-mannosidase II cannot act on bisected oligosaccharides (11). The reverse order of action leads to complex N-glycans. The relative abundance of GlcNAc-transferase III and α-mannosidase II in a particular tissue therefore controls the pathway towards hybrid or complex N-glycans.

GlcNAc-transferase II acts on the product of α-mannosidase II (10). The enzyme cannot act on $Man(\alpha1-6)[Man(\alpha1-3)]Man(\beta1-4)GlcNAc-R$, i.e., GlcNAc-transferase II requires the prior action of GlcNAc-transferase I. The addition of a bisecting GlcNAc residue or galactosylation of the $GlcNAc(\beta1-2)Man(\alpha1-3)Man\beta$-moiety prevents GlcNAc-transferase II action. The product of GlcNAc-transferase II is the entry point for the synthesis of all complex N-glycans.

The actions of GlcNAc-transferases III, IV, V and VI to form highly branched complex N-glycans have been reviewed [12–14] and will not be discussed in detail here.

3. Which fork in the road to follow?

The scheme for assembly of highly branched complex N-glycans is complicated [12–14] and involves many divergent branch points. The substrate level factors which route the metabolic flow [13,15] are discussed below.

3.1. Competition for a common substrate

The route taken at a divergent branch point is dictated primarily by the relative activities of glycosyltransferases which compete for a common substrate. Enzymes must be located in the same subcellular compartment if they are to compete. A major cross-roads in N-glycan synthesis is at GlcNAc-transferase I product ($GlcNAc_1Man_5GlcNAc_2Asn-X$) which can be acted on by β-1,4-Gal-transferase (non-bisected mono-antennary hybrid), GlcNAc-transferase III (bisected hybrid), GlcNAc-transferase IV (non-bisected di-antennary hybrid), α-mannosidase II (complex N-glycans), or core α-1,6-fucosyltransferase. Once a bisecting GlcNAc or a β-1,4-linked Gal residue has been incorporated, α-mannosidase II can no longer act and the pathway is committed to hybrid structures.

The product of GlcNAc-transferase II ($GlcNAc_2Man_3GlcNAc_2Asn-X$) is the major entry point for the synthesis of all complex N-glycans since it can be acted on by GlcNAc-transferases III, IV or V, β-1,4-Gal-transferase, β-1,4-GalNAc-transferase, or core α-1,6-fucosyltransferase.

The existence of these and other branch points implies that many different oligosaccharides can be made on a single assembly line depending on the relative activities of the competing enzymes. Competition explains the microheterogeneity of oligosaccharides at a single amino acid site of a glycoprotein.

3.2. GO and NO GO residues

3.2.1. The key role of UDP-GlcNAc:Man(α1–3)R [GlcNAc to Man(α1–3)] β-1,2-GlcNAc-transferase I as a GO signal

The structure $Man_5GlcNAc_2Asn$-X [10] is the physiological substrate for GlcNAc-transferase I. Prior GlcNAc-transferase I action is essential for the action of (i) α-mannosidase II, (ii) GlcNAc-transferases II and IV, enzymes that initiate N-glycan antennae, (iii) bisecting GlcNAc-transferase III, (iv) core α-1,6-fucosyltransferase and (v) indirectly for the actions of GlcNAc-transferases V and VI and the further addition of antennary Gal, Fuc and sialic acid residues. In the absence of GlcNAc-transferase I, the process stops at the high mannose stage and there is no formation of hybrid and complex N-glycans. Although cells lacking GlcNAc-transferase I (e.g., Lec 1 lectin-resistant Chinese hamster ovary cells [16]) can grow normally under tissue culture conditions, recent experiments have shown that in the transgenic mouse model the absence of this enzyme leads to the interruption of embryogenesis at 9–10 days [17].

GlcNAc-transferase I is an example of a GO signal, i.e., an enzyme which incorporates a carbohydrate residue that is an essential substrate component for succeeding enzymes. Other such GO signal glycosyl residues have been described [15].

It is to be noted [10,15,18–20] that (i) GlcNAc-transferases I, II, IV and V, α-mannosidase II and core α-1,6-fucosyltransferase cannot act on bisected substrates; (ii) GlcNAc-transferases II, III and IV, α-mannosidase II and core α-1,6-fucosyltransferase require the presence of the GlcNAc(β1–2)Man(α1–3)Manβ- moiety produced by GlcNAc-transferase I; (iii) GlcNAc-transferases II, III and IV, α-mannosidase II and core α-1,6-fucosyltransferase are inhibited by the presence of Gal(β1–4)GlcNAc- on the Man(α1–3)Manβ- arm; (iv) the bisecting GlcNAc does not block access to the point of reaction catalyzed by the above enzymes but either blocks binding (by steric hindrance) to the GlcNAc(β1–2)Man(α1–3)Manβ- moiety or prevents H-bonding with an essential OH group; (v) kinetic studies with GlcNAc-transferase II indicate that the GlcNAc(β1–2)Man(α1–3)Manβ- moiety is a substrate binding site for the enzyme. These findings have suggested that the GlcNAc(β1–2)Man(α1–3)Manβ- moiety is an essential binding site for all the above enzymes thereby explaining why GlcNAc-transferase I acts as a GO signal.

3.2.2. NO GO residues

Some key glycosyl residues act as STOP or NO GO signals. For example, it has been pointed out above that the insertion of a bisecting GlcNAc linked β-1,4 to the Manβ- residue of the N-glycan core prevents the actions of GlcNAc-transferases II, IV and V, α-mannosidase II and core α-1,6-fucosyltransferase. Although this reaction halts branching in the medial Golgi cisternae, it does not prevent movement to the *trans*-Golgi followed by addition of Gal, Fuc, sialic acid or other residues to the

284

antennae. The addition of Gal in β-1,4 linkage to the GlcNAc(β1–2)Man(α1–3)Manβ-arm acts as an analogous NO GO signal in the *trans*-Golgi. Both of these STOP signals may act by preventing binding of these enzymes to GlcNAc(β1–2)Man(α1–3)Manβ-but this has not been established.

3.3. The role of polypeptide

In view of the many possible routes available to the biosynthetic pathway, it is not surprising that there are often several different oligosaccharides at a particular Asn residue of a glycoprotein, a phenomenon which has been called microheterogeneity. On the other hand, there can be marked differences between the oligosaccharides on different Asn sites. On the basis of this and other evidence, it has been suggested that the polypeptide backbone must play a role in processing, either by interacting directly with the processing glycosidases and glycosyltransferases in the endomembrane system or by influencing the conformation of the oligosaccharide substrates of these enzymes (18, 19, 21–23). The oligosaccharyl-transferase which initiates N-glycosylation needs a specific sequon (Asn-X-Ser/Thr) but not all sequons are glycosylated either because of steric hindrance or other polypeptide-dependent control factors.

The term 'site-directed processing' has been coined [20] to describe the process whereby polypeptide influences oligosaccharide conformation and thereby oligosaccharide processing. For example, a human monoclonal IgG antibody has been described [24] with bisected complex N-glycans on Asn-107 of the L chain and mainly non-bisected complex N-glycans on Asn-297 of the H chain. Both sets of glycans have Fuc, Gal and sialyl residues indicating that the protein had entered the late Golgi compartments and that the glycans were sterically available to the processing enzymes. Why does GlcNAc-transferase III act at Asn-107 so much better than at Asn-297? The suggested explanation depends on the following hypotheses: (i) oligosaccharides in free solution can exist in a variety of different conformations; (ii) processing enzymes such as GlcNAc-transferase III can act only on a restricted number of these conformations; (iii) the interaction of polypeptide with co-valently bound N-glycan can stabilize the glycan into a restricted set of conformations. It is therefore possible that the glycan at Asn-297 is restricted to a conformation which is not a good substrate for GlcNAc-transferase III. Experiments with conformationally restricted synthetic substrate analogues for GlcNAc-transferase V [22,25], hybrid glycoproteins [26,27] and X-ray crystallography of carbohydrate–protein interactions [28,29] are compatible with this hypothesis but do not exclude other explanations. The role of glycan–peptide interactions in biosynthesis is discussed further by Cumming in this volume [30].

3.4. Conclusions

The surfaces of mammalian cells are covered with a variety of different protein- and lipid-bound complex carbohydrates. These structures differ from one tissue to another and also undergo time-dependent changes during many normal and pathological biological processes such as development and metastatic progression. The substrate specificities and subcellular localizations of processing enzymes play major roles in controlling the biosynthesis of complex N-glycans at the substrate level.

The processing enzymes are arranged along the endomembrane assembly line in the order in which they are needed. This arrangement minimizes the need for substrate to seek its proper enzyme and also exercises control over the synthetic process. For example, GlcNAc-transferase I and probably also the other GlcNAc-transferases which control branching are primarily residents of the medial-Golgi compartment whereas the sialyl- and galactosyl-transferases which add terminal sugars are in the *trans*-Golgi compartment and *trans*-Golgi network [1]. Thus, branching is determined before antennae are completed. This control may be important to the cell since there is a safe-guard mechanism which terminates branching once the glycoprotein leaves the medial-Golgi, i.e., addition of Gal residues in the *trans*-Golgi prevents the actions of GlcNAc-transferases II, III, IV and V. Even if there are small amounts of branching GlcNAc-transferases in the *trans*-Golgi, these enzymes are prevented from acting by galactosylation. The reason for this stringent control is not understood but may be related to the role of N-glycan antennae in cell-cell communication.

The interplay of assembly line mechanics and substrate level control of processing enzymes provides a sophisticated apparatus for glycan assembly. It is only by under-standing how this system operates that one can hope to understand why the cell makes so many diverse structures. Recent work on the gene organization and expression of some of the glycosyltransferases has indicated that there are also important transcrip-tional and translational controls [10,31]. Finally, the study of three-dimensional struc-tures in protein-carbohydrate interactions [28,29,32] is as yet in its infancy and may eventually supply important clues to the operation of the biosynthetic apparatus.

Acknowledgement

Supported by the Medical Research Council of Canada, and by the Canadian National Centres of Excellence Programme in "Protein Engineering: 3-D structure, function and design" (PENCE).

References

1. Roth, J. (1995) in: J. Montreuil, H. Schachter and J.F.G. Vliegenthart (Eds.), Glycoproteins. New Comprehensive Biochemistry, Vol. 29a. Elsevier, Amsterdam, pp. 287–312.
2. Haltiwanger, R.S., Kelly, W.G., Roquemore, E.P., Blomberg, M.A., Dong, L.Y., Kreppel, L., Chou, T.Y., Hart, G.W. (1992) Biochem. Soc. Trans. 20, 264–269.
3. Hart, G.W. (1992) Curr. Opin. Cell Biol. 4, 1017–1023.
4. Hart, G. (1995) in: J. Montreuil, H. Schachter and J.F.G. Vliegenthart (Eds.), Glycoproteins, New Comprehensive Biochemistry, Vol. 29b. Elsevier, Amsterdam. In prep.
5. Hemming, F. (1995) in: J. Montreuil, H. Schachter and J.F.G. Vliegenthart (Eds.), Glycoprote-ins. New Comprehensive Biochemistry, Vol. 29a. Elsevier, Amsterdam, pp. 127–143.
6. Verbert, A. (1995) in: J. Montreuil, H. Schachter and J.F.G. Vliegenthart (Eds.), Glycoproteins. New Comprehensive Biochemistry, Vol. 29a. Elsevier, Amsterdam, pp. 145–152.
7. Schachter, H. and Roden, L. (1973) in: W.H. Fishman (Ed.), Metabolic Conjugation and Metabolic Hydrolysis. Academic Press, New York, NY, Vol. III, pp. 1–149.

8. Milla, M., Capasso, J. and Hirschberg, C.B. (1989) Biochem. Soc. Trans. 17 (3), 447–448.
9. Abeijon, C. and Hirschberg, C.B. (1992) Trends Biochem. Sci. 17, 32–36.
10. Schachter, H. (1995) in: J. Montreuil, H. Schachter and J.F.G. Vliegenthart (Eds.), Glycoproteins. New Comprehensive Biochemistry, Vol. 29a. Elsevier, Amsterdam, pp. 153–199.
11. Harpaz, N. and Schachter, H. (1980) J. Biol. Chem. 255, 4894–4902.
12. Schachter, H., Brockhausen, I. and Hull, E. (1989) Meth. Enzymol. 179, 351–396.
13. Schachter, H. (1991) Glycobiology 1, 453–461.
14. Brockhausen, I., Carver, J. and Schachter, H. (1988) Biochem. Cell Biol. 66, 1134–1151.
15. Schachter, H. (1986) Biochem. Cell Biol. 64, 163–181.
16. Stanley, P. (1992) Glycobiology 2, 99–107.
17. Metzler, M., Gertz, A., Sarkar, M., Schachter, H., Schrader, J.W. and Marth, J.D. (1994) EMBO J. 13, 2056–2065..
18. Carver, J.P. and Brisson, J.-R. (1984) in: V. Ginsburg and P.W. Robbins (Eds.), Biology of Carbohydrates. John Wiley and Sons, New York, NY, Vol. 2, pp. 289–331.
19. Carver, J.P. (1984) Biochem. Soc. Trans. 12, 517–519.
20. Carver, J.P. and Cumming, D.A. (1987) Pure Appl. Chem. 59, 1465–1476.
21. Kornfeld, R. and Kornfeld, S. (1985) Annu. Rev. Biochem. 54, 631–664.
22. Lindh, I. and Hindsgaul, O. (1991) J. Am. Chem. Soc. 113, 216–223.
23. Snider, M.D. (1984) in: V. Ginsburg and P.W. Robbins (Eds.), Biology of Carbohydrates. John Wiley and Sons, New York, NY, Vol. 2, pp. 163–198.
24. Savvidou, G., Klein, M., Grey, A.A., Dorrington, K.J. and Carver, J.P. (1984) Biochemistry 23, 3736–3740.
25. Srivastava, O.P., Hindsgaul, O., Shoreibah, M. and Pierce, M. (1988) Carbohydrate Res. 179, 137–161.
26. Dahms, N.M. and Hart, G.W. (1986) J. Biol. Chem. 261, 13186–13196.
27. Lee, S.O., Connolly, J.M., Ramirez-Soto, D. and Poretz, R.D. (1990) J. Biol. Chem. 265, 5833–5839.
28. Sharon, N. (1993) TIBS 18, 221–226.
29. Lis, H. and Sharon, N. (1993) Eur. J. Biochem. 218, 1–27.
30. Cumming, D. (1995) in: J. Montreuil, H. Schachter and J.F.G. Vliegenthart (Eds.), Glycoproteins. New Comprehensive Biochemistry, Vol. 29a. Elsevier, Amsterdam, pp. 391–444.
31. Glick, M.C. (1995) in: J. Montreuil, H. Schachter and J.F.G. Vliegenthart (Eds.), Glycoproteins. New Comprehensive Biochemistry, Vol. 29a. Elsevier, Amsterdam, pp. 261–280.
32. Drickamer, K. (1993) Biochem. Soc. Trans. 21, 456–459.

J. Montreuil, H. Schachter and J.F.G. Vliegenthart (Eds.), *Glycoproteins*

287

Biosynthesis

4c. Compartmentation of glycoprotein biosynthesis

JÜRGEN ROTH

Division of Cell and Molecular Pathology, Department of Pathology, University of Zürich,
Schmelzbergstr. 12, CH-8091 Zürich, Switzerland

1. Introduction

From a simplistic viewpoint, each eukaryotic cell can be divided in two basic compartments which are the cytoplasmic matrix or the cytosol on one side and a variety of membrane-bounded organelles on the other side including those of the secretory pathway. The latter which is embedded in the cytoplasmic matrix is structurally and functionally both a highly complex and dynamic three-dimensional membranous network, the endomembrane system. Some of the parts of this endomembrane system are structurally continuous while others are functionally continuous by means of vesicular and tubular traffic. Further, quantitative aspects of the architecture of the endomembrane system may vary greatly depending on the cell type and the functional state of a given cell. Both, the cytoplasmic matrix and major parts of the endomembrane system, namely the endoplasmic reticulum and the Golgi apparatus, are sites of diverse steps in the *de novo* synthesis of proteins and lipids and subsequently occurring posttranslational modifications. A major posttranslational modification consists in the glycosylation leading to the construction of a myriad of oligosaccharide side chains of proteins and lipids which can greatly differ in their structure and composition. Glycosylation reactions proceed in distinct cellular organelles and these processes are intimately related to membrane traffic phenomena. It is beyond the scope of this review to cover the molecular basis of vesicular interorganelle traffic [1–5]. Various biochemical aspects of glycosylation have been the subject of previous reviews [6–8] and are covered in other chapters of this book. In this chapter the topographical aspects of the assembly of oligosaccharide side chains of glycoproteins are considered which have emerged by the combined use of morphological and biochemical techniques.

2. Architecture of cellular organelles involved in glycosylation

Various sugar nucleotides which are required as donor substrates in the biosynthesis of oligosaccharides are synthesized in the cytosol with the exception of CMP-sialic acid

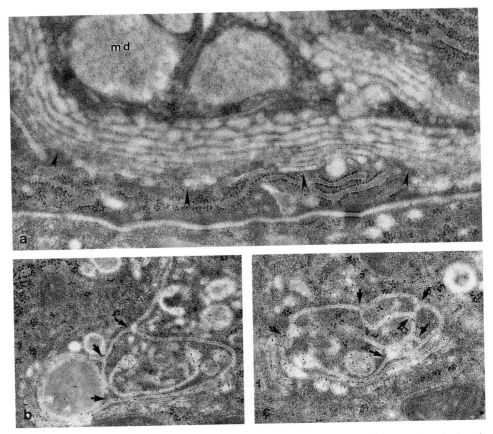

Fig. 1. (A): Part of a mucus producing goblet cell of rat colon. Cisternae of the rough endoplasmic reticulum (arrows) and a transitional element (open arrow) are adjacent to the *cis* side of the Golgi apparatus (arrowheads). Mucus droplets (md) are formed at the *trans* side of the Golgi apparatus. (B, C): Details of the Golgi apparatus in rat liver hepatocytes which exhibit well developed *trans* Golgi networks (arrows). The black spots represent gold particle immunolabeling to detect albumin reactivity. From Ref. [55].

which is synthesized in the nucleus. The other biosynthetic steps take place principally in two organelles: the endoplasmic reticulum and the Golgi apparatus (Figs. 1 and 2). The importance of the intermediate compartment, a pleiomorphic structure composed of vacuolar and tubulo-vesicular elements located between the endoplasmic reticulum and the Golgi apparatus as the possible site of glycosylation reactions remains to be elucidated. The aspects of structure and synthesis of nuclear and cytosolic glycoproteins characterized by the presence of an O-linked *N*-acetylglucosamine residue are covered in the chapter by G. Hart (Vol. 29b).

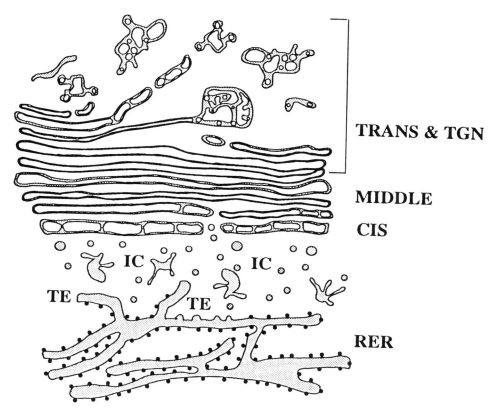

Fig. 2. Schematic representation of the rough endoplasmic reticulum (RER) and its transitional elements (TE), the intermediate compartment (IC) and the Golgi apparatus with its *cis*, *medial* and *trans* cisternae including the *trans* Golgi network (TGN).

2.1. The endoplasmic reticulum

The endoplasmic reticulum [9–11] extends through most of the cytoplasm and forms a network composed of membrane-bounded fenestrated cisternae and anastomosing tubules. Those cisternae which are covered on their cytosolic side by ribosomes constitute the rough endoplasmic reticulum and are the site of protein synthesis [12].

The rough endoplasmic reticulum is continuous with the nuclear envelope and there is evidence accumulating which suggests that the outer nuclear membrane which is covered with ribosomes is indeed part of the rough endoplasmic reticulum. Towards the Golgi apparatus the rough endoplasmic reticulum shows characteristic elements that are partly free of ribosomes and known as transitional elements [13,14]. Such transitional elements can also be found distant from the Golgi apparatus in peripheral regions of the cytoplasm. Naturally, the cisternae of the rough endoplasmic reticulum are most abundant in cell types with high synthetic activity and tend to be arranged in parallel arrays. The cisternal lumen is usually narrow but can be greatly distended in certain cell types such as plasma cells. Despite such quantitative variability, no obvious qualitative differences in the structural organization of the rough endoplasmic reticulum seem to exist in the various cell types.

In addition to the rough endoplasmic reticulum, a system of membrane-limited anastomosing and branching tubules devoid of ribosomes was observed by Fawcett [15] and was designated smooth endoplasmic reticulum. The tubular smooth endoplasmic reticulum is continuous with the cisternae of the rough endoplasmic reticulum, and with the exception of certain highly specialized cell types, is not as extensively developed as the latter.

2.2. The intermediate compartment

The existence of a compartment distinct from the endoplasmic reticulum and the Golgi apparatus has been demonstrated which is involved in the transport from the endoplasmic reticulum to the Golgi apparatus [16–22] Morphologically, it consists of a pleiomorphic structure composed of vacuolar and tubulo-vesicular elements located in the region of the Golgi apparatus and in peripheral parts of the cytoplasm distant from the Golgi apparatus. Biochemically, it has a protein composition distinct from that of the endoplasmic reticulum and the Golgi apparatus characterized by the presence of three membrane proteins of 53 kD, 58 kD and 63 kD. These proteins as well as newly synthesized viral glycoproteins accumulate in the intermediate compartment at 15°C. The function of the p53, p58 and p63 membrane proteins remains to be established. It also remains to be clarified if the intermediate compartment is a stable structure giving rise to transport vesicles or a dynamic transport structure that moves towards and finally fuses with the Golgi apparatus [23].

2.3. The Golgi apparatus

The Golgi apparatus is involved in many cellular functions such as the formation of secretory granules, posttranslational protein and lipid modifications including proteolytic cleavage, sulfation and glycosylation, membrane biogenesis and recycling as well as sorting [24]. Structurally, the Golgi apparatus is highly complex as evidenced from investigations of thick sections with high voltage electron microscopy [25,26] and consists of a continuous structure which is reminiscent of what was called originally by Golgi the "internal reticular apparatus". However, this network character of the Golgi apparatus is difficult to appreciate when single ultrathin sections are examined by electron microscopy. Rather, the Golgi apparatus appeared to be composed of separate units of twisting and curving stacks of smooth cisternae or sacculae. The number of cisternae forming the stack may differ significantly from cell type to cell type and additionally depending on the functional stage of a given cell. Therefore, as few as three or as many as twenty cisternea might be observed. The cisternal stack displays polarity. The side which faces the transitional elements of the rough endoplasmic reticulum and the intermediate compartment is termed *cis* side and is often but not always convex. The side which faces the centriole or may give rise to secretory granules is the *trans* side and is often concave. The identification of this polarity on purely morphological grounds can be difficult if not impossible. However, cytochemical techniques have provided useful means to identify and differentiate the *cis* from the *trans* side. Histochemically detectable thiamine pyrophosphatase and acid phosphatase activities were typically found in cisternae at the *trans* side or in the so-called *trans*-most cisterna, respectively [25–30]. Moreover, after prolonged osmification of the tissue, cisternae at the *cis* side of the Golgi apparatus became stained [31]. Nicoti-namide adenine dinucleotide phosphatase activity was found in cisternae situated in the middle of the stack [32]. However, the classical distribution of acid phosphatase activity can greatly vary depending on the cell type and its functional state [33–40]. Similarly, the distribution of thiamine pyrophosphatase in the Golgi apparatus may display great variability [38–40]. Collectively, these data prove the existence of cell-type specific and dynamic Golgi apparatus distributions of enzyme cytochemical markers.

The seemingly individual stacks of cisternae are bridged together by anastomosing tubules forming the intercisternal (or intersaccular) connecting regions [25,26,41–44]. At the two poles of the stacks complex networks exist, the *cis* Golgi network and the *trans* Golgi network [25,26]. In cross-sections the *cis* Golgi network appears in the form of 1 or 2 fenestrated cisternae. The *trans* Golgi network, in general, forms a much more elaborate system of interconnected tubules which is continuous with *trans* cisternae [26,45]. It appears to be a rather dynamic structure which can undergo changes in size and shape as a consequence of varying conditions [46–50]. There is now evidence that the acid phosphatase positive *trans* Golgi apparatus region, the so-called GERL [51–54] correspond to the *trans* Golgi network [45,55,56].

3. Techniques to study glycosylation in cells

3.1. Subcellular fractionation

An important advance in Golgi apparatus research resulted from the finding by Roth-man and coworkers [57,58] that in vesicular stomatitis virus infected CHO cells many of the glycosylation enzymes could be resolved and separated on analytical sucrose density gradients. The basic assumption for this approach was that a vectorial difference in density exists *in situ* across the stack of Golgi apparatus cisternae. Upon cell homogenization the Golgi apparatus cisternae or their remnants were assumed be torn apart providing the basis for their separation in subsequent gradient centrifugation. Experimentally, CHO cell membranes from a postnuclear supernatant could be resolved into different fractions in a six-step sucrose gradient where they accumulated in the sucrose layer corresponding to a density equal to their own. Measurement of specific enzymatic activities for different glycosylation enzymes revealed that kinetically early and late enzyme activities were concentrated in heavy and light membrane fractions, respectively. These data were the first biochemical evidence to suggest the existence of distinct Golgi apparatus glycosylation subcompartments and has been used by other investigators to study further aspects of glycosylation reactions in the Golgi apparatus. It should be noted, however, that although separation of enzymatic activities could be achieved, a certain degree of cross-contamination cannot be excluded as shown by the presence of broadly overlapping activities.

3.2. In situ labeling

3.2.1. Autoradiography
The technique of autoradiography applied to light and electron microscopy has provided one of the first direct in situ evidence of the involvement of the endoplasmic reticulum and the Golgi apparatus in glycosylation [59–62]. Injection of animals with ^{3}H-labeled monosaccharides serving as precursors in the synthesis of oligosaccharide side chains of glycoproteins is the first step in this technique. Thick or thin sections were then prepared from the different tissues and covered with a photographic emulsion. Following development, the localization of silver grains could be studied by light and electron microscopy to visualize the initial site of the incorporation of the radioactively labeled precursor and its subsequent transportation in the cells.

3.2.2. Immunolabeling
Electron microscopic gold labeling techniques have provided the most direct means for the high resolution in situ localization of enzymes involved in glycosylation and their products of action. For this purpose, ultrathin sections of low temperature Lowicryl K4M embedded cells and tissues [63] or frozen-thawed ultrathin sections [64,65] are incubated with specific antibodies or lectins. The binding of these reagents to subcellular structures is then visualized by the use of appropriate gold-labeled second step reagents [63,66–68]. The immunoelectron microscopic localization of enzymes involved in glycosylation reactions provides solely information about the distribution of the immunoreactive protein. Therefore, this information has to be combined with further data on the localization of the product of the action of the respective enzyme. For this purpose, lectins or anti-carbohydrate specific monoclonal antibodies recognizing specific moieties in oligosaccharide side chains have to be applied. Currently, these two independent approaches provide the most direct evidence for the subcellular site at which particular glycosylation reactions take place.

It should be mentioned that oligosaccharidic structures in tissue sections are not only accessible for interaction with lectins or antibodies but can be also subjected to various chemical modifications and enzymatical treatments involving the use of exo- and endoglycosidases [63,69–70]. Further, glycosylation reactions can be performed on ultrathin sections like in the reagent tube by using purified glycosyltransferases together with the appropriate donor substrate [71].

4. Topography of biosynthesis of asparagine-linked oligosaccharides

4.1. General

During the biosynthesis of asparagine-linked oligosaccharides distinct steps can be distinguished. These involve the synthesis of various donor nucleotide sugars and lipid-linked monosaccharides, the assembly of a lipid-linked oligosaccharide precursor and its transfer to the peptide, a number of subsequently occurring processing steps which involve different trimming and elongation reactions. These various steps occur in different cellular organelles. Beside such a level of interorganelle compartmentation, intraorganelle compartmentation has been shown to exist particularly in the Golgi apparatus. Depending on the cell type, however, the degree of intraorganelle compartmentation may exhibit considerable variation.

4.2. Reactions in the cytosol

The cytosol is the site of the localization and action of various nucleotide sugar synthetases involved in the synthesis of the precursor sugar donors needed for the synthesis of the lipid-linked oligosaccharide [72–76].

4.3. Reactions in the endoplasmic reticulum

The endoplasmic reticulum is the cellular site where the assembly of the lipid-linked oligosaccharide and its transfer to the peptide takes place. A number of subsequently occurring trimming reactions are also housed in the endoplasmic reticulum.

The question as to whether the synthesis of the lipid-linked oligosaccharide takes places entirely in the lumen of the endoplasmic reticulum, its cytosolic surface or partially in the former and partially on the latter has proven to be a challenging puzzle. The lipid-linked oligosaccharide is represented by the structure Dol-P-P-$(GlcNAc)_2$ $(Man)_9(Glc)_3$ and the current view of the topography of its synthesis [77] is schematically summarized in Fig. 3. Principally, the lipid-linked oligosaccharide is assembled by the stepwise addition of the various sugars. The sequential addition of GlcNAc-1-P and GlcNAc to dolichol-P to form Dol-P-P-$(GlcNAc)_2$ seems to occur on the cytosolic site (reaction 1 in Fig. 3) and not in the lumen of the endoplasmic reticulum. The evidence for this assumption comes from different and independent experimental data. The active site of both the dolichol-P:GlcNAc-1-P transferase and the dolichol-P-P-GlcNAc transferase faces the cytosolic side of the endoplasmic reticulum [73–75]. Further, by the use of the membrane-impermeable probes galactosyltransferase and UDP-galactose, GlcNAc could be galactosylated and therefore must be present on the cytosolic side of the endoplasmic reticulum [76]. These findings were somewhat unexpected since the donor substrate in these synthetic steps, UDP-GlcNAc, can be transported in a carrier-mediated manner from the cytosol into the lumen of the endoplasmic reticulum [77]. The next synthetic steps comprise the addition of five mannose residues from GDP-Man (reaction 2 in Fig. 3). The synthesis of GDP-Man takes place in the cytosol and since it cannot cross the endoplasmic reticulum membrane [77–79], these reactions also must occur on the cytosolic side of the endoplasmic reticulum. Indeed, studies employing the lectin Concanavalin A [80] demonstrated the presence of Dol-P-P-$(GlcNAc)_5(Man)_2$ on the cytosolic side of the endoplasmic reticulum [80]. By using the same experimental approach, evidence could be obtained that the product of the next synthetic steps, Dol-P-P-$(GlcNAc)_2$-$(Man)_{6-9}$, exists in the lumen of the cisternae of the endoplasmic reticulum [80–81]. This requires the prior translocation of the Dol-P-P(GlcNAc)$_2$-(Man)$_5$ across the endoplasmic reticulum membrane (reaction 3 in Fig. 3). Another translocation across the membrane must occur in addition. This involves the addition of the four last mannose residues since the immediate precursor for these reactions is not GDP-Man, which cannot traverse the membrane, but Dol-P-Man (reactions 4 and 5 in Fig. 3). The evidence that such translocation reactions of lipid-linked oligosaccharide and monosaccharide may occur is indirect and comes from studies with model membranes and in prokaryotes. It could be shown that polyisoprenols including dolichol-P and dolichol when added to liposomes cause a destabilization of the phospholipid bilayer and result in phosphatidylcholine exchange from the outer to the inner side of the vesicles [82–88]. Such changes which seem to involve perturbations of the lipid bilayer to a more non-bilayer or hexagonal conformation may facilitate the transmembrane movement of Dol-P-linked oligosaccharide. In prokaryotes, similar mechanisms involving the action of undecaprenol phosphate seem to be operating in the transmembrane movement of preassembled polysialic acid [89]. As was mentioned above, GDP-Man cannot traverse membranes and Dol-P-Man, the immediate precursor for the last five mannose residues to be added, is most probably synthesized on the cytosolic side of the endoplasmic reticulum and then transferred to the lumen (reactions 4 and 5 in Fig. 3) where it is used as donor substrate (reaction 6 in Fig. 3). This was concluded from data indicating that the Dol-P-Man synthetase may have an orientation in the membrane of the endoplasmic reticulum with its catalytic site facing the cytosol [75,90]. Other model studies have provided evidence that Dol-P-Man can traverse membranes [91]. The last step in the assembly of the lipid-linked oligosaccharide is the addition of three glucose residues. It has been shown that GDP-Glc can traverse the endoplasmic reticulum membrane by a carrier-assisted mechanism and thus, could be used in the lumen of the endoplasmic reticulum for the synthesis of Dol-P-Glc. However, there is evidence that the dolichol-P:UDP-Glc synthetase may be located on the cytosolic side of the endoplasmic reticulum [74,75; reaction 7 in Fig. 3]. This in turn would imply that a translocation reaction of the Dol-P-Glc must occur (reaction 8 in Fig. 3) before the glucosylation of the lipid-linked (GlcNAc)$_2$(Man)$_9$ can take place (reaction 9 in Fig. 3). Whatever mechanism operating in these synthetic steps is correct remains to be directly demonstrated. Irrespective of this missing direct evidence, the fully assembled lipid-linked oligosaccharide is finally transferred *en bloc* to nascent peptide chains and all available evidence indicates that this transfer reaction takes place in the lumen of the endoplasmic reticulum [80,81,92]. The transfer of the preassembled lipid-linked oligosaccharide to polypeptides is mediated by the N-oligosaccharyl transferase which is located in the endoplasmic reticulum [93–97]. The biochemical characterization of this enzyme is not yet completed and the molecular details of its membrane orientation are unknown. There is evidence that it forms a complex together with ribophorins I and II and a not further characterized 48 kDa protein [98]. In yeast, at least two proteins, Wbp1p and Swp1p, were found to form a complex with the N-oligosaccharyl transferase which was shown to be essential for the activity of the transferase [99]. These proteins are type I transmembrane proteins with their N-termini exposed to the lumen of the endoplasmic reticulum. The transfer of the preassembled lipid-linked oligosaccharide occurs to asparagine residues present in the sequence Asn-X-Ser/Thr where X can be

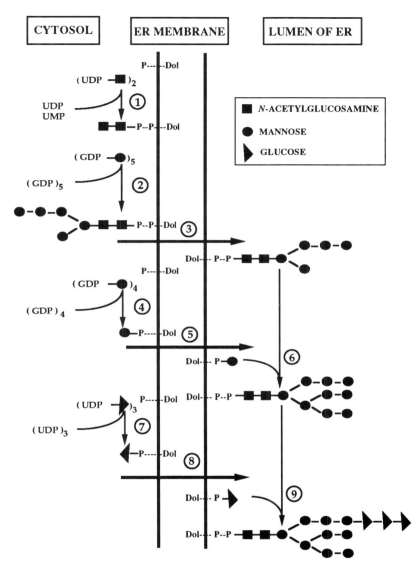

Fig. 3. Schema of the topography of the assembly of the lipid-linked oligosaccharide in the endoplasmic reticulum.

any amino acid except possibly Pro and Asn [100]. However, other currently not fully understood features of the polypeptide must have an influence as to whether such potential glycosylation sites become glycosylated or not [101–105].

The *en bloc* transfer of the preassembled oligosaccharide from Dol-P-P to asparagine in a polypeptide can occur cotranslationally when the polypeptide is still in the process of being vectorially discharged. This transfer reaction can be followed

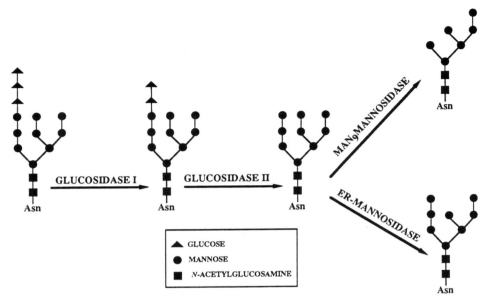

Fig. 4. Schema of the trimming reactions by glucosidases and mannosidases occurring on asparagine-linked oligosaccharides in the endoplasmic reticulum.

immediately by processing reactions which result in the trimming of all three glucose residues and a varying number of the mannose residues (Fig. 4). In addition a transient reglucosylation reaction may occur.

The first acting enzyme is the neutral processing α glucosidase I which removes the outer α-1,3-linked glucose residue [106–108]. The enzyme has not been localized *in situ* by immunoelectron microscopy. However, all the existing data unequivocally indicate that neutral processing α glucosidase I should by present and acting in the lumen of the endoplasmic reticulum. It could be demonstrated that glucosidase I acts on the G protein of vesicular stomatitis virus when it is still bound to ribosomes [106,109]. In general, the kinetics of removal of the single outer α-1,3-linked glucose residue show that hydrolysis occurs rapidly with a $t_{1/2} < 2$ min.

Following this first trimming reaction in the endoplasmic reticulum, the remaining two α-1,3-linked glucose residues are hydrolyzed by neutral processing α glucosidase II [110,111]. The outer α-1,3-linked glucose residue is hydrolyzed more rapidly ($t_{1/2} \sim$ 5 min) than the inner α-1,3-linked residue ($t_{1/2} \sim 20$–30 min) [106,109]. For example, studies on pulse-labeled glycoproteins have shown that after the rapid removal of the first of the two α-1,3-linked glucose residues, monoglucosylated species still exist after 20–25 min of chase. In hepatocytes, both biochemical and morphological evidence demonstrated the presence of glucosidase II in the lumen of the endoplasmic reticulum. The enzyme seems to carry one endo H-sensitive oligosaccharide side chain and apparently is only loosely membrane-associated and not a membrane integral protein [110,112]. By *in situ* immunoelectron microscopy, specific labeling for glucosidase II

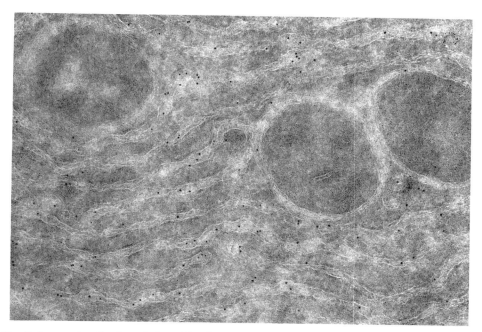

Fig. 5. Immunolocalization of glucosidase II in the rough endoplasmic reticulum of a hepatocyte in liver. Gold particle labeling is confined to the lumen of the cisternae. M: mitochondria. From Ref. [113].

was detected in the smooth and rough endoplasmic reticulum (Fig. 5) as well as in the nuclear envelope and not in the Golgi apparatus cisternal stack [113]. The intensity of gold particle labeling was similar over the rough and smooth endoplasmic reticulum and only about half of these values were found for the nuclear envelope. Interestingly, specific immunolabeling was also detectable over the transitional elements of the rough endoplasmic reticulum and smooth surfaced vesicular and tubular structures located in between the endoplasmic reticulum and the Golgi cisternal stack. Therefore, one may speculate that the latter structures may comprise the intermediate compartment or some parts of it. At the time when the glucosidase II immunolocalization was performed, the existence of the intermediate compartment was not yet proven. A specialized distribution pattern inside the glucosidase II immunolabeled structures as the possible basis for the observed differences in the time course of hydrolysis of the two α-1,3-linked glucose residues could not be detected. Therefore, differences in the rate of release of the two glucose residues by glucosidase II most probably account for the observed biochemical data. Collectively, the immunolocalization and biochemical studies on glucosidase II demonstrated that in hepatocytes, this enzyme represents a resident protein of the endoplasmic reticulum. Its diffuse distribution in this compartment is in agreement with its function in the bulk trimming reaction on asparagine-linked oligosaccharides. Subsequent studies on other cell types, however, have revealed striking differences from this pattern of distribution [114]. In various types of kidney epithelial cells, immunolabeling for glucosidase II was found additionally in the Golgi apparatus, the plasma membrane and a system of vesicular structures involved in exo- and endocytosis. This post-Golgi localization of glucosidase II was paralleled by the detection of enzymatically active, sialylated enzyme species in plasma membrane (brush border) fractions. Furthermore, evidence for the presence of a ligand for glucosidase II in this location could be obtained. Despite these findings, the actual function of glucosidase II in these post-Golgi locations remains to be shown. Furthermore, its possible role in these different intracellular sites in the deglucosylation of transiently reglucosylated glycoproteins is unclear. Such a transient reglucosylation of glycoproteins following the trimming of all three glucose residues has been shown to occur in the endoplasmic reticulum and may also take place after exit of glycoproteins from the endoplasmic reticulum [115–117]. In the endoplasmic reticulum, UDP-glucose represents the donor for this reaction which is executed by a specific glucosyltransferase which has been shown to be a soluble or loosely membrane attached protein which faces the lumen of the endoplasmic reticulum. The functional significance of the transient reglucosylation remains to be established.

Following the bulk trimming reactions catalyzed by the glucosidases I and II, which result in the formation of Asn-$(GlcNAc)_2(Man)_9$, glycoproteins may be transported from the endoplasmic reticulum to the Golgi apparatus for further processing. However, there is now unequivocal evidence that further trimming reactions can take place while a glycoprotein is still present in the endoplasmic reticulum. These reactions can result in the removal of a single or up to three mannose residues catalyzed by different neutral processing α-mannosidases (Fig. 4). Therefore, the presence of a $(GlcNAc)_2(Man)_6$ oligosaccharide structures can no longer be taken as unequivocal evidence of Golgi apparatus associated trimming.

The presence of mannosidases in the endoplasmic reticulum has been predicted from various studies analyzing the structure of oligosaccharide side chains of glycoproteins while present or experimentally retained in this compartment. Oligosaccharides of calf thyroglobulin in thyroid slices had been found to be processed to $(GlcNAc)_2(Man)_8$ in the presence of carbonyl cyanide N-chlorophenylhydrazone (CCCP), an inhibitor of transport from endoplasmic reticulum to Golgi apparatus [118]. Various viral glycoproteins have been shown to contain $(GlcNAc)_2(Man)_8$ oligosaccharides either during their passage in the endoplasmic reticulum or when retained by CCCP [119–121], as have endogenous endoplasmic reticulum glycoproteins [122–125]. Bischoff and Kornfeld [126,127] characterized a neutral α-mannosidase which was originally described as cytoplasmic α-mannosidase [128,129] and is highly enriched in endoplasmic reticulum membranes of rat liver. This presumptive endoplasmic reticulum α-mannosidase acts intracellularly on asparagine-linked $(GlcNAc)_2(Man)_9$ to yield a specific $(GlcNAc)_2(Man)_8$ isomer (Fig. 4). A specific antibody localized this enzyme in the endoplasmic reticulum of rat liver hepatocytes (Roth and Bischoff, unpublished data).

Oligosaccharide analyses have provided evidence for the presence of yet another endoplasmic reticulum α-mannosidase. The Z variant of human plasma α-1-antitrypsin which accumulates in the endoplasmic reticulum of hepatocytes was shown to possess a variety of high mannose oligosaccharides ranging from $(GlcNAc)_2(Man)_7$ to

$(GlcNAc)_2(Man)_6$ [130]. Similar observations were made for resident endoplasmic reticulum glycoproteins such as 3-hydroxy-3-methylglutaryl-CoA reductase [124], ribophorin I [125], endoplasmic reticulum protein ERp99 [123] and prolyl hydroxylase [122]. A mannosidase, named Man_9-mannosidase, which may be involved in this mannose trimming was reported by Bause and coworkers [131,132]. It acts immediately after the removal of the three glucose residues by glucosidases I and II and specifically cleaves three of the four α-1,2 mannose residues (Fig. 4). This Man_9-mannosidase is distinct therefore from the endoplasmic reticulum-mannosidase reported by Bishoff and Kornfeld [126,127] and the two Golgi mannosidases.

The immunoelectron microscopic localization of Man_9-mannosidase in pig liver with monospecific antibodies showed its presence in the smooth and rough endoplasmic reticulum as well as transitional elements of the latter and smooth surfaced membranes close to the Golgi apparatus. The mannosidase was undetectable in the nuclear envelope and the cisternal stack of the Golgi apparatus. When compared to glucosidase II by double immunoelectron microscopy, Man_9-mannosidase exhibited a more restricted and local distribution in the endoplasmic reticulum. The meaning of the differential distribution of the two resident endoplasmic reticulum trimming enzymes is unclear at present. It is tempting to speculate that it may be of functional significance. The diffuse distribution of glucosidase II can be seen in agreement with its function in the bulk trimming of asparagine-linked oligosaccharide intermediates [110,111,133]. On the other hand, the local distribution of Man_9-mannosidase could be related to its involvement in selective trimming reactions. Indeed there is evidence for selective trimming executed by the two different endoplasmic reticulum mannosidases. In their studies, Bischoff et al. [126,127] found that only one-third of the oligosaccharides on secreted glycoproteins and about one-half of the oligosaccharides on cellular glycoproteins in cultured hepatocytes were processed. In experiments using CCCP during the chase period to prevent transport out of the endoplasmic reticulum, the ratio of $(GlcNAc)_2(Man)_9$ to $(GlcNAc)_2(Man)_8$ was not altered. This was taken as indirect evidence that it is not the transportation rate, but rather some intrinsic properties of specific glycoproteins, which determines whether they become a substrate for this mannosidase. Another possibility, or an additional parameter in determining the occurrence of selective trimming reactions, could be the above reported specialized distribution of the trimming enzymes. Indeed, Kabcenell and Atkinson [121] have provided indirect evidence for the spatial separation of two different mannose trimming reactions occurring in the endoplasmic reticulum on VP7 glycoprotein of rotavirus SA11. In the presence of CCCP the processing of VP7 oligosaccharides was blocked at the level of $(GlcNAc)_2(Man)_8$ and it was concluded that further trimming to $(GlcNAc)_2(Man)_6$ required an energy-dependent transport process to a region of the endoplasmic reticulum containing another mannosidase. This study also provided evidence for the selective action of this other mannosidase since the oligosaccharides of NCPV5 rotavirus glycoprotein and the G protein of VSV virus were only trimmed to $(GlcNAc)_2(Man)_8$. Direct evidence for specialized domains in the rough endoplasmic reticulum exists for secretory proteins. Deschuyteneer et al. [134] reported the intermittent occurrence of apomucin in cisternae of the endoplasmic reticulum of submaxillary gland mucous cells and its absence in the nuclear envelope. Segregation of chondroitin sulfate glycoprotein precursor and its link protein has been found in chondrocyte endoplasmic reticulum [135].

As mentioned above, immunolabeling for both glucosidase II and Man9-mannosidase was detectable in smooth surfaced structures situated between transitional elements of the endoplasmic reticulum and *cis* Golgi cisternae and therefore may represent elements of the intermediate compartment. If this can be proven to be the case, then the intermediate compartment may be involved in the trimming of oligosaccharides.

A neutral, broad-specificity α-mannosidase was recently purified from rat liver microsomal membranes which catalyzes the removal of α-1,2, α-1,3 and α-1,6 linked mannose residues from $(GlcNAc)_2(Man)_{4-9}$ oligosaccharide structures [136]. This $Man_5GlcNAc$ mannosidase was detectable by immunoelecton microscopy throughout the rough and smooth endoplasmic reticulum and the entire Golgi apparatus cisternal stack [137]. Its physiological role remains to be elucidated.

4.4. Reactions in the Golgi apparatus

After initial trimming reactions in the endoplasmic reticulum, further trimming reactions as well as elongation reactions take place in the Golgi apparatus. A generally accepted principle is that the Golgi apparatus is divided in subcompartments: early glycosylation reactions are assumed to take place in the *cis* and medial part of the Golgi apparatus cisternal stack whereas the late one occur in the *trans* region of the organelle [8,45,57,138,139]. Although this principle seems to hold true for many cell types, cell-type specific variations in the extent of the subcompartmentation have been shown to exist. The first example was provided by observation of the highly differential distribution of sialyltransferase and blood group A transferase, both being terminal glycosyltransferases, in intestinal cells [140]. It is also necessary to state that of the many glycosylation enzymes which reside in the Golgi apparatus only a remarkably low number has been localized *in situ* by immunoelectron microscopy. In other words, the picture of the functional organization of the Golgi apparatus with regard to glycosylation is far from being complete.

Processing reactions on oligosaccharide side chains of glycoproteins can continue after their transport from the endoplasmic reticulum in the Golgi apparatus (Fig. 6). Removal of all remaining α-1,2-linked mannose residues takes place by the action of mannosidases IA and IB [141–143]. Biochemical evidence points to the *cis*/medial part of the Golgi apparatus as their site of location [144, 145]. Immunolocalization of the enzyme showed that it was principally detectable in the *medial* and undetectable in the *cis* Golgi apparatus [146]. The Golgi apparatus distribution of mannosidase I varied, however, depending on the cell type studied from a *medial* to a *medial* and *trans*, or *trans*, or diffuse localization. After removal of all α-1,2-linked mannose residues the glycosylation reactions leading to the formation of complex-type oligosaccharides can start. The initial reaction in this process is the addition of a single *N*-acetylglucosamine residue to the terminal α-1,3-linked mannose of the core structure $(GlcNAc)_2(Man)_5$ by the *N*-acetylglucosaminyl transferase I [147,148]. Monoclonal antibodies were prepared against this transferase and applied to detect immunoreactive protein in rabbit liver hepatocytes and LLC-RK1 cells [149]. A localization in the medial cisternae of

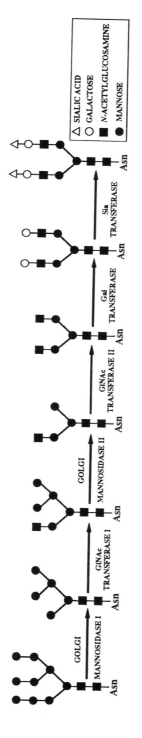

Fig. 6. Schema of processing reactions occurring on asparagine-linked oligosaccharides in the Golgi apparatus. The synthetic pathway yielding a bi-antennary complex-type oligosaccharide is depicted.

the Golgi stack was reported. From the published micrographs it seems that the only unlabeled cisternal element at the *cis* side corresponds to the fenestrated first *cis* element forming the *cis* Golgi network. Subsequent cytochemical detection of *N*-acetylglucosamine residues in the Golgi apparatus of hepatocytes became possible by the application of purified galactosyltransferase and UDP-gal [150]. This *in situ* galactosylation protocol revealed the presence of galactosylated *N*-acetylglucosamine in the *medial* and *cis* part of the Golgi apparatus including the first fenestrated element of the Golgi apparatus. Both results indicate that this glycosylation reaction is not restricted to the medial part of the Golgi apparatus in hepatocytes. None of the other known Golgi-associated *N*-acetylglucosaminyl transferases have been studied by immunoelectron microscopy until now. A similar broad distribution as detected for Golgi mannosidase I was observed for Golgi mannosidase II which catalyzes the removal of the non-core α 1,3- and α 1,6-linked mannose residues to yield the $(GlcNAc)_2(Man)_3GlcNAc$ structure [141,151–154]. The Golgi apparatus distribution of mannosidase II varied depending on the cell type studied from a *medial* to a *medial* and *trans* or *trans*, or diffuse localization [146]. In Chinese hamster ovary cells and mouse L929 cells, the use of the same antibody resulted in immunolabeling across the entire cisternal stack [155]. Thus, Golgi mannosidases I and II may exhibit, depending on the cell type studied, different distribution patterns in the Golgi apparatus as was shown for other glycosyltransferases earlier [140]. Of the families of galactosyl-, fucosyl- and sialyltransferases, *in situ* data for only one of the galactosyl- and one of the sialyltransferases are currently available. In HeLa cells, UDP-Gal: β-1,4-GlcNAc galactosyl transferase was localized to thiamine pyrophosphatase-positive *trans* cisternae [139]. Such a *trans* Golgi localization was also apparent in human hepatoma cell line HepG2 [156]. In another study, immunolocalization of CMP-Sia: Gal (β 1,4) GlcNAc α-2,6-sialyltransferase was performed [45]. In rat liver hepatocytes, immuno-label was present in both *trans* cisternae (positive for thiamine pyrophosphatase) and the *trans* Golgi network (positive for both thiamine pyrophosphatase and acid phosphatase) (Fig. 7). This labeling pattern coincided with the cytochemical localization of sialic acid residues as detected with the sialic acid specific *Limax flavus* lectin. In contrast to the restricted immunolabeling for α-2,6-sialyltransferase in hepatocyte and intestinal goblet cells was its diffuse distribution in the Golgi apparatus of intestinal absorptive cells [140]. With the exception of the first fenestrated element at the *cis* side all other cisternae exhibited immunolabeling (Fig. 8). This immunolabeling exhibited a gradient which increased in intensity from *cis* to *trans* cisternae. Cytochemically detectable sialic acid residues mirrored the distribution of sialyltransferase immunolabeling (Fig. 8). An equally broad distribution for sialic acid residues [157] and α-2,6-sialyltransferase immunolabel (Roth, Lee and Paulson, unpublished) existed in Chinese ovary hamster cells transfected with sialyltransferase expression vector. Thus, extreme patterns of distribution for the same glycosyltransferase can be observed in a cell-type specific manner. It may be significant that the broad distribution of sialyl-transferase in the enterocyte's Golgi apparatus is indeed a gradient. Thus, both distribution patterns may reflect solely different degrees of a gradient in the distribution of the glycosyltransferase in a Golgi cisternal stack. In other words, a steep gradient manifests itself in a more compartmentalized distribution than a more even gradient.

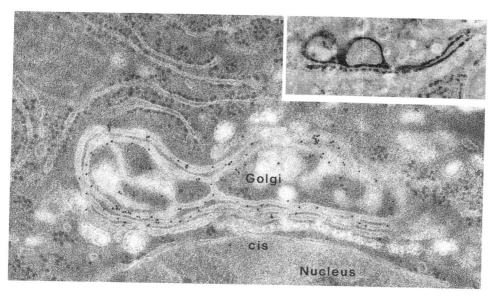

Fig. 7. Immunolocalization of α-2,6-sialyltransferase in the *trans* cisternae and the *trans* Golgi network of a hepatocyte from rat liver. The inset represents the histochemical demonstration of thiamine pyrophosphatase activity in *trans* cisternae and the *trans* Golgi network of this cell type. From Ref. [45].

A broad distribution of a terminal glycosyltransferase would imply that it would overlap in its distribution with other transferases present in the same Golgi apparatus. Indirect evidence for overlap between two glycosylation enzymes comes from the observed *cis* to *trans* distribution of both mannosidase II [155] and α-2,6-sialyltransferase [156] in Chinese hamster ovary cells. Direct evidence for the overlapping distribution of two glycosyltransferases was reported for HeLa cells in which galactosyltransferase had been demonstrated to reside in the *trans* Golgi apparatus [139, 157]. By double immunolabeling for galactosyltransferase and N-acetylglucosaminyl transferase I, the distribution of the two enzymes were found to partially overlap [158].

A glycosylation reaction which is characteristic for lysosomal enzymes may take place in the *cis*/medial part of the Golgi apparatus. This results in the synthesis of the recognition marker mannose-6-phosphate by two different enzymes. First, N-acetylglucosaminylphosphotransferase adds a GlcNAc-P moiety on position 6 of selected mannose residues of high mannose-type oligosaccharides [159–163]. Afterwards, N-acetylglucosamine-1-phosphodiester α-N-acetyl-glucosaminidase acts to form the mannosylphosphomonoester. Studies on liver Golgi apparatus subfractions or sucrose gradients prepared from mouse lymphoma and macrophage cell lines indicated that both enzymes seem to be grossly distributing together with Golgi mannosidase I [164–167].

304

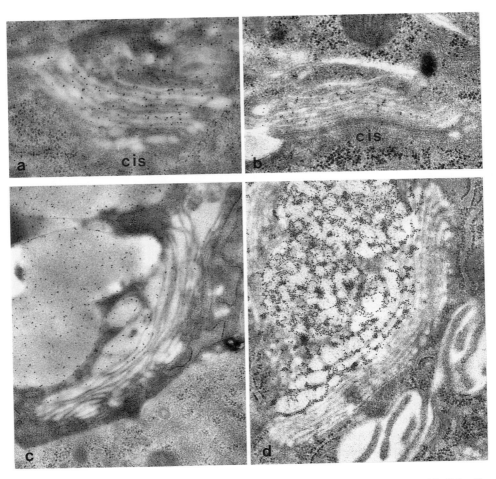

Fig. 8. Cell-type specific differential distribution of sialyltransferase in rat colonic epithelial cells. Both, immunogold labeling for sialyltransferase (A) and *Limax flavus* lectin-gold labeling for sialic acid residues (B) are diffusely distributed in the absorptive enterocyte Golgi apparatus whereas they exhibit a trans localization in the goblet cell apparatus (C, sialyltransferase; D, sialic acid moieties). From Ref. [140].

5. Topography of biosynthesis of threonine-/serine-linked oligosaccharides

5.1. General

Different types of Thr/Ser-linked oligosaccharides exist and only those characterized by the core structure Ser/Thr-GalNAc will be considered in the following paragraphs. Such oligosaccharides may be quite simple. The core glycosylation itself may remain the only reaction and gives rise to the so-called Tn antigen which is an onco-developmental antigen. The addition of a galactose moiety to the this core structure results in the formation of another onco-developmental antigen, the Thomsen–Friedenreich antigen. This structure is the precursor for the human blood group MN antigens and other more complex oligosaccharide side chains.

The synthetic pathway of Thr/Ser-linked oligosaccharides is much less complicated than the one for Asn-linked oligosaccharides. It does not involve the synthesis of a preassembled oligosaccharide and consists of a series of classical glycosyl transfer reactions.

5.2. Site of initial glycosylation reaction

The initial glycosylation reaction, the core O-glycosylation, involves the transfer of GalNAc from UDP-GalNAc to threonine or serine in a polypeptide. With respect to the subcellular site in which the core O-glycosylation takes place different sites such as the endoplasmic reticulum [168], intermediate compartment [169], *cis* Golgi apparatus [134, 170,171], *medial* Golgi apparatus [172, 173] and *trans* Golgi apparatus [174] have been suggested.

The corresponding enzyme, UDP-GalNAc:polypeptide *N*-acetylgalactosaminyl transferase has been purified to homogeneity from bovine colostrum [175] and porcine submaxillary glands [176]. Both the bovine and the porcine enzymes are membrane integral glycoproteins with a molecular mass of 69 kDa. Initial studies on the acceptor substrate specificity indicated that solely threonine residues become glycosylated [176, 177]. However, subsequent studies revealed that both threonine and serine are acceptor substrates and that the flanking amino acid sequences are critical for the occurrence of glycosylation on serine [178].

The immunolocalization of the UDP-GalNAc: threonine polypeptide *N*-acetylgalactosaminyl transferase has been achieved recently [179]. Specific polyclonal antibodies applied to thin sections of submaxillary glands resulted in labeling over the *cis* Golgi network and adjacent cisternae. No immunolabeling was detectable over *trans* Golgi cisternae and the rough endoplasmic reticulum including its transitional elements. The subcellular site at which immunolabeling for the transferase became first detectable was also the initial site at which the polypeptide-bound GalNAc could be detected cytochemically with the *Viccia villosa* isolectin B4 [179], as was reported in earlier studies [134,170]. These data are consistent with the previous immunolocalization of apomucin in porcine submaxillary gland (Fig. 9) which was found in the rough endoplasmic reticulum including its transitional elements but not more in the *cis* Golgi apparatus of the mucous cells [134]. In this study, the addition of the core GalNAc to apomucin was detected in the *cis* Golgi apparatus. Collectively, these data demonstrate that the initial reaction in the synthesis of O-linked oligosaccharides in the mucous cells of the porcine submaxillary gland and the intestinal goblet cells occurs posttranslationally in the *cis* Golgi apparatus and not in the endoplasmic reticulum. It remains to be clarified if this localization can be found in other cell types as well and to which extent the intermediate compartment is involved in the core O-glycosylation.

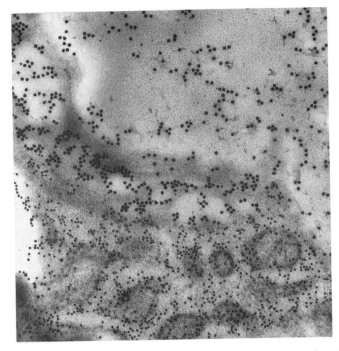

Fig. 9. Simultaneous detection of immunoreactivity for apomucin and lectin-gold labeling for serine/threonine-linked N-acetylgalactosamine in pig submaxillary gland for the identification of the site of core O-glycosylation. The smaller gold particles indicate the presence of apomucin which is restricted to the endoplasmic reticulum. The larger gold particles demonstrate the abrupt onset of labeling for N-acetylgalactosamine residues over smooth surfaced elements of the *cis* Golgi apparatus.

5.3. Site of elongation reactions

On the core GalNAc residue, a galactose residue may be added by a specific galactosyl transferase. Activity for this enzyme was found to co-sediment with the galactosyl-transferase acting on Asn-linked oligosaccharides [175]. Immunolabeling data on the localization of the enzyme are not available. However, the peanut lectin has been used for the cytochemical detection of the product of the action of this transferase. In gastrointestinal mucus-producing cells, the structure GalNAc (β1–3) Gal was detected in medial cisternae of the Golgi apparatus [180,181]. Evidence for the site of a terminal O-glycosylation reaction resulting in the formation of the blood group A determinant was obtained by the immunolocalization of the respective transferase and their reaction product [170,182]. The data pointed to the *trans* Golgi apparatus.

6. Conclusions

The synthesis of oligosaccharide side chains takes place in different intracellular compartments. The functional subcompartmentation may be of importance for the control and regulation of glycosylation. Other factors are important such as the levels of transcription of glycosyltransferase genes which in turn result in different levels of glycosyltransferases competing for a common donor substrate. Precedence for such situations exist in transfected cells in which the endogenous glycosylation pattern could be altered [156 and references therein]. Of primary importance for differential expression of glycosylation sequences during development, in differentiated cells and tissues and in malignant cells seems to be the cell-type specific expression of glycosyltransferase genes [183,184]. However, little is currently known about the molecular aspects of theses mechanisms. Probably, multiple promoters in glycosyltransferase genes with binding sites for tissue-specific transcription factors play a role in the regulation of cell-type specific glycosylation [185,186].

Abbreviations

Asn	asparagine
CCCP	carbonyl cyanide N-chlorophenyl hydrazone
CHO cells	Chinese hamster ovary cells
CMP	cytidine 5′-monophosphate
Dol-P-P	dolichol pyrophosphate
GalNAc	N-acetyl-D-galactosamine
GDP	Guanosine diphosphate
Glc	glucose
GlcNAc	N-acetyl-D-glucosamine
Man	mannose
Pro	proline
Ser	serine
Thr	threonine
UDP	uridine diphosphate

References

1. Sollner, T., Whiteheart, S. W., Brunner, M., Erjument-Bromage, H., Geromanos, S., Tempst, P. and Rothman, J.E. (1993) Nature 362, 318–324.
2. Bennet, M.K. and Scheller, R.H. (1993) Proc. Natl. Acad. Sci. USA 90, 2559–2563.
3. Schekman, R. (1992) Curr. Opin. Cell Biol. 4, 587–592.
4. Bennet, M.K., Calahos, N. and Scheller, R.H. (1992) Science 257, 255–259.
5. Rothmann, J.E. (1994) Nature 372, 55–63.
6. Hubbard, S.C. and Ivatt, R.J. (1981) Annu. Rev. Biochem. 50, 555–583.
7. Kornfeld, R. and Kornfeld, S. (1985) Annu. Rev. Biochem. 54, 631–664.
8. Roth, J. (1987) Biochim. Biophys. Acta 906, 405–436.
9. Porter, K.R. (1953) J. Exp. Med. 97, 727–750.
10. Palade, G.E. and Porter, K.R. (1954) J. Exp. Med. 100, 641–656.
11. Terasaki, M. (1989) In: Meth. Cell Biol. Vol. 29, pp. 125–136, Academic Press, New York.
12. Palade, G.E. (1955) J. Biophys. Biochem. Cytol. 1, 59–68.

308

13. Merisko, E.M., Fletcher, M. and Palade, G.E. (1986) Pancreas 1, 95–109.
14. Tartakoff, A.M. (1986) EMBO J. 5, 1477–1482.
15. Fawcett, D.W. (1955) J. Natl. Cancer Inst. 15, 1475.
16. Warren, G. (1987) Nature 327, 17–18.
17. Klausner, R.D. (1989) Cell 57, 703–706.
18. Saraste, J. and Svensson, K.(1991) J. Cell Sci. 100, 415–430.
19. Saraste, J. and Kuismanen, E. (1992) Semin. Cell Biol. 3, 343–355.
20. Schweizer, A., Matter, K., Ketcham, C.M. and Hauri, H.-P. (1991) J. Cell Biol. 113, 45–54.
21. Hauri, H.-P. and Schweizer, A. (1992) Curr. Opin. Cell Biol. 4, 600–608.
22. Lotti, L.V., Torrisi, M.R., Pascale, M.C. and Bonatti, S. (1992) J. Cell Biol. 118, 43–50.
23. Lippincott-Schwartz, J. (1993) Trends Cell Biol. 3, 81–88.
24. Farquhar, M.G. (1985) Annu. Rev. Cell Biol. 1, 447–488.
25. Rambourg, A., Clermont, Y. and Marraud, A. (1974) Am. J. Anat. 140, 27–46.
26. Rambourg, A. and Clermont, Y. (1990) Eur. J. Cell Biol. 51, 189–200.
27. Novikoff, A.B. (1964) Biol. Bull. 127, 358.
28. Novikoff, A.B. (1976) Proc. Natl. Acad. Sci. USA 73, 2781–2787.
29. Novikoff, A.B. and Novikoff, P.M. (1977) Histochem. J. 9, 525–551.
30. Novikoff, P.M., Novikoff, A.B., Quintana, N. and Hauw, J.J. (1971) J. Cell Biol. 50, 859–886.
31. Friend, D.S. and Murray, M.J. (1965) Am. J. Anat. 117, 135–150.
32. Smith, C.E. (1980) J. Histochem. Cytochem. 28, 16–26.
33. Boutry, J.M. and Novikoff, A.B. (1975) Proc. Natl. Acad. Sci. USA 72, 508–512.
34. Miller, J.S., Gavino, V.C., Ackerman, G.A., Sharma, H.M., Milo, G.E., Geer, J.C. and Cornwell, D.G. (1980) Lab. Invest. 42, 495–506.
35. Paavola, L.G. (1978) J. Cell Biol. 79, 45–58.
36. Paavola, L.G. (1978) J. Cell Biol. 79, 59–73.
37. Teichberg, S. and Holtzman, E. (1973) J. Cell Biol. 57, 88–108.
38. Pavelka, M. and Ellinger, A. (1982) J. Submicrosc. Cytol. 14, 577–585.
39. Pavelka, M. and Ellinger, A. (1983) Eur. J. Cell Biol. 29, 253–261.
40. Broadwell, R.D. and Oliver, C. (1981) J. Cell Biol. 90, 474–484.
41. Rambourg, A. and Chrétien, M. (1970) C.R. Acad. Sci. Paris, Sérié D 270, 981–983.
42. Carasso, N., Ovtracht, L. and Favard, P. (1971) C. R. Acad. Sci. Paris, Sérié D 273, 876–879.
43. Novikoff, P.M., Novikoff, A. B., Quintana, N. and Hauw, J.J. (1971) J. Cell Biol. 50, 859–886.
44. Rambourg, A., Clermont, Y. and Marraud, A. (1974) Am. J. Anat. 140, 27–46.
45. Roth, J., Taatjes, D.J., Lucocq, J.M., Weinstein, J. and Paulson, J.C. (1985) Cell 43, 287–295.
46. Doine, A.I., Oliver, C. and Hand, A.R. (1980) J. Histochem. Cytochem. 28, 601–602.
47. Hand, A.R. and Oliver, C. (1980) J. Cell Biol. 87, 304a.
48. Oliver, C., Auth, E.R. and Hand, A.R. (1980) Am. J. Anat. 158, 275–284.
49. Griffiths, G., Pfeiffer, S., Simons, K. and Matlin, K. (1985) J. Cell Biol. 101, 949–964.
50. Hand, A.R. and Oliver, C. (1984) In: M. Cantin (Ed.), Cell Biology of the Secretory Process. Karger, Basel, pp. 148–170.
51. Novikoff, A.B. (1964) Biol. Bull. 127, 358.
52. Novikoff, A.B. (1976) Proc. Natl. Acad. Sci. USA 73, 2781–2787.
53. Novikoff, A.B. and Novikoff, P.M. (1977) Histochem. J. 9, 525–551.
54. Novikoff, P.M., Novikoff, A.B., Quintana, N. and Hauw, J.J. (1971) J. Cell Biol. 50, 859–886.
55. Taatjes, D.J. and Roth, J. (1986) Eur. J. Cell Biol. 42, 344–350.
56. Griffiths, G. and Simons, K. (1986) Science 234, 438–443.
57. Dunphy, W.G., Fries, E., Urbani, L.J. and Rothman, J.E. (1981) Proc. Natl. Acad. Sci. USA 78, 7453–7457.
58. Dunphy, W.G. and Rothman, J.E. (1983) J. Cell Biol. 97, 270–275.

59. Neutra, M. and Leblond, C.P. (1966) J. Cell Biol. 30, 119–136.
60. Neutra, M. and Leblond, C.P. (1966) J. Cell Biol. 30, 137–150.
61. Bennett, G. and Leblond, C.P. (1971) J. Cell Biol. 46, 409–416.
62. Whur, P., Herscovics, A. and Leblond, C.P. (1969) J. Cell Biol. 43, 289–311.
63. Roth, J. (1989) In: A.M. Tartakoff (Ed.), Vesicular Transport, Part B. Meth. Cell Biol. Vol. 31 (part A). Academic Press, New York, pp. 513–551.
64. Griffiths, G., McDowall, A., Back, R. and Dubochet, J. (1984) J. Ultrastruct. Res. 8, 65–78.
65. Tokuyasu, K.T. (1989) Histochem. J. 21, 163–171.
66. Roth, J., Bendayan, M. and Orci, L (1978) J. Histochem. Cytochem. 26, 1074–1081.
67. Roth, J. (1983) J. Histochem. Cytochem. 31, 987–999.
68. Roth, J. (1983) In: G.R. Bullock and P. Petrusz (Eds.), Techniques in Immunocytochemistry, Vol. 2. Academic Press, London, pp. 217–284.
69. Deschuyteneer, M., Eckhardt, A.E., Roth, J. and Hill, R.L. (1988) J. Biol. Chem. 263, 2452–2459.
70. Roth, J., Wagner, Ph., Taatjes, D.J., Zuber, C., Weisgerber, C., Heitz, P.U., Goridis, C. and Bitter-Suermann, D. (1988) Proc. Natl. Acad. Sci. USA 85, 2999–300.
71. Lucocq, J.M., Berger, E.G. and Roth, J. (1987) J. Histochem. Cytochem. 35, 67–74.
72. Coates, S.W., Gurney Jr., T., Wyles Sommer, L., Yeh, M. and Hirschberg, C.B. (1980) J. Biol. Chem. 255, 9225–9229.
73. Kean, E.L. (1991) J. Biol. Chem. 266, 942–946.
74. Hannover, J.A. and Lennarz, W.J. (1982) J. Biol. Chem. 257, 2787–2794.
75. Snider, M.D., Sultzman, L.A. and Robbins, P.W. (1980) Cell 21, 385–392.
76. Abeijon, C. and Hirschberg, C.B. (1990) J. Biol. Chem. 265, 14691–14695.
77. Abeijon, C. and Hirschberg, C.B. (1992) Trends Biochem. Sci. 193, 32–36.
78. Perez, M. and Hirschberg, C.B. (1985) J. Biol. Chem. 260, 4671–4678.
79. Carey, D.J., W. Sommer, L. and Hirschberg, C.B. (1980) Cell 19, 597–605.
80. Snider, M.D. and Rogers, O.C. (1984) Cell 36, 753–761.
81. Snider, M.D. and Robbins, P.W. (1982) J. Biol. Chem. 257, 6796–6801.
82. McCkloskey, M.A. and Troy, F.A. (1980) Biochemistry 19, 2956–2060.
83. McCkloskey, M.A. and Troy, F.A. (1980) Biochemistry 19, 2061–2066.
84. Jensen, J.W. and Schutzbach, J.W. (1984) Biochemistry 23, 1115–1119.
85. Valtersson, C., Van Duijn, G., Verkleij, A.J., Chojnacki, T., De Kruijff, B. and Dallner, G. (1985) J. Biol. Chem. 260, 2742–2751.
86. Vigo, C., Grossman, C.H. and Drost-Hansen, W. (1980) Biochim. Biophys. Acta 774, 221–226.
87. Van Duijn, G., Valtersson, C., Chojnacki, T., Verkleij, A.J., Dallner, G. and De Kruijff, B. (1986) Biochim. Biophys. Acta 861, 211–223.
88. Knudsen, M.J. and Troy, F.A. (1990) Chem. Phys. Lipids 51, 205–212.
89. Troy, F.A., Cho, J.-W. and Ye, J. (1993) In: J. Roth, U. Rutishauser and F.A. Troy (Eds.), Polysialic Acid. Birkhäuser Verlag, Basel, pp. 93–111.
90. Beck, P.J., Orlean, P., Albright, C., Robbins, P.W., Gething, M.J. and Sambrook, J.F. (1990) J. Cell Biol. 111, 70a.
91. Haselbeck, A. and Tanner, W. (1982) Proc. Natl. Acad. Sci. USA 79, 1520–1524.
92. Hirschberg, C.B. and Snider, M.D. (1987) Annu. Rev. Biochem. 54, 63–87.
93. Czichi, U. and Lennarz, W.J. (1977) J. Biol. Chem. 252, 7901–7904.
94. Das, R.C. and Heath, E.C. (1980) Proc. Natl. Acad. Sci. USA 77, 3811–3815.
95. Welply, J.K., Kaplan, H.A., Shenbagamurthi, P., Naider, F. and Lennarz, W.J. (1986) Arch. Biochem. Biophys. 246, 808–819.
96. Tanner, W. and Lehle, L. (1987) Biochim. Biophys. Acta 906, 81–99.
97. Kaplan, H.A., Welply, J.K. and Lennarz, W.J. (1987) Biochim. Biophys. Acta 906, 161–173.

98. Kelleher, D.J., Kreibich, G. and Gilmore, R. (1992) Cell 69, 55–65.

99. Te Heesen, S., Knauer, R., Lehle, L. and Aebi, M. (1993) EMBO J. 12, 279–284.

100. Marshall, R.D. (1987) Annu. Rev. Biochem. 41, 673–702.

101. Aubert, J.P., Biserte, G. and Loucheux-Lefebvre, M.H. (1976) Arch. Biochem. Biophys. 175, 410–418.

102. Beeley, J.G. (1977) Biochem. Biophys. Res. Commun. 76, 1051–1055.

103. Pless, D.D. and Lennarz, W.J. (1977) Proc. Natl. Acad. Sci. USA 74, 134–138.

104. Bause, E. (1983) Biochem. J. 209, 331–336.

105. Mononen, I. and Karjalainen, E. (1984) Biochim. Biophys. Acta 788, 364–367.

106. Hubbard, S.C. and Robbins, P.W. (1979) J. Biol. Chem. 254, 4568–4576.

107. Elting, J.J., Chen, W.W. and Lennarz, W.J. (1980) J. Biol. Chem. 255, 2325–2331.

108. Hettkamp, H., Legler, G. and Bause, E. (1984) Eur. J. Biochem. 142, 85–90.

109. Atkinson, P.H. and Lee, J.T. (1984) J. Cell Biol. 98, 2245–2249.

110. Brada, D. and Dubach, U.C. (1984) Eur. J. Biochem. 141, 149–156.

111. Burns, D.M. and Touster, O. (1982) J. Biol. Chem. 257, 9991–10000.

112. Strous, G.J., van Kerkhof, P., Brok, R., Roth, J. and Brada, D. (1987) J. Biol. Chem. 262, 3620–3625.

113. Lucocq, J.M., Brada, D. and Roth, J. (1986) J. Cell Biol. 102, 2137–2146.

114. Brada, D., Kerjaschki, D. and Roth, J. (1990) J. Cell Biol. 110, 309–318.

115. Parodi, A.J., Mendelzon, D.H. and Lederkremer, G.Z. (1983) J. Biol. Chem. 258, 8260–8265.

116. Trombetta, S.E., Ganan, S.A. and Parodi, A.J. (1991) Glycobiology 1, 155–161.

117. Trombetta, S.E., Bosch, M. and Parodi, A.J. (1989) Biochemistry 28, 8108–8116.

118. Godelaine, D., Spiro, M.J. and Spiro, R.G. (1981) J. Biol. Chem. 256, 10161–10168.

119. Gabel, C.A. and Bergman, J.E. (1985) J. Cell Biol. 101, 460–469.

120. Hakimi, J. and Atkinson, P.H. (1982) Biochemistry 21, 2140–2145.

121. Kabcenell, A.K. and Atkinson, P.H. (1985) J. Cell Biol. 101, 1270–128.

122. Kedersha, N.L., Tkaz, S.J. and Berg, R.A. (1985) Biochemistry 24, 5952–5960.

123. Lewis, M.J., Turco, S.J. and Green, M. (1985) J. Biol. Chem. 260, 6926–6931.

124. Liscum, L., Cummings, R.D., Anderson, R.G.W., De Martino, G.N., Goldstein, J.L. and Brown, M.S. (1983) Proc. Natl. Acad. Sci. USA 80, 7165–7169.

125. Rosenfeld, M.G., Marcantonio, E.E., Hakimi, J., Ort, V.M., Atkinson, D.H., Sabatini, D. and Kreibich, G. (1984) J. Cell Biol. 99, 1076–1082.

126. Bischoff, J. and Kornfeld, R. (1983) J. Biol. Chem. 258, 7907–7910.

127. Bischoff, J. and Kornfeld, R. (1986) J. Biol. Chem. 261, 4758–4765.

128. Marsh, C.A. and Gourlay, G.C. (1971) Biochim. Biophys. Acta 235, 142–148.

129. Shoup, V.A. and Touster, O. (1976) J. Biol. Chem. 251, 3845–3852.

130. Hercz, A. and Harpaz, N. (1980) Can. J. Biochem. 58, 644–648.

131. Schweden, J., Legler, G. and Bause, E. (1986) Eur. J. Biochem. 157, 563–570.

132. Schweden, J. and Bause, E. (1990) Biochem. J. 264, 347–355.

133. Grinna, L.S. and Robbins, P.W. (1979) J. Biol. Chem. 254, 8814–8818.

134. Deschuyteneer, M., Eckhardt, A.E., Roth, J. and Hill, R.L. (1986) J. Biol. Chem. 263, 2452–2459.

135. Vertel, B., Velasco, A., LaFrance, S., Walters, L. and K. Kaczman-Daniel, K. (1989) J. Cell Biol. 109, 1827–1836.

136. Bonay, P. and Hughes, R.C. (1991) Eur. J. Biochem. 197, 229–238.

137. Bonay, P., Roth, J. and Hughes, R.C. (1992) Eur. J. Biochem. 205, 399–407.

138. Rothman, J.E. (1985) Sci. Am. 258, 84–95.

139. Roth, J. and Berger, E.G. (1982) J. Cell Biol. 92, 223–229.

140. Roth, J., Taatjes, D.J., Weinstein, J., Paulson, J.C., Greenwell, P. and Watkins, W.M. (1986) J. Biol. Chem. 261, 14307–14312.

141. Tulsiani, D.R.P., Opheim, D.J. and Touster, O. (1977) J. Biol. Chem. 252, 3227–3233.

142. Opheim, D.J. and Touster, O. (1978) J. Biol. Chem. 253, 1017–1023.

143. Tabas, I. and Kornfeld, S. (1979) J. Biol. Chem. 254, 11655–11663.

144. Dunphy, W.G., Fries, E., Urbani, L.J. and Rothman, J.E. (1981) Proc. Natl. Acad. Sci. USA 78, 7453–7457.

145. Dunphy, W.G. and Rothman, J.E. (1983) J. Cell Biol. 97, 270–275.

146. Velasco, A., Hendricks, L., Moremen, K.W., Tulsiani, D.R.P. and Farquhar, M.G. (1993) J. Cell Biol. 122, 39–51.

147. Harpaz, N. and Schachter, H. (1980) J. Biol. Chem. 255, 4885–4893.

148. Tabas, I. and Kornfeld, S. (1978) J. Biol. Chem. 253, 7779–7786.

149. Dunphy, W.G., Braands, R. and Rothman, J.E. (1985) Cell 40, 463–472.

150. Lucocq, J.M., Berger, E.G. and Roth, J. (1986) J. Histochem. Cytochem. 35, 67–74.

151. Tabas, I. and Kornfeld, S. (1978) J. Biol. Chem. 253, 7779–7786.

152. Harpaz, N. and Schachter, H. (1980) J. Biol. Chem. 255, 4894–4902.

153. Opheim, D.J. and Touster, O. (1978) J. Biol. Chem. 253, 1017–1023.

154. Tulsiani, D.R.P., Hubbard, S.C., Robbins, P.W. and Touster, O. (1982) J. Biol. Chem. 257, 3660–3668.

155. Thyberg, J. and Moskalewski, S. (1992) J. Submicrosc. Pathol. 24, 495–508.

156. Lee, E.U., Roth, J. and Paulson, J.C. (1989) J. Biol. Chem. 264, 13848–13855.

157. Strous, G.J., Van Kerkhof, P., Willemsen, R., Geuze, H.J. and Berger, E.G. (1983) J. Cell Biol. 97, 723–727.

158. Nilsson, T., Pypaert, M., Hoe, M.E., Slusarewicz, P., Berger, E.G. and Warren, G. (1993) J. Cell Biol. 120, 5–13.

159. Tabas, I. and Kornfeld, S. (1980) J. Biol. Chem. 255, 6633–6639.

160. Hasilik, A., Waheed, A. and v. Figura, K. (1981) Biochem. Biophys. Res Commun. 98, 761–767.

161. Reitman, S.L. and Kornfeld, S. (1981) J. Biol. Chem. 256, 4275–4281.

162. Varki, A. and Kornfeld, S. (1980) J. Biol. Chem. 255, 8398–8401.

163. Waheed, A., Hasilik, A. and v. Figura, K. (1981) J. Biol. Chem. 256, 5717–5721.

164. Pohlmann, R., Waheed, A., Hasilik, A. and v. Figura, K. (1982) J. Biol. Chem. 257, 5323–5325.

165. Goldberg, D. and Kornfeld, S. (1981) J. Biol. Chem. 256, 13060–13067.

166. Goldberg, D. and Kornfeld, S. (1983) J. Biol. Chem. 258, 3159–3165.

167. Deutscher, S.L., Creek, K.A., Merion, M. and Hirschberg, C.B. (1983) Proc. Natl. Acad. Sci. USA 80, 3938–3942.

168. Strous, G.J.A.M. (1979) Proc. Natl. Acad. Sci. USA 76, 2691–2698.

169. Tooze, S.A., Tooze, J. and Warren, G. (1988) J. Cell Biol. 106, 1475–1487.

170. Roth, J. (1984) J. Cell Biol. 98, 399–406.

171. Cummings, R.D., Kornfeld, S., Schneider, W.J., Hobgood, K.K., Tolleshaug, H., Brown, M.S. and Goldstein, J.S. (1983) J. Biol. Chem. 258, 15261–15273.

172. Elhammer, A. and Kornfeld, S. (1984) J. Cell Biol. 98, 327–331.

173. Niemann, H., Boschek, B., Evans, D., Rosing, M., Tamura, T., and Klenk, H.-D. (1982) EMBO J. 1, 1499–1504.

174. Johnson, D.C. and Spear, P.G. (1983) Cell 32, 987–997.

175. Elhammer, A. and Kornfeld, S. (1986) J. Biol. Chem. 261, 5249–5255.

176. Wang, Y., Abernethy, J.L., Eckhardt, A.E. and Hill, R.L. (1992) J. Biol. Chem. 276, 12709–12716.

177. O'Connel, B.C., Hagen, F.K. and Tabak, L.A. (1992) J. Biol. Chem. 267, 25010–15018.

178. Wang, Y., Agrwal, N., Eckhardt, A.E., Stevens, R.D. and Hill, R.L. (1993) J. Biol. Chem. 268, 22979–22983.

179. Roth, J., Wang, Y., Eckhardt, A.E. and Hill, R.H. (1994) Proc. Natl. Acad. Sci. USA, 51, 8935–8939.
180. Sato, A. and Spicer, S.S. (1982) Histochemistry 73, 607–624.
181. Sata, T., Zuber, C. and Roth, J. (1990) Histochemistry 94, 1–11.
182. Roth, J., Greenwell, P. and Watkins, W.M. (1988) Eur. J. Cell Biol. 46, 105–112.
183. Paulson, J.C., Weinstein, J. and Schauer, A. (1989). J. Biol. Chem. 264, 10931–10934.
184. Paulson, J.C. and Colley, K.J. (1989) J. Biol. Chem. 264, 17615–17618.
185. Svensson, E.C., Soreghan, B. and Paulson, J.C. (1990) J. Biol. Chem. 265, 20863–20868.
186. Van den Eijnden, D.H. and Joziasse, D.H. (1993) Current Op. Struct. Biol. 3, 711–721.

J. Montreuil, H. Schachter and J.F.G. Vliegenthart (Eds.), *Glycoproteins*
© 1995 Elsevier Science B.V. All rights reserved

Biosynthesis

5. Molecular basis of antigenic specificity in the ABO, H and Lewis blood-group systems

WINIFRED M. WATKINS

*Department of Haematology, Royal Postgraduate Medical School,
Hammersmith Hospital, Du Cane Road, London W12 0NN, UK*

1. Introduction

The classical ABO blood group system provided the first examples of single human polymorphic characters that were not associated with inherited diseases and understanding of the serological relationships in this system laid the basis for the safe transfusion of blood from one individual to another. The antigenic structures classified within the ABO blood-group system are now known to arise through the expression of genes encoding glycosyltransferase enzymes responsible for the terminal glycosylation of carbohydrate chains in glycoproteins, glycolipids and free oligosaccharides. In the 90 years between the discovery of the A and B antigens as serologically reactive entities on the human erythrocyte surface [1] and the cloning of the relevant glycosyltransferase genes [2] a wealth of information had accumulated on the serology, inheritance, formal genetics, tissue distribution, chemical nature and biosynthesis of these antigens and their interrelationships with other polymorphic blood group systems such as *Hh*, *Lele* and *Sese* [3–8]. Although the presence of the A and B antigens on the erythrocyte surface led them originally to be classified as "blood-group antigens" it was early recognised that substances exhibiting similar serological specificities could be found in human secretions [9,10] and in mucosal tissues of many animal species [11]. The wide tissue distribution of the antigens in humans and animals, and the fact that the expression of A and B antigens on erythrocytes appears to be of later evolutionary origin than the expression of the antigens on other tissues [11,6] has led to the suggestion that the ABH and Lewis antigens are more accurately defined as "histo-blood group" antigens [7].

Immunisation of rabbits with human erythrocytes in an attempt to identify other surface antigens led to the demonstration in the 1920s of the MN and P blood group systems [12]. However, the vast expansion in our knowledge of the antigenic structures on the erythrocyte surface came in the 1940s and 50s as a consequence of the greatly increased use of clinical blood transfusion that followed the outbreak of the Second

World War in 1939. Over 200 erythrocyte antigens have now been described and many of them have been accommodated into nineteen well defined blood group systems [13,14]. Recent work has shown that the specificity of a large number of these antigens resides in the expression of protein epitopes [15]; the antigens associated with ABO, Hh/Sese, Ii, Lewis, P and Sid (Cad) systems are the only ones definitely recognised as carbohydrate structures and hence known to be dependent on the inheritance and expression of genes encoding glycosyltransferases. The P [16] and Sid(Cad) [17] systems have recently been reviewed and will not be considered in this chapter. Although the *ABO*, *Hh*, *Sese*, *Ii* and Lewis *Lele* loci are genetically independent, the antigens arising from the action of the expressed glycosyltransferases are closely interrelated. The carbohydrate sequences designated Le^x, Le^y and sialyl-Le^x are not expressed on the erythrocyte membrane but the glycosyltransferases synthesising these structures, formerly known as *X* gene products (see [6]), act on the same precursor structures as the *H* gene encoded glycosyltransferase and, when expressed in certain tissues, form interaction products with the molecules carrying ABH determinants. Le^x and related determinants are loosely considered as part of the Lewis system although it is now recognised that they can be synthesised by various members of a family of closely related glycosyltransferases (see Sections 12 and 13).

2. Serology, inheritance and chromosomal location of blood-group genes

2.1. The ABO system

The ABO classification is based on the presence or absence of two antigens, A and B, on the erythrocyte surface and two antibodies anti-A and anti-B which always occur in the plasma when the corresponding antigen is missing. It is the presence of these naturally occurring antibodies that make knowledge of the ABO groups of crucial importance in blood transfusion. The relationships between the antigens on the erythrocytes and the antibodies in the plasma are shown in Table I. The antigens are inherited according to Mendelian laws [18,19] and on the basis of family and population studies a three allelic model for the inheritance of ABO characters was proposed in 1924 by Bernstein [20]. Although this model was challenged from time to time over the next six decades (reviewed in [21,22]) the recent cloning of the *ABO* locus [2] has triumphantly vindicated Bernstein's deductions. A child receives one of the three major *ABO* alleles from each parent giving six possible genotypes, AA, AO, BB, BO, AB and OO, and four main phenotypes, A, B, AB and O. Early serological studies led to the subdivision of group A into A$_1$ and A$_2$ [23] and detailed investigations over the years have revealed many rare variants of both A and B antigens [13,14]: these variants can be included in the Bernstein model by extending the number of alleles at the *ABO* locus. Family studies showed linkage of *ABO* to the loci for nail-patella syndrome, Np, [24] and adenylate kinase, AK$_1$ [25] and this linkage group was assigned to the distal end of the long arm of chromosome 9 [26]. Recently, the *ABO* locus has been precisely mapped to the region 9q34.2 on this chromosome [27].

TABLE I

The ABO system of antigens on erythrocytes and antibodies and glycosyltransferases in plasma

Red cells			Plasma	
Geno-type	Pheno-type	Minimal determinant structures	Antibodies	Glycosyl transferases
AA AO }	A	GalNAc(α1–3)Gal(β1-R) | Fuc(α1–2)	Anti-B	α-1,3-GalNAc-transferase
BB BO }	B	Gal(α1–3)Gal(β1-R) | Fuc(α1–2)	Anti-A	α-1,3-Gal-transferase
AB	AB	GalNAc(α1–3)Gal(β1-R) | Fuc(α1–2) and Gal(α1–3)Gal(β1-R) | Fuc(α1–2)	–	α-1,3-GalNAc- and α-1,3-Gal-transferase
OO	O	–	Anti-A and Anti-B	–

Abbreviations: Gal = D-galactopyranose; Fuc = L-fucopyranose; GlcNAc = N-acetyl-D-glucosami-nopyranose; GalNAc = N-acetyl-D-galactosaminopyranose; R = remainder of molecule.

2.2. The H system

The two structural alleles giving rise to glycosyltransferase enzymes at the *ABO* locus are *A* and *B*; the *O* allele does not give rise to an enzymic product. A number of reagents of human, animal and plant origin, that reacted preferentially with O cells, had earlier been thought to be detecting the product of the *O* gene. However, the demonstration that the antigen they recognised was detectable on the erythrocytes of group AB and homozygous AA and BB individuals, showed that this antigen could not be the product of the *O* gene if Bernstein's three allele model for the inheritance of *ABO* genes was correct. The specific substance detected by the so-called "anti-O" reagents was there-fore renamed H [28,29]. The H determinant was postulated to result from an inde-pendent genetic system *Hh* and, on the basis of biochemical observations, it was proposed that H is the substrate modified by the action of the enzymic products of the *A* and *B* genes [30,31]. With rare exceptions H is expressed on the cells of all group O individuals, but in persons belonging to groups A, B and AB there is complete or partial masking of H activity. The independence of the *H* gene from the *ABO* locus has been confirmed by the recent molecular cloning of a gene encoding a glycosyltransferase that completes the synthesis of the H structure [32]; the nucleotide sequence of this gene shows no resemblance to the cDNA sequences of the *ABO* genes (see Sections 9.2 and 11.8). Formal genetic evidence indicated that the *H* gene is located on the long arm of human chromosome 19 [33,34] and in somatic cell hybrid experiments the isolated cDNA was found to be syntenic with this chromosome [32].

2.3. The Sese (Secretor) system

The capacity to secrete ABH substances in a soluble form in saliva is a dimorphic character [35] determined by a pair of allelic genes, now referred to as *Sese*. On the basis of linkage studies the secretor gene locus was assigned to chromosome 19 [34] and is believed to be closely linked to the *H* gene locus [33]. Glycoproteins are the carriers of blood group determinants in saliva and the term 'secretor' refers only to the presence of the terminal oligosaccharide sequences that confer A, B or H antigenic specificities on the precursor glycoprotein molecules and not to the capacity of an individual to synthesise and secrete glycoprotein. Those carrying an *Se* allele, whether homozygous *SeSe* or heterozygous *Sese*, are 'secretors' whereas those homozygous for *sese* are 'non-secretors'.

Early biochemical studies on the blood-group-active glycoproteins from secretors and non-secretors demonstrated that the formation of H structures was the point in the biosynthetic pathways at which the substances made in ABH secretors diverged from those in non-secretors; it was therefore proposed that, in secretory tissues, the *Se* gene was a structural gene encoding an enzyme responsible for the formation of H structures [36]. However, in order to have a unifying hypothesis that covered the expression of H antigen on erythrocytes as well as in secretions, the idea was subsequently modified to indicate that the *Se* gene regulated the expression of the *H* gene in secretory tissues [30,31]. Individuals of the rare Bombay O$_h$ phenotype, first identified by Bhende et al. [37], fail to express H antigen on erythrocytes or in secretions, and were postulated to be homozygous for the silent allele *h* [29]. However, the discovery of other phenotypes in which individuals lacked (para-Bombay type 1), or showed very weak (para-Bombay type 2), ABH activity on erythrocytes but expressed normal patterns of ABH secretion in saliva [13,14], indicated that all the facts could not be explained by one regulator gene expressed in tissues secreting the blood-group-active glycoproteins and Solomon et al. [38] proposed a second independent regulatory locus, *Zz*, controlling the expression of H on erythrocytes. In the light of more recent structural [39] and conformational information [40] about H determinant structures, and the developing knowledge concerning the tissue specific expression of enzymes with closely related specificities, Oriol et al. [33] returned to the premise that *Se* is a structural gene and proposed that *H* and *Se* are two different, but closely linked genes on chromosome 19 [41] encoding glycosyltransferases that complete the synthesis of H structures in different cellular types: according to this model the *H* gene is expressed on erythrocytes and in some other tissues of mesodermal origin and the *Se* gene is expressed in cells of salivary glands and other tissues producing exocrine secretions. The biochemical and molecular genetic implications of these hypotheses are discussed in Sections 9 and 10.

2.4. The Lewis system

The Lewis blood group system was named after the donor of an antibody discovered by Mourant [42] that agglutinated the erythrocytes of about 20% of Europeans; the antigen detected by this antibody was subsequently called Le^a. In time it became apparent that this new blood group antibody was recognising an antigen on erythrocytes that was identical with an antigen earlier identified in saliva by Japanese workers [43]. A second human antibody [44] disclosed an apparently antithetical antigen, designated Le^b. On the basis of the reactivity of erythrocytes with the two types of antibody, adults are differentiated into three main groups, Le(a+b-), Le(a-b+) and Le(a-b-) with frequencies in Caucasians of approximately 20%, 75% and 5%, respectively. The incidence of the phenotype Le(a-b-) varies in different ethnic groups and rises to about 20% in some Black African populations [13]. The Lewis system differs from the ABO system in that the antigens are not synthesised in the erythrocyte precursor cells but are acquired from the plasma in which the cells circulate [45,46]. Grubb [47] made the striking observation that all persons whose erythrocytes group as Le(a+b-) are non-secretors of ABH substances; such individuals have Le^a-active glycoproteins in saliva. Le^a-activity is also detectable in the saliva of those whose erythrocytes group is Le(a-b+) but to a much lesser degree; these individuals are secretors of ABH and Le^b-active substances. Only those persons with erythrocytes grouping as Le(a-b-) have neither Le^a nor Le^b in their secretions: a small proportion of this group are non-secretors of ABH [47,48]. The distribution of ABH and Lewis (Le^a and Le^b) antigens on erythrocytes and in saliva of individuals with different combinations of *ABO*, *Hh*, *Sese*, and *Lele* genes are shown in Table II.

TABLE II

ABH and Lewis antigens expressed on erythrocytes and in secretions of individuals with different combinations of *ABO*, *Hh*, *Sese* and *Lele* genes

Group	Gene combination	Erythrocyte phenotype	Antigens					
			Erythrocytes			Secretions		
			ABH	Le^a	Le^b	ABH	Le^a	Le^b
1.	*ABO, H, Se, Le*	ABO,Le(a-b+)	+++	–	+++	+++	+	+++
2.	*ABO, H, sese, Le*	ABO,Le(a+b-)	+++	+++	–	–	+++	–
3.	*ABO, H, Se, lele*	ABO,Le(a-b-)	+++	–	–	+++	–	–
4.	*ABO, H, sese, lele*	ABO,Le(a-b-)	+++	–	–	–	–	–
5.	*ABO, hh, Se, Le*	^{a}O$_h$,Le(a-b+)	–	–	+++	+++	+	+++
6.	*ABO, hh, sese, Le*	O$_h$,Le(a+b-)	–	+++	–	–	+++	–
7.	*ABO, hh, Se, lele*	O$_h$,Le(a-b-)	–	–	–	+++	–	–
8.	*ABO, hh, sese, lele*	O$_h$,Le(a-b-)	–	–	–	–	–	–

^{a}O$_h$ = Erythrocytes deficient in ABH expression; groups 6 and 8 correspond to the classical Bombay phenotypes and groups 5 and 7 to para-Bombay phenotypes [13].
+++ Strong activity; + weak activity; – no activity.

TABLE III

Structures of H, Le^a, Le^b, sialyl-Le^a, Le^x, Le^y, sialyl-Le^x, VIM-2, dimeric Le^x, sialyl-dimeric-Le^x, Le^y-Le^x and trimeric Le^x determinants

Specificity	Structure
H	Fuc(α1–2)Gal(β1-R)

Le^a	Gal(β1–3)GlcNAc(β1-R)
	     |
	   Fuc(α1–4)

Le^b	Fuc(α1–2)Gal(β1–3)GlcNAc(β1-R)
	         |
	       Fuc(α1–4)

Sialyl-Le^a	NeuAc(α2–3)Gal(β1–3)GlcNAc(β1-R)
	          |
	        Fuc(α1–4)

Le^x	Gal(β1–4)GlcNAc(β1-R)
	     |
	   Fuc(α1–3)

Le^y	Fuc(α1–2)Gal(β1–4)GlcNAc(β1-R)
	         |
	       Fuc(α1–3)

Sialyl-Le^x	NeuAc(α2–3)Gal(β1–4)GlcNAc(β1-R)
	          |
	        Fuc(α1–3)

VIM-2	NeuAc(α2–3)Gal(β1–4)GlcNAc(β1–3)Gal(β1–4)GlcNAc(β1-R)
	                 |
	              Fuc(α1–3)

Dimeric-Le^x	Gal(β1–4)GlcNAc(β1–3)Gal(β1–4)GlcNAc(β1-R) \| \| Fuc(α1–3) Fuc(α1–3)

Sialyl-dimeric-Le^x	NeuAc(α2–3)Gal(β1–4)GlcNAc(β1–3)Gal(β1–4)GlcNAc(β1-R)
	          | |
	        Fuc(α1–3) Fuc(α1–3)

Le^y-Le^x	Fuc(α1–2)Gal(β1–4)GlcNAc(β1–3)Gal(β1–4)GlcNAc(β1-R)
	         | |
	       Fuc(α1–3) Fuc(α1–3)

Trimeric Le^x	Gal(β1–4)GlcNAc(β1–3)Gal(β1–4)GlcNAc(β1–3)Gal(β1–4)GlcNAc(β1-R) \| \| \| Fuc(α1–3) Fuc(α1–3) Fuc(α1–3)

Abbreviations: NeuAc = *N*-acetylneuraminic acid (sialic acid); others as in Table I.

Despite the failure to detect Le^a by normal serological methods on erythrocytes from Caucasians grouped as Le(a-b+), Cutbush et al. [49] early showed that some such cells reacted as though they had Le^a on their surface when injected into a recipient who had a circulating Le^a antibody. Donors whose erythrocytes group overtly as Le(a+b+) have been reported in Japanese [50], Aboriginal [51] and Polynesian blood donors [52]. Thus, although it is usual for the expression of Le^a and Le^b to be mutually exclusive on erythrocytes this is not an invariant rule.

Le^a and Le^b are not the products of allelic genes and the Lewis locus is believed to have only one functional allele, Le, and a null allele *le*. Ceppellini [53] observed that Le^b activity occurs only in the tissue fluids of ABH secretors and therefore suggested that its presence could be explained if Le^b is an interaction product of the *Le* gene and the secretor gene *Se*. Characterisation of the oligosaccharide structure responsible for Le^b specificity [54] (Table III) showed that it does arise from the combined action of two genes that, acting alone, would give an Le^a structure and an H structure.

On the basis of family linkage data the blood group Lewis locus was assigned to the short arm of human chromosome 19 in the region p13.3–p13.2 [55]. More recently, cDNA isolated by a gene transfer approach [56] has been tentatively identified as representing the *Lewis* blood-group locus (see Section 12.2); this cDNA cross-hybridises with sequences located on chromosome 19 [56].

Recognition of the genetic background of Le^x and related structures has emerged from biochemical and molecular genetic approaches rather than from family and linkage studies that were used to establish the inheritance of the ABO, Secretor and Lewis systems. Genes encoding glycosyltransferases with the specificity required for the synthesis of the Le^x structures have been mapped to both human chromosome 11 [57;58] and chromosome 19 [59]. At least one of these loci is known to be polymorphic (see Section 13.2) but the exact number of genes involved, and the possible existence of alleles at other loci has yet to be investigated.

2.5. The Ii system

The I and i specificities are defined by 'cold' agglutinins which most frequently are monoclonal antibodies associated with autoimmune disease. Those antibodies that react more strongly with normal adult erythrocytes than with cord cells are designated as anti-I [60] while those that react more strongly with cord cells, and rare adult i cells, are designated anti-i [61]. There is considerable heterogeneity among the antisera collected within the two definitions anti-I and anti-i [13]. The extreme rarity of the i adult phenotype precludes this blood-group system from being considered a useful polymorphism for anthropological or forensic purposes but interest in the I system stems from the changes that take place during mammalian development and differentiation [62–64]. In human erythrocytes the change from i to I specificity occurs gradually during the first year of life [65].

I and i antigens are not the products of allelic genes. The determinants detected by anti-i and anti-I reagents have been shown to react, respectively, with linear and branched poly-*N*-acetyllactosaminoglycans [66–68]. Therefore, in contrast to the ABH and Lewis genes which encode enzymes responsible for the addition of terminal sugars, the *I* gene, as originally suggested by Koscielak [69], can be considered as the gene encoding the glycosyltransferase that converts linear into branched polylactosamine chains. The gene for a branching enzyme with the requisite specificity has recently been cloned and shown to map to human chromosome 9q21 [70].

The relationship between Ii antigens and ABH antigens arises from the fact that the determinants may occur on the same molecules. The poly-*N*-acetyllactosaminoglycans contain peripheral core structures that, if unsubstituted, react with anti-I or -i reagents but are also potential acceptor sites for the H, and subsequently *A* or *B* transferases (see Section 11). The I/i determinants may therefore be present as cryptic internal sequences in chains bearing A, B or H determinants.

3. Chemical nature of the blood-group-determinant structures

3.1. Early approaches

The presence in secretions of substances with the capacity to inhibit the agglutination of erythrocytes by blood-group specific anti-sera was recognised long before anything was known about the chemical nature of the antigens. The work of Landsteiner and colleagues in the 1930s suggested that the soluble substances with blood group activity were some new type of carbohydrate–amino acid complex [71], but at that time, lipids and proteins were the only molecules recorded as components of the erythrocyte membrane. The indirect approaches that established the carbohydrate nature of the ABH and Lewis blood-group determinants have been reviewed in detail [72]. Major lines of attack were inhibition of lectin-mediated agglutination of human erythrocytes [73,74], or the inhibition of specific precipitation of soluble blood group substances [75], with simple sugars. These studies suggested a role for L-fucose in H-specificity, for *N*-acetyl-D-galactosamine in A specificity and of D-galactose in B-specificity. The inferences were tentative because plant lectins were not previously known to be carbo-hydrate-binding proteins combining specifically with monosaccharide units and, al-though the pioneering studies of Heidelberger, Avery, Goebel and Morgan (reviewed in [71]) had shown that the specificity of certain bacterial antigens is associated with carbohydrate structures, no other mammalian carbohydrate antigens had been identi-fied at that time.

In the 1930s a number of enzyme preparations of microbial or mollusc origin were described that destroyed the serological activity of water-soluble blood group sub-stances (reviewed in [72,76]) and inhibition by monosaccharides of the decomposition of A, B and H active structures in soluble blood-group substances [77] lent further credence to the idea that *N*-acetyl-D-galactosamine, D-galactose and L-fucose, respec-tively, constituted the immunodominant structures in A, B and H structures. The fact that the lectin-mediated agglutination of erythrocytes and the specific enzymic decom-position of water-soluble blood group substances were inhibited by identical sugars also supported the conclusion that carbohydrate units were responsible for ABH blood group activity both on the erythrocyte surface and in the secreted substances and that the same terminal sugars were involved in specificity in both tissues.

Application of the enzyme inhibition method to water-soluble Le[a]-active substances indicated a role for L-fucose in Le[a] specificity [78] and the fortunate coincidence of the isolation and structural identification of certain milk oligosaccharides by Kuhn and his colleagues in Heidelberg [79] made available low-molecular-weight fucose-containing compounds of established structure for antibody inhibition studies. These enabled the minimal Le[a] structure to be unequivocally identified (Table III) and certain important features of the Le[b] structure to be deduced [78,80].

3.2. Isolation and characterisation of minimal determinant structures

The chemical structures of the A, B, H and Lewis determinants were established in the 1950s and 1960s using secreted substances as the source materials. Earlier attempts to characterise the active substances on the erythrocyte surface had met with considerable difficulties (reviewed in [76]) because no methods for the isolation of membrane components present in very small amounts were available at that time. In consequence attention was turned to the active substances that had been demonstrated in animal mucosal tissues [76], and in soluble form in a number of normal human body fluids [10,81] and also in the pathological secretions from ovarian cysts [10]. The latter fluids were found to be a rich source of soluble blood group active substances [82] and were to provide the main starting materials for the isolation and characterisation of the A, B, H, Le[a], Le[b] and Le[x] determinants.

Irrespective of blood group activity the glycoproteins purified from ovarian cysts are composed of 85–90% carbohydrate, made up of D-galactose, L-fucose, N-acetyl-D-glucosamine and N-acetyl-D-galactosamine, and 15–20% amino acids [83,84,72]. The oligosaccharide chains are densely packed in the region of the peptide backbone rich in serine, threonine and proline, and a near 1:1 ratio of N-acetylgalactosamine to the hydroxy amino acids in H, B and Le[a] substances indicates a high degree of O-glycosidic substitution [85]. The results of serological precipitation experiments with monospecific anti-sera early demonstrated that the macromolecules in the purified preparations frequently carry more than one blood group antigenic specificity [86,31] and that, depending on the genetic endowment of the individual, A, B, H, Le[a] and Le[b] structures can be present on the same glycoprotein molecules (reviewed in [72]).

Many carbohydrate fragments were released and characterised from human ovarian cyst glycoproteins by acid hydrolysis and alkaline degradation procedures (reviewed in [72,84,87]). The failure of human anti-A or anti-B antibodies to be inhibited by the monosaccharides that neutralised the activity of the plant lectins demonstrated that the immunodominant sugars, although of major importance for specificity, did not constitute the complete determinants. The first serologically A- and B-active oligosaccharides isolated from ovarian cyst glycoproteins that contained the complete determinants were tetrasaccharides [88] or larger fragments [89] but binding experiments with monoclonal antibodies (reviewed in [90]) have disclosed that the minimal determinant sequences required for A and B blood group specificity are trisaccharides (Table I). The isolation of two H-active trisaccharides, each with a fucosyl residue in α-1,2-linkage to a β-galactosyl residue, and differing in the linkage of the galactose to the third sugar, N-acetylglucosamine [39], showed that the minimal H-structure is a disaccharide

(Table III). Chemical and serological characterisation of fragments isolated from Le[a]-active cyst materials [91,92] confirmed the trisaccharide structure for the Le[a] determinant (Table III) earlier deduced from the inhibitory activity of milk oligosaccharides with human anti-Le[a] serum [78]. Isolation and characterisation of Le[b] active fragments from an HLe[b]-active ovarian cyst glycoprotein [54] also confirmed the suggestion from serological inhibition tests that two fucose units were involved in this structure [78,80] and showed that the determinant comprised an H and an Le[a] structure (Table III). Each of the original specificities is masked in the composite structure.

The presence of fucose linked α-1,3 to internal N-acetylglucosamine residues in oligosaccharides isolated from A, B and H blood group active ovarian cyst glycoproteins was first noted by Lloyd et al. [89]. Fucose linked α-1,3 to glucose had been identified in milk oligosaccharides [79] but had not previously been found in glycoproteins or glycolipids. Subsequently, a trisaccharide, known from the method of isolation to be a terminal non-reducing structure, and characterised as Gal(β1–4)[Fuc(α1–3)]GlcNAc, was isolated from an Le[a] active glycoprotein [93]. This sequence was recognised as an isomer of Le[a] but at the time had no known immunological activity. Its apparent absence from the erythrocyte membrane means that in the strict sense it is not a blood group antigen, but in tissue antigens fucose α-1,3-linked to N-acetylglucosamine constitutes part of interaction products with ABH structures on Type 2 chains (Table IV) and the trisaccharide is now known to function as a determinant on white blood cells (see Section 16). The name X-hapten was applied to the trisaccharide structure when a glycosphingolipid with this same terminal sequence was isolated from human adenocarcinoma tissue [94]. Subsequently this sequence, and the antibodies that recognise it [95,63] have been assigned various names (X, SSEA1, CD15, Lewis-X, Le[c], Le[x]), but it is now most generally referred to as Le[x] (Table III). The structure isomeric with Le[b], with two fucose residues attached to adjacent sugars in the Type 2 peripheral core structure, is called Le[y] (Table III). The gene(s) encoding the fucosyltransferase(s) giving rise to Le[x] and Le[y] structures was earlier termed X, but it is now evident that the structure arises in different tissues from the expression of a family of genes encoding enzymes with closely related specificities (see Section 13). The use of the term Le[x] has caused some confusion amongst serologists because this nomenclature had

TABLE IV

Peripheral disaccharide core structures on which ABH and Lewis determinants are synthesised

Chain type	Structure	Determinants built on peripheral core structure
Type 1	Gal(β1–3)GlcNAc(β)	H, A, B, Le[a], Le[b] and sialyl-Le[a]
Type 2	Gal(β1–4)GlcNAc(β)	H, A, B, Le[x], Le[y] and sialyl-Le[x]
Type 3	Gal(β1–3)GalNAc(α)	H, A and B
Type 4	Gal(β1–3)GalNAc(β)	H, A and B
Type 6	Gal(β1–4)Glc	H, A and B

Abbreviations as in Table I.

earlier been used for an antibody that agglutinated both Le(a+) and Le(b+) adult erythrocytes and also cord bloods [13,96,97]; the structure detected by the latter anti-'Le^x' reagent, however, is not related to the Gal(β1–4)[Fuc(α1–3]GlcNAc structure [98].

3.3. Peripheral disaccharide core structures of A, B, H, Le^a, Le^b, Le^x and Le^y determinants

3.3.1. Type 1 and Type 2

The fragments isolated from acid hydrolysis products of ovarian cyst glycoproteins lacked acid-labile fucose or sialic acid residues but demonstrated that A, B and H determinants can be based on more than one peripheral carbohydrate core structure [39,99,100,101]. The two disaccharide sequences first identified as core structures were designated [101] Type 1 (Gal(β1–3)GlcNAc) and Type 2 (Gal(β1–4)GlcNAc) (Table IV). Unsubstituted Type 2 chain endings, occurring naturally, or as the result of removal of fucose under very mild acid conditions [102] are responsible for the reactivity of these substances with anti-Type XIV pneumococcal sera [103] and with certain anti-I sera [104]. The isolation from ovarian cyst glycoprotein of a reduced branched pentasaccharide [92], Gal(β1–3)GlcNAc(β1–3)[Gal(β1–4)GlcNAc(β1–6)] Gal-ol, containing both Type 1 and Type 2 chain endings indicated that H, A and B structures can occur in close proximity on branched chains.

Mono-fucosylated Le^a and di-fucosylated Le^b determinants occur only on Type 1 core structures and mono-fucosylated; Le^x and di-fucosylated Le^y determinants only on Type 2 core structures. Substitution of fucose in α-1,4-linkage onto N-acetylglucosamine in Type 1 A or B structures gives difucosyl ALe^b and BLe^b determinants and substitution of L-fucose in α-1,3-linkage on Type 2 structures gives ALe^y and BLe^y determinants (Table V). In poly-N-acetyllactosamine chains fucose residues may also be attached to the O-3 position of internal N-acetylglucosamine residues (Table III) to give di- and trimeric Le^x structures [105]. The advent of the hybridoma technique has allowed the production of monoclonal antibodies with the fine specificity needed to distinguish between these closely related structures. The Gal(β1–4)[Fuc(β1–3)]GlcNAc (Le^x) structure, for which specific antibodies were not previously available, was found to be highly immunogenic in mice and monoclonal antibodies raised against a variety of cell lines, as well as those raised against both normal and tumour tissues, were found to recognise this trisaccharide sequence [63,95,106]. Although many monoclonal anti-A reagents recognise only the terminal trisaccharide sequence common to all the A determinants, others are specific for structures which differ in the nature of the peripheral disaccharide core structure and the number of fucose residues (reviewed in [90,107]); immunostaining with these reagents is enabling the tissue and cell specific distribution of the various isoforms to be established.

3.3.2. Type 3

A third type of peripheral core structure, Gal(β1–3)GalNAc(α), occurring in glycoproteins with O-linked chains was designated Type 3 [108]. This core structure bears short ABH determinants that are joined via the N-acetylgalactosamine unit α-linked to serine or threonine residues in the peptide backbone (Table IV).

TABLE V

Mono- and di-fucosylated blood group A and B structures

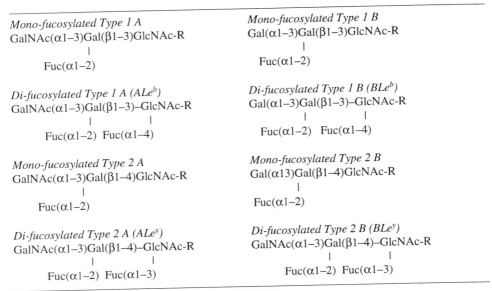

Mono-fucosylated Type 1 A GalNAc(α1–3)Gal(β1–3)GlcNAc-R | Fuc(α1–2)	*Mono-fucosylated Type 1 B* Gal(α1–3)Gal(β1–3)GlcNAc-R | Fuc(α1–2)
Di-fucosylated Type 1 A (ALe^b) GalNAc(α1–3)Gal(β1–3)–GlcNAc-R | | Fuc(α1–2) Fuc(α1–4)	*Di-fucosylated Type 1 B (BLe^b)* Gal(α1–3)Gal(β1–3)–GlcNAc-R | | Fuc(α1–2) Fuc(α1–4)
Mono-fucosylated Type 2 A GalNAc(α1–3)Gal(β1–4)GlcNAc-R | Fuc(α1–2)	*Mono-fucosylated Type 2 B* Gal(α13)Gal(β1–4)GlcNAc-R | Fuc(α1–2)
Di-fucosylated Type 2 A (ALe^y) GalNAc(α1–3)Gal(β1–4)–GlcNAc-R | | Fuc(α1–2) Fuc(α1–3)	*Di-fucosylated Type 2 B (BLe^y)* GalNAc(α1–3)Gal(β1–4)–GlcNAc-R | | Fuc(α1–2) Fuc(α1–3)

Abbreviations as in Table I.

3.3.3. Type 4
A fourth type of peripheral core sequence has been described in glycosphingolipids [109,110,111] (Type 4, Table IV). This structure is found as an extension of a globoside molecule (GalNAc(β1–3)Gal(α1–4)Gal(β1–4)Glc-Cer) in which the terminal non-reducing β-linked *N*-acetylgalactosamine unit of the glycolipid is first substituted with a β-1,3-linked galactosyl residue; this structure then forms a substrate for the biosynthesis of H, and subsequently, A or B structures.

3.3.4. Type 6
A fifth type of peripheral core structure that functions as a carrier of ABH determinants in naturally occurring oligosaccharides, Gal(β1–4)Glc, has been designated Type 6 (Table IV) (the designation Type 5 has been used for a chemically synthesised disaccharide, Gal(β1–3)Gal(β1–R), that functions *in vitro* as an acceptor substrate for the *H*-gene encoded glycosyltransferase [90]). Type 6 structures in which the fucose residue is α-2-linked to the β-galactosyl residue of lactose were first reported in the milk oligosaccharide, 2′-fucosyllactose (Fuc(α1–2)Gal(β1–4)Glc) [79] and blood-group-active oligosaccharides based on this core structure have been isolated from urine [112] and the faeces of breast-fed infants [113]. More recently glycolipids with, respectively, an A and an H determinant attached directly onto the β-galactosyl residue of lactosylceramide have been identified in small intestinal epithelial cells [114] and in renal vein tissue [115].

3 4. Molecules carrying blood-group-determinants

3.4.1. ABH structures

The terminal structures determined by the blood-group genes at the *ABO* and *H/Se* loci occur on glycoproteins, glycolipids and free oligosaccharides. Early attempts to isolate blood group antigens from erythrocytes by extraction with organic solvents suggested a possible lipid nature for the antigens (reviewed in [76]) and the pioneering studies of Yamakawa in Japan, Koscielak in Poland and Hakomori in the USA demonstrated the existence of families of glycosphingolipid molecules in lipid extracts of human erythrocytes that carried A, B and H activity. With the exception of the Type 4 structures, based on globoside, N-acetylgalactosamine is found only in blood-group A-active glycolipids [116–121]; L-fucose, D-galactose; N-acetylglucosamine and D-glucose are the other component sugars with D-glucose forming the linkage to the ceramide moiety. The glycolipid antigens integral to the erythrocyte membrane have Type 2 peripheral core structures, although small quantities of ABH glycolipids with Type 1 peripheral core sequences are adsorbed onto erythrocytes from the plasma in which the cells circulate [122,123]. Examples of blood group A-active glycolipids isolated from human group A erythrocyte membranes [116,124] are shown in Table VI. A series of H-active [117,119] and B-active [118,125] glycosphingolipids exhibiting a similar range of complexity were isolated from blood group O and B erythrocyte membranes, respectively. All contain the same minimal determinant structures (Tables I and III) earlier found in the glycoprotein carrier molecules.

Isolation of a hybrid Type 2 and "Type 3" chain A glycolipid from human erythrocytes [124,126] disclosed that, contrary to previous assumptions, the blood-group-active structures do not invariably occur at the terminal non-reducing ends of oligosaccharide chains. In these hybrid molecules the terminal N-acetylgalactosamine residue of a Type 2 A determinant is substituted with β-1,3-linked galactose which then constitutes the core structure for the addition of L-fucose and N-acetylgalactosamine to give a second A-trisaccharide sequence based on the Type 3 disaccharide core structure (Table IV). This extended sequence (Table VII) is sometimes referred to simply as a "Type 3" structure but the alternative designation of "Type 3 Repetitive A" or "Type 2/Type 3 A" avoids confusion with the short-chain Type 3 structures previously defined in glycoproteins [108]. Although undetected in earlier studies, the glycolipids with the extended A structure are now thought to constitute more than half the total A glycolipid on A$_1$ erythrocyte membranes [7]. A$_2$ erythrocytes, which virtually lack the Repetitive A structure, have a masked form of Type 2 A determinant in which the A structure is extended to terminate with an H determinant (Table VII) [127]. These structures with repetitive A and H determinants have not been described in glycoproteins.

Glycolipids carried on the Type 4 peripheral core sequence (Table VII) representing extended globo-series structures, occur in only small amounts in human erythrocyte membranes [109,110]. Globo-H (Table VII) represents only about 5–10% of the glycolipid isolated from O or A$_2$ erythrocyte membranes and has not been detected in glycolipids from A$_1$ membranes. Globo-A (Table VII), a minor component of the glycolipid from A$_1$ erythrocytes, has not been detected in the glycolipid fraction from

TABLE VI

Structures of glycosphingolipids of the *neo*-lacto-series isolated from human group A erythrocytes

Aa   GalNAc(α1–3)Gal(β1–4)GlcNAc(β1–3)Gal(β1–4)Glc(β1-Cer
     |
     Fuc(α1–2)

Ab   GalNAc(α1–3)Gal(β1–4)GlcNAc(β1–3)Gal(β1–4)GlcNAc(β1–3)Gal(β1–4)Glc(β1-Cer
     |
     Fuc(α1–2)

Ac   GalNAc(α1–3)Gal(β1–4)GlcNAc(β1–3)Gal(β1–4)GlcNAc(β1–3)Gal(β1–4)Glc(β1-Cer
     | |
     Fuc(α1–2)   GalNAc(α1–3)Gal(β1–4)GlcNAc(β1–6)
                              |
                      Fuc(α1–2)

norAc   GalNAc(α1–3)(Gal(β1–4)GlcNAc(β1–3)[Gal(β1–4)GlcNAc(β1–3)]₂Gal(β1–4)Glc(β1-Cer
        |
        Fuc(α1–2)

Ad   GalNAc(α1–3)Gal(β1–4)GlcNAc(β1–3)Gal(β1–4)GlcNAc(β1–3)Gal(β1–4)GlcNAc(β1–3)Gal(β1-Cer
     | |
     Fuc(α1–2)   GalNAc(α1–3)Gal(β1–4)GlcNAc(β1–6)
                                                      |
                                              Fuc(α1–2)

Compiled from Refs. [116,124].

TABLE VII

Extended *neo*-lacto (Type 2/Type 3 Repetitive) and extended globo-series A and H glycolipids

Extended *neo*-Lacto-series

H-active

Fuc(α1–2)Gal(β1–3)GalNAc(α1–3)Gal(β1–4)GlcNAc(β1–3)Gal(β1–4)Glc(β1–1)Cer

$\qquad$|

$\qquad$Fuc(α1–2)

A-active (Repetitive-A)

GalNAc(α1–3)Gal(β1–3)GalNAc(α1–3)Gal(β1–4)GlcNAc(β1–3)Gal(β1–4)Glc(β1–1)Cer

$\quad$| $\qquad\qquad\qquad$ |

Fuc(α1–2) $\qquad\qquad$ Fuc(α1–2)

Extended Globo-series

H-active (Globo-H)

Fuc(α1–2)Gal(β1–3)GalNAc(β1–3)Gal(α1–4)Gal(β1–4)Glc(β1–1)Cer

A-active (Globo-A)

GalNAc(α1–3)Gal(β1–3)GalNAc(β1–3)Gal(α1–4)Gal(β1–4)Glc(β1–1)Cer

$\quad$|

Fuc(α1–2)

Compiled from Refs. [126,127].

A_2 erythrocytes [109,110]. It has been reported in human meconium [128], teratocarcinoma cells [129] and in kidney glycosphingolipids [111], but the proportions of the total A-active glycolipid constituted by the globo-form varies in different tissues. In the kidney globo-A constitutes about half of the total A glycolipid [111] but, surprisingly, in contrast to this relative abundance, only minute amounts of Type 4 B glycolipids have been detected in group B kidneys [130].

The characterisation of the relatively simple range of glycolipids (Table VI) at first appeared to answer the question of the nature of the ABH blood group determinants on the erythrocyte surface. However, these glycolipids are now thought to account for only about 5% of the blood group activity [131–134]. More complex glycolipids, first reported by Gardas and Koscielak [135], with carbohydrate moieties containing 20–60 sugar residues per molecule, are believed to make a greater contribution (about 20%) to the total activity. These substances, designated polyglycosylceramides, are highly branched and, according to the inherited ABO phenotype, carry A, B and H determinants on peripheral Type 2 core structures [136–138].

The largest contribution to the ABH blood group activity of the red cell is now believed to come from determinants carried on glycoprotein molecules. Finne et al. [139,140] first reported blood group ABH antigens in human erythrocyte membranes carried on glycoprotein molecules in which the carbohydrate chains are joined to the protein via N-glycosidic linkages. Pronase digestion of these glycoproteins yielded glycopeptides in which the oligosaccharide chains comprised 50–60 sugars, and resembled the carbohydrate moieties of the polyglycosylceramides. Some of the minor erythrocyte glycoproteins carrying ABH determinants remain unidentified but the major carriers are Band 3 [133], the anion transporter protein [141,142], and Band 4.5, the glucose transporter protein [143,144]. Although, in contrast to the mucin type glycoproteins which carry multiple oligosaccharide chains, these glycoproteins each have only a single, N-linked poly-N-acetyllactosamine chain, the high degree of branching means that each chain may carry several Type 2 ABH determinants. There are approximately 1×10^6 Band 3, and 5×10^5 Band 4.5, molecules per cell [15] and some 75% of the ABH determinants on the erythrocyte surface are believed to be carried on these molecules [131–134].

Type 1 chains are widely expressed in both glycoproteins and glycolipids derived from lining epithelia and glandular epithelia and in secreted glycoproteins carrying blood group determinants (see [6]). Even in molecules terminating with the Type 1 core structure the inner part of the oligosaccharide chain normally consists of a Type 2 chain poly-N-acetyllactosamine structure [145] although exceptions have been reported in tumour tissues [146]. Blood-group A structures based on Type 1, Type 2 and Type 3 core structures were isolated from A_1 and A_2 ovarian cyst and salivary glycoproteins [108]. The yields of the three carrier isotypes did not necessarily give a precise reflection of the proportions in which they were present in the starting material because the different linkages may have had different susceptibilities to the isolation procedures used. However, in general, amounts of Type 2 structures were considerably smaller than the amounts of Type 1 or Type 3 structures in the specimens examined [108]. These results suggested that in these mucin-type glycoproteins perhaps up to one third of the oligosaccharide chains carrying blood group determinants are short chains built on the Type 3, Gal(β1–3)GalNAc(α1)-Ser/Thr core structures. The isolation of a branched hexasaccharide with the structure, Fuc(α1–2)Gal(β1–3) [Fuc(α1–2)Gal(β1–4)GlcNAc(β1–6)]GalNAc-ol, from the alkaline-borohydride degradation products of an H-active ovarian cyst preparation had in fact earlier demonstrated the presence of a Type 3 H determinant on the shorter branch of the hexasaccharide [147]. Immunostaining of tissue sections with monoclonal antibodies specific for Type 1 or Type 2 determinants have indicated a considerable predominance of Type 1 over Type 2 peripheral core structures in mucosal lining glycoproteins [6,148].

In addition to the well-documented presence of blood-group active determinants on the major gastric mucosal glycoproteins there is evidence that ABH structures occur on a wide range of small intestinal glycoproteins, including specific hydrolases [149,150]. The presence of ABH determinants has also been recorded on a number of diverse proteins including human Factor VIII/Von Willebrand Factor [151], on the epidermal growth factor receptor expressed on the surface of A431 cell line [152], on β-1,4-galactosyltransferase protein [153,154] and the blood group A-gene encoded α-1,3-N-acetylgalactosaminyl transferase [155].

3.4.2. Ii structures

Immunochemical inhibition experiments with a series of glycosphingolipids elucidated the basis of I/i specificity [66–68]. The i antigen present in foetal erythrocytes was identified as a linear repeating N-acetyllactosamine structure, represented by lacto-N-nor-hexaosylceramide while the I antigen present in adult erythrocytes was identified as a branched structure, represented by lacto-N-iso-octaosylceramide (Table VIII). Various monoclonal anti-i sera recognise different domains within the unbranched structure whereas monoclonal anti-I sera are directed towards different domains of the branched structure. On the adult erythrocyte, I activity is carried on the highly branched polygly-

TABLE VIII

Structures recognised by anti-sera directed towards blood group i and I determinants

	Ref.
i-active structure Gal(β1–4)GlcNAc(β1–3)Gal(β1–4)GlcNAc(β1–3)Gal(β)-R	66
I-active structure Gal(β1–4)GlcNAc(β1–6) 　　　　　| 　　　　　　Gal(β1–4)GlcNAc(β1–3)Gal(β)-R 　　　　　| Gal(β1–4)GlcNAc(β1–3)	67,68

cosylceramide molecules [156], on the oligosaccharide chains of the Band 3 and Band 4.5 glycoproteins [157,158] and the *neo*-lacto-series glycosphingolipids. From the structures of the i and I determinants it is apparent that the terminal non-reducing Type 2 sequences are potential substrates for further conversion to H, and A and/or B structures.

3.4.3. Le^a ,Le^b, sialyl-Le^a, Le^x, Le^y and sialyl-Le^x structures

3.4.3. Le^a ,Le^b, sialyl-Le^a, Le^x, Le^y and sialyl-Le^x structures
Marcus and Cass [159] demonstrated that the Le^a and Le^b antigens taken up onto erythrocytes are glycosphingolipids carried in the plasma by high- and low-density lipoproteins. The origin of the plasma glycolipids is still not entirely clear although the gastro-intestinal tract appears to be the most probable source. Le^a- and Le^b-active glycosphingolipids were first isolated from human adenocarcinoma tissue [160] and the determinant structures were shown to be identical to those characterised in the ovarian cyst glycoprotein fragments [54,91,92]. Extensive studies by Hanfland, Egge and colleagues [125,161–163] identified Le^a- and Le^b- blood-group-active glycolipids in the plasma from human group OLe(a-b+) individuals as mainly short pentaglycosyl- or hexaglycosyl-ceramides. An extended Le^b-active glycosphingolipid containing a repetitive Type 1 chain core structure was isolated from human colonic carcinoma cell line Colo 205 [146]; this structure was trifucosylated with an Le^b structure at the non-reducing terminal and an internal α-1,4-fucosylated *N*-acetylglucosamine residue. However, no evidence for extended Type 1 structures was found in a variety of normal tissues [164].

Compound structures with Le^b determinants underlying terminal Type 1 chain A or B structures (ALe^b and BLe^b, Table V) were first identified in glycosphingolipids isolated from erythrocytes and small intestine [165,166] but are now known to occur in both glycolipids and glycoproteins with O-linked chains, as are the difucosylated Type 2 chain ABH structures (Table V) first characterised in oligosaccharides isolated from ovarian cyst glycoproteins [89,92]. Recent reviews have given detailed analyses of the tissue distribution of blood-group-active glycosphingolipids based on the isolation and characterisation of the compounds [148] and on the detection of the various isoforms of the blood group determinants in different tissues by immunostaining methods [167].

Hybrid molecules with sialic acid and fucose attached to adjacent sugars in the core structure can be formed on Type 1 and Type 2 chain sequences in both glycoproteins and glycolipids to give, respectively, sialyl-Le^a and sialyl-Le^x determinants (Table III). The terminal sialic acid residues in these structures are always α-2,3-linked to the β-galactosyl residue. Glycoconjugate molecules bearing these structures have been isolated from both normal and tumour tissues [168–173]. Monoclonal antibodies specific for these hybrid structures have been raised by immunisation of mice with cancer cell membranes or with purified glycosphingolipids [170,172]. The presence of the sialic acid residue masks the specificity of the Le^a or Le^x structures in the same way as the presence of a second fucosyl residue on the terminal β-galactosyl unit masks the specificity of Le^a in Le^b structures [54].

4. Chemical synthesis of ABH, Lewis and Ii blood group determinants

Despite the widespread distribution of naturally occurring sources of blood group glycoproteins and glycolipids the macromolecules carrying the determinants are complex and heterogeneous. Isolation of blood-group-active oligosaccharide fragments from glycoproteins by controlled acid hydrolysis and alkaline degradation techniques gave yields which usually were sufficient only to characterise the oligosaccharides in serological and chemical terms. For more detailed studies on the conformation and antibody binding properties of the determinants chemical synthesis was evidently desirable but methods for synthesis in high yields of complex oligosaccharides composed of a mixture of sugars linked in both α- and β-anomeric configurations were not available until the mid 1970s when the pioneering work of Lemieux and his colleagues introduced the so-called "halide ion catalysed α-glycosidation reaction" [40,174]. The reaction is considered to proceed by way of the β-glycosyl halide which is brought into rapid equilibrium with the more stable α-anomer by way of ion-pair intermediates. The blood group Le^a and B trisaccharide determinants were first synthesised by this procedure [175,176]. These and other related blood group structures were prepared with a glycosidically-linked 8-methoxycarbonyl-octyl alcohol bridging arm which serves to link the determinant structure to a protein or an aminated solid support, thereby allowing the production of artificial antigens and solid immunoadsorbents [40,177,178].

An alternative methodology, the "imidate" procedure was introduced by Sinay and colleagues to chemically synthesise in good yield the Type 2 H determinant [179], the trisaccharide- and Type 2 tetrasaccharide- B determinants [180,181] and the Le^a [182] and Le^x [183] determinants. Chemical synthesis of the Type 1 tetrasaccharide B determinant was reported by Paulsen et al. [184]. Synthesis of the branched pentasaccharide carrying the Type 1 and Type 2 peripheral core structures, earlier isolated from a blood group active ovarian cyst glycoprotein [92], was achieved by Augé et al. [185] and oligosaccharides involved in I/i specific structures were synthesised by Alais and Veyrières [186].

5. Conformation of ABH and Lewis determinants

With the availability of synthetic determinant structures Lemieux and colleagues [187] undertook detailed studies of the conformations of oligosaccharides related to the ABH and Lewis determinants based on ^{13}C and ^{1}H NMR and computer assisted molecular modelling. Conformational analysis of the Type 1 and Type 2 disaccharides indicated that they are vastly different molecular structures (Fig. 1). In the Type 1 chain the 2-OH group of the β-D-galactosyl residue is in van der Waal contact with the carbonyl carbon residue of the acetamido group, whereas in the Type 2 chain this hydroxyl group is in near van der Waal contact with the methylene group at C-5 of the β-N-acetylglucosaminyl residue. It was therefore postulated that the two structures would present very different profiles to antibodies or enzymes. Conformational energy calculations, as well as proton NMR spectroscopy suggested that the non-reducing terminal blood-group determinants adopt single well defined low-energy conformations [187–189].

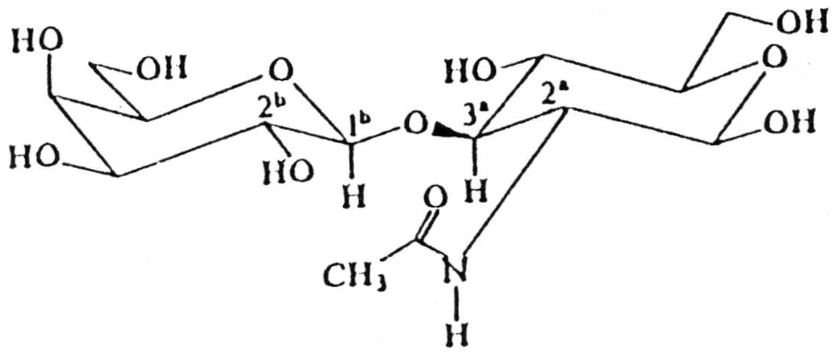

Fig. 1. Conformational formulae for the Type 1 (Gal(β1–3)GlcNAc) and Type 2 (Gal(β1–4)GlcNAc) di- saccharides predicted from hard sphere molecular modelling. Reproduced with permission from Lemieux et al. [187].

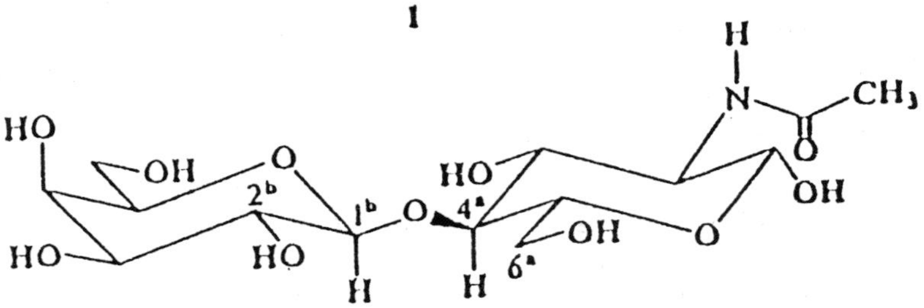

The projected conformational formulae for Type 1 human blood-group related oligosaccharides [187] is shown in Fig 2.

A conformational analysis of four different A-active glycosphingolipids (Type 1, Type 2, "Repetitive Type 3", and Type 4 "globo-A") revealed that in their minimum energy conformations the oligosaccharide chains are more or less curved [190] with the "Repetitive A" and Type 4 A having a strongly bent shape. The authors concluded that when the carbohydrate structures are linked to ceramide the A determinant of the Type 1 glycosphingolipid extends almost perpendicularly to the membrane plane whereas for Type 2, 3 and 4 the terminal parts of the oligosaccharide chains lie more or less parallel to the membrane. The fucose branch on "Repetitive Type 3" and Type 4 A structures appears directed towards the environment whereas in the Type 2 structure it would face the membrane. It was suggested that this peripheral core-dependent presentation of the A-determinant might explain the chain-type specificity observed for different monoclonal anti-A antibodies [190].

6. Lectin and antibody binding to ABH and Lewis determinant structures

Molecular recognition studies of the binding of the Type 2 H determinant by the anti-H lectin 1 of *Ulex europaeus* [191] using a wide range of chemically modified structures

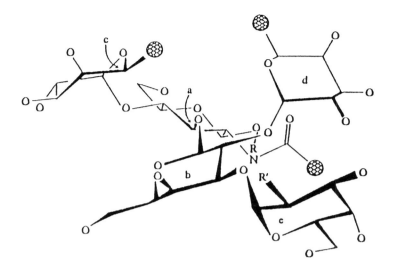

Structure	Residues
βDGal(1→3)βDGlcNAc (Type 1 core disaccharide, 1)	a, b
αL Fuc(1→4)βDGlcNAc (Le-disaccharide, 14)	a, c
βDGal(1→3)[αLFuc(1→4)]βDGlcNAc (Le*-trisaccharide, 15)	a, b, c
αL Fuc(1→2)βDGal(1→3)βDGlcNAc (Le^d-trisaccharide, 16)	a, b, d
αL Fuc(1→2)βDGal(1→3)[αLFuc(1→4)]βDGlcNAc (Le^b-tetrasaccharide, 17)	a, b, c, d
αL Fuc(1→2)βDGal (H-disaccharide, 18)	b, d
αDGalNAc(1→3)βDGal (A-disaccharide, 19)	b, e (R′ = NHAc)
αDGal(1→3)βDGal (B-disaccharide, 20)	b, e (R′ = OH)
αL Fuc(1→2)[αDGalNAc(1→3)]βDGal (A-trisaccharide, 21)	b, d, e (R′ = NHAc)
αL Fuc(1→2)[αDGal(1→3)]βDGal (B-trisaccharide, 22)	b, d, e (R′ = OH)
αL Fuc(1→2)[αDGalNAc(1→3)]βDGal(1→3)βDGlcNAc (A-tetrasaccharide, 23)	a, b, d, e (R′ = NHAc)
αL Fuc(1→2)[αDGal(1→3)]βDGal(1→3)βDGlcNAc (B-tetrasaccharide, 24)	a, b, d, e (R′ = OH)

Fig. 2. Projected conformational formulae of Type 1 human blood group related oligosaccharides (R = $(CH_2)_8COOCH_3$, $\phi^b = 50°$, $\psi^b = 20°$, $\phi^c = 55°$, $\psi^c = 25°$, $\phi^d = 40°$, $\psi^d = 20°$, $\phi^3 = -70°$, $\psi^e = -60°$). Reproduced with permission from Scheme 1 in Lemieux et al. [187].

related to Type 2 H, revealed that binding involves a wedge-shaped amphiphilic surface which extends over one side of the molecule. A cluster which involves OH-3, OH-4 and OH-2 of the fucose unit and OH-3 of the galactose unit provides the polar interactions but OH-3 and OH-4 of the α-L-fucose unit are regarded as providing the key polar interaction while OH-2 of fucose and OH-3 of galactose provide stability to the complex but are not essential. The groupings in the B-trisaccharide and the Le^b-tetrasaccharide determinants recognised, respectively, by hybridoma monoclonal anti-B and anti-Le^b reagents have been similarly defined [192,193].

Detailed investigations of the binding of the Le^b determinant by the lectin IV of *Griffonia simplicifolia* determined the area of the surface of the tetrasaccharide recognised by the lectin and led to the conclusion that the main driving force for the association of oligosaccharides and antibodies or lectins is the release to bulk of the water molecules at or near the interacting complementary surfaces [194]. Recently examination of the structures of this lectin and its complex with the Le^b determinant at high (2.0 Å) resolution revealed that several non-polar protein–carbohydrate interactions in the shallow binding site involved aromatic amino acids and that adjustments of positions of amino acids by approximately 1 to 2 Å took place in order to optimise the hydrophilic and hydrophobic interactions in the complex [195].

7. ABH and Lewis blood group antigen expression in embryonic development, cell maturation and adult tissues

Knowledge of the tissue specific expression of the human blood-group determinants is becoming increasingly important with the development of organ transplantation and production of reagents designed to target cell surface changes that accompany malignancy. Pioneering immunofluorescence studies carried out by Szulman in the 1960s disclosed the appearance and disappearance of ABH antigens at different stages of development of human embryos [196–200]. In early embryos, about five weeks post-fertilisation, ABH antigens are present on the surfaces of erythrocytes and are also strongly expressed on cell surfaces of the endothelium and epithelia of most early organs. In epithelial cells surface expression of ABH reaches a maximum at about nine weeks and thereafter decreases. In the digestive tract the decrease in cell surface expression coincides with the onset of secretion of mucus. Some cells, like the neurons of the central nervous system, muscle cells and bone cells completely lose the capacity to synthesise ABH antigens. The adult vestigial pattern of cell wall ABH antigens and the pattern of mucus secretions are established at 12–14 weeks of embryonic life [200].

In the foetal liver no ABH antigens are detectable except in the cells of the vascular endothelium [200]. In a study of the adult liver Rouger et al. [201] confirmed that normal hepatocytes lack blood-group antigens but demonstrated expression of Lewis and ABH antigens in biliary cells and i antigen in Kupfer cells. Comparison of the expression of ABH, Lewis and Ii antigens in the kidneys of three month embryos with the expression of these antigens in adult kidneys [202] demonstrated that, in both, ABH antigens were evident on the cells of the vascular endothelium in all specimens and were present in cells of the distal and collecting tubules in tissues from secretors. In the adult kidney, Lewis antigens were absent from the vascular structures but present in the cells of the distal and collecting tubules whereas in the fetal kidney virtually no Lewis antigens could be detected. Another difference was that in the adult kidney I and i antigens were observed on the membrane and in the cytoplasm of some cells whereas in the foetal kidney I antigen was absent and i antigen was present in the cells of the distal and collecting tubules. A recent study on H, Le^a, Le^b, Le^x, Le^y and sialylated-Lewis antigen expression in the developing human kidney [203] gives precise details of the cellular location and age at which the various determinants appear on Type 1 and Type 2 precursor core structures. Large amounts of blood group glycolipids are found in the adult pancreas [148,204]; Type 1 chain A, B, and Lewis compounds with six or seven sugar residues predominate with small amounts of Type 2 structures [148,205].

The pattern of ABH antigen expression during cell migration and maturation from the basal germinal layer to surface layers has been described for a number of tissues [167,206–208]. In some tissues, such as oral mucosa [209] and epidermis [210] there is a sequential appearance of precursor peripheral core determinants in the germinal layer, followed by monofucosylated H on the intermediate layers and later by A and B on more superficial layers but antigen expression does not always follow this sequential pattern [6,208].

Investigations of Le^a and Le^b antigens during embryonic development [211] confirmed that in salivary glands, stomachs and intestines of human embryos there is, in general, a reciprocal relationship between Le^a and ABH, whereas a parallel relationship is observed between Le^b and ABH. In normal adult urothelium the presence or absence of ABH antigens correlates with the secretor phenotype predicted from erythrocyte and saliva tests [208]. However, the appearance of Le^a and Le^b antigens in urothelium does not correlate with expression of these antigens in erythrocytes and saliva; saliva non-secretors with Le(a+b-) erythrocytes have Le^b in urothelium. More difficult to explain on the basis of classical explanations of Lewis blood-group genetics is the appearance of both Le^a and Le^b antigens in the urothelium of individuals grouped as Le(a-b-), and hence thought to be homozygous for the null allele *le* [212–215]. The identification of the antigens in this tissue is supported by the presence of the biosynthetic enzymes [216].

In acinar cells of salivary glands Le^b is expressed only in secretors and Le^a in non-secretors but in the epithelial cells of gland ducts and oral mucosae Le^b antigen is found in both secretors and non-secretors and has also been found in these structures from tissues derived from Le(a-b-) secretors [207]. The differential expression of Le^b in different cellular types of the same tissue or organ can be explained by the cell specific expression of either an *H* gene or an *Se* gene, either of which can synthesise a Type 1 H structure and thus form a substrate for an *Le* gene encoded transferase. However, the appearance of Le^a and Le^b in those grouped as Le(a-b-) suggests that there may be more than one gene locus responsible for the expression of the Le^a- and Le^b-positive phenotypes.

In haemopoietic tissue ABH antigens are both intrinsic and passively acquired on erythrocytes [122] and on platelets [217]. Granulocytes do not express ABH [218] and expression on lymphocytes comes from antigens passively acquired from the plasma [122,123]. Neither Le^a nor Le^b are expressed in haemopoietic cells. On neutrophils, the most abundant surface carbohydrate structures on glycoproteins and glycolipids are based on Type 2 polylactosamine chains [219,220] and the Type 2, monofucosyl, Le^x structure is expressed strongly on mature neutrophils and appears on immature cells from the promyelocytic stage onwards [221,222].

8. Biosynthesis of ABH and Lewis determinants

The evidence emerging in the 1950s on the carbohydrate nature of the blood group determinants coincided with the discoveries of Watson and Crick concerning DNA structure and the realisation that proteins were the primary translated products of genes [223]. It therefore became evident that the blood-group structures must be secondary gene products despite their clear-cut pattern of inheritance. A second discovery at about the same time was the characterisation by Leloir and colleagues (reviewed in [224]) of a new class of compound containing sugars esterified to nucleoside diphosphate which proved to be the natural donor substrates for the glycosyltransferase enzymes involved in oligosaccharide biosynthesis. In 1958/1959, in the light of the serological and bio-chemical evidence then available, schemes for the biosynthesis of ABH and Lewis determinants were put forward from our laboratory in which it was postulated that the determinants were built up by the sequential addition of sugar units to oligosaccharide chains in a precursor molecule. The blood-group-genes were envisaged as controlling the formation or function of the enzymes required for the attachment of those sugar residues that constitute the 'immunodominant' sugars in each determinant structure. The suggested interactions between *ABO*, *Hh*, *Sese* and *Lele* explained the different groups into which individuals can be placed on the basis of secretion of ABH, Le[a] and Le[b] antigens (Table II) [30,31,36,225]. An essentially similar scheme based on sero-logical and genetic considerations was advanced in 1959 by Ceppellini [226]. Experi-mental support for the idea that the gene products were glycosyltransferases came from many different laboratories (reviewed in [72,227–229], and has now been confirmed by the cloning of the relevant glycosyltransferase genes (see Sections 9.2, 11.8 and 12.2).

The glycosyltransferase encoded by the *H*-gene is a GDP-L-fucose: β-D-galactosyl α-1,2-fucosyltransferase (EC 2.4.1.69) that catalyses the reaction:

$$\text{GDP-Fuc} + \text{Gal}(\beta)–R \longrightarrow \text{Fuc}(\alpha1–2)\text{Gal}(\beta)–R + \text{GDP}$$

where R represents the remainder of a glycoprotein, a glycolipid or an oligosaccharide acceptor molecule.

Exoglycosidase degradation experiments established the precursor product relation-ship of H to A and B determinants (reviewed in [72]) and H structures are required substrates for the enzymes encoded by both the *A* and *B* genes. The glycosyltransferase encoded by the *A* gene is a UDP-N-acetyl-D-galactosamine: β-D-galactosyl α-1,3-N-acetyl-D-galactosaminyltransferase (EC 2.4.1.40) that catalyses the reaction:

$$\text{UDP-GalNAc} + \text{Fuc}(\alpha1–2)\text{Gal}(\beta)\text{-R} \longrightarrow \text{GalNAc}(\alpha1–3)\text{Gal}(\beta)\text{-R} + \text{UDP}$$
$$|$$
$$\text{Fuc}(\alpha1–2)$$

The transferase encoded by the *B* gene is a UDP-galactose: β-D-galactosyl α-1,3-D-galactosyltransferase (EC 2.4.1.37) that catalyses the reaction:

$$\text{UDP-Gal} + \text{Fuc}(\alpha1–2)\text{Gal}(\beta)\text{-R} \longrightarrow \text{Gal}(\alpha1–3)\text{Gal}(\beta)\text{-R} + \text{UDP}$$
$$|$$
$$\text{Fuc}(\alpha1–2)$$

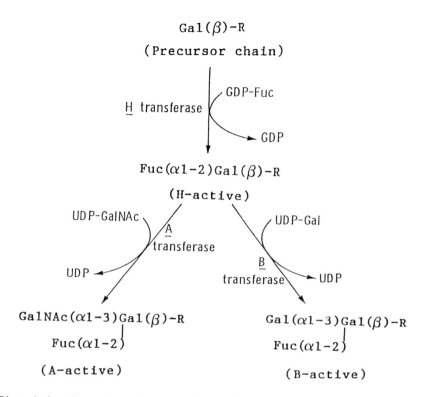

Fig. 3. Biosynthetic pathways for the formation of H, A and B blood group determinants. Abbreviations as in Table I.

The biosynthetic pathways for the formation of H, A and B determinant structures are summarised in Fig. 3. The third allele, O, at the ABO locus does not encode an active enzyme (see Section 11.8) and therefore in group O individuals the H structures on erythrocytes, tissue cell surfaces and soluble glycoconjugate molecules are not further changed.

The Le gene specifies a GDP-L-fucose: N-acetyl-D-glucosaminyl α-L-fucosyltransferase (EC 2.4.1.65) that catalyses the transfer of L-fucose to the subterminal N-acetyl-D-glucosamine residue of a Type 1 chain ending to form an Le^a structure:

$$GDP\text{-}Fuc + Gal(\beta1\text{--}3)GlcNAc(\beta)\text{-}R \longrightarrow Gal(\beta1\text{--}3)GlcNAc(\beta)\text{-}R + GDP$$
$$|$$
$$Fuc(\alpha1\text{--}4)$$

and to the same residue in a Type 1 H structure to form an Le^b structure:

$$GDP\text{-}Fuc + Gal(\beta1\text{--}3)GlcNAc(\beta)\text{-}R \longrightarrow Gal(\beta1\text{--}3)GlcNAc(\beta)\text{-}R + GDP$$
$$| \qquad\qquad\qquad | \qquad\quad |$$
$$Fuc(\alpha1\text{--}2) \qquad\qquad Fuc(\alpha1\text{--}2)\ Fuc(\alpha1\text{--}4)$$

338

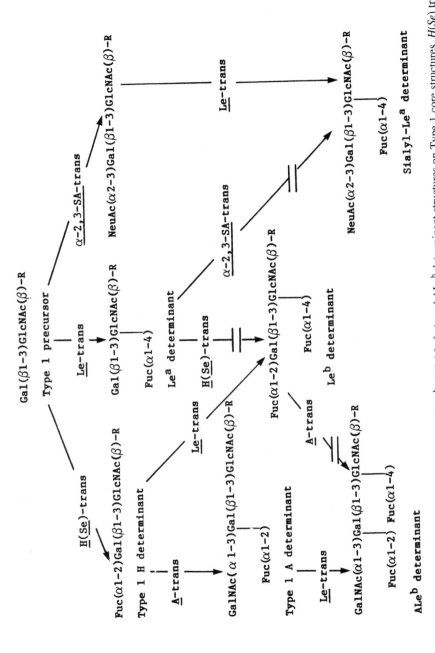

Fig. 4. Biosynthetic pathways for the formation of H, Le^a, Le^b, Sialyl-Le^a, A and ALe^b determinant structures on Type 1 core structures. H(Se) trans, α-1,2-fucosyltransferases encoded by H or Se genes; Le trans, α-3/4-fucosyltransferase encoded by the Le gene; A trans, α-1,3-N-acetylgalactosaminyl-α-1,2-fucosyltransferase encoded by the A-gene; α-2,3-SA trans, α-2,3-sialyltransferase. Breaks in arrows denote inadmissible pathways. Other abbreviations as in Tables I and III.

Inheritance of an *Le* gene has also been associated with the ability to fucosylate the O-3 position of terminal reducing glucose units in lactose-based oligosaccharides [230] according to the reaction:

$$GDP\text{-}Fuc + R\text{-}Gal(\beta 1\text{-}4)Glc \longrightarrow R\text{-}Gal(\beta 1\text{-}4)Glc + GDP$$
$$|$$
$$Fuc(\alpha 1\text{-}3)$$

and, to a considerably more limited extent [231,232], to transfer L-fucose to the O-3 position of *N*-acetylglucosamine in Type 2 chains (see Section 12.1). The biosynthetic pathways for the formation of Le^a, Le^b, ALe^b and sialyl-Le^a determinant structures are shown in Fig. 4.

9. The H-gene encoded α-1,2-L-fucosyltransferase

9.1. Specificity and purification of α-1,2-L-fucosyltransferase in human serum

The major part (90–95%) of the α-1,2-fucosyltransferase found in human plasma [233, 234] is believed to derive from haemopoietic tissue (evidence reviewed in [5]) and therefore to be the product of the blood group *H* gene, rather than the *Se* gene. The enzyme acts on low-molecular-weight aryl and alkyl-β-D-galactosides, on a range of di- and larger oligosaccharides terminating with β-D-galactosyl residues [235] and with high molecular weight glycoprotein and glycolipid substrates [5]. *In vitro* D-galactose is a weak acceptor and D-fucose (6-deoxy-D-galactose) is almost equally as effective, indicating that the OH-group at C-6 is not necessary for binding to the enzyme. Neither L-fucose nor L-galactose function as acceptors. An L-fucose substituent on the penulti-mate *N*-acetylglucosamine residue in a Type 1 or Type 2 acceptor, or a sialic acid residue on the terminal β-galactosyl unit in either α-2,3 or α-2,6-linkage, blocks the transfer of fucose to the terminal β-galactosyl unit [236]. Phenyl β-D-galactoside is a convenient substrate for measurement of levels of α-1,2-L-fucosyltransferase activity in plasma [235] and other sources in which there is also expression of α-1,3- or α-1,4-fucosyltransferases that transfer fucose to subterminal *N*-acetylglucosamine residues in, respectively, Type 2 or Type 1 acceptors. Results obtained with glycopro-tein substrates, such as asialo-fetuin, have to be interpreted with caution because the Type 2 *N*-acetyllactosamine structures at the ends of complex N-linked chains are substrates for both the α-1,2- and α-1,3-L-fucosyltransferases.

A finding for which no explanation has yet been given is that, while there is a spread of α-1,2-fucosyltransferase activity in serum or plasma for each ABO group, the average level from donors of groups A or AB is higher than for other groups [235,237]. A similar pattern of activity was also observed for platelet α-1,2-fucosyltransferase levels [238].

Purification of the enzyme in human serum with GDP-hexanolamine Sepharose as the main purification step [236,239,240] yielded preparations which each had an apparent $M_r \sim 150,000$ determined by gel filtration. The preparation described by

Yasawa and Furukawa [240], purified 20,000-fold, had the capacity to restore H activity to de-fucosylated group O erythrocyte membranes and was used to prepare a polyclonal antiserum that neutralised *H*-transferase activity. The highest purification (10×10^6-fold, specific activity 23.6 units/mg) was achieved by Sarnesto et al. [239] by subjecting the preparation obtained by GDP-hexanolamine chromatography to HPLC gel-filtration chromatography. This preparation was believed to be homogeneous as determined by non-denaturing and denaturing SDS-PAGE of ^{125}I-labelled protein. The apparent M_r 50,000 by SDS PAGE under reducing conditions, together with the susceptibility of the enzyme to β-mercaptoethanol, led the authors to conclude that the native enzyme exists in a polymeric form of co-valently bound sub-units. The yield from 10 litres of serum was estimated to be 5 μg. The lectin binding properties of the enzyme suggested that, in common with other glycosyltransferases [241], the α-1,2-fucosyltransferase is a glycoprotein.

The conformational studies of Lemieux and colleagues [40] on the Type 1 and Type 2 disaccharides (see Section 5 and Fig. 1) led them to predict that, because these structures assumed such different conformations, it was unlikely that the same α-1,2-L-fucosyltransferase could transfer fucose to both structures. Le Pendu et al. [242] examined the α-1,2-fucosyltransferase activity in native serum from a non-secretor individual and found that the enzyme acted on both Type 1 and Type 2 acceptors but possibly favoured the Type 2 oligosaccharides. Specificity studies by Kyprianou et al. [236] on the partially purified (~200,000-fold) enzyme failed to disclose a preference for one or other of the Type 1 and Type 2 disaccharides or tetrasaccharides. Although this latter preparation had not been purified to homogeneity competition experiments with Type 1 and Type 2 disaccharides and tetrasaccharides gave results close to those expected for a single enzyme. The Type 3 peripheral core structure [Gal(β1–3)Gal-NAc] was definitely a less effective acceptor for the plasma α-1,2-fucosyltransferase [236,242]. Only limited specificity studies were carried out by Sarnesto et al. [239] with the highly purified plasma enzyme but the results suggested some preference for Type 2 structures. The *H*-gene specified enzyme derived from haemopoietic tissue therefore appears to act on both Type 1 and Type 2 acceptors despite the conformational differences of these structures but whether there is a preference for one or the other of these peripheral core structures remains to be conclusively established.

α-1,2-L-Fucosyltransferase activity occurs in human bone marrow of both secretors and non-secretors [243]. Analysis of cells from the peripheral circulation shows that the enzyme is expressed in erythrocytes [237] and platelets but is absent from lymphocytes, neutrophils and monocytes [238]. Earlier reports of the presence of α-1,2-L-fucosyltransferase activity in lymphocytes [244,245] are almost certainly attributable to platelet contamination of lymphocytes isolated from blood that had not been defibrinated [238].

9.2. Molecular cloning of the H-gene encoded α-1,2-L-fucosyltransferase

The yields of purified α-1,2-L-fucosyltransferase from serum or plasma are very low and so far have proved insufficient for amino acid sequence studies on the enzyme protein. Molecular cloning of a candidate human *H* gene was achieved by Lowe and his colleagues who pioneered the use of a mammalian gene transfer approach to characterise fucosyltransferase genes [32,246,247]. The human epidermoid carcinoma cell line, A431, expresses blood-group ABH antigens [152] and genomic DNA from these cells was introduced into mouse L cells which express the necessary acceptor molecules for the biosynthesis of H structures but do not express detectable α-1,2-fucosyltransferase or cell surface H-antigenic activity. Transfectants that expressed H determinants were isolated by an immunological detection procedure using a monoclonal antibody specific for H structures. Two human DNA restriction fragments were identified by Southern blot analysis in the genome of each H-positive transfectant and the larger of the two fragments was shown to encode α-1,2-fucosyltransferase activity when expressed in COS-1 cells which, in the untransfected state, are deficient in this enzyme [246]. The apparent Michaelis constants for the acceptor, phenyl β-D-galactoside (2.0–4.4 mM) and donor, GDP-L-fucose (12.4–17.5 μM), substrates resembled the apparent K_m values (respectively 3.0–4.0 mM and 11–27 μM) recorded for these substrates with the H-gene associated α-1,2-fucosyltransferases purified from human plasma [236,239].

A cloned cDNA corresponding to this gene (EMBL/Genbank accession number M35531) was shown to identify a single 3.6 kb transcript in human cells, and to encode a protein with 365 amino acids and a calculated M_r 41,249 [32]. This protein is considered to have a Type II transmembrane topology typical of other mammalian glycosyltransferases [241] consisting of an NH_2-terminal cytosolic domain (8 amino acids), a hydrophobic transmembrane domain (17 amino acids) and a COOH terminal domain of 340 amino acids. The COOH-domain is predicted to reside within the lumenal compartments of the Golgi apparatus and the *trans*-Golgi network. Two potential N-glycosylation sites are present in the catalytic domain (Fig. 5). It is suggested that the soluble forms of the transferase that exist in plasma and other body fluids arise from proteolytic cleavage of the transmembrane segment [8]. Gene fusion experiments confirmed that the enzyme's COOH-terminal 333 amino acids are sufficient for catalytic activity [32]. Studies on the complete sequence of human genomic DNA corresponding to the α-1,2-fucosyltransferase cDNA have indicated the presence of four introns within the gene but the coding portion is encompassed within a single 1.1 kb exon [248,249]. Northern blot analyses have shown that the gene can generate multiple transcripts that differ at their 5′ ends. No similarities have been found between the coding sequence of this α-1,2-fucosyltransferase gene and sequences recorded for other glycosyltransferases or for other protein or DNA sequences so far lodged in data banks. Southern blot analysis of somatic cell hybrids mapped the gene to human chromosome 19 ([32], the chromosome to which the H locus had previously been mapped [41]. Recently by a combination of physical mapping and fluorescence *in situ* hybridisation the H gene locus has been shown to map distally to apolipoprotein loci E and C2 on 19q13.3 [250].

9.3. H deficient phenotypes

Individuals with Bombay and para-Bombay phenotypes are characterised by a deficiency of H activity on erythrocytes, the presence of anti-H antibodies in plasma [13] and a failure to express significant amounts of α-1,2-fucosyltransferase activity in

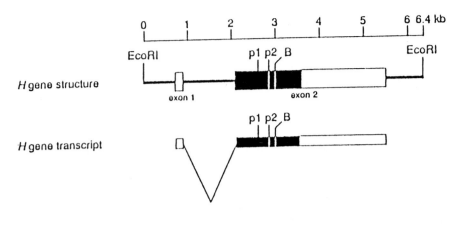

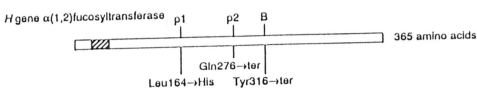

Fig. 5. Diagrammatic representation of the structure of the *H*-gene, the transcript and the encoded α-1,2-fucosyltransferase. The gene consists of two exons and the entire coding region (solid region) is contained in exon 2. The positions of the inactivating mutations giving rise to a Bombay (B) and two para-Bombay (pl and p2) phenotypes are indicated on the gene structure and in its transcript. The hatched portion of the α-1,2-fucosyltransferase represents the transmembrane segment. Reproduced with permission from Fig. 8 in Lowe [8].

plasma [233,234,237]. The absence of H acceptor substrate leads to an absence of A or B antigens on erythrocytes even though the *A* and *B* gene-associated glycosyltransferases are expressed normally in these phenotypes [251]. α-1,2-L-Fucosyltransferase activity is also absent from erythrocyte membranes of those with the Bombay O_h phenotype [237] but Schenkel-Brunner et al.[252] showed that membranes from these individuals can be made H-active by transfer of fucose mediated by an α-1,2-fucosyltransferase from human gastric mucosa; hence the membranes do not lack the appropriate precursors for the formation of H structures.

Recent sequence analysis of one allele at the *H* gene locus in an individual with the Bombay O_h phenotype identified six nucleotide sequence differences relative to the wild-type allele [8,249]. One of these differences, located within the coding portion of the gene, creates a termination codon corresponding to a tyrosine residue at amino acid 316 of the wild type allele (Fig 5); this change (Tyr 316 → ter) is predicted to yield a catalytically inactive mutant allele lacking 50 amino acids at the COOH terminal. The other five base pair changes were outside the coding region and gene transfer experiments showed that they were not functionally significant. When the Tyr 316 → ter mutation in the Bombay allele was changed back by mutagenesis to the wild type tyrosine codon, full enzyme activity was restored. PCR analyses of the genomic DNA from this Bombay O_h individual and his parents indicated that he was homozygous, and his parents heterozygous, for the mutated allele. Thus, this example provides confirmation of the prediction [29] that the Bombay O_h phenotype arises from the inheritance of a double dose of the null allele, h, at the Hh locus.

One distinction between Bombay O_h and para-Bombay phenotypes lies in the fact that, whereas those with the Bombay O_h phenotype lack ABH activity in both erythrocytes and secretions, some of those designated 'para-Bombay' are partially or completely deficient in ABH activity on erythrocytes but, if they carry a secretor Se gene, have normal amounts of ABH activity in saliva. The para-Bombay phenotype is heterogeneous and ranges from those expressing no, or only minute quantities of, erythrocyte A and B antigens, to those with A and B antigen levels only slightly lower than normal. All lack detectable H antigen on erythrocytes [13,14] but Mulet et al. [253] demonstrated that enzymic removal of galactose from para-Bombay cells exhibiting weak B activity exposed, as expected, small quantities of previously masked H structures. Nevertheless, because the H sequences are not exposed as terminal structures, these individuals may produce anti-H antibodies [13].

Examination of the coding region of both alleles at the H locus in a para-Bombay ABH-secretor individual revealed that each contained a single base sequence alteration relative to the wild-type α-1,2-fucosyltransferase gene. One of the sequence differences yields a termination codon at a position corresponding to amino acid residue 276 (Gln → ter) (Fig. 5) and is therefore predicted to generate a protein that has 90 amino acids deleted at the COOH terminal in comparison with the wild type gene [249,8]. The DNA sequence of the second allele has a missense mutation at codon 164 which substitutes a histidine residue for the leucine found in the wild type protein (Leu → His) (Fig. 5). When tested by transfection neither of these alleles yielded detectable α-1,2-fucosyltransferase activity [8]. These results clearly indicate that point mutations in the gene at the H locus are responsible for H antigenic deficiency in some Bombay and para-Bombay phenotypes. Different alleles responsible for expression of lesser amounts, or kinetically less efficient, α-1,2-fucosyltransferase enzymes may be expected to be found in the para-Bombay phenotypes in which the small amounts of H structures that are synthesised are completely converted into A and/or B structures.

An apparent para-Bombay phenotype that was not ascribable to deficiency of α-1,2-fucosyltransferase was observed by Herron et al. [254] in a person whose erythrocytes and saliva both lacked detectable H antigenic activity but showed weak A and B activity. A, B and H transferase levels were each normal in the serum of this person and it was suggested that the biochemical block must reside either in the availability of GDP-fucose or in the biosynthesis of the precursor acceptor oligosaccharide structures [254]. More recently two patients with recurrent severe infections resulting from a leukocyte adhesion deficiency were described who also had the Bombay phenotype [255]; the authors concluded that in these individuals the adhesion deficiency and the Bombay phenotype most probably resulted from a general defect in fucose metabolism.

10. The Se-gene encoded α-1,2-fucosyltransferase

Expression of H antigens on erythrocytes occurs whenever an individual inherits an *H* gene but in secretions expression of H, and hence of A and B, depends on the inheritance of a secretor *Se* gene which segregates independently of *ABO*. Examination of tissues where ABH activity is under the control of the *Se* gene, such as milk [256], submaxillary glands and gastric mucosal tissues [257], early established that significant amounts of α-1,2-fucosyltransferase activity were detectable only when the specimens came from ABH secretors. The suggested role of the *Se* gene as a regulatory gene implied that individuals of the Bombay phenotype could have genotypes *hh/sese*, *hh/Sese* or *hh/SeSe* since in the absence of an *H* gene the *Se* gene would have no function [258]. A family described in 1955 by Levine et al. [259] in which a woman with a Bombay O$_h$, non-secretor, phenotype, married to a non-secretor (*se/se*), had a child who was a secretor was for many years taken as support for the idea that individuals with the Bombay O$_h$ phenotype could carry unexpressed *Se* genes. However, the discovery of the para-Bombay phenotypes threw doubt on whether a regulatory role for the *Se* gene could explain all the facts [8]. An analysis of 44 published pedigrees from Bombay families [33] showed no other examples with evidence of suppression of Se: the author's concluded that the observations that had appeared to support the suppression theory [259] had most probably arisen from non-paternity. This finding, together with the ideas of Lemieux [40] relating to the conformational differences in the Type 1 and Type 2 core structures, and hence the possible existence of two different α-1,2-fucosyltransferases, led to the proposal that *H* and *Se* genes are two closely linked structural genes, each encoding α-1,2-fucosyltransferases [33]. According to this model the Bombay individuals who fail to secrete ABH are of the genotype *hh/sese* and the para-Bombay ABH secretors are those with the genotype *hh/SeSe* or *hh/Sese* (Table II).

The suggestion that *Se* is a structural gene led to a re-examination of the kinetics and acceptor specificities of known sources of α-1,2-fucosyltransferase and evidence soon emerged of species with differing properties. Salivas from non-secretors are not entirely devoid of ABH antigenic activity [81] and examination of human submaxillary glands from secretors and non-secretors demonstrated weak α-1,2-fucosyltransferase activity in the non-secretor glands amounting to about 5% of that found in secretors [260]. The enzymes from the two sources differed in solubility properties, charge and affinities for sugar donor substrate and, although the α-1,2-fucosyltransferase from 'secretor' glands reacted with both Type 1 and Type 2 acceptors, this enzyme had a distinct preference for Type 1 acceptors [260]. Stomach mucosal tissue from secretors contained a major species resembling the α-1,2-fucosyltransferase found in the secretor submaxillary glands and a minor species resembling the form in non-secretor glands [261]. Examination of α-1,2-fucosyltransferase expressed at different levels in the gastro-intestinal tract of secretors and non-secretors showed a marked difference between the activities with Type 1 and Type 2 acceptors in the upper regions of the digestive tract but the difference became much less marked in the samples from intermediate regions and in the caecum there was virtually no difference between the activities with the two substrates [150,262]. In the non-secretors, small amounts of α-1,2-fucosyltransferase activity were detectable in all levels of the tract but there was little discernable difference in the activity towards Type 1 and Type 2 substrates and the amount of activity measured with the Type 2 substrate was virtually the same in secretors and non-secretors. These results therefore suggested that small amounts of the *H*-gene encoded enzyme were present throughout the digestive tract and that the *Se* gene was expressed predominantly in the stomach, duodenum and jejunum [150,262] where the largest amounts of ABH antigenic activity are found [81,150].

The α-1,2-fucosyltransferase activity detected in the sera of individuals with the 'ABH-secretor, para-Bombay', phenotype (*hh/SeSe* or *hh/Sese*), amounting to 5–10% of the level found in normal group O individuals [263], is assumed to derive from secretory tissue and hence to be the product of the *Se* gene. A comparison of the properties of this enzyme with the α-1,2-fucosyltransferase in the serum from a normal group O non-secretor (*HH/sese* or *Hh/sese*) individual [242] revealed that the enzyme in the serum of the para-Bombay had a higher apparent K_m for GDP-fucose, a greater preference for Type 1 chain acceptors and an increased sensitivity to heat. A comparison of the α-1,2-fucosyltransferase in milk with that in serum of group O individuals led also to the conclusion that the enzyme derived from haemopoietic tissue differed from that secreted by epithelial tissues [264]. In the course of purification of the *H*-gene encoded α-1,2-fucosyltransferase from 10 litres of pooled serum Sarnesto et al. [239] separated a second species which they assumed to be the *Se* gene encoded α-1,2-fucosyltransferase [265]. A comparison of the kinetic parameters of the two purified species showed that they differed primarily in affinity for phenyl β-D-galactoside and GDP-fucose; Type 1 and Type 3 oligosaccharides were better acceptors than Type 2 structures for the presumed *Se*-gene encoded species [265].

The biochemical studies therefore support the proposal [33] that the *Se* gene encodes a second α-1,2-fucosyltransferase with different properties from the *H*-gene encoded enzyme, but these results did not exclude the possibility that the differences observed arose from some post-translational modification of the *H*-gene encoded enzyme brought about by the product of the *Se* gene. The most convincing evidence for the existence in the human genome of two independent α-1,2-fucosyltransferase genes has come from the recent analysis by Lowe and his colleagues of both alleles of the cloned *H*-gene encoded α-1,2-fucosyltransferase in a person with the para-Bombay phenotype [8,249]. The secretion of H substance by this individual, despite the fact that both his *H* alleles carry mutations that render them inactive, is strong support for the presence of a second gene in the human genome encoding an enzyme capable of adding fucose in α-1,2-linkage to β-D-galactosyl residues to form H structures [8,249].

The possibility of a third human gene encoding an α-1,2-fucosyltransferase has been suggested by the findings of Blaszczyk-Thurin et al. [266]. The biosynthesis of Le[b] and Le[y] structures (Table III) in normal epithelial tissues is believed to proceed by first the formation of Type 1 or Type 2 H structures, respectively, and then the addition of the second fucose to the subterminal *N*-acetylglucosamine residue to give the difucosylated Le[b] and Le[y] structures (Figs. 5 and 6); Le[a] and Le[x] do not normally function as acceptors for the α-1,2-fucosyltransferases [5,227,267]. However, an enzyme occurring in a gastric carcinoma cell line was reported to utilise Le[a] active glycolipid as a precursor for the biosynthesis of Le[b] [266]; Le[x] glycolipids were not converted to Le[y] by this

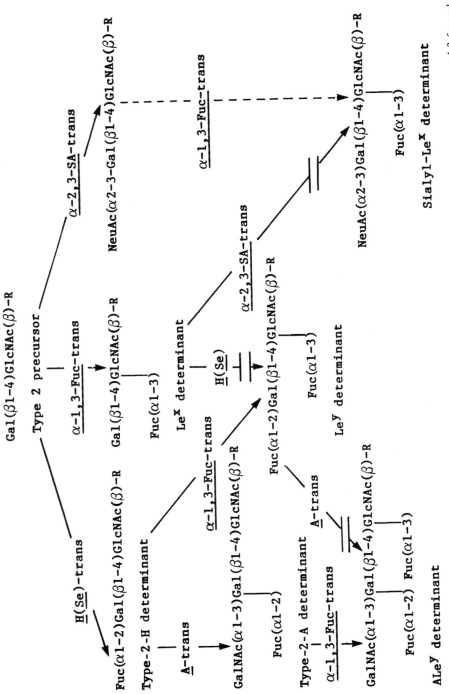

Fig. 6. Biosynthetic pathways for the formation of H, Le^x, Le^y, sialyl-Le^x, A and ALe^y determinant structures on Type 2 core structures. α-1,3-fucosyl-transferase(s), other abbreviations as in Fig. 4. Dotted line indicates a step catalysed by some, but not all, α-1,3-fucosyltransferases.

enzyme species. Subsequently enzymes with similar specificities were reported in a colon carcinoma cell line and in ovarian tumour tissue [268].

α-1,2-Fucosyltransferases have been described in a number of animal species [269–272]. The porcine enzyme, purified 124,000-fold to high specific activity from submaxillary glands, showed a preference for acceptors with Type 3 (Gal(β1–3)Gal-NAc(α)-R) or Type 1 terminal structures. Compounds with Type 2 peripheral core structures were very poor acceptors [269]; the porcine α-1,2-fucosyltransferase therefore bears a closer resemblance to the human *Se*-gene encoded transferase than to the human *H*-gene encoded α-1,2-fucosyltransferase.

11. The A- and B-gene encoded glycosyltransferases

11.1. Tissue distribution of A and B transferases

Activities associated with the *A*-gene encoded α-1,3-*N*-acetyl-D-galactosaminyltransferase and *B*-gene encoded α-1,3-D-galactosyltransferase have been reported in particle-bound form in human stomach mucosa and submaxillary glands, lung, kidney, bone marrow [5,273–276], erythrocyte membranes [277], ovarian cyst linings [278] and bladder urothelium [216] and in soluble form in milk [279,280], plasma [281–283], saliva [284], ovarian cyst fluids [278] and urine [285] from individuals of the appropriate A, B or AB blood groups. In contrast to the α-1,2-fucosyltransferase the distribution of these enzymes in tissues and body fluids is not dependent on secretor status and corresponds to the erythrocyte ABO phenotype of the individual.

Immuno-electron microscopy studies on the subcellular localisation of the *A*-gene encoded α-1,3-*N*-acetyl-galactosaminyltransferase revealed an unexpected difference in the localisation of the *A* transferase in adjacent cell types in intestinal mucus epithelium. In goblet cells the enzyme was restricted to the *trans*-cisternae of the Golgi apparatus whereas in adjacent absorptive cells the enzyme was present throughout the cisternal stack [286]. In addition this study showed the presence of the enzyme in the transtubular network and free, not membrane associated, transferase at the inner surface of the cisternal membranes and in early forming mucus droplets [287,288]. This observation is consistent with the suggestion that proteolytic cleavage can occur in the *trans* region of the Golgi apparatus [289] and with the presence of the soluble *A* (and *B*) transferase(s) in body fluids.

11.2. Specificity, cation requirements and pH optima

The *A* and *B*-gene encoded transferases both have an absolute requirement for a terminal H, Fuc(α1–2)Gal(β1-R), structure in the acceptor molecule. A second fucose substituent on a subterminal *N*-acetylglucosamine unit in a Type 1 or Type 2 structure prevents addition of *N*-acetyl-D-galactosamine or D-galactose to the β-D-galactosyl residue in the acceptor [5]. Although difucosyl structures occur (Table V) the second fucose residue is added after the addition of the terminal sugar (Figs. 4 and 6).

Both enzymes have a requirement for divalent cations and Mn^{2+} are the most effective. In serum Co^{2+} ions can replace Mn^{2+} ions in assays for the A^1 transferase but not for the A^2 transferase [290]. Differences are also found in pH optima; the pH optimum of the *B*-gene encoded α-1,3-galactosyltransferase is 6.5 [291] whereas the α-1,3-*N*-acetyl-galactosaminyltransferases in the sera of A_1 and A_2 individuals in the presence of Mn^{2+} ions have optima of 6.0 and 7.5–8.0, respectively [292–294]. The difference in pH optimum allows A_1 and A_2 phenotypes to be differentiated by testing the ratio of α-1,3-*N*-acetylgalactosaminyl transferase activity in sera at pH 6.0 and 8.0 [5].

11.3. Isoelectric points

The transferases in A_1 and A_2 sera differ in their isoelectric points as determined by isoelectric focusing [293,295,296]. Both give a major and a minor peak of α-1,3-*N*-acetyl-galactosaminyltransferase activity but in A_1 sera the major peak focuses in the pH range 8–9.5 whereas in A_2 sera the major peak focuses in the pH range 5–7. This difference in the isoelectric points of the A^1 and A^2 transferases enabled the presence of two distinct enzyme α-1,3-*N*-acetyl-D-galactosaminyl transferase species to be demonstrated in the serum of a donor of the genotype A^1A^2 [293,297]. The α-1,3-galactosyl-transferase in group B sera focuses in the same pH range as the A^1 transferase [293]. The differences in isoelectric points of A^1 and A^2 transferases appear to result from tissue-dependent post-translational modifications of the enzyme proteins since A^2 transferases in ovarian cyst fluids, stomach and kidney gave different focusing profiles from those obtained with the serum enzymes [296]. Treatment of A_2 serum with neuraminidase results in a shift in the pI of the transferase to a more basic pH [296] suggesting that glycosylation differences may account, at least in part, for the differences in isoelectric point.

11.4 Enzymic basis of A_1–A_2 differences

The question of whether the A_1 and A_2 subgroups arise from qualitative [298–301] or quantitative [302,303] differences in A antigen expression is one that has exercised workers in the blood group field for many years. Several groups of workers established that there are fewer A sites on A_2 cells than on A_1 cells [304–306] and the general properties of the A^2 transferase [292–294] are consistent with an enzyme that is less effective in transferring *N*-acetylgalactosamine from the donor substrate to H acceptor molecules. However, the discovery that Type 3 Repetitive A glycolipids and Type 4 Globo-A glycolipids (Table VII) occur almost exclusively on A_1 erythrocytes [110, 126], and that on A_1 erythrocytes Type 3 'Repetitive A' glycolipids possibly contribute about as much as Type 2 A glycolipids to the total A glycolipid component [7], has shown that qualitative differences do exist between the A antigens expressed on A_1 and A_2 cells.

The perception of the numbers of A receptors on A_1 and A_2 cells has also had to be changed in the light of these discoveries, since in both A_1 and A_2 cells Type 2 A structures on glycolipids may be masked by extension to Type 3 'Repetitive H' structures (Table VII). Enzyme analysis has indicated that the A^1 transferase is more efficient than the A^2 transferase in catalysing the addition of *N*-acetylgalactosamine to

H Type 3 determinants [307] and therefore in A_1 erythrocytes Type 3 'Repetitive H' structures are further converted to give more A antigen whereas in A_2 erythrocytes the Type 3 'Repetitive H' structures accumulate unchanged. In consequence the increased numbers of H determinants expressed on group A_2 erythrocytes in comparison with A_1 erythrocytes comprise masked Type 2 A structures as well as unconverted Type 2 H structures. Since Type 3 'Repetitive A' and Type 4 'Globo-A' determinants occur only on A_1 erythrocytes, monoclonal antibodies directed towards these structures behave as specific anti-A_1 reagents [7,126], but it is still not clear how much this qualitative distinction contributes to the A_1–A_2 differentiation detected by the classical human anti-A_1 reagents [13]. The lectin *Dolichos biflorus* which, appropriately diluted, is widely used as a reagent for determining the A sub-group is largely specific for terminal α-linked *N*-acetyl-D-galactosamine units [308] and therefore would be expected to react with both Type 2 A and Type 3 Repetitive A structures on glycolipids and also with Type 2 A determinants on glycoproteins. Viitala et al. [133] reported that 85% of the possible H acceptor sites were saturated with A determinants in A_1 polyglyco-sylpeptides whereas only 25% of H sites were filled in A_2 polyglycosylpeptides. Since the Type 3 'Repetitive A' structures are said not to occur on glycoproteins [7], and glycoprotein determinants are believed to account for 70–75% of the total erythrocyte ABH activity, differences in the numbers of Type 2 A structures would be expected to be the major contributors to A_1–A_2 distinctions established by the *Dolichos* reagent. Furukawa et al. [309] concluded that the qualitative versus quantitative views of A_1/A_2 were both correct depending on the reagent used to select the two cell types. Further support for quantitative differences in numbers of A determinants influencing A_1 or A_2 expression comes from investigations on rare blood group phenotypes where there is apparent competition between the *A* and *B* transferases for H-acceptor substrate. Although the *A* and *B* gene encoded enzymes are expressed co-dominantly, as ascer-tained by measurement of transferase levels in serum, in group AB individuals with very active *B*-gene encoded transferases an *A¹* gene product may be expressed on the erythrocyte surface as A_2 and an *A²*-gene product expressed as one of the weaker A variants [290,310–312].

In summary, the mutational changes in the *A* gene that give rise to the enzyme protein encoded by the *A²* allele result in both quantitative and qualitative variations in the complement of A antigenic determinants that are expressed on the erythrocyte surface.

11.5. Overlapping specificities of A and B transferases

The finding that serum from group B individuals had the capacity to transfer small amounts of *N*-acetylgalactosamine from UDP-*N*-acetylgalactosamine to H-acceptors to give A-active structures [313,314] raised the question of whether the *B*-gene encoded α-1,3-galactosyltransferase had an overlapping function with the A transferase. The properties of the enzyme bringing about this transfer were the same as those of the *B*-transferase and differed from the properties of *A¹* or *A²* transferases. Also, the level of activity with UDP-*N*-acetylgalactosamine could be directly correlated with the level of activity with the homologous UDP-D-galactose substrate [315]. These results there-fore strongly indicated that, although the *B*-transferase has not yet been purified to homogeneity, or the cloned gene expressed and tested for this overlapping function, the *B*-gene encoded enzyme can, to a limited extent, biosynthesise A determinants. Moreover, highly sensitive monoclonal anti-A blood grouping reagents have since revealed very weak A activity on some group B erythrocytes which have been designated B(A) [316–318]; the individuals exhibiting this weak A activity belonged to a subgroup [318] earlier reported by Badet et al. [319] in which unusually high levels of *B*-transferase activity are present in serum. The weak A reactivity on B cells was destroyed by treatment with α-*N*-acetyl-D-galactosaminidase and not by α-galactosidase [320] demonstrating that the reactivity is not attributable to recognition by the anti-A reagents of a structure common to the A and B determinants.

Utilisation of UDP-D-galactose by the *A* transferase to make B determinants has also been demonstrated [321] although the level of activity is even lower than for the reverse activity of the *B* transferase. This property was, however, retained by the α-1,3-*N*-acetylgalactosaminyltransferase purified to homogeneity from human gut mucosal tissue [322] and is therefore believed to be an inherent property of the *A* transferase. The presence of traces of B activity on group A erythrocytes has also been demonstrated with a powerful monoclonal anti-B reagent [323].

The capacity of the A and B transferases to utilise two different nucleotide sugar substrates is an apparent exception to the empirical rule of 'one enzyme–one glycosidic linkage' [324] but the fact that these two transferases are the products of alleles, and catalyse the transfer of sugars that structurally differ only in the nature of the substituents at C-2 in the sugar ring, makes it reasonable that the enzymes should have some capacity to recognise the alternative substrates. Of interest to note is the fact that the apparent K_m of the serum *B* transferase for UDP-*N*-acetylgalactosamine (285 μM) is far higher than the apparent K_m for UDP-galactose (11 μM) but the V_{max} for the two sugar donor substrates is approximately the same. In contrast, the apparent K_ms of the *A* transferase with the two donor substrates is approximately the same (~20 μM) but the V_{max} for UDP-*N*-acetylgalactosamine is 270-fold higher than that for UDP-galactose [315, 325]. Thus the *A* transferase appears to have a greater capacity to accommodate the 'wrong' substrate in its binding site than the *B* transferase but once the substrate is bound the *B* transferase has a greater capacity to catalyse the transfer of the sugar moiety.

11.6. Antibodies to the A- and B-transferases

Polyclonal antibodies have been raised in rabbits against the *A* transferase partially purified from plasma [326,327] and urine [328]. The authors each reported cross-reactivity of the antibodies with the *B* transferase in agreement with the allelic status of the *A* and *B* genes. Positive reactions were observed with *A* transferase partially purified from human kidney or stomach mucosal tissue but no reactivity was found with α-1,3-, or α-1,3/4-fucosyltransferases in human milk or saliva, or with α-1,3-fucosyl-, β-1,4-galactosyl- or β-1,3-*N*-acetylglucosaminyltransferases in human serum [327].

Mouse monoclonal antibodies to human *A* transferase were established by immunisation with enzyme purified from A plasma [155]. These antibodies also cross reacted with human *B*-transferase but did not react with *A* transferase from porcine submaxillary glands. The expression in the serum of group O individuals of an immunologically related protein cross reacting with the rabbit anti-*A*-transferase sera, reported by Yoshida et al. [326], was not detected with other polyclonal or monoclonal anti-*A*-transferase sera [155,329].

Of considerable interest is the observation, first made by Barbolla and colleagues [330,331], of the development of strong antibodies to *A*- and *B*-transferases in patients receiving ABO incompatible bone marrow or liver transplants. In the case of A recipients receiving bone marrow from O donors the antibody reacted with both *A* and *B* transferases and the patients exhibited severe graft-versus-host reactions. Formation of inhibitors of the *A* and *B* transferases following organ transplantation has been reported by others [332–334] and the antibody nature of these inhibitors was confirmed by the production of a monoclonal antibody to the *B*-transferase by hybrids constructed from mouse myeloma cells and Epstein-Barr-virus-transformed B lymphocytes from a blood-group B patient who had received group O bone marrow [335].

11.7. Purification of A- and B-transferases

Procedures for purification of the human plasma enzymes involving affinity adsorption onto erythrocyte membranes [291] and conventional multistep chromatography [336] are now known to have achieved only marginal purification of the *B*-gene encoded α-1,3-D-galactosyltransferase. The fortuitous binding of the *A* gene encoded α-1,3-*N*-acetylgalactosaminyltransferase onto agarose gel beads, and subsequent release from the gel by UDP [337] provides a valuable means of concentrating the *A*-transferase from large volumes of plasma and of separating the *A* from the *B* transferase in group AB plasma [315]. This absorbent was used by Nagai et al. [338] to purify the plasma *A*-transferase 70,000 to 100,000-fold. However, a membrane-bound α-1,3-*N*-acetyl-D-galactosaminyltransferase from porcine submaxillary gland [339], purified on UDP-hexanolamine agarose to a specific activity of 30 μmol/min/mg, was many thousand times more active than any of the previously described purified A enzymes and indicated the degree of purification that probably had to be achieved to obtain a homogeneous preparation.

Human α-1,3-*N*-acetylgalactosaminyltransferases have subsequently been purified to homogeneity from human lung [340–342], plasma [343] and gut mucosa [322] (Table IX). The lung and gut mucosal preparations, initially solubilised by extraction of the tissues with Triton-X 100, each had a M_r of ~40,000 and specific activities in the range 6–7 μmol/min/mg. Digestion of the lung preparation with N-glycanase yielded a reduction in M_r of ~6,000, estimated by SDS-PAGE, indicating that the *A*-transferase is a glycoprotein with N-linked carbohydrate chains [341]. The enzymes purified from lung [341] and gut mucosa [322] were subjected to N-terminal amino acid sequence determination; a 13 amino acid sequence obtained for the gut mucosal α-1,3-*N*-acetyl-galactosaminyltransferase was identical with the C-terminal 13 amino acids of a 23 N-terminal amino acid sequence obtained for the lung enzyme. The loss of 10 amino acids from the N-terminal sequence indicated that the gut mucosal preparation had probably undergone more extensive proteolysis in the process of solubilisation. Sufficient quantities of the lung enzyme were obtained for amino acid compositional analysis and the isolation and sequencing of peptides [341].

TABLE IX

Purification and properties of purified human blood group *A*-gene encoded α-1,3,-*N*-acetylgalac-
tosaminyltransferase preparations

Enzyme source	Main purification steps	Purification (fold)	Specific activity (μmol/min/mg)	M_r (kDa)	Ref.
Lung	1. Seph. 2. Ion-exchange FPLC	630,629	5.7	40,000	340,341
	1. UDP-Hex-Seph. 2. Octyl Seph.	100,000	6.1	40,000	342
Gut Mucosa	1. UDP-Hex-Seph. 2. Octyl-Seph.	124,175	7.0	40,000	322
Plasma	1. Seph. 2. IgG-Seph. 3. Blue Dextran-Seph.	27,000,000	16.0	35,000	343

Abbreviations: Seph. = Sepharose 4B, UDP-Hex-Seph. = UDP–Hexanolamine–Sepharose 4B; Octyl-
Seph. = Octyl Sepharose 4B; IgG-Seph. = rabbit IgG-Sepharose 4B; Blue Dextran Seph. = Blue
Dextran Sephadex G25.

11.8. Molecular cloning of the ABO locus

The partial amino acid sequence of the α-1,3-*N*-acetylgalactosaminyltransferase from
lung tissue enabled a series of corresponding deoxy-oligonucleotides to be constructed
which were used in conjunction with PCR to isolate human cDNA corresponding to
the blood-group *A* allele [344]. Nucleotide sequence analysis revealed a coding region
of 1062 base pairs encoding a protein of 41 kDa. Hydrophobicity plot analysis of the
predicted 353 amino acid sequence revealed a typical Type II transmembrane topology
with a short N-terminal domain (15 amino acids), a transmembranous hydrophobic
domain (24 amino acids) and a long C-terminal domain (314 amino acids). A single
N-glycosylation site was located at amino acid 112 in the C-terminal domain. The
N-terminus of the soluble form of the lung enzyme begins with an alanine residue at
amino acid 54 [344] whereas the N-terminus of the gut mucosal enzyme [322] was
apparently at amino acid 64. Southern hybridisation analysis indicated that the isolated
DNA did not represent a multigene family. Northern hybridisation analysis of mRNAs
from human colon adenocarcinoma cell lines originating from individuals with different ABO groups revealed multiple bands irrespective of A, B, AB or O phenotype
[344], thus showing that lack of enzyme activity of the *O* allele is due to structural
differences at the nucleotide level rather than to failure of expression of mRNA.
Isolation of this cDNA for the *A* transferase finally enabled the molecular basis for

the polymorphism at the *ABO* locus to be established in terms of the nucleotide sequences [2,344]. cDNAs were isolated from cell lines originating from people with different ABO blood groups by screening with the *A* transferase cDNA. Comparison of the DNA sequences revealed 99% homology between the coding regions of the *A* and *B* transferases and enabled seven nucleotide differences to be identified which yielded four consistent amino acid differences at residues 176 (*A*, Arg; *B*, Gly); 235 (*A*, Gly; *B*, Ser), 266 (*A*, Leu; *B*, Met) and 268 (*A*, Gly; *B*, Ala) (Fig. 7). Transfection experiments of blood-group O HeLa cells with reconstructed cDNAs encoding the crucial sequence for *A* and *B* transferase specificities resulted in expression of A and B antigenic expression. All the predicted *O* alleles sequenced in the first study [2] showed a single base deletion in the coding region at nucleotide 258 (later corrected to 261) which corresponds to one of the base pairs within the codon for amino acid 87 of the *A* transferase. This deletion results in a shift in the translational reading frame, and hence a different amino acid sequence distal to amino acid 86, and also leads to a termination codon at position 117 of the *A* transferase (Fig 7). The mRNA corresponding to the *O* allele would therefore be expected, if translated, to give rise to an altered, and shortened polypeptide lacking a functional catalytic domain. Since the *O* allele

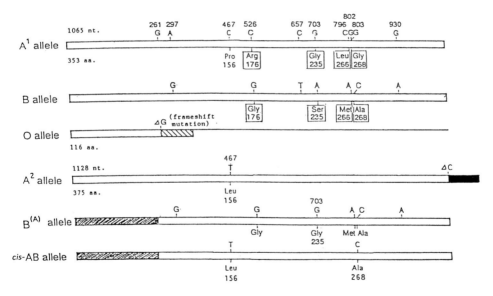

Fig. 7. Diagrammatic representation of alleles at the *ABO* locus. Only the positions where the other alleles differ from the *A¹* allele are shown except for nucleotide 703 in the *B⁽ᴬ⁾* allele where the substitution shown is the same as in the *A¹* allele. The boxed amino acids represent the four positions in the encoded proteins at which the *A¹* and *B* alleles differ. The cross-hatched region of the *O* allele indicates a different deduced amino acid sequence resulting from the frameshift mutation. The solid region of the *A2* allele indicates the extra domain that results from a single base deletion and a change in the reading frame. The hatched portions of the *B⁽ᴬ⁾* and *cis*-AB alleles represent undetermined portions of these sequences. Compiled from figures and data in Refs. [2,350,352,353].

does not give rise to an active enzyme it is to be anticipated that any missense mutations in the *A* or *B* alleles that drastically affect the binding or catalytic sites of the protein molecule will result in new *null* alleles at this locus and more recently Yamamoto et al. [345] described another type of *O* allele that did not have the single nucleotide deletion but had three nucleotide substitutions that resulted in a non-functional enzyme. Analysis of 132 Danish blood group O individuals by David et al. [346], using a set of restriction enzymes diagnostic for the deletion at nucleotide 261, revealed that, although 122 had this single deletion, ten of the phenotypically O individuals did not fit this pattern.

Johnson and Hopkinson [347] investigated the use of denaturing gradient gel electrophoresis (DGGE) on PCR amplified DNA for identification of genotypes within the ABO system and reported 4 different *O* alleles in 95 unrelated European individuals. The region of genomic DNA spanned included nucleotide 261, and the different *O* alleles could therefore have included ones with this single-nucleotide deletion plus additional different point mutations in regions flanking the deletion. Application of the DGGE method also allows *AO* and *BO* heterozygotes to be distinguished from *AA* and *BB* homozygotes, respectively, and has identified two different *B* alleles [347]. It is not yet clear whether these two *B* alleles correspond with the two groups into which Badet et al. [319] were able to classify Group B individuals on the basis of the levels of *B*-transferase activity in their sera. Application of an allele specific PCR method, that takes advantage of the single base deletion at nucleotide 261 and the substituents at nucleotides 532 and 700 which usually distinguish *B* from *A* and *O* alleles, has also been described by Ugozzoli and Wallace [348] for the direct determination of *ABO* genotypes.

The fact that the *A* and *B* alleles at the *ABO* locus encode glycosyltransferases that transfer different sugars makes them ideal candidates for the study of the extent of amino acid and structural differences in the enzyme protein required to change the utilisation from one sugar donor to another. Yamamoto and Hakomori [349] analysed the contributions made by the four different amino acid substitutions in the *A*-and *B*-transferases towards the ability of the enzymes to utilise UDP-*N*-acetyl-D-galactosamine or UDP-D-galactose. Chimeric cDNAs containing all possible combinations of the four polymorphic amino acid pairs were transfected into HeLa cells (expressing H, but not A or B antigens) and these cells were then tested for cell surface expression of A and/or B antigenic activity. These experiments suggested that leucine at position 266 and glycine at position 268 yields an A phenotype irrespective of the amino acids at the other two polymorphic positions whereas methionine at position 266 and alanine at position 268 yields a B phenotype irrespective of the amino acids at the other two positions. Therefore the amino acids at these two positions appear to be the most critical insofar as the utilisation of either UDP-*N*-acetyl-D-galactosamine or UDP-galactose is concerned. Computer analysis indicated that the amino acid substitutions at these two positions in the *B* transferase decreased the flexibility of the protein around these amino acids [349]; a factor which is in keeping with the decreased capacity of the *B*-transferase to accommodate the bulkier *N*-acetylgalactosamine residue [315] and may therefore be crucial in determining which sugar donor the enzyme can utilise. The amino acid at position 235 produces a weaker but still significant effect on specificity but the amino acid at 176 did not appear to influence the ability of the enzyme to discriminate between the two donor substrates.

11.9. Molecular genetic analysis of subgroups and rare variants of A and B blood groups

Cloning the ABO genes has opened up the possibility of determining the molecular basis of the subgroups and diverse rare variants known from serological and enzymic observations. At the time of writing the exon-intron boundaries and promoter sequences for these genes have yet to be completely defined but 91% (274 out of 301 amino acids) of the coding region of the soluble form of the transferases is contained in the last two coding exons of the gene and molecular genetic analysis of subgroups and rare ABO genotypes has focused on sequencing these two exons [350–353].

Enzymic and structural evidence has indicated that the A_2 subgroup arises from an α-1,3-*N*-acetylgalactosaminyltransferase that is less effective in converting H precursor substrates to A-active structures than is the product of the A_1 gene (see Section 11.4). Molecular genetic analysis of eight A^2 alleles showed that two differences in the nucleotide sequences distinguished them from A^1 alleles [350]. The first at nucleotide 467 results in substitution of leucine for proline at amino acid 156 and had previously been shown not to greatly alter enzymic activity or sugar-donor specificity [349]. The second single base deletion, found in one of the cytidine residues in nucleotide positions 1059–1061 (CCC) located near the C-terminal, results in a frame shift leading to an extra domain comprised of 21 amino acids (Fig. 7). Introduction of this deletion into an A^1 transferase cDNA expression construct results in drastically decreased α-1,3-*N*-acetylgalactos-aminyltransferase activity [350] and this single-base deletion is therefore believed to be the primary cause of the restricted substrate usage and weaker activity of the transferase encoded by the A^2 allele. The difference in isoelectric point of the A^2 transferase [296] remains unexplained as it is not possible to identify a new N-glycosylation site in the deduced amino acid sequence of the additional domain [350].

The A_3 phenotype is recognised serologically by an erythrocyte agglutination pattern in which small agglutinated clumps are outnumbered by unagglutinated cells [13]. Examination of the A-transferase in the serum of individuals with this phenotype indicated that the group was heterogeneous; in a few examples no A transferase activity was detectable and in others, although the enzyme levels were low, the properties of the transferase resembled those of an A^1 gene product whereas in others they were similar to A^2 gene products [5,277,290]. Molecular genetic analysis of A^3 alleles confirmed heterogeneity at the molecular level [351]. The nucleotide sequences of the last two coding exons of the genes revealed single base substitutions, leading to a change at amino acid 291, in two A_3 examples but in two other examples no differences were found in this region. No explanation has yet been advanced to account for the variable density of A-active sites on erythrocytes, and hence the mixed field agglutination pattern characteristic of this phenotype.

Two different weaker variants of A are A_x and A_m [13]. Powerful monoclonal reagents are now available that react with A_x erythrocytes [90,317] but previously they were frequently distinguished from group O erythrocytes only by their reactivity with anti-A,B sera from group O individuals, and from para-Bombay erythrocytes by the presence of H antigenic activity on erythrocytes and in saliva. Under normal assay conditions no A transferase activity is demonstrable in the serum of A_x individuals

[277,5] but concentration of the enzyme from large volumes of serum by absorption on to Sepharose 4B enables small amounts of α-1,3-*N*-acetylgalactosaminyltransferase activity to be demonstrated [354]. One A^x allele examined showed one nucleotide substitution resulting in a change from phenylalanine to leucine at amino acid 216 in the A^1 allele [353] but the effect of this change on the expression of α-1,3-*N*-acetyl-galactosaminyltransferase activity was not investigated. Erythrocytes of individuals with the A_m phenotype behave similarly to A_x cells but the saliva pattern is different in that, in A_m secretors, both A and H activities are present. Heterogeneity of the A_m phenotype was revealed by transferase studies which showed both varying levels of activity in serum and properties that were sometimes similar to A^1- and sometimes similar to A^2-transferases [5,277,290]. The molecular genetic basis of this phenotype has not been investigated but the serological and enzymic findings suggest a regulator gene acting to suppress the expression of A^1 and A^2 genes at the erythrocyte level. A possible candidate for a regulator gene acting on the A gene was described by Dorscheid et al. [355].

The *cis* AB phenotype [13,14] is rare, but genetically very interesting because individuals with this phenotype inherit both A and B characters from one parent, and therefore appear to challenge the accepted manner of inheritance of ABO blood groups. Transferase studies ([5] and references therein) on a number of such cases showed the presence of both *A* and *B* transferase activities in the sera of individuals with this phenotype but also revealed considerable heterogeneity with regard to the subgroup properties of the *A*-transferase and the level of activity of the *B*-transferase. In general the evidence supported the view that one enzyme was responsible for the two activities but was not conclusive. The interpretation that these individuals inherit a gene encoding an enzyme capable of transferring both *N*-acetyl-D-galactosamine and D-galactose, which had been mooted since the nature of the *A*- and *B*-transferases was established [356–358] has become a reasonable basis for the appearance of this phenotype since Yamamoto and Hakomori [349] demonstrated cell surface expression of both A and B activity resulting from chimaeric constructs incorporating elements of the nucleotide sequences of both the *A* and *B* alleles. Examination of the nucleotide sequence of the last two coding exons of ABO genes from two unrelated *cis*-AB individuals [352] revealed that they were identical and differed from the A^1 allele by two nucleotide substitutions which each resulted in an amino acid substitution. The first (nucleotide 467, amino acid 156) was identical with that found in A^2 alleles and the second occurred at the fourth position (nucleotide 803, amino acid 268) of the four amino acids that discriminate A^1 and *B* transferases. The change results in substitution of alanine for glycine at amino acid position 268 (Fig. 7), as is found in the *B* transferase, and led the authors to reassess their inference from the chimeric constructs that such a sequence gives rise only to A activity [349]. A *cis*-AB phenotype described by Yoshida et al. [359] did not fit the pattern of most other examples studied in that the serum contained separable *A*- and *B*-transferases; the authors suggested that in this case the phenotype resulted from unequal chromosomal crossing over: the molecular basis of this type of *cis*-AB has not yet been investigated.

Group B antigenic expression cannot be classified into well-defined subgroups corresponding to A_1, A_2, A_3 etc. but many weak B variants have been described [13,14].

The nucleotide sequences of the coding region of an allele classified as B^3 (although it is not clear whether the serological properties of the erythrocytes were analogous to those of A_3 cells) had a single base substitution resulting in a change at amino acid 352 (Arg → Trp) which discriminates this B^3 allele from normal B alleles [351].

Investigation of one example of the B(A) phenotype [316] revealed an allele with two nucleotide substitutions resulting in a change (serine to glycine) at amino acid 235 [353]. This substitution introduces into the B allele the amino acid that normally occupies this position in the A allele; previously cells transfected with a chimeric construct carrying this substitution had shown only B activity [349] but very powerful anti-A reagents are required to demonstrate A activity on B(A) erythrocytes and the experimental system used with the chimeric constructs may not have been sufficiently sensitive. On the basis of B-transferases levels in the serum of individuals with a normal B phenotype Badet et al. [360] described a sub-group that was clearly distinguished from the others by the high levels of expression of α-1,3-D-galactosyltransferase. Examples of this very active form of the enzyme were also observed in blood group AB subjects in which the expression of the A gene appeared to be suppressed by the unequal competition for the H precursor substrate [290,310–312]. In one case, in which the individual was grouped as $A_{weak}B$, but was shown by enzyme tests to have a normal A^2 transferase, a detailed study of the kinetic properties of the B transferase revealed an enzyme with a normal pH and apparent K_m for UDP-galactose and acceptor substrates but a maximum velocity several times higher than normal [290]. The examples of the B(A) phenotype examined for B transferase also had high levels of activity [318] and it will be of interest to know whether the high expressors of B transferase activity all have alleles with the substitution of glycine in place of serine at amino acid 235 in the encoded protein.

11.10. B-like antigen on animal cells and the relationship of bovine and mouse α-1,3-galactosyltransferase to human A and B transferases

B-like antigens similar, but not identical, in their serological properties to those of human B antigen were early reported in many animal species [11] and human anti-B sera were shown to contain mixtures of antibodies some specific for human erythrocyte B antigen and others for the B-like antigen [361]. The isolation by Eto et al. [362] of a pentaglycosylceramide from rabbit erythrocyte membranes with the structure: Gal(α1–3)Gal(β1–4)GlcNAc(β1–3)Gal(β1–4)Glc-Cer indicated that this straight chain compound, lacking the α-1,2-linked fucose on the subterminal β-galactosyl residue which is characteristic of the human group B structure, is responsible for the B-like activity of animal erythrocytes. This antigen has been evolutionarily conserved in many mammalian species and in New World monkeys but is absent from Old World monkeys, apes and man [11,363]. All humans have large amounts of anti-Gal(α-1–3)Gal antibodies that react with the B-like antigen [363,364] and these are important in relation to xenotransplantation of animal organs into man. Karlsson et al. [365] reported that the major target antigens in pig tissues for preformed IgG and IgM antibodies in human sera were endothelial glycoconjugates terminating in Gal(α1–3)Gal.

The presence of galactosyltransferases in particulate preparations from rabbit stomach mucosa with the capacity to transfer galactose in α-linkage to both Gal(β1–4)GlcNAc and Fuc(α1–2)Gal(β1–4)Glc [273], and the isolation from rabbit intestinal tissue of a hexglycosylceramide terminating with the fucosylated trisaccharide characteristic of a B determinant [366], showed that in this animal species both B and B-like structures could be synthesised. The demonstration that two distinct enzymes were responsible for these two activities indicated that separate genes were encoding the two transferases [367].

α-1,3-Galactosyltransferases (EC 2.4.1.151) catalysing the reaction:

$$\text{UDP-Gal} + \text{Gal}(\beta1\text{–}4)\text{GlcNAc-R} \longrightarrow \text{Gal}(\alpha1\text{–}3)\text{Gal}(\beta1\text{–}4)\text{Glc-R} + \text{UDP}$$

were purified to apparent homogeneity from calf thymus [368] and murine cells [369]. cDNA for a murine α-1,3-galactosyltransferase was cloned by a mammalian gene transfer system [370] and for a bovine α-1,3-galactosyltransferase using peptide sequence information and an anti-α-1,3-galactosyltransferase antiserum [371]. The sequences share substantial overall primary nucleic acid and derived protein similarity with the transferases encoded by the human blood-group ABO locus [2,344] in the regions corresponding to the C-terminal two thirds of the derived protein sequences, that is, in the regions comprising their respective catalytic domains. No corresponding functional gene encoding a α-1,3-galactosyltransferase acting on acceptor substrates containing terminal unsubstituted N-acetyllactosamine structures is believed to occur in humans but three distinct human genomic sequences that hybridised strongly with the murine or bovine α-1,3-galactosyltransferase cDNA probes have been isolated [372,373]. Although the human DNA segments showed substantial nucleotide sequence identity with the coding portions of the cDNAs they had many insertions and deletions with disruptions in the translational reading frame. Two of the fragments are thought to represent non-overlapping portions of a non-processed pseudogene that maps to human chromosome 9q33–q34 and the third segment to a processed pseudogene localised to human chromosome 12q14–q15 [374].

The proximity of the human pseudogene homologue of the α-1,3-galactosyltransferase genes to the ABO locus on chromosome 9, together with the significant degree of nucleotide sequence homology [344], supports the suggestion [373] that they are evolutionarily related. Shaper et al. [374] hypothesised that after duplication of an ancestral gene one copy might have evolved into the ABO genes and the other copy into an α-1,3-galactosyltransferase gene that, in higher primates and man, later became inactivated.

11.11. Animal genes cross hybridising with human ABO genes

Southern blot analysis of DNA from various species of animals with the human ABO gene probe, under conditions where the homologous α-1,3-galactosyltransferase pseudogene would not hybridise with human genomic DNA, revealed hybridisation with genomic DNA from marmoset, hamster, rat, mouse, sheep, cow, rabbit, cat and dog but not with other animals considered to be lower than mammals on the evolutionary tree

[375]. Portions of cross hybridising genes from several species of primates showed high conservation of the nucleotide, as well as the deduced amino acid, sequences during evolution.

Examination of the *A* and *B* alleles from the baboon for the four amino acid substitutions which discriminate human *A*- and *B*-transferases (a.a. 176, 235, 266 and 268) showed that the residues at the third and fourth amino acid substitutions (266 and 268) are conserved in the baboon alleles; thus giving further support to the conclusion that these amino acid substitutions are crucial for the different donor nucleotide-sugar specificities of the *A*- and *B*-transferases [349,375].

12. The Le-gene encoded α-1,3/1/4-fucosyltransferase

The genetics of the Lewis system, and the precise number and activities of the glycosyltransferase(s) involved in this system, are as yet less well defined than the genes and enzymes associated with the *ABO* locus.

12.1. Purification and properties of the α-1,3/1,4-fucosyltransferase

α-1,4-Fucosyltransferases (EC 2.4.1.65) acting on Type 1 chain sequences to give Le^a and Le^b determinants are detectable in milk [376], gastric mucosa and submaxillary glands [257], saliva [377], gall bladder and kidney [378] and bladder urothelium [208]. In contrast to the *A*, *B* and *H* transferases the α-1,4-fucosyltransferase does not occur in plasma [233] or cells of the haemopoietic system [244]. In milk [376], saliva [230], gastric mucosa and submaxillary glands [257] no expression of α-1,4-fucosyltransferase was detected in those individuals who, on the basis of erythrocyte and saliva tests for Le^a and Le^b activity, were grouped as Le(a-b-), and were therefore thought to have inherited the null allele, *le*. In bladder urothelium, however, the expression of Le^a and Le^b antigens, and of α-1,4-fucosyltransferase, does not correlate with the expression of these antigens on erythrocytes and saliva [216] and this raises the question of whether there are other tissue-specific genes encoding or regulating the expression of α-1,4-fucosyltransferases.

The acceptor specificity of the fucosyltransferase associated with the *Le* gene in saliva and milk is broader than for other glycosyltransferases so far described. Examination of fucosyltransferase activity in saliva of Le(a+b-), Le(a-b+) and Le(a-b-) individuals revealed that whereas all samples transferred fucose to the O-3 position of *N*-acetylglucosamine in Type 2 chains only those from individuals with an *Le* gene transferred L-fucose to the O-4 position of *N*-acetylglucosamine in Type 1 chain sequences and to the O-3 position of reducing glucose units in 2′-fucosyllactose [Fuc(α1–2)Gal(β1–4)Glc] and other milk oligosaccharides based on lactose [230]. Isoelectric focusing profiles of the saliva samples showed that the enzyme utilising Type 1 chains and 2′-fucosyllactose gave two major peaks of activity focusing at pH 5.1–5.4 and 9.5, whereas the major fractions utilising the Type 2 chain acceptor *N*-acetyllactosamine focused in the pH range 5.5–6.5. The enzyme in the Le(a-b-) saliva acting on Type 2 chains gave a single active peak in the range 5.5–6.5 [230].

Purification of the α-1,3-fucosyltransferase (500,000-fold) from pooled human milk by affinity chromatography on GDP-hexanolamine agarose [379] yielded an enzyme preparation that transferred fucose to both Type 1 and Type 2 acceptors to approximately the same extent and led to the suggestion that the two activities were due to a single α-1,3/1,4-fucosyltransferase that could utilise both Type 1 and Type 2 chains to synthesise Le[a], Le[b], Le[x], Le[y], sialyl-Le[a] and sialyl-Le[x] determinants. A second α-1,3-fucosyltransferase transferring fucose only onto N-acetylglucosamine in Type 2 acceptors is present in the saliva [230] and milk [380] of individuals with the Le(a-b-) phenotype. Chromatography of a milk preparation from an Le(a-b+) donor on CM-Sephadex C-50 removed a fraction which transferred fucose to the O-3 position of N-acetylglucosamine in Type 2 chains but had no activity with Type 1 sequences or with lactose [380]. These experiments showed the presence of two separable fucosyltransferases in human milk from Lewis-positive women and subsequent fractionation on Sephacryl S-200 of a preparation purified to approximately the same stage as the enzyme described by Prieels et al.[379] removed a large proportion of the α-1,3-fucosyltransferase responsible for transfer to Type 2 structures. The resultant α-1,3/1,4-fucosyltransferase added fucose primarily to the O-4 position of N-acetylglucosamine in Type 1 sequences and the O-3 position of D-glucose in lactose-based structures [381–384]. Conformational analysis [385] suggested that the steric environment of the 4-OH in the Type 1 disaccharide, Gal(β1–3)GlcNAc must be substantially different from the 3-OH of the Type 2 disaccharide, Gal(β1–4)GlcNAc, which is shielded by the rather large and rigidly held acetamido group (Fig. 2) whereas the environment of the 3-OH group of lactose (Gal(β1–4)Glc) can be expected to resemble that of the 4-OH of the Type 1 disaccharide. From a conformational point of view it is therefore not unreasonable that the α-1,4-fucosyltransferase should also react as an α-1,3-fucosyltransferase towards lactose but have limited capacity to utilise Type 2 sequences.

A preparation of α-1,3/1,4-fucosyltransferase isolated from human milk by affinity chromatography on GDP-agarose and HPLC, described by Eppenberger-Castori et al. [386], was purified 1.8×10^6-fold with respect to its ability to transfer fucose to the O-3 position of glucose in lactose. Unfortunately only limited specificity studies were reported on this highly purified enzyme preparation but it appeared to retain significant activity with Type 2 structures. Apparent molecular sizes based on gel filtration gave values of M_r 97,000 [386] and M_r 90,000 [383] for the milk α-1,3/1,4-fucosyltransferase and of M_r 47,000 [386] and M_r 53,000 [383] for the accompanying α-1,3-fucosyltransferase.

Specificity studies on the residual milk α-1,3/1,4-fucosyltransferase, remaining after removal of the contaminating enzyme acting on Type 2 structures, revealed a complex pattern of interactions [384]. Activity with the Type 2 disaccharide was only about one-tenth of that with the Type 1 disaccharide but the presence of fucose or sialic acid substituents on the terminal non-reducing β-galactosyl unit influenced the extent to which the substrate was utilised. Despite earlier reports to the contrary [379] oligosaccharides with terminal α-2,3-linked sialic acid residues, but not those with α-2,6-linked residues, are acceptors for both the fucosyltransferases in milk [387,384] but whereas the activity of the α-3-fucosyltransferase with Type 2 acceptors was enhanced by the presence of a terminal α-2,3-linked sialic acid the activity of the α-1,3/1,4-fucosyltransferase with Type 1 acceptors was decreased (Table X). Transfer of fucose took place to glycoproteins containing O-linked chains with terminal Type 1

TABLE X

Comparison of substrate specificities of α-1,3/1,4- and α-1,3-fucosyltransferases purified from natural sources with those expressed in COS-1 cells transfected with cloned fucosyltransferase genes

Substrate	Relative activities of purified enzymes[a]						
	α-1,3/1,4-Fuc-trans		α-1,3 Fuc-trans				
	Milk	A431	Plasma	Milk[b]	Liver	Neutrophils	
						Normal	CML
Gal(β1–4)GlcNAc	100	100	100	100	100	100	100
Gal(β1–3)GlcNAc	909	909	<1	<1	<1	<1	<1
Gal(β1–4)Glc	336	200	1	<1	–	–	–
NeuAc(α2–3)Gal(β1–4)GlcNAc	418	136	147	109	133	46	2
NeuAc(α2–6)Gal(β1–4)GlcNAc	<1	<1	<1	<1	<1	<1	<1
Fuc(α1–2)Gal(β1–4)GlcNAc	–	654	162	123	120	106	163
Fuc(α1–2)Gal(β1–4)Glc	827	1045	3	2	1	–	–

	Relative activities of enzymes in cells transfected with cloned fucosyltransferase genes[a]		
	Fuc-TIII	Fuc-TIV	Fuc-TV
Gal(β1–4)GlcNAc	100	100	100
Gal(β1–3)GlcNAc	420	<1	10
Gal(β1–4)Glc	145	3	11
NeuAc(α2–3)Gal(β1–4)GlcNAc	56	<1	115
Fuc(α1–2)Gal(β1–4)Glc	254	6	42

[a]Relative to activity with Gal(β1–4)GlcNAc.
[b]Purified from milk of Le(a-b-) donor.
Compiled from data in Refs. [381,384,388,393,408].

structures but no transfer was detectable to glycoproteins with N-linked oligosaccharide chains terminating in Type 2 structures irrespective of whether or not sialic acid had been removed [384]. A soluble, secreted, α-1,3/1,4-fucosyltransferase with a very similar specificity pattern to the further purified enzyme from milk was isolated from the culture medium of the human epidermoid carcinoma A431 cell line [388]. No other mammalian sources of α-1,4-fucosyltransferases have been described but a plant α-1,4-fucosyltransferase with the capacity to synthesise Le[a] structures on Type 1 chains [389] had no detectable activity against the Type 2 disaccharide and showed only very weak activity with 2′-fucosyllactose.

The acceptor specificity of the soluble, purified human α-1,3/1,4-fucosyltransferase towards Type 2 acceptors suggested that the Lewis gene-encoded enzyme would not have the capacity to synthesise Le^x or sialyl-Le^x to any appreciable extent [384,388]. However, the gene fragment coding for a putative Lewis enzyme, that has been cloned and expressed in COS-1 and CHO cells [56], gave rise to the appearance of both Le^a and Le^x determinants on the cell surfaces. The recombinant enzyme from COS cells also gave incorporation into both Type 1 and Type 2 chain acceptors, although the transfer to the Type 2 acceptor was appreciably lower. These results therefore suggested that the Lewis gene does in fact encode a broad specificity glycosyltransferase that breaches the empirical rule of one enzyme–one-glycosidic linkage [324]. In contrast, however, De Vries and Van den Eijnden [390] reported that in their hands the recombinant enzyme did not incorporate fucose into any Type 2 chain oligosaccharide whether free or linked onto a glycoprotein and whether sialylated or not, although the experimental evidence in support of this statement has yet to be published. The precise specificity of the putative human Lewis gene-encoded enzyme therefore remains to be completely clarified.

12.2. Molecular cloning of the Le-gene encoded transferase (Fuc-TIII)

Isolation of the cloned human cDNA tentatively assigned to the Lewis blood group gene locus [56] was achieved by a transient gene expression transfer approach similar to that used for cloning the *H*-gene [32]. The isolation was, in fact, based in the first instance, on the reactivity of the transfected recipient cells with antibody to the Le^x (Gal(β1–4)[Fuc(α-1–3)]GlcNAc) structure, that is, the procedure followed the α-1,3-fucosyltransferase activity of the expressed cDNA. A mammalian cDNA expression library prepared from the human epidermoid carcinoma A431 cell line was transfected into primate kidney COS-1 cells which are naturally deficient in endogenous α-1,3-fucosyltransferase activity but express cell surface Type 2 oligosaccharides that can potentially function as acceptors for this enzyme. The transfected COS-1 cells were screened first for Le^x expression and later for α-1,3-fucosyltransferase activity [56]. The sequence of the cloned cDNA isolated by this approach showed an open reading frame of 1083 bp which predicted a 361-amino acid long type II polypeptide, with a molecular mass of 42 kDa, consisting of a 15 residue NH$_2$ terminal cytosolic domain, a 19-amino acid long hydrophobic membrane spanning segment and a 327-residue COOH terminal catalytic domain (Fig. 8). The amino acid sequence revealed two potential N-glycosylation sites at amino acids 154 and 185 within the catalytic domain. Southern blot analysis localised this cDNA to human chromosome 19 to which the Lewis gene has been mapped [34,41]. Gene fusion experiments indicated that the cDNA encodes an enzyme that can synthesise a variety of α-1,3- and α-1,4-fucosylated molecules, including Le^a, Le^b, Le^x, sialyl-Le^a and sialyl-Le^x [391,392]. The apparent discrepancies between the characteristics of the cloned gene product and the specificity properties of the soluble milk and A431 α-1,3/1,4-fucosyltransferases [384,388] have still to be resolved.

The enzyme encoded by the gene isolated by Kukowska et al.[56] was later termed Fuc-TIII [393] to avoid calling it the *Lewis* gene until the assignment had been confirmed. This nomenclature followed the use of Fuc-TI and Fuc-TII to describe α-1,3-fucosyltransferases expressed by Chinese hamster ovary cell mutants [394] and

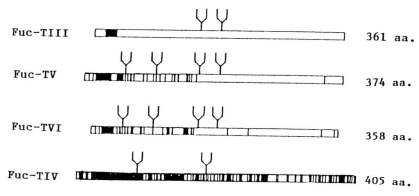

Fig. 8. Diagrammatic representation of the amino acid sequence homology of the cloned human α-1,3/1,4-fucosyltransferase (Fuc-TIII) and the three cloned α-1,3-fucosyltransferases (Fuc-TIV, Fuc-TV and Fuc-TVI). The putative transmembrane domains for each predicted protein are shaded and vertical bars denote amino acid residues differing from those in Fuc-TIII; the white areas thus represent areas of homology. The positions of potential glycosylation sites are indicated by the "Y"-shaped symbols.. The sizes of the predicted polypeptides are approximately to scale. Reproduced with permission from Part A, Fig. 2, Weston et al. [427].

the TIII is not intended to denote the linkage specificity of the fucosyltransferase. The nomenclature does, however, cause some confusion since the next α-1,3-fucosyltransferase to be cloned is now called Fuc-TIV [395–397] and this enzyme, in contrast to Fuc-TIII, has only α-1,3- and not α-1,4-fucosyltransferase activity.

12.3. Molecular genetics of the Lewis-negative phenotype

The isolation of the Fuc-TIII gene opened up the possibility of analysing this locus in Lewis-negative individuals believed to be homozygous for the null allele *le*. Two groups have shown single base changes at nucleotide 508 (A for G) which leads to an amino acid change from Gly to Ser at position 170 and observed that this mutation introduced a new PvuII restriction enzyme site [398,399]. Koda et al. [398] examined the mRNA from gastric mucosa of two Lewis-positive [Le(+)] and two Lewis-negative [Le(-)] individuals (grouped on the basis of red cell serology and immunostaining of gastric mucosa) with the cDNA isolated by Kukowska et al. [56] and demonstrated that the levels of mRNA were similar in both the Lewis-positives and negatives. Isolation of the protein coding region of the α-1,3/1,4-fucosyltransferase cDNA by PCR showed an identical sequence to that reported for the Le(+) cDNAs [56] but the sequence of one of the Le(a-) individuals contained two single base changes. Substitutions of G for T at nucleotide 59 and of A for G at nucleotide 508. Transfection of COS cells with chimeric cDNAs containing mutations at the two nucleotide positions revealed that the DNA with the mutation at nucleotide 508 did not lead to expression of Lewis antigen whereas the cells containing the cDNA with the mutation at nucleotide 59 did express this antigen. Use of the new PvuII restriction enzyme site enabled the authors to show that one of the Le(-) individuals was homozygous for the mutation whereas the other individual was heterozygous; thus suggesting the occurrence of other *le* alleles. Nishi-hara et al. [399] examined the genomic DNAs of seven Lewis-negative and five Lewis positive individuals. Southern blot analysis with the Fuc-TIII catalytic domain probe did not indicate large deletions in any of the Lewis-negative individuals. Restriction mapping revealed the additional PvuII site produced by the single base change at nucleotide 508 in the Fuc-TIII genes from Lewis-negative individuals; four were shown to be heterozygous and three homozygous for this site. Another single base substitution, resulting in a change from Asp to Ala at position 336, was observed in one Lewis-negative Fuc-TIII gene which did not have the PvuII site. All Fuc-TIII genes from Lewis-negative individuals, regardless of the PvuII site, had the single base substitution at nucleotide 59, also found by Koda et al. [398], resulting in the change of Leu to Arg at amino acid position 20, but this substitution did not alone lead to Fuc-TIII inactivation.

Elmgren et al. [400], using the PCR technique to analyse the coding sequence of Fuc-TIII in the genome of 40 individuals, found a specific base mutation in the coding sequence of Fuc-TIII in one Lewis-negative person; a C to T mutation at nucleotide 314 corresponding to a Thr to Met substitution at amino acid residue 105 in the transferase. This mutation creates a new NlaIII restriction enzyme cleavage site in the DNA and was found in both alleles in 5 of 18 Lewis-negative but in none of 22 Lewis-positive individuals. Heterozygous individuals for this substitution were found among both Lewis-negative (10 of 18) and Lewis-positive (4 of 22) individuals. The Lewis-negative phenotype thus appears to result from a number of different single base change mutations that render the protein encoded by the Fuc-TIII gene enzymically inactive. The molecular basis of the anomalous expression of Lewis antigens in normal bladder urothelium from Le(a-b-) individuals [212,208] has still to be explored.

13. α-1,3-Fucosyltransferases (Fuc-TIV, Fuc-TV, and Fuc-TVI)

13.1. Purification and properties of α-1,3-fucosyltransferases

Enzymes which catalyse the reaction:

GDP-Fuc + Gal(β1–4)GlcNAc-R $\longrightarrow$ Gal(β1–4)GlcNAc-R

$$| $$

Fuc(α1–3)

have been identified in human milk [256,380], submaxillary glands and stomach mu-cosa [257], plasma [233,401], saliva [230], amniotic fluid [402], kidney [378], liver [403], lymphocytes and granulocytes [244,238]. All sources that express α-1,3/1,4-fu-cosyltransferase activity also express an independent α-1,3-fucosyltransferase acting only on Type 2 structures and this renders more difficult purification of the α-1,3/1-4-enzyme, and assessment of its specificity towards Type 2 substrates. Plasma, liver and leukocytes, however, express α-1,3-fucosyltransferase in the absence of detectable α-1,3/1,4-enzyme. Biochemical studies have shown that although the enzymes in these tissues all synthesise the Le^x (Gal(β1–4)[Fuc(α1–3)]GlcNAc) structure they differ in other properties such as their ability to use sialylated acceptors, susceptibility to sulphydryl reagents and cation requirements (reviewed in [378,390,404,405]).

Purification of the α-1,3-fucosyltransferase in human plasma 600,000-fold by ammonium sulphate precipitation and chromatography on Phenyl Sepharose CL-4B and GDP-adipate-Sepharose yielded a preparation that transferred fucose preferentially to the N-acetylglucosamine residue in Fuc(α1–2)Gal(β1–4)GlcNAc and utilised α-2,3-sialylated Type 2 acceptors [387]. α-2,6-Sialyl compounds were not acceptors (Table X) and the presence of a second sialic acid residue as in NeuAc(α2–8)NeuAc(α-2,3)Gal(β1–4)GlcNAc, also abolished the activity. Similar initial purification steps involving ammonium sulphate precipitation and Phenyl Sepharose chromatography, followed by fractionation on S-Sepharose and affinity chromatography on GDP-hexanolamine Sepharose yielded a 600,000-fold purified preparation from plasma that had very similar specificity properties [406]. When this preparation was subjected to HPLC-gel filtration a 47×10^6-fold purified enzyme was obtained which on SDS-PAGE appeared to be a single molecular species with a M_r 45,000 [406]. The wheat-germ binding properties of the enzyme indicated that it was glycosylated.

An α-1,3-fucosyltransferase purified from a total homogenate of human liver to apparent homogeneity by a multi-stage process involving cation exchange chromatography, hydrophobic interaction chromatography and HPLC–gel exclusion chromatography yielded an enzyme protein with a M_r 40,000 Da and specificity properties very similar to the plasma enzyme [403] (Table X). Specificity studies with a hepatic cell line [378] and the Golgi fraction of isolated hepatocytes [407] revealed essentially the same pattern of activity with low molecular weight sialylated and non-sialylated substrates. With neutral or sialylated biantennary N-glycans the hepatocyte enzyme expressed highest affinity for the 3′-bisialylated derivative [407]. The α-1,3-fucosyl-transferase in milk from Le(a-b-) women is similar to the hepatic and plasma enzymes in its acceptor specificity requirements for terminal structures (Table X) [408]. A partially purified milk enzyme has also been shown to transfer fucose residues to internal N-acetylglucosamine units in polyl-N-acetyllactosamine chains; with a synthetic trimer of Gal(β1–4)GlcNAc this enzyme catalysed the addition of fucose residues in a given order to yield trimeric Le^x (Table III), with the N-acetylglucosamine nearest to the non-reducing end of the molecule being the last to be fucosylated [409]. The acceptor specificity properties of the α-1,3-fucosyltransferases occurring in plasma, milk and liver, collectively referred to as 'plasma-type' [378], resembles that found for the α-1,3-fucosyltransferase expressed by the CHO mutant cell line Lec 11 [394].

Partial purification of the α-1,3-fucosyltransferases in normal granulocytes showed that they utilised as substrates both low M_r compounds and glycoproteins (fetuin and α1-acid glycoprotein) containing terminal sialylated or unsialylated Type 2 structures (Table X), although, in contrast to the plasma enzyme, the presence of the sialic acid substituent decreased, rather than increased, the affinity of the enzyme for the Type 2 substrate [410]. In the author's laboratory the α-1,3-fucosyltransferase, isolated and partially purified from the 'leukocyte fraction' from three different patients with chronic myeloid leukaemia (CML) had very limited capacity to transfer fucose to low molecular weight sialylated compounds or glycoproteins with complex N-linked oligosaccharide chains unless the sialic acid had been removed [410]. Easton et al. [411] confirmed the activity of the enzyme in normal granulocytes with sialylated acceptors, although activity was very much higher with non-sialylated oligosaccharides, but reported that the enzymes in CML cells also had activity with sialylated acceptors and that the ratios of activity with the two types of substrate were similar for the normal granulocytes and CML cells. The suggestion that the α-1,3-fucosyltransferase in immature myeloid cells differs in its capacity to utilise sialylated acceptors from mature cells was, however, confirmed by examination of α-1,3-fucosyltransferase activity in normal granulocytes, acute myeloid leukaemic (AML) cells arrested at an early stage of development, and in the promyelocytic cell line HL60 which can be induced to differentiate along the granulocytic pathway [412]. The enzyme in mature granulocytes reacted almost equally well with α-2,3-sialylated and non-sialylated substrates whereas the enzyme in the leukaemic blasts showed a marked preference for non-sialylated substrates. Dimethylsulphoxide (DMSO) induced maturation of HL60 cells was associated with a decrease in α-1,3-fucosyltransferase activity, a change in substrate specificity towards increased activity with sialylated substrates and an increased cell surface expression of sialyl-Le^x. α-2,6-sialyltransferase activity, which was strong in the leukaemic blasts but absent from the mature cells, also decreased on maturation of the HL60 cells and it was suggested that this enzyme, which competes for the Type 2 acceptor substrate, might play a regulatory role in the expression of Le^x and sialyl-Le^x in differentiating myeloid cells [412].

Chou et al. [413] observed differences in the susceptibility of fucosyltransferases in plasma to inhibition with N-ethylmaleimide although the precise specificities of the enzymes were not clearly defined. Subsequent studies have shown that the α-1,3-fucosyltransferase in plasma, kidney, liver and gall bladder are much more sensitive to inhibition by this sulphydryl reagent than are the enzymes in myeloid cells or brain [378]. Enzymes resembling the α-1,3-fucosyltransferase in immature myeloid cells, by virtue of their failure to utilise sialylated acceptors, have been purified from the mouse teratocarcinoma stem cell F9 [414], a human neuroblastoma cell line [415], a wheat-germ-agglutinin resistant clone of mouse melanoma cells [416] and from a CHO mutant cell line LEC 12 [394]. Few studies have been carried out to determine the efficiency with which the human α-1,3-fucosyltransferases transfer L-fucose to internal N-acetylglucosamine residues on Type 2 chains substituted with α-2,3-linked sialic acid but Holmes et al. [417], employing glycolipid substrates, showed that the enzyme in a lung carcinoma cell line not only transferred fucose to the N-acetylglucosamine in the terminal trisaccharide sequence of α-2,3-sialylated substrates but also to internal N-acetylglucosamine residues in sialylated glycolipids with repeating N-acetyllactosamine structures.

13.2. Relationship between the Lewis-gene associated α-1,3/1,4-fucosyltransferase and the α-1,3-fucosyltransferase in plasma.

Examination of many hundreds of plasma samples for α-1,3-fucosyltransferase activity suggested that if there are null alleles at the locus encoding this enzyme they must be very rare [418]. However, in 1986 two unrelated individuals who lacked the enzyme in both plasma and saliva were described; both were of the Lewis phenotype Le(a-b-) and both were members of Black families [418]. Subsequently, plasma samples from two members of a different Black African family [419] and from 18 out of a randomly tested series of 200 Indonesians were found to be deficient in α-1,3-fucosyltransferase activity [420]; all belonged to the Lewis-negative Le(a-b-) phenotype. These results therefore suggested a relationship between the Lewis gene locus and the locus encoding the plasma α-1,3-fucosyltransferase despite the fact that the enzymes have different acceptor specificities and tissue specific expression. Furthermore, individuals showing the plasma α-1,3-fucosyltransferase deficiency expressed Le^x antigenic and α-1,3-fucosyltransferase activity in their white cells [419,421] supporting a different genetic origin for the enzyme in plasma from that expressed in haemopoietic tissue.

13.3. Molecular cloning of α-1,3-fucosyltransferase genes (Fuc-TIV, Fuc-TV and Fuc-TVI)

Biochemical and immunochemical analyses pointed to the existence of a family of related α-1,3-fucosyltransferases all of which can utilise Gal(β-1–4)GlcNAc acceptors but which differ in their capacities to synthesise Le^x, sialyl-Le^x, dimeric Le^x and VIM-2 determinants (Table III). These observations suggested to Lowe and co-workers that the primary sequences of the genes encoding these enzymes might be sufficiently similar to the Lewis α-1,3/1,4-fucosyltransferase to allow them to be cloned by cross-hybridisation with the Fuc-TIII cDNA. A human genomic DNA phage library, screened at low stringency with the coding portion of the Lewis cDNA, led first to the isolation of Fuc-TIV [395] which has been identified as a gene encoding an α-1,3-fucosyltransferase expressed in myeloid cells, although it is probably not exclusive to this tissue. Two other groups independently cloned this same cDNA using different isolation procedures [396,397]. Fuc-TIV has an intron-less coding region that predicts a protein 405 amino acids long, with a type II topology and a Golgi domain of 361 amino acid residues [395]. The amino acid sequence shares 60% homology with Fuc-TIII and is identical at 133 out of 231 residues in corresponding positions within the COOH terminal catalytic domains (Fig. 8). The gene yields several distinct transcripts in the myeloid lineage [395–397] which appear to encode the same protein and to derive from multiple distinct transcription initiation sites [396]. Discrepancies between the results of the three groups [395–397], which are not yet entirely resolved, relate to the capacity of the Fuc-TIV encoded enzyme to utilise sialylated Type 2 acceptors and hence to lead to the expression of sialyl-Le^x determinants (Table III and Fig. 6). Resolution of these inconsistencies has assumed considerable importance because sialyl-Le^x on neutrophils has been characterised as a ligand for the adhesion molecule E-selectin (formerly called ELAM-1) which mediates binding of neutrophils to the endothelium in the early stages of inflammation [422–425].

The gene associated with the expression of Le^x antigen on the surface of myeloid cells, and of α-1,3-fucosyltransferase in the cells, was earlier mapped to the long arm of human chromosome 11 [57,58]. Analysis of a panel of somatic cell hybrids carrying different rearrangements of chromosome 11 and high resolution hybrid mapping have recently enabled Fuc-TIV (called FUT4 for chromosomal localisation) to be assigned to chromosome band 11q21 between DIIS388 and D11s919 [426].

Two other genes encoding α-1,3-fucosyltransferases, Fuc-TV and Fuc-TVI have been cloned by cross-hybridisation with Fuc-TIII [393,427]. These are both single exon genes which are located on human chromosome 19. The DNA sequence of the Fuc-TIII catalytic domain probe is 96 and 95% identical, respectively, to corresponding positions in the Fuc-TV and Fuc-TVI genes and is 62% identical to the corresponding segment in the Fuc-TIV gene [427]. These genes are not alleles but Fuc-TVI has been mapped at only 13 kb 3′ from Fuc-TIII in tandem orientation on chromosome 19 [428]. The predicted protein sequences for both Fuc-TV and Fuc-TVI have each two new potential N-glycosylation sites as well as the conserved sites in the Fuc-TIII encoded protein [427] (Fig. 8).

The specificity of the enzyme encoded by Fuc-TV resembles that described for the purified plasma enzyme [387,406] in that it works efficiently with both N-acetyllactosamine and α-2,3-sialyl-N-acetyllactosamine to make, respectively, Le^x and sialyl-Le^x structures but differs in that it can also utilise 2′-fucosyllactose and lacto-N-biose 1 at a low but significant rate [393] (Table X). It is nevertheless frequently referred to as the gene encoding the 'plasma-type' α-1,3-fucosyltransferase. Expression of Fuc-TVI gives an enzyme that uses both N-acetyllactosamine and α-2,3-sialyl-N-acetyllactosamine and, according to Weston et al.[427], does not operate on 2′-fucosyllactose or lacto-N-biose 1; hence it more closely resembles the α-1,3-fucosyltransferase found in plasma than does the product expressed by Fuc-TV. However, Koszdin and Bowen [429], who independently cloned the same gene, reported activity of the Fuc-TVI encoded enzyme with 2′-fucosyllactose in addition to the other two substrates. The amounts of transfected cells available for assay, and differences in the methods used for estimating enzyme activity, may account for the discrepancies but the precise assignment of the cloned genes to tissues known to express α-1,3-fucosyltransferases in vivo still has to be clearly established.

On the basis of Southern blot analyses, Lowe and colleagues [8] believe that Fuc-TIV, -V and -VI genes fully represent all existing human genes with substantial DNA sequence similarity to the Fuc-III gene. However, these analyses have identified as yet uncloned sequences that cross-hybridise with the Fuc-TIV gene and which therefore may represent additional fucosyltransferase genes with sequences more closely related the Fuc-TIV gene on chromosome 11 than to Fuc-TIII, -TV and -TVI genes on chromosome 19.

14. Biosynthesis of i/I structures and cloning of the branching β-1,6-N-acetylglucosaminyltransferase

The I antigenic determinants are branched N-acetyllactosamino-glycans with one or several branch points at the 3- and 6-positions (see Section 3.4.2.). During development from foetal to adult erythrocytes linear i-active poly-N-acetyllactosamino-glycans are converted to branched structures having I activity [66–69]. β-1,3-N-acetylglucosaminyltransferases and β-1,6-N-acetylglucosaminyltransferases with the specificities required for the biosynthesis of i- and I-active structures have been identified in human serum [430,431], mouse T-lymphoma cells [432], hog gastric mucosa [433], Novikoff ascites tumour cells [434] and various rat tissues [435].

A cDNA encoding a β-1,6-N-acetylglucosaminyltransferase that converts linear into branched poly-N-acetyllactosamino-glycans [70] has been cloned by transfection of

CHO cells, that expressed i but not I antigen, with a cDNA library derived from PA-1 teratocarcinoma cells [436]. After transfection the cells were screened with an anti-I reagent and plasmid DNA was recovered from the I-positive cells. The cDNA sequence predicts a protein of 400 amino acids with a molecular mass of 45,860 and a type II membrane topology [436]. The amino acid sequence of the centre of the putative catalytic domain of this I-branching enzyme is strongly homologous with the corresponding region in another β-1,6-N-acetylglucosaminyltransferase that adds N-acetylglucosamine in β-1,6-linkage to Gal(β1–3)GalNAc in O-linked glycans [430]. The genes encoding both these β-1,6-N-acetylglucosaminyltransferases map to human chromosome 9, band q21, leading to the suggestion that they are possibly members of a gene family controlled by common regulatory elements that are responsible for their developmental pattern of expression [70].

15. Blood group antigens and disease

The earliest reports on ABO groups and disease related to slight, but statistically significant, associations linking group O with greater susceptibility to duodenal ulcer and rheumatoid arthritis, and group A with increased susceptibility to pernicious anaemia and carcinoma of salivary glands, stomach and colon (reviewed in [437]). The biochemical and genetic bases of these associations are still unclear.

The vast majority of publications on disease associations between ABO and Lewis groups concern changes in blood group antigen expression that occur in malignant tissues (reviewed in [438–440]). The precise relationships between gene expression, glycosyltransferase activity, cell surface antigenic expression and other possible regulatory mechanisms have still to be thoroughly investigated. However, since cell surface carbohydrate structures result from the concerted action of a number of different glycosyltransferases it is reasonable to assume that any changes in the expression of these enzymes will lead to a change in the nature of the structure synthesised. Lack of appearance of a blood group antigen may result from: (1) failure of expression of the glycosyltransferase that catalyses the addition of the terminal sugar that confers a particular antigenic specificity on the precursor structure; (2) from loss of one or more of the enzymes involved in the biosynthesis of the requisite precursor substrate; or (3) lack of nucleotide donor sugar, the appropriate divalent cations or other co-factor requirements. In addition,increased expression of an enzyme that glycosylates the same substrate as the blood-group-gene encoded glycosyltransferase, but which normally is a poor competitor for this substrate, may also lead to depression or complete loss of a normal cell surface antigen and to an abundance of an antigen that in normal tissues is expressed only to a very limited extent. Masking of the antigen by addition of a further sugar, or removal of sugars by exo- or endo-glycosidase action to expose usually cryptic structures, are other possibilities that could alter the balance of antigenic structures expressed on the cell surface.

Loss of A and B activity in human gastric carcinoma tissue, first reported in 1958 by Masamune et al. [441], was subsequently shown to reside in greatly reduced levels of A- and B-gene encoded transferases [442]. Similarly in bladder carcinoma A- and B-transferases, well expressed in normal tissues, are not detectable in tumour tissues which fail to show A or B antigen expression [208]. In the distal human colon blood group ABH activity can be detected during early foetal life but, disappears on development [443]; in malignancy the antigens re-appear [444,445]. The A- and B-gene encoded transferases are detectable in normal tissues from both secretors and non-secretors according to the ABO group of the individual but only very low levels of α-1,2-fucosyltransferase activity are found [445]; therefore in the normal distal colon the lack of A or B antigenic activity can be ascribed to paucity of H-acceptor substrate. In tumour tissues, however, there is a marked increase in α-1,2-fucosyltransferase activity in both secretors and non-secretors [444,445]. The re-expression of enzyme(s) synthesising H structures, together with normal presence of A (or B) transferase, thus provides an explanation for the re-appearance of A (or B) antigenic activity in carcinoma of the distal colon.

In addition to loss or gain of antigenic activity corresponding to the blood-group of the patient, aberrant expression of A and B antigens incompatible with the individual's ABO blood group occurs in some cancer tissues. Expression of A antigen in cancer tissues of group B or O individuals has been reported by several groups who have used antibodies or lectins for immunodetection [446–449]. Some of these observations have been attributed to cross reactivity of the anti-A reagents with Forsmann antigen (GalNAc(α1–3)GalNAc(β1–3)Gal(α1–4)Gal(β1–4)Glc-Cer) or Tn antigen (GalNAcα-Ser/Thr) [450,451]. However, in some instances glycolipids bearing normal A determinant structures have been isolated and characterised from cancer tissues of group O individuals [452–454] and A transferase activity has been demonstrated in adenocarcinoma tissues occurring in blood group O individuals [455,456]. The overlapping specificities of the A and B transferases (see Section 11.5) has been advanced as a possible explanation for the appearance of blood group B structures in tumour tissues of group A individuals and of A structures in tumour tissues of group B individuals [315]. Concentration of large volumes has also revealed the presence of an enzyme with the capacity to synthesise A determinants in the serum of group O individuals [354] and, although the activity was extremely weak, its existence demonstrated that individuals of all ABO groups may have the potential to make blood-group A structures. In view of our present knowledge of the truncated product of the O allele at the ABO locus, which if translated would lack the catalytic domain (see Section 11.8), the origin of this enzyme activity is unclear but its presence in normal group O individuals has been confirmed in another laboratory [457] and the up-regulation of this transferase activity in cancerous tissue could possibly account for the appearance of incompatible A structures.

A different explanation, invoking a repair mechanism to the O allele, has recently been advanced to account for the aberrant appearance of blood group A structures in group O individuals [456]. The authors suggest alternative splicing of the O gene product leading to loss of exon 2 and the restoration of the reading frame with the possible result of a protein containing a large part of the catalytic domain (see Section 11.8). Whether or not such a protein would express A transferase activity has yet to be established.

Changes in ABO blood group antigenic expression occur on the erythrocytes of some patients with acute leukaemia (reviewed in [458,459]). Deficiencies or complete disappearance of A,B or H antigens are sometimes observed but the expression of antigens incompatible with the patient's original blood-group has not been reported.

Examination of serum samples from patients with acute myeloid leukaemia (AML) revealed that all had low *H*-gene-associated α-1,2-fucosyltransferase activity [460]. Subsequent studies showed that depression of the *H*-transferase levels in the serum of these patients correlated with changes in platelet numbers and did not arise from abnormal expression of the α-1,2-fucosyltransferase in the leukaemic cells [461]. Similarly, elevated levels of the *H* transferase in the serum of patients with chronic myeloid leukaemia (CML) correlate with increases in platelet numbers [461]. In contrast, levels of α-1,3-fucosyltransferase, measured with asialo-substrates, are markedly elevated in leukaemic cells from patients with either AML or CML [412,461,462].

In solid tissues the α-1,3- and α-1,3/1,4-fucosyltransferases appear less susceptible to oncogenic change than the *ABH* gene-encoded transferases. The accumulated fucosylated products are classified as tumour-associated antigens [438] although they may occur normally in these same tissues in small amounts as overt cell surface structures or as cryptic structures masked by the further addition of sugar units. Sialylated and fucosylated Type 1 and Type 2 glycolipid and glycoprotein antigens accumulate in gastric, colonic, lung and bladder carcinomas, [208,463–466] and over-expression of α-1,3-fucosyltransferase has also been advanced to explain the strong expression of Le^x on glycoproteins observed in several hepatobiliary diseases [467].

Structures occurring in tumour tissues that are not detected to any extent in normal tissues are trifucosyl Le^y [468] and sialyl-dimeric Le^x [469,470] (Table III). The sialyl-dimeric-Le^x epitope has been implicated as a factor related to the metastatic potential of lung [471] and colon [472] carcinoma. Levels of α-1,3-fucosyltransferase have been reported to be higher in malignant gastrointestinal tissue than in normal mucosa [473] but it is not yet clear whether the fucosyltransferases responsible for the synthesis of these structures are different from those present in normal tissues, or whether the structures result from opportunistic activities of the resident fucosyltransferases when polylactosamine chains are formed, and there is an absence of other enzymes competing for terminal structures. De Vries et al. [409] showed that a partially purified α-1,3/1/4-fucosyltransferase from normal human milk had the potential to synthesise trimeric Le^x determinants and therefore that the accumulation of polyfucosylated structures may not indicate altered acceptor specificity of the transferase. On the other hand, Blaszczyk-Thurin et al. [266] and Yazawa et al. [474] have suggested that, in gastric and colorectal carcinoma cells, an aberrant α-1,2-fucosyltransferase can utilise Le^a structures to form Le^b determinants; a pathway that is not possible with the α-1,2-fucosyltransferases normally expressed in these tissues (see Section 9.1 and Fig. 4).

16. Function of ABO and Lewis blood group structures

Despite the vast accumulated knowledge of the serology, chemistry, and biosynthesis of the A and B blood group antigens, the identification of the glycosyltransferase products of the *A* and *B* genes and the crowning achievement of the cloning of these genes, it is still not possible to assign a clearly defined physiological function to the *ABO* locus. Similarly the nature of the selective pressures that maintain the ABO polymorphism are still open to speculation, although in this respect the relationship between the naturally occurring blood group antibodies and cell surface antigens of invading microorganisms has long been considered a possible factor in preserving the polymorphic balance [437,475]. The very existence of the polymorphism among healthy individuals is evidence that no one of the alleles has an essential physiological function, and the fact that group O individuals, who constitute a large proportion of many populations [13,14], completely lack the terminal sugars conferred on glyco-lipids and glycoproteins by the enzymic products of the A and B alleles makes it difficult to understand even what indirect advantage is conferred by the presence of this locus in the human genome. Taken in conjunction with the *Hh* locus, which controls the formation of the precursor structures necessary for the biosynthesis of A and B determinants, it can be argued that the terminal di- (H) or trisaccharide (A or B) sequences cap oligosaccharide chains and thereby mask structures that function as recognition signals in the course of differentiation and development. Koscielak [476] has referred to these as "singly-locked" and "doubly-locked" structures. The orderly appearance and disappearance of ABH determinants in the course of human embryo-genesis [199], lends credence to the idea that these antigenic structures are playing some significant role, but it would still appear to be one that could be accomplished by the activity of the *H* gene in the absence of *ABO* genes. Even absence of the *H* and *Se* genes in Bombay O_h non-secretor individuals does not have any apparent deleterious consequences although in view of the evidence of multi-gene fucosyltransferase fami-lies it is possible that different α-1,2-fucosyltransferase genes are expressed in the course of embryonic development. In erythrocytes from adult Bombay O_h individuals removal of sialic acid is required before addition of fucose can take place [252] and it is therefore feasible to assume that in the absence of the α-1,2-fucosyltransferases, α-2,3- or α-2,6-sialyltransferases take over the role of capping the oligosaccharide chains. Hakomori [107] has pointed out that deletion of A and B antigens, or the precursor H, Le^b or Le^y antigens, in human tumours is correlated with tumour invasion and metastasis; results that would accord with the role of these terminal structures as ones masking sequences that are important for binding to endogenous lectins involved in the migration of cells.

Equally difficult to assess is the function of the Lewis *Le* locus in relation to the blood group Le^a and Le^b characters, since absence of these determinants in those grouped as Le(a-b-) is compatible with normal life. However, the identification of Le^a and Le^b structures in normal bladder urothelium of Le(a-b-) individuals [208] suggests that other loci could be encoding genes giving rise to these determinants and that we still need to understand more of the genetic basis of this system. It is perhaps obvious that the determinants that are most likely to have an important function are those that arise through the activity of genes for which null alleles have not been detected, or from the activity of multigene families expressing enzymes with the same specificity. The Le^x, Le^y and sialyl-Le^x determinants, and α-1,3-fucosyltransferases involved in their biosynthesis fall into this category and are the ones for which a number of functions have now been ascribed. On leukocyte cell surfaces sialyl-Le^x structures are ligands for the adhesion molecule, E-selectin, which is transiently expressed on vascular endothe-lial cells in response to injury [422–425]. Binding of the carbohydrate ligand and the adhesion molecule are believed to result in the rolling and slowing down of the leukocytes along the vascular endothelium which is the first step in the inflammatory process and is followed by tighter protein–protein binding and leukocyte transmigration and extravasation (reviewed in [477]). E-selectin has also been shown to bind to sialyl-Le[x] structures expressed in some carcinoma tissues and it has been suggested that these ligands may promote E-selectin-dependent metastasis [478–480].

Roles have also been suggested for sialyl-Le[a] and sialyl-Le[x] as ligands for two other carbohydrate-binding adhesion molecules, P-selectin, expressed on endothelial cells and platelets [481] and L-selectin expressed on leukocytes [482]. These structures therefore have an important function in inflammatory reactions and deficiency of sialyl-Le[x] on the neutrophils of two patients was shown to result in extreme susceptibility to bacterial infections [255]. Although individuals with almost complete absence of α-1,3-fucosyltransferase in serum [418,419,421] did not exhibit this syndrome the deficiency in these cases appears to result from some failure of expression of the α-1,3-, and α-1,3/1,4-fucosyltransferase genes on chromosome 19 and not from absence of the products of the myeloid fucosyltransferase genes on chromosome 11. In animal experiments infusion of sialyl-Le[x] has been reported to drastically reduce P-selectin-dependent lung injury and to diminish accumulation of neutrophils [483]. The therapeutic possibilities of using sialyl-Le[x] for treatment of inflammatory diseases and cancers has led to considerable interest in the development of methods for the chemical and/or enzymic synthesis of this oligosaccharide and its analogues [484–487].

Le[x] is highly expressed at the morula stage in mice [488,489] and the observation that liposomes containing Le[x]-active glycolipids are self aggregating, whereas other related glycolipids are not, has indicated that carbohydrate–carbohydrate binding between Le[x] structures may be the mechanism underlying the part played by this determinant in embryo compaction [490–491]. Suggestions that a form of E-selectin recognising sialyl-Le[x] is involved in capillary morphogenesis [492] and that the localisation of Le[y], Le[x] and sialyl-Le[x] in guinea pig cochlea indicates a possible role for these structures in the physiology of hearing [493] have also been advanced. These claims will need further investigation but it is evident that of all the carbohydrate blood-group related determinants that have been studied these Type 2 fucosylated structures are emerging as the ones with the most obvious claims to be involved in important physiological functions.

17. Concluding remarks

The characteristics of the ABO system that make it difficult to ascribe a function to the locus are nevertheless ones that render it an ideal system for the analysis of many fundamental problems of glycoconjugate biosynthesis and expression. ABO was the first major human allo-antigen system to be identified [1] and the discovery of the carbohydrate nature of the antigenic determinants led first to structural studies to assign defined chemical structures to each blood group specificity [3–5] and then to investigations into how these well defined, and clearly inherited, carbohydrate structures, were controlled by the blood-group genes [30]. The characterisation of the glycosyltransferase products of the A and B genes provided the first examples in humans of a genetic polymorphism giving rise to enzymes with qualitatively different specificities and gave new genetic markers for the study of rare ABO variants [5]. The recent cloning of the *ABO* genes has identified the molecular basis of the polymorphism in terms of cDNA and deduced amino acid sequences [2] and has already yielded information on the extent of the changes in nucleotide sequences of DNA, and corresponding changes in amino acid sequence, necessary to alter the specificity of a glycosyltransferase from one donor sugar to another [349]. Moreover, since absence or weakness of one of the enzymically active gene products does not result in any apparent abnormality, or disease state, the system offers the possibility for investigating a wide range of mutant forms at a single genetic locus and for determining the changes in deduced amino-acid sequences of the encoded proteins that result in altered kinetic properties of the glycosyltransferases and hence in changes in the cell surface expression of the blood-group antigens. Expression of active recombinant enzymes, with the possibility of crystallisation of the proteins, should allow studies to be carried out on the conformation of the enzyme proteins and the nature of the interacting groups in the combining sites analogous to those being carried out with the carbohydrate-binding proteins of plant and mammalian origin; such studies have hitherto not been possible because of the very low abundance of these glycosyltransferases in natural sources.

The availability of cloned cDNAs for the ABO [2] and Lewis genes [8] has provided the necessary tools for more detailed investigations on the structural organisation of the genes and of the promoters and enhancers involved in their expression. Such studies should throw light on the mechanism underlying the tissue specific expression of ABH and Lewis antigens and of their ordered appearance and disappearance in the course of differentiation. Answers can now also be sought to other questions such as the relationship between mRNA expression, glycosyltransferase levels and cell surface antigen expression in normal tissues and the basis of the changes that take place in antigen expression in malignancy. The cloning of the genes at these blood group loci has thus ended the long pathway from the first serological observations on erythrocyte behaviour with certain anti-sera to the identification of the responsible DNA but, as has happened with each completed step in the pathway, the latest findings have opened up new areas of investigation which promise to throw further light not only on blood group antigen expression but on many related areas of glycobiology.

Abbreviations

Sugars

Gal	D-Galactose
GalNAc	*N*-Acetyl-D-galactosamine
Glc	Glucose
GlcNAc	*N*-Acetyl-D-glucosamine
Fuc	L-Fucose
NeuAc	*N*-Acetylneuraminic acid (Sialic acid)

Nucleic acids and nucleotides

G	Guanidine
A	Adenine
T	Thymidine

C	Cytidine
UDP	Uridine diphosphate
GDP	Guanosine diphosphate
DNA	Deoxyribonucleic acid
RNA	Ribonucleic acid
mRNA	messenger RNA
cDNA	complementary DNA
ter	termination codon
Southern blot	hybridisation of DNA on a nitrocellulose filter
Northern blot	hybridisation of mRNA on a nitrocellulose filter

Amino acids

Ser	Serine
Tyr	Tyrosine
Gly	Glycine
Gln	Glutamine
Leu	Leucine
His	Histidine
Met	Methionine
Arg	Arginine
Thr	Threonine
Try	Tryptophan
COOH-terminal	Carboxy terminal
N-terminal	Amino terminal
HPLC	High Performance Liquid Chromatography
PCR	Polymerase Chain Reaction
TLC	Thin Layer Chromatography
K_m	Michaelis constant
V_{max}	Maximum velocity
ϕ and ψ	Torsion angles for glycosidic bonds defined by neighbouring ^1H and ^{13}C atoms

Addendum

α-1,3-Fucosyltransferase genes
Two groups have recently described a fifth α-1,3-fucosyltransferase gene (Fuc-TVII) that is thought to encode the enzyme involved in the biosynthesis of sialyl-Le^x on human leukocytes [494,495]. Doubts concerning the interpretation of experiments showing that the lung inflammation process in a rat model is inhibited by perfusion of sialyl-Le^x oligosaccharide [483] have been raised by the demonstration that rodent polymorphonuclear leukocytes do not express sialyl-Le^x [496].

376

References

1. Landsteiner, K. (1900) Zbl. Bakt. 27, 357–362.
2. Yamamoto, F., Clausen, H., White, T., Marken, J. and Hakomori, S. (1990) Nature 345, 229–233.
3. Morgan, W.T.J. (1960) Proc. Roy. Soc. B, 151, 308–347.
4. Kabat, E.A. (1970) In: D. Aminoff (Ed.), Blood and Tissue Antigens. Academic Press, New York, pp. 187–198.
5. Watkins, W.M. (1980) Adv. Hum. Genet. 10, 1–136.
6. Oriol, R., Le Pendu, J. and Mollicone, R. (1986) Vox Sang. 51, 161–171.
7. Clausen, H. and Hakomori, S.-I. (1989) Vox Sang. 56, 1–20.
8. Lowe, J.B. (1993) In: G. Garratty (Ed.), Immunobiology of Transfusion Medicine. Marcel Dekker Inc., New York, pp. 1–36.
9. Yamakami, K. (1926) J. Immunol. 12, 185–189.
10. Yosida, K. (1928) Z. Ges. Exp. Med. 63, 331–339.
11. Wiener, A.S. (1943) Blood Groups and Transfusion. Thomas, Baltimore, pp. 332–60.
12. Landsteiner, K. and Levine, P. (1927) Proc. Soc. Exp. Biol. N.Y. 24, 941–942.
13. Race, R.R. and Sanger, R. (1975) Blood Groups in Man, 6th Edn.
14. Mollison, P.L., Engelfriet, C.P. and Contreras, M. (1993) Blood Transfusion in Clinical Medicine, 9th Edn. Blackwell, Oxford.
15. Anstee, D.J. (1990) Vox Sang. 58, 1–20.
16. Bailly, P., Piller, F., Gillard, B., Veyrières, A., Marcus, D.M. and Cartron, J.-P. (1992) Carbohydrate Res. 228, 277–287.
17. Watkins, W.M. (1994) In: J.-P. Cartron and P. Rouger (Eds.), Blood Cell Biochemistry, Vol. 16. Plenum Press, London. (In Press)
18. Epstein, A.A. and Ottenberg, R. (1908) Proc. N.Y. Pathol. Soc. 8, 117–123.
19. von Dungern, E. and Hirszfeld, L. (1910) Z. ImmunForsch. 6, 284–292.
20. Bernstein, F. (1924) Klin. Wochenschr. 3, 1495–1497.
21. Watkins, W.M., Greenwell, P. and Yates A.D. (1981) Immunol. Commun. 10, 83–100.
22. Yoshida, A. (1981) Acta Biol. Med. Germ. 40, 927–941.
23. Thomsen, O., Friedenreich, V. and Worsaee, E. (1930) Acta Path. Microbiol. Scand. 7, 157–190.
24. Renwick, J.H. and Lawler, S.D. (1955) Ann. Hum. Genet. 19, 312–320.
25. Rapley S.E., Robson E.B., Harris, H. and Maynard-Smith, S. (1968) Ann. Hum. Genet. 31, 237–242
26. Ferguson-Smith, M.A., Aitken, D.A., Turleau, C. and de Grouchy, J. (1976) Hum. Genet. 34, 35–43.
27. Kwaitkowski, D.J., Armour, J., Bale, A.E., Fourtain, J.W., Goudie, D., Haines, J.L., Knowles, M.A., Pilz, A., Slaugenhaupt, S. and Povey, S. (1993) Cytogenet. Cell Genet. 64, 94–103.
28. Morgan, W.T.J. and Watkins, W.M. (1948) Brit. J. Exp. Path. 29, 159–173.
29. Watkins, W.M. and Morgan, W.T.J. (1955) Vox Sang. (Old series) 5, 97–119.
30. Watkins, W.M. and Morgan, W.T.J. (1959) Vox Sang. 4, 97–119.
31. Watkins, W.M. (1959) In: G.E.W. Wolstenholme and C.M. O'Connor (Eds.), Ciba Foundation Symp. on Biochemistry of Human Genetics. Churchill, London, pp. 217–238.
32. Larsen, R.D., Ernst, L.K., Nair, R. and Lowe, J.B. (1990). Proc. Natl. Acad Sci. USA 87, 6674–6678.
33. Oriol, R., Danilovs, J. and Hawkins, B.R. (1981) Am. J. Hum. Genet. 33, 421–431.
34. Le Beau, M.M., Ryan, D. and Pericak-Vance, M.A. (1989) Cytogenet. Cell Genet. 51, 338–357.
35. Schiff, F. and Sasaki, H. (1932) Klin. Woch. 11, 1426–1429.
36. Watkins, W.M. (1958) Proc. 7th Congr. Int. Soc. Blood Transfus, Rome. Karger, Basle, pp. 692–696.

37. Bhende, Y.M., Deshpande, C.K., Bhatia, H.M., Sanger, R.S., Race, R.R., Morgan, W.T.J. and Watkins W.M. (1952) Lancet i, 903–904.
38. Solomon, J.M., Waggonner, R. and Leyshon, W.C. (1965) Blood 25, 470–485.
39. Rege, V.P., Painter, T.J., Watkins, W.M. and Morgan, W.T.J. (1964) Nature 203, 360–363.
40. Lemieux, R.U. (1978) Chem. Soc. Rev. 7, 423–452.
41 Ball, S.P., Tongue, N., Gibaud, A., Le Pendu, J., Mollicone, R., Gerard, G. and Oriol, R. (1991) Ann. Hum. Genet. 55, 225–233.
42. Mourant, A.E. (1946) Nature 158, 237.
43. Ueyama, R. (1939) Hanzaigaku-Zasshi 13, 51–64.
44. Andresen, P.H. (1948) Acta Path. Microbiol. Scand. 24, 616–618
45. Sneath, J.S. and Sneath, P.H.A. (1955) Nature 176, 172.
46. Makela, O. and Makela, P. (1956) Ann. Med. Exp. Fenn. 34, 157–162.
47. Grubb, R. (1948) Nature 162, 933.
48. Grubb, R. and Morgan, W.T.J. (1949) Brit. J. Exp. Path. 30, 198–208.
49. Cutbush, M., Giblett, E.R. and Mollison, P.L. (1956) Brit. J. Haemat. 2, 210–220
50. Sturgeon, P. and Arcilla, M.B. (1970) Vox Sang. 18, 301–322.
51. Boettcher, B. and Kenny, R. (1971) Human Hered. 21, 334–345.
52. Henry, S.M., Simpson, L.A. and Woodfield, D.G. (1988) Human Hered. 38, 111–116.
53. Ceppellini, R. (1955) Proc. 5th Congr. Int. Soc. Blood Transfus., Paris, 1954, p. 207.
54. Marr, A.M.S., Donald, A.S.R., Watkins, W.M. and Morgan, W.T.J. (1967) Nature 215, 1345–1349.
55. Sherman, S.L., Ball, S. and Robson, E.B (1985) Am. J. Hum. Genet. 49, 181–187.
56. Kukowska-Latallo, J.F., Larsen, R.D., Rajan, V.P. and Lowe, J.B. (1990) Genes Dev. 4, 1288–1303.
57. Tetteroo, P.A.T. and Geurts van Kessel, A. (1992) Histochemical J. 24, 777–782.
58. Couillin, P., Mollicone, R., Grisard, M.C., Gibaud, A., Ravise, N., Feingold, J. and Oriol, R. (1991) Cytogenet. Cell Genet. 56, 108–111.
59. Mollicone, R., Candelier, J.J., Reguine, L., Couillin, P., Fletcher, A. and Oriol, R. (1994) TCB 2, 91–97.
60. Wiener, A.S., Unger, L.J., Cohen, L. and Feldman, J. (1956) Ann. Intern. Med 44, 221–240.
61. Marsh, W.L. and Jenkins, W.J. (1960) Nature 188, 200–209.
62. Kapadia, A., Feizi, T. and Evans, M.J. (1981) Exp. Cell Res., 131, 185–195.
63. Feizi, T. (1985) Nature 314, 53–57.
64. Fukuda, M. (1985) Biochim. Biophys. Acta 780, 119–150.
65. Marsh, W.L. (1961) Brit J. Haematol. 7, 200–209.
66. Niemann, H., Watanabe, K., Hakomori, S., Childs, R.A. and Feizi, T. (1978) Biochem. Biophys. Res. Commun. 81, 1286–1293.
67. Feizi, T., Childs, R.A., Watanabe, K. and Hakomori, S. (1979) J. Exp. Med. 149, 975–980.
68. Watanabe, K., Hakomori, S., Childs, R.A. and Feizi, T. (1979) J. Biol. Chem. 254, 3221–3228.
69. Koscielak, J. (1977) In: J.F. Mohn, R.W. Plunkett, R.K. Cunningham and R.M. Lambert (Eds.), Human Blood Groups. Karger, Basle. pp. 143–169.
70. Bierhuizen, M.F.A., Mattei, M.G. and Fukuda, M. (1993) Genes Dev., 7, 468–478.
71. Landsteiner, K. (1945) The Specificity of Serological Reactions. Harvard University Press, Cambridge, MA, pp. 210–235.
72. Watkins, W.M. (1972) In: A. Gottschalk (Ed.), Glycoproteins: Their Composition, Structure and Function. Elsevier, Amsterdam, pp. 830–891.
73. Watkins, W.M. and Morgan, W.T.J. (1952) Nature 169, 825–826.
74. Morgan, W.T.J. and Watkins, W.M. (1953) Brit. J. Exp. Pathol. 34, 94–103.
75. Kabat, E.A. and Leskowitz, S. (1955) J. Am. Chem. Soc. 77, 5159–5164.

378

76. Kabat, E.A. (1956) Blood Group Substances. Academic Press, New York.
77. Watkins, W.M. and Morgan, W.T.J. (1955) Nature 175, 676–677.
78. Watkins, W.M. and Morgan, W.T.J. (1957) Nature 180, 1038–1040.
79. Kuhn, R. (1957) Angew. Chem. 60, 23.
80. Watkins, W.M. and Morgan, W.T.J. (1962) Vox Sang. 7, 129–150.
81. Hartmann, G. (1941) Group Antigens in Human Organs. Munksgaard, Copenhagen.
82. Morgan, W.T.J. and van Heyningen R. van (1944) Brit. J. Exp. Pathol., 25, 5–15.
83. Morgan, W.T.J. (1965) Methods in Carbohydrate Chemistry, 5, pp. 95–98.
84. Morgan, W.T.J. (1970) Ann. N.Y. Acad. Sci. 169, 118–130.
85. Donald, A.S.R. (1973) Biochem. Biophys. Acta. 317, 420–436.
86. Morgan, W.T.J. and Watkins, W.M. (1956) Nature, 177, 521–522.
87. Kabat, E.A. (1973) In: H. Isbell (Ed.), Chemistry of Carbohydrated in Solution. Adv. Chem. Ser. No. 117. American Chemical Society, Washington, DC, pp. 334–361.
88. Painter, T.J., Watkins, W.M. and Morgan, W.T.J. (1965) Nature 199, 282–283.
89. Lloyd, K.O., Kabat, E.A., Layug, E.J. and Gruezo, F. (1966) Biochemistry 5, 1489–1501.
90. Oriol, R., Samuelsson, B.E. and Messeter, L. (1990) J. Immunogenet. 17, 279–299.
91. Rege, V.P., Painter, T.J., Watkins, W.M. and Morgan, W.T.J. (1964) Nature 204, 740–742.
92. Lloyd, K.O., Kabat, E.A. and Licerio, E. (1968) Biochemistry 7, 2976–2990.
93. Marr, A.M.S., Donald, A.S.R. and Morgan, W.T.J. (1968) Biochem. J. 110, 789–791.
94. Yang, H. and Hakomori, S. (1971) J. Biol. Chem. 246, 1192–1200
95. Hakomori, S. (1984) In: R.H. Kennet, K.B. Bechtol and T.J. McKearn (Eds.), Monoclonal Antibodies and Functional Cell Lines. Plenum Press, New York, pp. 67–100.
96. Andresen, P.H. and Jordal, K. (1949) Acta Path. Microbiol. Scand. 26, 636–638.
97. Arcilla, M.B. and Sturgeon, P. (1974) Vox Sang. 26, 425–438.
98. Schenkel-Brunner and Hanfland, P (1981) Vox Sang. 40, 358–366.
99. Cheese, I.A.L. and Morgan, W.T.J. (1961) Nature 191, 149–150.
100. Painter, T.J., Watkins, W.M. and Morgan, W.T.J. Nature, 199, 282–283.
101. Rege, V.P., Painter, T.J., Watkins, W.M. and Morgan, W.T.J. (1963) Nature 200, 532–534.
102. Kabat, E.A., Baer, H., Bezer, A.E. and Knaub, V. (1948) J. Exp. Med. 88, 43–57.
103. Watkins, W.M. and Morgan, W.T.J. (1956) Nature 178, 1289–1290.
104. Feizi, T., Kabat, E.A., Vicari, G., Anderson, B. and Marsh, W.L. (1971) J. Exp. Med. 133, 39–52.
105. Holmes, E.H., Ostrander, G.K. and Hakomori, S. (1985) J. Biol. Chem. 260, 7619–7627.
106. Huang, L.C., Civin, C.K., Magnani, J.L., Shaper, J.H. and Ginsburg V. (1983) Blood 61, 1020–1023.
107. Hakomori, S. (1991) Bailliere's Clinical Haematology 4, 957–974.
108. Donald, A.S.R. (1981) Eur. J. Biochem. 120, 243–249.
109. Kannagi, R., Levery, S.B. and Hakomori, S (1984) FEBS Lett. 175, 397–401.
110. Clausen, H., Watanabe, K., Kannagi, R., Levery, S.B., Nudelman, E., Arao-Tomono, Y. and Hakomori, S. (1984) Biochem. Biophys. Res. Commun. 124, 523–529.
111. Breimer, M.E. and Jorvall, P.-A. (1985) FEBS Lett. 179, 165–172.
112. Lundblad, A (1978) Methods Enzymol. 50, 226–235.
113. Sabharwal, H., Chester, M.A., Sjoblad, S. and Lundblad, A. (1983) In: M.A. Chester, A. Heinegard, A. Lundblad and S. Svensson (Eds.), Proc. 7th Int. Symp. Glycoconjugates, Lund. Secretariat, Lund, p. 221.
114. Bjork, S., Breimer, M.E., Hansson, G.C., Karlsson, K.-A. and Leffler, H. (1987) J. Biol. Chem. 262, 6758–6765.
115. Holgersson, J., Clausen, H., Hakomori, S., Samuelsson, B.E. and Breimer, M.E. (1990) J. Biol. Chem. 265, 20790–20798.

116. Hakomori, S., Stellner, K. and Watanabe, K. (1972) Biochem. Biophys. Res. Commun. 49, 1061–1068.

117. Koscielak, J., Piasek, A., Gorniak, H., Gardas, A. and Gregor A. (1973) Eur. J. Biochem. 37, 214–225.

118. Hanfland, P. and Egge, H. (1975) Vox Sang. 28, 438–452.

119. Watanabe, K., Laine, R.A. and Hakomori, S. (1975) Biochemistry 14, 2725–2733.

120. Koscielak, J., Miller-Podraza, H., Krause, R. and Piasek, A. (1976) Eur. J. Biochem. 71, 9–18.

121. Hakomori, S. (1978) Methods Enzymol. 50, 207–211.

122. Crookston, M.C. and Tilley, C.A. (1977) In: J.F. Mohn, R.W. Plunkett, R.K. Cunningham and R.M. Lambert (Eds.), Human Blood Groups. Karger, Basle, pp. 246–256.

123. Oriol, R. (1980) Blood Transf. Immunohematol. 23, 517–524.

124. Clausen, H., Levery, S.B., Nudelman, E., Baldwin, M. and Hakomori, S. (1986) Biochemistry 25, 7075–7085.

125. Hanfland, P., Dabrowski, J. and Egge, H. (1983) In: J.-P. Cartron, P. Rouger and C. Salmon (Eds.), Red Cell Membrane Glycoconjugates and Related Genetic Markers. Libraire Arnette, Paris, pp. 107–117.

126. Clausen, H., Levery, S.B., Nudelman, E., Tsuchiya, S. and Hakomori, S. (1985) Proc. Natl. Acad. Sci. USA 82, 1199–1203.

127. Clausen, H., Levery, S.B., Kannagi, R. and Hakomori, S. (1986) J. Biol. Chem. 261, 1380–1387.

128. Karlsson, K.A. and Larson, G. (1981) J. Biol. Chem 256, 3512–3524.

129. Kannagi, R., Levery, S.B., Ishigami, F., Hakomori, S., Shevinsky, L.H., Knowles, B.B. and Solter, D. (1983) J. Biol. Chem. 258, 8934–8942.

130. Holgersson, J., Jorvall, P.-A., Samuelsson, B.E. and Breimer, M.E. (1991) Glycoconjugate J. 8, 424–433.

131. Wilzynska, Z., Miller-Podraza, H. and Koscielak, J. (1980) FEBS Lett. 112, 277–279.

132. Karhi, K.K. and Gahmberg, C.G. (1980) Biochem. Biophys. Res. Commun. 622, 344–354.

133. Viitala, J., Karhi, K.K., Gahmberg, C.G., Finne, J., Jarnefeld, J., Myllyla, G. and Krusius, T. (1981) Eur. J. Biochem. 113, 259–265.

134. Schenkel-Brunner, H. (1980) Eur. J. Biochem. 112, 529–534.

135. Gardas, A. and Koscielak, J. (1974) FEBS Lett. 42, 101–104.

136. Koscielak, J., Miller-Podraza, H., Krause, R. and Piasek, A. (1973) Eur. J. Biochem. 71, 9–18.

137. Dejter-Juszynski, M., Harpaz, N. and Flowers, H.M. (1978) Eur. J. Biochem. 83, 363–373.

138. Koscielak, J., Miller-Podraza, H. and Zdebska, E. (1978) Methods Enzymol. 50, 211–216.

139. Finne, J., Krusius, T., Rauvala, H., Kekomaki, R. and Myllyla, G. (1978) FEBS Lett. 89, 111–115.

140. Finne, J., Krusius, T., Rauvala, H. and Jarnefelt, J. (1980) Blood Transfus. Immunohematol. 23, 545–552.

141. Steck, T.L. (1974) J. Cell Biol. 62, 1–19.

142. Tanner, M.J.A. (1988) Biochem. J. 256, 703–712.

143. Mueckler, M., Caruso, C., Baldwin, S.A., Panico, M., Blench, I., Morris, H.E., Allard, W.J., Lienhard, G.E. and Lodish, H.F. (1985) Science, 229, 941–945.

144. Allard, W.J. and Lienhard, G.E. (1985) J. Biol. Chem. 160, 8668–8675.

145. Kannagi, R., Levery, S.B. and Hakomori, S., (1984) J. Biol. Chem. 260, 6410–6415.

146. Stroud, M.R., Levery, S.B., Salyan, M.E.K., Roberts, C.E. and Hakomori, S. (1992) Eur. J. Biochem. 203, 577–586.

147. Rovis, L., Anderson, B., Kabat, E.A., Gruezo, F. and Liao, J. (1973) Biochemistry 12, 5340–5354.

148. Holgersson, J., Breimer, M.E. and Samuelsson, B.E. (1992) APMIS Suppl. 27, 100, 18–27.

149. Triadou, N., Audran, E., Rousset, M., Zweibaum, A. and Oriol, R. (1983) Biochem. Biophys. Acta 761, 231–236.
150. Green, F.R., Greenwell, P., Dickson, L., Griffiths, B., Noades, J. and Swallow, D.M. (1988) In: J.R. Harris (Ed.), Subcellular Biochemistry, Vol. 12. Plenum Press, New York, pp. 119–153.
151. Sodetz, J.M., Paulson, J.P. and McKee, P.A. (1979) J. Biol. Chem. 254, 10754–10760.
152. Childs, R.A., Gregoriou, M., Scudder, P., Thorpe, S.J., Rees, A.R. and Feizi, T.. (1984) EMBO J 3, 2227–2233.
153. Feizi, T. and Childs, R.A. (1987) Biochem. J. 245, 1–11.
154. Amano, J., Straehl, P., Berger, E.G., Kochibe, N. and Kobata, A.J. (1991) J. Biol. Chem. 266, 11461–11477.
155. White, T., Mandel, U., Orntoft, T.E., Dabelsteen, E., Karkov, J., Kubeja, M., Hakomori, S. and Clausen, H. (1990) Biochemistry 29, 2740–2747.
156. Koscielak, J., Zdebska, E., Wilczynska, Z., Miller-Podraza, H. and Dzierzkowa-Borodej, W. (1979) Eur. J. Biochem. 96, 331–337.
157. Fukuda, M.N., Fukuda, M. and Hakomori, S. (1979) J. Biol. Chem. 254, 3700–3703.
158. Fukuda, M.N., Fukuda, M. and Hakomori, S. (1979) J. Biol. Chem. 254, 5458–5465.
159. Marcus, D.M. and Cass, L.E. (1969) Science 164, 553–555.
160. Hakomori, S. and Andrews, H.D. (1970) Biochem. Biophys. Acta 202, 225–228.
161. Hanfland, P. (1978) Eur. J. Biochem. 87, 161–170.
162. Hanfland, P. and Graham, H.A. (1981) Arch. Biochem. Biophys. 210, 383–395.
163. Egge, H. and Hanfland, P. (1981) Arch. Biochem. Biophys. 210, 396–404.
164. Stroud, M.R., Levery, S.B., Nudelman, E.D., Salyan, M.E.K., Towell, J.A., Roberts, C.E., Watanabe, M. and Hakomori, S. (1991) J. Biol. Chem. 266, 8439–8446.
165. Karlson, K.A. (1978) Proc. 15th Congr. Int. Soc. Blood Transfusion, Paris. Abstr. 641.
166. Smith, E.L., McKibbin, J.M., Karlsson, K.A., Pasher, I. and Samuelsson, B.E. (1975) J. Biol. Chem. 250, 6059–6064.
167. Oriol, R., Mollicone, R., Coullin, P., Dalix, A.-M. and Candelier, J.-J. (1992) APMIS Suppl. 27, 100, 28–38.
168. Rauvala, H. (1976) J. Biol. Chem. 251, 7517–7520.
169. Hanisch, F., Uhlenbruck, G. and Dienst, C. (1984) Eur J. Biochem. 144, 467–474.
170. Fukushima, K., Hirota, M., Terasaki, P., Wakishaka, A., Togashi, H., Chia, D., Suyama, N., Fukushi, Y., Nudelman, E. and Hakomori, S. (1984) Cancer Res. 44, 5279–5285.
171. Fukuda, M., Spooncer, E., Oates, J.E., Dell, A. and Glock, J.C. (1984) J. Biol. Chem. 259, 10925–10935.
172. Holmes, E.H., Ostrander, G.K. and Hakomori, S. (1986) J. Biol. Chem. 261, 3737–3743.
173. Kitigawa, H., Nakada, H., Fukui, S., Funakoshi, I., Kawasaki, T., Yamashima, I., Tate, S. and Inagaki, F. (1993) J. Biochem. 114, 504–508.
174. Lemieux, R.U., Hendriks, K.B., Stick, R.V. and James, K. (1975) J. Am. Chem. Soc. 97, 4056–4062.
175. Lemieux, R.U. and Driguez, H. (1975) J. Am. Chem. Soc. 97, 4063–4068.
176. Lemieux, R.U. and Driguez, H. (1975) J. Am. Chem. Soc. 97, 4069–4075.
177. Lemieux, R.U., Bundle, D.R. and Baker, D.A. (1975) J. Amer. Chem. Soc. 97, 4076–4083.
178. Lemieux, R.U., Baker, D.A. and Bundle, D.R. (1977) Can. J. Biochem. 55, 507–512.
179. Jacquinet, J.-C. and Sinay, P. (1976) Tetrahedron 32, 1693–1697.
180. Milat, M. and Sinay, P. (1979) Angewante Chemie Int, Ed. Eng. 18, 464–465.
181. Milat, M. and Sinay, P. (1981) Carbohydrate Res. 92, 183–189
182. Jacquinet, J.-C. and Sinay, P. (1979) Tetrahedron 35, 365–371
183. Jacquinet, J.-C. and Sinay, P. (1979) J. Chem. Soc. Perkins Trans. 1, 314–318.
184. Paulsen, H. and Kolar, C. (1978) Angew. Chem. 90, 823.

185. Augé, C., David, S. and Veyrières, A. (1977) J. Chem. Soc. Chem. Comm. pp. 449–450.

186. Alais, J. and Veyrières, A. (1981) Carbohydrate Res. 92, 310–313.

187. Lemieux, R.U., Bock, K., Delbaere, L.T.J., Koto, S. and Rao, V.S. (1980) Can. J. Chem. 58, 531–6653.

188. Thorgersen, H., Lemieux, R.U., Bock, K. and Meyer, B. (1981) Can. J. Biochem 60, 44–57.

189. Bush, C.A., Yan, Z.Y. and Rao, B.N.N. (1986) J. Am. Chem. Soc. 108, 6168–6173.

190. Nyholm, P., Samuelsson, B.E., Breimer, M. and Pascher, I. (1989) J. Mol. Recognition, 2, 103–113.

191. Hindsgaul, O., Khare, D.P., Bach, M. and Lemieux, R.U. (1985) Can. J. Chem. 63, 2653–2658.

192. Lemieux, R.U., Venot, A.P., Spohr, U., Bird, P., Mandal, G., Morishima, N. and Hindsgaul, O. (1985) Can. J. Chem. 63, 2664–2668.

193. Spohr, U., Hindsgaul, O. and Lemieux, R.U. (1985) Can. J. Chem. 63, 2644–2652.

194. Lemieux, R.U. (1993) In: P.J. Garegg and A.A. Lindberg (Eds.), Carbohydrate Antigens, ACS Symposium Series 519. American Chemical Society, Washington, pp. 5–18.

195. Delbaere, L.T.J., Vandonselaar, M. Prasad, L., Quail, J.W., Wilson, K.S. and Dauter, Z. (1993) J. Mol. Biol. 230, 950–965.

196. Szulman A. E. (1960) J. Exp. Med. 111, 1785–800.

197. Szulman, A.E. (1962) J. Exp. Med. 115, 977–996.

198. Szulman, A.E. (1964) J. Exp. Med. 119, 503–516.

199. Szulman, A.E. (1977) In: J.F. Mohn, R.W. Plunket and R.M. Lambert (Eds.), Human Blood Groups. Karger, Basle, pp. 426–436.

200. Szulman, A.E. (1980) Current Topics in Developmental Biology, 14, 127–145.

201. Rouger, P., Poupon, R., Gane, P., Mallissen, Darnis, F. and Salmon, C. (1986) Tissue Antigens, 27, 78–86.

202. Rouger R., Gane, P., Homberg, J.C. and Salmon, C. (1980) Blood Transfus. Immunohematol. 23, 553–562.

203. Candelier, J.-J., Mollicone, R., Mennessen, B., Bergemer, A.-M., Henry, S., Coullin, and Oriol, R. (1993) Lab. Invest. 69, 449–459.

204. Wherrett, J.R. and Hakomori, S. (1973) J. Biol. Chem. 248, 3046–3051.

205. Breimer, M.E. (1984) Arch. Biochem. Biophys. 228, 71–85.

206. Oriol, R., (1987) Biochem. Trans. 15, 596–599.

207. Mandel, U., Orntoft, T.F., Holmes, E.H., Sorensen, H., Clausen, H., Hakomori, S. and Dabelsteen, E. (1991) Vox Sang. 61, 205–214.

208. Orntoft, T.F. (1992) APMIS Suppl 27. 100, 181–187.

209. Mandel, U., Clausen, H., Vedtofte, P., Sorensen, H. and Dabelsteen, E. (1988) J. Oral Pathol. 17, 506–511.

210. Mollicone, R., Dalix, M.A., Jacobson, A. and Samuelsson, B.E. (1988) Glycoconjugate J. 5, 499–512.

211. Szulman, A.E. and Marcus, D.M. (1973) Lab. Invest. 28, 565–574.

212. Sakamato, J., Furukawa, C., Cordon-Cardo, Yin, B.W.T., Rettig, W.J., Oettgen, H.F., Old and Lloyd, K.O. (1986) Cancer Res. 46, 1553–1561.

213. Orntoft, T.F., Wolf, H., Clausen, H., Hakomori, S. and Dabelsteen, E. (1987) J. Urol. 138, 171–176.

214. Orntoft, T.F., Wolf, H., Clausen, H., Hakomori, S. and Dabelsteen, E. (1988) Lab. Invest. 58, 576–583.

215. Orntoft, T.F., Holmes, E., Johnson, P., Hakomori, S. and Clausen, H. (1991) Blood 77, 1389–1396.

216. Orntoft, T.E., Wolf, H. and Watkins, W.M. (1988) Cancer Res. 48, 4427–4433.

217. Dunstan, R.A., Simpson, M.B., Knowles, R.W. and Rosse, W. F. (1985) Blood 65, 615–619.

382

218. Kelton, J.G. and Bebenek, G. (1985) Transfusion 25, 567–569.
219. Spooncer, E., Fukuda, M., Klock, J.C., Oates, J.E. and Dell, A. (1984) J. Biol. Chem. 259, 4792–4801.
220. Fukuda, M.N., Dell, A., Oates, J.E., Wu, P., Klock, J.C. and Fukuda, M. (1985) J. Biol. Chem. 260, 1067–1082.
221. Civin, C.L., Mirro, J. and Banquerigo, M.L. (1981) Blood 57, 842–845.
222. Tetteroo, P.A.T., Mulder, A., Landsorp, P.M., Zola, H., Baker, D.A., Tisser, F.J. and von dem Borne, A.E.G.K. (1984) Eur. J. Immunol 14, 1089–1095.
223. Brenner, S. (1959) In: G.E.W. Wolstenholme and C.M. O'Connor (Eds.), Ciba Foundation Symp. Biochemistry of Human Genetics. Churchill, London, pp. 304–317.
224. Leloir, L.F. (1972) In: R. Pira and H.G. Pontis (Eds.), Biochemistry of the Glycosidic Linkage. Academic Press, New York, pp. 1–18.
225. Watkins, W.M. (1967) In: J.F. Crow and J.V. Neel (Eds.), Proc. 3rd Int. Congr. Human Genetics, Chicago, 1966. Johns Hopkins Press, Baltimore, pp. 171–187.
226. Ceppellini, R. (1959) In: G.E.W. Wolstenholme and C.M. O'Connor (Eds.), Ciba Foundation Symp. Biochemistry of Human Genetics pp. 242–263.
227. Ginsburg, V. (1972) Adv. Enzymol. 36, 131–149.
228. Schachter, H. and Tilley, C. (1978) In: D.J. Manners (Ed.), International Review of Biochemistry: Biochemistry of Carbohydrates II, Vol. 16. University Park Press, Baltimore, pp. 209–246.
229. Cartron, J.-P. (1978) Proc. 15th Congr. Int. Soc. Blood Transfus., Paris. Libraire Arnette, Paris, pp. 69–86.
230. Johnson, P.H., Yates A. D. and Watkins, W.M. (1981) Biochem. Biophys. Res. Commun. 100, 1611–1618.
231. Johnson, P.H., Donald, A.S.R., Feeney, J. and Watkins, W.M. (1992) Glycoconjugate J. 9, 251–264.
232. Johnson, P.H., Donald, A.S.R. and Watkins, W.M. (1993) Glycoconjugate J. 10, 152–164.
233. Schenkel-Brunner, H., Chester, M.A. and Watkins, W.M. (1972) Eur. J. Biochem. 30, 269–267.
234. Munro, J.R. and Schachter, H. (1973) Arch. Biochem. Biophys. 156, 534–542.
235. Chester, M.A., Yates, A.D. and Watkins, W.M. (1976) Eur. J. Biochem. 69, 583–592.
236. Kyprianou, P., Betteridge, A, Donald, A.S.R. and Watkins, W.M. (1990) Glycoconjugate J. (1990) 7, 573–588.
237. Mulet, C., Cartron, J.-P., Badet, J. and Salmon, C. (1977) FEBS Lett. 84, 74–78.
238. Skacel, P.O. and Watkins, W.M. (1987) Glycoconjugate J. 4, 267–272.
239. Sarnesto, A., Kohlin, T., Thurin, J. and Blaszczyk-Thurin, (1990) J. Biol. Chem. 265, 15067–15075.
240. Yazawa, S. and Furukawa, K. (1983) J. Immunogenetics 10, 349–360.
241. Paulson, J. and Colley, K.J. (1989) J. Biol. Chem. 264, 17615–17618.
242. Le Pendu, J., Cartron, J.-P., Lemieux, R.U. and Oriol, R. (1985) Am. J. Hum. Genet. 37, 749–760.
243. Pacuszka, T. and Koscielak, J. (1974) FEBS Lett. 41, 348–351.
244. Cartron, J.-P., Mulet, C., Bauvois, B., Rahuel, C. and Salmon, C. (1980) FEBS Lett 84, 74–78.
245. Greenwell, P., Ball, G.M. and Watkins, W.M. (1983) Febs Lett. 164, 314–317.
246. Ernst, L.K., Rajan, V.P., Larsen, R.D., Ruff, M.M. and Lowe, J.B. (1989) J. Biol. Chem. 264, 3436–3447.
247. Rajan, V.P., Larsen, R.D., Ajmera, S., Ernst, L.K. and Lowe, J.B. (1989) J. Biol. Chem. 264, 11158–11167.
248. Lowe, J.B., Kelly, R.J., Larsen, L.K. and Ernst, L.K. and Rajan, V.P. (1990) FASEB 1990 Glycoconjugates II Abstract 1372

249. Kelly, R.J., Ernst, L.K., Larsen, R.D., Bryant, J.G., Robinson, J.S. and Lowe, J.B. (1994) Proc. Natl. Acad. Sci. USA 91, 5843–5847.

250. Rouquier, S., Giorgi, D., Bergmann, A., Brandriff, B. and Lennon, G. (1994) Cytogenet. Cell Genet. 66, 70–71.

251. Race, C. and Watkins, W.M. (1972) FEBS Lett. 27, 125–130.

252. Schenkel-Brunner, H., Prohashka, R. and Tuppy, H. (1975) Eur. J. Biochem. 30, 269–277.

253. Mulet, C., Schenkel-Brunner, H., Cartron, J.-C. and Salmon, C. (1978) Proc. 15th Int. Congr. Blood Transfus., Paris. Abstract p. 689.

254. Herron, R., Greenwell, P., Westwood, M.C., Race, A.C., Smith, D.S. and Watkins, W.M. (1980) Vox Sang. 39, 186–194.

255. Etzioni, A., Frydman, M., Pollack, S., Avidor, I., Phillips, M.L., Paulson, J. and Gershoni-Baruch, R. (1992) New Eng. J. Med. 327, 1789–1792.

256. Shen, L., Grollman, E.F. and Ginsburg, V. (1968) Proc. Nat. Acad. Sci. USA 59, 224–230.

257. Chester, M.A. and Watkins, W.M. (1969) Biochem. Biophys. Res. Comm. 34, 835–842.

258. Watkins, W.M. (1966) Science 152, 172–181.

259. Levine, P., Robinson, E., Celano, M., Briggs, O. and Falkinburg, L. (1955) Blood 10, 1100–1108.

260. Betteridge, A. and Watkins, W.M. (1985) Glycoconjugate J. 2, 61–78.

261. Betteridge, A. and Watkins, W.M. (1985) Biochem. Soc. Trans. 13, 1126–1127.

262. Watkins, W.M. and Greenwell, P. (1987) Transplantation Proc. 19, 4413–4415.

263. Le Pendu, J., Oriol, R., Juszczak, Liberge, G., Rouger, P., Salmon, C. and Cartron, J.-P. (1983) Vox Sang. 44, 360–365.

264. Kumasaki, T. and Yoshida, A. (1984) Proc. Nat. Acad. Sci. USA 81, 4193–4197.

265. Sarnesto, A., Kohlin, T., Hindsgaul O., Thurin, J. and Blaszczyk-Thurin, M. (1992) J. Biol. Chem. 267, 2737–2744.

266. Blaszczyk-Thurin, M., Sarnesto, A., Thurin, J., Hindsgaul, O. and Koprowski, H. (1988) Biochem. Biophys. Res. Commun. 151, 100–108.

267. Beyer, T.A., Sadler, E. and Hill, R. (1980) J. Biol. Chem. 255, 5364–5372.

268. Jain, R.K., Pawar, S.M., Chandrasekaran, E.V., Piskorz, C.F. and Matta, K.L. (1993) Bioorgan. Medicinal Chem. Lett. 3, 1333–1338.

269. Beyer, T.A. and Hill, R.L. (1980) J. Biol. Chem. 255, 5373–5379.

270. Basu, S., Basu, M. and Chien, J.L. (1975) J. Biol. Chem. 250, 2956–2962.

271. Hoflack, B., Cacan, R. and Verbert, A. (1978) Eur. J. Biochem. 88, 1–6.

272. Sadler, J.E., Beyer, T.A., Oppenheimer, C.L., Paulson, J.C., Prieels J.-P., Rearick, J.I. and Hill, R.L. (1982) Methods Enzymol. 83, 460–470.

273. Ziderman, D., Gompertz, S., Smith, Z.G. and Watkins, W.M. (1967) Biochem. Biophys. Res. Commun. 29, 56–61.

274. Race, C., Ziderman, D. and Watkins, W.M. (1968) Biochem. J. 107, 733–735.

275. Hearn, V.M., Smith, Z.G. and Watkins, W.M. (1968) Biochem. J. 109, 315–317.

276. Tuppy, H. and Schenkel-Brunner, H. (1969) Vox Sang. 17, 138–142.

277. Cartron, J.-P., Badet, J., Mulet, C. and Salmon, C. (1978) J. Immunogenet. 5, 107–116.

278. Hearn, V.M., Race, C. and Watkins, W.M. (1972) Biochem. Biophys. Res. Commun. 46, 948–956.

279. Kobata A., Grollman, E.F. and Ginsburg, V. (1968) J. Biol. Chem. 245, 1484–1490.

280. Kobata, A., Grollman, E.F. and Ginsburg, V. (1968) Biochem. Biophys. Res. Commun. 32, 272–277.

281. Schachter, H., Michaels, M.A., Crookston, M.C., Tilley, C.A. and Crookston, J. (1971) Biochem. Biophys. Res. Chem. 45, 1010–1018.

282. Kim, Y.S., Perdoma, J., Bella, A. and Nordberg, J. (1971) Proc. Natl. Acad. Sci. USA 68,

384

1753–1756.
283. Sawicka, T. (1971) FEBS Lett. 16, 346–348.
284. Kogure, T. and Furukawa, K. (1976) J. Immunogen. 3, 147–154.
285. Takizawa, H., Kishi, K. and Iseki, S. (1980) Proc. Jpn. Acad. 56, 372–375.
286. Roth, J., Taatjes, D.J., Weinstein, J., Paulson, J., Greenwell, P. and Watkins, W.M. (1986) J. Biol. Chem. 261, 14307–14312.
287. Roth, J., Greenwell, P. and Watkins, W.M. (1988) J. Cell Biol. 46, 105–112.
288. Roth, J. (1995) In: J. Montreuil, H. Schachter and J.F.G. Vliegenthart (Eds.), Glycoproteins. New Comprehensive Biochemistry, 29a. Elsevier, Amsterdam, pp. xxx–xxx.
289. Orci, L., Ravazzola, M., Storch, M.J., Anderson, R.G.W., Vassalli, J.D. and Perrelet, A. (1987) Cell 49, 865–868.
290. Greenwell, P. (1983) Ph.D. Thesis, Council for National Academic Awards, U.K.
291. Carne, L.R. and Watkins, W.M. (1977) Biochem. Biophys. Res. Commun. 77, 700–707.
292. Schachter, H., Michaels, M.A., Crookston, M.C., Tilley, C.A. and Crookston, M.C. and Crookston, J. (1973) Proc. Nat. Acad. Sci. USA 70, 220–224.
293. Topping, M. and Watkins, W.M. (1975) Biochem. Biophys. Res. Commun. 64, 89–96.
294. Cartron, J.-P., Gerbal, A., Badet, J., Ropars, C. and Salmon, C. (1975) Vox Sang. 28, 347–365.
295. Greenwell, P. and Watkins, W.M. (1979) 9th Int. Congr. Biochem Toronto Abstract 03-1-S38, National Research Council of Canada, p. 156.
296. Greenwell, P. and Watkins, W.M. (1989) Biochem. Soc. Trans. 17, 134–135.
297. Watkins, W.M. (1977) In: J.F. Mohn, R.E. Plunkett, R.K. Cunningham and R.M. Lambert (Eds.), Human Blood Groups. Karger, Basle, pp. 134–142.
298. Moreno, C., Lundblad, A. and Kabat, E.A. (1971) J. Exp. Med. 134, 349–457.
299. Mohn, J.F., Cunningham, R.K. and Bates, J.E. (1977) In: J.F. Mohn, R.E. Plunkett, R.K. Cunningham and R.M. Lambert (Eds.), Human Blood Groups. Karger, Basle, pp. 316–325.
300. Kisailus, E.C. and Kabat, E.A. (1978) J. Exp. Med. 147, 830–843.
301. Hakomori, S., Watanabe K. and Laine R. A. (1977) In: J.F. Mohn, R.E. Plunkett, R.K. Cunningham and R.M. Lambert (Eds.), Human Blood Groups. Karger, Basle, 150–163.
302. Watkins, W.M. and Morgan, W.T.J. (1957) Acta Genet. Statist. Med. 6, 521–526.
303. Makela, O., Ruoslahti, E. and Ehnholm, C. (1969) J. Immunol. 102, 763–771.
304. Greenbury, C.L., Moore, D.H. and Nunn, L.A.C. (1963) Immunology 6, 421–433.
305. Economidou, J., Hughes-Jones, N.C. and Gardner, B. (1967) Vox Sang. 12, 321–328.
306. Williams, M.A. and Voak, D. (1972) Brit. J. Haematol. 23, 427–441.
307. Clausen, H., Holmes, E. and Hakomori, S. (1986) J. Biol. Chem. 261, 1388–1392.
308. Etzler, M.E. and Kabat, E.A. (1970) Biochemistry 9, 869–877.
309. Furukawa, K., Mattes, M.J. and Lloyd, K.O. (1985) J. Immunol. 135, 4090–4094
310. Watkins, W.M. and Greenwell, P. (1983) Proc 10th Congr. Soc. Forensic Haematol, Munich. pp. 67–75.
311. Yoshida, A. (1983) Am. J. Hum. Genet. 35, 1117–1125.
312. Fredrick, J., Hunter, J., Greenwell, P., Winter, K. and Gottshall, J.L. (1985) Transfusion 25, 30–33.
313. Greenwell, P., Yates, A.D. and Watkins, W.M. (1979) In: R. Schauer, P. Boer, E. Buddecke, M.-F. Kramer, J.F.G. Vliegenthart and H. Wiegandt (Eds.), Glycoconjugates. Theime, Stuttgart, pp. 268–269.
314. Yates, A.D., Feeney, J., Donald, A.S.R. and Watkins, W.M. (1984) Carbohydr. Res. 130, 251–260.
315. Greenwell, P., Yates, A.D. and Watkins, W.M. (1986) Carbohydr. Res. 149, 149–170.
316. Beck, M.L., Yates, A.D., Hardman, J.T. and Kowalski, M.A. (1987) Transfusion 27, 535.
317. Voak, D. (1987) Rev. Franc. Transfus. Immunohematol. 30, 363–367.

318. Yates, A.D. (1988) In: S.B. Moore (Ed.), Progress in Immunohaematology. American Association of Blood Banks, Arlington, VA. pp. 65–91.

319. Badet, J., Ropars, C., Cartron, J.-P., Doinel, C. and Salmon, C. (1976) Vox Sang. 30, 105–113.

320. Goldstein, J., Lenny, L., Davies, D. and Voak, D. (1989) Vox Sang. 57, 142–146.

321. Yates, A.D. and Watkins, W.M. (1982) Biochem. Biophys. Res. Commun. 109, 958–965.

322. Navaratnam, N., Findlay, J.B.C., Keen, J.N. and Watkins, W.M. (1990) Biochem. J. 271, 93–98.

323. Voak, D., Sonneborn, H. and Yates, A. (1992) Transfusion Med. 2, 119–127.

324. Hagopian, A. and Eylar, E.H. (1968) Arch. Biochem. Biophys. 128, 233–249.

325. Yates, A.D., Greenwell, P. and Watkins, W.M. (1983) Biochem. Soc. Trans. 11, 300–301.

326. Yoshida A., Yamaguchi, Y.F. and Dave, V. (1979) Blood 54, 344–350.

327. Cook, G.A., Greenwell, P., and Watkins, W.M. (1982) Biochem. Soc. Trans. 10, 446–447.

328. Takisawa, H. and Iseki, S. (1982) Proc. Jpn. Acad. 58, B, 65–68.

329. Greenwell, P. and Watkins, W.M. (1987) In: J. Montreuil J., A. Verbert, G. Spik and B. Fournet (Eds.), Proc. 11th Int. Symp Glycoconjugates, Lille, France. Abstract E34.

330. Barbolla, L., Mojena, M. and Bosca, L. (1988) Brit. J. Haematol. 70, 471–476.

331. Barbolla, L., Mojena, M., Cienfuegos, J.A. and Escartin, P. (1988) Brit. J. Haematol. 69, 93–96.

332. Matsue, K., Yasue, S, Matsuda, T., Iwabuchi, K., Ohtsuka, M., Ueda, M., Kondo, K., Shiobara, S., Mori, T. and Koizumi, S. (1989) Exp. Haematol. 17, 827–831.

333. Mojena, M. and Bosca, L. (1989) Blood 74, 1134–1138.

334. Rydberg, L. and Samuelsson, B.E. (1991) Transfusion Medicine 1, 177–182.

335. Kominata, Y., Fujikura, T., Shimada, I., Takisawa, H., Hayashi, K. and Mori, T. (1990) Vox Sang. 59, 116–118.

336. Nagai, N., Dave, V., Muensch, H. and Yoshida, A. (1978) J. Biol. Chem. 253, 380–381.

337. Whitehead, J.S., Bella, A. and Kim, Y.S. (1974) J. Biol. Chem. 249, 3442–3447.

338. Nagai, N., Dave, V., Kaplan, B.E. and Yoshida, A. (1978) J. Biol. Chem. 253, 377–379.

339. Schwyzer, M. and Hill, R.L. (1977) J. Biol. Chem. 252, 2346–2355.

340. Clausen, H., White, T., Stroud, M., Holmes, E., Takio, K., Titani, K., Thim, L. and Hakomori, S. (1989) N. Sharon, H. Lis, D. Duskin and I. Kahane (Eds.), Proc. 10th Int. Symp. Glycoconjugates Jerusalem, pp. 220–221.

341. Clausen, H., White, T., Takio, K., Titani, K., Stroud, M., Holmes, E., Karkov, J., Thim, L. aand Hakomori, S. (1990) J. Biol. Chem. 265, 1139–1145.

342. Navaratnam, N. and Watkins, W.M. (1989) In: N. Sharon, H. Lis, D. Duskin and I. Kahane (Eds.), Proc. 10th Int. Symp. Glycoconjugates, Jerusalem. pp. 225–226.

343. Takeya, A., Hosomi, O. and Ishiura, M. (1990) J. Biochem. Tokyo 107, 360–368.

344. Yamamoto, F., Marken, J., Tsuji, T., White, T., Clausen, H., and Hakomori, S. (1990) J. Biol. Chem. 265, 1146–1151.

345. Yamamoto, F., McNeill, P.D., Yamamoto, M., Hakomori, S., Bromilow, I.M. and Duguid, J.K.M. (1993) Vox Sang. 63, 175–178.

346. David, L., Leitao, D., Sobrinho-Simoes, M., Bennett, E.P., White, T., Mandel, U., Dabelsteen, E. and Clausen, H. (1993) Cancer Res. 53, 5494–5500.

347. Johnson, P.H. and Hopkinson, D.A. (1992) Human Mol. Genet. 1, 341–344.

348. Ugozzoli, L. and Wallace, R.B. (1992) Genomics 12, 670–674.

349. Yamamoto, F. and Hakomori, S. (1990) J. Biol. Chem. 265, 19257–19262.

350. Yamamoto, F., McNeill, P.D. and Hakomori, S. (1992) Biochem. Biophys. Res. Commun. 187, 366–374.

351. Yamamoto, F., McNeill, P.D., Yamamoto, M., Hakomori, S., Harris, H., Judd, W.J. and Davenport, P.D. (1993) Vox Sang. 64, 116–119.

352. Yamamoto, F., McNeill, P.D., Kominato Y., Yamamoto, M., Hakomori, S., Ishimoto, S., Nishida, S., Shima, M. and Fujimura, Y. (1993) Vox Sang. 64, 120–123.

353. Yamamoto, F., McNeill, P.D., Yamamoto, M., Hakomori, S. and Harris, T. (1993) Vox Sang. 64, 171–174.
354. Greenwell, P. and Watkins, W.M. (1987) In: J. Montreuil, A. Verbert, G. Spik and B. Fournet (Eds.), Proc. 11th Int. Symp. Glycoconjugates, Lille. Abstract E. 34.
355. Dorscheid, D., Friedlander, P. and Price, G. (1991) (Abstract 4. 19) Glycoconjugate J. 8, 151.
356. Watkins, W.M. (1968) 21st John Gibson II Lecture, College of Physicians and Surgeons, Columbia University, New York.
357. Badet J., Ropars, C. and Salmon, C. (1978) J. Immunogenet. (1978) 5, 221–231.
358. Hirschfield, J. (1977) Vox Sang. 33, 286.
359. Yoshida, A., Yamaguchi, H. and Okubo, Y. (1980) Am. J. Hum. Genet. 32, 332–338.
360. Badet, J., Ropars, C., Cartron, J.-P. and Salmon, C. (1974) Biomedicine 21, 230–232.
361. Landsteiner, K. and Miller, C.P. (1925) J. Exp. Med. 42, 863–872.
362. Eto, T., Itchikawa, Y., Nishimura, K., Ando, S. and Yamakawa, T. (1968) J. Biochem. Tokyo. 64, 205–212.
363. Galili, U. (1988) Transf. Med. Rev. 2, 112–121.
364. Galili, U., Rachmilewich, E.A. and Peleg, A. (1984) J. Exp. Med. 160, 1519–1531.
365. Karlsson, E., Cairns, T., Holgersson, J., Welsh, K. and Samuelsson, B. (1993) Glycoconjugate J. 10, 297.
366. Breimer, E., Hanson, G.C., Karlson, K.-A., Leffler, T., Pimlott, W. and Samuelsson, B.E. (1979) Biomed. Mass Spectrosc. 6, 231–241.
367. Betteridge, A. and Watkins, W.M. (1983) Eur. J. Biochem. 132, 29–35.
368. Blanken, W.M. and van den Eijnden, D.H. (1985) J. Biol. Chem. 260, 12927–12934.
369. Elices, M.J., Blake, D.A. and Goldstein, I.J. (1986) J. Biol. Chem. 261, 6064–6072.
370. Larsen, R.D., Rajan, V.P., Ruff, M.M., Kukowska-Latallo, J., Cummings, R.D. and Lowe, J.B. (1989) Proc. Nat. Acad. Sci. USA 86, 8227–8231.
371. Joziasse, D.H., Shaper, J.H., van den Eijnden, D.H., Van Tunen, A.J. and Shaper, N.L. (1989) J. Biol. Chem. 264, 14290–14297.
372. Larsen, R.D., Rivera-Marrero, C.A., Ernst, L., Cummings, R. and Lowe, J.B. (1990) J. Biol. Chem. 265, 7055–7061.
373. Joziasse, D.H., Shaper, J.H., Jabs, E.W. and Shaper, N.L. (1991) J. Biol. Chem. 266, 6991–6998.
374. Shaper, N.L., Lin, S.-P., Joziasse, D.H., Kim, D.Y. and Yang-Feng, T. (1992) Genomics 12, 613–615.
375. Kominato, Y., McNeill, P.D., Yamamoto, M., Russel, M. Hakomori, S. and Yamamoto, F. (1992) Biochem. Biophys. Res. Commun. 189, 154–164.
376. Grollman, E.F., Kobata, A. and Ginsburg, V. (1969) J. Clin. Invest. 48, 1489–1494.
377. Yazawa, S. (1976) Kitakanto Igato 26, 203–214.
378. Mollicone, R., Gibaud, A., Francis, A., Ratcliffe, M. and Oriol, R. (1990) Eur. J. Biochem. (1990) 191, 169–176.
379. Prieels, J.-P., Monnom, D., Dolmans, M., Beyer, T.A. and Hill, R.L. (1981) 256, 10456–10463.
380. Johnson, P.H. and Watkins, W.M. (1982) Biochem. Soc. Trans. 10, 1119–1120.
381. Johnson, P.H. and Watkins, W.M. (1987) In: J. Montreuil, A. Verbert, G. Spik and B. Fournet (Eds.), Proc. 9th Int. Symp. Glycoconjugates, Lille, France Abstract E 107.
382. Watkins, W.M., Greenwell, P., Yates, A.D. and Johnson, P.H. (1988) Biochemie 70, 1597–1641.
383. Johnson, P.H. and Watkins, W.M. (1992) Glycoconjugate J. 9, 241–249.
384. Johnson, P.H., Donald, A.S.R., Feeney, J. and Watkins, W.M. (1992) Glycoconjugate J. 9, 251–264.
385. Khare, D.P., Hindsgaul, O. and Lemieux, R.U. (1985) Carbohydrate Res. 136, 285–308.

386. Eppenberger-Castori, S., Lotscher, H. and Finne, J. (1989) Glycoconjugate J. 6, 101–114.
387. Johnson, P.H. and Watkins, W.M. (1985) Biochem. Soc. Trans 13, 1119–1120.
388. Johnson, P.H., Donald, A.S.R. and Watkins, W.M. (1993) Glycoconjugate J. 10, 152–164.
389. Crawley, S.C., Hindsgaul, O., Ratcliffe, R.M., Lamontagne, L.R. and Palcic, M.M. (1989) Carbohydr. Res. 193, 249–256.
390. De Vries, T.H. and van den Eijnden, D.H. (1992) Histochem. J. 24, 761–770.
391. Lowe, J.B., Stoolman, L.M., Nair, R.P., Larsen, R.D., Berhend, T.L. and Marks, R.M. (1990) Cell 63, 475–484.
392. Dumas, D.P., Itchikawa, Y., Wong, C.H., Lowe, J.B. and Nair, R.P. (1991) Bioorg. Med. Chem. Lett. 8, 25–428.
393. Weston, B.W., Nair, R.P., Larsen, R.D. and Lowe, J.B. (1992) J. Biol. Chem. 267, 4152–4160.
394. Howard, D.R., Fukuda, M., Fukuda, M.N. and Stanley, P. (1987) J. Biol. Chem. 262, 16830–16837.
395. Lowe, J.B., Kukowska-Latallo, J.F., Nair, R.P., Larsen, R.D. and Marks, R.M. (1991) J. Biol. Chem. 266, 17467–16477.
396. Kumar R., Potvin, B., Muller, W.A. and Stanley, P. (1991) J. Biol. Chem. 266, 21777–21783.
397. Goeltz, S.E., Hession, C., Goff, D., Griffiths, Tizard, R., Newman, B., Chi-Rosso, G. and Lobb, R. (1990) Cell 63, 1349–1356.
398. Koda, Y., Kimura, H. and Mekada, E. (1993) Blood 82, 2915–2919.
399. Nishihara, S., Yasawa, S., Iwasaki, H., Nakazano, M., Kudo, T., Ando, T. and Narimatsu, H. (1993) Biochem. Biophys. Res. Commun. 196, 624–631.
400. Elmgren, A., Rydberg, L. and Larson, G. (1993) Biochem. Biophys. Res Commun. 196, 515–520.
401. Packuska, T. and Koscielak, J. (1976) Eur. J. Biochem. 64, 499–506.
402. Misakos, A. and Hanisch, F.G. (1989) Biol. Chem. Hoppe-Seyler 370, 239–243.
403. Johnson, P.H. and Watkins, W.M. (1989) In: N. Sharon, H. Lis, D. Duksin and I. Kahane (Eds.), Proc. 10th Int. Symp. Glycoconjugates, Jerusalem pp. 214–215.
404. Macher, B., Holmes, E., Swieldler, S.J., Stults, C.L.M. and Srnka, C.A. (1991) Glycobiology 1, 577–584.
405. Watkins, W.M., Skacel, P.O. and Johnson, P.H. (1993) In: P.J. Garegg and A.A. Lindberg (Eds.), Carbohydrate Antigens, ACS Symposium Series 519. American Chemical Society, Washington, pp. 34–63.
406. Sarnesto, A., Kohlin, T., Hindsgaul, O., Vogele, K., Blaszczyk-Thurin, M. and Thurin, J. (1992) J. Biol. Chem. 267, 2745–2752.
407. Jezequel-Cuer, M., N'Guyen-Cong, H., Biou, D. and Durand, G. (1993) Biochim. Biophys. Acta 1157, 252–258.
408. Johnson, P.H. (1988) PhD Thesis, Council for National Academic Awards, U. K.
409. De Vries, T., Norberg, T., Lonn, H. and van den Eijnden, D.H., (1993) Eur. J. Biochem. 216, 769–777.
410. Johnson, P.H. and Watkins, W.M. (1987) Biochem Soc. Trans. 15, 396.
411. Easton, E.W., Schiphorst, W.E.C.M., van Drunen, E., van der Schoot, C.E. and van den Eijnden, D.H. (1993) Blood 81, 425–428.
412. Skacel, P.O., Edwards, A.J., Harrison, C.T. and Watkins, W.M. (1991) Blood 78, 1452–1460.
413. Chou, T.H., Murphy, C. and Kessel, D. (1977) Biochem. Biophys Res. Commun. 74, 1001–1006.
414. Muramatsu, H., Kamada, Y. and Murmatsu, T. (1980) Eur. J. Biochem. 157, 71–75.
415. Foster, C.S., Gillies, D.R.B. and Glick, M.C. (1991) J. Biol. Chem. 266, 3526–3531.
416. Prieels, J.-P., Monnom, D., Perraudin, J.-P., Finne, J. and Burger, M. (1983) Eur. J. Biochem. 130, 347–351.
417. Holmes, E.H., Ostrander, G.K. and Hakomori, S. (1986) J. Biol. Chem. 261, 3737–3743.

418. Greenwell, P., Johnson, P.H., Edwards, J.H., Reed, R.M., Moores, P.P., Bird, A., Graham, H.A. and Watkins, W.M. (1986) Blood Transfus. Imunnohaemtol. 29, 233–249.
419. Caillard, T., le Pendu, J., Ventura, M., Mada, M., Rault, G., Mannoni, P. and Oriol, R. (1988) Exp. Clin. Immunol. 5, 15–23.
420. Oriol, R., Mollicone, R., Masri, R., Whan, I. and Lovric, V.A. (1991) Glycoconjugate J. Abstract 4. 7, 8, 147.
421. Johnson, P.H., Skacel, P.O., Greenwell, P. and Watkins, W.M. (1989) Biochem. Soc. Trans. 17, 133.
422. Philips, M.L., Nudelman, E., Gaeta, F.C.A., Perez, M., Singhal, A.K., Hakomori, S. and Paulson, J.C. (1990) Science 250, 1130–1132.
423. Walz, G., Arrufo, A., Kolanus, W., Bevilaqua, M. and Seed, B. (1990) Science 250, 1132–1135.
424. Tiemeyer, M., Swiedler, S.J., Ishihara, M., Moreland, M., Schweingruber, H., Hirtzer, P. and Brandley, B.K. (1991) Proc. Natl. Acad. Sci. USA 88, 1138–1142.
425. Springer, T.A. and Lasky, L.A. (1991) Nature 349, 196–197.
426. Reguine, I., James, M.R., Richard III, C.W., Mollicone, Seawright, A., Lowe, J.B., Oriol, R. and Coullin, P. (1994) Cytogenet. Cell Genet, 66, 104–106.
427. Weston, B.W., Smith, P.L., Kelly, R.J. and Lowe, J.B. (1992) J. Biol. Chem. 267, 24575–24584.
428. Nishihara, S., Nakazato, M., Kudo, T., Kimura, H., Ando, T. and Narimatsu, H. (1993) Biochem. Biophys. Res. Commun. 190, 42–46.
429. Kosdin, K.L. and Bowen, B.R. (1992) Biochem. Biophys. Res. Commun. 187, 152–157.
430. Piller, F. and Cartron, J.-P. (1983) J. Biol. Chem. 258, 12293–12299.
431. Yates, A.D. and Watkins, W.M. (1983) Carbohydr. Res. 120, 251–268.
432. Basu, M. and Basu, S. (1984) J. Biol. Chem. 259, 12557–12562.
433. Piller, F., Cartron, J.-P., Maranduba, A., Veyrières, A., Leroy, Y. and Fournet, B. (1984) J. Biol. Chem. 259, 13385–13390.
434. Van den Eijnden, D.H., Koederman, A.H.L. and Schiphorst, W.E.C.M. (1988) J. Biol. Chem. 263, 12461–12471.
435. Gu, J., Nishikawa, A, Fujii, Gasa, S. and Taniguchi, N. (1992) J. Biol. Chem. 267, 2994–2999.
436. Bierhuizen, M.F.A. and Fukuda, M. (1992) Proc. Natl. Acad. Sci. USA 89, 9326–9330.
437. Mourant, A.E., Kopec, A.C. and Donaniewska-Sobczar, K. (1978) Blood Groups and Diseases. Oxford University Press, Oxford.
438. Hakomori, S. (1985) Cancer Res. 45, 2405–2414.
439. Kuhns, W.J. (1980) In: Contemporary Haematology/Oncology, Vol. 1. Plenum, New York, pp. 149–200.
440. Fukuda, M. (1985) Biochem. Biophys. Acta 780, 119–150.
441. Masamune, H., Kawasaki, H., Abe, S., Oyama, K. and Yamaguchi, Y. (1958) Tohoku J. Med. 68, 81–91.
442. Stellner, K., Hakomori, S. and Wagner, G.A. (1973) Biochem. Biophys. Res. Commun. 55, 439–445.
443. Szulman, A.E. (1966) Ann. Rev. Med. 17, 307–322.
444. Piller, F., Schenkel-Brunner, H., Tuppy, H. (1979) In: R. Schauer, M.F. Kramer, J.F.G. Vliegenthart and H. Wiegandt (Eds.), Proc. 5th Int Symp. Glycoconjugates, Kiel. Thieme, Stuttgart, pp. 639–640.
445. Orntoft, T.F., Greenwell, P., Clausen, H. and Watkins, W.M. (1991) Gut 32, 287–293.
446. Hakkinen, I. (1970) J. Natl. Cancer Inst. 44, 1183–1193.
447. Denk, H., Tappeiner, G., Davidovits, A., Eckerstorfer, R. and Holzner, J.H. (1974) J. Natl. Cancer Inst. 53, 933–942.
448. Kapadia, A., Feizi, T., Jewell, D., Keeling, J. and Slavin, G. (1981) J. Clin. Pathol. 34, 320–337.
449. Finan, P.J., Wight, D.G.D., Lennox, E.S., Sacks, S.H. and Bleehen, N.M. (1983) J. Natl. Cancer

Inst. 70, 679–685.

450. Hakomori, S., Wang, S.M. and Young, W.W. (1977) Proc. Natl. Acad. Sci. USA 74, 3023–3027.

451. Yokota, M., Warner, G. and Hakomori, S. (1981) Cancer Res. 41, 4185–4190.

452. Breimer, M.E. (1980) Cancer Res. 40, 897–908.

453. Hattori, H., Uemura, K., Ogata, H., Katsuyama, T. and Kanfer, J.N. (1987) Cancer Res. 47, 1968–1972.

454. Hattori, H., Uemura, K., Ishihara, H. and Ogata, H. (1992) Biochim. Biophys. Acta 1125, 21–27.

455. Clausen, H., Hakomori, S., Graem, N. and Dabelsteen, F. (1986) J. Immunol. 136, 326–330.

456. David, L., Leitao, D., Sobrinho-Somoes, M., Bennett, E.P., White, T., Mandel, U., Dabelsteen, E. and Clausen, H. (1993) Cancer Res. 53, 5494–5500.

457. Clausen, H., Personal communication.

458. Salmon, C. (1978) In: Plenary sessions of 15th Congr. Int. Soc. Blood Transfus. Paris. Libraire Arnette, Paris, pp. 37–50.

459. Kuhns, W.J. and Primus, F.J. (1985) In: Progress in Clinical Biochemistry and Medicine Vol. 2. Springer-Verlag, Berlin, pp. 51–95.

460. Kuhns, W.J., Oliver, R., Watkins, W.M, and Greenwell, P. (1980) Cancer Res. 40, 268–275.

461. Skacel, P. and Watkins, W.M. (1988) Cancer Res 48, 3998–4001.

462. Kessel, D., Shah-Reddy, L., Mirchandani, I., Khilanani, P. and Chou, T. (1980) Cancer Res. 40, 3576–3578.

463. Hakomori, S., Nudelman, E., Levery, S. and Kannagi, R. (1984) J. Biol. Chem. 259, 4672–4680.

464. Fukushima, K., Hirota, M., Terasaki, P.L., Wakisaka, A. Togashi, H., Chia, D., Suyama, N., Fukushi, Y., Nudelman, E. and Hakomori, S. (1984) Cancer Res. 44, 5279–5285.

465. Itzkowitz, S.H., Yuan, M., Fukushi, Y., Palekar, A., Phelps, P. C., Shamsuddin, A.M., Trump, B.F. and Hakomori, S. and Kim, Y.S. (1986) Cancer Res. 46, 2627–2632.

466. Nilsson, O. (1992) APMIS Suppl. 27, 100, 149–161.

467. Jezequel-Cuer, M., Dalix, A.M., Flejou, J.-F., and Durand, G. (1992) Liver 12, 140–146.

468. Nudelman, E., Levery, S., Kaizu, T. and Hakomori, S. (1986) J. Biol. Chem. 261, 11247–11253.

469. Fukushi, Y., Nudelman, E., Levery, S.B., Hakomori, S. and Rauvala, H. (1984) J. Biol. Chem. 259, 10511–10517.

470. Matsushita, Y., Cleary, K.R., Ota, D.M., Hoff, S.D. and Irimura, T. (1990) Lab. Invest. 63, 780–791.

471. Infusa, H., Kojima, N., Yasutomi, M. and Hakomori, S. (1991) Clin. Exp. Metastasis 9, 245–257.

472. Matsushita, Y., Nakamori, S., Seftor, E.A., Hendrix, M.J.C. and Irimura, T. (1991) Exp. Cell Res. 196, 20–25.

473. Dohi, T., Hashigawa, M., Yamamoto, S., Morita, H. and Oshima, M. (1994) Cancer 73, 1552–1561.

474. Yasawa, S., Nakamura, J., Asao, T., Nagamachhi, Y., Sagi, M., Matta, K.L., Tachikawa, T. and Akamatsu, M. (1993) Jpn. J. Cancer Res. 84, 989–995.

475. Giblett, E.R. (1969) Genetic Markers of Human Blood. Blackwells Scientific Publications, Oxford, p. 314.

476. Koscielak, J. (1986) Glycoconjugate J. 3, 95–108.

477. Feizi, T. (1993) Curr. Opin. Struct. Biol. 3, 701–710.

478. Lowe, J.B., Stoolman, L.M., Nair, R.P., Berhend, T.L. and Marks, R.M. (1990) Cell 63, 474–484.

479. Berg, E.L., Robinson, M.K., Mansson, O., Butcher, E.C. and Magnani, J.L. (1991) J. Biol.

Chem. 266, 14869–14872.

480. Takada, A., Ohmori, K., Takahashi, N., Tsuyoka, K., Yago, A., Zenita, K., Hasegawa, A. and Kannagi, R. (1991) Biochem. Biophys. Res. Commun. 179, 713–719.

481. Zhou, Q., Moore, K.L., Smith, D.F., Varki, A., McEver, R.P. and Cummings, R.D. (1991) J. Cell Biol. 115, 557–564.

482. Brandley, B.K., Sweidler, S.J. and Robbins, P.W. (1990) Cell 63, 861–863.

483. Mulligan, M.S., Paulson, J.C., De Frees, S., Zheng, Z.-L., Lowe, J.B. and Ward, P.A. (1993) Nature 384, 149–151.

484. De Vries, T., van den Eijnden, D.H., Schultz, J.E. and O'Neill, R.A. (1993) FEBS Lett. 330, 243–248.

485. Ichikawa, Y., Halcomb, R.L. and Wong, C. (1994) Chem. Brit. Feb. 1994, 117–121.

486. Kashem, M.A., Wlasichuk, K.B., Gregson, J.M. and Venot, A.P. (1993) Carbohydrate Res. 250, 129–144.

487. Nikrad, P.V., Kashem, M.A., Wlasichuk, K.B., Alton, G. and Venot, A.P. (1993) Carbohydrate Res. 250, 145–160.

488. Solter, D. and Knowles, B.B. (1978) Proc. Nat. Acad. Sci. USA. 75, 5565–5569.

489. Gooi, H.C., Feizi, T., Kapadia, A., Knowles, B.B., Solter, D. and Evans, M.J. (1981) Nature 292, 156–158.

490. Eggens, I., Fenderson, B.A., Toyokuni, T., Dean, B., Stroud, M.R. and Hakomori, S. (1989) J. Biol. Chem. 264, 9476–9484.

491. Fenderson, B.A., Eddy, E.M. and Hakomori, S. (1990) BioEssays 12, 173–179.

492. Nguyen, M., Strubel, N.A. and Bishoff, J. (1993) Nature 365, 267–269.

493. Howzawa, K., Wataya, H., Takasaka, T., Fenderson, B.A. and Hakomori, S. (1993) Glycobology J. 3, 47–55.

494. Sasaki, K., Kurata, K., Funayama, K., Nagata, M., Watanabe, E., Ohta, S., Hanai, N. and Nishi (1994) J. Biol. Chem. 269, 14730–14737.

495. Natsuka, S., Gersten, K.M., Zenita, K., Kannagi, R. and Lowe, J.B. (1994) J. Biol. Chem. 269, 16789–16794.

496. Ito, K., Handa, K. and Hakomori, S. (1994) Glycoconjugate J. 11, 232–237.

J. Montreuil, H. Schachter and J.F.G. Vliegenthart (Eds.), *Glycoproteins*
© 1995 Elsevier Science B.V. All rights reserved

Biosynthesis

6. The role of polypeptide in the biosynthesis of protein-linked oligosaccharides

RAYMOND T. CAMPHAUSEN, HSIANG-AI YU and DALE A. CUMMING

Small Molecule Drug Discovery, Genetics Institute, 87 Cambridge Park Drive, Cambridge, MA 02140, USA

1. Introduction

There is now ample evidence demonstrating that glycoconjugates are messengers bearing information important to a number of fundamental biological events [1–3] and clear expectation that continued study will elucidate new biological roles for this class of biopolymers. As our understanding of the biology of glycoconjugates has grown, considerable attention has been directed to questions focused upon the regulation of glycoconjugate biosynthesis. This has been especially true for glycoproteins where three essential elements have been identified in defining and regulating the structure of oligosaccharides covalently linked to proteins. In her seminal paper on the regulation of protein glycosylation, Hubbard [4] delineates, with eloquence and simplicity, these elements which together comprise a system capable of complex regulation: "... not only protein structure and cell type, but also growth status can affect the course of N-glycosylation." The present discussion is focused upon the central role of one of these elements, the polypeptide chain itself, in regulating its glycosylation.

It is our central tenet that information is encoded within the polypeptide sequence, typically in a nonlinear fashion, that specifies which class(es) of oligosaccharide structures are found at a given glycosylation site. This information is 'decoded' during biosynthesis through mechanisms involving protein–protein and protein–oligosaccharide interactions between the glycosylation processing apparatus of the host cell and the nascent glycoprotein. We will review the experimental data which supports this tenet and discuss several hypotheses which have appeared in the literature as possible 'decoding' mechanisms. We will conclude by suggesting that all of these hypotheses are but elements describing a spectrum of interaction modes, all unified by the hypothesis that the local 3D structure about a glycosylation site regulates the interaction with the host processing apparatus.

1.1. Historical perspective

Protein glycosylation is a complex biosynthetic process that is strictly regulated. The simple observation that the oligosaccharide structures observed on any given protein are but a subset of all possible oligosaccharides that could be synthesized by the cell is, perhaps, the clearest demonstration of this point. However, it was not many years ago that the prevailing view of protein glycosylation was that of a random, perhaps kinetically driven process without significant functional consequence and certainly capable of generating a dazzling degree of glycoprotein heterogeneity. At the time, various suggestions were put forward to explain this apparent behavior including differential transit times of nascent glycoproteins through various subcellular compartments of the secretory pathway.

Extensive research has produced a dramatic change in the prevailing view. This change was catalyzed not only by advances in the techniques used to isolate, fractionate, and characterize protein-linked glycans, but also by the application of techniques from fields such as molecular and cellular biology. As a result, it is now clear that the glycosylation of a protein is, under defined conditions, a non-random and reproducible process. At the same time, it is also a process capable of great plasticity in response to, for example, hormonal changes or the onset of disease. This plasticity no doubt relates to the functional roles of protein-linked glycans but it can (and did) shroud just how this pathway is regulated.

Examination of the biosynthetic pathway for N-glycosylation [5] shows that there is a considerable investment by the cell in regulating this process. For example, the pathway is ordered and thus deterministic. In its most elaborated form, the pathway involves disassembly of a preassembled oligosaccharide (cotranslationally linked to the nascent protein) and subsequent cell type-specific reconstruction of that oligosaccharide following protein-specific instructions. The removal or addition of specific carbohydrate residues can affect subsequent processing steps. Thus, the number of different glycan structures possible at the end of the pathway is strongly dependent upon the nature and occurrence of intermediate processing steps.

Consideration of the molecular components of the processing apparatus (glycohydrolases and glycosyltransferases) also illustrates the extensive cellular investment in the pathway. For example, the addition of antennal GlcNAc residues in N-linked glycans is catalyzed by no fewer than five GlcNAc-transferases, each a separate gene product despite the fact that the glycosidic linkage formed (e.g. GlcNAc[β1–2]) by these transferases is identical for some members of the set. Moreover, different processing enzymes possess distinct subcellular localization sites, allowing physical and chronological segregation of processing events. The specificity, maintenance, and regulation of the processing apparatus support the contention that protein glycosylation is a highly structured and ordered process.

1.2. Terms and concepts

The definition of a few terms employed in this review is warranted. A 'structural class' of an oligosaccharide refers to the structural subsets of both O- and N-linked oligosaccharides, which are themselves most frequently defined by the branching pattern emanating from 'core' structures. For example, structural classes of complex-type N-glycans include diantennary, triantennary, tetraantennary (and so forth) oligosaccharides, these distinctions made without regard to the specific monosaccharide residues that adorn each antenna.

There are a set of characteristics associated with protein glycosylation which, as originally detailed by Dwek and colleagues [6], deserve some mention here. First, different proteins synthesized by the same cell can be glycosylated very differently. Individual polypeptide chains frequently contain multiple sites of protein glycosylation and, at any given site, multiple oligosaccharide structures are often observed. This latter feature is known as 'site heterogeneity'. Most importantly, the site heterogeneity observed at a specific locale on a given protein is, under constant conditions, defined and reproducible.

Finally, two other terms are worthy of note. 'Site occupancy' refers to the actual utilization of a potential site of glycosylation within a population of a glycoprotein. Site occupancy can vary dramatically and is a powerful means of modulating the functional and physiochemical attributes of a glycoprotein [7]. The term 'site-specific glycosylation' derives from the observation that, within a single protein with multiple sites of glycosylation, the type and distribution of attached oligosaccharide structures at each site can vary dramatically and can even be mutually exclusive. The potential significance of site-specific glycosylation has been discussed elsewhere [8].

1.3 Other participating factors

In addition to the polypeptide chain of the nascent glycoprotein, two other factors are important determinants in specifying the structural class of the resulting glycan chains. These factors are the glycosylation phenotype of the host cell and environmental (physiological) factors [9,10]. The glycosylation phenotype of the host is defined by the complement of glycosylation processing enzymes (glycohydrolases and glycosyltransferases) expressed by the cell. It is well known that different cell types possess distinct sets of processing enzymes and that the same polypeptide chain synthesized in different cells can bear glycans with distinct primary structures. Thus, CHO cells do not normally attach sialic acid to N-linked glycoconjugates in an α-2,6-linkage, since the necessary sialyltransferase(s) is not expressed, while BHK cells are capable of forming this linkage [11]. Similarly, recombinant proteins produced in murine L cells bear glycan chains absent the 'bisecting' GlcNAc observed in the native protein [12].

It should be borne in mind that the glycosylation phenotype of a cell is also subject to sometimes dramatic changes due to environmental (physiological) factors. Those factors demonstrated to impact the glycosylation of proteins include hormones, pathogens, metabolites, and growth factors. This area has been extensively reviewed by others and the reader is referred to two monographs on the subject [9,13]. The sensitivity of protein glycosylation to environmental factors may provide an efficient cellular mechanism for modulating the biological attributes of a protein in response to an altered environment, or at a minimum, serve as a convenient and sensitive indicator of changes in cellular phenotype. Most importantly, it is clear that it is the interplay between a specific polypeptide sequence and the glycosylation phenotype of a particular cell type, as modulated by any environmental factors, which specifies the identity and spectrum of attached glycans in a mature protein.

2. Evidence that the polypeptide chain specifies attached oligosaccharide structure

It is now clear that primary, secondary, tertiary, and quaternary structural characteristics of proteins are involved in specifying the structures of the glycans covalently linked to proteins (for example, see Dahms and Hart, 1986 [14]; Troesch et al., 1990 [15]). The following discussion seeks to illustrate the available evidence which establishes the role of the protein chain in regulating attached glycan structures.

2.1 Protein structural features

Primary amino acid consensus sequences are the simplest example of instructions encoded within the polypeptide chain which can specify glycosylation features. Perhaps the most well known of these is the tripeptide consensus sequence Asn-X-Ser/Thr which is a necessary, but not sufficient, condition for N-glycosylation of proteins [5,16]. Within this sequon, 'X' can be any amino acid save Pro. In addition, it appears that Cys residues can substitute for Ser/Thr in certain cases [17]. It is believed that the Ser and Thr hydroxyl side chains are important determinants of the consensus tripeptide as recognized by the ER-resident oligosaccharide transferase which catalyzes the addition of a precursor glycan to the nascent peptide chain [18].

In contrast, there is no known consensus sequence for O-linked glycans, the other major class of protein-linked oligosaccharides, although several studies have indicated that sequences proximal to O-glycosylation sites do impact site utilization [19,20]. Recently, a transferase catalyzing the first step in O-glycan biosynthesis was purified to homogeneity from porcine submaxillary glands [21]. Utilizing mucin-like peptide acceptor sequences (modeling the addition of GalNAc to Thr) and a peptide fragment of EPO (to model N-acetylgalactosaminylation of Ser), Wang et al. [21] demonstrated a significant effect upon glycosylation by amino acid substitutions around the site of glycosylation. In particular, the utilization of serine residues in model peptides was found to be sensitive to the length or identity of sequences three or more residues away coupled with a high degree of dependency upon the presence of flanking Ala-Ala sequences on both sides of the Ser residue. In contrast, the utilization of Thr residues was far less sensitive to flanking residues with X-X-Thr-Thr-X-X containing peptides being the best substrates.

Few other consensus sequences for protein glycosylation are known. The differential glycosylation of the pituitary gonadotropic hormones (i.e. LH, FSH, TSH) appears to be regulated by a polypeptide-resident recognition sequence present in some members. Thus, these heterodimeric glycoproteins can bear N-linked glycan chains terminally substituted with 4-O-sulfated GalNAc residues or, in the absence of this polypeptide recognition sequence, the more typical sialylated Gal residues. Baenziger and colleagues [22] have demonstrated that the biosynthesis of sulfated N-linked chains requires the recognition by a GalNAc transferase and the subsequent action of a specific sulfotransferase [GalNAc(β1–4)-GlcNAc(β1–2)-Man(α4)-sulfotransferase]. A tripeptide sequence (Pro-X-Arg/Lys) located 6–9 residues upstream of the N-glycosylation site appears to constitute the required GalNAc-transferase recognition sequon.

At the other end of the structure spectrum, protein quaternary structure has been demonstrated to influence site-specific β-chain glycosylation of two sets of integrins [14,15]. Integrins are heterodimeric cell adhesion molecules which can be grouped into families of related molecules on the basis of common β-subunits paired with distinct α-chains. Formation of stable, noncovalent α/β-heterodimers is a relatively early biosynthetic event, occurring prior to any golgi-mediated glycan processing. In both studies [14,15], the glycosylation of the β-chain of two members of a specific integrin family were evaluated. To illustrate the essential points, we shall focus our attention on one of these reports.

In the study from Dahms and Hart [14], the site-specific glycosylation of two β_1 integrins, LFA-1 ($\alpha_L\beta_1$) and MAC-1 ($\alpha_M\beta_1$), was investigated. The macrophage-like line P388D$_1$ synthesizes both of these β_1 integrins concurrently, thus defining a system which allows evaluation of β-chain glycosylation when the polypeptide chain and the cellular glycosylation apparatus are identical. At least five sites of glycosylation were identified on the β-chains and, employing a variety of biochemical and structural techniques, at least four of these sites were shown to be differentially glycosylated between LFA-1 and MAC-1 β-chains. Therefore, differential association of subunits can influence site-specific glycosylation of proteins even when the peptide chain and cell type remain the same.

2.2. Comparison of identical glycosylation sites for proteins expressed in heterologous systems

Application of recombinant techniques allows the synthesis of a specific protein in a wide variety of cell types. This allows, in combination with analytical techniques capable of unambiguous characterization of oligosaccharide primary structure (e.g. NMR spectroscopy and mass spectrometry) [23], the assessment and identification of any role the polypeptide chain plays in directing glycosylation at a given site. Thus, while cell-type specific glycosylation features will certainly be observed (and hence, the precise primary structure of the glycan will likely be different) for a given glycoprotein, any conserved structural features can reasonably be attributed to the common polypeptide chain. To date, a large number of glycoproteins have been expressed in heterologous cell lines and detailed structural characterization of the glycan structures at each site made. Table I summarizes some of these data.

The principal finding from these studies is that the site-specific glycan structure class observed at a given glycosylation site is preserved regardless of the cell type of expression. For example, the predominant oligosaccharide observed at the N-glycosylation site at Asn-83 in recombinant erythropoietin (rEPO) is a tetraantennary complex structure regardless of whether the protein is expressed in CHO or BHK cells (Table I). Indeed, 'native' EPO purified from human urine also retains this structural feature

TABLE I

Influence of cell type on site-specific glycan structure class[a]

Protein	Cell line/source			
	A	B	C	D
EPO[b] (e.g. Asn-83)	Tetraantennary	Tetraantennary	Tetraantennary	–
tPA[c] (Asn-117)	High mannose	High mannose	High mannose	High mannose
Ovalbumin[d] (Asn-293)	Hybrid/high mannose	Hybrid/high mannose	–	–
Interferon β1[e] (Asn-80)	Diantennary	Diantennary	Diantennary	Diantennary
IL-2[f] (Thr-3)	S3GS6GalNAc	S3GS6GalNAc	S3GS6GalNAc	

[a] Table designates glycan structural class (i.e. high mannose, complex diantennary, complex triantennary, etc.) identified. See [1] for detailed references. Reprinted from Ref. [1] by permission of Oxford University Press.

[b] Cell lines/sources used were: A, urinary EPO; B, BKH cell rEPO; C, CHO cell rEPO.

[c] Asn-117 is located in the first Kringle domain. Cell lines/sources used were: A, Bowes melanoma; B, colonic fibroblasts; C, CHO cell rtPA; D, murine C127 rtPA.

[d] Asn-293 is the single N-glycosylation site utilized. Chicken ovalbumin contains roughly equal amounts of hybrid- and high-mannose-type oligosaccharides. Both structural classes are present in ovalbumin from both sources. Cell lines/sources used were: A, oviduct; B, murine L cells.

[e] Asn-80 is the single site of N-glycosylation. Cell lines/sources used were: A, human fibroblasts; B, human lung tumor-derived PC8 rIFN-β; C, CHO cell rIFN-β; D, murine C127 rIFN-β.

[f] IL-2 contains a single O-glycosylation site. Natural IL-2 exists as two glycosylated forms containing NANA(α2–3)-Gal(β1–3)-GalNAc(α) and NANA(α2–3)-Gal-(β1–3)-[NANA(α2–6)]GalNAc(α), respectively. The notation is shorthand for any type of O-linked glycan structures. Cell lines/sources used were: A, peripheral blood lymphocytes; B, Jurket cell IL-2; C, CHO rIL-2.

at Asn-83. Moreover, EPO is also unusual among circulating serum glycoproteins in that its N-linked glycans can be 'decorated' with polylactosamine [Gal(β1–4)-GlcNAc(β1–3)-] moieties. This characteristic is retained in both recombinant forms of the molecule. These data suggest that the polypeptide chain of EPO contains information necessary to specify the presence of these carbohydrate structural features.

Comparable observations have been made for a number of other recombinant glycoproteins (Table I) for both N- and O-linked glycan chains. Thus, the first kringle domain of both natural and recombinant tissue plasminogen activator (tPA) is invariably glycosylated at Asn-117 with high mannose oligosaccharides [24,25]. Again, where known, the glycans observed at specific glycosylation sites are of the same structural class in both natural and recombinant forms. Of course, these observations are made for systems where each host cell contains the necessary enzymes to enable the synthesis of the conserved glycan structural class. The expression of proteins in cell types genetically or phenotypically incapable of synthesizing a particular glycan structural class obviously cannot satisfy this rule. For example, expression in yeast of a protein normally bearing complex-type oligosaccharides in mammalian cells results in the presence of high mannose-type glycans since yeast cells do not express the requisite enzymes to synthesize complex-type glycans [26].

2.3. Glycosylation and protein folding

Protein glycosylation and folding are posttranslational events which share a dependency upon instructions encoded within the primary amino acid sequence. In both events, conformational modulation by the peptide chain of either itself or of its attached carbohydrate chain is an important and common conceptual tenet. A number of experimental observations link the two events as well.

The first of these observations is that improper glycosylation leads to abnormal protein folding. In the simplest examples of this, either inhibition of N-glycosylation with tunicamycin or elimination of N-glycosylation sites by site-directed mutagenesis leads to aberrant folding [27,28]. In the latter case, restoration of a subset of the N-glycosylation sites is sufficient to restore normal processing and secretion in a number of model systems, including the G protein of vesicular stomatitis virus (VSV), H-2 protein, and the membrane anchored form of human chorionic gonadotropin [29,30]. There is a complementarity to this relationship as well in that improper protein folding leads to abnormal protein glycosylation. For example, a temperature-sensitive folding mutant of the VSV G protein was observed to bear incompletely processed glycan chains at the nonpermissive temperature [31,32].

N-glycosylation, in particular, appears to be intimately tied to proper disulfide bond formation. Within the endoplasmic reticulum there exists a complex 'editing' apparatus which monitors the integrity of nascent proteins and which has been likened to a cellular 'quality control' system [33]. Two principal functions for this system are to monitor both protein glycosylation and early protein folding events. One of the critical components of this cellular 'QC' system is protein disulfide isomerase (PDI) which catalyzes the formation of appropriate disulfide bonds in nascent proteins [34]. Additional 'editing' activities have also been ascribed to PDI (or closely related proteins based on amino acid sequence homologies) including prolyl-4-hydroxylase, a lipid transfer protein, and a component of the ER N-oligosaccharyltransferase complex that recognizes the tripeptide consensus sequence for N-glycosylation. While the precise relationship between PDI and these other editing activities requires further clarification, additional studies do suggest that PDI may play a role in regulating the extent of protein glycosylation by modulating the rate of protein folding [35]. An interesting example of this is provided by protein C, a component of plasma, which bears one unusual site of potential N-glycosylation at a Asn-X-Cys tripeptide beginning at Asn-329. Plasma protein C exists in two forms which differ in glycosylation site occupancy at this Asn. The study of Miletich and Broze [36] suggests that the extent of glycosylation at this site is dependent upon disulfide bond formation catalyzed by PDI. Thus, an intracellular competition between protein N-glycosylation and disulfide bond formation appears to exist. Since protein N-glycosylation is a cotranslational event, these observations further imply that disulfide bond formation may also be dependent upon the translation rate.

2.4. Molecular manipulations

A number of studies involving molecular alterations of the protein backbone also provide evidence for the role of the polypeptide chain in defining the structures of attached glycans. For example, Wilhelm et al. [37] observed that structural alterations to tPA made amino terminal to the glycosylation site on Asn-117, which typically bears high-mannose oligosaccharides (see Table I), resulted in complex type glycans at this site. Ashford et al. [38] observed a greater degree of glycan processing at a specific site in a truncated form of soluble rat CD4 containing only the first two, instead of the usual four, Ig-like domains suggesting an attenuation of processing occurs in the full length form.

The precise position of N-glycosylation sites within the peptide sequence is also critical for the folding, processing, and secretion of nascent glycoproteins. Using the VSV G protein as a paradigm, Rose and colleagues [39,40] have studied the impact of introducing novel sites of N-glycosylation within the peptide backbone both in the presence or absence of the two wild-type glycosylation sites. In all cases, mutant constructs exhibited impaired transport to the cell surface, relative to wild-type. Many mutant constructs which include both novel and wild-type sites of N-glycosylation exhibit a temperature-sensitive phenotype, implying an effect on protein folding, while all constructs missing the normal sites of N-glycosylation exhibited improper disulfide pairing.

3. Mechanistic models

If glycosylation is protein specific, how does the polypeptide chain influence or specify the final attached glycan structures? Several hypotheses have been put forward, including differential accessibility [41,42], a kinetic rate model [43], and site-directed processing [44]. The following discussion will describe each model and highlight its strengths and weaknesses. While evidence exists to support all three mechanistic models, no single model is capable of explaining all of the available experimental data. Our discussion will conclude with a hypothesis that attempts to unite all three current models.

3.1. Steric accessibility

The potential for simple steric blockage of glycan processing sites is both conceptually simple and appealing. This possibility was originally predicated upon the observation that a 'gradient' of glycan processing can be observed in some proteins with multiple glycosylation sites such that the more highly processed structures are disposed towards the N-terminus while less processed oligosaccharides are disposed C-terminally [45]. Ignoring possible steric considerations for the proximity of C-terminal glycosylation sites to the membrane, such a gradient could occur if co-translational glycosylation during the later part of peptide chain synthesis could result in glycan chains on a partially folded protein scaffolding and thus arresting further processing. On the other hand, N-terminal glycosylation events would occur on a peptide chain largely unfolded and presumably without any immediate impediment to glycan processing.

The involvement of steric effects in the processing of protein-linked glycans has been suggested by a number of studies. For example, using Sindbis virus-infected wild-type and mutant CHO cells and an enzymic probe, Hsieh et al. [41] demonstrated that glycosylation sites originally bearing high-mannose-type glycans, but destined to undergo processing to complex-type glycans, were more susceptible to enzymatic digestion than sites destined to remain high-mannose. Specifically, glycosylation sites on the Sindbis virion E1 and E2 glycoproteins, after infection of wild-type CHO cells, were identified as either complex or high-mannose [41]. A mutant CHO cell line, CHO 15B, was then employed as a viral host. CHO 15B cells, which lack the processing enzyme N-acetylglucosaminyltransferase I, cannot convert high mannose type to complex type glycans and yield E1 and E2 glycoproteins with only high mannose structures. An enzyme probe (endo-β-N-acetylglucosaminidase or endo H), which liberates high mannose oligosaccharides from glycoproteins and glycopeptides [46]) was used to assess the steric accessibility as determined by the extent of cleavage at each glycosylation site within the E1 and E2 glycoproteins from infected CHO 15B cells. It was determined that the E1 and E2 glycosylation sites which normally bear complex type glycans in wild-type CHO cells were more susceptible to endo H than those normally bearing high mannose glycans. This differential susceptibility to endo H was abolished upon detergent or pronase treatment of the glycoproteins. In a similar study, Trimble et al. [42] observed a correlation between select glycosylation sites on native yeast invertase and carboxypeptidase Y that are resistant to digestion with both endo H and α-mannosidase and N-glycans that are relatively small in size and lack peripheral phosphate residues. While both resistant and non-resistant classes of yeast N-glycans are composed only of mannose peripheral to the invariant pentasaccharide core present in all N-glycans, the resistant class is comparable to the high mannose N-glycans of higher eukaryotes while the larger, phosphorylated (and hence more highly processed) and non-resistant N-glycans could be taken as comparable to the complex type of N-glycans of mammalian cells. As put forward by the authors, the simplest explanation of these results is that at certain sites of glycosylation, the attached oligosaccharides become inaccessible to processing enzymes due to folding of the peptide chain.

While certainly an attractive interpretation of these data, the steric accessibility model also seems unsatisfactory in certain respects. Although these data could account for site-specific differences between high mannose and more processed oligosaccharides (i.e. complex type glycans in mammalian cells or high molecular weight, phosphorylated glycans in yeast), the model can not explain the restricted subsets of complex or high mannose oligosaccharides observed at individual sites (e.g. the identity or distribution of complex-type glycans observed at a site destined to be of the complex-type). Other limitations upon these studies also exist. For example, there is no evidence that the tertiary structures of the E1 and E2 proteins from CHO 15B infected cells are similar to those from native CHO cells or to the various folded states of biosynthetic intermediates of these proteins which may change during intracellular processing and trafficking. Questions also arise about the significance of endo H as the paradigm for processing enzymes, especially in that the site of cleavage by endo H is more proximal to the point of attachment of the oligosaccharide to the polypeptide than where most enzymes of the processing apparatus act and thus may exhibit an exaggerated sensitivity to 'steric' effects. Certainly the use of additional enzymatic probes, as employed by Trimble et al. [42], greatly substantiate conclusions drawn from these data. Nonetheless, there can be little doubt that physical accessibility of the protein-linked glycan processing apparatus towards a given site of oligosaccharide attachment is an important condition for glycan processing.

3.2. Kinetic efficiency model

Given the experimental observation that the polypeptide of a nascent glycoprotein can influence attached glycan structures, a simple and defined model system has been established by Wold and colleagues [43,47–50] which allows assessment of the impact of a local protein environment ('matrix') upon the ability of various glycan processing enzymes to act on oligosaccharide substrates.

The model system (Fig. 1) employs biotinylated oligosaccharide substrates bound to avidin. This complex is then employed as a substrate for various processing enzymes. Kinetic parameters are then extracted and compared to those obtained with the oligosaccharide substrate free in solution. The effect of the protein matrix upon the kinetic parameters of each processing enzyme can be probed further by varying the proximity of the bound oligosaccharide to avidin through the insertion of 'spacer' groups between the biotinyl and oligosaccharide moieties. Employing this system, the kinetic parameters of a number of different glycan processing enzymes were evaluated (Table II).

Two different effects of the protein matrix upon glycan processing, termed short and long range effects, were observed in these studies. Short range effects were those where a strong inhibition of enzyme activity was observed with the biotinylglycan-avidin complex (compared to the activity towards free oligosaccharide substrate) that was attenuated when the spacer group was introduced. Inhibitory effects independent of spacer length were denoted long-range effects. In general, 'commitment steps' (e.g., where the glycan is either to remain unprocessed as a high-mannose glycan or to be further processed to a complex-type oligosaccharide) exhibit short range effects whereas processing steps involving addition of peripheral monosaccharides exhibit long-range effects.

Opposite: Fig. 1. Models of biotinylated glycans with the 6-amino-hexanoic acid spacer (a) Man(α1–6)[Man(α1–3)]Man(β1–4)GlcNAc(β1–4)GlcNAc(β1-N)-Asn-CO-(CH₂)₅-NH-Biotin, and without the spacer (b) Man(α1–6)[Man(α1–3)]Man(β1–4)GlcNAc(β1–4)GlcNAc(β1-N)-Asn-Biotin. The biotin/streptavidin complex structure is from Protein Data Bank 1STP [51]. The glycan structure was generated with the program CCM [70] with preferred glycosidic torsion angles. The molecular modeling package QUANTA was used to generate the 6-amino-hexanoic acid spacer as well as linking and docking the glycans to the biotin/streptavidin complex. The docked structures were energy minimized using standard QUANTA/CHARMM parameters [71,72].

(a)

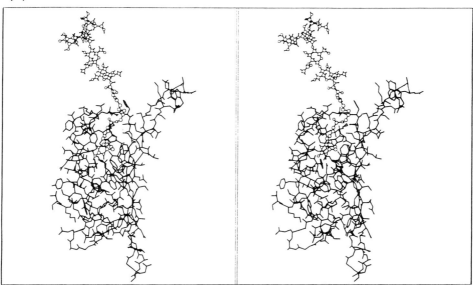

(b)

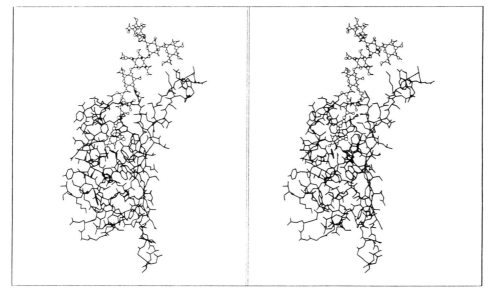

TABLE II

Protein matrix effects upon glycosylation processing enzymes[a]

Processing enzyme	Enzyme source[b]	Effect[c]	Strength[d]	Comments
Mannosidase I	O	Long	Potent	–
GlcNAc-T-I	O/L	Short	Potent	Primarily Affects V_{max}
Mannosidase II	O/L	Short	Potent	–
GlcNAc-T-II	L	Short	Weak	–
GlcNAc-T-III	O	Short	Weak	–
Gal-T	O/L	Long	Weak	Addition of 2nd Gal residue in hybrid structures with α-1,3-Man branch disubstituted with GlcNAc is abolished
GlcNAc-T-IV	O	None	–	–
GlcNAc-T-V	O	None	–	–
α-2,6-Sialyl-T	L	Long	Weak	3-arm substituted first in a diantennary substrate
α-1,6-Fuc-T	L	Long	Potent	No enzyme activity towards proximal complexes

[a] Data taken from the published reports of Wold and colleagues [43,47–50]. Matrix effects are measured by comparing activity of the specified processing enzyme towards two biotinylated forms of an appropriate glycan substrate measured either free in solution or bound to avidin. The two forms are the biotinyl- or the 6-biotinamidohexanoyl derivatives.
[b] Enzyme sources are either from hen oviduct (O) or rat liver (L).
[c] Effects are categorized as either short range (Short), long range (Long), or no (None) effect. Short range effects are those where significant matrix effects are observed only with the biotinylated (proximal) not the biotinamidohexanoyl (distal) derivatives. Long range effects are those of comparable magnitude with both the proximal and distal substrates.
[d] The magnitude of the effects are classified as either weak or potent. Weak effects are those that result in a < 5-fold retardation of the observed rate, relative to free glycan substrate.

While this model has the advantages of a defined system and allows qualitative assessment of protein matrix effect upon glycan processing, it does suffer from several limitations. Most notably, potential protein–oligosaccharide or protein–protein interactions which affect the kinetic properties of processing enzymes are by definition artificial and may inadequately represent the local environment occurring at actual sites of protein glycosylation. In addition, oligosaccharide characterization frequently was performed by mass spectrometry and thus positional isomers of reaction products were not distinguishable. As pointed out by the authors, the most serious deficiency resulting from this analysis is the inability to elucidate the contribution of the protein matrix to the generation of specific isomeric products. Finally, this approach is dependent upon the availability of processing enzymes free of potentially competing hydrolases, a situation that is difficult to guarantee when utilizing impure extracts of enzymes. Nonetheless, judicious interpretation of experimental results supports the overall conclusion that the generation of different glycan structures, even within a given protein, results from processing events regulated in part by the protein matrix.

3.3. Site-directed processing

In contrast to models where there are steric or kinetic effects upon the glycan processing enzymes by the attached peptide, the hypothesis of site-directed processing [44] embodies the concept that the protein matrix specifies the glycan structures observed at a glycosylation site by conformational modulation of the oligosaccharide substrate, yielding differential oligosaccharide processing within the same protein molecule. This model is so named because it relies upon protein–oligosaccharide interactions dictated by the local 3D structure about a given site of glycosylation to direct the biosynthetic processing of oligosaccharides.

The hypothesis of site-directed processing has three fundamental tenets. The first is that the biosynthetic intermediates in protein-linked glycan processing frequently are found in more than one conformational state. This concept has been amply demonstrated for isolated N-linked and O-linked oligosaccharides [52,53] and it seems likely that at least a limited set of glycan conformers are produced on the nascent protein. Secondly, there are processing enzymes for which only a limited subset of the available oligosaccharide three-dimensional structures are substrates or where the kinetic properties of the processing enzyme are influenced by the conformational state of the oligosaccharide substrate. A number of studies have suggested that this indeed is the case, at least for certain glycan processing enzymes [44]. And finally, protein–oligosaccharide interactions can preferentially stabilize select conformations which either are not substrates for particular processing enzymes, or engender altered enzymatic utilization. Such conformational modulation of glycan substrates could itself be temporally modulated depending upon the folded state of the nascent glycoprotein. In addition to protein–oligosaccharide interactions, it has been suggested that oligosaccharide–oligosaccharide interactions between the glycans at different sites of glycosylation might also play a role in specifying glycan structure [54,55].

There are numerous examples where the hypothesis of site-directed processing can explain the observed differential processing of protein-linked glycans within the same protein. For example, we can consider the case of the glycosylation of a human myeloma IgG$_1$ [56]. This myeloma IgG contains two sites of N-glycosylation, one at Asn-107 of the light chain and the other at the site within the heavy chain at Asn-297. Despite the fact that both chains assemble in the same cell and share the same intracellular routing, differential glycosylation was observed at the two sites. Structural analysis of the glycans from Asn-107 indicated the presence of sialylated, bisected diantennary oligosaccharides, while those at Asn-297 were essentially similar but all unbisected.

To understand how the concept of site-directed processing could explain these observations, we need to describe a conformational feature of N-linked glycans. The glycosidic linkages connecting the component monosaccharides of oligosaccharides possess varying degrees of flexibility [57,58]. This conformational flexibility can be simply described in terms of the torsion angles which constitute a given glycosidic

linkage. For most glycosidic linkages, the conformation describing the relative disposition between two connected monosaccharides can be described in terms of two
torsion angles, ϕ and ψ (Fig. 2a). However, glycosidic linkages involving substitution
at C6 possess an additional degree of freedom and require three torsion angles to define
the linkage: ϕ, ψ, and ω (Fig. 2a). As first predicted by Montreuil [59], the 1,6-linkages
of N-glycans, including those found in IgG's, are a biologically significant source of

(a)

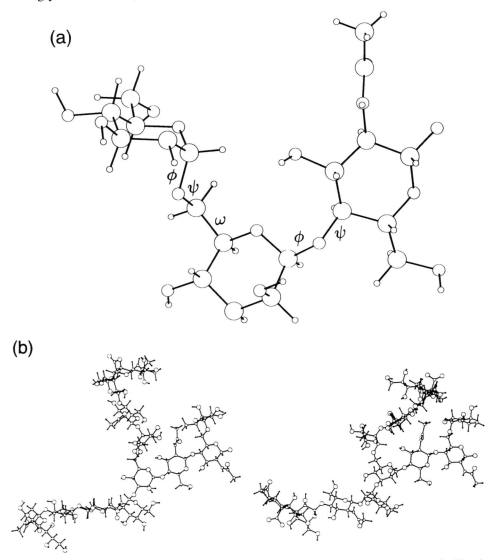

(b)

Fig. 2. Definition of the glycosidic torsion angles. (a) ϕ: H1-C1-O1-Cx, ψ: C1-O1-Cx-Hx; ϕ:
H1-C1-O1-C6, ψ: C1-O1-C6-C5, ω: O1-C6-C5-H5. (b) A complex diantennary glycan at the $\omega = -60°$
(right), and $\omega = 180°$ conformations (see Fig. 4).

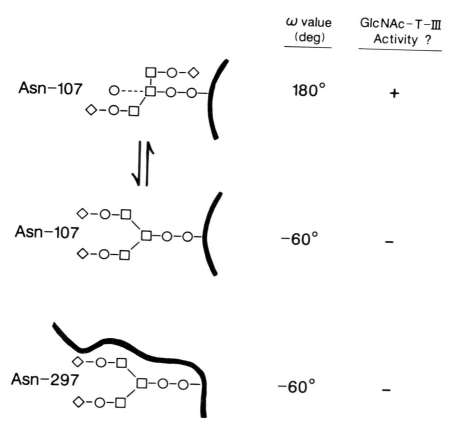

	ω value (deg)	GlcNAc-T-Ⅲ Activity ?
Asn-107	180°	+
Asn-107	-60°	-
Asn-297	-60°	-

Fig. 3. Site-directed processing of IgG N-linked glycans. Schematic representation of site-directed processing based (in this example) upon the data of Savvidou et al. [56] and Deisenhofer [60] as put forward by Carver and Cumming [44]. Reprinted from Ref. [1] by permission of Oxford University Press.

conformational flexibility, particularly in defining the gross shape of the oligosaccharide. This is especially true in terms of ω torsion angle which can adopt two low energy values of –60° and 180°. This is illustrated in Fig. 2b, where the overall conformation of a simple diantennary glycan is shown with these two values of ω.

The mechanism put forward to explain the differential glycan structures observed with myeloma IgG is summarized in Fig. 3. The enzyme which catalyzes the addition of the 'bisecting' GlcNAc residue, GlcNAc-T-III, requires a particular conformational orientation for one of the antennae of the oligosaccharide substrate ($\omega = 180°$ for the α-1,6-Man antenna, but not the $\omega = -60°$ orientation). While this conformation is accessible at Asn-107, protein-oligosaccharide interactions at Asn-297 preclude the preferred orientation, excluding GlcNAc-T-III action at this site (Fig. 3). Direct experimental support for this mechanism is obtained from X-ray crystallographic analysis of the human myeloma Fc fragment where the oligosaccharide at Asn-297 is observed in the $\omega = -60°$ orientation [60].

The major limitation of this hypothesis is the relative lack of direct experimental evidence for a conformational dependency of many glycosyltransferases for their oligosaccharide substrates. There are several reasons for this lack of understanding. First, relatively few glycosyltransferases can be obtained in sufficient quantity to conduct the detailed enzymological studies necessary to establish a conformational preference for oligosaccharide substrates. A similar comment can be made for many substrates as well since detailed conformational analysis of many protein-derived oligosaccharides has yet to be performed. It is also likely that clear delineation of a conformational preference by a transferase would entail the synthesis of oligosaccharide derivatives that are 'locked' in specific conformational states, a situation that presents significant synthetic challenges. Finally, the number of X-ray crystallography-derived glycoprotein structures having sufficient glycan density to warrant interpretation, has not increased significantly in recent decades. Nonetheless, recent advances in glycoprotein crystallography, carbohydrate synthetic chemistry and in the application of recombinant techniques to glycosyltransferases and glycoproteins should facilitate the continued evaluation of the hypothesis of 'site-directed' processing.

4. A unifying synthesis: 'site-specific' topological modulation

Even cursory consideration of the preceding discussion clearly indicates that delineation between the three current mechanisms for protein-directed glycosylation is often merely semantics, or at least a matter of perspective. For example, steric hinderance can be viewed as an extreme case of modulation of the effective K_m of a processing enzyme for its glycoprotein substrate. On the other hand, presentation of an oligosaccharide substrate to a processing enzyme in a conformationally restricted state could also alter the catalytic efficiency of the reaction, though not necessarily to zero. All three of the existing mechanisms of protein-directed glycosylation can be taken as segments of a continuum defined by a unifying operational feature: modulation of the action of processing enzymes by protein–protein or protein–oligosaccharide interactions.

We would suggest that the underlying mechanism by which the polypeptide chain directs the oligosaccharide structures at a given glycosylation site is by modulation of the 'local' three dimensional topology at a glycosylation site and we would term this hypothesis 'site-specific topological modulation'. It can be appreciated readily that this hypothesis is consistent with all of the existing data demonstrating that the characteristics of protein structure (primary through quaternary) and the conformational state of the glycan on the nascent glycoprotein can be important determinants in directing the maturation of oligosaccharide chains. In this scheme protein–protein, protein–oligosaccharide, and oligosaccharide–oligosaccharide interactions would provide the mechanisms for topological alterations about a glycosylation site. In turn, the precise topology at a nascent glycosylation site would influence, for example, the effective K_m and/or V_{max} of the processing enzyme. All three of the previously discussed mechanisms for protein-directed regulation of glycan structure are clearly special examples of 'site-specific topological modulation'.

In addition, several other features of protein glycosylation are also consistent with this hypothesis. It has been demonstrated that several glycosyltransferases exhibit a catalytic preference for specific antennae of N-linked oligosaccharides [61,62]. For example, the glycosyltransferase which catalyzes the α-1,3-linked galactosylation of the non-reducing termini of isolated diantennary glycans shows a marked preference for galactose addition to the α-1,6-Man antenna [61]. Topological modulation about protein-linked oligosaccharides could likely affect these arm preferences, perhaps by altering conformational equilibria of glycan antennae. Similar observations can be made for the 'ordered addition' observed for many glycosyltransferases, especially where the action of one transferase precludes or enables the subsequent action of a different transferase [63]. For example the action of GlcNAc-transferase V, which catalyzes the formation of triantennary from diantennary glycans is completely blocked by the prior action of β-1,4-Gal transferase [63]. Since many transferases have over-lapping subcellular sites of distribution [64], catalytic modulation of β-1,4-Gal-trans-ferase at a local site could enable the formation of tri- and tetraantennary oligo-saccharide chains.

Finally, it should be noted that this hypothesis allows for alterations in the local 3D structure as a function of time and subcellular localization, an important allowance given the observation that protein quaternary structure, for example, can play a role in specifying glycosylation. Indeed, numerous protein folding, assembly, and other proc-essing events occur in concert with various elements of the glycosylation apparatus and evidence for interplay between these two processes has been given above. Such interplay is all the more intriguing since protein folding and glycosylation both share a dependency upon instruction sets encrypted with the protein sequence. Thus, the hypothesis of site-specific topological modulation should provide a useful conceptual framework for further studies into the means by which a polypeptide chain directs its own glycosylation.

5. Conclusions and future directions

While it is clear that the polypeptide chain of a protein plays a significant role in specifying the structural class of glycan observed at a given site of glycosylation, it should be evident from the above discussion that many fundamental questions remain regarding the mechanism(s) by which this occurs. An example of the challenges to be faced in future studies and of the required evolution of our current set of mechanistic models can be seen in the glycosylation of macrophage colony-stimulating factor (M-CSF), a cytokine that induces the proliferation and differentiation of hematopoietic precursors of the mononuclear phagocytic lineage [65].

M-CSF is comprised of two disulfide-linked monomers each containing two sites of N-linked glycosylation, at Asn-122 and Asn-140. Structural characterization by NMR spectroscopy of the glycans liberated from a recombinant form of human M-CSF produced in CHO cells has shown that the glycans attached to both sites are diantennary complex-type glycans with lesser amounts of tri- and tetraantennary oligosaccharides (Camphausen and Cumming, manuscript in preparation). Only glycans attached to

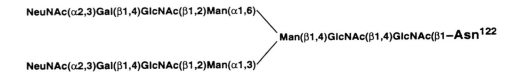

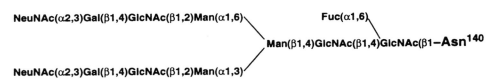

Fig. 4. The dominant complex biantennary N-linked glycans found in M-CSF at Asn-122 and Asn-140.

Asn-140, however, contain a fucose residue linked to the 'core' *N*-acetylglucosaminyl residue most proximal to the polypeptide chain (Fig. 4). Since it has been amply demonstrated that CHO cells are highly adept at fucosylating complex-type N-glycans in many recombinant proteins [24,66], the absence of core fucose at Asn-122 demonstrates that some mechanism of site-specific processing is occurring during the biosynthesis of M-CSF.

The availability of the X-ray crystal structure for recombinant human M-CSF [67] allows at least facile assessment of the three mechanistic models. The α-carbon trace of M-CSF indicates that the dimer forms approximately a rectangular slab of dimension 26×38×76 Å. Both Asn-122 and Asn-140 glycosylation sites (unoccupied by N-glycans in this recombinant form of M-CSF) are found in surface-exposed regions, well removed from the monomer–monomer interface and are not proximal to any crystal contact points that might produce secondary structure artifacts. The most obvious difference between the two glycosylation sites pertains to the nature of secondary structures in which they are found: Asn-140 is found in a loop region while Asn-122 is found within the middle of a six-turn α-helix. Thus, a simple explanation for the glycosylation differences observed between the two sites is that core fucosylation of the Asn-140 glycan is allowed to proceed unhindered by secondary structure-related steric constraints while, in contrast, accessibility of the fucosyltransferase to the α-helix-resident Asn-122 glycan is sterically restricted. A close examination of the M-CSF structure using molecular modeling techniques, however, does not permit this interpretation.

For these studies, ten models were generated via simulated annealing by applying Cα–Cα distances (from the published X-ray structure of M-CSF; A. Bohm, personal communication) as fictitious NOE constraints in the program XPLOR [68]. The models include all heavy atoms (N, C, O, etc.) and polar hydrogens. Additionally, two structures were generated for a complex, core fucosylated diantennary glycan (Fig. 4) using the CCM program [69] and preferred glycosidic torsion angles for all residues.

One glycan structure was generated for each of the two preferred conformers ($\omega = -60°$ and $\omega = 180°$) about the α-1,6-Man linkage. The different glycan structures were attached to the appropriate Asn residues of a representative M-CSF model and the intact 'neo' glycoproteins subjected to energy minimization using standard QUANTA/CHARMM parameters [70,71]. The resultant structures are shown in Fig. 5.

The ten M-CSF models generated by simulated annealing show little variation in structure at Asn-122 while somewhat larger differences are observed between models near Asn-140. These differences are largely due to variation in location of the side chain of a neighboring residue (Tyr-95) that, in some models, approaches Asn-140 quite closely. Ironically, this potentially inhibitory interaction is occurring not at the site where core fucosylation is restricted but rather where core fucose is found quantitatively on the glycan. An 'extreme case' model that contained the Tyr-95 residue in closest proximity to the Asn-140 site was selected to illustrate this point. Evaluation of the glycan-attached models (Fig. 5) indicates that core fucose can be accommodated at either Asn residue regardless of which α-1,6-Man torsion angle ($\omega = 180°$ or $\omega = -60°$) is used. Additionally, there appear to be no obvious obstructions at or surrounding the glycosylation sites that might restrict accessibility of the core fucosyltransferase to the Asn-122 glycan.

If steric accessibility issues are at play, perhaps a more important determinant in the differential fucosylation at the two glycosylation sites is the location of the sites within the dimeric protein. The Asn-122 residues of the two subunits are found on opposing large surfaces (76×38 Å) of the aforementioned rectangular slab while the Asn-140 sites are found on opposing sides of the smallest rectangular surface (26×38 Å). Hence, in this particular case, the fucosyltransferase may be insensitive to the local sterics at the individual glycosylation sites but it may be sensitive to the dimensions of the protein surface it approaches due to steric interactions well removed from the catalytic site.

An interpretation of the differential fucosylation of the M-CSF glycans using the kinetic efficiency model is likewise fraught with difficulties. While the specificity and relative catalytic efficiency of the α-1,6-fucosyltransferase has been described [72], nothing is known regarding the conformational sensitivity of the enzyme. Wold and colleagues [50], however, have embarked on a preliminary investigation of the role of the protein matrix in modulating core α-1,6-fucosylation using their biotinylglycan-streptavidin system and the known specificity of the enzyme. It has been demonstrated [72] that the preferred substrate for the enzyme is the processing intermediate GlcNAcMan$_3$GlcNAc$_2$-Asn which contains the essential GlcNAc(β1–2)Man(α1–3) substitution. Intermediates containing Man(α1,6) disubstituted with Man(α1–6)[Manα1–3] or monosubstituted with GlcNAc(β1–2) are poorer, yet potential, substrates. Characterization of the α-1,6-fucosyltransferase activity towards these substrates in the biotinylglycan-streptavidin system indicates that, by criteria established for the kinetic rate model [43], this processing event is 'short range' in nature: incorporation of a hexanoyl spacer group between the biotin and glycan moiety ('distal display' mode) exposes in part the fucosylation site that is fully 'masked' in the 'proximal display' mode [50].

The applicability of the above kinetic rate observation to the glycosylation of M-CSF is questionable owing in large part to the validity of streptavidin as a general

(a)

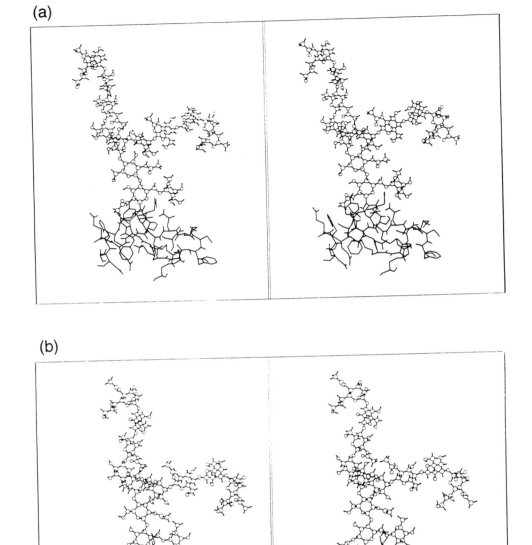

(b)

Fig. 5. Models of N-glycosylated human M-CSF. The glycan is fucosylated at the core GlcNAc. (a) gly-can-Asn-140 at $\omega = -60°$, (b) glycan-Asn-122 at $\omega = -60°$, and (c) glycan-Asn-122 at $\omega = 180°$.

(c)

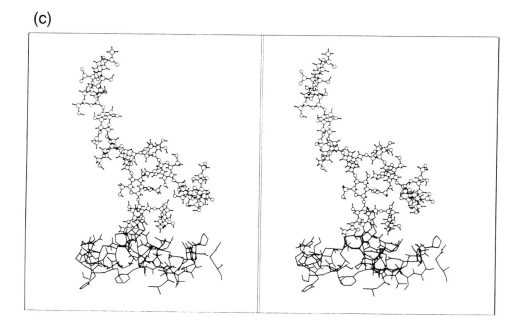

Fig 5. (c) Caption opposite.

protein matrix effector. This concern is most immediately apparent when distance measurements are made of proximity of the 6-OH group of the glycan GlcNAc-**1**, the position to which the fucose is added, to the surrounding protein surface in both the biotinylglycan-streptavidin complex (see Fig. 1) and the glycan-added M-CSF models (Fig. 5). Rotation of the Asn-linked GlcNAc-**1** about the glycosidic linkage in the biotinylglycan–streptavidin model produces a maximal distance between the 6-OH and the protein surface of 7 Å for the 'proximal display' biotinyl derivative and, consistent with the addition of a hexanoyl spacer, 12 Å for the 'distal display' biotinyl form. An analogous evaluation of the M-CSF glycosylation sites establishes a distance of 6 Å for both Asn-122 and Asn-140 linked glycans. Therefore, the application of criteria established from the biotinylglycan–streptavidin system would result in the prediction that the site for core fucosylation of M-CSF Asn-140 would be fully 'masked' by the protein matrix, a prediction that is in direct opposition to that which is observed.

The absence of an X-ray crystal structure for glycosylated M-CSF, either as the mature complex glycan-containing form or as an immature biosynthetic intermediate, which might provide critical information regarding the disposition of the glycans relative to the protein surface, precludes direct evaluation of the site-directed process-ing model as a mechanism for producing differential glycan processing in M-CSF. However, as noted above, the glycosylation sites within M-CSF are found in different global environments: Asn-122 is located on a protein surface significantly larger in dimension than that in which Asn-140 is found. Therefore, the possibility exists that

412

differential protein–carbohydrate interactions at the two sites stabilize, or alternatively destabilize, glycan conformers that are in turn substrates, or not substrates, for α-1,6-fucosyltransferase. Whether these interactions ultimately result in processing events that differ in their kinetic efficiency or are dependant upon site-specific stabilization by the protein of preferred or forbidden substrate conformers remains to be determined.

References

1. Cumming, D.A. (1991) Glycobiology 1, 115130.
2. Drickamer, K. and Carver, J. (1992) Curr. Opin. Struct. Biol. 2, 653–654.
3. Varki, A. (1993) Glycobiology 3, 97–130.
4. Hubbard, S.C. (1988) J. Biol. Chem. 263, 19303–19317.
5. Kornfeld, R. and Kornfeld, S. (1985) Ann. Rev. Biochem. 54, 631–664.
6. Rademacher, T.W., Parekh, R.B. and Dwek, R.A. (1988) Ann. Rev. Biochem. 57, 785–838.
7. Cebon, J., Nicola, N., Ward, M., Gardner, I., Dempsey, P., Layton, J., Dursen, U., Burgess, A.W., Nice, E., and Morstyn, G. (1990) J. Biol. Chem. 265, 4483–4491.
8. Cumming, D.A., (1992) Develop. Biol. Standard. 76, 83–94.
9. Goochee, C.F., Gramer, M.J., Andersen, D.C., Bahr, J.B. and Rasmussen, J.R. (1992) Frontiers Bioproc. II, 199–240.
10. Goochee, C.F and Monica, T. (1990) Bio/Technology 8, 421–427.
11. Tsuda, E., Goto, M., Murakami, A., Akai, K., Ueda, M. and Kawanishi, G. (1988) Biochemistry 27, 5646–5654.
12. Sheares, B.T. and Robbins, P.W. (1986) Proc. Natl. Acad. Sci. USA 83, 1993–1997.
13. Cumming, D.A. (1992) Improper Glycosylation and the Cellular Editing of Nascent Proteins. In: T. Ahern and M. Manning (Eds.), Stability of Protein Pharmaceuticals: In Vivo Pathways of Degradation and Strategies for Protein Stabilization. Plenum Publishing, New York.
14. Dahms, N.M. and Hart, G.W. (1986) J. Biol. Chem. 261, 13186–13196.
15. Troesch, A., Duperray, A., Polack, B. and Marguerie, G. (1990) Biochem. J. 268, 129–133.
16. Gavel, Y. and von Heijne, G. (1990) Protein Engineering 3, 433–442.
17. Yan, S.C.B., Razzano, P., Chao, Y.B., Walls, J.D., Berg, D.T., McClure, D.B. and Grinnell, B.W. (1990) Bio/Technology 8, 655–661.
18. Bause, E. (1983) Biochem. J. 209, 331–336.
19. Wilson, I.B.H., Gavel, Y. and von Heijne, G. (1991) Biochem. J. 275, 529–534.
20. O'Connell, B.C., Hagen, F.K. and Tabak, L.A. (1992) J. Biol. Chem. 267, 25010–25018.
21. Wang, Y., Agrwal, N., Eckhardt, A.E., Stevens, R.D. and Hill, R.L. (1993) J. Biol. Chem. 268, 22979–22983.
22. Smith, P.L. and Baenziger, J.U. (1992) Proc. Natl. Acad. Sci. USA 89, 329–333.
23. Dwek, R. A., Edge, C.J., Harvey, D.J., Wormald, M.R. and Parekh, R.B. (1993) Annu. Rev. Biochem. 62, 65–100.
24. Spellman, M.W., Basa, L.J., Leonard, C.K., Chakel, J.A., O'Connor, J.V., Wilson, S. and van Halbeek, H. (1989) J. Biol. Chem. 264, 14100–14111.
25. Parekh, R.B., Dwek, R.A., Thomas, J.R., Opdenakker, G. and Rademacher, T.W. (1989) Biochemistry 28, 7644–7662.
26. Bergh, M.L.E., Hubbard, S.C. and Robbins, P.W. (1988) Banbury Report 29, 59–69.
27. Gibson, R., Leavitt, R., Kornfeld, S. and Schlesinger, S. (1978) Cell 13, 671–679.
28. Machamer, C.E. and Rose, J.K. (1985) Mol. Cell Biol. 5, 3074–3083.
29. Miyazaki, J.-I., Appella, E., Zhoa, H., Forman, J. and Ozato, K., (1986) J. Exp. Med. 163, 856–871.

30. Guan, J.-L., Cao, H. and Rose, J.K., (1988) J. Biol. Chem. 263, 5306–5313.
31. Suh, K., Bergmann, J.E. and Gabel, C.A. (1989) J. Cell Biol. 108, 811–819.
32. Machamer, C.E. and Rose, J.K. (1990) J. Biol. Chem. 265, 6879–6883.
33. Hurtley, S.M. and Helenius, A. (1989) Annu. Rev. Cell Biol. 5,277–307.
34. Rose, J.K. and Doms, R.W. (1988) Annu. Rev. Cell Biol. 4, 257–288.
35. Bulleid, N.J. and Freedman, R.B. (1990) EMBO J. 9, 3527–3532.
36. Miletich, J.P. and Broze, G.J., Jr. (1990) J. Biol. Chem. 265, 11397–11404.
37. Wilhelm, J., Lee, S.G., Kalyan, N.K., Cheng, S.M., Wiener, F., Pierzchala, W. and Hung, P.P. (1990) Bio/Technology 8, 321–325.
38. Ashford, D.A., Alafi, C.D., Gamble, V.M., Mackay, D.J.G., Rademacher, T.W., Williams, P.J., Dwek, R.A., Barclay, A.N., Davis, S.J., Somoza, C., Ward, H.A. and Williams, A.F. (1993) J. Biol. Chem. 268, 3260–3267.
39. Machamer, C.E. and Rose, J.K. (1988) J. Biol. Chem. 263, 5948–5954.
40. Machamer, C.E. and Rose, J.K. (1988) J. Biol. Chem. 263, 5955–5960.
41. Hsieh, P., Rosner, M.R. and Robbins, P.W. (1983) J. Biol. Chem. 258, 2555–2561.
42. Trimble, R.B., Maley, F. and Chu, F.K. (1983) J. Biol. Chem. 258, 2562–2567.
43. Yet, M.-G., Shao, M.-C., and Wold, F. (1988) FASEB J. 2, 22–31.
44. Carver, J.P. and Cumming, D.A. (1987) Pure Appl. Chem., 59, 1465–1476.
45. Pollack, L. and Atkinson, P.H. (1983) J. Cell Biol. 97, 293–300.
46. Tarentino, A.L. and Maley, F. (1974) J. Biol. Chem. 249, 811–817.
47. Shao, M.-C. and Wold, F. (1988) J. Biol. Chem. 263, 5771–5774.
48. Shao, M.-C. and Wold, F. (1988) J. Biol. Chem. 264, 6245–6251.
49. Shao, M.-C., Krudy, G., Rosevear, P.R. and Wold, F. (1989) Biochemistry 28 4077–4083.
50. Shao, M.-C., Sokolik, C.W. and Wold, F. (1994) Carbohydr. Res. 251, 163–173.
51. Weber, P.C., Ohlendorf, D.H., Wendoloski, J.J. and Salemme, F.R. (1989) Science 243, 85–88.
52. Cumming, D.A. and Carver, J.P. (1987) Biochemistry 26, 6664–6676.
53. Carver, J.P. (1991) Curr. Opin. Struct. Biol. 1, 716–720.
54. Parekh, R.B., Tse, A.G.D., Dwek, R.A., Williams, A.F. and Rademacher, T.W. (1987) EMBO J. 6, 1233–1244.
55. Sasaki, H., Ochi, N., Dell, A. and Fukuda, M. (1988) Biochemistry 27, 8618–8626.
56. Savvidou, G., Klein, M., Grey, A., Dorrington, K.J. and Carver, J.P. (1984) Biochemistry 23, 3736–3740.
57. Rutherford, T.J., Partridge, J., Weller, C.T. and Homans, S.W. (1993) Biochemistry 32, 12715–12724.
58. Mukhopadhyay, C., Miller, K.E. and Bish, C.A. (1994) Biopolymers 34, 21–29.
59. Montreuil, J. (1984) Pure Appl. Chem. 56, 859–877.
60. Deisenhofer, J. (1981) Biochemistry 20, 2361–2370.
61. Joziasse, D.H., Schiphorst, W.E.C.M., van den Eijden, D.H., van Kuik, J.A., van Halbeek, H. and Vliegenthart, J.F.G. (1987) J. Biol. Chem. 252, 2025–2033.
62. Blanken, W.M. and Van den Eijden, D.H. (1985) J. Biol. Chem. 260, 12927–12934.
63. Schachter, H. (1986) Biochem. Cell Biol. 64, 163–181.
64. Nilsson, T., Pypaert, M., Hoe, M.H., Slusarewicz, P., Berger, E.G. and Warren, G. (1993) J. Cell Biol. 120, 5–13.
65. Rettenmier, C.W. and Sherr, C.J. (1989) Hematology/Oncology Clinics of North America 3, 479–493.
66. Sasaki, H., Bothner, B., Dell, A. and Fukuda, M. (1987) 262, 12059–12076.
67. Pandit, J., Bohm, A., Jancarik, J., Halenbeck, R., Koths, K. and Kim, S.-H. (1992) Science 258, 1358–1362.
68. Brunger, A.T. (1992) X-PLOR, Version 3.1. Yale University Press, New Haven.

414

69. Brisson, J.-R. and Carver, J.P. (1983) Biochemistry 22, 1362–1368.
70. Brooks, B.R., Bruccoleri, R.E., Olafson, B.D., States, D.J., Swaminathan, S. and Karplus, M. (1983) J. Comp. Chem. 4, 187–217.
71. Momany, F.A. and Rone, R. (1992) J. Comp. Chem. 13, 888–900.
72. Longmore, G.D. and Schachter, H. (1982) Carbohydr. Res. 100, 365–392.

J. Montreuil, H. Schachter and J.F.G. Vliegenthart (Eds.), *Glycoproteins*
© 1995 Elsevier Science B.V. All rights reserved

Biosynthesis

7. How can N-linked glycosylation and processing inhibitors be used to study carbohydrate synthesis and function

Y.T. PAN and ALAN D. ELBEIN

*Department of Biochemistry and Molecular Biology,
The University of Arkansas for Medical Sciences, Little Rock, AR, USA*

1. Introduction

Glycoproteins are widely distributed in nature, not only in animal cells but also in plants, microorganisms and viruses. The oligosaccharide chains of the N-linked glycoproteins are believed to be involved in a wide variety of biological functions [1,2]. Using various inhibitors to prevent synthesis of or to modify the carbohydrate portion of the glycoprotein has provided a useful way to determine the function of the oligosaccharide portion of a given glycoprotein. Such inhibitors are also useful for studies on the mechanism of biosynthesis of the oligosaccharide chains, since they may give rise to various intermediates that can provide important structural information.

Several reviews have covered different aspects of the biosynthesis of the oligosaccharide chains of the N-linked glycoproteins [3,4]. In order to describe the effects of inhibitors in their proper context, a brief overview of N-linked glycoprotein biosynthesis is described below.

The assembly of the various types of N-linked oligosaccharides involves two distinct series of reactions. In the first stage of synthesis, a precursor oligosaccharide is synthesized on a lipid carrier by the stepwise addition of sugars from their nucleoside diphosphate derivatives or from a lipid-linked monosaccharide intermediate to the lipid oligosaccharide, and then this oligosaccharide is transferred 'en bloc' to the polypeptide chain. In the second series of reactions, this newly transferred oligosaccharide is modified by a variety of glycosidases that remove some sugars from the oligosaccharide, and glycosyltranferases that add other sugars, to produce a large variety of different carbohydrate structures.

The biosynthesis of the precursor lipid-linked oligosaccharide, generally referred to as the dolichol pathway, involves the sequential addition of sugars to the lipid carrier, dolichol-phosphate, to form the common intermediate having the structure, $Glc_3Man_9(GlcNAc)_2$-pyrophosphoryl-dolichol [3], as shown in Fig. 1. This series of reactions occurs in the endoplasmic reticulum (ER) via a number of membrane-bound glycosyltransferases. The first reaction in the biosynthetic path-

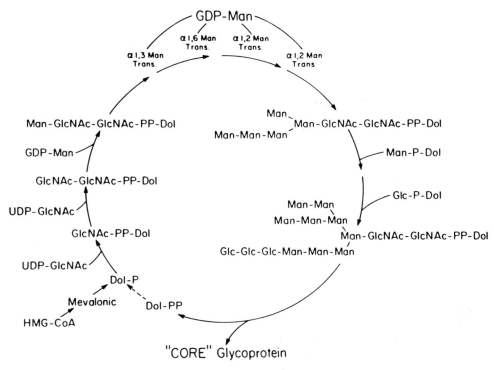

Fig. 1. Pathway for the formation of lipid-linked saccharides that produces the common intermediate for N-linked glycosylation.

way involves the transfer of a GlcNAc-1-P from the sugar nucleotide, UDP-GlcNAc, to dolichyl-P to form the first lipid intermediate, dolichyl-PP-GlcNAc (or GlcNAc-PP-dolichol). Dolichols are a family of long chain polyprenols that range in size from C_{80} to C_{100}, and which have a saturated α-isoprene unit. Like other polyprenols such as cholesterol, they are biosynthesized from mevalonic acid, probably either in the endoplasmic reticulum or in the Golgi [5]. A second GlcNAc is then added from UDP-GlcNAc to the GlcNAc-PP-dolichol to form GlcNAcβ1→4 GlcNAc-PP-dolichol [3,6]. The first mannose is then added in a β1,4-linkage from GDP-mannose to the disaccharide lipid to produce the trisaccharide lipid, Manβ1→4GlcNAcβ1→4GlcNAc-PP-dolichol [3,6,7]. Elongation of the trisaccharide-lipid by the addition of eight α-linked mannose residues and three α-linked glucose units leads to the final product, $Glc_3Man_9(GlcNAc)_2$-PP-dolichol [3]. The first four α-linked mannose residues come directly from GDP-mannose whereas the last four mannose residues are donated from dolichyl-P-mannose. Likewise, the donor of the three glucose units to the oligosaccharide is dolichyl-P-glucose. The final step in the lipid-linked oligosaccharide pathway is transfer of the oligosaccharide to the protein while the peptide chain is being synthesized on membrane-bound ribosomes [4,8,9]. This oligosaccharide transfer is catalyzed by a protein-oligosaccharyl transferase that recognizes the amino acid asparagine as the acceptor in the tripeptide sequence, Asn-X-Thr(Ser), where X can be any amino acid except a proline residue [10,11].

Once the $Glc_3Man_9(GlcNAc)_2$ has been transferred to the protein, various processing reactions occur that result in a number of modifications of the N-linked oligosaccharide chain to produce many different types of structures. The initial processing begins in the ER, probably before the synthesis of the protein is complete, and processing continues as the protein is transferred through the various Golgi stacks to its ultimate goal [12]. These processing reactions are outlined in Fig. 2. The initial reactions involve removal of all three glucose residues by two different membrane-bound glucosidases. Glucosidase I removes the outer most α-1,2-linked glucose while glucosidase II removes the next two α-1,3-linked glucoses [13]. The reactions of glucosidase I and glucosidase II are likely to occur on the cisternal surface of the rough and smooth endoplasmic reticulum [12]. A novel processing enzyme has been found in liver and has endomannosidase activity [14]. That enzyme can cleave a $Glc\alpha$-1,3-Man disaccharide from the $Glc_1Man_9(GlcNAc)_2$-protein to give a $Man_8(GlcNAc)_2$-protein. After the three glucose residues have been removed, the resulting glycoprotein would have a high mannose structure. The $Man_9(GlcNAc)_2$ (or $Man_8(GlcNAc)_2$) structure can then be further processed to form smaller-sized high mannose type oligosaccharides or processing may continue as the glycoprotein is transformed through the Golgi compartments to form complex types of oligosaccharides. Removal of all the α-1,2-linked mannose residues from the $Man_9(GlcNAc)_2$ to give a $Man_5(GlcNAc)_2$ structure is apparently the next stage of processing. The removal of the four α-1,2-linked mannose residues is catalyzed by the cis-Golgi associated α-mannosidase I [15–17]. One α-mannosidase, originally reported to be located in rat liver endoplasmic reticulum, apparently removes only one or two of the α-1,2-mannose residues from the $Man_9(GlcNAc)_2$ structure to give mostly a $Man_8(GlcNAc)_2$ structure, and smaller amounts of $Man_7(GlcNAc)_2$ [18]. The role of this ER α-mannosidase is not clear, but a similar antibody-reactive enzyme is also found in the cytoplasm of the cell. Once the $Man_5(GlcNAc)_2$ structure has been produced, further processing involves the removal of the α-1,3-and α-1,6-linked mannoses from the α-1,6-branch. This processing cannot occur without the addition of an N-acetylglucosamine (from UDP-GlcNAc) to the mannose residue that is linked in α-1,3-linkage to the β-linked mannose [19,20]. This reaction is catalyzed by GlcNAc transferase I, which is located in the medial Golgi region and is an essential enzyme in the processing of the $Man_5(GlcNAc)_2$ structure to complex or hybrid oligosaccharides. The next enzyme to act in the synthesis of complex-types of oligosaccharides is the Golgi-associated mannosidase II which rapidly removes the α-1,3 and α-1,6-linked mannosyl residues which are attached to the 6-linked core mannose [19,21]. The substrate specificity of this enzyme is quite strict, and it will not remove mannosyl residues from a $Man_5(GlcNAc)_2$ structure or from a $GlcNAc-(GlcNAc)Man_5(GlcNAc)_2$ structure that contains a "bisecting" GlcNAc linked to the β-mannosyl residue of the core [19,20]. Once the $GlcNAcMan_3(GlcNAc)_2$, has been formed, GlcNAc transferase II can catalyze the addition of a β-1,2-linked GlcNAc residue from UDP-GlcNAc to the terminal $Man(\alpha$-1,6)-residue [20,22]. The resulting product is then transported into the trans-Golgi region for the formation of complex structures with two or more outer branches. Complex and

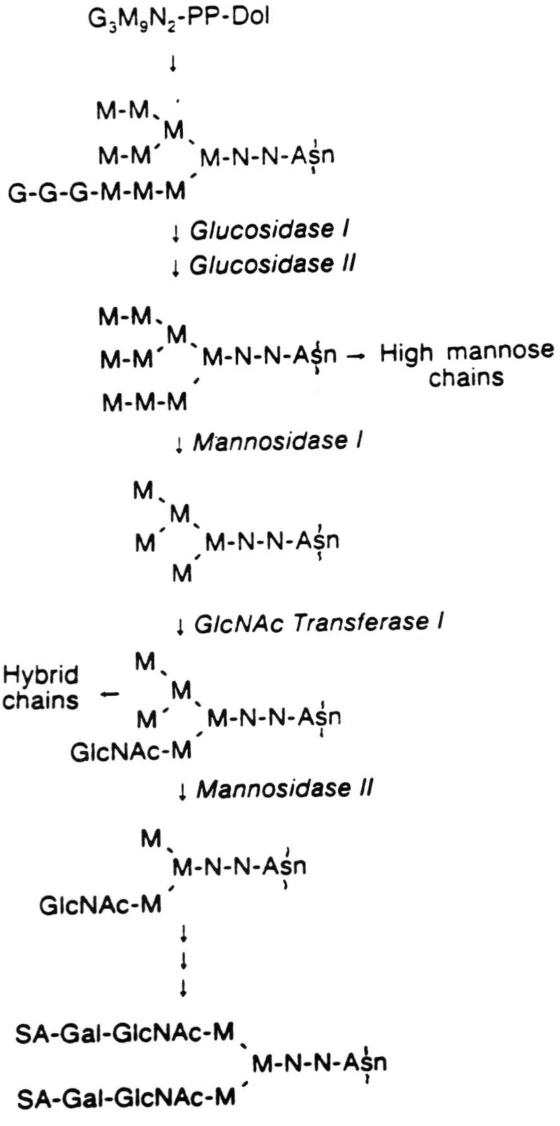

G$_3$M$_9$N$_2$-PP-Dol

↓

```
M-M
    M
M-M'   M-N-N-Asn
G-G-G-M-M-M
```

↓ *Glucosidase I*

↓ *Glucosidase II*

```
M-M
    M
M-M'   M-N-N-Asn  →  High mannose
                          chains
M-M-M
```

↓ *Mannosidase I*

```
M
  M
M'  M-N-N-Asn
M
```

↓ *GlcNAc Transferase I*

```
          M
Hybrid      M
chains  —
         M'   M-N-N-Asn
      GlcNAc-M
```

↓ *Mannosidase II*

```
M
  M-N-N-Asn
GlcNAc-M
```

↓
↓
↓

```
SA-Gal-GlcNAc-M
                 M-N-N-Asn
SA-Gal-GlcNAc-M
```

Fig. 2. Pathway of processing of the N-linked oligosaccharide (i.e., Glc$_3$Man$_9$(GlcNAc)$_2$) to produce the various complex chains.

hybrid-type oligosaccharide chains are completed by the addition of galactose, fucose, and sialic acid residues to their non-reducing termini from the appropriate nucleoside diphosphate sugar. Each of these sugars can be present in a variety of different glycosidic linkages.

2. Inhibitors of lipid-linked saccharide formation

A number of compounds have been described that inhibit glycosylation of N-linked glycoproteins either by preventing the formation of the lipid-linked oligosaccharide intermediate, or by causing alterations in the structure of the oligosaccharide portion of the lipid-linked oligosaccharides [23,24]. In the former case (tunicamycin and derivatives), glycosylation does not occur since no N-linked oligosaccharide donor can be produced. In the latter cases, various alterations in the structure of the oligosaccharide portion of the lipid-linked oligosaccharides may be caused by the inhibitors, but generally these altered oligosaccharides can still be transferred to protein and processing of the oligosaccharide may still occur. These various glycosylation inhibitors are described below.

2.1. Tunicamycin and related antibiotics

One approach to studying the role of carbohydrate in the function of the glycoprotein is to prevent the attachment of sugars during synthesis of the glycoprotein by using inhibitors of glycosylation. Tunicamycin has been widely used in many systems to study the significance of N-linked glycosylation on function of certain proteins. Tamura and co-workers first isolated this inhibitor produced by *Streptomyces lysosuperificus* in the early 1970s and found it to be inhibitory toward gram-positive bacteria [25]. The antibiotic was later shown to inhibit replication in yeast, fungi, protozoa, enveloped virus and mammalian cell lines in culture [26,27].

Tunicamycin is an uracil-containing, nucleoside antibiotic that also contains *N*-acetylglucosamine, an 11-carbon amino-deoxydialdose called tunicamine and a long-chain fatty acid [28,29]. The structure of tunicamycin is shown in Fig. 3. This antibiotic is produced as a mixture of several closely related compounds which differ from each other in the size and structure of the fatty acid chain. The total chemical synthesis of tunicamycin V, a major component of the tunicamycin homologs, has been described [30]. In cell-free extracts of several different animal tissues, tunicamycin was shown to inhibit the first enzyme in the synthesis of lipid-linked oligosaccharides, the GlcNAc-1-P transferase, thereby inhibiting the formation of dolichyl-pyrophosphoryl-GlcNAc [31–33]. Tunicamycin also inhibited the synthesis of GlcNAc-pyrophosphoryl-polyprenol in cell-free extracts of several plant tissues [34] and in yeast [35].

Tunicamycin did not affect the GlcNAc transferase that adds the second GlcNAc from UDP-GlcNAc to form GlcNAc-GlcNAc-PP-dolichol either in yeast [35], animals or plants [36]. Furthermore, tunicamycin did not inhibit the GlcNAc transferases that add terminal GlcNAc residues to the N-linked oligosaccharide chains [33], nor did it affect the activity of the phospho-N-acetylglucosaminyl transferase that adds GlcNAc-1-P from UDP-GlcNAc to terminal mannose residues on the high mannose oligosaccharides of lysosomal hydrolases [37].

The mechanism of action of tunicamycin on the solubilized GlcNAc-1-P transferase was studied [38–40], but it was not possible to demonstrate competitive inhibition because of the very strong affinity of this enzyme for tunicamycin [38]. That is, the

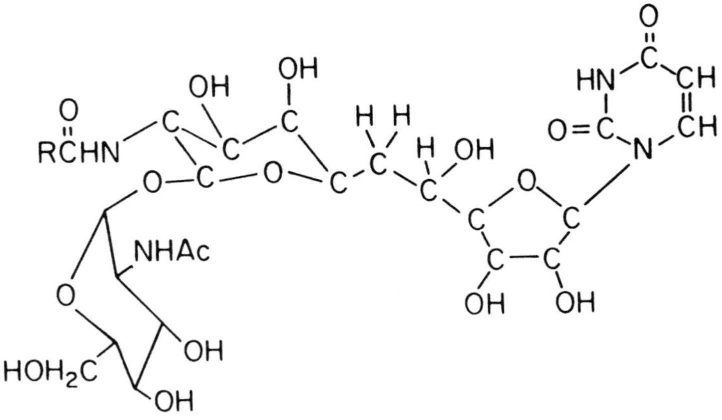

I : R = CH$_3$)$_2$CH(CH$_2$)$_7$CH=CH-   VI : R = (CH$_3$)$_2$CH(CH$_2$)$_{11}$ -

II :   (CH$_3$)$_2$CH(CH$_2$)$_8$CH=CH-   VII :   (CH$_3$)$_2$CH(CH$_2$)$_{10}$CH=CH-

III :   CH$_3$(CH$_2$)$_{10}$CH=CH-   VIII :   CH$_3$(CH$_2$)$_{12}$CH=CH-

IV :   CH$_3$(CH$_2$)$_{11}$CH=CH-   IX :   CH$_3$(CH$_2$)$_{13}$CH=CH-

V :   (CH$_3$)$_2$CH(CH$_2$)$_9$CH=CH-   X :   (CH$_3$)$_2$CH(CH$_2$)$_{11}$CH=CH-

Fig. 3. Structure of the various tunicamycin components. R refers to the different fatty acids indicated as I to X that may be attached to tunicamycin.

binding affinity of the transferase for tunicamycin was estimated to be about 5×10^{-8} M, whereas the K_m for UDP-GlcNAc was about 3×10^{-6} M [40]. Since the enzyme binds tunicamycin at least 100 fold better than it binds its substrate, it is obviously difficult to demonstrate competitive inhibition.

Most tunicamycin preparations do have some inhibitory activity toward protein synthesis, but that activity is usually considerably less (on a molar basis) than the inhibition of protein glycosylation [41–44]. Much of the inhibitory activity toward protein synthesis can be removed from tunicamycin preparations when they are purified by high performance liquid chromatography [45].

The inhibition of the first reaction of the dolichol pathway by tunicamycin affects the glycosylation of asparagine-linked glycoproteins. The *in vivo* studies with tunicamycin in various cell types have shown a wide range of effects as well as great variations in the K_i. Most of these effects are probably due to the absence of glycosylation of N-linked glycoproteins in the treated cells. However, the effects of tunicamycin also depended on the specific cell type and also on the protein in question. Thus, tunicamycin did not have any effect on the secretion of procollagen from chick embryo [46], or from cranial bones in organ culture [42]; the secretion of the serum proteins, transferrin, the apo-β component of very low density lipoprotein or α-fetoprotein by primary cultures of hepatocytes [47,48]; interferon secretion by leukocytes [49], or interferon in L-cells induced by Newcastle virus [50], β-N-acetylhexosaminidase by fibroblasts from normal individuals [51]; secretion of glycoprotein α-subunit by carcinoma cells [52]; secretion of pre-opiomelanocontin or the cleavage products ACTH and endorphin by cultured pituitary cells [53]; secretion of prothrombin from the livers of treated rats [54]. The above studies suggest that N-linked oligosaccharides are not essential in the synthesis, transport or secretion of these extracellular proteins. However, the immunoglobulins IgA, IgM and IgG were inefficiently secreted when these proteins were synthesized in the presence of tunicamycin [41,55]. Tunicamycin inhibited the secretion of IgM and IgA by 81% and 64%, respectively, whereas the inhibition of secretion of IgG was 28% in normal B cell. The hierarchy of inhibition of secretion by tunicamycin mirrors the degree of glycosylation of the heavy chains involved. However, in similar experiments using hybridoma cell lines, the secretion of IgD, which is heavily glycosylated as IgM, was not inhibited by tunicamycin whereas this antibiotic had same effect on IgM and IgG as in B cells. Therefore, the single degree of immunoglobulin heavy chain glycosylation does not determine the extent of the requirement for secretion. Treatment with tunicamycin prevented the secretion of invertase and acid phosphatase in yeast and the nonglycosylated proteins were entrapped in intracellular membranes [56]. In plants, this antibiotic caused a 60–80% inhibition in the secretion of α-amylase by aleurone layers of barley seeds [57].

Tunicamycin has been extensively used to study the role of glycosylation in the function of different kinds of receptors. Inhibition of glycosylation by tunicamycin caused a rapid depletion of insulin binding at the surface of 373-L1 adipocytes [58]. The non-glycosylated insulin receptor was not proteolytically processed to the functional insulin receptor, and was not translocated to the plasma membrane [59,60]. Treatment of cultured calf-aortic smooth muscle cells with tunicamycin resulted in the progressive loss of receptors for epidermal growth factor [61], and glycosylation was found to be essential for epidermal growth factor receptor to function in terms of ligand binding and phosphorylation [62]. The expression of acetylcholine receptor on the surface of muscle cells that were treated with tunicamycin was diminished to about 10% of normal, but the non-glycosylated receptor still retained normal kinetics for α-bungarotoxin binding. Glycosylation was found to be necessary to stabilize the receptors against proteolytic degradation [63]. Other membrane receptors also show greatly diminished function when they are synthesized in the presence of tunicamycin. Among those receptors affected are the following: the β-adrenergic receptor in astrocytoma cells [64]; gonadotropin receptor [65]; nerve growth factor receptor of pheochromocytoma cells [66]; low density lipoprotein receptor of fibroblasts or smooth muscle cells [67,68]; opiate receptors in neurotumor cells [69]. The adherence of group B streptococci to influenza A virus-infected MDCK cells was also blocked by treatment of MDCK cells with tunicamycin. In this study, it appeared that the non-glycosylated viral proteins were not inserted into the mammalian cell membrane, and therefore no adhesion occurred [70].

Inhibition of glycosylation in virus infected cells usually has dramatic effects on virus multiplication and infectivity of the newly-formed virus particles. In the presence of tunicamycin, virus particles that lacked infectivity and were devoid of envelope glycoproteins were found in a deletion mutant of Rous sarcoma virus and in a temperature-sensitive mutant of VSV [71–74]. Tunicamycin treatment also caused decreases in the infectivity of other viruses, including herpes virus [75], Semliki forest virus (76), Sindbis virus [77], Newcastle disease virus [78], mouse leukemia virus [79], a number of strains of influenza virus [76], Hantaan virus [80], and rotavirus [81]. In some cases, the non-infectious virus particles were unusually susceptible to proteolytic digestion *in vivo*. Thus in cells infected with fowl plaque virus and incubated in the presence of tunicamycin, the non-glycosylated hemagglutinin could only be found when a protease inhibitor such as TPCK(N-α-P-tosyl-L-lysine-chloromethyl ketone) was included in the medium [76].

Many other proteins have been found to be quite susceptible to proteolytic digestion when they are synthesized in the presence of tunicamycin and are therefore not properly glycosylated. Such has been shown to be case for fibronectin [82], thyroid stimulating hormone [83], the acetylcholine receptor [63], the ACTH-endorphin common precursor [84], various immunoglobulins [85] and alkaline phosphatase [86]. The addition of protease inhibitors such as TPCK or leupeptin, may prevent the tunicamycin-induced disappearance of the non-glycosylated protein. From these various studies, it was concluded that glycosylation was required to protect against non-specific proteolysis during intracellular transport.

The lack of carbohydrate affected the physicochemical properties of some glycoproteins. The non-glycosylated G protein of VSV synthesized at 38°C in the presence of tunicamycin pelleted to the bottom of the tube upon sucrose density gradient centrifugation. In contrast, the non-glycosylated G protein of VSV that was synthesized at 30°C stayed at the top of the gradient, as did the normally glycosylated protein [87,88]. This means that at the lower temperature, the non-glycosylated G protein had a conformation similar to that of the glycosylated species. These studies suggest that oligosaccharides can affect the conformation or stability of a protein, and that proteins with different amino acid sequences would be expected to have different requirements for carbohydrate. One study using two different strains of vesicular stomatitis virus (San Juan strain or Orsay strain) showed that these strains were differentially affected by the absence of glycosylation. These two strains of VSV have related but distinct G proteins. In the presence of tunicamycin, virus yields for both strains of VSV were severely depressed at 38°C, while at 30°C the growth of VSV (San Juan) was still inhibited by tunicamycin but the yield of VSV (Orsay) was nearly equal to that of the controls [88].

Non-glycosylated interferon retained its antiviral activity, but showed a decreased thermal stability and decreased affinity for those antibodies that were directed against its fully glycosylated protein [89,90]. The transport of nutrients in cultured cells is also affected by tunicamycin and this is probably due to the fact that the transport proteins are N-linked glycoproteins. Thus, glucose transport in a number of different cell lines, such as chick embryo fibroblasts [91], Swiss 3T3 cells [92], erythrocytes [93] and malignant and non-malignant cells [94] is inhibited, as are some of the amino acid transport systems in hepatocytes [95]. However, other amino acid carrier systems are not affected by this drug.

Several antibiotics which are chemically and biologically related to tunicamycin have been described. In general, these compounds have the same structure as tunicamycin, and only differ in the nature of the fatty acid components. Thus, the ones with longer chain length fatty acids appear to be more active when they are tested in cell culture systems. It seems likely that those structures with longer chain length fatty acids are able to pass through the cell membrane more easily. These various antibiotics are known as streptovirudin [96,97], mycospocidin [98], antibiotic 24010 [99], antibiotic MM19290 [100] and corynetoxin [101,102]. Streptovirudin and antibiotic 24010 were shown to have the same mechanism of action as tunicamycin in inhibiting *in vitro* transfer of GlcNAc from UDP-[³H]GlcNAc to dolichyl phosphate in various eukaryotic tissue. These antibiotics also inhibited the formation of dolichyl phosphoryl glucose, but this inhibition required much higher concentrations of the antibiotics [103].

2.2. Other antibiotics that inhibit lipid-linked saccharide formation

Amphomycin, isolated from *Streptomyces canus*, is a lipopeptide antibiotic [104], containing either 3-isododecanoic or 3-anteisododecanoic acid attached to an N-terminal aspartic acid in an amide linkage [105] (Fig. 4). Amphomycin was initially shown to inhibit cell wall synthesis in bacteria [106] which made it a logical choice to test as an inhibitor of N-linked glycoprotein biosynthesis. This antibiotic inhibited the transfer of mannose, GlcNAc and glucose from their nucleoside diphosphate sugar donors to dolichyl-phosphate, thereby preventing the formation of dolichyl-P-mannose, dolichyl-PP-GlcNAc and dolichyl-P-glucose in plant and animal tissues [107,108]. On the other hand, amphomycin had no effect on the transfer of mannose from exogenous

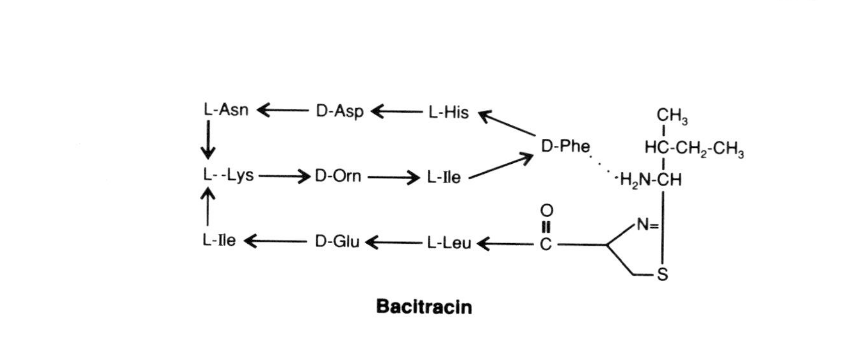

Amphomycin

Bacitracin

Fig. 4. Structure of the peptide antibiotics, amphomycin and bacitracin.

dolichyl-P-mannose to endogenous oligosaccharide acceptor, or the transfer of mannose from GDP-mannose to exogenous dolichyl-PP-(GlcNAc)$_2$ [109]. At amphomycin concentrations which completely inhibited the formation of dolichyl-P-mannose, mannose was still transferred to linked-liked oligosaccharides, and the major lipid-linked oligosaccharide species synthesized under these conditions was identified as Man$_5$(GlcNAc)$_2$-PP-dolichol. These studies demonstrated that without participation of dolichyl-P-mannose, the mannose residues in Man$_5$(GlcNAc)$_2$-PP-dolichol must come directly from GDP-mannose. Similar results were reported in hen oviduct tissue [110] and in embryonic liver [111]. These results have received strong support by the isolation of a mutant CHO cell line that is defective in dolichyl-P-mannose formation, but that still is able to synthesize Man$_5$(GlcNAc)$_2$-PP-dolichol [112]. Furthermore, a partially purified mannosyltransferase that catalyzes the transfer of mannose from GDP-mannose to form Man$_5$(GlcNAc)$_2$-PP-dolichol was resistant to inhibition by amphomycin [113] indicating the absence of dolichyl-P-mannose in this reaction. The inhibition by amphomycin on the formation of dolichyl-P-mannose provides good evidence for the role of this intermediate as the mannosyl donor in the elongation of Man$_5$(GlcNAc)$_2$-PP-dolichol to Man$_9$(GlcNAc)$_2$-PP-dolichol. The mechanism of amphomycin inhibition on the glycosylation of dolichyl-P apparently involves the formation of a complex between amphomycin and dolichyl-P at the specific reaction site for the acceptor lipid [109]. Tsushimycin, another antibiotic that is similar to amphomycin in structure, also inhibits the formation of dolichyl-P-mannose, dolichyl-P-glucose and dolichyl-PP-GlcNAc using an enzyme preparation from pig aorta [114].

Bacitracin is a cyclic peptide antibiotic isolated from *Bacillus licheniformis* [115] and the structure of this compound is shown in Fig. 4. This antibiotic was shown to inhibit bacterial cell wall biosynthesis by blocking the enzymatic dephosphorylation of the C$_{55}$-polyisoprenyl-pyrophosphate. Bacitracin was reported to cause an accumulation of Man(GlcNAc)$_2$-polyisoprenoid lipid in oviduct membranes [116], and to inhibit the transfer of GlcNAc from UDP-GlcNAc to polyisoprenyl-P in calf pancreas microsomes [117]. However, it had no effect on the synthesis of dolichyl-PP-(GlcNAc)$_2$, dolichyl-P-mannose or dolichyl-P-glucose in these various extracts. On the other hand, in studies with the aorta enzyme system, bacitracin inhibited the formation of dolichyl-P-mannose and dolichyl-PP-GlcNAc, as well as the transfer of mannose from GDP-mannose to lipid-linked oligosaccharides and glycoprotein [118]. The inhibition of dolichyl-P-mannose formation could be overcome by the addition of high concentrations of dolichyl-phosphate, but this lipid was not able to reverse the inhibition of dolichyl-PP-GlcNAc formation. In yeast membrane preparations, bacitracin inhibited the transfer of GlcNAc from UDP-GlcNAc to form dolichyl-PP-GlcNAc, but it did not inhibit addition of a second GlcNAc to produce dolichyl-PP-(GlcNAc)$_2$ [119].

The structure of diumycin has not yet been elucidated, but its main components are known to be glucosamine, glucose, ammonia, acetic acid, phosphate and a C$_{25}$ fatty acid. It is produced by *Streptomyces umbrinus* [120], and it has been shown to inhibit cell wall biosynthesis in *Staphylococcus aureus* [121]. Using yeast membrane preparations as the enzyme source, diumycin effectively inhibited the formation of dolichyl-P-mannose but did not inhibit the glycosyl transfer reaction from GDP-mannose to dolichyl-PP-(ClcNAc)$_2$. The inhibitory action of diumycin on dolichyl-P-mannose

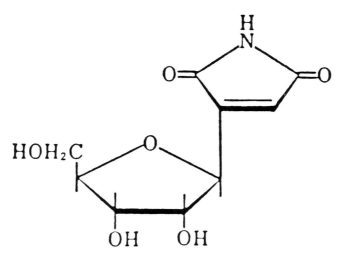

Fig. 5. Structure of the nucleoside antibiotic, showdomycin.

formation did not seem to be competitive with respect to dolichyl-phosphate [122]. When enzyme preparations isolated from *Acanthamoeba* were used, diumycin was found to inhibit the transfer of both mannose and acetylglucosamine from their nucleoside diphosphate sugars [123]. Diumycin showed a greater inhibitory effect on the transfer of the second GlcNAc to dolichyl-PP-GlcNAc than the transfer of GlcNAc-1-P to dolichyl-P. The formation of dolichyl-P-glucose was not inhibited.

Flavomycin is another phosphoglycolipid antibiotic of the moenomycin group that is produced by various species of streptomycetes [124]. This antibiotic blocked the formation of dolichyl-P-mannose and the transfer of lipid-linked saccharides to exogenous acceptor in particulate membrane fractions from pig brain [125]. The formation of dolichyl-P-mannose was hardly affected by this antibiotic.

Showdomycin is a nucleoside antibiotic isolated from *Streptomyces showdoensis* [126] and the structure of this compound is shown in Fig. 5. This antibiotic was found to inhibit the synthesis of certain lipid-linked saccharides in enzyme preparations of aorta [127]. Using a particulate enzyme fraction, the antibiotic was found to be quite effective in inhibiting the formation of dolichyl-P-glucose. Showdomycin also inhibited the transfer of mannose to both dolichyl-P-mannose and lipid-linked oligosaccharides. This inhibition was much more pronounced in the presence of NP-40. In the green alga *Volvox carteri*, showdomycin was found to inhibit the formation of both dolichyl-P-glucose and dolichyl-PP-GlcNAc. The inhibitory effect was lost when the antibiotic was preincubated in the presence of high concentrations of dithiothreitol. It was suggested that the inhibitory effect of showdomycin was caused by an irreversible reaction of the maleimide structure of the antibiotic with an essential thiol group near, or at, the active site of the glycosyltransferase [128].

2.3. Sugar analogues and amino sugars

Sugar analogs of glucose and mannose were previously detected as inhibitors of protein glycosylation by virtue of their anti-viral properties [129]. A number of sugar analogs such as 2-deoxy-D-glucose, 2-deoxy-2-fluoro-D-glucose(2-deoxy-2-fluoro-D-mannose), 4-deoxy-4-fluoro-D-mannose,6-deoxy-6-fluoro-D-glucose, glucosamine and mannosamine, have been synthesized, and results with these compounds have been extensively reviewed [130,131]. Sugar analogs have been shown to affect protein glycosylation *in vivo* by inhibiting the lipid-linked oligosaccharide pathway. Apparently the sugar analogs need to be metabolized to their respective guanosine and/or uridine nucleoside diphosphate derivative in order to exert their inhibitory effect. Thus, the respective nucleoside diphosphate sugar prevents the elongation of the oligosaccharide chain either because the sugar analog is transferred to the lipid-acceptors (i.e. trapping of lipid acceptors), or because the normal transfer of sugar from its sugar nucleotide is blocked by the analog.

When 2-deoxyglucose was incubated with cultured chicken cells, the sugar was converted to both UDP-2-deoxy-glucose and GDP-2-deoxy-glucose, and then to dolichyl-P-2-deoxy-glucose. The inhibition of protein glycosylation by these inhibitors appears not to be due to depletion of sugar nucleotide pools since the levels of GDP-mannose and UDP-GlcNAc in the inhibited cell are actually increased. The major compound involved in the inhibition appears to be GDP-2-deoxymannose (glucose), since the addition of mannose to the inhibited cells reversed the inhibition, and decreased the level of GDP-2-deoxyglucose in the cells [132]. GDP-2-deoxyglucose also serves as a sugar donor in the formation of the trisaccharide-lipid dolichyl-PP-GlcNAc-GlcNAc-2-deoxyglucose, but this oligosaccharide could not be further elongated nor could it be transferred to protein [133]. GDP-2-deoxy-glucose also caused the depletion of the dolichyl-P pool in the cell by competing with the natural nucleoside diphosphate sugar for dolichyl-P to form the dolichyl-derivative of the sugar analog. GDP-deoxyglucose and UDP-deoxyglucose also prevent the attachment of the peripheral glucose residues in the formation of the Glc_3Man_9-$(GlcNAc)_2$-PP-dolichol intermediate [133].

When chicken embryo cells were treated with fluoroglucose or fluoromannose (2-deoxy-2-fluoro-D-glucose or 2-deoxy-2-fluoro-D-mannose), lipid-linked oligosaccharides containing decreased amounts of glucose and mannose were produced. This inhibition was reversed by the addition of glucose or mannose to the culture medium [134]. The synthesis of β-Man$(GlcNAc)_2$-PP-dolichol was inhibited when GDP-2-deoxy-2-fluoro-mannose was added to the microsomal enzyme fraction of chicken embryo cells suggesting that the production of this abnormal nucleoside diphosphate sugar is responsible for the inhibition [135].

Earlier studies demonstrated that glucosamine inhibited the multiplication of a variety of enveloped virus by interfering with the production of the viral glycoproteins [136]. These studies suggested that animal cells grown in glucosamine were abnormal in terms of protein glycosylation but the specific site of action was not determined [137]. The removal of glucosamine from the medium led to a decrease in the concentration of intracellular glucosamine and a reversal of the inhibition, suggesting that this amino sugar was responsible. It appeared likely that glucosamine was inhibiting at an early step in the assembly of the lipid-linked oligosaccharides, but the inhibition required intact cells suggesting that glucosamine had to be metabolized in order to cause the inhibition [138]. When female Wistar rats were treated with D-galactosamine, high tissue contents of UDP-glucosamine and UDP-galactosamine were detected [139,140]. UDP-glucosamine was found to function as a sugar donor with microsomal preparations of both chicken embryo cells and rat liver, yielding GlcN-P-dolichol [141]. However, glucosamine-P-dolichol did not serve as substrate for the transfer of glucosamine to dolichol-linked oligosaccharides. Interestingly enough, the synthesis of dolichyl-P-[^{3}H]glucose from UDP-[^{3}H]-glucose was depressed by 56–75% in the presence of UDP-GlcN. These results suggest that, dolichyl-P-GlcN interfered with dolichyl-P-dependent reactions.

Glucosamine was also found to cause alternations in the lipid-linked oligosaccharide pathway in MDCK cells. Thus, when MDCK cells were incubated in the presence of a 1 mM concentration of glucosamine, there was a complete shift in the size of the major lipid-linked oligosaccharide, from the normal Glc$_3$Man$_9$(GlcNAc)$_2$-PP-dolichol to a Man$_{6-8}$(GlcNAc)$_2$-PP-dolichol. Furthermore, when the concentration of glucosamine was raised to 10 mM, the major oligosaccharide associated with the dolichol was a Man$_3$(GlcNAc)$_2$. The effect of glucosamine was reversible and following removal of glucosamine, the cells could renew their synthesis of Glc$_3$Man$_9$(GlcNAc)$_2$-PP-dolichol [142]. The exact site of inhibition by glucosamine that caused this alternation has not yet been identified.

Mannosamine also caused alternations in the lipid-linked oligosaccharide profile in MDCK cells, but the intermediates that accumulated were different from those seen in the presence of glucosamine [143]. When MDCK cells were incubated in the presence of mM concentrations of mannosamine and labeled with [2-^{3}H]mannose, they accumulated various lipid-linked oligosaccharides that had smaller sized oligosaccharides with unusual structures, and the normal Glc$_3$Man$_9$(GlcNAc)$_2$-pyrophosphoryl-dolichol was not observed [144]. More than 80% of the oligosaccharides released from the lipid-linked oligosaccharides were eluted from a column of concanavalin A-sepharose with 10 mM α-methylglucoside indicating that they were not the normal high-mannose types of oligosaccharides. In addition, 20-40% of these oligosaccharides bound to a column of Dowex 50-H$^+$ indicating that they had a positive charge. These abnormal oligosaccharides were still transferred to protein. The behavior of the mannosamine-induced oligosaccharides on columns of Bio-Gel P-4 after chemical N-acetylation and gel filtration at high pH strongly suggested that some mannosamine had been incorporated into the lipid-linked oligosaccharides [144]. Moreover, MDCK cells did incorporate label from [^{3}H]mannosamine into lipid-linked oligosaccharides. Mannosamine was also found to be an inhibitor of the formation of the glycan portion of the glycosylphosphatidylinositol anchor [145]. It also blocked the expression of a recombinant GPI-anchored protein in MDCK cells and converted this protein to an unpolarized secretory protein. The effect of mannosamine on the GPI-anchor precursor showed that mannosamine itself was incorporated into the glycan, probably in the second mannose position and thereby blocked the further addition of mannose into the anchor components [146].

2.4. Glucose starvation and energy charge

When Chinese hamster ovary (CHO) cells were placed in a medium devoid of glucose, there was a rapid cessation of the synthesis of the usual $Glc_3Man_9(GlcNAc)_2$ lipid-linked oligosaccharide and the accumulation of a $Man_5(GlcNAc)_2$ lipid linked oligosaccharide. The latter compound was glucosylated and transferred to protein and then was subsequently processed [147,148]. This effect of glucose starvation was found at low to moderate cell densities. At high cell densities, the effect was not evident and addition of glucose or mannose, but not glycerol, glutamine, galactose, inositol or glycine, prevented the accumulation of $Man_5(GlcNAc)_2$-PP-dolichol [148]. The time required for the glucose starvation effect ranged from 20 minutes in CHO [147,148] to 6 hours in BHK cells [149]. In BHK cells that were infected with VSV and starved for glucose, the $Glc_3Man_5(GlcNAc)_2$PP-dolichol was formed and transferred to the G-protein [150]. Rat hepatoma cells under glucose free culture conditions produced glycoproteins with the small molecular weight-form of α_1-acid glycoprotein. This reduction in molecular weight was due not only to a truncated oligosaccharide but also to a reduced number of oligosaccharide chains on the protein [151].

In the presence of carbonyl cyanide m-chlorophenyhydrazone (CCCP), an uncoupler of oxidative phosphorylation, the formation of dolichyl-P-mannose in chicken embryo cells was inhibited. However, CCCP did not affect the formation of $(GlcNAc)_2$-PP-dolichol or the pool of dolichyl-P-glucose and affected only to slight extent the pool size of GDP-mannose. Therefore, in the energy-depleted cells, the cells only produced a $Glc_3Man_5(GlcNAc)_2$-PP-dolichol or $Glc_3Man_4(GlcNAc)_2$-PP-dolichol. These oligosaccharides were able to be transferred to protein [152]. Different alterations in oligosaccharide-lipid patterns were observed when energy deficit was induced in thyroid slices by CCCP or respiratory inhibitors, such as N_2 or antimycin A. Energy deprivation leads to impairment in the glycosylation of lipid-linked oligosaccharide with an attendant accumulation of predominantly $Man_3(GlcNAc)_2$ and to a lesser extent, $Man_5(GlcNAc)_2$ [153]. Some uncouplers of oxidative phosphorylation have been reported to disrupt the recycling of the glucose transport carrier [154] which limited the entry of glucose into cells. It is possible that the effect of CCCP may be due to the same situation as that of glucose starvation.

3. Inhibitors of processing glycosidases

The initial reactions in the N-linked oligosaccharide processing pathway involve the removal of all three glucose residues by 2 membrane-bound ER α-glucosidases (glucosidase I and glucosidase II), followed by the removal of up to 6 α-linked mannose residues. It is still not clear exactly how many α-mannosidases are actually involved in glycoprotein processing, but at least 2 (mannosidase I and mannosidase II) are necessary to remove all of the α-1,2-linked mannoses (mannosidose I), and the α-1,3 and α-1,6 mannoses from the α-1,6-branched mannose (mannosidase II). In recent years,

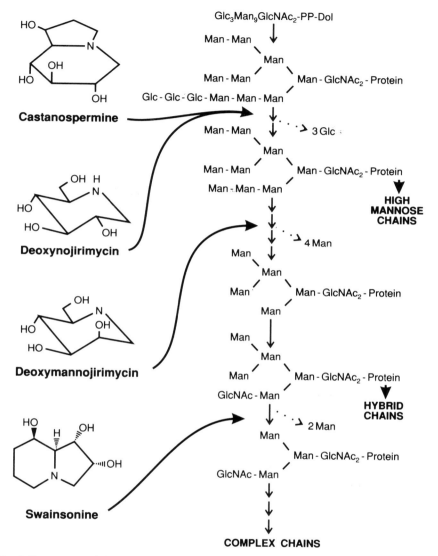

Fig. 6. Structures and sites of action of some of the inhibitors of glycoprotein processing.

a number of plant indolizidine or pyrrolizidine alkaloids have been shown to be potent and reasonably specific inhibitors of the processing glycosidases (Fig. 6). These compounds are discussed below and their site of action and utilization in biological systems are described.

3.1. Inhibitors of glucosidase I and glucosidase II

3.1.1. Castanospermine, Deoxynojirimycin and DMDP

Castanospermine [(1S,6S,7R,8αR)-1,6,7,8-tetrahydroxy-indolizidine] is a plant alkaloid that was isolated from the seeds of the Australian tree, *Castanospermum australe*, also called the Moreton Bay chestnut [155]. The structure of castanospermine is shown in Fig. 7. This alkaloid is toxic to sheep and other animals that eat these seeds, resulting

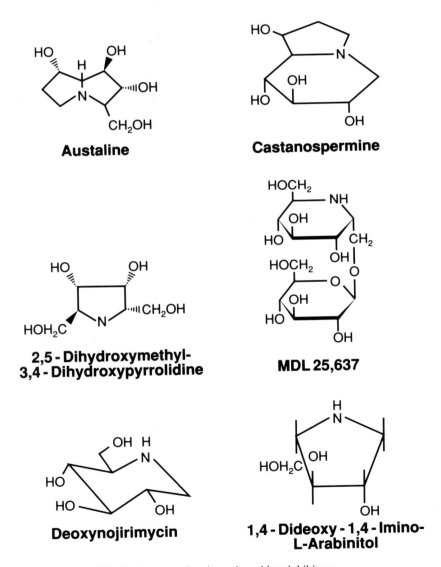

Austaline

Castanospermine

2,5 - Dihydroxymethyl-3,4 - Dihydroxypyrrolidine

MDL 25,637

Deoxynojirimycin

1,4 - Dideoxy - 1,4 - Imino-L-Arabinitol

Fig. 7. Structure of various glucosidase inhibitors.

in severe gastrointestinal upset, possibly leading to death. These symptoms are related to the fact that castanospermine is a potent inhibitor of α-glucosidases, including sucrase, maltase and lysosomal α- and β-glucosidases [156,157]. When rats are given daily injections of castanospermine over a 3-day period, the activity of the disaccharidases, sucrase and maltase, are completely inhibited. Moreover, the animals display symptoms that are similar to humans suffering from Pompe's disease, a lysosomal storage disease where individuals are missing the lysosomal α-glucosidase, and therefore accumulate glycogen storage products in their lysosomes. Thus, in these castanospermine-treated animals there is almost a complete absence of α-glucosidase activity, and the lysosomes of liver and skeletal muscle contain large deposits of glycogen [158].

Castanospermine appears to be non-toxic to eucaryotic cells in culture, and does not affect the formation of lipid-linked oligosaccharides or the synthesis of protein [159, 160]. However, castanospermine was shown to inhibit the processing of the influenza viral envelope glycoprotein, i.e., hemagglutinin [159], as well as the glycoproteins of suspension-cultured soybean cells [161]. This inhibition was due to a block in the activities of glucosidase I and glucosidase II, causing the production of N-linked glycoproteins having oligosaccharides mostly of the $Glc_3Man_{7-9}(GlcNAc)_2$ structure in both the plant cells and in various animal cells.

Deoxynojirimycin is another compound that inhibits the processing glucosidases. This compound was isolated from certain bacteria of the *Bacillus* spp. [162]. Deoxynojirimycin is a nitrogen containing analog of glucose and its structure is shown in Fig. 7. It was initially found to inhibit intestinal α-glucosidase, pancreatic α-amylase [162,163] and human liver lysosomal α-glucosidase [164]. In intact cells, deoxynojirimycin blocks the formation of complex-types of oligosaccharides and results in the accumulation of glycoproteins with $Glc_{1-3}Man_{7-9}(GlcNAc)_2$ structures [165]. Thus, oligosaccharides containing 3 glucose residues accounted for only 20% of the mixture, suggesting that *in vivo*, glucosidase II was preferentially inhibited [166]. Deoxynojirimycin also inhibited the formation of lipid-linked oligosaccharides in cultured cells [167]. The major lipid-linked oligosaccharide found in IEC-6 cells, incubated in the presence of deoxynojirimycin, was $Man_9(GlcNAc)_2$-PP-dolichol whereas in control cells it was $Glc_3Man_9(GlcNAc)_2$-PP-dolichol. *In vitro*, the sensitivity of glucosidases to deoxynojirimycin varied depending upon the enzyme source. Thus, with the enzymes from yeast, 50% inhibition of activity was obtained at deoxynojirimycin concentrations of 20 μM for glucosidase I and 5 μM for glucosidase II [165]. However, with calf liver microsomes, 50% inhibition was obtained at 3 μM deoxynojirimycin for glucosidase I and 20 μM for glucosidase II [168].

N-Methyldeoxynojirimycin is obtained by methylation of the ring nitrogen of deoxynojirimycin [169]. This derivative is a more potent inhibitor of glucosidase I than is the parent compound [170]. In the presence of N-methyldeoxynojirimycin, 70% of the N-linked oligosaccharides contained 3 glucose residues, in contrast to only 20% in the deoxynojirimycin treated-cells [166]. In addition, this methylated derivative did not affect the formation of the lipid-linked oligosaccharides.

DMDP (2,5-dihydroxymethyl-3,4-dihydroxypyrrolidine) is an analog of β-D-fructofuranose, and is an inhibitor of glycoprotein processing. The structure of DMDP is shown in Fig. 7. DMDP was first isolated from the leaves of *Derris elliptica* [171], and was subsequently synthesized from glucose [172]. It proved to be a potent inhibitor of both α-and β-glucosidases, and was shown to be 10-to-60-fold as effective as deoxynojirimycin [173]. In influenza virus-infected MDCK cells, DMDP inhibited the synthesis of the complex types of N-linked oligosaccharides and gave rise to new structures of the $Glc_3Man_{7-9}(GlcNAc)_2$ type, indicating that DMDP was an inhibitor of glucosidase I [174]. However, in cultured intestinal epithelial cells, DMDP was reported to cause the accumulation of $Man_{7-9}(GlcNAc)_2$ oligosaccharides on the glycoproteins, indicating that DMDP was acting on the α-1,2-mannosidase [175]. It is not clear why DMDP shows different effects in MDCK cells from those seen in intestinal epithelial cells.

The inability of cells to remove glucose from their N-linked oligosaccharides may have dramatic effects on the synthesis, transport and secretion of various glycoproteins. In the case of the biosynthesis of the α-subunits of the ion-channel of rat brain neurons, post translational processing includes incorporation of palmityl residues in thioester linkage to the protein, as well as sulfate residues on an internal GlcNAc of the N-linked oligosaccharides. The incorporation of [³H]palmitate into the α-subunit is inhibited by tunicamycin indicating that this addition occurs at an early step of biosynthesis, but only after addition of the N-linked oligosaccharide has occurred. However, the α-subunit synthesized in the presence of castanospermine appears to be of lower molecular weight. Castanospermine prevented the incorporation of most of the sialic acid into the α-subunit, and it also inhibited sulfation of the molecule, but not palmitylation. Interesting enough, although castanospermine caused extensive changes in the structure of this glycoprotein, it did not prevent the covalent assembly of the α- and β2-subunits, or the transfer of the αβ2-complexes to the cell surface. Furthermore, the sodium channels synthesized under these conditions showed the normal affinity for binding saxitoxin [176]. Several other studies have demonstrated that certain post translational modifications are inhibited by castanospermine, presumably because the compound prevents the formation of the necessary oligosaccharide acceptor structure. Thus, when influenza virus-infected MDCK cells are incubated in the presence of castanospermine, the viral glycoprotein fails to become fucosylated [177] or sulfated [178]. On the other hand, when the cells are incubated in swainsonine, the viral glycoproteins are of the hybrid structure and can be both fucosylated and sulfated.

Some indication was obtained in several systems that glucose removal may cause changes in the sensitivity of the protein toward temperature changes, probably because of temperature effects on protein folding. Deoxynojirimycin and castanospermine inhibited the formation of Sindbis virus in BHK cells, and in several other cell lines. The growth of Sindbis was inhibited to a much greater extent at 37 than at 30°C in BHK cells [179]. Studies with the San Juan strain of VSV indicated that the formation of the virus becomes temperature sensitive when glucose residues are retained on the oligosaccharide chain of the G protein [180]. As indicated earlier, temperature effects on VSV proteins were also seen in the presence of tunicamycin.

Deoxynojirimycin affects the secretion of some, but not all cellular glycoproteins. In HEP G-2 cells grown in the presence of deoxynojirimycin, the rate of secretion of $α_1$-antitrypsin and $α_1$-antichymotrypsin decreased significantly, but only marginal effects were observed on other liver glycoproteins such as complement, C3, transferrin and albumin [181]. The secretion of IgD, but not that of IgA, was inhibited by deoxynojirimycin [182]. Deoxynojirimycin also prevented the secretion of α_1-proteinase inhibitor [183]. It was suggested that the presence of glucose on the oligosaccharide chain might retard the transport of the glycoprotein from the endoplasmic reticulum to the Golgi. Thus, α_1-antitrypsin and α_1-antichymotrypsin were found in the rough endoplasmic reticulum [181]. However, since a number of other serum glycoproteins with similar N-linked oligosaccharides are not retained by this inhibitor, this explanation is not completely satisfactory. Perhaps the effect has to do with fact that deoxynojirimycin causes an under-glycosylation of some proteins as outlined above, and this could result in a slower transport of these proteins.

Glucosidase inhibitors also show pronounced effects on membrane receptors. Thus, the epidermal growth factor receptor in deoxynojirimycin-treated human A431 cells showed a delay in the acquisition of both ligand binding and resistance to endoglycosidase H [184]. In the presence of deoxynojirimycin, there was a decrease in the number of nicotine acetylcholine receptors in BC3H-1 cells because of an increase in the receptor degradation rate [185]. In treated cells, approximately the same amount of the α-subunit of the receptor was translated as in control cells, but the α-subunits were degraded more rapidly and only 25% of them acquired the capacity to bind α-bungarotoxin. These results suggest that oligosaccharide processing may aid in protecting the primary translation product of the α-subunit from degradation.

Cultured IM-9 lymphocytes, treated with either castanospermine or deoxynojirimycin also showed a 50% reduction in the number of surface insulin receptors. In this case, the degradation rates of the surface receptors were similar in both control and inhibitor-treated cells, indicating that accelerated receptor degradation could not account for this reduction of receptor numbers. Biosynthetic labeling experiments with [^{3}H]leucine and [^{3}H]mannose showed that glucose removal from the core oligosaccharide was not necessary for the cleavage of the insulin pre-receptor, but it did delay the processing of this precursor. This probably accounts for the reduction in the number of cell-surface receptors [186].

Similar results have been observed with regard to the low-density lipoprotein (LDL) receptor [187] and the IL-2 receptor [188]. Thus, when aortic smooth muscle cells were grown in the presence of castanospermine, there was a considerable decrease in the ability of these cells to internalize and degrade exogenous ^{125}I-low-density lipoprotein. This reduction in LDL internalization was due to the fact that there were fewer LDL receptors present at the cell surface of these inhibited cells. Since castanospermine-grown cells had the same number of total receptors (surface and internal), but there were fewer receptors at the cell surface, it seems likely that the inhibitor slows down the transport of receptor from the ER to the surface [187]. The specific step in targeting that is inhibited by castanospermine is not known, but it is likely to be the transport of glycoprotein from ER to Golgi. In the presence of castanospermine, T-lymphocytes are not activated by B-cells [188]. In this case, castanospermine causes a suppression of T-lymphocyte activation, apparently due to a diminished number of IL-2 receptors at the cell surface.

Castanospermine and N-methyl-1-deoxynojirimycin were used to study the transport of the glycoprotein E2, and the intracellular maturation of mouse hepatitis virus A59. Both inhibitors selectively inhibited the processing glucosidase, caused a drop in virion formation by two log steps, and a drastic delay in the surface expression of glycoprotein E2. The E2 species synthesized under such conditions was acylated but accumulated intracellularly in a compartment distinct from the Golgi. Concomitantly synthesis of the matrix glycoprotein E1 of mouse hepatitis virus A59 was impaired [189].

The use of processing inhibitors to prevent the formation of the aberrant asparagine-linked glycoprotein and to inhibit catabolic glycosidases is being actively pursued with castanospermine as a therapeutic agent for cancer [190]. *In vitro* studies with transformed cells have shown that the induced changes in glycosylation caused by processing inhibitors result in a variety of effects on the properties of the transformed cells. Thus, the inhibition of processing may increase the rate of growth of tumors [191] and the expression of non-glycosylated oncogene products [192]. Castanospermine was effective in reducing the rate of tumor growth in animals when it was placed in the diet. Thus, when mice were injected with 10^6 feline sarcoma viral particles (SM-FME-cells) and then fed various amounts of castanospermine, those mice receiving 0.84 mg castanospermine/g body weight/day had much slower growing tumors that were 2.6 times smaller than those of control animals [191]. When murine melanoma B16-F10 cells were pretreated with castanospermine prior to intravenous injection into mice, pulmonary colonization was decreased ($\geq$80%) [193]. The complex types of oligosaccharides have been implicated in metastasis [194]. The inhibitor treated transformed cells had endoglycosidase H-sensitive and concanavalin-A binding glycans at their cell surface. In two different tumor cell lines, a mastocytoma and a murine T-cell lymphoma, growth in the presence of 1 to 3 mM deoxynojirimycin for 24 h resulted in cells that were considerably less sensitive to lysis by interferon-activated macrophages than were control cells. At 3 mM deoxynojirimycin, there was 71% inhibition in the macrophage induced cell lysis, suggesting that N-linked oligosaccharides may participate in the recognition and/or lysis [194].

The inhibitors of the processing glucosidases did decrease the infectivity of the human immunodeficiency virus (HIV) responsible for the acquired immunodeficiency syndrome (AIDS) [195–197], as well as other retroviruses [198] such as the feline equivalent of HIV [199]. The cellular basis of this activity is not very well understood. An interaction between the heavily glycosylated viral envelope protein gp 120 and the membrane glycoprotein CD4 on the surface of T-lymphocytes is essential for infection by HIV. This interaction leads to syncytium formation or a fusion of the cells. The decrease in syncytium formation caused by processing inhibitors could be attributed to inhibition of the processing of the envelope precursor gp 160, and then decreased cell surface expression of the mature envelope glycoprotein gp 120. Both castanospermine and deoxynojirimycin affected HIV infectivity and were not cytotoxic to lymphocytes, whereas the mannosidase inhibitors such as swainsonine and deoxymannojirimycin had no effect on HIV infectivity. The most effective anti-HIV agents have been reported to be N-butyldeoxynojirimycin [200] and 6-O-butanoyl-castanospermine [201].

3.1.2. Australine

Australine [(1R, 2R, 3R, 7S, 7αS)-3-hydroxymethyl-1-2,7-trihydroxypyrrolizidine] is a polyhydroxylated pyrrolizidine alkaloid that was isolated from the seeds of the Australian tree, *Castanospermum australe* [202]. The structure of australine is shown in Fig. 7. *In vitro*, australine inhibited the glycoprotein processing enzyme glucosidase I, but had only slight activity against glucosidase II. When incubated with cultured cells, this alkaloid inhibited glycoprotein processing at the glucosidase I step and caused the accumulation of glycoproteins with $Glc_3Man_{7-9}(GlcNAc)_2$ oligosaccharides [203]. The inhibitory activity toward glucosidase I was less potent relative to castanospermine, but australine is the first example of a compound inhibiting glucosidase I without also inhibiting glucosidase II (even at concentrations of up to 500 μM).

3.1.3. Alexine

Several other bicyclic pyrrolizidine alkaloids have been isolated and tested as inhibitors as glycosidases. Alexine [IR, 2R, 3R, 7S, 8S)-3-hydroxymethyl-1,2,7-trihydroxypyrrolizidine] was isolated from *Alexa leiopetala* [204] whereas 3,8-diepialexine was isolated from the seeds of *Castanospermum australe* [205]. Alexine and its 3- and 7-epimers have also been chemically synthesized from a protected 2-azido-2-deoxymannose [206]. In contrast to australine, alexine and its derivatives have been found to be very poor inhibitors of glycosidases and almost millimolar concentrations of these compounds are necessary to achieve 50% inhibition.

3.1.4. DIA

1,4-Dideoxy-1,4-imino-D-arabinitol (DIA), a chemically synthesized aminopentitol (Fig. 7), is a potent competitive inhibitor of yeast α-glucosidase (50% inhibition at 0.18 μM) [207]. It is also an effective inhibitor of glucosidase I in both chick embryo cells and in BHK cells. Treatment of influenza virus-infected chick embryo cells with DIA in medium containing 10 mM glucose resulted in an inhibition in the formation of complex types of oligosaccharides but without the accumulation of glucosylated high-mannose oligosaccharides. However, when the glucose concentration in the medium was reduced to 2 mM, the accumulation of $Glc_3Man_{8-9}(GlcNAc)_2$ occurred [208].

3.1.5. MDL

2,6-Dideoxy-2,6-imino-7-0-(β-D-glucopyranosyl)-D-glycero-L-guloheptitol (MDL) is a novel compound that was chemically designed to mimic a disaccharide to act as a transition state inhibitor of the intestinal glucohydrolases, such as sucrase, maltase, isomaltase, glucoamylase and trehalase [209]. The structure of MDL is shown in Fig. 7. When this compound was tested as an inhibitor of the processing of the influenza viral N-linked glycoprotein, it was found to preferentially affect glucosidase II. In cell culture, MDL caused the accumulation of glycoproteins having mostly $Glc_2Man_{7-9}(GlcNAc)_2$ structures [210]. However, this inhibitor did not appear to affect the formation of lipid-linked saccharides, or the synthesis of protein or nucleic acid.

3.1.6. Bromoconduritol

Bromoconduritol (6-bromo-3,4,5-trihydroxycyclohex-1-ene), obtained by chemical synthesis from myoinositol, is an active site-directed covalent inhibitor of glucosidases [211]. It inhibited the formation of complex oligosaccharides on the viral hemagglutinin and neuraminidase in chick embryo cells infected with influenza virus [212]. In a cell-free glycosidase preparation from rat liver, bromoconduritol inhibited the release of the innermost glucose residue from the oligosaccharide, $Glc_3Man_9GlcNAc$. Therefore, it was proposed to be a glucosidase II inhibitor. However, glucosidase II is supposed to remove the last two glucoses, so inhibition of this enzyme should give $Glc_2Man_{7-9}(GlcNAc)_2$ structures. Bromoconduritol also inhibited the release of viral particles from infected chick embryo cells.

3.2. Inhibitors of Mannosidase I and Mannosidase II

3.2.1. Deoxymannojirimycin
Deoxymannojirimycin (1,5-dideoxy-1,5-imino-D-mannitol) is the synthetic mannose analog of 1-deoxynojirimycin [213] (Fig. 8). It was the first inhibitor to be described that prevented the *in vivo* conversion of high-mannose types of oligosaccharides to complex chains by inhibiting the Golgi mannosidase IA/B activity. It also caused the accumulation of high mannose oligosaccharides of the $Man_{8-9}(GlcNAc)_2$ structure in cultured cells [214,215]. Deoxymannojirimycin (DMJ) did not prevent the secretion of either IgD or IgM [214] in contrast to deoxynojirimycin which does inhibit IgD secretion. DMJ had no effect on the surface appearance of several membrane glycoproteins, such as the VSV-G protein, influenza viral hemagglutinin and HLA-A, -B, and C antigens [216], as well as α_1-acid glycoprotein and α_1-proteinase inhibitor [217]. DMJ was used as a tool to study the recycling of membrane glycoproteins through those regions of the Golgi that contain mannosidase I. In this experiment, membrane glycoproteins were synthesized in the presence of DMJ and [2-^{3}H]-mannose in order to produce glycoproteins with $Man_{8-9}(GlcNAc)_2$ structures. After growth of the cells in the presence of inhibitor and label, the media was removed and replaced with fresh media, and the change in the structure of the oligosaccharide chains of the transferrin receptor was followed with time. The data indicated that this receptor and other surface glycoproteins were endocytosed and transported to the mannosidase I compartment, and some of this protein was further processed during recycling [218].

The rat liver endoplasmic reticulum α-mannosidase is not sensitive to deoxymannojirimycin. Therefore DMJ was used to determine the role of the ER α-mannosidase in the processing of the HMG CoA reductase of UT-1 cells. In the absence of inhibitor, the predominant oligosaccharides on the reductase were single isomers of $Man_6(GlcNAc)_2$. However, in the presence of DMJ, the $Man_8(GlcNAc)_2$ accumulated, indicating that the ER-α-mannosidase is responsible for the initial mannose processing [219]. However, not all hepatocyte glycoproteins were found to be substrates for the ER α-mannosidase.

3.2.2. Kifunensine
Kifunensine, produced by the actinomycete *Kitasatosporia kifunensine* 9482, is an alkaloid that corresponds to the cyclic oxamide derivative of 1-amino-mannojirimycin (Fig. 8). Kifunensine was a weak inhibitor of jack bean α-mannosidase [220], but a

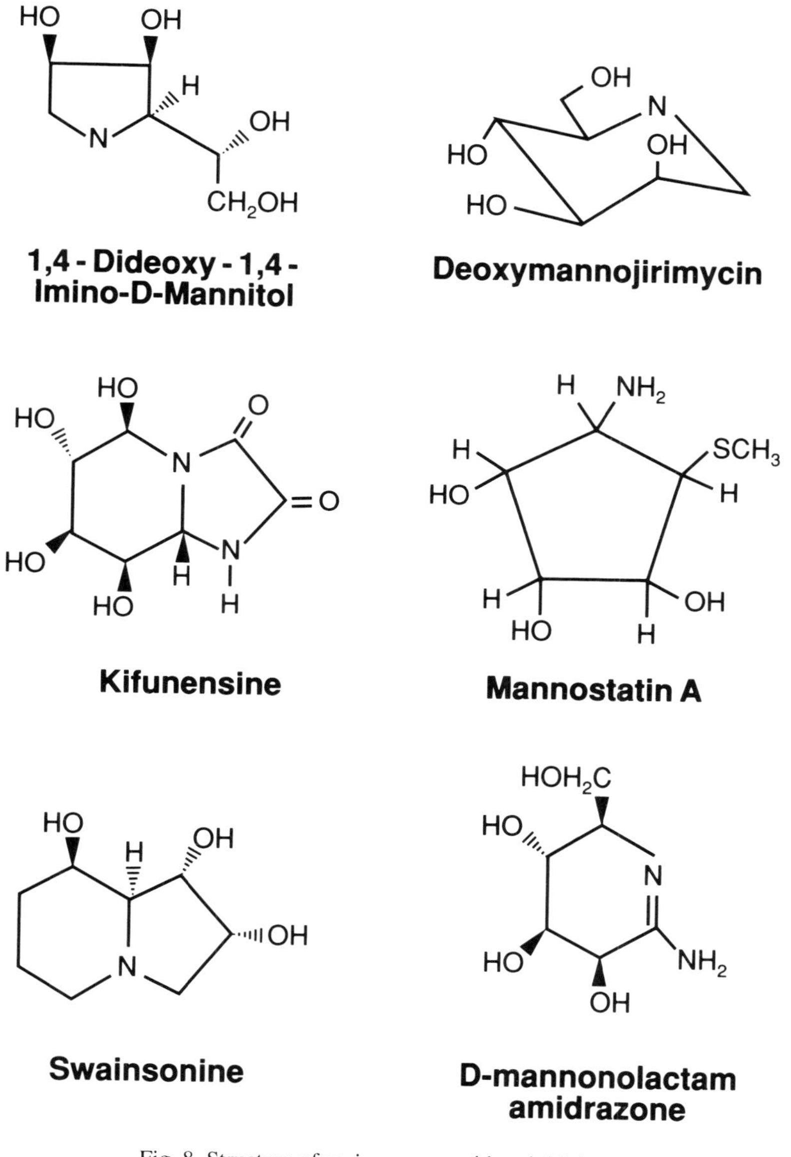

1,4-Dideoxy-1,4-Imino-D-Mannitol

Deoxymannojirimycin

Kifunensine

Mannostatin A

Swainsonine

D-mannonolactam amidrazone

Fig. 8. Structure of various mannosidase inhibitors.

very potent inhibitor of the glycoprotein processing enzyme, mannosidase I [221,222]. However, this alkaloid was inactive toward mannosidase II, or the ER mannosidase. When kifunensine was tested in cell culture at concentrations of 1 μg/ml, it caused almost complete inhibition of complex chain formation, with the accumulation of N-linked glycoproteins having $Man_9(GlcNAc)_2$ structures. Kifunensine is probably one of the most effective glycoprotein processing inhibitors observed thus far. Its inhibitory capacity towards mannosidase I was 50 to 100 times more potent than that of deoxymannojirimycin [221].

3.2.3. D-Mannonolactam amidrazone

D-Mannonolactam amidrazone was synthesized chemically in order to mimic the mannopyranosyl cation which is believed to be the active intermediate in mannosidase hydrolysis [223]. The structure of this compound is shown in Fig. 8. This inhibitor proved to be a more general mannosidase inhibitor than other known compounds. D-mannonalactam amidrazone not only inhibited the Golgi mannosidase I (IC_{50} = 4 μM), but it also inhibited mannosidase II (IC_{50} = 100 nm), the ER mannosidase (IC_{50} = 1 μM) and aryl α-mannosidases (IC_{50} = 40 nM) [224]. It was tested in MDCK cells and was found to almost completely prevent the formation of complex types of N-linked oligosaccharides, with the formation of about equal amounts of $Man_9(GlcNAc)_2$ and $Man_8(GlcNAc)_2$ structures.

3.2.4. DIM

DIM (1,4-dideoxy-1,4-imino-D-mannitol) was synthesized chemically from benzyl α-D-mannopyranoside and was an effective inhibitor of jack bean α-mannosidase and mannosidase I [225]. The structure of DIM is shown in Fig. 8. In both cases, the inhibition was of the competitive type and the inhibition was better at higher pH values. It was suggested that DIM was more effective when the nitrogen in the ring was in the unprotonated form. This compound also inhibited the glycoprotein processing enzyme, mannosidase I, and prevented the formation of complex types of oligosaccharides in influenza virus-infected MDCK cells. In the presence of DIM, most of the viral N-linked oligosaccharides were of the $Man_9(GlcNAc)_2$ structure, but almost 15% of the endo-N-acetylglucosaminidase H-released oligosaccharides appeared to be hybrid types of structures. These results suggested that DIM may also inhibit mannosidase II [225].

3.2.5. MMNT

A number of the glycosyl triazenes have been studied and found to be irreversible active site-directed "suicide" inhibitors of their respective glycosidases [226] and some of these compounds have been used to measure the turnover time of these glycosidases [227]. MMNT (α-D-mannopyranosyl-methyl-para-nitrophenyltriazene) was found to be a mannosidase I inhibitor. It had no effect on Golgi mannosidase II or endoplasmic reticulum α-mannosidase [228], but it did inhibit the α-mannosidase isolated from lysosomes as well as that from the cytosol [229].

3.2.6. Mannostatin A

Mannostatin A was isolated from the culture filtrate of the microorganism *Streptoverticillium verticillus* [230] and its structure is shown in Fig. 8. It was shown to be a potent competitive inhibitor of rat epididymal α-mannosidase. Mannostatin A was also a potent inhibitor of lysosomal α-mannosidase. The inhibition was competitive in nature, as was the inhibition of the glycoprotein-processing enzyme, mannosidase II. However mannostatin was inactive toward the glycoprotein processing mannosidase I. In cell culture studies, mannostatin caused an increase in the amount of hybrid types of oligosaccharides [231].

3.2.7. Swainsonine

Swainsonine is an indolizine alkaloid [(1S,2R,8R,8αR)-1,2,8-trihydroxyoctahydroin-dolizidine], and was the first compound reported to be a glycoprotein processing inhibitor. The structure of this mannosidase II inhibitor is shown in Fig. 8. This compound was first isolated from *Swainsona canescens* [232], an Australian plant that is toxic to animals and produces symptoms resembling those of human α-mannosidosis upon prolonged ingestion of this plant [233]. This alkaloid has also been isolated from the fungus *Rhizoctonia leguminicola* [234] and from spotted locoweed (*Astragalus lentiginosus*) [235,236], and has also been chemically synthesized [237–240]. Swainsonine was found to be a potent inhibitor of lysosomal [241] and jack bean α-mannosidase with 50% inhibition of these enzymes requiring about a 1×10^{-7} M concentration of the inhibitor [242]. It was suggested that the inhibitory activity of swainsonine results from the structural similarity to the protonated form of the mannosyl cation, since the hydrolytic intermediate is probably formed during normal catalysis by glycosidases [243].

Swainsonine was subsequently found to prevent the conversion of high mannose oligosaccharides into those of the complex type in various mammalian cell lines [244,245], and also to block normal processing of the influenza viral hemagglutinin [246]. Although the specific point of inhibition was not determined in those early studies, a direct test of the effect of swainsonine on the partially purified mannosidase I and mannosidase II showed that only the latter enzyme was inhibited [247]. Mannosidase II is the enzyme that specifically removes the α-1,3 and α-1,6-linked mannose residues from the GlcNAcMan$_5$(GlcNAc)$_2$ structure. Swainsonine did not show any effect on the ER α-mannosidase, nor on the soluble α-mannosidase [248]. Swainsonine has also been shown to have the same effect on glycoprotein processing of other proteins, such as the G protein of VSV [249], fibronectin [250] and the glycoproteins of BHK cells [251].

Swainsonine has been used in a number of studies in order to determine whether changes in the structure of the N-linked oligosaccharides affect the function of a given glycoprotein. In most cases, swainsonine had little effect on the glycoprotein in question, which may indicate that a partial complex chain (i.e., hybrid structure) is sufficient for activity and that this change in structure (from complex to hybrid) does not affect protein conformation. For example, swainsonine did not impair the synthesis or export of thyroglobulin in porcine thyroid cells [252], nor did it affect surfactant glycoprotein A from type II epithelial cell [253]. H2-DK histocompatibility antigens from macrophages [254] and Von Willebrand protein in epithelial cells [255] were likewise not affected by synthesis in the presence of this alkaloid. In one study, the secretion of α$_1$-antitrypsin from primary rat hepatocytes was not altered by growth of the cells in swainsonine [256], but in another case using hepatoma cells, the alkaloid increased the rate of secretion of transferrin, ceruloplasmin, α$_1$-antitrypsin and α$_2$-macroglobulin [257]. The authors suggested that these proteins traversed the Golgi more rapidly than their normal counterparts. Swainsonine also did not affect the insertion or function of the insulin receptor [258], the epidermal growth factor receptor [259], or the receptor for asialoglycoproteins [260]. The inhibitor, however, did block the receptor-mediated uptake of mannose-terminated glycoproteins by macrophages. This inhibition appeared to be due to the formation of hybrid types of oligosaccharides on the glycoprotein at the macrophage surface, which then reacted with and tied up the mannose receptor [261]. Viral proteins are still assembled and processed in the presence of this alkaloid, and infectious particles are still formed [262].

On the other hand, some functions are affected by swainsonine. For example, this inhibitor caused a time-dependent loss in rat intestinal sucrase activity when it was injected into rats, even though swainsonine has no direct affect on sucrase activity *in vitro* [157]. It is likely that sucrase inhibition is caused by an alteration in the oligosaccharide structure of the enzyme which may cause mistargeting of the protein, or result in an enzyme with an altered conformation which is then inactive. The glucocorticoid stimulation of resorptive cells, which probably involves the attachment of cells such as osteoclasts to bone, is blocked by swainsonine [263]. This alkaloid also reduced the interaction of *Trypanosoma cruzi* with peritoneal macrophages, when it was used to treat either the host or the parasite [264]. Swainsonine treatment of either B16-F10 murine melanoma cells [265] or MDAY-D2 murine lymphoreticular tumor cells [266] resulted in a substantial impairment of tumorigenic activity. Tumor cells cultured in the presence of swainsonine for 24–48 h showed reduced organ colonization potential when injected into mice [266,267]. This alkaloid has also been shown to alleviate both chemically-induced and tumor-associated immune suppression [268], to increase natural killer cell activity [265] and to increase IL-2 production by lymphocytes [269]. In addition, swainsonine may also affect ras gene expression since NIH 3T3 cells transfected with human tumor DNA lose their metastatic phenotype when treated with the drug [270].

4. Inhibitors of dolichyl-P synthesis

Since dolichyl-phosphate is an essential cofactor in N-linked glycosylation and acts as a carrier of sugars to form a large-sized oligosaccharyl-PP-dolichol, a block in the formation of dolichol would be expected to affect glycosylation of N-linked glycoproteins. Dolichol is synthesized in the cell via a pathway involving condensation of acetate units to form hydroxymethylglutaryl-CoA, conversion to mevalonic acid, and production of isoprene units. These reactions are identical to those involved in the initial steps of cholesterol biosynthesis and the biosynthesis of ubiquinones. The isoprene units are then condensed (and may be cyclized) to form dolichol, other polyisoprenols, cholesterol, and so on.

In this pathway, HMG-CoA reductase is considered to be the major rate-limiting and control enzyme in cholesterol and probably in dolichol synthesis [271]. The inhibitors of this enzyme (compactin and its analogs) should prevent the synthesis of dolichol and block N-glycosylation *in vivo*.

4.1. 25-Hydroxycholesterol

In aortic smooth muscle cells, 25-hydroxycholesterol strongly inhibited the incorporation of acetate into dolichol and cholesterol by 91% and 81% respectively, and also diminished dolichol-dependent glycoprotein synthesis. On the other hand, the incorporation of mevalonate into these lipids was not affected by 25-hydroxycholesterol. In fact, the addition of mevalonate to the cells reverses the inhibition of glycoprotein synthesis [272]. When L cells and MOPK104E cells were treated with 25-hydroxycholesterol, the incorporation of acetate into both cholesterol and dolichol was also inhibited. However, the relationship between the concentration of 25-hydroxycholesterol and the level of inhibition differed for the two lipids. When comparing the rates of cholesterol and dolichol synthesis from acetate, large fluctuations in cholesterol synthesis were observed, while the rate of dolichol synthesis was only slightly affected. These results suggested to these workers that HMG-CoA reductase affects the rate of dolichol synthesis by altering the concentration of a substrate for an enzyme which catalyzes a rate-limiting reaction peculiar to dolichol synthesis [273]. The rate of synthesis of dolichyl-phosphate was further shown to be controlled during the condensation of isopentenyl-pyrophosphate with farnesyl-pyrophosphate. When animals were fed diets supplemented with cholestyramine, a compound that stimulates HMG-CoA reductase, the incorporation of acetate into cholesterol was greatly increased, but acetate into dolichyl-phosphate was not affected. These results indicated that the rate of dolichyl-phosphate biosynthesis in animals was not regulated by the HMG-CoA reductase, but was probably regulated by the level of dolichyl phosphate synthetase [274]. When animal are fed a diet supplemented with high amounts of cholesterol, the biosynthesis of cholesterol is inhibited, but the incorporation of mevalonate into dolichol and dolichol-phosphate-dependent mannosyl transferase was stimulated [275]. Thus, the inhibition of the cholesterol branch of the biosynthetic pathway could result in the accumulation of various intermediates (farnesyl-pyrophosphate and isopentenyl-pyrophosphate), and could therefore stimulate the synthesis of dolichyl-phosphate.

4.2. Compactin

A specific nonsteroidal inhibitor of cholesterol synthesis, called compactin or ML-236B, was isolated. This compound is produced by several fungal strains, including *Penicillium brevicompactum* [276,277], and its structure includes a lactonized ring that resembles the lactone form of mevalonic acid (Fig. 9). As a result, compactin acts as a potent, reversible competitive inhibitor of HMG-CoA reductase [276]. In CHO cells in culture, compactin inhibits cholesterol synthesis at concentrations of 1–10 µM [278]. In developing sea urchin embryos, compactin induced abnormal gastrulation by inhibiting the synthesis of dolichol, glycolipids and N-linked glycoproteins [279]. The inhibition of development and glycoprotein formation by compactin could be overcome by supplementing the embryos with exogenous dolichol or dolichyl-phosphate, but not by adding cholesterol or coenzyme Q [280]. In monolayers of rat hepatocytes cultured in the presence of compactin for 24 hours, the synthesis of dolichyl-phosphate was inhibited by 91% as estimated by the incorporation of [3H] acetate, and by 77% as

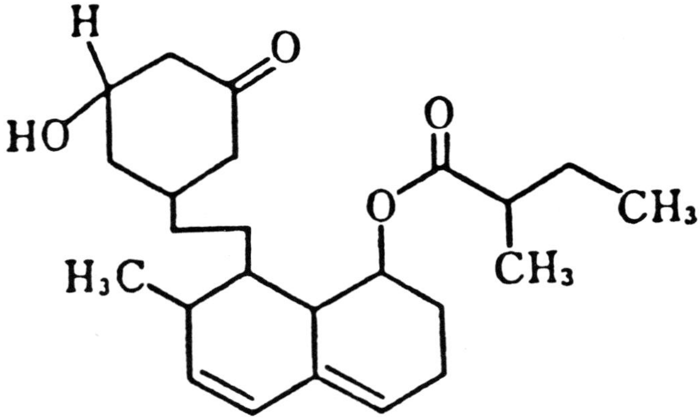

Fig. 9. Structure of compactin.

estimated by the incorporation of ^{32}P. These results indicate that dolichyl-phosphate is mainly synthesized through a *de novo* pathway, whereas phosphorylation through the CTP-mediated kinase is of limited functional importance [281]. Similar results were obtained with mouse embryos where development was arrested by either compactin or an oxygenated sterol at the 32-cell stage, leaving the blastomeres decompacted [280]. In this case also, N-linked glycosylation was inhibited and mevalonate could reverse this inhibition.

5. Inhibition of protein targeting or movement

During the synthesis and processing of N-linked oligosaccharides, the asparagine-linked glycoproteins are transported from their site of synthesis in the ER through the various Golgi stacks (from *cis* to *trans*) to their ultimate location. Although the signals involved in targeting these proteins to specific locations are just becoming known, a number of compounds have been found to perturb the movement of glycoproteins. Most of these perturbants are ionophores which affect the concentrations of various ions in certain cellular compartments, and may also affect the internal pH of some of these compartments.

5.1. Monensin

Monensin is an ionophore that binds monovalent cations and has been shown to interfere with the intracellular transport of newly synthesized secretory proteins in a variety of systems. It also inhibits incorporation of glycoproteins into the plasma membrane [282–287]. The principle site of arrest of intracellular transport has been studied by subcellular fractionation, electron microscopic autoradiography, and morphological and immunofluorescent methods. In all cases the principal site of blockage has proven to be within the Golgi complex [288,289]. Treatment of mammalian cells with monensin results in modification of the post-translational processing of glycoproteins by inhibiting the conversion of high-mannose oligosaccharides to their complex forms [287,290–292]. Recently, the effect of monensin on human immunodeficiency virus type 1 (HIV-1) was studied and the infectivity of this virus was found to be reduced by monensin treatment [293]. The endoproteolytic cleavage of gp 160 and the glycosylation of gp 120 are required for infectivity of HIV-1 and for binding of virions to CD4 [293–295], but processing and transport of gp 160 and secondary glycosylation of gp 120 are blocked by monensin [296].

5.2. Brefeldin A

Brefeldin A (BFA), an antifungal metabolite, inhibits the intracellular transport of viral envelope proteins and various secretory proteins, including the G protein of stomatitis virus that contains endoglycosidase H sensitive oligosaccharides [297]. Interestingly, while BFA blocked G protein transport to the plasma membrane, the movement of cholesterol to the plasma membrane was unaffected. These data are consistent with the hypothesis that vesicles containing the viral G protein traverse the Golgi, whereas cholesterol bypasses this organelle en route to the cell surface. BFA also inhibited the proteolytic conversion of proalbumin to albumin and blocked the terminal glycosylation of α_1-protease inhibitor and haptoglobin in rat hepatocytes [298]. In mouse pituitary thyrotropic tumor tissue, BFA inhibited the carbohydrate processing of thyroid stimulating hormone, resulting in the accumulation of $Man_{5-8}(GlcNAc)_2$ [299]. BFA caused different effects on the sialylation of low density lipoprotein receptor and epidermal growth factor receptor depending upon the linkage types of their oligosaccharides [300]. This sialylation of N-linked oligosaccharides was blocked, but not that of O-linked oligosaccharides. The blockage by BFA of intracellular transport and processing of various glycoproteins was accompanied by a dissembly of the Golgi complex, and these effects were reversible [298,301–303]. Biochemical and immunocytochemical studies have shown an accumulation of Golgi-related enzymes in the endoplasmic reticulum of various cells [304–306]. The redistribution of the Golgi-specific enzymes, mannosidase II and galactosyltransferase, into the ER suggests that BFA disrupts the normal membrane flow and allows Golgi proteins to mix with ER proteins. BFA has been shown to prevent the assembly of the coats of non-clathrin-coated Golgi vesicles that are responsible for anterograde transport from ER to Golgi [305–307].

6. Compounds that modify protein structure or synthesis

6.1. β-Hydroxynorvaline and fluoroasparagine

The tripeptide sequence Asn-X-Thr (Ser), where X represents any amino acid except proline or aspartic acid, serves as the recognition site for the attachment of the $Glc_3Man_9(GlcNAc)_2$ oligosaccharide that is transferred from membrane-bound dolichyl-pyrophosphate [308,309].

The importance of threonine in the Asn-X-Thr recognition sequence for asparagine-linked glycosylation was tested by examining the effect of a threonine analog, β-hydroxynorvaline, on co-translational glycosylation in Krebs' II ascites tumor lysates [310]. β-Hydroxynorvaline inhibited the glycosylation of the α-subunit of human chorionic gonadotropin, indicating that this threonine analog acted via its incorporation into protein. The substitution of β-hydroxynorvaline for threonine caused steric hindrance for the attachment of the oligosaccharide chain. In cultured fibroblasts, β-hydroxynorvaline caused the formation of a variety of cathepsin D molecules having two, one, or no oligosaccharide chains. The non-glycosylated form was a minor species and was degraded within 45 minutes of its synthesis, as previously demonstrated with tunicamycin. Cathepsins with two or one oligosaccharide chains were normally segregated into lysosomes, and their proteolytic maturation was not affected [311]. In the study of rat hepatocytes, β-hydroxynorvaline not only blocks glycosylation of α_1-acid glycoprotein but also shows secretion of this protein with or without a concomitant effect on glycosylation [312]. Rat hepatocytes synthesized α_1-acid glycoprotein with zero to six oligosaccharide chains in the presence of β-hydroxynorvaline. Partially glycosylated (with one to five oligosaccharide chains) molecules and unglycosylated (tunicamycin-inhibited) molecules exited the cells more slowly than native α_1-acid glycoprotein, but secretion of fully glycosylated (with six oligosaccharide chains) glycoprotein was retarded in β-hydroxynorvaline-treated cells when compared to untreated cells. It was suggested that incorporation of β-hydroxynorvaline into the peptide also changed its structure in such a way as to reduce its rate of transport.

Threo-β-fluoroasparagine was selectively toxic to asparagine-requiring Jensen sarcoma cells when aspartic acid was included in the culture medium [313]. The cytotoxicity in mammalian cells in culture may involve inhibition of asparagine-linked glycosylation of proteins. In cell free translation systems, the threo-isomer of β-fluoroasparagine inhibited glycosylation at a concentration of 1 mM, and this effect was blocked by L-asparagine. However, the erythro-isomer was not incorporated into protein and had no effect on glycosylation [314]. Substitution of β-fluoroasparagine for asparagine in the short synthetic peptides resulted in decreased N-glycosylation in a cell free system [315]. The results implied an inhibition of N-linked glycosylation by producing protein substrates that were ineffective for glycosylation.

6.2. Inhibitors of protein synthesis

Several inhibitors of protein synthesis and RNA synthesis have been found to affect the synthesis of lipid-linked oligosaccharides [316–319]. In cell culture, when protein synthesis is blocked by cycloheximide or puromycin or when RNA synthesis is inhibited by actinomycin D, the incorporation of mannose into lipid-linked oligosaccharides is also inhibited. However, mannose incorporation into dolichyl-P-mannose is not inhibited under these circumstances [316,317,319]. The limitation in the amount of dolichyl-P available to serve as a carrier for the oligosaccharide was suggested as being the factor responsible for the inhibition. Using actinomycin D to depress the levels of mRNA or cycloheximide to inhibit protein synthesis, experiments showed that the synthesis of lipid-linked oligosaccharide was proportional to the rate of protein synthesis.

The regulated step appeared to be that prior to the formation of Man$_5$(GlcNAc)$_2$-PP-dolichol. These results suggested that a likely control point was the availability of dolichyl-P [318]. However, other studies indicated that the availability of dolichyl-P in the inhibited cells was not limiting since the formation of dolichyl-P-mannose did not stop in MDCK cells under these inhibitory conditions [317]. The addition of exogenous dolichyl-P to LM cells or to MDCK cells that were inhibited with cycloheximide did not overcome the inhibition of mannose incorporation into lipid-linked oligosaccharide, even though this addition of dolichyl-P to control cells greatly stimulated the incorporation of mannose into lipid-linked oligosaccharide [319,320]. Furthermore, the addition of the tripeptide acceptor, N-acetyl-Asn-Try-Thr, to the inhibited MDCK cells did overcome the puromycin inhibition to some extent [320]. This result suggests that the accumulation of some intermediates such as lipid-linked oligosaccharides might be involved in the inhibition. Such a feedback mechanism might play an important role in the control of lipid-linked oligosaccharide synthesis.

Abbreviations

Glc	glucose
Man	mannose
GlcNAc	N-acetylglucosamine
GlcN	glucosamine
GlcNAc-1-P	N-acetylglucosamine-1-phosphate
UDP-GlcNAc	UDP-acetylglucosamine
dolichyl-P	dolichyl phosphate
Asn	asparagine
Thr	threonine
Ser	serine
ACTH	adrenocorticotropic hormone
MDCK	Madin–Darby canine kidney
BHK	baby hamster kidney
VSV	vesicular stomatitis virus
GPI	glycosylphosphatidylinositol
HMG-CoA reductase	3-hydroxy-3-methylglutaryl CoA reductase

References

1. Olden, K., Parent, B. and White, S.L. (1982) Biochim. Biophys. Acta 652, 209–232.
2. Schwarz, R.T. and Datema, R. (1982) Adv. Carbohyd. Chem. and Biochem. 40, 287–379.
3. Hubbard, S.C. and Ivatt, R.J. (1981) Ann. Rev. Biochem. 50, 555–583.
4. Kornfeld, R. and Kornfeld, S. (1985) Am. Rev. Biochem. 54, 631–664.
5. Wong, T.K. and Lennarz, W.J. (1981) J. Biol. Chem. 257, 6619–6624.
6. Elbein, A.D. (1979) Ann. Rev. Plant Physical 30, 229–272.
7. Parodi, A.J. and Leloir, L.F. (1979) Biochim. Biophys. Acta, 599, 1–37.

8. Waechter, C.J. and Lennarz, W.J. (1976) Ann. Rev. Biochem. 45, 95–112.
9. Struck, D.K. and Lennarz, W.J. (1980) In: W.J. Lennarz (Ed.), The Biochemistry of Glycoproteins and Proteoglycans, Plenum Press, New York, pp. 35–83.
10. Bause, E. and Hettkamp, H. (1979) FEBS Lett. 108, 341–344.
11. Pless, D.D. and Lennarz, W.J. (1977) Proc. Natl. Acad. Sci. USA 74, 134–138.
12. Grinna, L.S. and Robbins, P.W. (1979) J. Biol. Chem. 254, 8814–8818.
13. Grinna, L.S. and Robbins, P.W. (1980) J. Biol. Chem. 255, 2255–2258.
14. Lubas, W.A. and Spiro, R.G. (1988) J. Biol. Chem. 263, 3990–3998.
15. Opheim, D.J. and Touster, O. (1978) J. Biol. Chem. 253, 1017–1023.
16. Tabas, I. and Kornfeld, S. (1979) J. Biol. Chem. 254, 11655–11663.
17. Forsee, W.T. and Schutzbach, J.S. (1981) J. Biol. Chem. 256, 6577–6583.
18. Bischoff, J. and Kornfeld, R. (1983) J. Biol. Chem. 258, 7907–7910.
19. Tabas, I. and Kornfeld, S. (1978) J. Biol. Chem. 253, 7779–7786.
20. Harpaz, N. and Schachter, H. (1980) J. Biol. Chem. 255, 4885–4893.
21. Harpaz, N. and Schachter, H. (1980) J. Biol. Chem. 255, 4894–4902.
22. Oppenheimer, C.L., Eckhardt, A.E. and Hill, R.L. (1981) J. Biol. Chem. 256, 11477–1182.
23. Elbein, A.D. (1987) Ann. Rev. Biochem. 56, 497–534.
24. Elbein, A.D. (1991) FASEB J., 5, 3055–3063.
25. Takatsuki, A., Arima, K. and Tamura, G. (1971) J. Antibiot. 24, 215–223.
26. Takatsuki, A. and Tamura, G. (1971) J. Antibiot. 24, 224–230.
27. Takatsuki, A., Shimizu, K.I. and Tamura, G. (1972) J. Antibiot. 25, 75–85.
28. Takatsuki, A., Kawamura, K., Okina, M., Kodama, Y., Ito, T. and Tamura, G. (1977) Agric. Biol. Chem. 41(11), 2307–2309.
29. Ito, T., Takasuki, A., Kawamura, K., Sata, K. and Tamura, G. (1980) Agric. Biol. Chem. 44(3), 695–698.
30. Suami, T., Sasai, H., Matsumo, K. and Suzuki, N. (1985) Carbohydr. Res. 143, 85–96.
31. Tkacz, J.S. and Lampen, J.O. (1975) Biochem. Biophys. Res. Commun. 65, 248–257.
32. Takatsuki, A., Kohno, K. and Tamura, G. (1975) Agric. Biol. Chem. 39, 2089–2091.
33. Struck, D.K. and Lennarz, W.J. (1977) J. Biol. Chem. 252, 1007–1013.
34. Erison, M.C., Gafford, J. and Elbein, A.D. (1977) J. Biol. Chem. 252, 7431–7433.
35. Lehle, L. and Tanner, W. (1976) FEBS Lett. 71, 167–170.
36. Kaushal, G.P. and Elbein, A.D. (1986) Plant Physiol. 81, 1086–1091.
37. Reitman, M.L. and Kornfeld, S. (1981) J Biol. Chem. 256, 4275–4281.
38. Heifetz, S., Keenan, R.W. and Elbein, A.D. (1979) Biochemistry 18, 2186–2192.
39. Keller, R.K., Boon, D.Y. and Crum, F.C. (1979) Biochemistry 18, 36–64.
40. Takatsuki, A. and Tamura, G. (1982) In: G. Tamura (Ed.), Tunicamycins, Japan Sci. Soc. Press, Tokyo, pp. 37–64.
41. Hickman, S., Kulczyki, A., Jr., Lynch, R.G., and Kornfeld, S. (1977) J. Biol. Chem. 252, 4402–4408.
42. Duskin, D. and Bornstein, P. (1977) J. Biol. Chem. 252, 959–962.
43. Struck, D.K., Suita, P.B., Lane, M.D. and Lennarz, W.J. (1978) J. Biol. Chem. 253, 5332–5337.
44. Olden, K., Pratt, R.M. and Yamada, K.M. (1978) Cell 13, 461–473.
45. Mahoney, W.C. and Duskin, D. (1979) J. Biol. Chem. 254, 6572–6576.
46. Tanzer, M.L., Rowland, I.N., Murray, L.W. and Kaplan, J. (1977) Biochim. Biophys. Acta 506, 187–196.
47. Ledford, B.E. and Davis, D.F. (1983) J. Biol. Chem. 258, 3304–3308.
48. Katz, N.R., Goldfarb, V., Liem, H. and Müller-Eberhard, U. (1985). Eur J. Biochem., 146, 155–159.
49. Mizrahi, S., O'Malley, J.A., Carter, W.A., Takatsuki, A., Tamura, G. and Sulkowski, E. (1978)

J. Biol. Chem. 253, 7612–7615.

50. Fujisawa, J., Iwakura, Y. and Kawade, Y. (1978) J. Biol. Chem. 253, 8677–8679.
51. Miller, A.L., Kress, B.C., Lewis, L., Stein, R. and Kinnon, C. (1980) Biochem. J. 186, 971–975.
52. Cox, G.S. (1981) Biochemistry 20, 4893–4900.
53. Budarf, M.L. and Herkert, E. (1982) J. Biol. Chem. 257, 10128–10135.
54. Swanson, J.C. and Suttie, J.W. (1985) Biochemistry 24, 3890–3897.
55. Hickman, S. and Kornfeld, S. (1978) J. Immunol. 121, 990–996.
56. Onishi, H.R., Tkacz, J.S. and Lampen, J.O. (1979) J. Biol. Chem. 254, 11943–11949.
57. Schwaiger, H. and Tanner, W. (1979) Eur. J. Biochem. 102, 375–387.
58. Rosen, O.M., Chia, G.H. Fung, C. and Rubin, C.S. (1979) J. Cell Physiol. 99, 37–42.
59. Reed, B.C., Ronnett, G.V. and Lane, M.D. (1981) Proc. Natl. Acad. Sci. USA 78, 2908–2912.
60. Ronnett, G.V., Knutson, V.P., Kohanski, R.A., Simpson, T.L. and Lane, M.D. (1984) J. Biol. Chem. 259, 4566–4575.
61. Bhargava, G. and Makman, M.H. (1980) Biochim. Biophys. Acta 629, 107–112.
62. Soderquist, A.M. and Carpenter, G. (1984) J. Biol. Chem. 259, 12586–12594.
63. Prives, J.M. and Olden, K. (1980) Proc. Natl. Acad. Sci. USA 77, 5263–5267.
64. Perkins, J.P., Toews, M.L. and Harden, T.K. (1984) Adv. Cyclic Nucleotide Protein Phosphoryl. Res. 17, 37–46.
65. Schwartz, I. and Hazum, E. (1985) Endocrinology 116, 2341–2346.
66. Baribault, T.J. and Neet, K.E. (1985) J. Neurosci. Res. 14, 49–60.
67. Chattergee, S., Kwiterovich, P.O. Jr., and Scherke, C.S. (1979) J. Biol. Chem. 254, 3704–3707.
68. Filipovic, I. and Von Figura, K. (1980) Biochem. J. 181, 373–375.
69. Law, P.Y., Umgar, H.G., Horn, D.S. and Loh, H.H. (1985) Biochem. Pharmacol. 34, 9–17.
70. Pan, Y.T., Schmidt, J.W., Sanford, B.A. and Elbein, A.D. (1979) J. Bacteriol. 139, 507–514.
71. Schule, C.M. and Hanafusa, H. (1971) Virology 45, 401–410.
72. Ogura, H. and Tris, R.R. (1975) J. Virol. 16, 443–446.
73. Halpern, M.S., Bologenes, D.P. and Frus, R.R. (1976) J. Virol. 18, 504–510.
74. Schnitzer, T.J. and Lodish, H.F. (1979) J. Virol. 29, 443–447.
75. Courtney, R.J., Steiner, S.M. and Benyesh-Melnick, M. (1973) Virology 52, 447–455.
76. Schwarz, R.T., Rohrschneider, J.M. and Schmidt, M.F.G. (1976) J. Virol. 19, 782–791.
77. Leavitt, R., Schlesinger, S. and Kornfeld, S. (1977) J. Biol. Chem. 252, 9018–9025.
78. Takatsuki, A. and Tamura, G. (1978) Agric. Biol. Chem. 42, 275–278.
79. Polonoff, E., Machida, C.A. and Kabat, D.A. (1982) J. Biol. Chem. 257, 14023–14028.
80. Schmaljohn, C.S., Hasty, S.E., Rasmussen, L and Dalrymple, J.M. (1986) J. Gen. Virol. 67, 707–717.
81. Petrie, B.L., Estes, M.K. and Graham, D.Y. (1983) J. Virol. 46, 270–274.
82. Olden, K., Parent, J.B. and White, S.L. (1982) Biochim. Biophys. Acta 650, 209–232.
83. Weintraub, B.D., Stannard, B.S. and Meyers, L. (1983) Endocrinology 112, 1331–1335.
84. Loh, Y.P. and Gainer, H. (1979) Endocrinology 105, 474–487.
85. Dulis, B.H., Kloppel, T.M., Grey, H. and Kubo, R.T. (1982) J. Biol. Chem. 257, 4369–4374.
86. Firestone, G.L. and Heath, E.C. (1981) J. Biol. Chem. 256, 1404–1411.
87. Gibson, R., Leavitt, R., Kornfeld, S. and Schlesinger, S. (1978) Cell 13, 671–679.
88. Gibson, R., Schlesinger, S. and Kornfeld, S. (1979) J. Biol. Chem. 254, 3600–36007.
89. Fujisawa, J., Iwakura, Y. and Kawade, Y. (1978) J. Biol. Chem. 253, 8677–8679.
90. Havell, E.S., Vileck, J., Falcoff, E. and Bermann, B. (1975) Virology 63, 475–483.
91. Olden, K., Pratt, R.M., Jaworski, C. and Yamada, K.M. (1979) Proc. Natl. Acad. Sci. USA 76, 791–795.
92. Kitagawa, K., Nishino, M. and Iwashima, S. (1985) Biochim. Biophys. Acta 821, 67–71.
93. Haspel, H.C., Birnbaum, M.J., Wilk, E.W. and Rosen, O.M. (1985) J. Biol. Chem. 260,

448

7219–7225.
94. White, M.K., Bramwell, M.E. and Harris, H. (1984) J. Cell Sci. 68, 257–270.
95. Barber, E.F., Handlogten, M.E. and Kilberg, M.S. (1983) J. Biol. Chem. 258, 11851–11855.
96. Eckardt, K. Wetzstein, H., Thrum, H. and Ihn, W. (1980) J. Antibiot. 33, 908–910.
97. Elbein, A.D., Occolowitz, J.L., Hamill, R.L. and Eckardt, K. (1987) Biochemistry 20, 4210–4216.
98. Tkacz, J. and Wong, A. (1978) Fed. Proc. 37, 1766.
99. Murazumi, N., Yamamori, S., Araki, Y. and Ito, E. (1979) J. Biol. Chem. 254, 11791–11793.
100. Kenig, M. and Reading, C. (1979) J. Antibiot. 32, 549–554.
101. Vogel, P., Stynes, B.A., Cookley, D., Yoeh, G.T. and Patterson, D.S. (1982) Biochem. Biophys. Res. Commun. 105, 835–840.
102. Frahn, J.L., Edgar, J.A., Jones, A.J., Cockrum, P.A., Anderton, N. and Culvenor, C.C.J. (1984) Aust. J. Chem. 37, 165–182.
103. Elbein, A.D., Gafford, J. and Kang, M.S. (1979) Arch. Biochem. Biophys. 196, 311–318.
104. Heinemann, B., Kaplan, M.A., Muir, R.D. and Hooper, I.R. (1953) Antibiot. Chemother. 3, 1239–1242.
105. Bodansky, M., Sigler, G.B. and Bodansky, A. (1953) J. Am. Chem. Soc. 95, 2353–2357.
106. Tanaka, H., Iwai, Y., Oiwa, R., Dinohara, S., Shimizer, S., Oha, T. and Bmura, S. (1972) Biochim. Biophys. Acta 497, 633–640.
107. Kang, M.S., Spencer, J.P. and Elbein, A.D. (1978) J. Biol. Chem. 253, 8860–8866.
108. Ericson, M.C., Gafford, J. and Elbein, A.D. (1978) Arch. Biochem. Biophys. 191, 698–704.
109. Banerjee, D.K., Scher, M.G. and Waechter, C.J. (1981) Biochemistry 20, 1561–1568.
110. Hayes, G.R. and Lucas, J.J. (1983) J. Biol. Chem. 258, 15095–15100.
111. Kyosseva, S.V. and Zhivkov, V.I. (1985) Int. J. Biochem. 17, 813–817.
112. Chapman, A., Fujimoto, K. and Kornfeld, S. (1980) J. Biol. Chem. 255, 4441–4446.
113. Jensen, J.W. and Schutzbach, J.S. (1981) J. Biol. Chem. 256, 12899–12904.
114. Elbein, A.D. (1981) Biochem. J. 193, 477–484.
115. Stone, K.J. and Strominger, J.L. (1971) Proc. Natl. Acad. Sci. USA 68, 3223–3227.
116. Chen, W.W. and Lennarz, W.J. (1976) J. Biol. Chem. 251, 7802–7809.
117. Herscovics, A., Bugge, B. and Jeanloz, R.W. (1971) FEBS Lett. 82, 215–218.
118. Spencer, J.P., Kang, M.S. and Elbein, A.D. (1978) Arch. Biochem. Biophys. 190, 829–837.
119. Reuvers, F., Boer, P. and Steyn-Parve, E.P. (1978) Biochem. Biophys. Res. Commun. 82, 800–805.
120. Meyers, E., Slusarchy, D.S., Bouchard, J.L. and Weisenborn, F.L. (1989) J. Antibiot. 22, 490–493.
121. Lugtenberg, E.J.J., Hellings, J.A. and Van de Berg, G.J. (1972) Antimicrobiol. Agents Chemother. 2, 485–490.
122. Babazinski, P. (1980) Eur. J. Biochem. 112, 53–58.
123. Villemez, C.L. and Carlo, P.L. (1980) J. Biol. Chem. 255, 8174–8178.
124. Welzel, P., Witteler, F.T., Müller, D. and Riemer, W. (1981) Angew Chem. Int. Ed. Engl. 93, 130–131.
125. Bause, E. and Legler, G. (1982) Biochem. J. 201, 481–487.
126. Nishimura, H., Mayama, M., Komatsu, Y., Kato, H., Shimaoka, N. and Tanaka, Y. (1964) Antibiotics. 17A, 148–154.
127. Kang, M.S., Spencer, J.P. and Elbein, A.D. (1979) J. Biol. Chem. 254, 10037–10043.
128. Müller, T., Bause, E. and Jaenicke, L. (1981) FEBS Lett. 128, 208–212.
129. Datema, R., Olofsson, S. and Romero, P.A. (1981) Pharmacol Ther. 33, 221–286.
130. Schwarz, R.T. and Datema, R. (1982) Adv. Carbohyd. Chem. Biochem. 40, 287–379.
131. McDowell, W. and Schwarz, R.T. (1988) Biochimie 70, 1535–1549.

132. Datema, R. and Schwarz, R.T. (1978) Eur. J. Biochem. 90, 505–516.
133. Datema, R., Pont-Lezica, R., Robbins, P.W. and Schwarz, R.T. (1981) Arch Biochem. Biophys. 206, 65–71.
134. Datema, R., Schwarz, R.T. and Jankowski, A.W. (1980) Eur. J. Biochem. 109, 331–341.
135. McDowell, W., Datema, R., Romero, P.A. and Schwarz, R.T. (1985) Biochemistry 24, 8145–8152.
136. Scholtissek, C. (1975) Curr. Top. Microbiol. Immunol. 70, 101–119.
137. Koch, H.U., Schwarz, R.T. and Scholtissik, C. (1979) Eur. J. Biochem. 94, 515–522.
138. Datema, R. and Schwarz, R.T. (1979) Biochem. J. 184, 113–123.
139. Keppler, D.O.R., Rudigier, J.F.M., Bischoff, E. and Decker, K. (1970) Eur. J. Biochem. 17, 246–253.
140. Weckbecker, G. and Keppler, D.O.R. (1982) Eur. J. Biochem. 128, 163–168.
141. McDowell, W., Weckbecker, G., Keppler, D.O.R. and Schwarz, R.T. (1986) Biochem. J. 233, 749–754.
142. Pan, Y.T. and Elbein, A.D. (1982) J. Biol. Chem. 257, 2795–2801.
143. Pan, Y.T. and Elbein, A.D. (1985) Arach. Biochem. Biophys. 242, 447–456.
144. Pan, Y.T., DeGespari, R., Warren, C.D., and Elbein, A.D. (1992) J. Biol. Chem. 267, 8991–8999.
145. Lisanti, M.P., Field, M.C., Caras, I.W., Menon, A.K. and Rodiguez-Boulan, E. (1991) EMBO J. 10, 1969–1977.
146. Pan, Y.T., Kamitani, T., Bhuvaneswaran, C., Hallaq, Y., Warren, C.D., Yeh, E.T.H. and Elbein, A.D. (1992) J. Biol. Chem. 267, 21250–21255.
147. Rearick, J.I., Chapman, A. and Kornfeld, S. (1981) J. Biol. Chem. 256, 6255–6261.
148. Gershman, H. and Robbins, P.W. (1981) J. Biol. Chem. 256, 7774–7780.
149. Turco, S.J. (1980) Arch, Biochem. Biophys. 205, 330–339.
150. Turco, S.J. and Picard, J.L. (1982) J. Biol. Chem. 257, 8674–8679.
151. Baumann, H. and Jahreis, G.P. (1983) J. Biol. Chem. 258, 3942–3949.
152. Datema, R. and Schwarz, R.T. (1981) J. Biol. Chem. 256, 11191–11198.
153. Spiro, R.G., Spiro, M.J. and Bhoyroo, V.D. (1983) J. Biol. Chem. 258, 9469–9476.
154. Kono, T., Suzuki, D., Dansey, L.E., Robinson, F.W. and Blevins, T.L. (1981) J. Biol. Chem. 256, 6400–6407.
155. Hohenschutz, L.D., Bell, E.A., Jewess, P.J., Leworthy, P., Pryce, R.J., Arnold, E. and Clardy, J. (1981) Phytochemistry 20, 811–814.
156. Saul, R., Molyneux, R.J. and Elbein, A.D. (1984) Arch. Biochem. Biophys. 230, 668–675.
157. Pan, Y.T., Ghidoni, J. and Elbein, A.D. (1993) Arch. Biochem. Biophys. 134–144, 303.
158. Saul, R., Ghidoni, J., Molyneux, R.J. and Elbein, A.D. (1985) Proc. Natl. Acad. Sci. USA 82, 93–97.
159. Pan, Y.T., Hori, H., Saul, R., Sanford, B.A., Molyneux, R.J. and Elbein, A.D. (1983) Biochemistry 22, 3975–3984.
160. Elbein, A.D. (1987) Methods Enzymol. 138, 661–709.
161. Hori, H., Pan, Y.T., Molyneux, R.J. and Elbein, A.D. (1984) Arch. Biochem. Biophys. 228, 525–533.
162. Frommer, W., Junge, B., Mueller, L., Schmidt, D. and Truscheit, E. (1979) Planta Med. Phytother. 35, 195–207.
163. Schmidt, D., Frommer, W., Mueller, L. and Truscheit, E. (1979) Naturwissenschaften 66, 504–585.
164. Chambers, J.P., Elbein, A.D. and Williams, J.C. (1982) Biochem. Biophys. Res. Commun. 107, 1490–1496.
165. Saunier, B., Kilker, R.D., Tkacz, J.S., Quaroni, A. and Herscovics, A. (1982) J. Biol. Chem.

257, 14155–14161.

166. Romero, P.A., Saunier, B. and Herscovics, A. (1985) Biochem. J. 226, 733–740.

167. Romero, P.A., Friendlander, P. and Herscovics, A. (1985) FEBS Lett. 183, 29–32.

168. Hettkamp, H., Bause, E. and Legler, G. (1982) Biosci. Rep. 2, 899–906.

169. Murai, H., Enomoto, H., Aoyagi, Y., Yoshikuri, Y., Masahiro, J. and Sirahase, I. (1977) Deutsche Offenlegungsschrift 2824781 Anmelder Nippon Shiriyaku Co., Ltd. Kyoto (Japan) Vol. 4-1-1979.

170. Romero, P.A., Datema, R. and Schwarz, R.T. (1983) Virology 130, 238–242.

171. Welter, A., Jadot, J., Dardenne, G., Marlier, M. and Casimic, J. (1976) Phytochemistry 15, 747–749.

172. Fleet, G.W.J. and Smith, P.W. (1985) Tetrahedron Lett. 26, 1469–1472.

173. Evans, S.V., Fellows, L.E., Shing, T.K.M. and Fleet, G.W.J. (1985) Phytochemistry 24, 1953–1955.

174. Elbein, A.D., Mitchell, M., Sanford, B.A., Fellows, L.E. and Evans, S.V. (1984). J. Biol. Chem. 259, 12409–12413.

175. Romero, P.A., Friendlauder, P., Fellows, L., Evans, S.V. and Herscovics, A. (1985) FEBS Lett. 184, 197–201.

176. Schmidt, J.W. and Catterall, W.A. (1987) J. Biol. Chem. 262, 13713–13723.

177. Schwartz, P. and Elbein, A.D. (1985) J. Biol. Chem. 260, 14452–14458.

178. Merkle, R., Elbein, A.D. and Heifetz, A. (1985) J. Biol. Chem. 260, 1083–1089.

179. Schlesinger, S., Koyama, A.H., Malfer, C., Gee, S.L. and Schlesinger, M.J. 1985, Virus Res. 2, 139–149.

180. Schlesinger, S., Malfer, C. and Schlesinger, M.J. (1984) J. Biol. Chem. 259, 7597–7601.

181. Lodish, H.F. and Kong, N. (1984) J. Cell Biol. 98, 1720–1729.

182. Peyrieras, N., Bause, E., Legler, G., Vasilov, R., Claesson, L., Peterson, P. and Ploegh, H. (1983) EMBO J. 2, 823–832.

183. Gross, V., Andus, T., Tran-Thi, T-A., Schwarz, R.T., Decker, K. and Heinrich, P.C. (1983) J. Biol. Chem. 258, 12203–12209.

184. Slicker, L.J., Martenson, T.M. and Lane, M.D. (1986) J. Biol. Chem. 261, 15233–15241.

185. Smith,M.M., Schlesinger, S., Lindstrom, J. and Merlie, J.P. (1986) 261, 14825–14832.

186. Arakai, R.F., Hedo, J.A., Collier, E. and Gordon, P. (1987) J. Biol. Chem. 262, 11886–11892.

187. Edwards, E., Sprague, E.A., Kelley, J.L., Kerbacher, J.J., Schwartz, C.J. and Elbein, A.D. (1989) Biochemistry 28, 7676–7687.

188. Wall, K.A., Pierce, J.D. and Elbein, A.D. (1988) Proc. Natl. Acad. Sci. USA 85, 5644–5648.

189. Repp, R., Tamura, T., Boschek, C.B., Wege, H., Schwarz, R.T. and Niemann, H. (1985) J. Biol. Chem. 260, 15873–15879.

190. Oden, K., Breton, P., Grzegorzewski, K., Yasuda, Y., Gause, B.L., Oredipe, O.A., Newton, S.A. and White, S.L. (1991) Pharmacol. Ther. 50, 285–290.

191. Ostrander, G.K., Scribner, N.K. and Rohrschneider, L.R. (1988) Cancer Res. 48, 1091–1094.

192. Desantis, R., Santer, V.V. and Glick, M.C. (1987) Biochem. Biophys. Res. Commun. 142, 348–353.

193. Humphris, M.J., Matsumoto, K., White, S.L. and Olden, K. (1986) Cancer Res. 46, 5215–5222.

194. Mercurio, A.M. (1986) Proc. Natl. Acad. Sci. USA 83, 2609–2613.

195. Gruters, R.A., Neefjes, J.J., Tersmette, M., deGoede, R.E.Y., Tulp, A., Huisman, H.G., Miedema, F. and Ploegh, H.L. (1987) Nature 330, 74–77.

196. Walker, B.D., Kowalski, M., Goh, W.C., Kozarsky, K., Krieger, M., Rosen, C., Rohrschneider, L., Haseltine, W.A. and Sodorski, J. (1987) Proc. Natl. Acad. Sci. USA 84, 8120–8124.

197. Fleet, G.W., Karpas, A., Dwek, R.A., Fellows, L.E., Tyms, A.S., Petursson, S., Namgoog, S.K., Ramsden, N.G., Smith, P.W., Son, J.C, Wilson, F., Witty, D.R., Jacob, G.S. and Rademacher,

T.W. (1988) FEBS Lett. 237, 128–132.

198. Ruprecht, R.M., Mullaney, S., Anderson, J. and Bronson, R. (1989) J. Acquir. Immun. Defic. Syndr. 2, 149–157.

199. Stephens, E.B., Monck, E., Reppas, K. and Butfiloski, E.J. (1991) J. Virol. 65, 1114–1123.

200. Karpas, A. Fleet, G.W.J., Dwek, R.A., Petursson, S., Namgoong, S.K., Ramsden, N.G., Jacob, G.S. and Rademacher, T.W. (1988) Proc. Natl. Acad. Sci. USA 85, 9229–9233.

201. Taylor, D.L., Sunkara, P.S., Liu, P.S., Kang, M.S., Bowlin, T.L., Tyms, A.S. (1991) AIDS 5, 693–698.

202. Molyneux, R.J., Benson, M., Wong, R.Y., Tropea, J. and Elbein, A.D. (1988) J. Nat. Prod. 51, 1198–1206.

203. Tropea, J., Molyneux, R.J., Kaushal, G.P., Pan, Y.T., Mitchell, M. and Elbein, A.D. (1989) Biochemistry 28, 2027–2034.

204. Nash, R.J., Fellows, L.E., Dring, J.V., Fleet, G.W.J., Derome, A.E., Hamor, T.A., Scofield, A.M. and Watkins, D.J. (1988) Tetrahedron Lett. 29, 2487–2490.

205. Nash, R.J., Fellows, L.E., Plant, A.C., Fleet, G.W.J., Derome, A.E., Baird, P.D., Hegarty, M.P. and Scofield, A.M. (1988) Tetrahedron 44, 5959–5964.

206. Fleet, G.W.J., Haraldson, M., Nash, R.J. and Fellows, L.E. (1988) Tetrahedron Lett. 29, 5441–5444.

207. Fleet, G.W.J., Nicholas, S.J., Smith, P.W., Evans, S.V., Fellows, L.E. and Nash, R.J. (1985) Tetrahedron Lett. 26, 3127–3130.

208. McDowell, W., Fleet, G.W.J., Fellows, L.E. and Schwarz, R.T. (1988) In Swainsonine and Related Glycosidases Inhibitors (James L.F. et al., eds.) pp. 220–230, Iowa State University Press, Ames.

209. Rhinehart, B.L., Robinson, K.M., Liu, P.S., Payne, A.J., Wheatley, M.E. and Wagner, S.R. (1987) J. Pharmacol. Exp. Ther. 241, 915–921.

210. Kaushal, G.P., Pan, Y.T., Tropea, J.E., Mitchell, M., Liu, P. and Elbein, A.D. (1988) J. Biol. Chem. 263, 17278–17283.

211. Legler, G. (1977) Methods Enzymol. 46, 368–381.

212. Detama, R., Romer, P.A., Legler, G. and Schwarz, R.T. (1982) Proc. Natl. Acad. Sci. USA 79, 6786–6791.

213. Legler, G. and Julich, E. (1984) Carbohyd. Res. 128, 61–72.

214. Fuhrmann, U., Bause, E., Legler, G. and Ploegh, H. (1984) Nature 307, 755–758.

215. Elbein, A.D., Legler, G., Tlusty, A., McDowell, W. and Schwarz, R.T. (1984) Arch. Biochem. Biophys. 235, 579–588.

216. Burke, B., Matlin, K., Bause, E., Legler, G., Peyrieras, N. and Ploegh, H. (1984) EMBO J. 3, 551–556.

217. Gross, V., Steuse, K., Tran-Thi, T-A., McDowell, W., Schwarz, R.T., Decker, K., Gerok, W. and Heinrich, P.C. (1985) Eur. J. Biochem. 150, 41–46.

218. Snider, M.D. and Rogers, O.C. (1986) J. Cell Biol. 103, 265–275.

219. Bischoff, J., Liscum, L. and Kornfeld, R. (1986) J. Biol. Chem. 261, 4766–4774.

220. Kayakiri, H., Takese, S., Shibata, T., Okamoto, M., Terano, H., Hashimoto, M., Tada, T. and Koda, S. (1989) J. Org. Chem. 54, 4015–4016.

221. Elbein, A.D., Tropea, J.E., Mitchell, M. and Kaushal, G.P. (1990) J. Biol. Chem. 265, 15599–15605.

222. Elbein, A.D., Kerbacher, J.K., Schwartz, C.J. and Sprague, E.A. (1991) Arch. Biochem. Biophys. 288, 177–184.

223. Papandreau, G. and Ganem, B. (1991) J. Am. Chem. Soc. 113, 8984–8985.

224. Pan, Y.T., Kaushal, G.P., Papandreou, G., Genem, B. and Elbein, A.D. (1992) J. Biol. Chem. 267, 8313–8318.

452

225. Palamarczyk, G., Mitchell, M., Smith, P.W., Fleet, G.W.J. and Elbein, A.D. (1985) Arch. Biochem. Biophys. 243, 35–45.

226. Coxon, B. and Fletcher, H.G., Jr. (1964) J. Am. Chem. Soc. 86, 922–926.

227. Van Diggelen, O.P., Galjaard, H., Sinnott, M.L. and Smith, P.J. (1980) Biochem. J. 188, 337–343.

228. Docherty, P.A., Kuranda, M.J., Aronson, N.N., Jr., BeMiller, J.N., Myers, R.W. and Bohn, J.A. (1986) J. Biol. Chem. 261, 3457–3463.

229. Docherty, P.A. and Aronson, N.N., Jr. (1987) Biochim. Biophys. Acta 914, 283–288.

230. Aoyagi, T., Yamamoto, T., Kojiri, K., Morishima, H., Nagai, M., Hamada, M., Takeuchi, T. and Umezawa, H. (1989) J. Antibiot. 42, 883–889.

231. Tropea, J.E., Kaushal, G.P., Pastuszak, I., Mitchell, M., Aoyagi, T., Molyneux, R.T. and Elbein, A.D. (1990) Biochemistry 29, 10062–10069.

232. Colegate, S.M., Dorling, P.R. and Huxtable, C.R. (1979) Aust. J. Chem. 32, 2257–2264.

233. Dorling, P.R., Huxtable, C.R. and Colegate, S.M. (1978) Appl. Neuobiol. 4, 285–295.

234. Schneider, M.J., Ungemach, F.S., Broquist, H.P. and Harris, T.M. (1983) Tetrahedron 39, 29–32.

235. Molyneux, R.J. and James, F. (1982) Science 216, 190–191.

236. Davis, D., Schwarz, P., Hernander, T., Mitchell, M., Warnock, B. and Elbein, A.D. (1984) Plant Physiol. 76, 972–975.

237. Suami, T., Kinichi, T. and Youichi, I. (1984) Chem. Lett. 4, 513–516.

238. Ali, M.H., Hough, L. and Richardson, A.C. (1984) J. Chem. Soc. Chem. Commun. 447–448.

239. Adams, C.E., Walker, F.J. and Sharpless, K.B. (1985) J. Org. Chem. 50, 420.

240. Yasada, N., Tsutsumi, H. and Takaya, T. (1984) Chem. Lett. 1984, 1201–1204.

241. Dorling, P.R., Huxtable, C.R. and Colegate, S.M. (1980) Biochem. J. 191, 649–651.

242. Kang, M.S. and Elbein, A.D. (1983) Plant Physiol. 71, 551–554.

243. Broquist, H.P. (1985) Ann. Rev. Nutr. 5, 391–409.

244. Elbein, A.D., Solf, R., Dorling, P.R. and Vosbeck, K. (1981) Proc. Natl. Acad. Sci. USA 78, 7393–7397.

245. Elbein, A.D., Pan, Y.T., Solf, R. and Vosbeck, K. (1983) J. Cell Physiol. 115, 265–275.

246. Elbein, A.D., Dorling, P.R., Vosbeck, K. and Horisberger, M. (1982) J. Biol. Chem. 257, 1573–1576.

247. Tulsiani, D.R.P., Hubbard, S.C., Robbins, P.W. and Touster, O. (1982) J. Biol. Chem. 257, 3660–3668.

248. Bischoff, J., Liscum, L. and Kornfeld, R. (1986) J. Biol. Chem. 261, 4766–4774.

249. Kang, M.S. and Elbein, A.D. (1983) J. Virol. 46, 60–69.

250. Arumugham, R.G. and Tanzer, M.L. (1983) J. Biol. Chem. 258, 11883–11889.

251. Foddy, L., Feeney, J. and Hughes, R.C. (1986) Biochem. J. 233, 697–706.

252. Franc, J.L., Housepian, S., Fayet, G. and Bouchilloux, S. (1986) Eur. J. Biochem. 157, 225–232.

253. Whitsett, J.A., Ross, G., Weaver, T., Rice, W., Dion, C. and Hull, W. (1985) J. Biol. Chem. 260, 15273–15279.

254. Lee, A.Y. and Doyle, D. (1985) Biochemistry 24, 6238–2645.

255. Wagner, D.D., Mayadas, T., Urban-Pickering, M., Lewis, B.H. and Marder, V.J. (1985) J. Cell. Biol. 101, 112–120.

256. Gross, V., Tran-Thi, A.A., Vosbeck, K. and Heinrich, R.C. (1983) J. Biol. Chem. 258, 4032–4036.

257. Yeo, T.K., Yeo, K.T., Parent, J.B. and Olden, K. (1985) J. Biol. Chem. 260, 2565–2569.

258. Duronio, V., Jacobs, S. and Cuatracasas, P. (1986) J. Biol. Chem. 259, 12586–12594.

259. Soderquist, A.M. and Carpenter, G. (1984) J. Biol. Chem. 259, 12586–12594.

260. Breitfeld, P.P., Rup, D. and Schwartz, A.L. (1984) J. Biol. Chem. 259, 10414–10421.

261. Chung, K.M., Shepard, V.L. and Stahl, P. (1984) J. Biol. Chem. 259, 14637–14641.

262. Hadwiger, A., Niemann, H., Kablisch, A., Bauer, H. and Tamura, T. (1986) EMBO J. 5, 689–694.

263. Bar-Scavit, Z., Kahn, A.J., Pegg, L.E., Stone, R.R., and Teitelbaum, S.L. (1984) J. Clin. Invest. 73, 1277–1283.

264. Villalta, F. and Kierszenbaum, F. (1985) Mol. Biochem. Parasitol. 16, 1–10.

265. Humphries, M.J., Matsumoto, K., White, S.L., Molyneux, R.J. and Olden, K. (1988) Cancer Res. 48, 1410–1415.

266. Dennis, J.W. (1986) Cancer Res. 46, 5131–5136.

267. Humphries, M.J., Matsumoto, K., White, S.L. and Olden, K. (1986) Proc. Natl. Acad. Sci. USA 83, 1752–1756.

268. Hino, M., Nakayama, O., Tsurami, Y., Adachi, K. and Shibata, T. (1985) J. Antiobiot. 38, 926–935.

269. Bowlin, T.L., McKown, B.J., Kang, M.S. and Sunkara, P. (1989) Cancer Res. 49, 4109–4113.

270. De Santis, R., Santer, U.V. and Glick, M.C. (1987) Biochem. Biophys. Res. Commun. 142, 348–353.

271. Brown, M.S. and Goldstein, J.L. (1980) J. Lipid Res. 21, 505–517.

272. Mills, J.T. and Adamany, A.M. (1978) J. Biol. Chem. 253, 5270–5273.

273. James, M.J. and Kandutsch, A.A. (1979) J. Biol. Chem. 254, 8442–8446.

274. Keller, R.K., Adair, W.J.,Jr., and Ness, G.C. (1979) J. Biol. Chem. 254, 9966–9969.

275. White, D.A., Middleton, B., Pawson, S., Bradshaw, J.P., Clegg, R.J., Hemming, F.W. and Bell, G.D. (1981) Arch. Biochem. Biophys. 208, 31–36.

276. Endo, AA., Kuroda, M. and Tanzawa, K. (1976) FEBS Lett. 72, 323–326.

277. Brown, A.G., Smale, T.C., King, T.J., Hasenkamp, R. and Thompson, R.H. (1976) J. Chem. Soc. Perkin Trans. 1, 1165–1170.

278. Goldstein, J.L., Helgeson, J.A.S. and Brown, M.S. (1979) J. Biol. Chem. 254, 5403–5409.

279. Carson, D. and Lennarz, W.J. (1979) Proc. Natl. Acad. Sci. USA 76, 5709–5713.

280. Surani, M.A., Kimber, S.J. and Osborn, J.C. (1983) J. Embryol. Exp. Morphol. 75, 205–223.

281. Astrand, I.M., Fries, E., Chojnacki, T. and Dallner, G. (1986) Eur. J. Biochem. 155, 447–452.

282. Tartakoff, A.M. and Vassalli, P. and Detraz, M. (1977) J. Exp. Med. 41, 1332–1345.

283. Tartakoff, A.M. and Vassalli, P. (1978) J. Cell Biol. 79, 694–707.

284. Tartakoff, A.M. (1983) Cell 32, 1026–1028.

285. Strous, G.J.A.M. and Lodish, H.F. (1980) Cell 22, 709–717.

286. Uchida, N., Smilowitz, H. and Tanaez, M.L. (1979) Proc. Natl. Acad. Sci. USA 76, 1868–1872.

287. Kuhn, L.J., Hadman, M. and Sabban, E.L. (1986) J. Biol. Chem. 261, 3816–3825.

288. Kaariainen, L., Hashimoto, K., Saraste, J., Virtanen, I. and Penttinen, K. (1980) J. Cell Biol. 87, 783–791.

289. Tartakoff, A.M. (1982) Trends Biochem. Sci. 7, 174–176.

290. Danielsen, E.M., Cowell, G.M. and Poulsen, S.S. (1983) Biochem. J. 216, 37–42.

291. Stewart, J.R. and Kenny, A.J. (1984) Biochem. J. 224, 559–568.

292. Pesonen, M. and Kaariainen, L. (1982) J. Mol. Biol. 158, 213–230.

293. Gruters, R.A., Nufjes, J.J., Tersmette, M., De Gaede, R.E.Y., Tulp, A., Huisman, H.G., Miedema, F. and Ploegh, H.L. (1987) Nature 330, 74–77.

294. Matthews, T.J., Weinhold, K.J., Lyerly, H.K., Langlois, A.J., Wigzell, H. and Bolognesi, D.P. (1987) Proc. Natl. Acad. Sci. USA 84, 5424–5428.

295. McCune, J.M., Rabin, L.B., Feidberg, M.B., Lieberman, M., Kosek, J.C., Reyes, G.R. and Weissmann, I.L. (1988) Cell 53, 55–67.

296. Dewar, R.L., Vasudevachari, M.B., Natarajan, V. and Salzman, N.P. (1989) J. Virol. 63, 2452–2456.

297. Takatsuki, A. and Tamura, G. (1985) Agric. Biol. Chem. 49, 899–902.

298. Misumi, Y., Miki, K., Takatsuki, A., Tamura, G. and Ikehera, Y. (1986) J. Biol. Chem. 261, 11398–11403.

299. Perkel, V.S., Liu, A.Y., Miura, Y. and Magner, J.A. (1988) Endocrinology 123, 310–318.

300. Shiti, S., Seguchi, T., Mizoguchi, H., Ono, M. and Kuwano, M. (1990) J. Biol. Chem. 265, 17385–17388.

301. Fujiwara, T., Oda, K., Yokota, S., Takatsuki, A. and Ikehara, Y. (1988) J. Biol. Chem. 263, 18545–18552.

302. Doms, R.W., Russ, G. and Yewdell, J.W. (1989) J. Cell Biol. 109, 61–72.

303. Nuchtern, J.G., Bonifacino, J.S., Biddison, W.E. and Klausner, R.D. (1989) Nature 339, 223–226.

304. Fujiwara, T., Oda, K. and Ikehara, Y. (1989) Cell Struct. Funct. 14, 605–616.

305. Lippincott-Schwartz, J., Yuan, L.C., Bonifacino, J.S. and Klausner, R.D. (1989) Cell 56, 801–813.

306. Lippincott-Schwartz, J., Donaldson, J.G., Schweizer, A., Berger, E.G., Hauri, H., Yuan, L.C. and Klausner, R.D. (1990) Cell 60, 821–836.

307. Ulmer, J.B. and Palade, G.E. (1989) Proc. Natl. Acad. Sci. USA 86, 6992–6996.

308. Marshall, R.D. (1974) Biochem. Soc. Symp. 40, 17–26.

309. Hart, G.W., Brew, K., Grant, G.A., Bradshaw, R.A. and Lennarz, W.J. (1979) J. Biol. Chem. 254, 9747–9754.

310. Hortin, G. and Boime, I. (1980) J. Biol. Chem. 255, 8007–8010.

311. Hentez, M., Masilik, A. and Von Figura, K. (1984) Arach. Biochem. Biophsy. 230, 375–382.

312. Docherty, P.A. and Aronson, N.N., Jr. (1985) J. Biol. Chem. 260, 10847–10855.

313. Stern, A.M., Foxman, B.M., Tashjian, A.H., Jr. and Abeles, R.H. (1982) J. Med. Chem. 25, 544–550.

314. Hortin, G., Stern, A.M., Miller, B., Abeles, R.H. and Boime, I. (1983) J. Biol. Chem. 258, 4047–4050.

315. Rothod, P.K., Tashjian, A.H., Jr. and Abeles, R.H. (1986) J. Biol. Chem. 261, 6461–6469.

316. Spiro, M.J., Spiro, R.G. and Bhoyroo, V.D. (1976) J. Biol. Chem. 251, 6400–6408.

317. Schmitt, J.W. and Elbein, A.D. (1979) J. Biol. Chem. 254, 12291–12294.

318. Hubbard, S.C. and Robbins, P.W. (1980) J. Biol. Chem. 255, 11782–11793.

319. Grant, S.R. and Lennarz, W.J. (1983) Eur. J. Biochem. 134, 575–583.

320. Pan, Y.T. and Elbein, A.D. (1990) Biochemistry 29, 8077–8084.

J. Montreuil, H. Schachter and J.F.G. Vliegenthart (Eds.), *Glycoproteins*

455

Bacterial glycoproteins

MANFRED SUMPER[1] and FELIX T. WIELAND[2]

[1]*Universität Regensburg, Lehrstühl Biochemie I, Postfach 397,*
93053 Regensburg, Germany
[2]*Institut für Biochemie I, Im Neuenheimer Feld 328, 69120 Heidelberg, Germany*

1. Introduction

Since the first demonstration of a covalent linkage between carbohydrate and protein by the groups of Neuberger and Jovins in 1958, glycoproteins were believed to be restricted to the eukaryotic kingdom. About twenty years later, first reports appeared on glycoproteins in Archaebacteria, and consequently, glycosylation in prokaryotes was taken as a speciality of Archaea, and used to emphasize the non-eubacterial character of Archaebacteria. Another decade later, glycoproteins were established as constituents of Eubacteria as well. Thus, the ability to glycosylate proteins is ubiquitous in all kingdoms of life.

During the past ten years or so, quite a variety of glycoconjugate structures as well as of linkage units between carbohydrate and protein have been isolated from prokaryotes and characterized, and none of them was known to occur in eukaryotes. Here, we would like to present an overview on structures and linkages of archaebacterial as well as eubacterial glycoproteins known to date, and describe in some detail and compare with the eukaryotic system what we know about the molecular mechanisms of their biosynthesis in bacteria. Many of the data given here and more details are found in reviews that either focus on eubacterial S-layer glycoproteins [1,2,22] or on archaebacterial glycoproteins [3,4].

2. Glycoproteins as constituents of bacterial S-layers

The first prokaryotic glycoprotein was discovered in the surface layer (denoted as S-layer according to Sleytr [5]) of the archaebacterium *Halobacterium salinarium* (which is identical to *Halobacterium halobium* used for most of the structural work on this glycoprotein, see below) by Mescher and Strominger [6]. S-layers surround many eubacterial and archaebacterial cells as the outermost component of the cell wall. They consist of a single protein species that is able to form a two-dimensional crystalline layer (for a recent review see Ref. [2]). Halobacteria of the genus *Halobacterium* live in saturated sodium chloride solutions. Their cell wall only consists of an S-layer with hexagonally arranged protein subunits and this S-layer is very tightly joined to the plasma membrane. The original finding was that this halobacterial S-layer is composed of a single glycoprotein species. This glycoprotein species was reported to have a molecular mass of about 200 kDa and to contain 10–12% (w/w) of carbohydrate, attached to the polypeptide via N- and O-glycosyl linkages, analogous to those found in eukaryotes [6]. Subsequent work on the detailed chemical structure of this halobacterial glycoprotein has corrected most of these early reports.

In the meantime, evidence has accumulated that S-layer subunits from many different species of archaebacteria and eubacteria represent glycoproteins as well.

2.1. Archaebacteria

2.1.1. Halobacteria

The most detailed chemical and genetic characterization of S-layer glycoproteins was performed on the extreme halophile *Halobacterium halobium* and on the moderate halophile *Haloferax volcanii*.

The S-layer glycoprotein from *H. halobium* is extremely acidic [7]. More than 50 sulfate residues were found to be covalently attached per glycoprotein molecule [7]. This high degree of sulfation has allowed for an easy and selective labelling of the glycoprotein: halobacteria do not reduce sulfate, but rather they incorporate radioactive sulfate exclusively into a few sulfated macromolecules. This selective labeling has greatly facilitated studies on the structure and the biosynthesis of the S-layer glycoprotein.

Exhaustive pronase digestion of the purified glycoprotein and subsequent chromatography on a Biogel P-10 column resulted in the separation of three glycopeptide fractions [8]. Work over several years gave a detailed picture of the chemical structures involved. To summarize these data, a schematic representation of the S-layer glycoprotein of *H. halobium* is given in Fig. 1.

A high molecular mass saccharide is attached to the asparagine residue in position 2 of the mature glycoprotein and this saccharide consists of repeating units of a pentasaccharide [8,9]. Remarkable features of this saccharide are the occurrence of a methylated galacturonic acid and a furanosidic galactose. Both these unusual sugars are linked peripherally to the backbone of the polysaccharide that consists of a linear chain of GalNAc-GalA-GlcNAc repeats. The overall chain length of this glycosaminoglycan-like polysaccharide ranges between 10 and 20 repeats of the pentasaccharide unit. Each pentasaccharide bears two sulfate ester residues, one attached to the 4-position of GalNAc. Unlike the eukaryotic glycosaminoglycans, this halobacterial glycosaminoglycan does not contain a special linker region, but the GalNAc residue at the reducing end is linked directly to an asparagine residue of the polypeptide as was proven by the isolation and chemical characterization of the novel linkage unit asparaginyl-GalNAc [10]. This asparagine residue is located within the typical N-glycosyl acceptor sequence Asn-X-Thr (Ser) that has been established as the minimal peptide structure common to all eukaryotic N-glycosidically linked saccharides investigated so far. One repeating unit saccharide is present per glycoprotein molecule.

Another type of sulfated saccharides with a low molecular mass is found to be present in about 10 copies per protein molecule [8,11]. These sulfated oligosaccharides are again linked to the polypeptide in N-glycosyl bond, and asparaginylglucose was

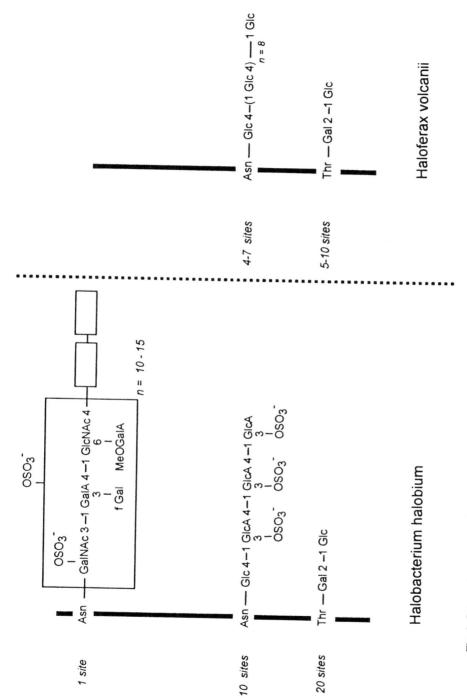

Fig. 1. Schematic representation of the S-layer glycoproteins of *H. halobium* and *Haloferax volcanii* according to Refs. [8–12,19].

characterized as another novel linkage unit [12]. Again, the asparagine involved is a constituent of the Asn-X-Thr (Ser) sequence. This linkage unit is extended by a chain of two or three β-1,4-linked glucuronic acids. About one third of these glucuronic acid residues are found to be replaced by iduronic acid [13]. A sulfate residue in ester linkage to the 3-position of each glucuronic acid further increases the negative charge density of these oliogosaccharides. Strikingly, the N-glycosyl linkage unit is not restricted to bacteria, but has most recently been isolated from the mammalian basement membrane protein laminin [14]. This protein forms an extracellular matrix together with other components like collagen type IV, fibronectin and heparansulfate proteoglycan.

Finally, about 20 copies of a neutral disaccharide Glc-(α1-2)Gal are attached to threonine residues in O-glycosyl linkage [8]. These disaccharide units occur in a highly clustered arrangement within the polypeptide chain as was concluded from amino acid sequence data. Again, the same type of disaccharide is known to occur in the eukaryotic protein collagen [15]. With these features and the occurrence of a glycosaminoglycan, this prokaryotic S-layer glycoprotein structurally resembles the eukaryotic proteoglycan–collagen complexes.

The primary structure of the cell surface glycoprotein was elucidated by the identification and complete characterization of the corresponding gene [16]. Some of the features of the deduced amino acid sequence are summarized as follows:

1. The mature glycoprotein consists of a polypeptide chain with 818 amino acids.
2. The open reading frame encodes an N-terminal leader peptide of 34 amino acid residues reminiscent of a typical signal peptide. The sequence preceding the cleavage site is Ala-Ala-Ala.
3. Whereas the complete polypeptide chain mainly consists of polar and negatively charged amino acids there is a single exception at the C-terminal end. Only three amino acid positions away from the C-terminus a stretch of 21 amino acid residues exclusively consists of hydrophobic amino acids and this hydrophobic peptide most probably serves as a membrane anchor. Close to this membrane binding domain, a cluster of 14 threonine residues could be identified as the binding site of all the above-mentioned neutral disaccharides. Perhaps this structural element located immediately above the membrane binding domain serves as a spacer to the extracellular domain of the glycoprotein [17]. A total of 12 potential N-glycosylation ('sequon') sites Asn-X-Thr/Ser are found throughout the polypeptide chain. As mentioned above, the very first sequon structure is found at position 2 of the mature polypeptide chain, and it is this Asn-residue to which the repetitive pentasaccharide structure is linked occurring once per glycoprotein molecule. With a single exception, all the other sequon structures distributed along the polypeptide appear indeed to be linked to the sulfated oligosaccharides via the novel linkage Glc(β1–N)Asn. This glycoprotein offers the unique situation of being equipped with two different types of N-glycosyl bonds at defined positions of the polypeptide chain. Therefore additional and as yet unknown recognition signals are required to denote individual glycosylation sites.

S-layer proteins represent the outermost component of the cell envelope. The extreme habitat of *H. halobium* imposes particularly serious problems with respect to the stabilization of the protein structure of the exposed S-layer glycoprotein. As other members of the halobacterial family only tolerate moderately halophilic conditions (e.g. *Haloferax volcanii* requires 2 M NaCl for optimal growth), a comparison of the corresponding S-layer protein structures should give hints on how adaptation to high salt conditions is achieved. Recently, also the primary structure and the glycosylation of the S-layer glycoprotein from the moderate halophile has become available [18,19]. The amino acid sequence deduced from the cloned cDNA gene indicated that the mature polypeptide contains 793 amino acids that is 25 residues less than found for the *H. halobium* protein. A comparison of the amino acid sequences of the S-layer glycoproteins from *H. halobium* and *H. volcanii* shows that the proteins share 40.5% identity. Remarkably, there are regions of high homology and these are arranged in a regular pattern: Stretches of nearly complete homology are interrupted by stretches of unrelated sequences, and the degree of homology strikingly drops towards the N-terminal end, i.e. towards the most extracellular domain. Possibly, the regions of highly conserved sequences indicate sites of essential protein-protein interactions. Common to both polypeptides is the location and sequence of the putative membrane binding domain with the adjacent clusters of threonine residues. Thus, the adaptation of *H. halobium* to high salt conditions does not reflect any obvious structural features within the polypeptide chain of the S-layer glycoprotein.

Unexpectedly, however, it is the structure of the N-glycosidically bound saccharides that turned out to be completely different in the moderate and extreme halophile [19]. Only glucose and galactose were found as the main sugar components of the S-layer glycoprotein from *Haloferax volcanii*. Neither amino sugars nor uronic acids nor sulfate residues could be detected in the corresponding saccharides. This fact immediately implies a completely different set of N-glycosidically bound saccharides compared with the extreme halophilic glycoprotein. The absence of amino sugars excludes the existence of the above mentioned repeating unit saccharide. After tryptic digestion, glycopeptides covering 4 out of the total of 7 N-glycosylation sites were isolated and the saccharide structures analyzed by permethylation analysis as well as enzymatic degradation studies. Again, Glc(β1–N)Asn was identified as the linkage unit, but instead of being linked with a few sulfated uronic acids, a chain of 9 or 10 β-1,4-linked glucose residues follows. Thus, completely uncharged saccharides replace the highly charged sulfated oligosaccharides found in the extremely halophilic glycoprotein (Fig. 1). With respect to the O-glycosidically bound disaccharides no differences between both glycoproteins were found.

The marked differences in the glycosylation pattern of otherwise related (40.5% overall identity) S-layer glycoproteins lead to a drastically changed net surface charge, and most probably this reflects the adaptation from a moderately to an extremely halophilic environment. The ratio of negatively versus positively charged amino acids is about the same for both polypeptide chains and the calculated isoelectric points are as low as 3.2 and 3.4. In the case of the moderate halophile no additional charges are introduced by the neutral saccharide structures. In sharp contrast, the saccharides of the extreme halophile contribute at least 120 additional negative charges. This is nearly a doubling of the negative surface charges. S-layer glycoprotein isolated from *H. halobium* grown in the presence of bacitracin selectively lacks the repeating unit saccharide

[7] and therefore its surface charge is reduced by about 60 negative charges [9]. The biological consequence is quite spectacular. Bacitracin-treated bacteria are no longer rods but grow as spheres indicating the collapsing of the native S-layer structure. This is good evidence for a functional role of the glycosaminoglycan chain in maintaining the protein structure of the S-layer glycoprotein under high salt conditions.

2.2.1. Methanogens

The extreme thermophilic archaebacterium *Methanothermus fervidus* has a double-layered cell envelope. The inner pseudomurein sacculus is covered by a hexagonally arranged S-layer that consists of glycoprotein subunits. Recently, this glycoprotein was characterized in some detail [20]. The primary structure of the polypeptide chain was deduced from the nucleotide sequence of the cloned gene. The mature polypeptide consists of 593 amino acid residues. Again, a typical leader peptide was identified, and the mature polypeptide exhibits a total of 20 potential N-glycosylation sites. A heterosaccharide derived from the glycoprotein contains mannose, 3-*O*-methylmannose, (3-*O*-methylglucose), and N-acetylgalactosamine. In addition, galactose and N-acetylglucosamine were identified as constituents of the intact glycoprotein. The heterosaccharide was reported to consist of mannose and 3-*O*-methylmannose (partly replaced by 3-*O*-methylglucose) residues [21]: α-D-3-O-MetManp-(1→6)-α-D-3-O-MetManp-[(1→2)-α-D-Manp]₃-(1→4)-D-GalNAc.

2.3. Eubacteria

In recent years, experimental evidence has accumulated that procaryotic glycoproteins are not restricted to archaebacteria but are of widespread occurrence in eubacterial S-layers as well [22]. Chemical analysis of some of these glycoproteins revealed novel structural features, most remarkably another type of N-glycosyl linkage unit, asparaginyl–rhamnose (RhaAsn) as well as a novel O-glycosyl linkage unit (tyrosyl–glucose) [23–26]. RhaAsn was identified in a glycopeptide derived from the S-layer glycoprotein from *Bacillus stearothermophilus* NRS 2004/3a [23]. The saccharide moiety of this glycopeptide is a rhamnan with a trisccharide repeating unit [24]. Tyrosyl–glucose as a linkage unit was identified in the S-layer glycoproteins from two different species, namely *Clostridium thermohydrosulfuricum* S102-70 and *Acetogenium kivui* [25,26]. The glycan structures of the S-layer glycoproteins from five strains of *Clostridium thermohydrosulfuricum* were analyzed in some detail using ¹H- and ¹³C-NMR measurements [27–30]. The purified S-layer proteins were exhaustively digested with pronase and the resulting peptides subjected to gel permeation chromatography. Glycopeptides with a high molecular mass [20–25 kD) were obtained and these fractions were further analyzed. All these glycans studied so far were found to be polymers of linear or branched repeating unit saccharides containing 2–6 monosaccharide residues. Remarkably, an unexpected diversity of glycan structures was reported to exist even among strains of the same species. Some of the repeating unit saccharides found in these eubacterial species are summarized in Table I. By the same experimental approach, the glycan structure of the S-layer glycoprotein from *Bacillus alvei* was identified as a repeating trisaccharide unit [31].

TABLE I

Some structures of repeating unit saccharides isolated from eubacterial S-layer glycoproteins

---

Bacillus stearothermophilus NRS 2004/3

[→4)-ManA2,3(NAc)$_2$-β-(1→3)-GlcNAc-α-(1→4)ManA2,3(NAc)$_2$-β-(1→6) Glc-α-(1→]
[→2)-Rha-α-(1→3)-Rha-β-(1→]

Bacillus alvei CCM 2051

[→3)-Gal-β-(1→4)-ManNAc-β-(1→]
          ↑α-(1→6)
          Glc

Clostridium thermosaccharolyticum D 120-70

[→3)-Man-β-(1→3)-Glc-α-(1→4)Rha-α-(1→]
   ↑β-(1→6)   ↑α-(1→2)
   Glc   Gal

[→4)-GlcNAc-β-(1→3)-ManNAc-β-(1→]
            ↑β-(1→6) ↑α-(1→2)
            Gal

Clostridium thermohydrosulfuricum L 111-69

[→4)-Man-α-(1→3)-Rha-α-(1→]

Clostridium thermohydrosulfuricum L 77-66

[→3)-GalNAc-α-(1→)-GalNAc-α-(1→]
         ↑β-(1→4)
         Man
         ↑α-(1→2)
         GlcNAc

---

3. Glycoproteins as constituents of flagellins

3.1. Halobacteria

Besides the S-layer protein, a set of proteins with molecular masses within the range of 26–36 kD was observed to be glycosylated in the archaebacterium *Halobacterium halobium*, as well. These proteins were identified as their flagellins, the first demonstration of glycosylated flagellins [32]. Flagellar bundles of *Halobacterium halobium* consists of 5–10 filaments. Halobacteria swim forward by clockwise and backward by counterclockwise rotation of their right handed flagellar bundles [33]. The halobacterial flagellins are encoded in five different but highly homologous genes [34]. Two of the genes are arranged in tandem at one locus (flg A1 and A2), and the other three in a tandem arrangement at a different locus (flg B1, B2 and B3). All five gene products are expressed and integrated into the flagellar bundle [35]. The sulfated oligosaccharides attached to these flagellins turned out to be identical to the sulfated oligosaccharides of the S-layer glycoprotein described above [32]. Furthermore, these saccharides are attached to the flagellins by the same linkage unit Glc(β1–N)Asn. No other type of saccharides has been detected in the flagellins.

3.2. Methanogens

The discovery of glycosylated flagellins in halobacteria suggested that flagellins may be candidates for glycosylation in other families of archaebacteria as well. In fact, the flagellins of *Methanospirillum hungatei*, *Methanococcus deltae* and *Metanothermus fervidus* were reported to be glycosylated [36–39].

Glycosylation of archaebacterial flagellins is an unexpected posttranslational modification obviously excluding the mechanism of flagellum biogenesis established for eubacteria. Glycosylation of archaebacterial glycoproteins takes place at the extracellular surface of the cell membrane [3,4]. As a consequence, the flagellin polypeptide has to be transported across the cell membrane before glycosylation can be initiated. This, however, seems to exclude transport of flagellin molecules through the central channel of the hook to the tip of the flagellum. Thus, an alternative pathway of flagellum assembly appears to be involved. This view is further supported by the fact that archaebacterial flagellin genes encode leader peptides [38,39].

4. Glycosylated exoproteins

All glycoproteins mentioned so far are constituents of the bacterial cell surface or of flagella. However, there is a report on a soluble secreted glycoprotein from the extremely halophilic archaebacterium *Haloarcula marismortui* [40]. This organism, upon starvation of inorganic phosphate, produces an alkaline phosphatase and secretes it into the medium. The enzyme has an apparent subunit molecular weight of 160 kDa, and its glycoprotein nature has been assessed by hydrolysis and chemical quantification. It contains 3% (w/w) of carbohydrate, and metabolic labelling studies indicated the presence of glucosamine in the glycoconjugate.

Additional examples of glycosylated exoproteins are the cellulase complex of *Clostridium thermocellum* [41], the glycoprotein toxin of *Bacillus thuringiensis* [42] (toxic for mosquito larvae), and an autolysin of *Streptococcus faecium* [43]. This protein is attached to bacterial wall, and after autodigestion of the walls it is a soluble active enzyme.

5. Biosynthetic aspects

Most of the studies on bacterial glycoprotein biosynthesis have been performed on archaebacteria. Only recently more detailed studies on eubacterial glycoproteins have been published.

5.1. Biosynthesis of the cell surface glycoprotein and the flagella of halobacteria

First attempts to investigate the biosynthesis of the halobacterial cell surface glycoprotein have been made by Mescher and Strominger, who described an activity in cell homogenates for the formation of polyisoprenyl phosphoglucose, polyisoprenyl phosphomannose, and polyisoprenyl diphospho-GlcNAc [44]. From these findings, the authors concluded that lipid-linked intermediates are involved in the biosynthesis of the amino sugar-containing saccharide. Later investigations have not confirmed the presence of mannose in the amino sugar-containing saccharide, which turned out to be a glucosaminoglycan, as described above. It would be of interest, therefore, to determine if a lipid-linked mannose residue could serve as a precursor of mannose-containing membrane glycolipid sulfates in halobacteria [45]. The activity of formation of lipid-linked N-acetyl-D-glucosamine is dependent on a high concentration of potassium ions, which predominate inside the halobacterial cells, and therefore transfer of GlcNAc has been suggested to occur inside the cell. In the meantime, the linkage unit of the amino sugar-containing glycosaminoglycan moiety has been shown to consist of GalNAc-Asn rather than GlcNAc [10], and therefore it will be of interest to learn about the role of the lipid diphosphate GlcNAc.

5.1.1. Sulfated glucosaminoglycan

According to recent results, the repeating unit saccharide strand is completed in a lipid-linked state including sulfation, and thereafter transferred 'en bloc' to the nascent protein chain [46]. After *in vivo* pulse labeling with $^{35}SO_4^{2-}$, the lipid-linked intermediates analyzed by SDS-PAGE appear as a regular pattern of 10–15 bands, which are thought to represent different chain lengths of the growing saccharide. Most likely the distance between adjacent bands is due to one repeating unit of the glycosaminoglycan. This implies that the glycosaminoglycan is assembled on the lipid carrier by polymerization of preformed pentasaccharides, a mechanism similar to the one described for the biosynthesis of the *Salmonella* O antigen [47]. The above-mentioned inhibition by bacitracin of this transfer implies two conclusions: (a) the lipid involved is likely to be a diphosphate, as bacitracin complexes prenyl diphosphate compounds, thus inhibiting hydrolytic regeneration of the corresponding monophosphate [48]; and (b) transfer to protein of the glycosaminoglycan most likely takes place at the cell surface, as bacitracin does not penetrate the plasma membrane of halobacteria [44,49]. Thus, biosynthesis of the halobacterial glycosaminoglycan differs from the synthesis of animal glycosaminoglycans [50] in that an already sulfated lipid precursor occurs in halobacteria, whereas in animals the carbohydrate chain is established at the protein-linked level through step-by-step addition of monosaccharides (from their nucleotide-activated derivatives), and sulfation takes place only on the protein-linked saccharide strand. On the other hand, as mentioned before, sequence analysis has revealed a 'sequon' acceptor peptide Asn-Ala-Ser for this transfer that results in the linkage unit GalNAc-Asn [10]. Additional principal structural similarities with the animal glycosaminoglycans exist: both classes are linear strands composed of uronic acids, amino sugars, and sulfate, and although many animal glycosaminoglycans are O-glycosidically linked to protein via xylose, corneal keratan sulfate is bound to protein via a true N-glycosyl linkage [51].

5.1.2. Sulfated oligosaccharides

Studies on the biosynthesis of the second novel type of N-glycosyl linkage unit in halobacteria, β-GlcAsn, have revealed some unexpected features: pulse-chase labeling experiments with $^{35}SO_4^{2-}$ demonstrated the existence of already sulfated precursors of this type of glycoconjugate [11]. The protein acceptors for this sulfated precursor material are the S-layer glycoprotein and the set of proteins of lower molecular weight that have been shown to represent the halobacterial flagellins [32]. Thus, two different types of glycoproteins share the same pool of precursors.

The radioactive label has facilitated purification to homogeneity of a group of lipid-linked oligosaccharides that make up this pool [11,52]. Detailed chemical analysis of these compounds has revealed that, as in the case of the glycosaminoglycan, completely sulfated lipid-linked precursors are established before transfer of the oligosaccharides to protein. The lipid is a C_{60}-polyprenol of the eukaryotic dolichyl- rather than of the bacterial undecaprenyl-type. The reducing-end glucose residue of the oligosaccharide part is linked to the dolichol via a monophosphate rather than a diphosphate bridge. Furthermore, the occurrence of iduronic acid in a lipid-activated saccharide has first been demonstrated in these precursors [13]. This hexuronic acid is a typical constituent of eukaryotic extracellular glycoconjugates and is reported to result from epimerization of glucuronic acid residues within the completed, protein-linked saccharide chain [50].

To prove a precursor-product relationship of these lipid-linked sulfated oligosaccharides with the sulfated oligosaccharides from the S-layer glycoprotein, their molecular structure has been determined by permethylation analysis [52]. Typical structures are given in Fig. 2, upper panel. Composition, linkage pattern, and sequence of the

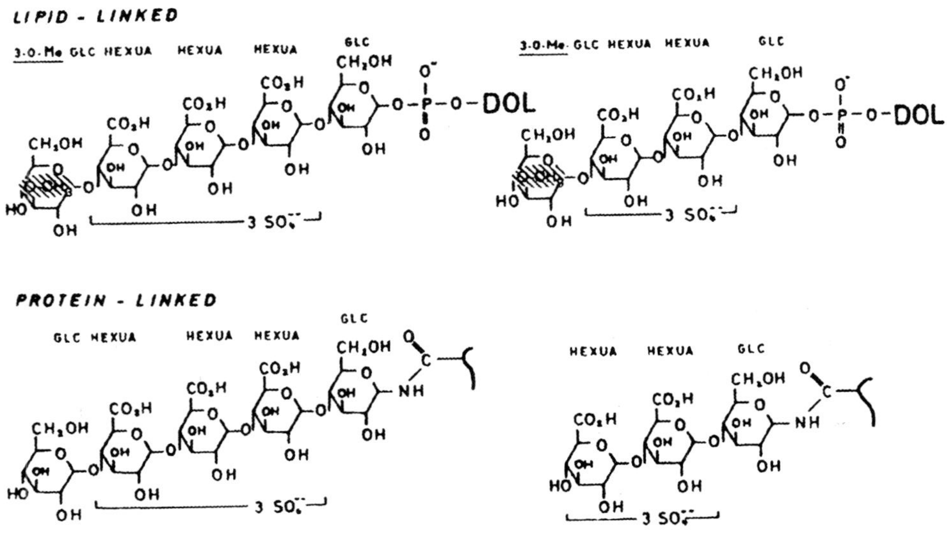

Fig. 2. Structures of lipid-linked and protein-linked oligosaccharides in the cell surface glycoprotein of *H. Halobium* according to Ref. [52].

lipid-linked oligosaccharides turned out to be identical to those obtained from glycopeptides (Fig. 2, lower panel) with one surprising exception: the lipid-linked saccharides contained a chemical modification that could not be found in the protein-linked saccharides. Specifically, position 3 of the oligosaccharides' peripheral, nonreducing-end glucose residue carries a methyl group. Except for this methylation, all structural details were identical in the lipid-linked and protein-linked sulfated oligosaccharides, and the pool of $^{35}SO_4^{2-}$-labeled lipid-linked precursors could quantitatively be chased into glycoprotein. Therefore, a true precursor–product relationship exists between the two species, and the methylation observed represents a transient modification of the lipid-linked oligosaccharides [52].

What might be the role of this transient methylation? To investigate this, sulfated glycoprotein biosynthesis was analyzed under conditions where S-adenosylmethionine-dependent methylation is inhibited. Surprisingly, sulfated glycoprotein biosynthesis is depressed greatly, although (a) general protein biosynthesis (as assessed by incorporation of ^{35}S-methionine) is not altered, and (b) synthesis of an unmethylated pool of sulfated lipid-linked precursors is not inhibited, but rather this pool remains stable in pulse-chase experiments performed in the presence of the inhibitors of methylation.

Thus, a transient methylation is involved at some stage in the biosynthesis of the novel N-glycosyl linkage β-1,N-Asn.

Inhibition of this methylation could only be achieved by use of a combination of adenosine and homocysteine thiolactone, and not with the actual direct inhibitor S-adenosylhomocysteine. The two former substances do permeate cell membranes to a sufficient extent, whereas the latter does not [53]. Therefore, it is assumed that construction of the dolichol monophosphate-linked oligosaccharides, including methylation of their peripheral glucose residues, occurs at the cytosolic face of the plasma membrane [52,54].

Where then does transfer to protein occur? To investigate this, advantage was taken of the finding that typical N-glycosyl 'sequon' sequences are present around the N-glycosyl linkage units. Accordingly, a synthetic glycosyl acceptor peptide was employed *in vivo* that has successfully been used before in *in vitro* glycosylation studies [55]. The hexapeptide Tyr-Asn-Leu-Thr-Ser-Val contains two ionic charges and cannot therefore permeate membranes. Interestingly, this acceptor-peptide becomes glycosylated with sulfated oligosaccharides when added to the medium of stirred suspensions of halobacteria [11]. Thus, transfer to protein of the sulfated oligosaccharides occurs on the surface of the halobacterial cell. Partial characterization of the sulfated glycopeptides obtained after exogenous addition of acceptor peptide revealed that only the sulfated oligosaccharides are transferred to this acceptor, but no glycosaminoglycan. This finding indicates that for the transfer of the glycosaminoglycan, structural information may be needed in addition to a general N-glycosyl acceptor sequence.

5.1.3. Model for the biosynthesis of N-linked saccharides in halobacteria

A concluding scheme reflecting a hypothesis on the biosynthesis of the sulfated S-layer glycoprotein of halobacteria is depicted in Fig. 3. Two different N-glycosyl linkages are synthesized within this complex glycoprotein, and the two biosynthetic pathways

466

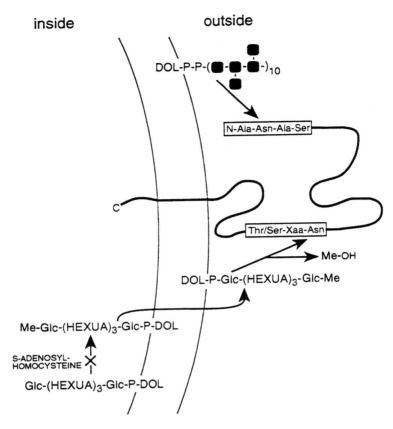

inside   outside

DOL-P-P-(■-■-■-)₁₀

N-Ala-Asn-Ala-Ser

C

Thr/Ser-Xaa-Asn

Me-OH

DOL-P-Glc-(HEXUA)₃-Glc-Me

Me-Glc-(HEXUA)₃-Glc-P-DOL

S-ADENOSYL-
HOMOCYSTEINE

Glc-(HEXUA)₃-Glc-P-DOL

Fig. 3. Biosynthesis of N-linked glycoproteins in *halobacteria* according to Ref. [4].

differ in the type of lipid phosphate involved. A dolichol monophosphate is used for the Glc(β1–N)Asn, and a lipid (most likely a dolichol) -diphosphate for the Gal-NAc(β1–N)Asn-type. All three types of N-glycosyl linkages established so far have in common sequences that fit the general formula Asn-X-Thr(Ser), known as the eukaryotic N-glycosylacceptor sequence. This strongly supports the notion that the hydroxyl group of the amino acid next but one to Asn-carbohydrate is involved in the process of glycosyl transfer in a catalytic manner, and is not merely a structural recognition site for the transferase [56,57]. In addition to the hitherto unknown types of N-glycosyl linkages found in the S-layer glycoprotein in halobacteria, differences exist in details of the biosynthetic pathways leading to sulfated glycoconjugates in halobacteria and eukaryotes. In eukaryotes, glycoconjugates are covalently modified at the protein-linked level in the Golgi apparatus, where trimming, further glycosylation, sulfation, or epimerization occurs. Halobacteria lack organelles like the Golgi apparatus and therefore must follow a different biosynthetic pathway leading to their glycoproteins. Thus, the oligosaccharides are completed and sulfated while still attached to dolichol on the cytosolic side of the cell membrane. Thereafter, they are translocated to the cell surface, possibly by a mechanism that involves a transient methylation of their peripheral glucose residues. Finally, transfer to the protein occurs at the cell surface, and with this generation of N-glycosyl linkages, the halobacterial cell surface is functionally equivalent to the luminal side of the endoplasmatic reticulum membrane in eukaryotic cells. With the mechanism described, an unglycosylated protein core can be translocated through the cell membrane, and the corresponding glycoconjugates may pass this membrane in their lipid-linked stage. Thus, despite the lack of compartmentalization in prokaryotes, halobacterial glycoprotein biosynthesis in essence resembles the mechanism of glycoprotein biosynthesis in eukaryotes.

A variety of questions remains to be answered: how is the glycosaminoglycan translocated to the cell surface? Is it transported block by block, similar to the biosynthesis of the bacterial O-antigens [47], or is it completed inside the cell and then translocated through the membrane? What exactly is the role of the transient methylation of the sulfated dolichol monophophate oligosaccharides?

Further problems wait to be resolved: How are the O-glycosyl linkage units established, from nucleotide or from lipid-activated saccharide precursors, inside or outside the cell?

In summary, even the seemingly simple halobacterial system has turned out to be a far more complicated model for the investigation of structure, biosynthesis, and function of glycoproteins than had been anticipated at the onset.

5.2. Methanothermus fervidus glycoprotein

Recently, the composition of potential precursors of protein-glycoconjugates from this extreme thermophilic archaebacterium was published [58]. Although the structures of these saccharides have not yet been completely established, some unusual features of glycoprotein biosynthesis in this bacterium have been described:

1. Besides nucleoside diphosphate-activated monosaccharides, oligosaccharides with rather complex compositions have been found to occur as UDP-derivatives, as well. Oligosaccharides with this type of activation have been found to be involved in the biosynthesis of a methanobacterial pseudomurein [59], of methanochondroitin [60], and of the S-layer glycoprotein of *Bacillus alvei* [61].
2. Processing (methylation) of oligosaccharides seems to occur at the UDP-linked level. Interestingly, as in halobacteria, 3-O-methylglucose is generated by methylation of the oligosaccharide at the precursor level, but unlike in halobacteria, the modification is stable and found as a stoichiometric component of the glycoprotein.
3. Like in halobacteria, lipid-activated oligosaccharides occur with short chain (C_{55}) dolichol rather than undecaprenol. Unlike in halobacteria, exclusively dolichyl diphosphate oligosaccharides were found, with no dolichyl monophosphate oligosaccharide. Probably the oligosaccharides are transferred from UDP to dolichyl phosphate. On the other hand, undecaprenol precursors have been identified for methanobacterial pseudomurein and for methanochondroitin [59,60].
4. Finally, as is the case in eukaryotes and halobacteria, a transient modification of the precursor oligosaccharides seems to occur. Rather than a transient methyla-

tion, here a transient glycosylation takes place with up to 6 or 8 glucose residues that were found in lipid-linked oligosaccharides, and are absent in the surface glycoprotein. This is reminiscent of the transient glycosylation with three glucose units of the unique dolichol-linked precursor saccharide for N-glycosylation in eukaryotes [62]. Final insight into the mechanism of this transient modification depends on the knowledge of the molecular structures of the glycoconjugates in their precursor-linked as well as their protein-bound state. Mysteriously, neither UDP- nor lipid diphosphate-activated glucose or 3-O-methylmannose and 3-O-methylglucose could be detected in *Methanothermus fervidus*. On the other hand, these monosaccharides are constituents of UDP-activated and dolichol-activated oligosaccharides. Therefore it will be of interest to elucidate the process of activation for transfer of glucose in this organism.

5.3. S-layer glycoprotein of Bacillus alvei

This glycoprotein covers the murein sacculus of the eubacterium *B. alvei*. The glycoconjugate is a repeating unit chain that contains ManNAc, Gal and Glc as constituents. From analytical studies of activated saccharide compounds isolated from bacterial cell extracts a sequence of events is proposed to be involved in the construction of this glycan chain, as depicted in Scheme 1 [61]. Monosaccharides are activated as their UDP-derivatives (Glc, Gal and GlcNAc), and as GDP-ManNAc. Strikingly, from these activated monosaccharides activated oligosaccharides are formed with

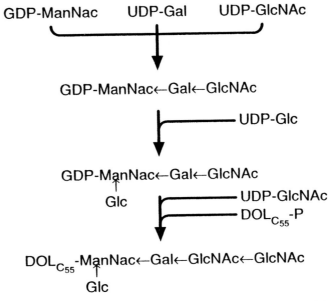

Scheme 1. Hypothetical steps in the biosynthesis of glycoconjugate precursors for the S-layer glycoprotein in *Bacillus alvei* according to Ref. [61].

guanosinediphosphate and ManNAc at the reducing end. Stepwise addition of monosaccharides results in a branched GDP-oligosaccharide which is transferred to a C_{55}-Dol-P, yielding the corresponding Dol-PP-oligosaccharide (short chain Dol-P and -Dol-PP oligosaccharides are also involved in halobacterial glycoprotein biosynthesis, cf. 5.1.). Further addition of a GlcNAc unit results in a hexasaccharide-structure with a peripheral chain of three GlcNAc residues. Note that none of these GlcNAc moieties is found in the mature S-layer glycoprotein. Thus, the three GlcNAc residues must be removed after transfer to protein of the carbohydrate chain. This is again reminiscent of the mechanism of transfer of the unique oligosaccharide during N-glycosylation in eukaryotes, where a triglucosylstructure is hydrolyzed immediately after transfer from the lipid to protein of the oligosaccharide chain.

6. Conclusions

Although only a few structures have been solved to date of bacterial protein-glyco-conjugates, it is a strong prediction that bacterial protein-linked carbohydrates will turn out to represent a far greater variety of structures than those found in eukaryotic glycoconjugates. Likewise, with only relatively few bacterial species analyzed so far a couple of new types of linkage units between protein and carbohydrate have already been found. At the level of our present knowledge, the most striking structural similarities between bacterial and mammalian glycoproteins are found in the cell surface glycoprotein of halobacteria and the basement membrane proteoglycan/collagen/laminin complexes. Particularly, collagen type IV shares disaccharides of the structure O-Gal(α1-2)Glc [15] with the cell surface glycoprotein, and the sulfated saccharide repeats in halobacteria resemble the sulfated repeating units in mammalian proteoglycans. In addition, the linkage unit Glc(β1–N)Asn, first isolated from halobacterial glycoproteins, has now been demonstrated to exist in the mammalian protein laminin [14]. Thus, laminin contains two types of N-glycosyl linkage units, and it will be of interest to learn about the biosynthesis of the novel unit. More is known about the biosynthesis of the two types of N-glycosyl-units in halobacteria; GalNAc(β1–N)Asn and Glc(β1–N)Asn. A completed, sulfated glycosaminoglycan chain is transferred en bloc from a C_{60} Dol-PP-carrier to the Asn-residue in position 2 of the mature protein to yield the β-GalNAc-Asn bond. In contrast, sulfated oligosaccharides are activated by Dol-monophosphates, with transfer to Asn-residues again occurring at the cell surface, yielding the β-Glc-Asn-linkage. In both cases, the Asn residues are constituents of typical N-glycosyl acceptor sequences (sequons). From these findings one might predict that the β-Glc-Asn in the mammalian protein laminin will turn out to be part of a sequon structure, as well. It is not clear how the decision is made as to which carbohydrate is transferred to which sequon. Most likely, additional signals need to exist for this discrimination. In any case, functionally the cell surface of halobacteria resembles the endoplasmic reticulum of eukaryotes: proteins are first translocated through a membrane and thereafter N-glycosylated.

A novel type of activated sugars begins to emerge from the work on the biosynthesis of prokaryotic glycoproteins: for the first time, nucleoside diphosphate activated

470

oligosaccharides have been described. These soluble precursors have in common with the Dol-linked precursors the existence of transient chemical modifications: a triglucosyl structure on the eukaryotic unique lipid linked oligosaccharide, 3-O-methylglucose on Dol-P-linked oligosaccharides in halobacteria, and a variety of homooligosaccharides on the nucleoside-diphosphate activated sugars. The sites and mechanisms of carbohydrate transfer remain to be established in the eubacterial system.

A variety of individual functions have recently emerged for individual structures in eukaryotic glycoproteins. Due to the much younger history of their bacterial counterparts our knowledge of functions of prokaryotic glycoconjugates is still modest: ionic charges of uronic acids and sulfate were convincingly attributed to high salt tolerance in halobacteria, and a glycosaminoglycan structure is responsible for forming or maintaining the rod shape of halobacteria. Most likely, as investigations proceed prokaryotic protein linked carbohydrates will disclose a wild variety of individual functions attributable to an infinitely increasing number of individual glycoconjugate structures.

Acknowledgements

We thank Dieter Jeckel for critically reading the manuscript. Mathias Neufeld's help with the figures is gratefully acknowledged.

References

1. Sleytr, U.B. and Messner, P. (1983) Crystalline surface layers on bacteria, Annu. Rev. Microbiol. 37, 311–339.
2. Messner, P. and Sleytr, U.B. (1992) Crystalline Bacterial Cell-surface layers, in: A.H. Rose and D.W. Tempest (Eds.), Advances in Microbial Physiology, Vol. 33, 213–275.
3. Sumper, M. (1987) Halobacterial glycoprotein biosynthesis, Biochim. Biophys. Acta 906, 69–79.
4. Lechner, J. and Wieland, F. (1989) Structure and biosynthesis of prokaryotic glycoproteins, Annu. Rev. Biochem. 58, 173–194.
5. Sleytr, U.B. (1978) Regular arrays of macromolecules on bacterial cell walls: structure chemistry, assembly, and function. Int. Rev. Cytol. 53, 1–64.
6. Mescher, M.F. and Strominger, J.L. (1976) Purification and characterization of a procaryotic glycoprotein from the cell envelope of *Halobacterium salinarium*, J. Biol. Chem. 251, 2005–2014.
7. Wieland, F., Dompert, W., Bernhardt, G. and Sumper, M. (1980) Halobacterial glycoprotein saccharides contain covalently linked sulphate, FEBS Lett. 120, 110–114.
8. Wieland, F., Lechner, J. and Sumper, M. (1982) The cell wall glycoprotein of Halobacteria: structural, functional and biosynthetic aspects, Zbl. Bakt. Hyg., 1. Abt. Orig. C3, 161–170.
9. Paul, G. and Wieland, F. (1987) Sequence of the halobacterial glucosaminoglycan, J. Biol. Chem. 262, 9587–9593.
10. Paul, G., Lottspeich, F. and Wieland, F. (1986) Asparaginyl-N-acetylgalactosamine: linkage unit of halobacterial glycosaminoglycan, J. Biol. Chem. 261, 1020–1024.

11. Lechner, J., Wieland, F. and Sumper, M. (1985) Biosynthesis of sulfated oligosaccharides N-glycosidically linked to the protein via glucose. J. Biol. Chem. 260, 860–866.

12. Wieland, F., Heitzer, R. and Schaefer, W. (1983) Asparaginylglucose: novel type of carbohydrate linkage, Proc. Natl. Acad. Sci. USA 80, 5470–5474.

13. Wieland, F., Lechner, I. and Sumper, M. (1986) Iduronic acid: Constituent of sulphated dolichyl phosphate oligosaccharides in halobacteria. FEBS Lett. 195, 77–81.

14. Schreiner, R., Schnabel, E. and Wieland, F. (1994) Novel N-Glycosylation in Mammals: Laminin contains the linkage unit β-Glucosylasparagine. J. Cell Biol. 124, 1071–1081.

15. Spiro, G.T.R. (1967) Studies on the renal glomerular basement membrane. J. Biol. Chem. 242, 1923–1932.

16. Lechner, J. and Sumper, M. (1987) The primary structure of a prokaryotic glycoprotein, J. Biol. Chem. 262, 9724.

17. Kessel, M., Wildhaber, I., Cohen, S. and Baumeister, W. (1988) Three-dimensional structure of the regular surface glycoprotein layer of *Halobacterium volcanii* from the Dead Sea, EMBO J. 7, 1549–1554.

18. Sumper, M., Berg, E., Mengele, R. and Strobel, I. (1990) Primary structure and glycosylation of the S-layer protein of *Haloferax volcanii*, J. Bacteriol. 172, 7111–7118.

19. Mengele, R. and Sumper, M. (1992) Drastic differences in glycosylation of related S-layer glycoproteins from moderate and estreme halophiles, J. Biol. Chem., 267, 8182–8185.

20. Bröckl, G., Behr, M., Fabry, S., Hensel, R., Kaudewitz, M., Biendl, E. and König, H. (1991) Analysis and nucleotide sequence of the genes encoding the surface-layer glycoproteins of the hyperthermophilic methanogens *Methanothermus fervidus* and *Methanothermus sociabilis*. Eur. J. Biochem. 199, 147–152.

21. Kärcher, U., Schröder, H., Haslinger, E., Allmeier, G., Schreiner, R., Wieland, F., Haselbeck, R. and König, H. (1993) Primary structure of the heterosaccharide of the surface glycoprotein of *Methanothermus fervidus*. J. Biol. Chem. 268, 26821–26826.

22. Messner, P. and Sleytr, U.B. (1991) Bacterial surface layer glycoproteins, Glycobiol. 1, 545–551.

23. Messner, P. and Sleytr, U.B. (1988) Asparaginyl-rhamnose: a novel type of protein–carbohydrate linkage in a eubacterial surface-layer glycoprotein, FEBS Lett. 228, 317–320.

24. Christian, R., Schulz, G., Unger, F.M., Messner, P., Küpcü, Z. and Sleytr, U.B. (1986) Stucture of a rhamnan from the surface-layer glycoprotein of *Bacillus stearothermophilus* strain NRS 2004/3a, Carbohydr. Res. 150, 265–272.

25. Messner, P., Christian, R., Kolbe, I., Schulz, G. and Sleytr, U.B. (1992) Analysis of a novel linkage unit of O-linked carbohydrates from the crystalline surface layer glycoprotein of *Clostridium thermohydrosulfuricum* S 102-70, J. Bacteriol. 174, 2236–2240.

26. Peters, J., Rudolf, S., Oschkinat, H., Mengele, R., Sumper, M., Kellermann, J., Lottspeich, F. and Baumeister, W. (1992) Evidence for tyrosine-linked glycosaminoglycan in a bacterial surface protein, Biol. Chem. Hoppe-Seyler 373, 171–176.

27. Christian, R., Messner, P., Weiner, C., Sleytr, U.B. and Schulz, G. (1988) Structure of a glycan from the surface-layer glycoprotein of *Clostridium thermohydrosulfuricum* strain L111-69, Carbohydr. Res. 176, 160–163.

28. Altmann, E., Brisson, J.-R., Messner, P. and Sleytr, U.B. (1990) Chemical characterization of the regularly arranged surface layer glycoprotein of *Clostridium thermosaccharolyticum* D120-70, Eur. J. Biochem. 188, 73–82.

29. Messner, P., Bock, K., Christian, R., Schulz, G. and Sleytr, U.B. (1990) Characterization of the surface layer glycoprotein of *Clostridium symbiosum* HB25, J. Bacteriol. 172, 2576–2583.

30. Altmann, E., Brisson, J.-R., Gagne, S.M., Kolbe, J., Messner, P. and Sleytr, U.B. (1992) Structure of the glycan chain from the surface layer glycoprotein of *Clostridium thermohydro-*

472

sulfuricum L77-66, Biochim. Biophys. Acta 1117, 71–77.

31. Altmann, E., Brisson, J.-R., Messner, P. and Sleytr., U.B. (1991) Structure of glycan chain from the surface layer glycoprotein of *Bacillus alvei* CCM 2051, Biochem. Cell Biol., 69, 72–78.

32. Wieland, F., Paul, G. and Sumper, M. (1985) Halobacterial flagellins are sulfated glycoproteins, J. Biol. Chem. 260, 15180–15185.

33. Alam, M. and Oesterhelt, D. 1984, Morphology, function, and isolation of Halobacterial flagella. Mol. Biol. 176, 459–475.

34. Gerl, L. and Sumper, M. (1988) Halobacterial flagellins are encoded by a multigene family, J. Biol. Chem. 263, 13246–13251.

35. Gerl, L., Deutzmann, R. and Sumper, M. (1989) Halobacterial flagellins are encoded by a multigene family: Identification of all five gene products, FEBS Lett. 244, 137–140.

36. Southam, G., Kalmokoff, M.L., Jarrell, K.F., Koval, S.F. and Beveridge, T.J. (1990) Isolation, characterization, and cellular insertion of the flagella from two strains of *Methanospirillum hungatei*, J. Bacteriol. 172, 3221–3228.

37. Faguy, D.M., Koval, S.F. and Jarrell, K.F. (1992) Correlation between glycosylation of flagellin proteins and sensitivity of flagellar filaments to Triton X-100 in methanogens, FEMS Microbiol. Lett. 90, 129–134.

38. Kalmokoff, M.L. and Jarrell, K.F. (1991) Cloning and sequencing of a multigene family encoding the flagellins of *Methanococcus detae*, J. Bacteriol. 173, 7113–7125.

39. Kalmokoff, M.L., Koval, S.F. and Jarrell, K.F. (1992) Relatedness of the flagellins from methanogens, Arch. Microbiol. 157, 481–487.

40. Goldman, S., Hecht, K., Eisenberg, H. and Mevarech, M. (1990) Extracellular Ca^{2+}-dependent inducible alkaline phosphatase from extremely halophilic archaebacterium *Haloarcula marismortui*. J. Bacteriol. 172, 7065–7070.

41. Gerwig, G. J., de Waard, P., Kamerling, J.P., Vliegenthart, J. F. G., Morgenstern, E., Lamed, R. and Bayer, E. A. (1989) Novel O-linked carbohydrate chains in the cellulase complex (Cellulosome) of Clostridrium thermocellum. 3-*O*-Methyl-*N*-Acetylglucosamine as a constituent of a glycoprotein, J. Biol. Chem. 264, 1027–1035.

42. Muthukamar, G. and Nickerson, K.W. (1987) The glycoprotein toxin of Bacillus thuringiensis subsp. israelensis indicates a lectin like receptor in the larval mosquito gut. Appl. Env. Microbiol. 417, 2650–2655.

43. Kawamura, T. and Schokman, G. D. (1983) Purification and some properties of the endogenous, autolytic *N*-Acetylmuramoylhydrolase of streptococcus faecium, a bacterial glycoenzyme, J. Biol. Chem. 258, 9514–9521.

44. Mescher, M.F. and Strominger, J.-L. (1978) In: Energetics and structure of halophilic microorganisms, eds. S. R. Caplan, M. Ginzburg, Elsevier/North-Holland Biomedical, 503–512.

45. Kates, M. and Küshwaha, S.C. (1976) Halobacterial Lipids, in Lipids, ed. Porcellati, G. and Jacini, G., New York, Raven, 1, 276–294.

46. Wieland, F., Lechner, J., Bernhardt, G. and Sumper, M. (1981) Sulphation of a repetitive saccharide in Halobacterial cell wall glycoprotein. FEBS Lett. 132, 319–323.

47. Robbins, P.W., Bray, D., Dankert, M. and Wright, A. (1967) Direction of chain growth in polysaccharide synthesis. Science, 158, 1536–1543.

48. Siewert, G. and Strominger, J.L. (1967) Bacitracin: An inhibitor of the dephosphorylation of lipid pyrophosphate, intermediate in biosynthesis of the peptidoglycan of bacterial cell walls. Proc. Natl. Acad. Sci. USA 57, 767–771.

49. Mescher, M.F. and Strominger, J.L. (1978) Glycosylation of the surface glycoprotein of halobacterium salinarium via a cyclic pathway of lipid-linked intermediates. FEBS Lett. 89, 37–41.

50. Rodén, L. (1980) Structure and metabolism of connective tissue proteoglycans. In: The Biochemistry of glycoproteins and proteoglycans, ed. Lennarz, W.J. New York, Plenium, 267–371.

51. Stein, T. Keller, R., Stuhlsatz, H.W., Greilling, H., Ost, E. et al. (1982) Structure of the linkage-region between polysaccharide chain and core protein in bovine corneal proteokeratan sulfate. Hoppe Seyler's Z. Physiol. Chem. 363, 825–833.

52. Lechner, J. Wieland, F. and Sumper, M. (1985) Transient methylation of dolichyl oligosaccharides is an obligatory step in halobacterial sulfated glycoprotein synthesis. J. Biol. Chem. 260, 8984–8989.

53. Barber, J.R. and Clarke, S. (1984) Inhibition of protein carboxyl methylation by S-adenosyl-L-homocysteine in intact erythrocytes. J. Biol. Chem., 259, 7115–7122.

54. Lechner, J., Wieland, F. and Sumper, M. (1986) Sulphated dolicholphosphate oligosaccharides are transiently methylated during biosynthesis of halobacterial glycoproteins. Syst. Appl. Microbiol. 7, 286–292.

55. Lehle, L. and Bause, E. (1984) Primary structural requirements for N- and O-glycosylation of yeast-mannoproteins. Biochim. Biophys. Acta, 799, 246–251.

56. Bause, E. and Legler, G. (1981) The role of the hydroxy amino acid in the triplet sequence Asn-Xaa-Thr(Ser) for the N-glycosylation step during glycoprotein biosynthesis. Biochem. J. 195, 639–644.

57. Bause, E. (1984) Model studies on N-glycosylation of proteins. Biochem. Soc. Trans., 12, 514–517.

58. Hartmann, E. and König, H. (1989) Uridine and dolichyl diphosphate activated oligosaccharide intermediates are involved in the biosynthesis of the surface layer glycoprotein of Methanothermus fervidus. Arch. Microbiol., 151, 274–281.

59. Hartmann, E. and König, H. (1990) Comparison of the biosynthesis of the methanobacterial pseudomurein and the eubacterial murein. Naturwissenschaften, 77, 472–475.

60. Hartmann, E. and König, H. (1991) Nucleotide-activated oligosacchardies are intermediates of the cell wall polysaccharide of Methanosarcina barkeri. Biol. Chem. Hoppe Seyler, 372, 971–974.

61. Hartmann, E., Messner, P., Allmeier, G. and König, H. (1993) Proposed pathway for biosynthesis of the S-layer glycoprotein of Bacillus alvei. J. Bacteriol. 175, 4515–4519.

62. Turco, S.J., Stetron, B. and Robbins, P.W. (1977) Comparative rates of transfer of lipid-linked oligosaccharides to endogenous glycoprotein acceptors in vitro. Proc. Natl. Acad. Sci. USA, 74, 4411–4414.

J. Montreuil, H. Schachter and J.F.G. Vliegenthart (Eds.), *Glycoproteins*

475

Protein glycosylation in yeast

L. LEHLE and W. TANNER

*Lehrstuhl für Zellbiologie und Pflanzenphysiologie,
Universität Regensburg, 93040 Regensburg, Germany*

1. Introduction

Glycosylated proteins are found in all eukaryotes, in many archaebacteria and exceptionally in eubacteria [1–6]. The glycosylation of proteins is the most complex type of protein modification known in nature. Ten to twenty percent by weight of glycoproteins normally consist of saccharide moieties and sometimes these even amount to more than 80%. Since it is obviously very costly for a cell to invest in the synthesis of these carbohydrate chains, they must have an important function. This assumption is strengthened by the observation that the initial rather evolved and complicated reaction sequence responsible for protein N-glycosylation, the dolichol cycle, has been conserved in evolution from yeast to man.

In recent years a number of interesting results related to functional aspects of protein-bound saccharides have been reported. Leucocyte function and selection [7], embryonic development and stage specific cell surface oligosaccharides [8], coordination of cellular activation and the family of glycoprotein hormones [9] are some of the relevant examples and they all are covered in this series. These examples clearly demonstrate that glycoproteins are generally involved in cellular information transfer, in communication of cells of the immune system with those of the endothelium, in cell–cell interaction determining in part embryogenesis, and in the crosstalk of various cells and tissues with each other via hormones. All these phenomena typically are restricted to multicellular organisms, to higher eukaryotes. The question therefore arises: why do *uni*cellular organisms, such as baker's yeast, glycosylate proteins?

Two functional levels have to be distinguished when considering protein glycosylation. A cellular, most likely intracellular, one that corresponds from an evolutionary point of view to the primary function which should be present in *all* organisms, and secondary functions that have arisen later during evolution; the latter are restricted to higher eukaryotes and obviously are represented by the examples referred to above. The ER-located reactions of N-glycosylation are conserved from yeast to man, *because* they must have been and still must be of importance in any organism. Late in evolution additional, secondary functions became associated with saccharide chains that are synthesized late on the biosynthetic route (for example in the Golgi) by reactions which have *not* been conserved in evolution.

To find out why unicellular eukaryotes glycosylate proteins might help to uncover those primary functions. Of course yeasts — and we will mainly concentrate in this article on *Saccharomyces cerevisiae* — have a number of additional properties that are responsible for the extended studies on protein glycosylation over the years. Thus *S. cerevisiae* can be inexpensively grown in large amounts and is therefore well suited for biochemistry. The organism is extremely well characterized genetically; it can be transformed and mutants can easily be obtained by gene disruption. Finally, its biotechnological use for heterologous gene expression has aroused interest in yeast specific protein modifications.

In this review we will try to summarize what is known about typical protein-bound saccharide structures in *S. cerevisiae*, their N- and O-glycosylation pathways and discuss possible functions of protein glycosylation. Finally a short survey will be given on protein glycosylation in other fungi as well as on aspects of glycosylation of heterologously expressed proteins. Reviews on the topic have last appeared in 1987 [2,3] and in 1993 [10].

2. Yeast mannoproteins and cell wall architecture

The yeast cell wall may be considered as a massive extracellular organelle. It consists of three classes of polysaccharides making up about 90% of the wall [11]: an external layer of *mannoproteins*, previously termed mannan, is responsible for the surface properties like immunogenicity, sexual adhesion and flocculation; an internal β-*glucan* layer forms the structural network and is interwoven with a recently discovered *glucomanno*protein constituent; the third polymer, *chitin*, is a minor constituent present predominantly on bud scars, but also in a diffuse manner all over the cell envelope. Chitin synthesis occurs in different phases of the yeast life cycle and at least four structural genes encode chitin synthetases, *CHS1*, *CHS2*, *CSD2/CAL1/DIT101* and *CSD4* [12–17]. Chitin extensively cross links the glucan fraction to make it alkali-insoluble. A defect in chitin synthase 3 activity responsible for cell wall chitin not found in the primary septum [15] rendered all β-glucan alkali-soluble. The β-glucan fraction is predominantly a β-1,3-glucan with some β-1,6-branching and a minor β-1,6-glucan portion [18–21]. Important new insights concerning the formation and assembly of glucans have come from genetic analysis of several *kre* mutants defective in glucan synthesis [22,23]. In addition antibodies have been raised recently against β-1,6-glucan, so that an immunochemical probe for its localization is available now [24].

The mannan, originally thought to be exclusively a carbohydrate polymer, turned out to be the prototype of high mannose type glycoproteins (although extensively processed) found in all eukaryotes. The carbohydrate part of different mannoproteins may vary between 5 and 90%; the higher amounts usually occur on extracellular mannoproteins. The overall structure of the carbohydrate component of bulk cell wall mannoprotein we mainly owe to Ballou's laboratory; the structure is shown in Figs. 1 and 2. Notable features are the differentiation of the N-linked chains into an *inner core* and an *outer chain* as well as the occurrence of linear short O-linked oligosaccharides; they will be discussed below in more detail. The N-linked structure has been revised

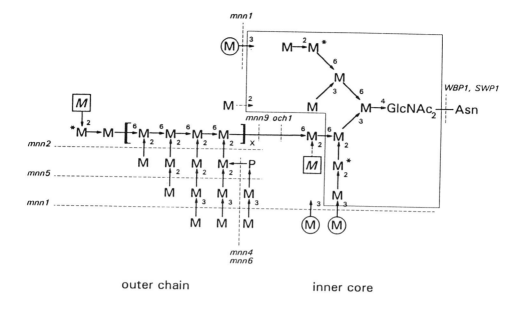

outer chain · inner core

x = ~ 10; wild type
x = 1-4; *mnn7, mnn8*
mnn9 outer chain shortened to one α - 1,6-mannose
mnn10 outer chain 10-12 mannoses
[M] the italicized mannose acts as a stop signal for outer chain formation
* phosphorylation sites

Fig. 1. Structure of N-linked carbohydrate component. The trimmed $Man_8GlcNAc_2$ core originating from Dol-PP-$GlcNAc_2Man_9Glc_3$ is boxed. It is modified to a minor extent by α-1,6-, α-1,2- and α-1,3 -linked mannoses (circled M) to give rise to the heterogenous *inner core* ranging from $Man_{8-15}GlcNAc_2$. A minority of *inner core* oligosaccharides retain the α-1,2 mannose (dashed arrow) on the inner α1,3 branch; this mannose is not removed by the *MNS1* encoded α-mannosidase. Particular oligosaccharides are further elongated by extensive α-1,6- followed by α-1,2- and α-1,3 mannosylation to build up the outer chain. Various *mnn* mutations affecting the mannoprotein structure are indicated. In the *mnn2* mutation the *boxed* mannose (dashed arrow) on the *inner core* is absent and located at the non reducing end of the outer chain. It is not regulated by the *MNN2* locus and is postulated to act as a stop signal for α1,6-backbone elongation. Phosphorylation sites are indicated by asterisks. Outer chain phosphorylation is reduced in *mnn4* and *mnn6* mutants. The *mnn1* mutation affects outer chain, inner core (circled mannose) as well as O-linked α-1,3 terminal mannose linkages.

and differs from previous proposals in the attachment site of the outer chain [25–27]. Most of the structural work has been carried out on bulk material obtained by extraction of cells with hot citrate buffer followed by precipitation of the mannan as a cetyl-trimethylammonium–borate complex. Studies using defined glycoproteins, like invertase, are in essential agreement with that deduced from the bulk mannoprotein fraction [28]. It should be emphasized, however, that the literature before 1989 assigns the outer

478

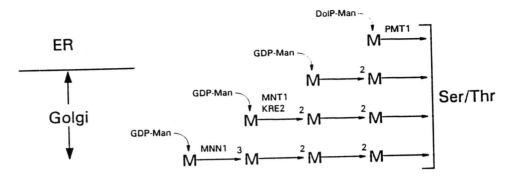

Fig. 2. Structure and assembly of alkaline labile O-glycan chains. Dol-P-Man is the donor for the first mannose residue; GDP-Man for the subsequent mannoses. *PMT1*, *MNT1/KRE2* and *MNN1* encode genes affecting mannosyl transferases.

chain to the α-1,6-mannose branch; therefore, some of the oligosaccharide structures have to be reconsidered.

Investigations carried out to get more insight into the cell wall architecture and into the interaction of the various components showed that two types of cell wall mannoproteins can be distinguished. One part can be extracted by hot SDS and comprises 80% of the wall proteins; the other type can only be liberated by digestion of the glucan layer with β-1,3-glucanase [29–32] indicating that this mannoprotein fraction is tightly linked to glucan. Approximately 60 different SDS extractable mannoproteins of molecular masses <100 kDa can be distinguished by SDS-polyacrylamide gel electrophoresis. A dominant and well-characterized N-glycosylated component of this fraction is a 29/33.5 kDa species [29–31,35,36] that seems to occur also in *Candida* [33,34]. The corresponding *S. cerevisiae* gene, *BGL2*, was cloned [35] and shown to encode an *endo* β-1,3-glucanase [36]. The remaining glucanase extractable mannoproteins are mainly of high molecular weight [29,30,37,38]. A recent intriguing observation is that some of these glycoproteins carry a novel type of carbohydrate chain characterized by the presence of linear β-1,6-linked glucoses and α-1,6-linked mannoses [24,39]. The glucose containing chains cannot be released by endo H or N-glycosidase F and are also resistant to β-elimination. This finding of a novel *glucomanno*protein fraction and earlier observations supporting a covalent linkage between glucan and mannoproteins [32,40,41] led to the suggestion that the glucose containing chains are interwoven with the glucan layer thereby anchoring these proteins into the cell wall [24]. Since there is also a glucan–chitin complex [42], which may interact with the glucomannoproteins, a picture of the three-dimensional organization of the cell wall gradually takes shape. It is noteworthy to mention that SDS-extractable mannoproteins are incorporated continuously during the cell cycle, whereas the incorporation of glucanase extractable mannoproteins, which affect cell wall porosity [43,44], show cyclic oscillation with a maximum after nuclear migration [45].

3. N-linked carbohydrate structures

Among the N-linked chains two types can be distinguished. There is a family of small oligosaccharides ranging from $Man_{8-15}GlcNAc_2$ [46–49] which is called *inner core* of mannan [50]. It can be considered to result from minor modifications of the lipid-bound $Glc_3Man_9GlcNAc_2$ precursor core after its transfer to the nascent polypeptide. On secreted glycoproteins, some of these oligosaccharides may become elongated by an *outer chain* variable in size (>50 mannose residues) by extension of an α-1,6-linked polymannose backbone and of short α-1,2-linked mannose side chains (one to two mannoses). Some of the side branches both in the *outer chain* and the *inner core* are 'capped' with a terminal α-1,3-linked mannose residue or they are substituted with mannosylphosphate and mannobiosylphosphate residues, respectively. The *outer chain* is highly immunogenic. Several so-called *mnn* mutants (for <u>m</u>an<u>n</u>an *defective*) with an altered carbohydrate structure helped to dissect the immunochemical properties of mannoprotein carbohydrate, but also contributed to elucidate the mannosylation pathway [50]. None of the affected genes besides *MNN1* [51,52] and perhaps *MNN3* seem to account for a structural gene encoding the corresponding mannosyltransferase. The presence of an *outer chain* seems to be a characteristic of extracellular mannoproteins, whereas intracellular ones lack this modification. A possible function for the *outer chain* cannot be given at the moment. However, a clumpy morphology, reduced viability and distortion of the cell wall arise when *outer chain* elongation is prevented [53].

Outer chain attachment site. Recent investigations of core oligosaccharide structures of several *mnn* mutants lead to a revision of the carbohydrate structure originally proposed. It was believed previously that the α-1,6-linked terminal mannose of the $Man_8GlcNAc_2$ core is the site of initiation of the outer chain. However, Ballou and coworkers found that in mutants which lack outer chains, this terminal α1,6-mannose was phosphorylated at position 6 [25–27]. Further examination of the *mnn9* structure using negative ion LSIMS indicated that the first α-1,6 *outer chain* mannose is attached to the α-1,3-branch of the β-1,4-linked mannose (Fig. 1). Moreover, it was shown that carboxypeptidase Y from wild type yeast has the same branching, indicating that the novel structural assignment is not restricted to the mutants [54]. The new data are consistent with the methylation, acetolysis and ¹H-NMR data published previously [46,55] and the specificity of the α-1,6-mannosyltransferase initiating *outer chain* formation [56].

Heterogeneity of 'inner core' oligosaccharides. The core portion is identical with the high mannose oligosaccharides of higher eukaryotes, except that it is modified in yeast by the addition of several α-1,3-mannose units [27] resulting in a series of oligosaccharides differing by one mannose. Heterogeneity on glycoproteins is a common structural feature. Isolated $Man_{8-14}GlcNAc$ oligosaccharides from highly purified invertase were shown to constitute a homologous series of nearly homogeneous compounds, which appear to define biosynthetic intermediates [28]. In contrast, studying the N-glycan chains from the total soluble glycoprotein pool of log-phase cells considerable structural heterogeneity in all but $Man_8GlcNAc$ was seen [57]. The principle isomer in each *inner core* oligosaccharide was that found in invertase.

Heterogeneity appeared to arise from degradative rather than from synthetic events: it results from metabolic degradation in the vacuole and to a lesser extent from ER trimming and core extension in the Golgi. A powerful method to separate oligosaccharide isomers is high performance anion-exchange chromatography [58] which has also been applied to high-mannose oligosaccharides from yeast [59,60].

4. N-glycosylation pathway

Similar to higher eukaryotes, the initial steps of the N-glycosylation pathway involve the stepwise assembly of the unique precursor oligosaccharide $Glc_3Man_9GlcNAc_2$ on the dolichyl pyrophosphate carrier lipid and its *en bloc* transfer to the amide group of an Asn-Xaa-Thr/Ser sequon of nascent polypeptide chains. Subsequently the precursor is modified in the ER and Golgi by processing glycosidases and glycosyltransferases. Whereas the initial processing is highly conserved in all eukaryotes, yeast Golgi modifications are very different from those in mammalian cells. No complex or hybrid type structures are synthesized in yeast.

4.1. Lipid-linked oligosaccharide assembly

The core oligosaccharide is assembled in a stepwise fashion on dolichyl phosphate and several of the reactions have been studied in detail both *in vitro* and *in vivo* (see Refs. [2,3,10]). In yeast the dolichols are a mixture of 14–18 isoprene units, somewhat shorter than in mammalian cells [61]. Besides a minimum of around seven isoprene units no preference for a chain length was observed *in vitro* [62,63]. The glycosyl donors leading to $Man_5GlcNAc_2$-PP-Dol are UDP-GlcNAc and GDP-Man [2,3,10], respectively, and these reactions are believed to take place on the cytosolic face of the ER. Core-oligosaccharide completion occurs on the lumenal side with lipid intermediates Dol-P-Man and Dol-P-Glc as donors (for review see Ref. [64]).

Complementation of the temperature sensitive mutant *sec59*, which accumulates truncated glycoproteins in the ER [65] revealed that the *SEC59* gene encodes a CTP-dependent dolichol kinase [66]. Other temperature sensitive mutants that have defects in the steps of core assembly are the so-called *alg*-mutants (*asparagine linked glycosylation*) [67,68]. *alg* mutants defective in the formation of the $Man_5GlcNAc_2$ oligosaccharides are not viable at the restrictive temperature, whereas mutations affected in later stages of oligosacharide assembly are not lethal. This clearly shows that protein N-glycosylation is a vital reaction. Several of the *ALG* genes have been cloned. *ALG1* encodes the structural gene for the β-1,4-mannosyltransferase forming Dol-PP-$GlcNAc_2$Man [69]. The enzyme is a type II membrane protein with a cytosolic orientation [70]. *ALG7* codes for the first gene product required in the pathway, the UDP-GlcNAc: Dol-P GlcNAc phosphate transferase. *ALG7* has been cloned on the basis of its ability to rescue cells from tunicamycin toxicity [71]. A transcript heterogeneity was observed that seems to be subject to differential expression [72]. Alg7p has structural homology (up to 60% identity in several portions constituting one-third of the protein) to the corresponding protein from hamster [73]. The essential *ALG2* gene most likely encodes the α-1,3-mannosyltransferase involved in the conversion of Man$_2$GlcNAc$_2$-PP-Dol to Man$_3$GlcNAc$_2$-PP-Dol. The amino acid sequence predicts a membrane-bound protein with a mol. mass of 60 kDa [72a]. *ALG3*, although not cloned yet, seems to encode a DolP-Man:Man$_5$GlcNAc$_2$ α-1,3-mannosyltransferase. The *alg3* mutant accumulates Man$_5$GlcNAc$_2$ both on dolichol and glycoprotein [74]. In special mutants of mammalian cells a similar phenotype results from a defect in the formation of Dol-P-Man [75]. *ALG5* is a transmembrane protein of 38 kDa and encodes the structural gene for Dol-P-Glc synthase [76]. Using 5-azido-UDP-glucose as an active site directed photoaffinity analogue a protein of 35 kDa was identified, what is in good agreement with the above value [76a].

4.2. Core glycosylation

Although the Glc$_3$Man$_9$GlcNAc$_2$ is the preferred donor for saccharide transfer to nascent polypeptide chains, transfer of smaller oligosaccharides has been demonstrated both *in vitro* [77] and *in vivo* [78–80]. Studies with a solubilized extract from yeast membranes showed that the glucosylated lipid-oligosaccharide lowers the apparent K_m-value for the peptide acceptor by a factor of about ten as compared with the donor Dol-PP-GlcNAc$_2$ [77]. Genetic evidence concerning a role of glucose residues suggests that they are neither necessary for viability of yeast cells under normal laboratory growth conditions nor for outer chain elongation [48,68]. In the *alg3* mutant structural analysis of oligosaccharides on invertase showed that >75% of the N-linked glycosylation occurs by transfer of Man$_5$GlcNAc$_2$ without prior addition of the three glucoses [79,80]. However, a lack of glucose residues may result in underglycosylation of mannoproteins as shown for yeast invertase [68,81].

Substantial progress has been made recently both in yeast and mammals in the characterization of the *N*-oligosaccharyltransferase (OTase) catalyzing the saccharide transfer to asparagine both from yeast and mammals. Earlier attempts to purify this ER membrane-bound enzyme have failed due to its lability upon solubilization [77,82,83] and, as it turns out, probably also because the transferase is an enzyme complex consisting of at least 4 subunits. In yeast two essential genes, *WBP1* (wheat germ agglutinin binding protein) and *SWP1* (suppressor of wbp1) have been isolated and characterized and shown to be essential for *N*-oligosaccharyl transferase activity *in vivo* and *in vitro* [84,85]. Chemical crosslinking experiments and genetic data indicate that the two proteins form a complex. *Wbp*1p was originally purified due to its binding to WGA [86] and subsequently found to be a component of the OTase. It is a mainly luminally oriented type I ER transmembrane protein with a 32 amino acid long hydrophobic C-terminal anchor domain. The protein sequence predicts two potential N-glycosylation sites, which are glycosylated (Lehle and Aebi, unpublished) giving rise to a molecular mass of 48 kDa. Swp1p is a 30 kDa type I transmembrane protein. Its gene was isolated as an allele specific suppressor of a *wbp1* ts-mutant [85]. Since overexpression of both genes does not increase OTase activity, it was postulated that both are non-limiting components of a larger protein complex. Using an anti-wbp1p-antibody, which is able to precipitate OTase activity, a third 60/63 kDa and a fourth 34 kDA component of the complex was isolated (Knauer and Lehle, unpublished). The doublet differs in the degree of glycosylation. Recently it was shown that the mammalian oligosaccharyl transferase activity cofractionated with a complex of ribophorins I and II and a 48 kDa protein (OST48) [87]. Sequence comparisons of the yeast and mammalian components display interesting homologies. Wbp1p is 25% identical to the OST48 protein [88] and may be the homologous gene product. *SWP1* originally showed no significant homology to other known sequences in several databases [85]. However, comparing the C-terminal half of ribophorin II (MW 67 kDa) with Swp1p (MW 30kDa) an overall identity of 22% and a similarity of 46% was detected. Interestingly, ribophorin II also reveals 2–3 potential membrane spanning domains close to the C-terminus. So far the specific function of each of the various components is not known. The proposed dolichol recognition motif deduced from sequence comparison of the *ALG1*, *ALG2*, *ALG7* and *DPM1* genes [70,72a] is also present in ribophorin I [87]. However, more recent studies have questioned such a role. It was found that deletion of this element is not essential for Dol-P-Man synthase activity [88a]. It is also not present in the Dol-P-Glc synthase [76].

A 57 kDa luminal ER protein from chicken called GSBP (glycosylation site binding protein) was originally believed to recognize the sequon motif and to be involved in N-oligosaccharyl transfer [89,89a]. However, the yeast homologous gene, *TRG1* (for thioredoxin related glycoprotein) encoding a 72 kDa essential glycoprotein with two typical thioredoxin motifs, does not prevent N-glycosylation under the control of a regulatory promotor, rather it causes a maturation defect of vacuolar carboxypeptidase [90]. Other yeast lumenal ER protein homologues of *TRG1* were also reported. The yeast gene *PDI1* [91] is almost identical to *TRG1*. *EUG1* (a soluble resident ER protein unnecessary for growth, which is not essential and contains two thioredoxin boxes with only one cysteine residue) has 43% identity with *TRG1* [92]. Its synthesis is increased in response to the accumulation of proteins in the ER. Eug1p overproduction allows growth in the absence of *PDI1* but does not suppress CPY accumulation. The precise *in vivo* role and relationship of these three proteins is not clear at the moment.

4.3. N-glycosylation sites

Not all potential N-glycosylation sites of the type Asn-X-Ser/Thr (X can be any one of the 20 amino acids apart from proline) in a polypeptide are used. Among various factors, the importance of the appropriate secondary structure in the protein backbone has been pointed out. Earlier work predicted used acceptor sites as being present in β-turns [93,94]. This has been questioned in recent computer surveys and statistical analysis of known sequences of glycoproteins: both glycosylated and nonglycosylated acceptor sites occur in similar predicted conformations [95,96]. On the other hand, *in vitro* studies using cotranslational peptide acceptor substrates in an extended β-turn or Asx-turn motif (a local conformation topologically equivalent to a β-turn) indicate that adoption of an Asx-turn is the favourable conformation [97,98]. Further work is needed to answer the question of whether there is a role for peptide conformation in Asn-linked glycosylation. *In vitro* experiments with yeast membranes using synthetic peptides [77,99] agreed with data obtained from animals [100] that the tripeptide sequence is sufficient to function as glycosyl acceptor, if both the α-amino group of asparagine and the carboxyl group of the hydroxyamino acid are blocked. More recently, the ability of potential acceptor peptides to interfere with glycosylation of nascent invertase chains was tested in a cell free translation/glycosylation system prepared from reticulocytes and dog pancreas microsomes [101]. It was observed that peptides which are acceptors for N-oligosaccharyltransferaseinhibitcotranslationalglycosylationofinvertase,whereas non-acceptors have no effect. Results obtained with proline-containing peptides are compatible with the notion that proline residues at the C-terminus or within the tripeptide sequence protein prevent glycosylation, but allow oligosaccharyl transfer when present in an N-terminal position ([101]; see also Bause, 1983). This was further substantiated by *in vivo* studies using site-directed mutagenesis to introduce a proline residue at the C-terminus of a selected glycosylation site of invertase. Expression of this mutation in three different systems, yeast cells, oocytes from *Xenopus laevis*, and in cell free translation/glycosylation using reticulocyte lysates in the presence of dog pancreas membranes, demonstrated an inhibition of glycosylation, although with qualitative and quantitative differences [102]. These data are in agreement with the statistical study of Gavel and van Heijne [96] indicating that proline residues in positions X or Y of the consensus sequence Asn-X-Thr/Ser-Y strongly reduce the likelihood of N-linked glycosylation.

4.4. Maturation to inner core and outer chain biosynthesis

As mentioned earlier, the protein-bound core oligosaccharide can be processed to two types of N-linked chains depending on individual glycosylation sites and on the glycoprotein. Minor modifications give rise to the family of slightly enlarged *inner core* oligosaccharides $Man_8-Man_{15}GlcNAc_2$. In the second type of modification extensive mannosylation elongates the *inner core* to large polymannose chains. Both types of chains can be present on one and the same protein [103–105]. Heterologous expression of yeast invertase in *Xenopus laevis* oocytes showed that the number of chains acquired is the same as in yeast, but the polymannose chains are modified to complex type structures [102]; yeast lacks the glycosylation machinery for complex type structures. Therefore, *outer chain* formation and complex type processing may be considered as analogous types of processing reactions: both seem to occur on saccharide chains still accessible after folding [106,107].

4.4.1. Inner core modificaton

Based on the observation that $Man_8GlcNAc_2$ is the smallest N-linked oligosaccharide in pulse-chase studies of mannose-labeled yeast cells [47] and in isolated invertase [28,49] the early processing steps were deduced to consist of removal of 3 glucoses and a single specific α-1,2-linked mannose on the inner α-1,3-branch of the core, identical to the reactions in mammalian cells.

Glucosidase I (cleaving terminal α-1,2-linked glucose) and glucosidase II (cleaving the two α-1,3-linked glucoses) have been partially purified and characterized [108,109]. The specific $Man_9GlcNAc_2$ α-mannosidase forming the unique isomer of $Man_8GlcNAc_2$ has been purified [110,111] and the corresponding gene, *MNS1*, has been cloned [112]. The yeast trimming mannosidase is a type II membrane glycoprotein [111,113] with a catalytic C-terminal domain. The calculated MM is 63 kDa. On nonreducing SDS gels the isolated enzyme gives a sharp band at 66 kDa, while in the presence of β-mercaptoethanol two peptides of 44 and 23 kDa are obtained. The enzyme requires low Ca^{2+} and works on $Glc_3Man_9GlcNAc_2$ as well as $Man_9GlcNAc_2$ [111]. This is in agreement with the observation that glucose removal is not a prerequisite for mannose removal [114]. For maximal enzyme activity the α-1,2-linked mannose on the α-1,6-branch was shown to be important. Gene disruption of *MNS1* shows that it is not essential for survival nor for outer chain modification [115]. On the other hand it is unclear, why in almost all *inner core* oligosaccharides the inner α-1,3 branch is trimmed, if there were no function.

The heterogeneity of the *inner core* is caused both by a size and a structural heterogeneity. Structural investigation of the oligosaccharides and their relative abundance from a defined glycoprotein, like invertase, showed that they are homologous isomers resulting from a single-step mannose addition to the trimmed $Man_8GlcNAc$ [28]. The data suggest that there is a core filling major pathway of *inner core* formation rather than a random synthesis. As mentioned earlier, the structural heterogeneity *within* a particular size class of the oligosaccharides from bulk soluble glycoprotein pool can be assigned both to minor alternate biosynthetic routes (altered processing, partially mannan elongation) and mainly to α-mannosidase degradation reactions [57].

4.4.2. Outer chain attachment

There is biochemical and genetic evidence that the synthesis of the *outer chain* is carried out in the Golgi by a separate set of mannosyltransferases and that exclusively GDP-Man is used as sugar donor [117,118]. Selective assays for mannosyltransferases involved in the formation of α-1,6-, α-1,2- and α-1,3-linkages have been devised using small manno-oligosaccharides as acceptors [119–124]. Formation of *outer chain* occurs in a structural, and also topological (see below), controlled and coordinated elongation process rather than in a random fashion [125–127].

The pathway has been mainly inferred from the known saccharide structures and from effects of various *mnn* mutants on these structures. At least 7 steps appear to require distinct mannosyltransferases. Mutation *mnn1* prevents terminal α-1,3-mannose addition in all parts of the N-linked glycan (and also in the O-linked chains), while *mnn2* and *mnn5* mutants lack the first and second α-1,2-mannose residue, respectively. It was observed that in these mutants α-1,2-branching still occurs at the nonreducing terminus of the backbone implying that an additional *mnn2/mnn5* independent α-1,2-mannosyltransferase exists [25,122,125,126]. Studies of the *mnn9* mutant have led to the identification of two α-1,6-mannosyltransferases, one that initiates *outer chain* synthesis [128] and one that acts to elongate the α-1,6-backbone [125,127]. As pointed out earlier, only the *mnn1* mutant shows a reduced α-1,3-mannosyltransferase activity [51,121], whereas *mnn2* and *mnn9* strains are not deficient in enzyme activities that would explain the phenotype. The *mnn9* gene [52] and also *mnn2* seem to affect glycosylation indirectly and may encode functions of the secretory pathway. The *mnn 8,9,10* mutants were shown to be resistant to vanadate, and in turn other vanadate resistant mutants isolated independently and belonging to different complementation groups were also defective in protein glycosylation [129]. Finally a mannosylphosphate transferase reaction introduces phosphate residues to some of the side branches, a reaction type originally described in *Hansenula* [130,131]. There are at least four mannosylphosphorylation sites in *S. cerevisiae* mannoproteins: side chains in the outer chain part [132], two sites in the core [26,133] and on the non-reducing terminus in the *mnn10* mutant [26]. Phosphorylation of the O-linked oligosaccharides has not been demonstrated. A mannosylphosphate transferase has partially been purified [124]. Since it does not act on large exogeneous mannan acceptors, the phosphate branch seems to be introduced while the outer chain is being synthesized. It is not clear whether introduction of mannosylphosphate in the core is catalyzed by a different enzyme. The *mnn4* and *mnn6* mutations prevent phosphorylation, but are not structural genes for mannosylphosphate transferase. In any case mannose phosphorylation in yeast is not involved in sorting vacuolar proteins (see also below) [134–137].

In summary (Fig. 3) it seems that an initiator α-1,6 transferase [56] introduces the first mannose residue. The further steps appear to involve the coordinated action of an α-1,6-elongating transferase and an α-1,2-transferase introducing the side chains on the backbone. If the terminal α-1,6-mannose in the growing chain obtains an α-1,2-mannose before the next α-1,6-elongation then *outer chain* synthesis is terminated. A 'capping' α-1,3-transferase completes the side chains.

Several of the outer chain enzymes have been characterized and purified to some extent. The initiator α-1,6-mannosyltransferase has been partially purified [128]. It acts both on $Man_8GlcNAc$ and $Man_9GlcNAc$ and the structural analysis of the products formed is in agreement with the revised attachment site on the outer α-1,3-branch. The observation that $Man_9GlcNAc$ also acts as an acceptor is consistent with the observation of the minor core elongation pathway in which no mannose is removed from the core [57]. Whether this enzyme activity is encoded by *OCH1* is not clear at present. *OCH1* was cloned by complementation of a *ts*-mutant defective in the *outer chain* [138]. Its sequence predicts a 55 kDa type II protein, which is variably glycosylated to give four forms of 58–66 kDa. Originally it was assumed that Och1p is functional in the elongation [138]. Subsequent studies of the carbohydrate structure of invertase from a *och1 mnn1* double mutant reveal that *OCH1* is rather functional in the initiation of the α-1,6-backbone, since predominantly a single $Man_8GlcNAc_2$ species accumulates on this glycoprotein [139]. Two α-1,2-mannosyltransferase activities from *mnn1* microsomes were characterized [122]. One, the M_2MT-I transferase, was novel and has the properties of the postulated α-1,6-backbone terminase. The other α-1,2-transferase, M_1MT-I, was described earlier [119]; it transfers mannose to free mannose or methylmannoside. The reduction of this activity in *mnn3* extracts suggests that it may function in branching the outer chain; but it also cannot be ruled out that it corresponds to Mnt1p involved in O-mannosylation (see below). As mentioned earlier, neither removal of the 3 glucoses, nor of one mannose residue is a prerequisite for elongation; and also $Man_5GlcNAc_2$ oligosaccharide in the *alg3* mutant gets extended *in vivo*. From structural oligosaccharide studies on $Man_5GlcNAc_2$ elongation *in vivo* in the *alg3* mutant [79,80] and the observation that the oligosaccharide-lipid form of $Man_5GlcNAc_2$ acts as an acceptor for the α-1,6-mannosyltransfer, it appears that *outer chain* initiation is dependent upon the presence of α-1,2-linked mannoses at the 3-branch.

486

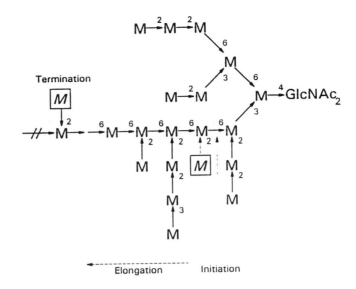

Structurally controlled process, not random

Initiation   1 ⟶ 6 transferase

Elongation   1 ⟶ 6 transferase
                      1 ⟶ 2 transferase
                      1 ⟶ 3 transferase   'capping'

Termination 1 ⟶ 2 transferase

Fig. 3. Outer chain formation. The assembly of the *outer chain* is a structurally controlled process, established by the coordinated action of distinct, compartmentalized mannosyltransferases. The elongation reaction of the backbone stays ahead the formation of the side branches. Addition of an α-1,2-linked mannose to the core (*italicized, boxed mannose M*) prior to an α-1,6-elongation leads to termination. Thus the boxed mannose with the dashed arrow is not present on *inner cores* that are extended.

5. The O-glycosylation pathway

5.1. Biosynthesis, molecular biology

Whereas the initial reactions of protein N-glycosylation proceed identically in all eukaryotic cells as mentioned above, protein O-glycosylation differs considerably among organisms (Table I). Serine and threonine residues of proteins are glycosylated

TABLE I

Comparison of protein-O-glycosylation in yeast and mammals

	Yeast	Mammals
Protein-linked sugar	Man[ab]	GalNAc[a]
Sugar-donor	Dol-P-Man	UDP-GalNAc
Cellular site of transfer reaction		
(a) First sugar	ER	Golgi[c]
(b) Chain extension	Golgi	Golgi

[a] O-glycosylated proteins containing only one GlcNAc residue have been found in higher eukaryotes [180] as well as in *S. cerevisiae* [180; Hilgarth and Tanner, unpublished].
[b] An exception is Galβ1,3-GalNAc in the ascomycetes *Cordyceps ophioglossoides* [181].
[c] Roth [182].

in fungal cells for example via Dol-P-Man as an intermediate which so far has not been observed in higher eukaryotes [140–146]. The following reaction sequence (see also Fig. 2) has been established for *S. cerevisiae* [140,141]:

I) GDP-Man + Dol-P $\xrightleftharpoons{Mg^{++}}$ Dol-P-Man + GDP

II) Dol-P-Man + Protein(Ser/Thr) $\xrightarrow{Mg^{++}}$ Protein(Ser/Thr)Man + Dol-P

III) Protein(Ser/Thr)Man + nGDP-Man $\xrightarrow{Mg^{++}}$ Protein(Ser/Thr)Man$_{1+n}$ + nGDP

Evidence has been presented that reactions I and II proceed in the ER [147–150], whereas those reactions summed up in III (mannosyltransferases for α-1,2 and α-1,3 linked mannoses) most likely take place in the Golgi [123,148,150], although the attachment of the second mannose in the ER cannot be excluded [150–152]. Strong *in vivo* evidence has been presented by Orlean [153] that reactions I and II represent the only pathway for protein O-mannosylation in *S. cerevisiae* (see below).

The enzyme catalyzing reaction I, a membrane protein of 30 kDa, has been purified [154,155]. Since the anomeric configuration of the Ser/Thr-linkage was shown to be α [99], the enzyme transfers mannose from Dol-P-β-mannose with inversion of the mannose configuration. The corresponding gene, *DPM1*, has been cloned [156]. It codes for a protein 267 amino acids in length with one potential transmembrane anchor. The protein seems to be attached to the membrane both with the C- and N-terminus. The main part of the protein either faces the lumenal or more likely the cytoplasmic side of the ER [155,217]. Disruption of the *DPM1* gene resulted in a lethal phenotype [156]. The reason is not known. Mammalian cells do not require this reaction for growth in culture [75], which may indicate that protein O-glycosylation — a reaction sequence not present in mammalian cells — is essential in yeast. With a *ts* mutant of

DPM1 it was shown that protein O-glycosylation was completely switched off when cells were shifted to non-permissive temperature [153]; therefore, no alternative O-glycosylation pathway seems to exist. The yeast *DPM1* gene is functional in mammalian cells and complements defects in Dol-P-Man, GPI anchor and partly N-linked saccharide synthesis [157,158].

The enzyme catalyzing reaction II has been purified [159,160] and an antibody precipitating the enzyme activity was shown to react with a membrane glycoprotein of 92 kDa [159]. With the help of an antibody affinity column followed by SDS-PAGE the protein has been purified to homogeneity and via peptide sequences a positive clone was obtained from a genomic *S. cerevisiae* library [161]. An open reading frame of 2451 bp codes for an 817 amino acid protein with three potential N-glycosylation sites. The gene has been called *PMT1* for protein mannosyl transferase. The hydropathy plot indicates a tripartite structure of the predicted protein: an N-terminal one third with 4–6 potential transmembrane helices, a C-terminal one third with probably 4 membrane spanning domains, and a central hydrophylic part. Since all three glycosylation sites most likely are glycosylated [159] the central part of the protein as well as the C-terminal end probably face the ER lumen. When the clone was expressed in *E. coli* the bacterial extract catalyzed reaction II. Gene disruption led to a complete loss of *in vitro* mannosyltransfer activity from Dol-P-Man to a hexapeptide used as acceptor in the enzymatic assay. *In vivo*, however, protein-O-mannosylation had only decreased to about 50% in the disruptant, indicating the existence of a second gene for a transferase not detectable by the enzyme assay [159]. In the meantime a test for the second transferase has been set up: the enzyme requires a tenfold higher concentration of peptide and its pH optimum (6.4) is about one pH-unit lower than that for Pmt1p (M. Gentzsch, unpublished).

A membrane-bound α-1,2-mannosyltransferase from *S. cerevisiae* has been purified, cloned and sequenced [162]. The *in vitro* specificity of the enzyme is similar to that described from this laboratory earlier [119]. The *MNT1* gene codes for a 41 kDa Golgi protein which is required for the attachment of the third mannosyl residue of O-linked saccharides [123]. The same gene has independently been cloned by selecting for killer toxin resistent strains [163].

In the field of protein O-glycosylation it is still a wide open question which serine and threonine residues of a secreted protein will be glycosylated as compared to many others that will not be. Neither for yeast cells nor for higher eukaryotes is a sequon sequence analogous to that for N-glycosylation in sight. In yeast cells a requirement for one or more prolines preceeding the hydroxy amino acid could not be demonstrated [159,164]; a glycine N-terminally neighbouring serine/threonine seems to have a strong negative influence on O-glycosylation as has a negative charge in the immediate neighbourhood of the hydroxy amino acid [159]. From *in vivo* studies with cloned human parathyroid hormone a positive effect of neighbouring leucine and lysine residues has been proposed [164]. The transferases coded for by *PMT1* and *PMT2* (see above) both transfer mannosyl residues to serine as well as to threonine [159, and unpublished results]. An extensive investigation with the *Candida albicans* protein mannosyltransferase demonstrated that within all peptides tested threonine residues were better acceptors than serines [165]. The peptide serving best as mannosyl acceptor was biotin-YATAV-NH$_2$ for which the *Candida* enzyme possessed a K$_m$-value of 0.075 mM.

It has been shown that depending on the proteins, sequences which get O-glycosylated in mammalian and fungal cells may either differ or may be identical [164,166–168]. In the case of the human granulocyte-macrophage colony-stimulating-factor it was observed that the *in vivo* O-glycosylation pattern was similar, whereas a peptide of the factor used for enzymatic glycosylation *in vitro* yielded different results depending on whether yeast or rat liver extracts were the enzyme source [168,169]. This may well be related to the finding that *in vitro* assays not necessarily measure the various mannosyl transferases (see above).

5.2. Defined O-mannosylated proteins

Originally Sentandreu and Northcote described mannosyl-residues linked O-glycosidically to hydroxy amino acids in yeast [170]. Although cell walls contain significant amounts of O-linked saccharides only few, well-defined proteins are known so far with this type of covalent modification. A highly N- and O-glycosylated cell wall protein with a protein moiety of approximately 22 kDa and a carbohydrate part which consists of about 50% O-linked saccharides and amounts to more than 88% of the total molecular mass has been described. It was characterized from an outer chain mutant of *S. cerevisiae* [37]. A smaller cell surface component, the a-agglutinin induced by α-factor in a-cells has an apparent molecular mass of 18 kDa and consists of 30% O-linked carbohydrate [171,172]. The molecular mass of the serine/threonine rich apoprotein amounts to 7.5 kDa; the corresponding gene has been cloned and sequenced [173]. Another pheromone-induced cell surface protein (80 kDa) which is highly O-glycosylated is the 58 kDa *FUS1* gene product [152]. Chitinase, an exclusively O-glycosylated yeast protein is secreted into the medium [151], although a small part also remains cell wall associated. The apparent molecular mass of chitinase shifts from 150 to 60 kDa, when the protein is not glycosylated in a temperature sensitive *dmp1* mutant under non-permissive conditions [153]. Secretory glycoproteins, the formation of which is stimulated by heat shock, have been reported to be highly O-glycosylated [174,175]. The gene of a 150 kDa protein has been cloned (*HSP150*) and the non-glycosylated protein amounts to 47 kDa [174]. Finally the product of the *KEX2* gene is an O-glycosylated protein located in the Golgi [176] and also the ER located *SEC20* gene product is thought to contain O-linked saccharides [177].

The function of protein O-glycosylation is still not fully understood. However two consequences of protein O-glycosylation seem more generally accepted: a relative resistance towards proteases and a conformational effect resulting in a stiff and extended structure [178]. For an O-mannosylated yeast protein corresponding data are not available so far, however the unusual running behaviour of a-agglutinin in SDS gels (18 instead of 11 kDa) as well as its resistance to boiling and SDS treatment [172,173] may indicate an extended and stiff conformation. O-mannosylation of certain proteins also positively affects their secretion [179; Strahl-Bolsinger, unpublished results].

6. Protein modification by GPI-anchors

Many eukaryotic plasma membrane surface proteins are modified in the ER at their C-terminus by a glycosylphosphatidylinositol (GPI) anchor. This topic has been reviewed by Ferguson [183] and Herscovics and Orlean [10]. The primary function of the GPI anchors is to guarantee the stable association of the protein with the membrane. The existence of GPI-anchored proteins in yeast was first shown by Conzelman and coworkers [184–186]. The most abundant GPI glycoprotein was studied by two different groups and described as a 125 kDa [184,186] and 115 kDa glycoprotein [187], respectively. Isolation of the corresponding genes (GAS1/GGP1) revealed that both glycoproteins are identical [188, 189]. gp 125 is a cell cycle modulated protein [190], but not essential for cell viability. A gas1 mutant lacking C-terminal hydrophobic sequences showed that this domain is necessary for anchor addition; the truncated forms failed to become membrane attached and were secreted into the medium [188]. The sequence requirements for peptide cleavage and glycolipid addition were also studied with this protein. The findings that only a subset of amino acids with small side chains are tolerated at the attachment site are consistent with the situation in animals [191,192].

Other candidates for GPI-anchored proteins as predicted from the C-terminal protein sequences are α-agglutinin [193] involved in mating, the KRE1 protein [22] involved in cell wall synthesis and the YAP3 protein [194] encoding an aspartyl protease. For α-agglutinin GPI modification has been proven experimentally [193]. The conserved core-region of mammalian GPI anchors — EtN-PO$_4$-6Man(α1–2)Man(α1–6) Man(α1–4)GlcNH$_2$(α1–6) myoinositol phosphate lipid — seems to be the same also in yeast (Fig. 4). With regard to the lipid moiety two different components occur in yeast. Besides the usual diacylglycerol residue evidence exist for a base resistant ceramide-like lipid [195]. The relative amounts of the two lipids in the GPI moiety depends on the proteins. From in vivo pulse-chase studies it seems that GPI takes part in a lipid remodelling past the sec 18 block. A ceramide-like lipid has also been detected in another case in the contact site A protein of Dictyostelium discoideum [196]. It is not known whether GPI modification is essential for yeast. Mutants have been isolated that

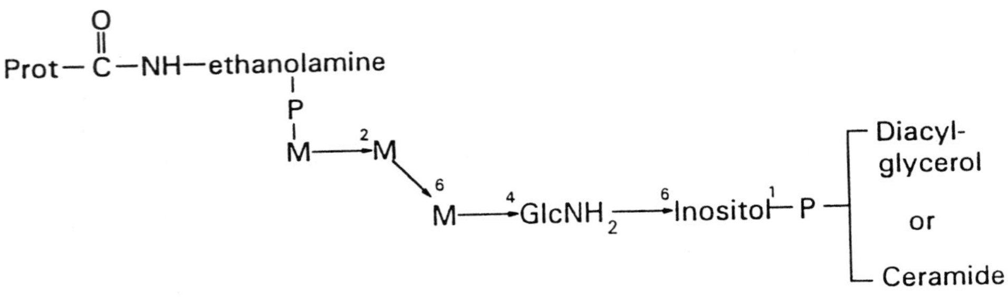

Fig. 4. Yeast GPI anchor. The lipid portion may constitute either a diacylglycerol or a ceramide [196] residue. The inositol may be acylated leading to phospholipase C resistance (not shown).

are able to survive without sphingolipids [197]; thus anchors with ceramide-lipid are not essential for yeast.

Biosynthesis of GPI anchors was studied both *in vivo* and *in vitro* [153,198] and the yeast pathway is identical to that of higher eukaryotes. A transfer of GlcNAc from UDP-GlcNAc to phosphatidylinositol is followed by deacetylation to glucosamine. Three mannose residues are added, at least one coming from Dol-P-Man. This is inferred from the observation that in the *dpm1* mutant, defective in the synthesis of Dol-P-Man, GPI anchoring is stopped [153]. A glycolipid accumulates with the structure GlcNH$_2$-PI, that is fatty acylated on the inositol residue. The acyl group is introduced, therefore, prior to mannosylation; acyl-CoA was shown to be the donor [198]. It is probably removed again during later processing of the GPI, since most of the GPI anchored proteins in yeast can be released by phospholipase C [185,198], which can act only when there is no extra fatty acid attached to the inositol ring [199]. The significance of this modification is unknown.

7. Topology and compartmental organization of glycosylation reactions and the secretory pathway

Protein glycosylation is intimately associated with aspects of translocation of polypeptides and hydrophilic sugar moieties across the lipid bilayer, with vesicular transport through the secretory pathway and the problem of targeting to specific compartments. Recent progress has been achieved in this issue by the convergence of biochemical and genetic approaches. In yeast several mutants (*sec* mutants) defective in individual steps of the secretory pathway have been isolated and the topic has been reviewed in detail [200,201]. Core glycosylation occurs while the nascent polypeptide chain is translocated through the ER membrane. Candidates for constituents of the translocation apparatus in yeast include *SEC 61,62* and *63* genes that have been found by genetic screening; certain mutations of these genes lead to a defect of protein translocation [202–204]. A mammalian homologue of Sec61p was isolated and found to be located in the immediate vicinity of nascent chains during their membrane passage; it is tightly associated with ribosomes and also cofractionates with three proteins identified earlier as the *N*-oligosaccharyltransferase complex [205]. Thus the OTase may be part of the machinery that translocates polypeptides into the ER lumen. The functional advantage of such a location is obvious, since glycosylation occurs prior to folding. It was observed that potential glycosylation sites are most efficiently glycosylated on unfolded polypeptides [206]. By *in vitro* translation of model proteins in the presence of dog pancreas membranes the distance between the *N*-oligosaccharyltransferase active site and the ER membrane was determined. It was found that a precise distance constraint of 12–14 residues corresponding to a distance of 40–45 Å above the lumenal membrane surface allows glycosylation [207]. This observation is compatible with the dimensions of the carrier dolichol, which is about 80–100 Å for dolichol-19 [208] of which 35–60 Å could be buried in the membrane. It should be recalled that both Wbp1p and Swp1p have the majority of the protein portion lumenally oriented. Attempts to reconstitute protein translocation from solubilized yeast membranes have been successful. However, so far no N-glycosylation was achieved [209], also not in similar reconstitution experiments with membranes from mammalian origin [210]. It has been shown that lumenal N-glycosylation, likewise proper protein folding, can prevent retrograde movement of nascent polypeptide chains across the ER membrane *in vitro* [211]. Along the same theme, experiments with acceptor peptide for OT indicate that they freely entered and could be washed from the ER but got trapped after their glycosylation [212]. Thus an indirect function of OTase could be to assist in protein translocation across the ER membrane.

The topological complexity and the problems of glycosylation reactions at the ER have already been mentioned: whereas the various sugar nucleotide donors are synthesized in the cytoplasm, glycosyl transfer to proteins occurs in the ER lumen. Although transporters for UDP-GlcNAc, UDP-Glc (but not GDP-Man) into the ER have been reported, the current experimental evidence suggests stepwise assembly up to Dol-PP-GlcNAc$_2$Man$_5$ on the cytosolic face [2,3,64,213]. Translocation of Man$_5$GlcNAc$_2$-PP-Dol has not been demonstrated yet and remains to be established. The possibility of a flip-flop mechanism is rather unlikely. Movement of polyprenol phosphates in artificial bilayers is exceedingly slow: $T_{1/2} > 5$ h as compared to saccharide translocation, which is calculated to be $T_{1/2} < 1$ s [214]. However, dolichol phosphate has an effect on membrane fluidity. It induces disorder and formation of hexagonal II phase in artificial lipids, which may facilitate transmembrane movement in natural membranes [215]. It has been proposed that oligosaccharide transfer is mediated by a specific 'flippase' protein. In the case of Dol-P-Man, which also has to be transferred across the ER membrane in order to elongate Man$_5$GlcNAc$_2$ to Man$_9$GlcNAc$_2$, it was shown that partially purified yeast Dol-P-Man synthetase acts as a translocase in sealed phospholipid vesicles [154]. Since a recombinant synthetase did not catalyze this translocation [216], the possibility exists that a 'contaminating' activity was responsible for translocation rather than the mannosyl transferase itself.

The bulk of Dol-P-Man synthetase probably faces the cytosol [217; see however 155]. Indirect evidence exists that this may also be true for the enzyme forming Dol-P-Glc (inactivation of the enzyme activity by mild protease treatment in intact microsomes was observed [64]). This would mean that Dol-P-Glc is formed in the cytosol and must also be translocated. On the other hand a transmembrane ER-transport for UDP-Glc was found; however, the transported UDP-Glc may be used for a different purpose, for example transient reglucosylation of high mannose-linked chains. This type of reaction may occur subsequent to the loss of the three glucoses and needs UDP-Glc directly and not Dol-P-Glc. It was originally discovered in trypanosomes but subsequently found also in microsomes from mammals, plants and protozoa. Among fungi it was observed in *Mucor rouxii*, not however in *S. cerevisiae* [218,219].

Outer chain modification, as already mentioned, occurs in the Golgi apparatus. Although typical stacks of Golgi cisternae are rarely detectable in wild type yeast, there is growing evidence from immuno-EM [220], immunofluorescence microscopy [221, 222], biochemical studies combined with classical and molecular genetic methods [116,223–225] that yeast has a functional Golgi compartment. The non random outer chain synthesis takes place in the Golgi: glycosyltransferases reside within distinct compartments. A model derived from results described in [223,224] is depicted in Fig. 5. Using carbohydrate specific antibodies as probes, glycoprotein products were analyzed

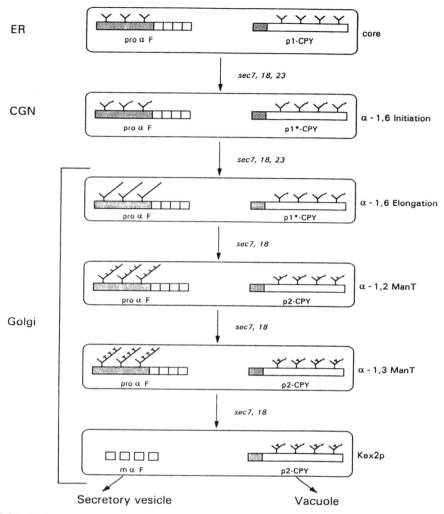

Fig. 5. Model for compartmental organization of N-glycosylation. Data presented in [223,224] reveal that within secretory compartments functionally different modifying glycosylation activities are localized, leading to a sequential assembly of the N-linked carbohydrate along the secretory pathway. In the case of CPY the core glycosylated p1 form, p1CPY, is partially modified by outer chain α1,6 mannoses p1*CPY; p2CPY has received all its mannoses. The various forms can be distinguished by their different mobility on SDS-PAGE. The mature form, mCPY, gets its pro-peptide cleaved off upon arrival in the vacuole. In the case of α-factor a pre-pro-form is core glycosylated that receives more extensive mannose addition (as is the case for most of the externally secreted mannoproteins) due to α-1,6- elongation and α-1,2- and α-1,3-branching (*outer chain*). This hyperglycosylated product migrates on SDS-PAGE as a heterogenous diffuse band. The αf-propeptide (proαF) contains four copies of the 13 aa mature peptide. The excision occurs in a late Golgi compartment characterized by endopeptidase Kex2p. The boundaries need not be interpreted as representing physically distinct compartments. Relevant functions affecting secretory transport are also depicted. ER, endoplasmic reticulum; CGN, cis Golgi network; ManT, mannosyl transferase; CPY, carboxypeptidase Y.

deriving from a cell free ER to Golgi transport reaction or from metabolically labelled *sec* mutants that are affected in vesicular traffic. By this approach it was possible to distinguish different compartments by their different glycosyltransferase contents [223]. Additional evidence comes from density gradient centrifugation experiments. Separation of enzyme activities was achieved for $\alpha 1,3$-mannosyl transferase and GDPase (intermediate Golgi reactions) from fractions that contain Kex2 protease activity (late Golgi) [225,226]. The Golgi glycosylation reactions occur on the lumenal side. A sugar nucleotide transport system for GDP-Man and a GDPase activity has been described in yeast Golgi vesicles [226]. In the case of a corresponding transport system in the mammalian Golgi, GDP-Man entry is coupled to the exit of GMP, arising from cleavage of GDP after sugar transfer. The GDPase from yeast has been purified [227] and the corresponding gene, *GDA1*, isolated. Its disruption is not lethal, but displays underglycosylation of glycoproteins [227a].

8. Specific topics

8.1. Glycoproteins and cell/cell recognition

Saccharide moieties of glycoproteins are frequently thought to be involved in cell–cell recognition phenomena [228]. Although this is the case, for example in the leukocyte/endothelium interaction, such examples are still rare.

In *S. cerevisiae* two mating type specific cell surface glycoproteins are the agglutinins. Their formation is induced by the peptide pheromones α- and a-factor in the corresponding haploid mating partners, the a- and α-cell, respectively [193,229]. The α-agglutinin is a highly N-glycosylated protein with an apparent molecular mass of >250 kDa [38, 230]. The *AGα1* gene coding for a 68.2 kDa protein moiety has been cloned. From the DNA sequence a protein 650 amino acids long with a 19 amino acid signal peptide, with a GPI anchor and with 12 potential N-glycosylation sites is predicted [38,231,232]. The completely deglycosylated protein still shows full biological activity [38,230] demonstrating that for mating type specific cell–cell interaction the protein moiety of α-agglutinin recognizes its a-agglutinin counterpart exposed on a-cells. This interaction is caused by a 1:1 complex formation between the two agglutinins [173].

The a-agglutinin consists of two parts: a small highly O-glycosylated protein with an apparent molecular weight of 18 kDa which is attached via S-S linkage(s) to a serine/threonine rich cell wall protein potentially 725 amino acids in length [172,173, 233]. Also the a-agglutinin retains most of its biological activity when deglycosylated by HF treatment [173]. This has recently been supported by expressing the a-agglutinin gene in *E. coli* as a fusion protein. A C-terminal 30 amino acid long peptide obtained from this carbohydrate-free agglutinin showed biological activity at a concentration of 6×10^{-8} M; the corresponding glycosylated peptide from authentic a-agglutinin was active when applied at 1.5×10^{-8} M (Cappellaro, unpublished results). Therefore, it seems most likely that cell/cell recognition and mating type specific agglutination in *S. cerevisiae* is based mainly on a protein/protein interaction. The O-linked sugars of the a-agglutinin may only have a supportive, auxiliary function.

Thus for baker's yeast, a major function of O- or N-linked saccharides in cell–cell interaction cannot be demonstrated.

8.2. Cell wall glycoproteins and killer toxin action

S. cerevisiae killer strains contain a double stranded RNA virus encoding a toxic protein which is secreted and kills sensitive strains of the same species [234]. Three killer toxins have been well defined: K1, K2 and KT28 [235]. To act as toxin the killer proteins have first to bind to a cell wall component and subsequently they interfere with membrane functions [236]. Killer toxin resistant mutants have been obtained which show defects in mannoprotein synthesis; as spheroplasts these mutants remain killer sensitive [136]. The *KRE2* gene product, which confers sensitivity to killer toxin K1, was shown to be identical to the *MNT1* gene product, a Golgi mannosyltransferase involved in O-mannosylation [123] (see above). The sensitivity of yeast cells towards the killer toxin K1, therefore, depends on the extension of O-linked cell wall saccharides beyond the disaccharide. At the moment it is not clear, however, whether the *kre2/mnt1* mutation may indirectly be responsible for K1 resistance, since binding of K1 toxin to β1,6-glucan has also been claimed to be required for K1 toxicity [163,236]. Killer toxin KT28 binds to the *outer chain* of cell wall mannoproteins and mutants in *outer chain* synthesis are resistent to KT28 [237,238].

8.3. Relevance of N- and O-glycosylation in general

As pointed out before the strong conservation of the dolichol cycle reactions of N-glycosylation certainly suggests that this protein modification has an essential function in all eukaryotes. In *S. cerevisiae* early experiments with tunicamycin [239], with lethal *alg* mutations [240] as well as the recent knock-out experiments concerning the oligosaccharyl transferase [84,85] all show that protein N-glycosylation is a vital reaction. When this reaction is prevented in a growing culture, all cells complete their division cycle, daughter and mother cells separate, and they all get arrested as unbudded cells [240]. Since the nucleus in such cells proceeds through a second S-phase and stops in G2 [241], the arrested unbudded cells are 'pseudo-G1' cells. Thus during two instances of the yeast cell cycle, at bud initiation and at the G2/M transition, N-glycosylation of proteins is obviously required. The reason is obscure at the moment.

It has been shown conclusively that carbohydrate moieties influence the formation of oligomeric structures of invertase and acid phophatase [105,242–244]. These two secretory proteins exist in the cell periplasm as octamers and this form — possibly simply due to its size — may be responsible in retaining the enzyme at the cell surface in association with the cell wall. It has been demonstrated repeatedly that glycosylated proteins *in vitro* are more stable against proteolytic degradation, against thermal or chemical denaturation as compared to their nonglycosylated counterparts. This, as well as the lack of *in vivo* data has been discussed previously [2]. On the other hand the increased tendency of some proteins to aggregate when not glycosylated has been shown *in vitro* [245] and *in vivo* [246–248]. Due to the fact that protein folding is much in vogue nowadays, these latter effects have been taken up again and studied in more detail. Indeed, for viral proteins it has been shown that lack of glycosylation leads to non-productive protein aggregation and to an inhibited secretion [249] confirming in a sense the old data of Gibson et al. [246]. However, is this a general phenomenon or a problem of a few specific proteins? The latter is not unlikely, since firstly the secretion of a number of glycoproteins is *not* affected when N-glycosylation is prevented [2,248]. And secondly, if the feature discussed were a general one, then prevention of N-glycosylation should result in the same phenotype as *sec* mutations: when shifted to non-permissive temperature *sec* mutants stop in the cell cycle wherever they are and remain a completely unsynchronized culture (Tanner et al., unpublished results). When N-glycosylation is switched off, on the other hand, young buds grow to normal size daughter cells, complete their cell cycle and are highly synchronized (see above).

How about protein-targeting? The mannose-6-phosphate signal pathway for lysosomal enzymes in mammalian cells is a well-documented example [250]. In *S. cerevisiae*, however, although the mannose-6-phosphate group exists on various glycoproteins, it does not seem to be functional as a signal which targets proteins to the yeast's lysosome, its vacuole [134–137,251].

Whereas there is no doubt that N-glycosylation is a vital reaction for yeast, this is not yet known for O-glycosylation. Since the knock out experiment of the Dol-P-Man:protein mannosyltransferase (see above) has uncovered the existence of a second enzyme, the answer to this question has to await the cloning of the second gene.

Thus, it certainly holds for yeasts what Hakomori recently stated: "The function of glycoconjugates remains essentially a great enigma." [252].

9. Protein glycosylation in other fungi

The biosynthesis of glycoproteins in species other than *S. cerevisiae* has been studied less extensively. Again the better characterized examples are mannoproteins of the cell envelope. They have structural features closely related to 'baker's yeast mannan' but with species specific differences. The relevant literature on the biosynthesis up to 1981 has been summarized previously [253]; structural aspects were discussed by Gorin and Spencer [254,255] and Ballou and Raschke [256].

Hansenula species contain a phosphomannan besides a neutral mannan [257,258]. A well-characterized surface mannoprotein is the 5-agglutinin, which is involved in cell–cell recognition between 5- and 21-cells [259–261]. A mannosylphosphate transferase as well as a mannosyl transferase activity involved in O-mannosylation has been described [131,143,257].

Kluyveromyces lactis mannan is very similar to that of *S. cerevisiae* but has an immunodominant *N*-acetylglucosamine residue added in α1,2-linkage to a mannotetraose side branch instead of mannosylphosphate groups [262]. The α-*N*-acetylglucosamine transferase was purified and characterized [263,264]. Mutants were isolated with a defect in this enzyme activity [263–265].

Schizosaccharomyces pombe, the most frequently investigated yeast besides *S. cerevisiae*, is known to galactosylate both cell wall mannoproteins [266,267] and to secrete glycoproteins like invertase [268] or acid phosphatase [269,270]. A galactosyltransferase activity was described that is encoded by the *gma12* gene [270,271].

Deletion of the gene results in loss of this activity, but does not abolish the ability to synthesize galactomannan.

Several *Candida* species are human pathogens. The mannans are more complex with side branches up to six mannose residues [272–274]. Mannan determines the antigenic specificity of *Candida* species [275–277]. Antimannan antibodies increase during invasive candidiasis. Mannose incorporation from GDP-Man into endogenous protein acceptor using membranes from *Candida albicans* was reported [278]. A $Man_6GlcNAc_2$-Asn oligosaccharide could be elongated up to 7,8,9,10 mannose residues with GDP-Man as donor and a crude membrane fraction [279]. The same acceptor was processed to about 25 mannoses in *S. cerevisiae* [127].

The filamentous fungi *Aspergillus* contains galactomannans. Their structures were studied in *A. niger* [280]. A well-characterized example for a secretory glycoprotein in this species is glucoamylase. Its gene has been cloned [281–283] and its carbohydrate structure has been characterized [284,285]. It contains both N- and O-linked mannose oligosaccharides. The latter are clustered in a Ser/Thr-rich linker region of the protein and, unusual as compared to *S. cerevisiae*, they are branched and contain also glucose residues. The O-glycan chains were reported to stabilize the catalytic and binding domains of glycoamylase. A smooth membrane fraction of *A. niger* was shown to catalyze transfer of mannose from GDP-Man to endogenous lipid and protein acceptor. The mannolipid appears to be a polyprenol derivative, the oligosaccharide is almost identical to the saccharide chain from ovalbumin [286].

10. Heterologous expression and glycosylation of proteins

Saccharomyces cerevisiae has gained popularity as a host to express and secrete foreign proteins for research, industrial and medical use. This topic has been reviewed recently in detail [287]. In this chapter only aspects of glycosylation and secretion will be discussed. When trying to express and secrete foreign proteins in yeast a major drawback is that both the N- and O-glycosylation apparatus differs from that of higher eukaryotic cells. Due to hyperglycosylation of proteins by outer chain attachment or due to the lack of complex type glycosylation the biological function and/or immunogenicity of the particular protein may be affected, although not necessarily. For example the envelope glycoprotein gp350 from Epstein Barr virus, lost reactivity with antibodies when expressed in yeast, because of extensive glycosylation [288]. On the other hand yeast specific addition of α-1,3-linked mannose units can be highly immunogenic [50] by creating new antigenic epitopes on a protein. No conclusive predictions can be made for successful expression of a protein in yeast and each case has to be investigated by considering inherent features of the protein and with respect to its intended use. Nevertheless many heterologous glycoproteins have been produced in yeast including tissue plasminogen activator [289], somatostatin [290], human erythropoetin [291], glucoamylase from *Aspergillus* [292], human α1-antitrypsin [294], porcine urokinase [295], gamma-interferon [296], interleukin-6 [297], human parathyroid hormone [164], wheat germ agglutinin [298] or human granulocyte-macrophage colony stimulating factor (h GM-CSF) [299]; the list is by far not complete. In the case of the parathyroid hormone the protein was O-glycosylated in a yeast specific manner but in spite of that fully active.

TABLE II

Glycosylation mutants and genes

Gene	Mutant	Phenotype	Gene product	Localization	Lethality	Ref.
ALG7			Dol-PP-GlcNAc synthase	ER	+	71
ALG1		accumulates Dol-PP-GlcNAc$_2$	β-ManT	ER	+	69
ALG2		accumulates Dol-PP-GlcNAc$_2$Man$_2$	GDP-Man: Dol-PP-GlcNAc$_2$Man$_2$ ManT ?	ER	+	68,72a
	alg3	accumulates Dol-PP-GlcNAc$_2$Man$_5$	Dol-P-Man: Dol-PP-GlcNAc$_2$Man$_5$ ManT ?	ER	−	68,74
SEC53	alg4/sec53	accumulates Dol-PP-GlcNAc$_{1-8}$	Phosphomannomutase	ER	+	68,309
ALG5		accumulates Dol-PP-GlcNAc$_2$Man$_9$	Dol-P-Glc synthase	ER	−	76,78
	dpg1	suppressor of gls1	Dol-P-Glc synthase ?	ER	−	81
	alg6	accumulates Dol-PP-GlcNAc$_2$Man$_9$	Dol-P-Glc: Dol-PP-GlcNAc$_2$Man$_9$ GlcT	ER	−	78
	alg8	Dol-PP-GlcNAc$_2$Man$_9$Glc$_1$	Dol-P-Glc: Dol-PP-GlcNAc$_2$Man$_9$Glc$_1$ GlcT	ER	−	310
WBP1			N-oligosaccharylT subunit	ER	+	84,86
SWP1			N-oligosaccharylT subunit	ER	+	85
	gls1		α-glucosidase I ?	ER	−	48
MNS1			Man$_9$GlcNAc$_2$ α-mannosidase	ER	−	110–112

SEC59	Dolichol kinase		ER	+	66
DPM1	Dol-P-Man synthase		ER	+	156
PMT1	Dol-P-Man: Protein(Ser/Thr) ManT		ER	–	161
OCH1	α-1,6-ManT(?)	outerchain	GA	–	138,139
MNN1	α-1,3-ManT		GA	–	51,52,60
mnn2		outerchain mannosylation	GA	–	60
mnn3		outerchain mannosylation	GA	–	60
mnn4		outerchain phosphorylation	GA	–	60
mnn5		outerchain mannosylation	GA	–	60
mnn6		outerchain phosphorylation	GA	–	60
mnn9		outerchain mannosylation	GA	–	52,60
mnn7/8		outerchain mannosylation	GA	–	60
mnn10		outerchain mannosylation	GA	–	60
MNT1/KRE2	α-1,2-ManT		–	–	123,162,163
GDA1	GA	GDPase	–	–	227a

T = transferase.

Several possibilities are conceivable to overcome problems with yeast glycosylation. The use of mutants, defective in *outer chain* biosynthesis and/or addition of the α-1,3-linked mannose residues will lead to more mammalian like high mannose oligosaccharide glycans. Mutations leading to a Golgi bypass, as assumed to be the case in the *pmr1* mutant, also give rise to more core like glycosylated proteins [300]. Eliminating potential glycosylation sites by site directed mutagenesis may be a further alternative, in case glycosylation is not important for a particular protein. Thus expression of a mutated plasminogen activator was no longer glycosylated but fully active and non-immunogenic [301].

The secreted products are either delivered to the medium or into the periplasmic space of the wall. Various examples reveal that this is not a simple size problem. Thus αIFN with 20 kDa, a relatively small protein, is mainly localized in the cell wall, whereas various large glycoproteins e.g. cellobiohydrolase (~200 kDa) [302], viral hemagglutinin (~250 kDa) [303], Eppstein Barr virus with ~400 kDa are released into the medium. Although there are reports that a M_r of 760 is a threshold for molecules to diffuse freely through the yeast cell wall, it is clear that molecules up to 200–400 kDa can still diffuse [see review 304]. Several factors seem to influence permeability, including the yeast strain, growth phase and nutrition. Also mutations leading to an altered cell wall structure affect wall porosity. In *mnn9* mutants periplasmically localized invertase is released in its octameric state [243]. Mutants with an altered β-1,6-glucan, like *kre1*, oversecrete yeast proteins [305]. On the other hand, it is also possible to specifically target and incorporate heterologous proteins to the cell wall. This may be relevant for several interesting biotechnological processes. It was found that fusion of the C-terminal half of yeast α-agglutinin and guar α-galactosidase as a reporter gene anchors the enzyme to the cell wall [306].

Finally to circumvent drawbacks of *S. cerevisiae* as a host for foreign protein expression, alternative systems have been developed and may be used, such as expression in *Kluyveromyces lactis* or *Schizosaccharomyces pombe* (see Ref. [287]). In the case of *Pichia*, invertase from *S. cerevisiae* was expressed and the oligosaccharide structure was analyzed [307,308]. It was found that this invertase is more homogeneous than that homologously expressed in *S. cerevisiae* with over 90% of the saccharides of the size $Man_{8-11}GlcNAc_2$. Unexpectedly, the oligosaccharides did not have terminal α-1,3- linked mannose. It is not clear at the moment, whether *Pichia* can carry out this reaction at all.

Note added in proof:

Recently the active oligosaccharyltransferase complex from S. cerevisiae was isolated independently by two groups [311,312].

Acknowledgments

We would like to thank V. Mrosek for her care and patience in preparing the manuscript and R. Knauer for drawing the figures. The original work referred to from this institute has been supported by Deutsche Forschungsgemeinschaft (SFB 43) and by Fonds der Chemischen Industrie.

Abbreviations

mnn	mannan
alg	asparagine linked glycosylation
sec	secretory
GPI	glycosylphosphatidyl isnositol
OTase	N-oligosaccharyltransferase
Glc	glucose
Man	mannose
GlcNAc	N-acetylglucosamine
Dol	dolichol
Dol-P	dolichylphosphate
UDP-GlcNAc	uridine diphospho-N-acetylglucosamine
UDP-Glc	uridine diphospho glucose
GDP-Man	guanosine diphospho mannose

References

1. Kornfeld, R. and Kornfeld, S. (1985) Annu. Rev. Biochem. 54, 631–634.
2. Tanner, W. and Lehle, L. (1987) Biochim. Biophys. Acta 906, 81–99.
3. Kukuruzinska, M.A., Bergh, M.L.E. and Jackson, B.J. (1987) Annu. Rev. Biochem. 56, 915–944.
4. Schachter, H. (1991) Glycobiology 1, 453–461.
5. Lechner, J. and Wieland, F. (1989) Annu. Rev. Biochem. 58, 173–194.
6. Messner, P., Sleyter, U.B. (1991) Glycobiology 1, 545–551.
7. Springer, T.A. (1990) Nature 346, 425–434.
8. Feizi, T. (1985) Nature 314, 53–57.
9. Stockel Hartree, A. and Renwick, A.G.C. (1992) Biochem. J. 287, 665–579.
10. Herscovics, A. and Orlean, P. (1993) FASEB J. 7, 540–550.
11. Cabib, E., Roberts, R. and Bowers, B. (1982) Annu. Rev. Biochem. 51, 763–793.
12. Bulawa, C.E., Slater, M., Cabib, E., Au-Young, J., Sburlati, A., Adair, W.L. Jr. and Robbins, P.W. (1986) Cell 46, 213–225.
13. Cabib, E., Sburlati, A., Bowers, B. and Silverman, S.J. (1989) J. Cell Biol. 108, 1665–1672.
14. Silverman, S.J., Sburlati, A., Slater, M.L. and Cabib, E. (1988) Proc. Natl. Acad. Sci. USA 85, 4735–4739.
15. Shaw, J.A., Mol, P.C., Bowers, B., Silverman, S.J., Valdivieso, M.H., Durán, A. and Cabib, E. (1991) J. Cell Biol. 114, 111–123.
16. Valdivieso, M.H., Mol, P.C., Shaw, J.A., Cabib, E. and Durán, A. (1991) J. Cell Biol. 114, 101–109.
17. Bulawa, C.E. (1992) Mol. Cell Biol. 17, 1764–1776.
18. Fleet, G.H. and Phaff, H.F. (1981) In: W. Tanner and F.A. Loewus (Eds.), Encyclopedia of Plant Physiology, New Series, Vol. 13B, Springer Verlag, Berlin, pp. 416–440.
19. Wessels, J.G.H. and Sietsma, J.H. (1981) In: W. Tanner and F.A. Loewus (Eds.), Encyclopedia of Plant Physiology, New Series, Vol. 13B, Springer-Verlag, Berlin, pp. 352–394.
20. Manners, D.J., Masson, A.J. and Patterson, J.C. (1973) Biochem. J. 135, 19–30.

502

21. Manners, D.J., Masson, A.J., Patterson, J.C., Bjorndal, H. and Lindberg, B. (1973) Biochem. J. 135, 31–36.
22. Boone, C., Sommer, S.S., Hensel, A. and Bussey, H. (1990) J. Cell. Biol. 110, 1833–1843.
23. Brown, J.L. Kossaczka, Z., Jiang, B. and Bussey, H. (1993) Genetics 133, 837–849.
24. Montijn, R.C., van Rinsum, J. and Klis. F.M. (1993) 16th International Specialized Symposium on Yeast, abstract.
25. Hernandez, L.M., Ballou, L., Alvarado, E., Gillece-Castro, B.L., Burlingame, A.L. and Ballou, C.E. (1989) J. Biol. Chem. 264, 11849–11856.
26. Hernandez, L.M., Ballou, L., Alvarado, E., Tsai, P.K. and Ballou, C.E. (1989) J. Biol. Chem. 264, 13648–13659.
27. Alvarado, E., Ballou, L., Hernandez, L.M. and Ballou, C.E. (1990) Biochemistry 29, 2471–2482.
28. Trimble, R.B. and Atkinson, P.H. (1986) J. Biol. Chem. 261, 9815–9824.
29. Valentin, E., Herrero, E., Pastor, F.I.J. and Sentandreu, R. (1984) J. Gen. Microbiol. 130, 1419–1428.
30. Pastor, F.I.J., Valentin, E., Herrero, E. and Sentandreu, R. (1984) Biochim. Biophys. Acta 802, 292–300.
31. Sanz, P., Herrero, E. and Sentandreu, R. (1989) FEMS Microbiol. Lett. 57, 265–270.
32. Shibata, N., Mizugami, K., Takano, K. and Suzuki, S. (1983) J. Bacteriol. 156, 552–558.
33. Elorza, M.V., Murgui, A., Sentandreu, R. (1985) J. Gen. Microbiol. 131, 2209–2216.
34. Elorza, M.V., Marcilla, A. and Sentandreu, R. (1988) J. Gen. Microbiol. 134, 2393–2403.
35. Klebl, F. and Tanner, T. (1989) J. Bacteriol. 171, 6259–6264.
36. Mrsa, V., Klebl, F. and Tanner, W. (1993) J. Bacteriol. 175, 2102–2106.
37. Frevert, J. and Ballou, C.E. (1985) Biochemistry 24, 753–759.
38. Hauser, K. and Tanner, W. (1989) FEBS Lett. 255, 290–294.
39. Van Rinsum, J., Klis, F.M. and van den Ende, H. (1991) Yeast 7, 717–726.
40. Fleet, G.H. and Manners, D.J. (1976) J. Gen. Microbiol. 94, 180–192.
41. Fleet, G.H. and Manners, D.J. (1977) J. Gen. Microbiol. 98, 315–327.
42. Mol, P.C. and Wessels, J.G.H. (1987) FEMS Microbiol. Lett. 41, 95–99.
43. De Nobel, J.G., Klis, F.M., Priem, J., Munnik, T. and van den Ende, H. (1990) Yeast 6, 491–499.
44. De Nobel, J.G., Klis, F.M., Munnik, T., Priem, J. and van den Ende H. (1990) Yeast 6, 483–490.
45. De Nobel, J.G., Klis, F.M., Ram, A., van Unen, H., Priem, J., Munnik, T. and van den Ende, H. (1991) Yeast 7, 589–598.
46. Cohen, R.E., Zhang, W. and Ballou, C.E. (1982) J. Biol. Chem. 257, 5730–5737.
47. Byrd, J.C., Tarentino, A.L., Maley, F., Atkinson, P.H. and Trimble, R.B. (1982) J. Biol. Chem. 257, 14657–14666.
48. Tsai, P.K., Ballou, L., Esmon, B., Schekman, R. and Ballou, C.E. (1984) Proc. Natl. Acad. Sci. USA 81, 6340–6343.
49. Lehle, L., Roitsch, T., Strahl, S. and Tanner, W. (1991) In: J.P. Latgé and D. Boucias (Eds.), NATO ASI Series, Vol. H.53, Springer-Verlag, Berlin, pp. 69–80.
50. Ballou, C.E. (1990) Meths. Enzymol. 185, 440–470.
51. Graham, T.R. Verostek, M.F., MacKay, V., Trimble, R. and Emr, S.D. (1992) Yeast 8, p. 458.
52. Yip, C.L., Welch, S.K., Klebl, F., Gilbert, T., Seidel, P., Grant, F.J., O'Hara, P.J. and MacKay, V.L. (1994) Proc. Natl. Acad. Sci. USA (accepted for publication)
53. Ballou, L., Cohen, R.E. and Ballou, C.E. (1980) J. Biol. Chem. 255, 5986–5991.
54. Ballou, L., Hernandez, L.M., Alvarado, E. and Ballou, C.E. (1990) Proc. Natl. Acad. Sci. USA 87, 3368–3372.
55. Nakajima, T. and Ballou, C.E. (1974) J. Biol. Chem. 249, 7679–7684.
56. Reason, A.J., Dell, A., Romero, P.A. and Herscovics, A. (1991) Glycobiology 1, 387–391.

503

57. Trimble, R.B. and Atkinson, P.H. (1992) Glycobiology 2, 57–75.

58. Townsend, R.R. and Hardy, M.R. (1991) Glycobiology 1, 139–147.

59. Hernandez, L.M., Ballou, L. and Ballou, C.E. (1990) Carbohydrate Res. 203, 1–11.

60. Townsend, R.R., Atkinson, P.H. and Trimble, R.B. (1991) Carbohydrate Res. 215, 211–217.

61. Jung, P. and Tanner, W. (1973) Eur. J. Biochem. 37, 1–6.

62. Palamarczyk, G., Lehle, L., Mankowski, T., Chojnacki, T. and Tanner, W. (1980) Eur. J. Biochem. 105, 517–523.

63. Reuvers, F., Boer, P. and Hemming, F. (1978) Biochem. J. 169, 505–508.

64. Abeijon, C. and Hirschberg, C.B. (1992) Trends Biochem. Sci. 17, 32–36.

65. Bernstein, M., Kepes, F. and Schekman, R. (1989) Mol. Cell. Biol. 9, 1191–1199.

66. Heller, L., Orlean, P. and Adair, W.L., Jr. (1992) Proc. Natl. Acad. Sci. USA 89, 7013–7016.

67. Huffaker, T.C. and Robbins, P.W. (1982) J. Biol. Chem. 257, 3203–3210.

68. Huffaker, T.C. and Robbins, P.W. (1983) Proc. Natl. Acad. Sci. USA 80, 7466–7470.

69. Couto, J.R., Huffaker, T.C. and Robbins, P.W. (1984) J. Biol. Chem. 259, 378–382.

70. Albright, C.F., Orlean, P. and Robbins, P.W. (1989) Proc. Natl. Acad. Sci. USA 86, 7366–7369.

71. Rine, J., Hansen, W., Hardeman, E. and Davis, R.W. (1983) Proc. Natl. Acad. Sci. USA 80, 6750–6754.

72. Kukuruzinska, M.A. and Robbins, P.W. (1987) Proc. Natl. Acad. Sci. USA 84, 2145–2149.

72a. Jackson, B.J., Kukuruzinska, M.A. and Robbins, P. (1993) Glycobiology 3, 357–365.

73. Zhu, X. and Lehrman, M.A. (1990) J. Biol. Chem. 265, 14250–14255.

74. Verostek, M.F., Atkinson, P.H. and Trimble, R.B. (1991) J. Biol. Chem. 266, 5547–5551.

75. Chapman, A., Fujimoto, K. and Kornfeld, S. (1980) J. Biol. Chem. 255, 4441–4446.

76. Te Heesen, S., Lehle, L., Weissmann, A. and Aebi, M. (1994) Eur. J. Biol. Chem. 224, 71–79.

77. Sharma, C.B., Lehle, L. and Tanner, W. (1981) Eur. J. Biochem. 116, 101–108.

78. Runge, K.W., Huffaker, T.C. and Robbins, P.W. (1984) J. Biol. Chem. 259, 412–417.

79. Verostek, M.F., Atkinson, P.H. and Trimble, R.B. (1993) J. Biol. Chem. 268, 12095–12103.

80. Verostek, M.F., Atkinson, P.H. and Trimble, R.B. (1993) J. Biol. Chem. 268, 12104–12115.

81. Ballou, L., Gopal, P., Krummel, B., Tammi, M. and Ballou, C.E. (1986) Proc. Natl. Acad. Sci. USA 83, 3081–3085.

82. Das, R.C. and Heath, E.C. (1980) Proc. Natl. Acad. Sci. USA 77, 3811–3815.

83. Kaplan, H.A., Welply, J.K. and Lennarz, W.J. (1987) Biochim. Biophys. Acta 906, 161–173.

84. Te Heesen, S., Janetzky, B., Lehle, L. and Aebi, M. (1992) EMBO J. 11, 2071–2075.

85. Te Heesen, S., Knauer, R., Lehle, L. and Aebi, M. (1992) EMBO J. 12, 279–284.

86. Te Heesen, S., Rauhut, R., Aebersold, R., Abelson, J., Aebi, M. and Clark, M.W. (1991) Eur. J. Cell Biol. 56, 8–18.

87. Kelleher, D.J., Kreibich, G. and Gilmore, R. (1992) Cell 69, 55–65.

88. Silberstein, S., Kelleher, D.J. and Gilmore, R. (1992) J. Biol. Chem. 267, 23658–23663.

88a. Zimmermann, J.W. and Robbins, P.W. (1993) J. Biol. Chem. 268, 16746–16753

89. Geetha-Habib, M., Noiva, R., Kaplan, H.A. and Lennarz, W.J. (1988) Cell 54, 1053–1060.

89a. Noiva, R., Kimura, H., Roos, J. and Lennarz, W.J. (1991) J. Biol. Chem. 266, 19645–19649

90. Günther, R., Bräuer, C., Janetzky, B., Förster, H.-H., Ehbrecht, I.-M., Lehle, L. and Küntzel, H. (1991) J. Biol. Chem. 266, 24557–24563.

91. LaMantia, M.L., Miura, T., Tachikawa, H., Kaplan, H.A., Lennarz, W.J. and Mizunaga, T. (1991) Proc. Natl. Acad. Sci. USA 88, 4453–4457.

92. Tachibana, C. and Stevens, T.H. (1992) Mol. Cell. Biol. 12, 4601–4611.

93. Albert, J.P, Biserte, G. and Loucheux-Lefebvre, M.H. (1976) Arch. Biochem. Biophys. 175, 410–418.

94. Beely, J.G. (1977) Biochem. Biophys. Res. Commun. 76, 1051–1055.

95. Mononen, I. and Karjalainen, E. (1984) Biochim. Biophys. Acta 788, 364–367.

96. Gavel, Y. and von Heijne, G. (1990) Protein Engineering 3, 433–442.
97. Imperiali, B., Shannon, K.L. and Rickert, K.W. (1992) J. Am. Chem. Soc. 114, 7942–7944.
98. Imperiali, B., Shannon, K.L., Unno, M. and Rickert, K.W. (1992) J. Am. Chem. Soc. 114, 7944–7945.
99. Bause, E. and Lehle, L. (1979) Eur. J. Biochem. 101, 531–540.
100. Bause, E., Legler, G. (1981) Biochem. J. 195, 639–644.
101. Roitsch, T. and Lehle, L. (1989a) Eur. J. Biochem. 181, 525–529.
102. Roitsch, T. and Lehle, L. (1989b) Eur. J. Biochem. 181, 733–739.
103. Lehle, L., Cohen, R.E. and Ballou, C.E. (1979) J. Biol. Chem. 254, 12209–12218.
104. Reddy, A.V., Johnson, R.S., Biemann, K., Williams, R.S., Ziegler, F.D., Trimble, R.B. and Maley, F. (1988) J. Biol. Chem. 263, 6978–6985.
105. Reddy, A.V., MacColl, R. and Maley, F. (1990) Biochemistry 29, 2482–2487.
106. Trimble, R.B., Maley, F. and Chu, F.K. (1983) J. Biol. Chem. 258, 2562–2567.
107. Hsieh, P., Rosner, M.R. and Robbins, P.W. (1983) J. Biol. Chem. 258, 2548–2554.
108. Kilker Jr., R.D., Saunier, B., Tkacz, J.S. and Herscovics, A. (1981) J. Biol. Chem. 256, 5299–5303.
109. Saunier, B., Kilker, R.D. Jr., Tkacz, J.S., Quaroni, A. and Herscovics, A. (1982) J. Biol. Chem. 257, 14155–14161.
110. Jelinek-Kelly, S. and Herscovics, A. (1988) J. Biol. Chem. 263, 14757–14763.
111. Ziegler, F.D. and Trimble, R.B. (1991) Glycobiology 1, 605–614.
112. Camirand, A., Heysen, A., Grondin, B. and Herscovics, A. (1991) J. Biol. Chem. 266, 15120–15127.
113. Grondin, B. and Herscovics, A. (1992) Glycobiology 2, 369–372.
114. Esmon, B., Esmon, P. and Schekman, R. (1984) J. Biol. Chem. 359, 10322–10327.
115. Puccia, R., Grondin, B. and Herscovics, A. (1993) Biochem. J. 290, 21–26.
116. Esmon, B., Novick, P. and Schekman, R. (1981) Cell 25, 451–460.
117. Parodi, A.J. (1978) Eur. J. Biochem. 83, 253–259.
118. Lehle, L. (1980) Eur. J. Biochem. 109, 589–601.
119. Lehle, L. and Tanner, W. (1974) Biochim. Biophys. Acta 350, 225–235.
120. Farkas, V., Vagabov, V.M. and Bauer, S. (1976) Biochim. Biophys. Acta 428, 573–582.
121. Nakajima, T. and Ballou, C.E. (1975) Proc. Natl. Acad. Sci. USA 72, 3912–3916.
122. Lewis, M.S. and Ballou, C.E. (1991) J. Biol. Chem. 266, 8255–8261.
123. Häusler, A., Ballou, L., Ballou, C.E. and Robbins, P.W. (1992) Proc. Natl. Acad. Sci. USA 89, 6846–6850.
124. Karson, E.M. and Ballou, C.E. (1978) J. Biol. Chem. 253, 6484–6492.
125. Gopal, P.K. and Ballou, C.E. (1987) Proc. Natl. Acad. Sci. USA 84, 8824–8828.
126. Ballou, L., Alvarado, E., Tsai, P., Dell, A. and Ballou, C.E. (1989) J. Biol. Chem. 264, 11857–11864.
127. Flores-Carreón, A., Hixson, S.H., Gomez, A., Shao, M.C., Krudy, G., Rosevear, P.R. and Wold, F. (1990) J. Biol. Chem. 265, 754–759.
128. Romero, P.A. and Herscovics, A. (1989) J. Biol. Chem. 264, 1946–1950.
129. Ballou, L., Hitzeman, R.A., Lewis, M.S. and Ballou, C.E. (1991) Proc. Natl. Acad. Sci. USA 88, 3209–3212.
130. Bretthauer, R.K., Kozak, L.P. and Irwin, W.E. (1969) Biochem. Biophys. Res. Commun. 37, 820–827.
131. Bretthauer, R.K., Wu, S. and Irwin, W.E. (1973) Biochim. Biophys. Acta 304, 736–747.
132. Thieme, T.R. and Ballou, C.E. (1971) Biochem. Biophys. Res. Commun. 39, 621–625.
133. Hashimoto, C., Cohen, R.E. Zhang, W.-j. and Ballou, C.E. (1981) Proc. Natl. Acad. Sci. USA 78, 2244–2248.

134. Schwaiger, H., Hasilik, A., von Figura, K., Wiemken, A. and Tanner, W. (1982) Biochem. Biophys. Res. Commun. 104, 950–956.

135. Valls, L.A., Hunter, C.P., Rothman, J.H. and Stevens, T.H. (1987) Cell 48, 887–897.

136. Johnson, L.M., Bankaitis, V.A. and Emr, S.D. (1987) Cell 48, 875–885.

137. Clark, D.W., Tkacz, J.S. and Lampen, J.O. (1982) J. Bacteriol. 152, 865–873.

138. Nakayama, K., Nagasu, T., Shimma, Y., Kuromitsu, J. and Jigami, Y. (1992) EMBO J. 11, 2511–2519.

139. Nakanishi-Shindo, Y., Nakayama, K.-I., Tanaka, A., Toda, Y. and Jigami, Y. (1994) J. Biol. Chem. 268, 26338–26345

140. Babczinski, P. and Tanner, W. (1973) Biochem. Biophys. Res. Commun. 54, 1119–1124.

141. Sharma, C.B., Babczinski, P., Lehle, L. and Tanner, W. (1974) Eur. J. Biochem. 46, 35–41.

142. Letoublon, R. and Got, R. (1974) FEBS Lett. 46, 214–217.

143. Bretthauer, R.K. and Wu, S. (1975) Arch. Biochem. Biophys. 167, 151–160.

144. Gold, M.H. and Hahn, H.J. (1976) Biochemistry 15, 1808–1814.

145. Soliday, C.L. and Kolattukudy, P.E. (1979) Arch. Biochem. Biophys. 197, 367–378.

146. Kruszewska, J., Messner, R., Kubicek, C.P. and Palamarczyk, G. (1989) J. Gen. Microbiol. 135, 310–307.

147. Larriba, G., Elorza, M.V., Villanueva, J.R. and Sentandreu, R. (1976) FEBS Lett. 71, 316–320.

148. Lehle, L., Bauer, F. and Tanner, W. (1977) Arch. Microbiol. 114, 77–81.

149. Marriott, M. and Tanner, W. (1979) J. Bacteriol. 139, 565–572.

150. Haselbeck, A. and Tanner, W. (1983) FEBS Lett. 158, 335–338.

151. Kuranda, M.J. and Robbins, P.W. (1991) J. Biol. Chem. 266, 19758–19767.

152. Truehart, J. and Fink, G.R. (1989) Proc. Natl. Acad. Sci. USA 86, 9916–9920.

153. Orlean, P. (1990) Mol. Cell. Biol. 10, 5796–5805.

154. Haselbeck, A. and Tanner, W. (1982) Proc. Natl. Acad. Sci. USA 79, 1520–1524.

155. Haselbeck, A. (1989) Eur. J. Biochem. 181, 663–668.

156. Orlean, P., Albright, C. and Robbins, P.W. (1988) J. Biol. Chem. 263, 17499–17507.

157. Beck, P.J., Orlean, P., Albright, C., Robbins, P.W., Gething, M.-J. and Sambrook, J.F. (1990) Mol. Cell. Biol. 10, 4612–4622.

158. Thomas, L.J., De Gaspari, R., Sugiyama, E., Chang, H.-M., Beck, P.J., Orlean, P., Urakaze, M., Kamitani, T., Sambrook, J.F., Warren, C.D. Yeh, E.T. (1991) J. Biol. Chem. 266, 23175–13184.

159. Strahl-Bolsinger, S. and Tanner, W. (1991) Eur. J. Biochem. 196, 185–190.

160. Sharma, C.B., D'Souza, C. and Elbein, A.D. (1991) Glycobiology 1, 367–373.

161. Strahl-Bolsinger, S., Immervoll, T., Deutzmann, R. and Tanner, W. (1993) Proc. Natl. Acad. Sci. USA 90, 8164–8168.

162. Häusler, A. and Robbins, P.W. (1992) Glycobiology 2, 77–84.

163. Hill, K., Boone, C., Goebl, M., Puccia, R., Sdicu, A.-M. and Bussey, H. (1992) Genetics 130, 273–283.

164. Olstad, O.K., Reppe, S., Gabrielsen, O.S., Hartmanis, M., Blingsmo, O.R. Gautvik, V.T., Haflan, A.K., Christensen, T.B. Øyen, T.B. and Gantvik, K.M. (1992) Eur. J. Biochem. 205, 311–319.

165. Weston, A., Nassau, P.M., Henly, C. and Marriott, M.S. (1994) Eur. J. Biochem., in press.

166. Gellerfors, P., Axelsson, K., Helander, A., Johansson, S., Kenne, L., Lindqvist, S., Pavlu, B., Skottner, A. and Fryklund, L. (1989) J. Biol. Chem. 264, 11444–11449.

167. Ernst, J.F. and Richman, L.H. (1989) Bio/Technology 7, 716–720.

168. Ernst, J.F., Mermod, J.-J. and Richman, L. (1992) Eur. J. Biochem. 203, 663–667.

169. Lorenz, C., Strahl-Bolsinger, S. and Ernst, J.F. (1992) Eur. J. Biochem. 205, 1163–1167.

170. Sentandreu, R. and Northcote, D.H. (1969) Carbohydr. Res. 10, 584–585.

171. Orlean, P., Ammer, H., Watzele, M. and Tanner, W. (1986) Proc. Natl. Acad. Sci. USA 83,

6263–6266.

172. Watzele, M., Klis, F. and Tanner, W. (1988) EMBO J. 7, 1483–1488.

173. Cappellaro, C., Hauser, K., Mrsa, V., Watzele, M., Watzele, G., Gruber, C. and Tanner W. (1991) EMBO J. 10, 4081–4088.

174. Russo, P., Kalkkinen, N., Sareneva, H., Paakkola, J. and Makarow, M. (1992) Proc. Natl. Acad. Sci USA 89, 3671–3675.

175. Lupeshin, V.V., Kononova, S.V., Ratner, Y.N., Tsiomenko, A.B. and Kulaev, I.S. (1992) Yeast 8, 157–169.

176. Wilcox, C.A. and Fuller, R.S. (1991) J. Cell Biol. 115, 297–307.

177. Sweet, D.J. and Pelham, H.R.B. (1992) EMBO J 11, 423–432.

178. Jentoft, N. (1990) TIBS 15, 291–294.

179. Kubicek, C.P., Panda, T., Schreferl-Kunar, G., Gruber, F. and Messner, R. (1987) Can. J. Microbiol. 33, 698–703.

180. Hart, G.W., Haltiwanger, R.S., Holt, G.D. and Kell, W.G. (1989) Annu. Rev. Biochem. 58, 841–874.

181. Kawaguchi, N., Ohmori, T., Takeshita, Y., Kawanishi, G., Katayama, S. and Yamada, H. (1986) Biochem. Biophys. Res. Commun. 140, 350–356.

182. Roth, J. (1984) J. Cell Biol. 98, 399–406.

183. Ferguson, M.A.J. (1991) Biochem. Soc. Trans. 20, 243–256.

184. Conzelmann, A., Riezman, H., Desponds, C. and Bron, C. (1988) EMBO J. 7, 2233–2240.

185. Conzelmann, A., Fankhauser, C. and Desponds, C. (1990) EMBO J. 9, 653–661.

186. Fankhauser, C. and Conzelmann, A. (1991) Eur. J. Biochem. 195, 439–448.

187. Vai, M., Popolo, L., Grandori, R., Lacanà, E. and Alberghina, L. (1990) Biochim. Biophys. Acta 1038, 277–285.

188. Nuoffer, C., Jenö, P., Conzelmann, A. and Riezman, H. (1991) Mol. Cell. Biol. 11, 27–37.

189. Vai, M., Gatti, E., Lacana, E., Popolo, L. and Alberghina, L. (1991) J. Biol. Chem. 266, 12242–12248.

190. Popolo, L. and Alberghina, L. (1984) Proc. Natl. Acad. Sci. USA 81, 120–124.

191. Caras, I.W.,Weddell, G.N. and Williams, S.R. (1989) J. Cell Biol. 108, 1387–1396.

192. Gerber, L.D., Kodukula, K. and Udenfriend, S. (1992) J. Biol. Chem. 267, 12168–12173.

193. Lipke, P.N. and Kurjan, J. (1992) Microbiol. Rev. 56, 180–194.

194. Egel-Mitani, M., Flygenring, H.P. and Hansen, M.T. (1990) Yeast 6, 127–137.

195. Conzelmann, A., Puoti, A., Lester, R.L. and Desponds, C. (1992) EMBO J. 11, 457–466.

196. Stadler, J., Keenan, T.W., Bauer, G. and Gerisch, G. (1989) EMBO J. 8, 371–377.

197. Dickson, R.C., Wells, G.B., Schmidt, A. and Lester, R.L. (1990) Mol. Cell. Biol. 10, 2176–2181.

198. Costello, L.C. and Orlean, P. (1992) J. Biol. Chem. 267, 8599–8603.

199. Roberts, W.L., Myher, J.J., Kuksis, A., Low, M.G. and Rosenberry, T.L. (1988) J. Biol. Chem. 263, 18766–18775.

200. Schekman, R. (1985) Annu. Rev. Cell. Biol. 1, 115–143.

201. Pryer, N.K., Wuestehube, L.J. and Schekman, R. (1992) Annu. Rev. Biochem. 61, 471–516.

202. Deshais, R.J. and Schekman, R. (1987) J. Cell. Biol. 105, 633–645.

203. Deshais, R.J. and Schekman, R. (1989) J. Cell. Biol. 109, 2653–2664.

204. Deshais, R.J., Sanders, S.L., Feldheim, D.A. and Schekman, R. (1991) Nature 349, 806–808.

205. Görlich, D., Prehn, S., Hartmann, E., Kalies, K.U. and Rapoport, T.A. (1992) Cell 71, 489–503.

206. Pless, D.D. and Lennarz, W.J. (1977) Proc. Natl. Acad. Sci. USA 74, 134–138.

207. Nilsson, I. and von Heijne G. (1993) J. Biol. Chem. 268, 5798–5801.

208. Murgolo, N.J., Patel, A., Stivala, S.S. and Wong, T.K. (1989) Biochemistry 28, 253–260.

209. Brodsky, J.L., Hamamoto, S., Feldheim, D. and Schekman, R. (1993) J. Cell Biol. 120, 95–102.

210. Nicchitta, C.V., Migliaccio, G. and Blobel, G. (1991) Cell 65, 587–598.

211. Ooi, C.E. and Weiss, J. (1992) Cell 71, 87–96.

212. Welply, J.K., Shenbagamurthi, P., Lennarz, W.J. and Naider, F. (1983) J. Biol. Chem. 258, 11856–11863.

213. Snider, M.D. and Robbins, P.W. (1982) J. Biol. Chem. 257, 6796–6801.

214. McCloskey, M.A. and Troy, F.A. (1980) Biochemistry 19, 2061–2066.

215. Valtersson, C., van Duyn, G., Verkleij, A.J., Chojnacki, I., de Kruijff, B. and Dallner, G. (1985) J. Biol. Chem. 260, 2742–2751.

216. Schutzbach, J.S. and Zimmerman, J.W. (1992) Biochem. Cell Biol. 70, 460–465.

217. Beck, P.J., Orlean, P., Albright, C., Robbins, P.W., Gething, M.J. and Sambrook, J.F. (1990) J. Cell Biol. 111, abstract 378.

218. Parodi, A.J., Mendelzon, D.H., Lederkremer, G.Z. and Martin-Barrientos, J. (1984) J. Biol. Chem. 259, 6351–6357.

219. Trombetta, S.E., Ganan, A. and Parodi, A.J. (1991) Glycobiology 1, 155–161.

220. Preuss, D., Mulholland, J., Franzusoff, A., Segev, N. and Botstein, D. (1992) Mol. Biol. Cell 3, 789–803.

221. Franzusoff, A., Redding, K., Crosby, J., Fuller, R.S. and Schekman, R. (1991) J. Cell Biol. 112, 27–37.

222. Redding, K., Holcomb, C. and Fuller, R.S. (1991) J. Cell Biol. 113, 527–538.

223. Franzusoff, A. and Schekman, R. (1989) EMBO J. 8, 2695–2702.

224. Graham, T.R. and Emr, S.D. (1991) J. Cell Biol. 114, 207–218.

225. Cunningham, K.W. and Wickner, W.T. (1989) Yeast 5, 25–33.

226. Abeijon, C., Orlean, P., Robbins, P.W. and Hirschberg, C.B. (1989) Proc. Natl. Acad. Sci. USA 86, 6935–6939.

227. Yanagisawa, K., Resnick, D., Abeijon, C., Robbins, P.W. and Hirschberg, C.B. (1990) J. Biol. Chem. 265, 19351–19355.

227a Abeijon, C., Yanagisawa, K., Mandon, E.C., Häusler, A., Moremen, K., Hirschberg, C.B. and Robbins, P.W. (1993) J. Cell Biol. 122, 307–323.

228. Sharon, N. and Lis, H. (1993) Sci. Am. 268, 74–81.

229. Yanagishima, N. (1984) In: H.F. Linskens and J. Heslop-Harrison (Eds.), Encyclopedia of Plant Physiology: Cellular Interactions, New Series, Vol. 17, Springer-Verlag, Berlin, pp. 402–423.

230. Terrance, K., Heller, P., Wu, Y.-S. and Lipke, P.N. (1987) J. Bacteriol. 169, 475–482.

231. Lipke, P.N., Wojciechowicz, D. and Kurjan, J. (1989) Mol. Cell. Biol. 9, 3155–3165.

232. Wojciechowicz, D., Lu, C.-F., Kurjan, J. and Lipke, P.N. (1993) Mol. Cell. Biol. 13, in press.

233. Roy, A., Lu, C.F., Marykwas, D.L., Lipke, P.N. and Kurjan, J. (1991) Mol. Cell. Biol. 11, 4196–4206.

234. Wickner, R.B. (1986) Annu. Rev. Biochem. 55, 373–395.

235. Tipper, D.J. and Schmitt, M.J. (1991) Mol. Microbiol. 5, 2331–1338.

236. Zhu, H. and Bussey, H. (1991) Mol. Cell. Biol. 11, 175–181.

237. Schmitt, M. and Radler, F. (1987) J. Gen. Microbiol. 133, 3347–3354.

238. Schmitt, M. and Radler, R. (1988) J. Bacteriol. 170, 2192–2196.

239. Arnold, E. and Tanner, W. (1982) FEBS Lett. 148, 49–53.

240. Klebl, F., Huffaker, T.C. and Tanner, W. (1984) Exp. Cell Res. 150, 309–313.

241. Vai, M., Popolo, L. and Alberghina, L. (1987) Exp. Cell Res. 17, 448–459.

242. Esmon, P.C., Esmon, B.E., Schauer, I.E., Taylor, A. and Schekman, R. (1987) J. Biol. Chem. 262, 4387–4394.

243. Tammi, M., Ballou, L., Taylor, A. and Ballou, C.E. (1987) J. Biol. Chem. 262, 4395–4401.

244. Mrsa, V., Barberic, S., Ries, B. and Mildner, P. (1989) Arch. Biochem. Biophys. 273, 121–127.

245. Kern, G., Schülke, N., Schmid, F.X. and Jaenicke, R. (1992) Protein Science 1, 120–131.

508

246. Gibson, R., Schlesinger, S. and Kornfeld, S. (1979) J. Biol. Chem. 254, 3600–3607.
247. Hickman, S., Kulczycki, A. Jr., Lynch, R.G. and Kornfeld, S. (1977) J. Biol. Chem. 252, 4402–4408.
248. Mizunaga, T., Izawa, M., Ikeda, K. and Maruyama, Y. (1988) J. Biochem. 103, 321–326.
249. Marquardt, T. and Helenius, A. (1992) J. Cell Biol. 117, 505–513.
250. Kornfeld, S. and Mellman, J. (1989) Annu. Rev. Cell. Biol. 5, 483–525.
251. Winther, J.R., Stevens, T.H. and Kielland-Brandt, M.C. (1991) Eur. J. Biochem. 197, 681–689.
252. Hakomori, S. (1991) Trends Glycosci. Glycotechnol. 3, 1–3.
253. Lehle, L. (1981) In: W. Tanner and F.A. Loewus (Eds.), Encyclopedia of Plant Physiology, New Series, Vol. 13 B Plant Carbohydrates II, Springer Verlag, Berlin, pp. 458–483.
254. Gorin, P.A.J., Spencer, J.F.T. and Bhattacharjee, S.S. (1968) Can. J. Chem. 47, 1499–1507.
255. Gorin, P.A.J. and Spencer, J.F.T.(1970) Advances in Applied Microbiology 13, 25–89.
256. Ballou, C.E. and Raschke, W.C. (1974) Science 184, 127–134.
257. Kozak, I.P. and Bretthauer, R.K. (1970) Biochemistry 9, 1115–1122
258. Slodki, M.E. (1963) Biochim. Biophys. Acta 69, 96–102.
259. Crandall, M.A. and Brock, T.D. (1968) Bacteriol. Rev. 32, 139–163.
260. Yen, P.H. and Ballou, C.E. (1974) Biochemistry 13, 2420.
261. Yen, P.H. and Ballou, C.E. (1974) Biochemistry 13, 2428.
262. Raschke, W.C. and Ballou, C.E. (1972) Biochemistry 11, 3807–3816
263. Douglas, R.H. and Ballou, C.E. (1980) J. Biol. Chem. 255, 5979–5985.
264. Douglas, R.H. and Ballou, C.E. (1982) Biochemistry 21, 1570–1574.
265. Smith, W.L., Nakajima, T. and Ballou, C.E. (1975) J. Biol. Chem. 250, 3426–3435.
266. Bush, D.A., Horisberger, M., Horman, I. and Wursch, P. (1974) J. Gen. Microbiol. 81, 199–206.
267. Horisberger, M., Vonlanthen, M. and Rosset, J. (1978) Arch. Microbiol. 119, 107–111.
268. Moreno, S., Ruiz, T., Sanchez, V., Villanueva, J.R. and Rodriguez, L. (1985) Arch. Microbiol. 142, 370–374.
269. Dibenedetto, G. and Cozzani, I. (1975) Biochemistry 14, 2847–2852.
270. Chappel, T.G. and Warren, G. (1989) 109, 2693–2702
271. Chappel, T.G. (1991) In: J. Pringle, S. Reed and S. Emr (Eds.), Yeast Cell Biology. Cold Spring Harbor, New York, p. 134.
272. Kocourek, J. and Ballou, C.E. (1969) J. Bacteriol. 100, 1175–1181.
273. Shibata, N., Ichikawa, T., Tojo, M., Takahashi, M., Ito, N., Okuo, Y. and Suzuki, S. (1985) Arch. Biochem. Biophys. 243, 338–348.
274. Shibata, N., Kobayashi, II., Tojo, M. and Suzuku, S. (1986) Biophys. 251, 697–708.
275. Tsuchiya, T., Fukazawa, Y., Tagushi, M., Kase, T. and Shinoda, T. (1974) Mycol. Appl. 53, 77–85.
276. Suzuki, M. and Fuzakawa, Y. (1982) Microbiol. Immunol. 26, 387–409.
277. Hayette, M.P., Strecker, G., Faille, C., Dive, D., Camus, D., Mackenzie, D.W.R. and Poulain, D. (1992) J. Clin. Microbiol. 30, 411–417.
278. Marriott, M.S. (1977) J. Gen. Microbiol. 103, 673–702.
279. Flores-Carreón, A. and Balcázar-Orozco, R. (1993) FEMS Microbiol. Lett. 110, 121–126.
280. Bardalaye, P.C. and Nordin, J.H. (1977) J. Biol. Chem. 252, 2584–2591.
281. Boel, E., Hansen, M.T., Hjort, I., Høegh, I. and Fiil, N.P. (1984) EMBO J. 3, 1581–1585.
282. Boel, E., Hjort, I., Svensson, B., Norris, F., Norris, K.E. and Fiil, N.P. (1984) EMBO J. 3, 1097–1102.
283. Nunberg, J.H., Meade, J.H., Cole, G., Lawyer, F.C., McCabe, P., Schweickart, V., Taì, R., Wittman, V.P., Flatgaard, J.E. and Innis, M.A. (1984) Mol. Cell. Biol. 4, 2306–2315.
284. Gunnarsson, A., Svensson, B., Nilsson, B. and Svensson, S. (1984) Eur. J. Biochem. 145, 463–467.

285. Svensson, B., Larsen, K. and Gunnarsson, A. (1986) Eur. J. Biochem. 154, 497–502.

286. Rudick, M.J. (1979) J. Bacteriol. 137, 301–308.

287. Romanos, M.A., Scorer, C.A. and Clare J.J. (1992) Yeast 8, 423–488.

288. Schultz, L.D., Tanner, J., Hofmann, K.J., Emini, E.A., Condra, J.H., Jones, R.E., Kieff, E. and Ellis, R.W. (1987) Gene 54, 113–123.

289. Meyhack, B. and Hinnen, A. (1984) Eur. Pat. Appl. 84810564.9

290. Green, R., Schaber, M.D., Schields, D. and Kramer, R. (1986) J. Biol. Chem. 261, 7558–7565.

291. Elliott, S., Giffin, J., Suggs, S., Lau, E.P. and Banks, A.R. (1989) Gene 79, 167–180.

292. Innis, M.A., Holland, M.J., McCabe, P.C., Cole, G.E., Wittman, V.P., Tal, R., Watt, K.W.K., Gelfand, D.H., Holland, J.P. and Meade, J.H. (1985) Science 228, 21–26.

293. Williamson, G., Belshwa, N.J., Noel, T.R., Ring, S.G. and Williamson, M.P. (1992) Eur. J. Biochem. 20, 661–670.

294. Moir, D.T. and Dumais, D.R. (1987) Gene 56, 209–217.

295. Zaworski, P.G., Marotti, K.R., MacKay, V., Yip, C. and Gill, G.S. (1989) Gene 85, 545–551.

296. Hitzeman, R.A., Leung, D.W., Perry, L.J., Kohr, W.J., Levine, H.C. and Goeddel, D.V. (1983) Science 219, 620–625.

297. Guisez, Y., Tison, B., Vandekerckhove, J., Demolder, J., Bauw, G., Haegeman, G., Fiers, W. and Contreras, R. (1991) Eur. J. Biochem. 198, 217–222.

298. Nagahora, H., Ishikawa, K., Niwa, Y., Muraki, M. and Jigami Y. (1992) Eur. J. Biochem. 210, 989–997.

299. Ernst, J.F. (1988) DNA 7, 355–360.

300. Rudolph, H.K., Antebi, A., Fink, G.R., Buckley, C.M., Dorman, T.E., LeVitre, J., Davidow, L.S., Mao, J. and Moir, D.T. (1989) Cell 58, 133–145.

301. Melnick, L.M., Turner, B.G., Puma, P., Price-Tillotson, B., Salvato, K.A., Dumais, D.R., Moir, D.T., Broze, R.J. and Avgerinos, G.C. (1990) J. Biol. Chem. 265, 801–807.

302. Pentillä, M.E., André, L., Lehtovaara, P., Bailey, M., Teeri, T.T. and Knowles, J.K.C. (1988) Gene 63, 103–112

303. Jabbar, M.A. and Nayak, D.P. (1987) Mol. Cell. Biol. 7, 1476–1485.

304. De Nobel, J.G. and Barnett, J.A. (1991) Yeast 7, 313–323.

305. Bussey, H., Steinmetz, O. and Sville, D. (1983) Curr. Genet. 7, 449–456.

306. Schreuder, M.P., Brekelmans, S., van den Ende, H. and Klis, F.M. (1993) Yeast 9, 399–409.

307. Grinna, L.S. and Tschopp, J.F. (1989) Yeast 5, 107–115.

308. Trimble, R.B., Atkinson, P.H., Tschopp, J.F., Townsend, R.R. and Maley, F. (1991) J. BioBiol. Chem. 266, 22807–22817.

309. Kepes, F. and Schekman, R. (1988) J. Biol. Chem. 263, 9155–9161.

310. Runge, K.W. and Robbins, P.W. (1986) J. Biol. Chem. 261, 15582–15590.

311. Knauer, R. and Lehle, L. (1994) FEBS Lett. 344, 83–86.

312. Kelleher, D.J. and Gilmore, R. (1994) J. Biol. chem. 269, 12908–12917.

J. Montreuil, H. Schachter and J.F.G. Vliegenthart (Eds.), *Glycoproteins*

511

O-glycosylation in plants

FRANS M. KLIS

*Institute of Molecular Cell Biology, University of Amsterdam, BioCentrum Amsterdam,
Kruislaan 318, 1098 SM Amsterdam, The Netherlands*

1. Introduction

The extracellular matrix of plants does not only consist of polysaccharides, but also contains proteins. It is now clear that both enzymes and structural proteins are present at the cell surface and that they play an important role in many processes ranging from cell wall assembly, differentiation, somatic embryogenesis, and defence against invading micro-organisms [1,2]. From a relatively neglected area opened up by the pioneering efforts of Lamport [3,4], it has developed into a rapidly expanding field worthwhile the attention of glycobiologists and molecular biologists alike. This review is focused on a particular category of cell surface proteins, which are mainly or exclusively glycosylated through hydroxyamino acids (hydroxyproline, serine, and threonine) and are therefore designated as hydroxyproline-rich glycoproteins (HRGPs) or, more correctly, O-glycosylproteins (OGPs).

2. O-glycosylproteins in higher plants

Three main categories of OGPs are found in higher plants: (a) extensins; insoluble cell wall proteins, which derive their name from their supposed role in cell extension; (b) Solanaceous lectins; soluble proteins found only in the Solanaceae family, which bind specifically to chito-oligosaccharides [5] and of which potato lectin is the best characterized example; (c) arabinogalactan proteins; soluble proteins present in the cell wall, in the intercellular spaces of plant tissues, and in plant exudates [6]; they are also found in association with the plasma membrane [7,39]. They are often categorized as proteoglycans because of their high degree of glycosylation.

2.1. Extensins

Plant cell walls contain an insoluble protein network consisting of one or more OGPs called extensins, which are particularly rich in hydroxyproline and serine (Table I). They are secreted into the cell wall as soluble precursors, but rapidly become insoluble presumably by forming cross-links to each other and/or to yet unknown wall components [9]. The presence of an extensin network in the cell wall becomes evident, when cell walls are treated with HF [10], which cleaves glycosyl linkages but leaves the peptide linkages intact. About 95% of the wall is solubilized and thin cell wall-like structures remain consisting of about equal amounts of extensin and another substance. This observation further suggests that extensins could play a role in cell wall assembly. The expression of extensins is strongly regulated and tissue-specific [2]. In addition, their synthesis is induced by stress conditions such as wounding, fungal infection, and heavy metal ions [11,12].

Hydroxyproline and serine are often present in the form of the repeating amino acid motif SO_4:Ser-Hyp-Hyp-Hyp-Hyp [2] and may represent about 60% of the polypeptide chain (Table I). Extensins are heavily glycosylated and carry short side-chains consisting of arabinose residues and linked to hydroxyproline, and some galactose α-linked to serine. Under the electron microscope [13–15] they look thin and rod-like. Their length is consistent with a polyproline II helix, an extended left-handed helix with three residues per turn and a pitch of 0.94 nm [13]. This has been confirmed by circular

TABLE I

Characteristics of representative plant O-glycosylproteins

Feature	Higher plants			Volvocales	
	Extensins	Solanaceous lectins	Arabinogalactan proteins	Volvox extracellular matrix	Sexual agglutinins
Mr (kDa)	86	50	140	185	1300
Carbohydrate (%)	65	50	90	70	50
Principal amino	Hyp 46	Hyp 21	Hyp 26	Hyp 17	Hyp 10
acids (%)	Ser 14	Ser 11	Ser 18	Ser 8	Ser 13
	Lys 7	Gly 11	Ala 16		Gly 12
Peptide motifs	SO_4[a]		AOA		
Main sugars (%)	Ara 97	Ara 92	Ara 22	Ara 47	Ara 37
	Gal 3	Gal 8	Gal 60	Gal 36	Gal 24
			UA 14	Man 11	Glc 14
Sulfated sugars	ND[b]	ND	ND	yes	yes
Glycopeptide	Ara-Hyp	Ara-Hyp	Ara-Hyp	ND	ND
linkages	Gal-Ser	Gal-Ser	Gal-Hyp		
			Glc-Hyp (?)		
Polyproline II	100	35	30	likely	ND
conformation (%)					
Length (nm)	84	ND	ND	28[c]	345
Ref.	13,14,19 20,22,23	24–26	23,31,38,71	44,54	43,50,57

[a] SO_4: Ser-Hyp4.

[b] ND: Not determined.

[c] Hydroxyproline-rich domain.

$$\beta\text{-L-Ara}f\text{-}(1\to2)\text{-}\beta\text{-L-Ara}f\text{-}(1\to2)\text{-}\beta\text{-L-Ara}f\text{-}(1\to4)\text{-Hyp}$$
$$\alpha\text{-L-Ara}f\text{-}(1\to3)\text{-}\beta\text{-L-Ara}f\text{-}(1\to2)\text{-}\beta\text{-L-Ara}f\text{-}(1\to2)\text{-}\beta\text{-L-Ara}f\text{-}(1\to4)\text{-Hyp}$$

Fig. 1. The structure of Araf_3-Hyp and Araf_4-Hyp in extensins.

dichrometry showing that extensins are completely in the polyproline II conformation [13]. The carbohydrate side-chains are essential in maintaining the elongated structure of extensins. When extensins are deglycosylated, they loose the polyproline II conformation and their elongated form [13,15].

The carbohydrate side-chains in extensins consist either of arabinose or of galactose. The arabinosyl side-chains are β-linked to hydroxyproline and consist of 1 to 4 arabinofuranosyl residues mostly in the form of Araf_3-Hyp and Araf_4-Hyp [16]. The glycosidic linkage between arabinose and hydroxyproline is extremely acid-labile as is characteristic for arabinofuranosyl linkages, but is on the other hand alkali-stable [16]. The latter property allows the isolation of hydroxyproline-arabinosides — after cleavage of the peptide linkages by heating in saturated barium hydroxide by either gel filtration [17] or by ion exchange chromatography [3,18]. The structures of Araf_3-Hyp and Araf_4-Hyp have been elucidated by ^{13}C-NMR [19] and are identical except for the terminal, nonreducing α-linked arabinosyl residue in Araf_4-Hyp (Fig. 1). The galactosyl side-chains consist of one [20] or two [21] galactosyl residues α-linked [22] to serine.

2.2. Solanaceous lectins

As their name already indicates, they occur only in the Solanaceae (potato, tomato, *Datura stramonium*) and they are capable of agglutinating blood cells [23]. The best studied example is the lectin from potatoes [5,24–26] (Table I). It has long been thought that potato lectin was mainly cell wall-bound because, after homogenisation, most of the lectin was associated with the cell wall fraction [27]. However, in a recent study, it has been shown by immunogold-labelling that the cell wall is almost completely devoid of potato lectin. Instead, the lectin was found in the vacuole, and in the cytoplasm, presumably, associated with membranous structures [28]. As potato lectin is a positively charged protein, it is very well possible that the earlier observations were an artefact of homogenisation due to the binding of lectin molecules to the polygalacturonic acid fraction in the cell wall.

Potato lectin consists of at least two domains [24,26]: (a) a short stem-like domain rich in hydroxyproline and serine with a predicted length of about 30 nm [26], and (b) a chitin-binding domain [5] rich in disulfide bridges and glycine [26]. The stem-like domain, which is obtained by extensive pronase treatment of the lectin, contains all the hydroxyproline and essentially all the sugars (arabinose and galactose) and shows all the known characteristics of extensin. It has the polyproline II conformation [26] and carries short arabinosyl side-chains β-linked to hydroxyproline in the form of Araf_3-Hyp and Araf_4-Hyp (Fig. 1) and identical in structure to that determined in extensins [25]. After deglycosylation, the polyproline II conformation is lost [26] as has also been observed in extensins. Finally, the galactose residues are α-linked to serine residues in the form of Gal*p*-α1→Ser in potato [24] and in the form of Gal*p*-α-1→Ser and Gal*p*-α-1→3Gal*p*→Ser in *Datura stramonium* seeds [25]. The presence of an exten-sin-like stem domain separate from the recognition domain in the Solanaceous lectins indicates that such domains are the plant equivalent of the spacer domains found in mammalian cells and in fungi, where regions rich in glycosylated seryl and threonyl residues function as such [29,30].

The role of the Solanaceous lectins is unknown, but in view of their chitin-binding domain it seems possible that they play a role in plant defense mechanisms.

2.3. Arabinogalactan proteins

2.3.1. General description and occurrence

Arabinogalactan proteins (AGPs) are extracellular plant proteoglycans with a protein content usually varying between 2 and 10% [2,6,23]. Their peptide moiety is rich in hydroxyproline and serine, and both amino acids function as attachment sites for O-glycosyl side-chains [31]. As these plant proteoglycans contain a very prominent type of side-chain formed by an arabinogalactan linked to hydroxyproline [32], they have been called AGPs. They are present in both Dicots and Monocots in all species investigated [6,33].

AGPs are often called 'β-lectins' because they can be precipitated by Yariv's reagent, a β-glycosylated compound derived from phloroglucinol, with which they form a red-orange complex [23]. Although the exact nature of this interaction is not known, it has greatly helped the identification, localisation and purification of AGPs. AGPs occur as soluble molecules in cell walls, in the intercellular space of plant tissues, in exudates and in the medium of suspension cultures [6,34–36]; in addition, they are also found in association with the plasma membrane [7,8]. Using crossed electropho-resis, it has been demonstrated that AGPs isolated from tissues and organs are hetero-geneous in net charge [37].

2.3.2. Composition and properties of AGPs

AGPs are generally soluble in saturated ammonium sulfate and in 5% trichloroacetic acid, which greatly facilitates their purification. The major amino acids are hydroxy-proline, serine, and alanine [6,71] (Table I). A recurring amino acid motif in AGPs is the sequence Ala-Hyp-Ala (AOA) [23]. The major monosaccharides in AGPs are D-galactopyranose and L-arabinofuranose. Other sugars often present are uronic acids (D-glucuronic acid and its 4-O-methyl derivative, and D-galacturonic acid and its 4-O-methyl derivative), L-rhamnopyranose, D-mannopyranose, and D-glucopyranose [6]. The presence of uronic acids probably explains the low isoelectric points of AGPs, but it cannot be excluded that sulfated sugars are also present as in OGPs in algae (see below).

Hydroxyproline is linked to both short oligosaccharide side-chains and an arabi-nogalactan [31]. The arabinogalactan is linked to hydroxyproline through galactose [32,35] and, possibly, also through glucose [35]. It consists of a branched β-galac-topyranose framework having predominantly 1,3-linkages with some 1,6-linkages; this framework is often substituted with terminal Ara*f* residues or uronic acids (or their 4-O-methylated derivatives) [6]. As expected in view of their amino acid composition,

AGPs are partially in the polyproline II conformation [38] (Table I). It is not yet known if the polyproline II conformation is limited to a specific domain of the molecule as in the Solanaceous lectins, but it is attractive to assume that the β-lectin activity of AGPs is located in a domain distinct from a purely structural O-glycosylated region. The situation may, however, be more complicated because in contrast to the Solanaceous lectins, AGPs seem to be resistant to proteolytic attack.

2.3.3. The function of AGPs

The function of AGPs has long been a matter of discussion. Recently, some intriguing data indicating a role for AGPs in tissue differentiation and embryogenesis have been obtained [36]. Carrot seeds and embryogenic lines of carrot cell suspension cultures, which form somatic embryos under certain conditions, produce similar sets of AGPs as determined by crossed electrophoresis with the specific AGP-binding Yariv reagent. Non-embryogenic cell lines form, however, a different set of AGPs. When AGPs isolated from seed were added to a non-embryogenic cell line, the cell line became embryogenic again indicating that specific AGPs are essential for somatic embryogenesis. Seed AGPs also increased the embryogenic potential of embryogenic cell lines. In the case of plasma membrane-associated AGPs, it has been proposed that they might function as attachment sites for cell wall polysaccharides [39].

3. O-glycosylproteins in green algae

The extracellular matrix of the *Volvocales*, an order of the green algae, consists almost entirely of OGPs generally rich in glycosylated hydroxyproline and serine residues. The best-studied examples are found in the unicellular algae *Chlamydomonas reinhardtii* and *C. eugametos*, and in the multicellular alga *Volvox carteri* that forms spheroids consisting of several thousand cells. The OGPs in the extracellular matrix of *Chlamydomonas* and *Volvox* show some resemblance in their glycosylation to OGPs in higher plants. Alkaline hydrolysis in barium hydroxide of the extracellular matrix has demonstrated the presence of hydroxyproline-arabinose, hydroxyproline-galactose and hydroxyproline-linked hetero-oligosaccharides consisting of arabinose, galactose, and glucose [40]. Many of their sugar residues are sulfated [41–43]. On electron micrographs, the OGPs in the *Volvocales* generally show rod-like domains indicating that these domains are in the polyproline II conformation [44–50] and, in the case of *Chlamydomonas* wall protein, it has been demonstrated directly by circular dichroism [51]. The OGPs of the *Volvocales* have, therefore, much in common with their counterparts in higher plants raising the question in how far they are evolutionary related. O-glycosylproteins of the *Volvocales* have two functions: most OGPs seem to be involved in the organisation and assembly of the extracellular matrix, whereas other OGPs play a role in sexual adhesion (Section 3.2).

3.1. O-glycosylproteins in the assembly of the extracellular matrix

After extraction of *Chlamydomonas* cells or *Volvox* spheroids with hot SDS, colourless 'ghosts' remain [44,52]. This reflects the fact that the extracellular matrix of the

Volvocales consists of a network of OGPs insoluble in chaotropic agents and hot SDS [53,54], and a set of chaotrope-soluble OGPs that is capable of recrystallizing and of self-assembling in the presence of chaotrope-extracted cells [52]. The precursor of the insoluble network has best been characterised in *Volvox* (Table I). It consists of a sulfated 185-kDa OGP with a protein core of about 50 kDa [44]. Its most remarkable feature is a central domain of about 80 amino acids that almost entirely consists of arabinosylated hydroxyproline residues interspersed with a few serine residues [44,54] and that probably is in the polyproline II conformation. This is supported by electron microscopic pictures [44] showing a fibrous domain of 29 nm long, which agrees well with the predicted length. The 185-kDa OGP contains about 70% carbohydrate (Table I) and some of its sugars are sulfated [54]. Surprisingly, all sulfated sugars are present in a single sulfated 28-kDa polysaccharide that can be released by mercaptolysis and is, according to electron microscopic pictures, attached close to the middle of the central rod-like domain [44]. Chemical analysis has shown that it consists of a 1,3-linked mannose backbone to which di-arabinosyl side-chains are attached, and that both mannose and arabinose are substituted with sulfate groups [44]. Indirect immunofluorescence in combination with electron microscopy has shown that after polymerisation the 185-kDa OGP forms chambers enclosing the individual cells leading to a honeycomb-like structure [44]. The exact nature of the cross-links between the monomers is not known, but, recently, a phosphodiester bridge between the C-5 atoms of two arabinose residues in the 185-kDa OGP has been identified [55] that fulfils the requirements for such a cross-link. Interestingly, after deglycosylation of the polymeric network with anhydrous hydrogen fluoride, the network is dissolved and a single polypeptide chain of about 50 kDa is obtained indicating that carbohydrate side-chains are indeed involved in cross-linking.

The chaotrope-soluble OGPs in the extracellular matrix of both *Chlamydomonas* and *Volvox* are just as their insoluble counterparts sulfated glycoproteins [41,42], and are capable of self-assembly [52]. After removal of matrix OGPs with perchlorate, the extracted cells can function as nucleating agents. When they are combined with biotinylated OGPs in the presence of perchlorate, and the concentration of perchlorate is subsequently lowered by dialysis, reconstitution of the wall can be monitored by fluorescence microscopy after adding FITC-streptavidin [52]. The molecular interactions that bring this orderly reconstitution of the external wall layers about are not known but it seems possible that sulfated sugars might play a role in this.

3.2. O-glycosylproteins in sexual adhesion

The unicellular alga *Chlamydomonas* has two flagella at the anterior side of the cell body, which are normally used for swimming. When the cells are starved, they differentiate into gametes and a mating type-specific sexual adhesion protein appears on the flagellar surface [46, 56]. The sexual adhesion proteins of both mating types directly interact with each other [43] resulting in flagellar adhesion by cells of the opposite mating type and, finally, in cell fusion. The sexual adhesion proteins are unusually long, fibrous molecules of several hundred nanometers in length and they often, but not always, carry a terminal, globular domain that is believed to be involved in recognition of the complementary adhesion protein [46,49,50]; their molecular masses can be as high as 1300 kDa [57]. The protein part is rich in hydroxyproline, serine, and glycine [57–59] and their main sugars are arabinose, galactose, and glucose; N-acetylglucosamine is virtually absent [50] indicating that N-glycans do not play a significant role in the glycosylation of the sexual adhesion proteins. Versluis et al. [43] have shown that the sexual adhesion proteins of *Chlamydomonas eugametos* are extensively sulfated through their sugar residues. Beta-elimination released about 25% of the sulfated side-chains indicating that these were either bound to serine or threonine; in view of the virtual or total absence of N-linked glycans, it was concluded that the remainder was probably bound to hydroxyproline [43]. Although firm evidence is lacking, it seems likely, taking into account their form and their amino acid and sugar compositions, that the sexual adhesion proteins are at least partially in the polyproline II conformation.

Recognition between the sexual adhesion proteins is at least partially based on ionic interactions as flagellar adhesion is inhibited by low salt concentrations [60,61]. This suggests that sulfated sugar residues might play a role in this process; this is supported by the observation that isolated side-chains from the sexual adhesion proteins are capable of inhibiting sexual adhesion at micromolar concentrations (Versluis R., personal communication).

4. O-glycosyltransferases

Compared to mammalian systems, our knowledge about the glycosyl transferases involved in the synthesis of O-glycosyl side-chains of OGPs is almost negligible. Only arabinosyltransferases involved in the formation of the Hyp-Ara linkage [62] and in the elongation of Hyp-arabinoside side-chains [63,64] have been studied in some detail. However, even in these cases the interpretation of the findings is generally hindered by the lack of defined acceptors.

4.1. Properties of arabinosyltransferases

Arabinosyltransferases have both been demonstrated in higher plants [63–66] and in the *Volvocales* [62, 67]. The synthesis of the tetra-arabinoside attached to hydroxyproline [α-L-Araf-(1→3)-β-L-Araf-(1→2)-β-L-Araf-(1→2)-β-L-Araf-(1→4)-Hyp] in extensins and in the Solanaceous lectins requires at least three and probably four different enzymes. Karr [63] was the first to show that a crude membrane preparation from cultured sycamore cells could be used both to initiate the synthesis of hydroxyproline-arabinosides by transferring arabinose from UDP-arabinose to peptidyl hydroxyproline, and to elongate existing hydroxyproline-arabinosides. The pH optimum was 6.5, and the reaction was stimulated by divalent cations and by the addition of de-arabinosylated trypsin fragments of extensin. Owens and Northcote [64] obtained similar results with a crude membrane preparation from potatoes.

Günther et al. [62] were the first to use defined acceptors. Using a membrane preparation from *Volvox* and synthetic oligopeptides such as Tyr-(Ser)-Hyp$_{1-4}$-Lys containing one or more internal hydroxyproline residues as potential sugar acceptor sites, they could show that arabinosyl residues were attached. The glycopeptide bond that was formed was resistant to conditions used for β-elimination as expected for an arabinosyl-hydroxyproline linkage. The reaction product was also resistant to an α-L-arabinofuranosidase suggesting that the Hyp-Ara linkage has a β-configuration similar to extensins and potato lectin [16,25]. Peptides with the acceptor hydroxyproline residue at their C-terminus were not glycosylated. When the number of internal hydroxyproline residues was increased to four, the acceptor activity of the peptide became significantly better suggesting that such clusters of hydroxyproline, which are often seen in OGPs, form a more natural substrate for the arabinosyltransferase. The optimal pH was close to 7, and as observed for arabinosyltransferases in higher plants [63,64] divalent cations were required for optimal activity.

4.2. Intracellular location of arabinosyltransferases

Arabinosyltransferase activity has been found both in the ER and in the Golgi apparatus [64–69]. This has led to a long-standing controversy about the intracellular location of arabinosyltransferases, which has still not been completely resolved. There are several possible explanations for this. First, all localisation studies so far have been carried out with undefined acceptors present in the crude membrane preparations used, so it is unclear if different investigators have studied the same enzyme. This is particularly relevant because, for the synthesis of the tetra-arabinoside attached to hydroxyproline in extensins and potato lectin, several different enzymes are probably needed. It seems also possible that the plant systems used in these studies were actively engaged in the biosynthesis of more than one OGP at the same time — for example extensins plus AGPs — each with their own requirement for arabinosyl transferases. Finally, it is useful to remember that the O-glycosylation of serine in yeast starts in the ER with the addition of a single mannose residue to serine and that the O-chain is elongated in the Golgi [70]. A similar situation might hold for the glycosylation of hydroxyproline. In this connection the observations by Moore et al. [66] might be relevant. Using an anti-extensin antibody that recognises the terminal α-linked arabinosyl residue of the tetra-arabinoside linked to hydroxyproline, they could show by immunogold labelling that the cis-cisternae of the Golgi apparatus contain completely glycosylated extensin indicating that either all arabinosyl transferases needed for the glycosylation of extensin are present in the cis-cisternae as the authors believe or that chain initiation occurs in the ER and is followed by chain elongation in the Golgi. .

Abbreviations

AGPs	arabinogalactan proteins
HRGPs	hydroxyproline-rich glycoproteins
OGPs	O-glycosylproteins

References

1. Cassab, G.I. and Varner, J.E. (1988) Annu. Rev. Plant Physiol. and Plant Mol. Biol. 39, 321–353.
2. Showalter, A.M. (1993) Plant Cell 5, 9–23.
3. Lamport, D.T.A. (1967) Nature 216, 1322–1324.
4. Lamport, D.T.A. (1970) Annu. Rev. Plant Physiol. 21, 235–270.
5. Allen, A.K. and Neuberger, A. (1973) Biochem. J. 135, 307–314.
6. Fincher, G.B., Stone, B.A. and Clarke, A.E. (1983) Annu. Rev. Plant Physiol. 34, 47–70.
7. Samson, M., Klis, F.M., Sigon, C.A.M. and Stegwee, D. (1983) Planta 159, 322–328.
8. Pennell, R.I. (1992) In: J.A. Callow and J.R. Green (Eds.), Society for Experimental Biology Seminar Series 48: Perspectives in Plant Cell Recognition. Cambridge University Press, Cambridge, pp. 105–121.
9. Cooper, J.B. and Varner, J.E. (1983) Biochem. Biophys. Res. Commun. 112, 161–167
10. Mort, A.J. and Lamport, D.T.A. (1977) Anal. Biochem. 82, 289–309.
11. Klis, F.M., Rootjes, M., Groen, S. and Stegwee, D. (1983) Z. Pflanzenphysiol. 110, 301–307
12. Esquerré-Tugayé, M.T., Mazau, D., Pélissier, B., Roby, D., Rumeau, D. and Toppan, A. (1985) In: J.L. Key and T. Kosuge (Eds.), Cellular and Molecular Biology of Plant Stress. Alan R. Liss, New York, pp. 459–473.
13. Van Holst, G.-J. and Varner, J.E. (1984) Plant Physiol. 74, 247–251.
14. Stafstrom, J.P. and Staehelin, L.A. (1986) Plant Physiol. 81, 234–241.
15. Stafstrom, J.P. and Staehelin, L.A. (1986) Plant Physiol. 81, 242–246.
16. Lamport, D.T.A. (1980) In: J. Preiss (Ed.), The Biochemistry of Plants, Vol. 3. Academic Press, New York, Ch. 13 pp. 501–540.
17. Klis, F.M. and Eeltink, H. (1979) Planta 144, 479–484.
18. Lamport, D.T.A. and Miller, D.H. (1971) Plant Physiol. 48, 454–456.
19. Akiyama, Y., Mori, M. and Kato, K. (1980) Agric. Biol. Chem. 44, 2487–2489.
20. Lamport, D.T.A., Katona, L. and Roerig, S. (1973) Biochem. J. 133, 125–131.
21. Cho, Y.P. and Chrispeels, M.J. (1976) Phytochem. 15, 165–169.
22. O'Neill, M.A. and Selvendran, R.R. (1980) Biochem. J. 187, 53–63.
23. Showalter, A.M. and Varner, J.E. (1989) In: A. Marcus (Ed.), The Biochemistry of Plants. Vol. 15, Molecular Biology. Academic Press, New York, pp. 485–520.
24. Allen, A.K., Desai, N.N., Neuberger, A. and Creeth, J.M. (1978) Biochem. J. 171, 665–674.
25. Ashford, D., Desai, N.N., Allen, A.K., Neuberger, A., O'Neill, M.A. and Selvendran, R.R. (1982) Biochem. J. 201, 199–208.
26. Van Holst, G.-J., Martin, S.R., Allen, A.K., Ashford, D. and Desai, N.N. (1986) Biochem. J. 233, 731–736.
27. Casalongué, C. and Pont Lezica, R. (1985) Plant Cell Physiol. 26, 1533–1539.
28. Millar, D.J., Allen, A.K., Smith, C.G., Sidebottom, C., Slabas, A.R. and Bolwell, G.P. (1992) Biochem. J. 283, 813–821.
29. Jentoft, N. (1990) TIBS 15, 291–294.
30. Lipke, P.N., Wojciechowicz, D. and Kurjan, J. (1989) Mol. Cell. Biol. 9, 3155–3165.
31. Van Holst, G.-J. and Klis, F.M. (1981) Plant Physiol. 68, 979–980.
32. Strahm, A., Amado, R. and Neukom, H. (1981) Phytochem. 20, 1061–1063.
33. Jermyn, M.A. and May Yeow, Y. (1975) Aust. J. Plant Physiol. 2, 501–531.
34. Samson, M.R., Jongeneel, R. and Klis, F.M. (1984) Phytochem. 23, 493–496.
35. Pope, D.G. (1977) Plant Physiol. 59, 894–900.
36. Kreuger, M. and Van Holst, G.-J. (1993) Planta 189, 243–248.
37. Van Holst, G.-J. and Clarke, A.E. (1986) Plant Physiol. 80, 786–789.

38. Van Holst, G.-J. and Fincher, G.B. (1984) Plant Physiol. 75, 1163–1164.
39. Pennell, R.I., Knox, J.P., Scofield, G.N., Selvendran, R.R. and Roberts, K. (1989) J. Cell Biol. 108, 1967–1977.
40. Miller, D.H., Lamport, D.T.A. and Miller, M. (1972) Science 176, 918–920.
41. Roberts, K., Gay, M.R. and Hills, G.J. (1980) Physiol. Plant. 49, 421–424.
42. Wenzl, S. and Sumper, M. (1982) FEBS Lett. 143, 311–315.
43. Versluis, M., Klis, F.M., Van Egmond, P. and Van Den Ende, H. (1993) J. Gen. Microbiol. 139, 763–767.
44. Ertle, H., Wenzl, S., Engel, J. and Sumper, M.J. (1989) J. Cell Biol. 109, 3493–3501.
45. Ertl, H., Hallmann, A., Wenzl, H. and Sumper, M. (1992) EMBO J. 11, 2055–2062.
46. Goodenough, U.W., Adair, W.S., Collin-Osdoby, P. and Heuser, J. (1985) J. Cell Biol. 101, 924–941.
47. Goodenough, U.W., Gebhart, B., Mecham, R.P. and Heuser, J.E. (1986) J. Cell Biol. 103, 405–417.
48. Goodenough, U.W. and Heuser, J.E. (1988) J. Cell Sci. 90, 717–733.
49. Crabbendam, K.J., Klis, F.M., Musgrave, A. and van Den Ende, H. (1986) J. Ultrastr. Molec. Struct. Res. 96, 151–159.
50. Klis, F.M., Crabbendam, K., Van Egmond, P. and Van Den Ende, H. (1989) Sex Plant Reprod. 2, 213–218.
51. Homer, R.B. and Roberts, K. (1979) Planta 146, 217–222.
52. Adair, W.S., Steinmetz, S.A., Mattson, D.M., Goodenough, U.W. and Heuser, J.E. (1987) J. Cell Biol. 105, 2373–2382
53. Hills, G.J. (1973) Planta 115, 17–23.
54. Wenzl, S., Thym, D. and Sumper, M. (1984) EMBO J. 3, 739–744.
55. Holst, O., Christoffel, V., Fründ, R., Moll, H. and Sumper, M. (1989) Eur. J. Biochem. 181, 345–350.
56. Van Den Ende, H., Musgrave, A. and Klis, F.M. (1990) In: R.A. Bloodgood (Ed.), Ciliary and Flagellar Membranes. Plenum Publ. Corp., London and New York, pp. 129–147.
57. Samson, M.R., Klis, F.M., Homan, W.L., Van Egmond, P., Musgrave, A. and Van Den Ende, H. (1987) Planta 170, 314–321.
58. Samson, M.R., Klis, F. M., Crabbendam, K. J., Van Egmond, P. and Van Den Ende, H. (1987) J. Gen. Microbiol. 133, 3183–3191.
59. Collin-Osdoby, P. and Adair, W.S. (1985) J. Cell Biol. 101, 1144–1152.
60. Versluis, R., Schuring, F., Klis, F.M., Van Egmond, P. and Van Den Ende, H. (1992) FEMS Microbiol. Lett. 97, 101–105.
61. Goodenough, U. (1986) Exp. Cell Res. 166, 237–246.
62. Günther, R., Bause, E. and Jaenicke, L. (1987) FEBS Lett. 221, 293–298.
63. Karr, A.L. (1972) Plant Physiol. 50, 275–282.
64. Owens, R.J. and Northcote, D.H. (1981) Biochem. J. 195, 661–667.
65. Jones, R.L. and Robinson, D.G. (1989) New Phytol. 111, 567–597.
66. Moore, P.J., Swords, K.M.M., Lynch, M.A. and Staehelin, L.A. (1991) J. Cell Biol. 112, 589–602.
67. Zhang, Y.-H, Lang, W.C. and Robinson, D.G. (1989) Plant Cell Physiol. 30, 617–622.
68. Andreae, M., Blankenstein, P., Zhang, Y.-H. and Robinson, D.G. (1988) Eur. J. Cell Biol. 47, 181–192.
69. Gardiner, M.C. and Chrispeels, M.J. (1975) Plant Physiol. 55, 536–541.
70. Haselbeck, A. and Tanner, W. (1983) FEBS Lett. 158, 335–338.
71. Van Holst, G.J., Klis, F.M., De Wildt, P.J.M., Hazenberg, C.A.M., Buijs, J. and Stegwee, D. (1981) Plant Physiol. 68, 910–913.

J. Montreuil, H. Schachter and J.F.G. Vliegenthart (Eds.), *Glycoproteins*

N-Glycosylation of plant proteins

ARND STURM

Friedrich Miescher-Institut, Postfach 2543, CH-4002 Basel, Switzerland

1. Introduction

In plants, the majority of proteins of the extracellular space and the endomembrane system (the endoplasmic reticulum and its continuum the nuclear envelope, the Golgi apparatus, the vacuole and its surrounding membrane the tonoplast and the plasma membrane) are N-glycosylated. In the N-linked glycoproteins, oligosaccharides (glycans) are covalently linked to the protein backbone via the amide nitrogen of asparagine (Asn). The only glycan acceptors are the Asn residues in the tripeptide Asn-Xaa-Ser/Thr (glycosylation recognition sequence, glycosylation sequon), where Xaa can be any amino acid other than proline. The ring structure formed by this tripeptide is specifically recognized by the oligosaccharyltransferase, which transfers precursor oligosaccharides to newly synthesized proteins [1]. In addition, to function as an N-glycosylation signal, the tripeptide must be located in a protein domain that favors the formation of β-turns [2].

The N-linked oligosaccharides of plant glycoproteins have 5–13 glycosyl units and can be divided into two main classes: high-mannose-type glycans and complex-type glycans (Fig. 1). Both classes of glycans have a core of two N-acetylglucosamine residues (N,N'-diacetylchitobiose) linked at the reducing end to Asn and at the non-reducing end to mannose. The high-mannose-type glycans have up to nine Man residues with the typical Man-Man branching pattern shown in Fig. 1, whereas the complex glycans have less Man residues but additional sugars such as fucose, xylose, and galactose. With the exception of β-1,2-linked Xyl in the complex glycans of some glycoproteins from invertebrates [3,4], and α-1,3-linked fucose in the complex glycans of some glycoproteins from honey bee venom [5], Xyl β-1,2-linked to the core Man residue and Fuc α-1,3-linked to the proximal GlcNAc residue are plant-specific glycan residues. In animal glycoproteins, Fuc is α-1,6-linked to the proximal GlcNAc residue [1]. Furthermore, sialic acid, a common terminal sugar of complex glycans from animal glycoproteins, has not been found in plant glycoproteins.

The N-linked oligosaccharide side-chains are bulky residues which probably have a great impact on the physicochemical properties of glycoproteins due to their pronounced hydrophilicity. Y-, T-, bird- and broken wing-conformations have been proposed for diantennary glycans, whereas glycans with more antennae seem to favor an umbrella-conformation with the N,N'-diacetyl-chitobiose core corresponding to the

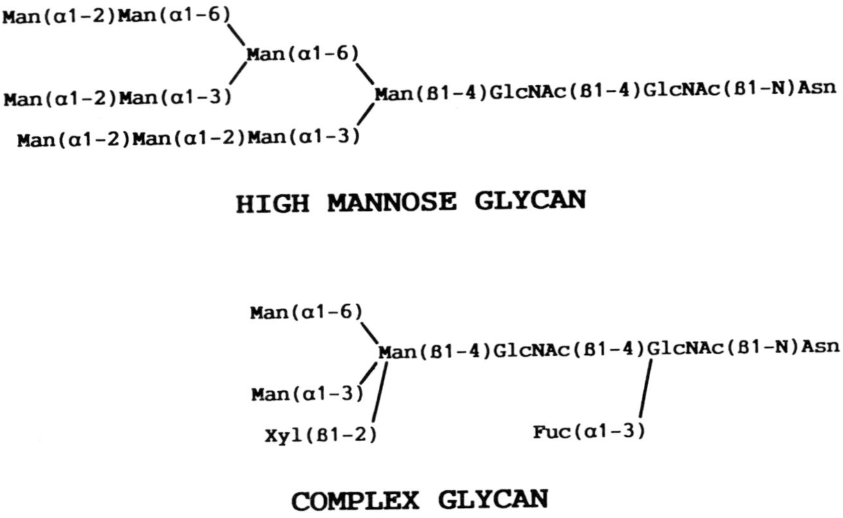

```
Man(α1-2)Man(α1-6)
                    \
                     Man(α1-6)
                    /         \
Man(α1-2)Man(α1-3)             Man(β1-4)GlcNAc(β1-4)GlcNAc(β1-N)Asn
Man(α1-2)Man(α1-2)Man(α1-3)
```

HIGH MANNOSE GLYCAN

```
Man(α1-6)
         \
          Man(β1-4)GlcNAc(β1-4)GlcNAc(β1-N)Asn
         /                     |
Man(α1-3)                      |
        /                      |
Xyl(β1-2)          Fuc(α1-3)
```

COMPLEX GLYCAN

Fig. 1. Structures of a representative high-mannose- and complex-type glycan from plant N-linked glycoproteins.

stem and the antennae forming the umbrella [2]. Multiantennary glycans extruding from a globular polypeptide will cover large areas of the protein and may also have a significant impact on its surface properties.

2. *The biosynthesis and processing of N-linked glycoproteins*

Our knowledge about the biosynthesis and processing of N-linked glycans is based primarily on studies with mammalian cells and yeast. The biosynthetic pathway unraveled by these studies [6–8] has been verified by plant glycobiologists (see reviews by Elbein [9] and Faye et al. [10]). Both high-mannose- and complex-type N-linked glycans originate from a common oligosaccharide precursor, a lipid-linked glycan of the high-mannose-type with nine Man residues and an additional three glucose residues (Fig. 2) [11,12], which is synthesized at and in the endoplasmic reticulum (ER). The lipid dolichol [13] anchors the oligosaccharide precursor in the ER membrane with the carbohydrate facing the lumen of the ER. During protein biosynthesis on membrane-bound ribosomes, the oligosaccharide precursor is transferred *en bloc* from the lipid to the growing polypeptide chain. The three Glc residues are removed immediately after protein glycosylation to give high-mannose-type glycans with nine Man residues. During maturation of glycoproteins, the N-linked glycans either remain in the high-mannose form or are modified to the complex-type glycans. Glycans that remain in their high-mannose form have the common structure $Man_{9-5}GlcNAc_2$, which results from limited processing of the $Man_9GlcNAc_2$ structure by α-mannosidases. Conversion of high-mannose-type glycans to complex-type glycans may take place as the

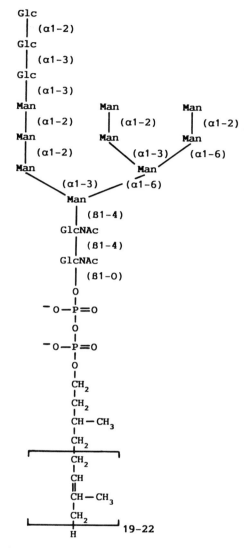

Fig. 2. Structure of the dolichol-linked oligosaccharide precursor of the plant N-linked glycans.

proteins move from the rough ER via the Golgi to their ultimate destinations, such as the extracellular matrix, or the vacuoles. These modifications are catalyzed by glycosidases and glycosyltransferases. As a result of this complex biosynthesis and processing pathway, which is schematically shown in Fig. 3, specific oligosaccharide structures are attached to specific glycosylation sites.

The reason why some oligosaccharide side-chains remain in the high-mannose form whereas others are extensively modified seems to be differences in their accessibility

524

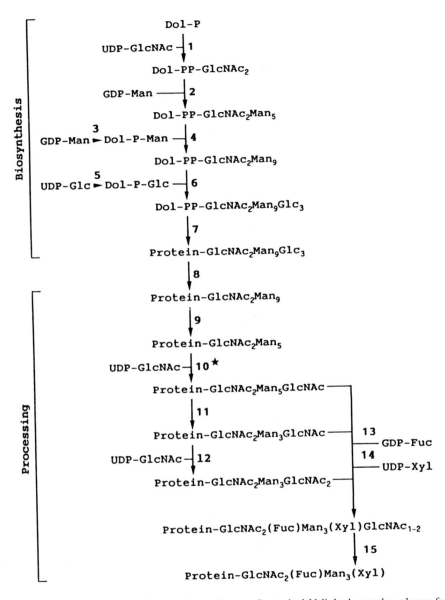

Fig. 3. Proposed biosynthesis and processing pathway of a typical N-linked complex glycan from plants. The enzymes involved are GlcNAc-1-P-transferase and GlcNAc-transferase (**1**), mannosyl-transferase (**2**), Dol-P-mannose synthase (**3**), Dol-P-mannosyl-transferase (**4**), Dol-P-glucose synthase (**5**), Dol-P-glucosyl-transferase (**6**), oligosaccharide-transferase (**7**), glucosidase I and II (**8**), mannosidase I (**9**), *N*-acetylglucosaminyltransferase I (**10**), mannosidase II (**11**), *N*-acetylglucos-aminyltransferase II (**12**), fucosyl-transferase (**13**), xylosyltransferase (**14**), *N*-acetyl glucosaminidase (**15**). The enzyme activity **10** (*N*-acetyl-glucosaminyltransferase I) which is missing in the *A. thaliana* mutants C5 and C6 [109] is marked by an asterisk.

to glycan-processing enzymes: glycans located on the surface of folded polypeptides are accessible to the processing enzymes and become converted to complex-type glycans. In contrast, glycans completely or partly buried in the folded polypeptide remain in their high-mannose form [14]. Structural differences in the complex glycans from different plant glycoproteins indicate the presence of additional signals directing the formation of the different structures. These signals may be encoded in the primary sequence or in the surface structure of the glycoproteins. Furthermore, spatial and temporal differences in the expression of glycosidases and glycosyltransferases may have a profound impact on the resulting glycan structures.

3. N-Glycan structures of plant glycoproteins

To date, several oligosaccharide structures from vacuolar plant glycoproteins [15–30] but only a few from extracellular plant glycoproteins [31–37] have been characterized. High-mannose-type oligosaccharides with 9–5 Man residues have been found attached to glycoproteins from both subcellular compartments. Analysis of the high-mannose-type glycans at individual glycosylation sites usually reveals some structural heterogeneity, e.g. a protein with predominantly $Man_7GlcNAc_2$ attached to a specific glycosylation site also has some glycans with six and eight Man residues at the same Asn residue.

The complex glycans from vacuolar glycoproteins are very similar, showing very little heterogeneity at individual glycosylation sites. A consensus 'vacuole-type' complex glycan structure [36], a β-1,2-xylosylated α-1,3-fucosylated mannotriosyl N,N'-diacetyl-chitobiose glycan, is shown in Fig. 4A. Most of the complex glycans from secreted glycoproteins are fairly large and display pronounced structural heterogeneity at individual glycosylation sites. The smallest of these oligosaccharide structures is identical to the 'vacuole-type' complex glycan. The larger glycans have additional sugars such as GlcNAc, Gal, and Fuc attached to the non-reducing ends of the molecules [31,36]. Figure 4B shows the structures of the complex glycans from laccase [31], a glycoprotein secreted by sycamore (*Acer pseudoplatanus*) cells. Interestingly, in these large complex-type glycans, Fuc is α-1,3-linked to the proximal GlcNAc residue of the N,N'-diacetyl-chitobiose core and α-1,6-linked to the terminal GlcNAc residues ('terminal' Fuc), suggesting the existence of two different fucosyltransferases. The data summarized above show that the known complex N-linked glycan structures of extracellular glycoproteins differ from the complex N-linked glycan structures of vacuolar glycoproteins by the additional terminal sugar residues. Whether this finding can be generalized is not yet clear because the glycoproteins compared were from different plants.

Up to now, only five plant N-linked glycoproteins have been studied in sufficient detail to allow the assignment of specific structures to defined glycosylation sites (Fig. 5).

Ricin D, a glycoprotein from the seeds of castor plants (*Ricinus communis*), has two non-identical polypeptide chains (A- and B-chain) held together by a disulfide linkage. The amino acid sequences of the two polypeptides contain one potential glycosylation site for the A-chain (Asn10) and two for the B-chain (Asn95 and Asn135). Xyl-

526

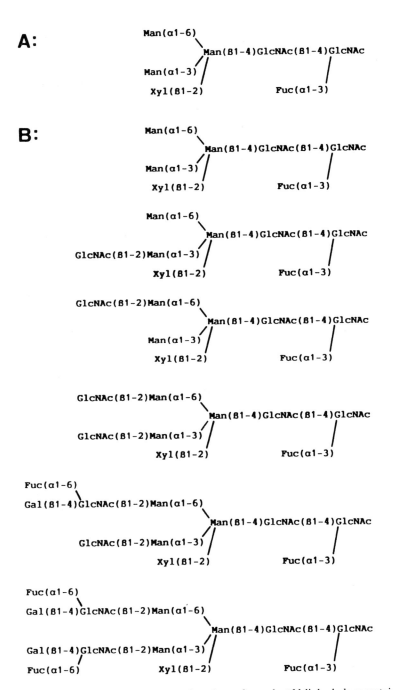

Fig. 4. A: Structure of the 'vacuole-type' complex glycan from plant N-linked glycoproteins [36]. B: Proposed structures of the complex-type N-linked glycans from laccase excreted by sycamore cells [31].

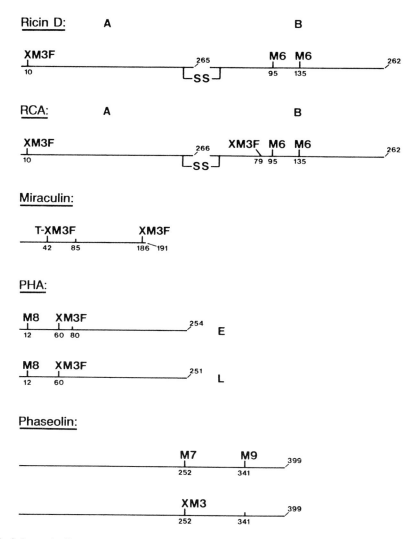

Figure 5. Schematic linear polypeptide structures of five well-characterized plant glycoproteins showing the attachment points and the formula of their N-linked oligosaccharides (M9, M8, M7, and M6 indicate high-mannose-type glycans with 9–6 Man residues; XM3F, $XylMan_3FucGlcNAc_2$; T-XM3F, $XylMan_3FucGlcNAc_2$ with terminal sugars such as GlcNAc, Gal, and Fuc linked to the α-1,3- and α-1-6-Man residues; XM3, $XylMan_3GlcNAc_2$). The numbers of the Asn residues to which glycans are attached are indicated below the linear protein structures. Only the formula of the most abundant glycan attached to a specific glycosylation site is shown above the linear protein structures. Information on structural heterogeneity at individual glycosylation sites can be found in the present review and in the original manuscripts on the A-chain and B-chain of ricin D [22], the A-chain and B-chain of RCA [23], miraculin [35], the E- and L-subunit of PHA [28], and the polypeptides of phaseolin [21]. The length of the polypeptides (number of amino acid residues) is indicated at the end of each linear protein structure.

Man$_3$FucGlcNAc$_2$ is predominantly attached to Asn10 of the A-chain. A high-mannose-type glycan with six Man residues has been found attached to Asn95 and heterogenous high-mannose glycans with 5–7 Man residues and some Xyl-Man$_4$GlcNAc$_2$ have been found attached to Asn135 of the B-chain [22]. The extent of glycan processing on the three glycosylation sites differs markedly, indicating that the sugar chains on Asn10 are more accessible to the processing enzymes than the chains on Asn95 and Asn135. This interpretation is consistent with the results of X-ray structural analyses of ricin D [38], which show that Asn10 of the A-chain is situated completely outside of the polypeptide, while Asn95 and Asn135 are situated near the contact region of the A- and B-chains.

  R. communis agglutinin (RCA) has a protein structure similar to ricin D with an A- and a B-chain linked by a disulfide bridge. The B-chain has an additional glycosylation site at Asn79 to which XylMan$_3$FucGlcNAc$_2$ is attached. Glycosylation of Asn10 of the A-chain and Asn95 and Asn135 of the B-chain is very similar to ricin D [23].

  Miraculin, a glycoprotein with 191 amino acid residues from the red berries of *Richadella dulcifica*, has three glycosylation sites at Asn42, Asn85 and Asn186 [39]. Complex oligosaccharides are attached to Asn42 and Asn186 and the glycosylation site at Asn85 is not used. XylMan$_3$FucGlcNAc$_2$ is the prominent glycan on Asn186, whereas the same structure with varying numbers of additional terminal sugars (GlcNAc, Gal, and Fuc) is linked to Asn42 [35]. These large and heterogeneous N-glycan structures are identical to those found attached to laccase [31].

  Phytohemagglutinin (PHA), the lectin from the seeds of the common bean (*Phaseolus vulgaris*), consists of two subunits (PHA-E and PHA-L) with 82% identity in their amino acid sequences. PHA-E has three glycosylation sites (Asn12, Asn60, and Asn80), whereas PHA-L has only two (Asn12 and Asn60) [40]. The positions and sequences of the glycosylation sites at Asn12 and Asn60 are conserved in both polypeptides and only these sites are used. A high-mannose glycan with eight Man residues was found attached to Asn12 and XylMan$_3$FucGlcNAc$_2$ attached to Asn60 [28,41]. Modification of the glycan attached to Asn60 seems to be controlled by its accessibility to the Golgi-localized processing enzymes. The second glycan attached to Asn12 is shielded by the folded polypeptide and remains in its high-mannose form [42].

  Phaseolin, the storage protein from seeds of *P. vulgaris*, consists of four size classes of polypeptides. The two more abundant classes have two oligosaccharides each, whereas the others have one oligosaccharide each [43]. The amino acid sequences of the phaseolin polypeptides are very homologous and contain two conserved glycosylation sites at Asn252 and Asn341 [44]. Polypeptides with two glycans have Man$_7$GlcNAc$_2$ attached to Asn252 and Man$_9$GlcNAc$_2$ attached to Asn341. Polypeptides with only one glycan have XylMan$_3$GlcNAc$_2$ attached to Asn252 [21]. The extent of glycan processing on the different glycosylation sites differs markedly, and again correlates with differences in the accessibility of the different glycans to the processing enzymes. It is of particular interest that the glycan at Asn252, which is accessible to glycosidases and glycosyltransferases, is in a hydrophilic region of the polypeptide, whereas the inaccessible glycan at Asn341 is in a hydrophobic region. Furthermore, the presence of XylMan$_3$GlcNAc$_2$ and Man$_7$GlcNAc$_2$ at the same Asn residue (Asn252) of different polypeptides seems to be controlled by the glycosylation status of Asn341. When Asn341 is unoccupied, the glycan at Asn252 is complex. When Asn341 is occupied, the glycan at Asn252 is only modified to the extent that two mannosyl residues are removed [21].

A comparison of the five plant glycoproteins analyzed in detail (Fig. 5) does not reveal any correlation between glycan classes and position in the amino acid sequence. In contrast, a survey of 50 animal glycoproteins showed an asymmetric distribution of complex and high-mannose glycans: the majority of complex glycans were found in the N-terminal domain, whereas the majority of high-mannose glycans were located in the C-terminal domain [45].

4. Glycosidases and glycosyltransferases that catalyze the synthesis of the lipid-linked oligosaccharide precursor

The enzymes GlcNAc-1-P-transferase [46] and GlcNAc-transferase [47], which successively transfer the first and the second GlcNAc residues to dolichol-P to form $GlcNAc_2$-PP-dolichol (Fig. 3, step 1) have been enzymatically characterized. The donor for both GlcNAc residues is UDP-GlcNAc. The addition of Man residues to the N,N'-diacetylchitobiose core to form $Man_9GlcNAc_2$-PP-dolichol is, as in mammalian cells, a two-step process. Man residues, which make up the heptasaccharide core ($Man_5GlcNAc_2$), are transferred directly from GDP-Man to the disaccharide lipid (Fig. 3, step 2). The four outer Man residues are donated by Man-P-dolichol (Fig. 3, step 4), synthesized from GDP-Man and dolichol-P (Fig. 3, step 3). These findings indicate that in both plants and animals the biosynthesis of the oligosaccharide precursor starts at the cytoplasmic side of the ER and is completed in the ER lumen [48]. A detailed characterization of the mannosyltransferases in plant cells is still lacking but it is likely that formation of the Man(α1–2)Man, Man(α1–3)Man, Man(α1–6)Man and Man(β1–4)GlcNAc linkages is catalyzed by different mannosyl-transferase isoenzymes. Glucosyltransferase, which transfers Glc to dolichol-P (Fig. 3, step 5), has been characterized [49] but the enzyme(s) that transfers three Glc residues from Glc-P-dolichol to $Man_9GlcNAc_2$-PP-dolichol (Fig. 3, step 6) has not yet been studied. Again, the formation of Glc(α1–2)Glc, Glc(α1–3)Glc and Glc(α1–3)Man may require different glucosyl-transferase isoenzymes. Plant oligosaccharyltransferase, which catalyzes the transfer of the oligosaccharide precursor to the nascent polypeptide chain (Fig. 3, step 7), has not yet been characterized.

None of the enzymes for the biosynthesis of the lipid-linked precursor have been purified in sufficient quantities to allow protein sequencing and the generation of antibodies. Furthermore, there is no information on the genes encoding glycosyltransferases or their regulation.

5. Glycosidases and glycosyltransferases that catalyze the processing of N-linked glycans

Immediately after N-glycosylation of plant proteins by the oligosaccharide precursor $Glc_3Man_9GlcNAc_2$, the three Glc residues are removed by glucosidases (Fig. 3, step 8), whereby glucosidase I hydrolyses Glc(α1–2)Glc and glucosidase II Glc(α1–3)Glc and

Glc(α1–3)Man. Removal of Glc takes place in the ER [50] and, in general, seems to be very efficient [10]. Most likely, the unusual $Glc_1Man_9GlcNAc_2$ glycan structure on jack bean α-mannosidase [51] arises due to the inaccessibility of the precursor oligosaccharide to glucosidase II resulting in the cleavage of only the outer α-1,3-linkage. Removal of the four Man residues with α-1,2-linkages from $Man_9GlcNAc_2$ is catalyzed by mannosidase I (Fig. 3, step 9) and takes place in the Golgi apparatus [50]. The action of mannosidase I is a prerequisite for the addition of a GlcNAc residue by *N*-acetylglucosaminyl-transferase I yielding the structure $GlcNAcMan_5$ $GlcNAc_2$ (Fig. 3, step 10). The removal of two more α-1,3- and α-1,6-linked Man residues by mannosidase II (Fig. 3, step 11) is a prerequisite for the addition of a second GlcNAc residue by *N*-acetyl-glucosaminyltransferase II, resulting in the formation of $GlcNAc_2Man_3GlcNAc_2$ (Fig. 3, step 12). Glucosidases I and II [52,53], α-mannosidases I and II [54,55], and N-acetyl-glucosaminyltransferases I and II [56,57] have been characterized from mung bean seedlings but only glucosidase II and mannosidase II were purified to homogeneity. On Coomassie-stained SDS polyacrylamide gels, the purified enzymes migrated as single polypeptides with molecular masses of 110 kDa and 125 kDa [53,55].

The formation of the highly branched complex glycans structures of some mammalian glycoproteins is initiated by further *N*-acetylglucosaminyltransferases (*N*-acetylglucosaminyl-transferases IV to VI) [1]. These transferases and the *N*-acetylglucosaminyltransferase that links the bisecting GlcNAc residue β-1,4 to the core β-Man residue (*N*-acetylglucosaminyltransferase III) [1] are apparently absent in plants. The place of the bisecting GlcNAc may be taken by the β-1,2-linked Xyl residue. Xylosyltransferase and fucosyltransferase are required for the formation of the 'vacuolar-type' [36] complex glycan structure (Fig. 3, steps 13 and 14). Galactosyltransferase and the 'terminal' fucosyltransferase catalyze synthesis of the large, complex, N-linked glycan structures of the secreted plant N-glycosylproteins [31,35]. No work has been done on these terminal glycosyltransferases.

The enzymes for the processing of N-linked glycans in plants have been characterized with respect to their substrate specificity [58,59]. In bean cotyledons, the formation of $GlcNAcMan_5GlcNAc_2$ is an ordered process and requires the sequential action of mannosidase I and *N*-acetylglucosaminyl-transferase I. The substrate specificities of the two enzymes are very tight and the octasaccharide formed is essential for the completion of glycan processing. The 'vacuole-type' complex glycan Xyl-$Man_3FucGlcNAc_2$ is formed from the octasaccharide by an overlapping sequence of glycan modifications. Fucosyltransferase can transfer Fuc to $GlcNAcMan_5GlcNAc_2$, $GlcNAcMan_3GlcNAc_2$, and $GlcNAc_2Man_3GlcNAc_2$, whereas none of the mannosyl-only glycans showed detectable acceptor activity. The broad substrate specificity of fucosyltransferase indicates that it probably begins to act right after *N*-acetylglucosaminyltransferase I *in vivo*. Xylosyltransferase exhibits significant activity towards $GlcNAcMan_3GlcNAc_2$ and $GlcNAc_2Man_3GlcNAc_2$ only and, therefore, probably acts after fucosyltransferase and α-mannosidase II *in vivo*. Since diantennary glycans with terminal Gal residues were inactive in the fucosyl- and xylosyltransferase assays, the transfer of Gal residues to the complex glycans of some secreted glycoproteins may succeed the addition of Fuc and Xyl.

The processing pathway for glycoproteins of *R. communis* seeds was deduced from a detailed analysis of the glycan structures linked to Ricin D and *R. communis* agglutinin [60]. Xyl was found in the glycans XylMan$_4$GlcNAc$_2$ and XylMan$_3$GlcNAc$_2$, whereas Fuc was only detected in XylMan$_3$FucGlcNAc$_2$. The authors concluded that xylosylation precedes fucosylation in *R. communis* but not in *P. vulgaris*.

In sycamore cells, the sequence of processing events was determined by analyzing the substrate specificity of the processing enzymes and the kinetics of the formation of glycan intermediates and mature glycan structures [59]. The pyridylaminated high-mannose glycan with five Man residues (Man$_5$-PA) was converted by *N*-acetylglu-cosaminyltransferase I to GlcNAcMan$_5$-PA. GlcNAcMan$_5$-PA was rapidly converted to GlcNAcMan$_3$-PA, which was formed *via* GlcNAcMan$_4$-PA by the action of α-man-nosidase II. Assays of *N*-acetylglucosaminyltransferase II and xylosyltransferase showed that GlcNAcMan$_3$-PA is the substrate of both enzymes. However, the reaction rate of *N*-acetylglucosaminyl-transferase is much higher than that of xylosyltransferase when both enzymes are active in the reaction mixture. This result suggests that *N*-acetylglucosaminyltransferase II acts initially on GlcNAcMan$_3$-PA to produce GlcNAc$_2$Man$_3$-PA and xylosyltransferase then produces the Xyl-containing oligosac-charide GlcNAc$_2$XylMan$_3$-PA, which is found in the plant glycoproteins. Since Xyl-containing oligosaccharides without an α-1,3-Fuc residue are linked to rice α-amylase [34], ascorbic acid oxidase of Zucchini [32], and phaseolin of the common bean [21], xylosyltransferase seems to act independently of the existence of an α-1,3-Fuc residue. The addition of an α-1,3-Fuc residue to the proximal GlcNAc residue of the *N,N'*-diacetylchitobiose core could take place before or after addition of Xyl. However, there is insufficient experimental support for this suggestion because of the low activity of fucosyltransferase in the *in vitro* assays. Furthermore, just when terminal Gal and terminal Fuc are added to the complex glycans of sycamore glycoproteins such as laccase [31] is also unclear.

Enzymes catalyzing the formation of complex glycans from Man$_9$GlcNAc$_2$ have been localized in the Golgi apparatus, but attempts to biochemically dissect the Golgi into *cis*, *medial*, and *trans* cisternae and assign the different glycosidases and glycosyl-transferases to specific Golgi sub-compartments failed [50]. Recent studies on the localization of glycan intermediates using anti-carbohydrate antibodies and the Golgi apparatus of sycamore cells showed that Xyl-containing complex glycans were pre-dominantly located in the *medial* Golgi, and Fuc-containing glycans in the *trans* Golgi [61–63]. These findings suggest that xylosylation of glycans precedes fucosylation in cells of sycamore, which is consistent with the proposed processing pathway in *R. communis* but contradicts the pathway postulated for *P. vulgaris*.

6. The function of protein glycosylation in plants

The conservation of the N-glycosylation pathway in all plant species and the large number of highly specific enzymes involved suggest an important function for protein N-glycosylation in plants. However, numerous studies have failed to reveal a general function, and the idea has developed that the function of N-glycosylation is specific for each protein and must be studied case by case [64].

Useful tools for these studies are specific inhibitors of glycan biosynthesis and processing [9]; these may help in understanding the role of N-linked glycans in general and the significance of complex glycans in particular. In recent years, the study of individual glycoproteins with for example, mutated glycosylation recognition signals has become possible using recombinant DNA technology and methods for plant transformation.

6.1. Protein targeting

In plants as in animals and yeast, secretion is by bulk flow (default pathway) and requires no information beyond the presence of a signal peptide necessary for entry into the ER [65]. Therefore, N-linked glycans have no important role in targeting proteins to the outside of the cell [66].

Based on the role of phosphorylated high-mannose-type glycans as targeting determinants for the transport of glycoproteins into animal lysosomes [67], the function of N-linked glycans in the targeting of proteins into the vacuole, the plant equivalent of lysosomes, has been extensively studied. Man-6-phosphate residues on high-mannose glycans have not been detected in plants [68]. The transport of plant glycoproteins with complex N-linked glycans from the Golgi to the vacuole is characterized by transient terminal GlcNAc residues [69]. The addition of these sugars in the Golgi and their removal in the vacuole resembles the phosphorylation of high-mannose glycans of lysosomal glycoproteins in the Golgi and their subsequent dephosphorylation in the lytic compartment, suggesting a role for these terminal GlcNAc residues in vacuolar targeting in plants. Despite this analogy, a number of studies on the transport of glycoproteins in the presence of tunicamycin, which inhibits biosynthesis of the lipid-linked oligosaccharide precursor, have shown that plant N-linked glycans contain no vacuolar targeting information [70–72]. Furthermore, mutagenesis of sequences encoding N-glycosylation sites of several plant glycoproteins and their expression in transgenic plants confirmed these findings [73–76].

The transport of plant membrane glycoproteins was also studied in the presence of tunicamycin [77]. The lack of carbohydrate on the newly synthesized membrane polypeptides did not prevent their appearance in the Golgi or the plasma membrane, suggesting that the intracellular transport and intercalation of newly synthesized glycoproteins into plant cell membranes also does not require the presence of N-linked glycans.

6.2. Protein processing

The removal of a C-terminal propeptide from some vacuolar glycoproteins during or directly after transport into the target organelle seems to be a common feature of proteins with a C-terminal targeting signal [78]. In the lectins of barley [74] and wheat [79], and in β-1,3-glucanase from tobacco [80], the C-terminal propeptide is a glycopeptide with a high-mannose-type glycan. Inhibition of glycosylation with tunicamycin [72,79] had no effect on the proteolytic cleavage, suggesting that glycosylation has no role in C-terminal processing. When the glycosylation site from the C-terminal propeptide of barley lectin was mutated, pulse-chase labeling experiments demonstrated that the mutated proprotein was processed to the mature protein at least twice as fast as the glycosylated protein. In addition, the mutated proprotein was transported from the Golgi complex faster than the glycosylated protein [74]. These results point to an indirect functional role for the glycan in post-translational processing and transport of barley lectin to vacuoles.

6.3. Protein activity

Concanavalin A (ConA), the lectin from seeds of jack beans (*Canavalia ensiformis*), is also synthesized as a glycosylated precursor (pro-ConA) [81]. The mature ConA is generated by the excision of an internal glycopeptide from pro-ConA and subsequent ligation of the two resultant polypeptides [82,83]. The glycosylated pro-ConA has no sugar-binding activity but can be converted to an active lectin by deglycosylation with N-glycanase [84,85]. Jack beans are known to contain an N-glycanase that could be responsible for the observed processing event [86,87]. The conversion of a protein with no carbohydrate-binding activity to an active lectin by the removal of an oligosaccharide side-chain indicates a role for N-glycans in the regulation of protein activity. Inhibition of N-linked glycosylation with tunicamycin significantly impedes transport of pro-ConA from the ER to vacuoles [88], suggesting that only the non-lectin ConA precursor is a transport competent. A lectin with strong affinity to high-mannose-type glycans will most likely bind to N-linked glycoproteins of the ER thereby preventing the exit of pro-ConA from this compartment.

6.4. Protein folding

A function for N-linked glycans in the folding and assembly of nascent polypeptides has been suggested for mammalian and yeast glycoproteins (reviewed in [89]). The role of the intramolecular high-mannose glycans in the folding and assembly of soybean lectin polypeptides has recently been investigated [89]. Active soybean lectin is a homooligomer with a molecular mass of 120 kDa made up of four inactive subunits with molecular masses of 28 kDa and a single high-mannose oligosaccharide chain, $Man_9GlcNAc_2$, per polypeptide [15]. The purified lectin dissociated by 6 M guanidine hydrochloride into its inactive subunits was quantitatively reconstituted to the active tetrameric structure by simple dilution. However, neither the activity nor the tetrameric structure was regained when the lectin was chemically deglycosylated or when the glycosylated lectin was reconstituted in the presence of 100 μM $Man_9GlcNAc_2$. In fact, the glycan at this concentration gradually dissociated the native lectin into subunits. A high-mannose glycan with five Man residues was less efficient in the inhibition of lectin reassembly, and a diantennary complex glycan with terminal *N*-acetyl lactosamine residues had no inhibitory activity. These findings are consistent with the idea that soybean lectin polypeptides possess particular sites with affinity for the intramolecular high-mannose oligosaccharide chains, and that binding of the sugar chains to these sites is essential both for the proper folding of the subunit polypeptides and for subunit assembly [89].

A role of the complex-type N-linked glycan of the *Erythrina corallodendron* lectin (EcorL) in polypeptide folding has been studied by Shaanan et al. [90]. Legume lectins, which in general are highly homologous in their primary amino acid sequences, are also usually very similar in their three-dimensional structures. Surprisingly, despite these similarities, most lectins have different carbohydrate-binding specificities. The three-dimensional structure of EcorL, which is specific for Gal and its derivatives, was determined crystallographically [90] and was shown to be similar to the known structures of concanavalin A (ConA) and pea lectin (PL); these are both specific for Man and Glu. EcorL differs from ConA and PL by having a β-1,2-xylosylated and α-1,3-fucosylated complex glycan attached to Asn17 of each EcorL monomer. In contrast, mature ConA and PL are not glycosylated. The crystal structure analysis of EcorL shows that the bulky, complex heptasaccharides force the EcorL dimers into a drastically different quaternary structure, altering the structure of the carbohydrate binding site so that it becomes specific for Gal instead of for Man and Glc.

6.5. Protein stability

When suspension-cultured cells of sycamore (*Acer pseudoplatanus*) were treated with inhibitors of glycoprotein processing, such as castanospermine and deoxymanno-jirimycin, the formation of complex glycans was prevented but only the secretion of glycoproteins with high-mannose-type oligosaccharides was unchanged, indicating that complex glycans are not necessary for protein stability [91]. In contrast, when the sycamore cells were treated with tunicamycin, the accumulation of newly synthesized non-glycosylated proteins in the culture medium was inhibited [92]. The apparent inhibition of secretion suggests that attachment of high-mannose-type oligosaccharides is required for the accumulation of secreted proteins in the extracellular compartment.

When suspension-cultured cells of carrot (*Daucus carota*) were treated with tuni-camycin, the accumulation of cell wall β-fructosidase, which is usually glycosylated with one high-mannose-type and two complex-type oligosaccharides [36], was also inhibited. In the presence of tunicamycin, non-glycosylated β-fructosidase is synthe-sized in the ER. The results of pulse-chase experiments suggested that the protein is secreted in the same way as glycosylated β-fructosidase, but is degraded during the last stage of secretion. In analogy to the fate of non-glycosylated cell wall β-fructosidase, it is likely that the non-glycosylated sycamore proteins are also degraded during their intracellular transport prior to secretion [91].

Expression of an intron less phaseolin gene in transgenic tobacco led to the accumu-lation of the N-glycosylated bean storage protein in tobacco seeds. A gene from which the two glycosylation sites had been removed was also transformed into tobacco. Analysis of seeds containing non-glycosylated phaseolin revealed a substantial reduc-tion in the total amount of intact phaseolin and an increase in processing products, suggesting a higher rate of proteolysis of the mutated proteins [76].

6.6. Plant development

The effect of tunicamycin on growth and the synthesis of DNA and protein was studied in a suspension culture of *Catharanthus rosea* [93]. In the presence of 0.1–1 μg/ml of the inhibitor of protein N-glycosylation, cell division and DNA synthesis stopped in cells which had been proliferating logarithmically, but protein synthesis continued. Cytophotometric determination of the nuclear DNA content showed that the cell cycle arrest had occurred in G1. Metabolic labeling of cells with glucosamine or mannose was also inhibited. When tunicamycin-blocked cells were resuspended in new medium without inhibitor, the cell cycle arrest was released. The results indicate that one or more N-linked glycoproteins are needed for the plant cell to pass through the G1 phase, as was postulated for animal [94,95] and yeast cells [96].

Carrot somatic embryogenesis is a well-characterized model system of early plant development [97]. A subset of proliferating suspension-cultured cells (preembryonic masses) cultivated in the presence of the synthetic plant auxin 2,4-dichlorophenoxy-acetic acid (2,4-D) undergo somatic embryogenesis when the auxin is omitted. Carrot somatic embryogenesis is inhibited by tunicamycin at a very early stage, before globular stage embryos are formed. Somatic embryogenesis of tunicamycin-blocked cultures was completely rescued by addition of extracellular, correctly glycosylated proteins from untreated embryo cultures [98]. These results provide evidence that the hormone-controlled release of extracellular N-linked glycoproteins is essential for carrot somatic embryogenesis. Fractionation of the extracellular proteins identified an isoenzyme of cationic peroxidase that could restore embryogenesis of tunicamycin-treated carrot cells to more than 50% of that in untreated controls [99]. Peroxidases may reduce cell wall plasticity [100] by oxidative cross-linking of cell wall components, and this seems to be essential for morphogenesis [101]. The peroxidase identified is an N-linked glyco-protein which is either not active or is degraded when secreted without glycans.

At non-permissive temperature (32°C), the temperature-sensitive carrot cell variant ts11 is arrested at the globular stage [102]. The defect in ts11 can be fully comple-mented by the addition of extracellular wild-type proteins, which have much more Fuc than the extracellular proteins synthesized under non-permissive conditions. Bio-chemical characterization show that ts11 is not able to perform proper glycosylation at the non-permissive temperature and suggests that the activity of certain extracellular proteins essential for the transition of globular- to heart-stage somatic embryos depends on correct modification of their oligosaccharide side-chains. Recently, the rescue activity was identified as an N-glycosylated endochitinase, which may play an impor-tant role in protoderm formation [103].

6.7. Signal molecules

Free glycans of the high-mannose- and the complex-type have been isolated from the medium of suspension-cultured cells of white campion (*Silene alba*) [104] and from ripening tomato fruits [105]. These free glycans were interpreted to be possible break-down products of N-linked glycoproteins. The 'vacuole-type' complex glycan Xyl-$Man_3FucGlcNAc_2$ isolated from white campion cells acted as a growth factor that stimulated the elongation of flax hypocotyls at lower concentrations than the synthetic auxin 2,4-D [106]. At higher concentrations, the glycan showed an anti-auxin effect. The same glycan or a glycan of the high-mannose-type with five Man residues coinfil-trated into mature green tomato fruit at concentrations of about 1 ng per gram fresh weight with 40 µg per gram fresh weight Gal, a level of Gal insufficient to promote ripening, stimulated ripening as measured by red coloration and ethylene production [107]. Thus, some of the plant N-linked glycans appear to function as signal molecules.

6.8. Analysis of a plant N-glycosylation mutant

A polyclonal antibody against Xyl β-1,2-linked to the core β-Man residue of complex N-linked glycans [108], which is a hall mark of complex glycans from different plant species, was used to screen for N-glycosylation mutants in a plant population derived from chemically mutagenized seeds of *Arabidopsis thaliana*. Two plants were identified with polypeptides completely lacking Xyl residues [109]. Genetic and biochemical analyses showed that the two mutants were allelic. The mutant plants synthesize N-linked glycoproteins with high-mannose-type oligosaccharides only. The biosynthesis of the oligosaccharide precursor $Glc_3Man_9GlcNAc_2$ and its transfer to polypeptides appears to be normal, whereas the conversion to complex-type glycans is blocked at the level of high-mannose-type glycans with five Man residues. This finding, which indicates a mutation in *N*-acetylglucosaminyltransferase I, was confirmed by direct enzyme analysis. The mutant plants are able to complete normal development, suggesting that the complex glycans are not essential for developmental processes under optimal growth conditions. The complex glycans may only be needed on a small subset of glycoproteins which are synthesized when plants are grown under non-laboratory conditions and subjected to particular biotic or abiotic stresses. The analysis of mutant plants grown under non-laboratory conditions is in progress. Preliminary results confirm the hypothesis and indicate that plants lacking complex N-linked glycans are less resistant to specific stresses than wild-type plants. Leaves of mutant plants grown at 8°C have a slightly reduced protein content and a sick appearance. Mutant seeds germinating on non-sterilized soil have a reduced survival rate compared to wild-type seeds. Furthermore, only 66–75% of the mutant seedlings transferred from sterile agar into soil survive, whereas 100% of the wild-type seedlings develop to mature plants. Leaves of mutant plants grown at 37°C appeared to have necrotic spots ([109]; Sturm, von Schaewen and Chrispeels, unpublished results).

7. Concluding remarks

The past 10 years of research have greatly increased our knowledge about N-glycosylation of plant proteins and at the same time has generated numerous challenging questions to be answered in the future. Because nothing is known about the glycosylation of plant membrane proteins, it is quite obvious that more structural work has to be done. In addition, to support the hypothesis that glycoproteins from different subcellular compartments have different complex glycan structures, glycoproteins from different compartments of the same cell type must be characterized. The effect of N-linked glycans on the physicochemical properties of glycoproteins is only poorly understood and needs to be studied in more detail. Furthermore, the role of N-linked glycans in protein folding in the ER and thereby the recognition of these proteins as correctly folded or misfolded should be analyzed.

A challenging task for the future is the purification of the glycosyltransferases and glycosidases involved in the biosynthesis and modification of N-linked glycans, which will allow the generation of specific antibodies and partial amino acid sequencing. The antibodies and the partial amino acid sequences could lead to the isolation of the genes and the study of their regulation. Antibodies against the purified enzymes will lead to their subcellular localization, e.g. in different Golgi cisternae. These localization data, together with the amino acid sequences deduced from the genes, will enable us to study how the different enzymes are targeted to specific endomembrane subcompartments. Furthermore, detailed enzymatic characterization of the glycosyltransferases and glycosidases may give us insight how a complex pathway such as protein glycosylation operates.

The search for the function of protein glycosylation has indicated that N-linked glycans are essential for plants, but has not yet demonstrated a role for specific complex glycan structures. The absence of a phenotype of the *Arabidopsis* mutant lacking complex glycans when grown under normal laboratory conditions suggests that in most cases high-mannose-type oligosaccharides are sufficient for fulfilling the role of N-linked glycans. Therefore, it will be an interesting task to identify the minimal structure satisfying the role of N-linked glycans by screening for more glycosylation mutants.

In general, protein glycosylation seems to be vital for the stability of secreted plant proteins. N-glycans also seem to have a role in the activation of proteins. In addition, N-linked glycoproteins appear to be essential for important cellular processes such as the maintenance of the cell cycle and the formation of embryos. The identification of the glycoproteins involved is an important task for the future. The degradation of the plant N-linked glycoproteins is also only poorly understood. The finding that glycans released from their proteins may function as signal molecules in plant development needs to be followed up in the future. Furthermore, the possible function of the highly antigenic and allergic plant-specific residues of the complex glycans in the protection of plants and their seeds against herbivores should be considered.

Acknowledgment

I am grateful to my collaborators Maarten J. Chrispeels, Alan D. Elbein, Loic Faye, Kenneth D. Johnson, Antje von Schaewen, Alessandro Vitale, and Johannes F.G. Vliegenthart for their valuable experimental contributions and the countless stimulating discussions. I also thank Roland Beffa, Jürg Bilang, Maarten Chrispeels, and Pat King for critically reading the manuscript.

Abbreviations

Asn	asparagine
ConA	concanavalin A
ER	endoplasmic reticulum
Gal	galactose
GlcNAc	*N*-acetylglucosamine

Man	mannose
Ser	serine
Thr	threonine
ts	temperature sensitive
Xyl	xylose

References

1. Kobata, A. (1992) Eur. J. Biochem. 209, 483–501.

2. Montreuil, J. (1984) Biol. Cell 51, 115–131.

3. Van Kuik, J.A., Van Halbeek, H., Kamerling, J.P. and Vliegenthart, J.F.G. (1985) J. Biol. Chem. 260, 13984–13988.

4. Van Kuik, J.A., Sijbesma, R.P., Kamerling, J.P., Vliegenthart, J.F.G. and Wood, E.J. (1987) Eur. J. Biochem. 169, 399–411.

5. Staudacher, E., Altmann, F., Glössl, J., März, L., Schachter, H., Kamerling, J.P., Hard, K. and Vliegenthart, J.F.G. (1991) Eur. J. Biochem. 199, 745–751.

6. Hubbard, S.C. and Ivatt, R.J. (1981) Annu. Rev. Biochem. 50, 555–583.

7. Kornfeld, R. and Kornfeld, S. (1985) Annu. Rev. Biochem. 54, 631–664.

8. Kukuruzinska, M.A., Bergh, M.L.E. and Jackson, B.J. (1987) Annu. Rev. Biochem. 56, 915–944.

9. Elbein, A.D. (1988) Plant Physiol. 87, 291–295.

10. Faye, L., Johnson, K.D., Sturm, A. and Chrispeels, M.J. (1989) Physiol. Plant. 75, 309–314.

11. Lehle, L. (1981) FEBS Lett. 123, 63–66.

12. Hori, H., James, D.W.,Jr. and Elbein, A.D. (1982) Arch. Biochem. Biophys. 215, 12–21.

13. Lehle, L. and Tanner, W. (1983) Biochem. Soc. Trans. 11, 568–574.

14. Faye, L., Johnson, K.D. and Chrispeels, M.J. (1986) Plant Physiol. 81, 206–211.

15. Dorland, L., Van Halbeek, H., Vliegenthart, J.F.G., Lis, H. and Sharon, N. (1981) J. Biol. Chem. 256, 7708–7711.

16. Neeser, J.-R., Del Vedovo, S., Mutsaers, J.H.G.M. and Vliegenthart, J.F.G. (1985) Glycoconjugate J. 2, 355–364.

17. Hase, S., Koyama, S., Daiyasu, H., Takemoto, H., Hara, S., Kobayashi, Y., Kyogoku, Y. and Ikenaka, T. (1986) J. Biochem. 100, 1–10.

18. Kitagaki-Ogawa, H., Matsumoto, I., Seno, N., Takahashi, N., Endo, S. and Arata, Y. (1986) Eur. J. Biochem. 161, 779–785.

19. Ashford, D., Dwek, R.A., Welpy, J.K., Amatayakul, S., Homans, S.W., Lis, H., Taylor, G.N., Sharon, N. and Rademacher, T.W. (1987) Eur. J. Biochem. 166, 311–320.

20. Fournet, B., Leroy, Y., Wieruszeski, J.-M., Montreuil, J. and Poretz, R.D. (1987) Eur. J. Biochem. 166, 321–324.

21. Sturm, A., Van Kuik, J.A., Vliegenthart, J.F.G. and Chrispeels, M.J. (1987) J. Biol. Chem. 262, 13392–13403.

22. Kimura, Y., Hase, S., Kobayashi, Y., Kyogoku, Y., Ikenaka, T. and Funatsu, G. (1988) J. Biochem. 103, 944–949.

23. Kimura, Y., Hase, S., Kobayashi, Y., Kyogoku, Y., Ikenaka, T. and Funatsu, G. (1988) Biochim. Biophys. Acta 966, 248–256.

24. Bouwstra, J.B., Spoelstra, E.C., De Waard, P., Leeglang, B.R., Kamerling, J.P. and Vliegenthart, J.F.G. (1990) Eur. J. Biochem. 190, 113–122.

25. Capon, C., Piller, F., Wieruszeski, J.-M., Leroy, Y. and Fournet, B. (1990) Carbohydr. Res. 199, 121–127.

26. Ramires-Soto, D. and Poretz, R.D. (1991) Carbohydr. Res. 213, 27–36.
27. Yamaguchi, H., Funaoka, H. and Iwamoto, H. (1992) J. Biochem. 111, 388–395.
28. Sturm, A., Bergwerff, A.A. and Vliegenthart, J.F.G. (1992) Eur. J. Biochem. 204, 313–316.
29. Debray, H., Wieruszeski, J.-M., Strecker, G. and Franz, H. (1992) Carbohydr. Res. 236, 135–143.
30. Wantyghem, J., Platzer, N., Giner, M., Derappe, C. and Goussault, Y. (1992) Carbohydr. Res. 236, 181–193.
31. Takahashi, N., Hotta, T., Ishihara, H., Mori, M., Tejima, S., Bligny, R., Akazawa, T., Endo, S. and Arata, Y. (1986) Biochem. 25, 388–395.
32. D'Andrea, G., Bouwstra, J.B., Kamerling, J.P. and Vliegenthart, J.F.G. (1988) Glycoconjugate J. 5, 151–157.
33. Takayama, S., Isogai, A., Tsukamoto, C., Shiozawa, H., Ueda, Y., Hinata, K., Okazaki, K., Koseki, K. and Suzuki, A. (1989) Agric. Biol. Chem. 53, 713–722.
34. Hayashi, M., Tsuru, A., Mitsui, T., Takahashi, N., Hanzawa, H., Arata, Y. and Akazawa, T. (1990) Eur. J. Biochem. 191, 287–295.
35. Takahashi, N., Hitotsuya, H., Hanzawa, H., Arata, Y. and Kurihara, Y. (1990) J. Biol. Chem. 265, 7793–7798.
36. Sturm, A. (1991) Eur. J. Biochem. 199, 169–179.
37. Woodward, J.R., Craik, D., Dell, A., Khoo, K.-H., Munro, S.L.A., Clarke, A.E. and Bacic, A. (1992) Glycobiology 2, 241–250.
38. Montfort, W., Villafranca, J.E., Monzingo, A.F., Ernst, S.R., Katzin, B., Rutenber, E., Xuong, N.H., Hamlin, R. and Robertus, J.D. (1987) J. Biol. Chem. 262, 5398–5403.
39. Theerasilp, S., Hitotsuya, H., Nakajo, S., Nakaya, K., Nakamura, Y. and Kurihara, Y. (1989) J. Biol. Chem. 264, 6655–6659.
40. Hoffman, L.M. and Donaldson, D.D. (1985) EMBO J. 4, 883–889.
41. Sturm, A. and Chrispeels, M.J. (1986) Plant Physiol. 81, 320–322.
42. Faye, L., Sturm, A., Bollini, R., Vitale, A. and Chrispeels, M.J. (1986) Eur. J. Biochem. 158, 655–661.
43. Bollini, R., Vitale, A. and Chrispeels, M.J. (1983) J. Cell Biol. 96, 999–1007.
44. Slightom, J.L., Sun, S.M. and Hall, T.C. (1983) Proc. Natl. Acad. Sci. USA 80, 1897–1901.
45. Pollack, L. and Atkinson, P.H. (1983) J. Cell Biol. 97, 293–300.
46. Kaushal, G.P. and Elbein, A.D. (1986) Plant Physiol. 82, 748–752.
47. Kaushal, G.P. and Elbein, A.D. (1986) Plant Physiol. 81, 1086–1091.
48. Hori, H., Kaushal, G.P. and Elbein, A.D. (1985) Plant Physiol. 77, 840–846.
49. Miernyk, J.A. and Riedell, W.E. (1991) Phytochem. 30, 2865–2867.
50. Sturm, A., Johnson, K.D., Szumilo, T., Elbein, A.D. and Chrispeels, M.J. (1987) Plant Physiol. 85, 741–745.
51. Sturm, A., Chrispeels, M.J., Wieruszeski, J.M., Strecker, G. and Montreuil, J. (1987) in: J. Montreuil, A. Verbert, G. Spik and B. Fournet (Eds.), Glycoconjugates — Proceedings of the 9th International Symposium on Glycoconjugates, Lille, France, p. A107.
52. Szumilo, T., Kaushal, G.P. and Elbein, A.D. (1986) Arch. Biochem. Biophys. 247, 261–271.
53. Kaushal, G.P., Pastuszak, I., Hatanaka, K. and Elbein, A.D. (1990) J. Biol. Chem. 265, 16271–16279.
54. Szumilo, T., Kaushal, G.P., Hori, H. and Elbein, A.D. (1986) Plant Physiol. 81, 383–389.
55. Kaushal, G.P., Szumilo, T., Pastuszak, I. and Elbein, A.D. (1990) Biochemistry 29, 2168–2176.
56. Szumilo, T., Kaushal, G.P. and Elbein, A.D. (1986) Biochem. Biophys. Res. Commun. 134, 1395–1403.
57. Szumilo, T., Kaushal, G.P. and Elbein, A.D. (1987) Biochemistry 26, 5498–5505.
58. Johnson, K.D. and Chrispeels, M.J. (1987) Plant Physiol. 84, 1301–1308.

540

59. Tezuka, K., Hayashi, M., Ishihara, H., Akazawa, T. and Takahashi, N. (1992) Eur. J. Biochem. 203, 401–413.
60. Kimura, Y., Hase, S., Kobayashi, Y., Kyogoku, Y., Funatsu, G. and Ikenaka, T. (1987) J. Biochem. 101, 1051–1054.
61. Lainé, A.-C., Gomord, V. and Faye, L. (1991) FEBS Lett. 295, 179–184.
62. Zhang, G.F. and Staehelin, L.A. (1992) Plant Physiol. 99, 1070–1083.
63. Staehelin, L.A., Driouich, A., Giddings, T.H. and Zhang, G.F. (1993) J. Cell Biol. Supplement 17A, 4.
64. Olden, K., Parent, J.B. and White, S.L. (1982) Biochim. Biophys. Acta 650, 209–232.
65. Vitale, A. and Chrispeels, M.J. (1992) Bioassays 14, 151–160.
66. Kurosaki, F., Tokitoh, Y., Morita, M. and Nishi, A. (1989) Plant Sci. 65, 39–43.
67. Griffiths, G., Hoflack, B., Simons, K., Mellman, I. and Kornfeld, S. (1988) Cell 52, 329–341.
68. Gaudreault, P.-R. and Beevers, L. (1984) Plant Physiol. 76, 228–232.
69. Vitale, A. and Chrispeels, M.J. (1984) J. Cell Biol. 99, 133–140.
70. Bollini, R., Ceriotti, A., Daminati, M.G. and Vitale, A. (1985) Physiol. Plant. 65, 15–22.
71. Faye, L., Greenwood, J.S., Herman, E.M., Sturm, A. and Chrispeels, M.J. (1988) Planta 174, 271–282.
72. Sticher, L., Hinz, U., Meyer, A.D. and Meins, F.,Jr. (1992) Planta 188, 559–565.
73. Voelker, T.A., Herman, E.M. and Chrispeels, M.J. (1989) Plant Cell 1, 95–104.
74. Wilkins, T.A., Bednarek, S.Y. and Raikhel, N.V. (1990) Plant Cell 2, 301–313.
75. Sonnewald, U., von Schaewen, A. and Willmitzer, L. (1990) Plant Cell 2, 345–355.
76. Bustos, M.M., Kalkan, F.A., VandenBosch, K.A. and Hall, T.C. (1991) Plant Mol. Biol. 16, 381–395.
77. LaFayette, P.R. and Travis, R.L. (1989) Plant Physiol. 89, 299–304.
78. Nakamura, K. and Matsuoka, K. (1993) Plant Physiol. 101, 1–5.
79. Mansfield, M.A., Peumans, W.J. and Raikhel, N.V. (1988) Planta 173, 482–489.
80. Shinshi, H., Wenzel, H., Neuhaus, J.-M., Felix, G., Hofsteenge, J. and Meins, F.,Jr. (1988) Proc. Natl. Acad. Sci. USA 85, 5541–5545.
81. Herman, E.M., Shannon, L.M. and Chrispeels, M.J. (1985) Planta 165, 23–29.
82. Bowles, D.J., Marcus, S.E., Pappin, D.J.C., Findlay, J.B.C., Eliopoulos, E., Maycox, P.R. and Burgess, J. (1986) J. Cell Biol. 102, 1284–1297.
83. Chrispeels, M.J., Hartl, P.M., Sturm, A. and Faye, L. (1986) J. Biol. Chem. 261, 10021–10024.
84. Sheldon, P.S. and Bowles, D.J. (1992) EMBO J. 11, 1297–1301.
85. Min, W., Dunn, A.J. and Jones, D.H. (1992) EMBO J. 11, 1303–1307.
86. Sugiyama, K., Ishihara, H., Tejima, S. and Takahashi, N. (1983) Biochem. Biophys. Res. Commun. 112, 155–160.
87. Yet, M.-G. and World, F. (1988) J. Biol. Chem. 263, 118–122.
88. Faye, L. and Chrispeels, M.J. (1987) Planta 170, 217–224.
89. Nagai, K. and Yamaguchi, H. (1993) J. Biochem. 113, 123–125.
90. Shaanan, B., Lis, H. and Sharon, N. (1991) Science 254, 862–866.
91. Driouich, A., Gonnet, P., Makkie, M., Lainé, A.-C. and Faye, L. (1989) Planta 180, 96–104.
92. Faye, L. and Chrispeels, M.J. (1989) Plant Physiol. 89, 845–851.
93. Ettlinger, C., Schindler, J. and Lehle, L. (1986) Planta 168, 101–105.
94. Nishikawa, Y., Yamamoto, Y., Kaji, K. and Mitsui, H. (1980) Biochem. Biophys. Res. Commun. 97, 1296–1303.
95. Savage, K.E. and Baur, P.S. (1983) J. Cell Sci. 64, 295–306.
96. Arnold, E. and Tanner, W. (1982) FEBS Lett. 148, 49–53.
97. Sung, Z.R., Fienberg, A., Chorneau, R., Borkird, C., Furner, I. and Smith, J. (1984) Plant Mol. Biol. Rep. 2, 3–14.

98. de Vries, S.C., Booij, H., Janssens, R., Vogels, R., Saris, L., Lo Schiavo, F., Terzi, M. and van Kammen, A. (1988) Genes Dev. 2, 462–576.

99. Cordewener, J., Booij, H., van der Zandt, H., van Engelen, F., van Kammen, A. and de Vries, S.C. (1991) Planta 184, 478–486.

100. Goldberg, R., Imberty, A., Liberman, M. and Prat, R. (1986) in: H. Greppin, C. Penel and T. Gaspar (Eds.), Molecular and Physiological Aspects of Plant Peroxidases. University of Geneva, Switzerland, pp. 209–220.

101. Fry, S.C. (1990) in: H.J.J. Nijkamp, L.H.W. van der Plas and J. van Aartrijk (Eds.), Progress in Plant Cellular and Molecular Biology. Kluwer Academic Publishers, Dordrecht, pp. 504–513.

102. Lo Schiavo, F., Giuliano, G., de Vries, S.C., Genga, A., Bollini, R., Pitto, L., Cozzani, F., Nuti-Ronchi, V. and Terzi, M. (1990) Mol. Gen. Genet. 223, 385–393.

103. De Jong, A., Cordewener, J., Lo Schiavo, F., Terzi, M., Vandekerckhove, J., van Kammen, A. and de Vries, S.C. (1992) Plant Cell 4, 425–433.

104. Priem, B., Solokwan, J., Wieruszeski, J.-M., Strecker, G., Nazih, H. and Morvan, H. (1990) Glycoconjugate J. 7, 121–131.

105. Priem, B., Gitti, R., Bush, C.A. and Gross, K.C. (1993) Plant Physiol. in press.

106. Priem, B., Morvan, H. and Hafez, A.M.A. (1990) C. R. Acad. Sci. Paris 311, 411–416.

107. Priem, B. and Gross, K.C. (1992) Plant Physiol. 98, 399–401.

108. Laurière, M., Laurière, C., Chrispeels, M.J., Johnson, K.D. and Sturm, A. (1989) Plant Physiol. 90, 1182–1188.

109. von Schaewen, A., Sturm, A., O'Neill, J. and Chrispeels, M.J. (1993) Plant Physiol. 102, 1103–1118.

J. Montreuil, H. Schachter and J.F.G. Vliegenthart (Eds.), *Glycoproteins*

Protein glycosylation in insects

LEOPOLD MÄRZ, FRIEDRICH ALTMANN, ERIKA STAUDACHER and
VIKTORIA KUBELKA

*Institut für Chemie, Universität für Bodenkultur,
Gregor-Mendelstrasse 33, A-1180 Vienna, Austria*

1. Introduction

While a wealth of information exists on the structures, the biosynthesis and catabolism, and the function of mammalian glycoprotein glycans — these volumes provide an impressive documentation — our knowledge of those from invertebrate sources, in particular from insects, is at best fragmentary. This is indeed surprising, as insects have over decades served as classical biochemical models; they provided important insights which often proved applicable to mammalian systems [1]. Among other new developments, it was especially the availability of the now well-established baculovirus vector system in combination with the technology of insect cell culture, which has stimulated the quest for their protein glycosylation capacity. How important this aspect has become, can be deduced from the ever growing list of glycoproteins, which have in the meantime been expressed in insect cells.

2. From early data to current views

2.1. Bits and pieces

In a very early investigation, Warren [2] had searched for the occurrence of sialic acids in insects and found no indication. In a series of pioneering studies, Butters et al. [3–5] drew a first picture of the overall protein glycosylation potential of cultured insect cells, Mos 20A from mosquito (*Aedes aegypti*) larvae. Lectin binding studies indicated the presence of Man, Fuc and GalNAc residues and again the absence of sialic acids. Isolated plasma membrane glycoprotein fractions were investigated for their affinity towards immobilised Concanavalin A and Soybean Agglutinin, analysed for their monosaccharide compositions and treated with endo- and exoglycosidases. The authors concluded that in their basic structural features, especially in the presence of the $Man_3GlcNAc_2$ core pentasaccharide, these glycans resembled those of mammalian origin. The fact that the incorporation of radioactive GalNAc, in contrast to that of GlcNAc, was not affected by tunicamycin [5], together with the susceptibility of the

GalNAc content to β-elimination, suggested this monosaccharide to be part of O-linked carbohydrate. No evidence could be found for the presence of sialic acids in the GalNAc-containing glycans [4]. As only 'insignificant levels' of sialyl-, galactosyl-, and N-acetylglucosaminyltransferases were detectable in the mosquito cells, it was proposed that they do not possess the capacity to synthesise complex type N-glycans [5].

2.2. Do insects make high mannose N-glycans only and are they able to O-glycosylate?

2.2.1. N-Glycans

The 'high-mannose only' hypothesis was subsequently adapted by the scientific community and still represents a widely held belief. Indeed, Hsieh and Robbins [6] demonstrated that the complex type N-glycan structures of vertebrate cell-derived Sindbis Virus were replaced by $Man_3GlcNAc_2$, when mosquito (Aedes albopictus) C6/36 cells were used for virus propagation; earlier, Stollar et al. [7] had not been able to detect sialic acids in a Sindbis Virus preparation derived from Aedes albopictus cells. Subsequently, Ryan et al. isolated a glycopeptide from the major hemolymph protein of Manduca sexta larvae and identified, by endoglycosidase treatment and 250 MHz ^{1}H-NMR spectroscopy, the well-known $Man_9GlcNAc_2$ structure [8].

Recently, Williams et al. [9] released oligosaccharides from Drosophila melanogaster larval membrane glycoproteins and a purified larval serum protein (LSP2) by hydrazinolysis. High-resolution gel permeation chromatography, together with exoglycosidase treatment and controlled acetolysis, identified high-mannose type structures only, ranging from Man_9- to $Man_2GlcNAc_2$. Fuc was found α-1,6-linked to the asparagine-bound GlcNAc of Man_3- and Man_2-oligosaccharides. Despite the carbohydrate-based immunological cross-reactivity between Drosophila tissue and horseradish peroxidase [10,11], neither α-1,3-fucosylation of this residue nor the presence of Xyl, two substitutions typical for plants [12] could be detected in the Drosophila glycoprotein glycans.

2.2.2. O-Glycans

Apart from the very preliminary findings by Butters et al. [5], additional evidence for the occurrence of O-glycans on insect glycoproteins was presented by Kress [13], who conducted studies with tunicamycin and β-hydroxynorvaline on Drosophila virilis salivary gland glue proteins to suggest that glycosylation occurs at threonine residues, and by Campbell et al. [14]. Papilin, a proteoglycan-like glycoprotein isolated from culture media of Drosophila K_c cells, carries a large number of sulfated O-glycans, which could be β-eliminated. Neither the linkage sugar nor the composition of these side chains were reported [14].

In addition to the identification of partly fucosylated high-mannose type N-glycans, Williams et al. also presented preliminary evidence for the occurrence of small anionic, partially neuraminidase-sensitive oligosaccharide species in the hydrazinolysate of D. melanogaster embryonic preparations [9]; although no speculations as to the possible linkage of these oligosaccharides to protein were presented in this report, it is obvious that they are likely candidates for O-glycans.

The presence of sialylated, O-linked oligosaccharides was also indicated for locust (*Locusta migratoria*) vitellogenin [15]. As these observations were entirely based on lectin blotting and digestions with endo- and exoglycosidases, no structural conclusions could be drawn. In contrast to previously published work, Roth et al. [16] were able to detect sialic acids throughout embryonic and post-embryonic development of *Drosophila* using *Limax flavus* lectin-gold histochemical labeling, immunoblotting with a polysialic acid-specific monoclonal antibody and gas–liquid chromatography–mass spectrometric identification of N-acetylneuraminic acid after isolation from embryos. Evidence for the occurrence of sialic acid as a constituent of insect glycoproteins was also obtained in two other studies [17,18]. No attempts were made in any of these studies to define, whether sialylation occurs on N- and/or O-glycans.

Non-sialylated O-glycans made up of GalNAc or Gal-GalNAc disaccharides, in place of sialylated O-glycans of the native counterparts, were identified in recombinant, insect cell-produced glycoproteins by Thomsen et al. [19], Chen et al. [20], Wathen et al. [21], Grabenhorst et al. [22] and Sugyiama et al. [23].

As cytosolic and nucleoplasmic membrane proteins of many eukaryotic organisms are known to contain O-linked GlcNAc residues, Kelly and Hart [24] used FITC-WGA and bovine milk galactosyltransferase together with UDP-[^{3}H]Gal to detect large amounts of terminal GlcNAc residues on *Drosophila* polytene chromosomes, where they exist as single, protein-linked, monosaccharides. Their function is still subject to speculation [25].

2.3. Biosynthetic aspects

2.3.1. Lipid intermediates

The evidence so far available suggests that the precursor oligosaccharide assembly in insect cells, although not yet fully elucidated, proceeds along the same routes as has been worked out for mammalian organisms, involving dolichyl phosphates as lipid carriers.

Quesada Allue and Belocopitow [26] used extracts from the fruit fly (*Ceratitis capitata*) to transfer Glc, Man and GlcNAc from their respective nucleotide precursors to endogenous polyprenyl phosphate. The formation of a radiolabeled chitobiosyl lipid was demonstrated as well as the incorporation of [^{14}C]-Man into lipid-linked oligosaccharides. The labeled monosaccharides were finally found in a protein fraction, indicating a role of lipid-bound oligosaccharide precursors in insect protein N-glycosylation.

Another piece of evidence was added by Butters et al. [5] during their already mentioned work with *Aedes aegypti* larval cell extracts; they detected Man- and GlcNAc-P-Dol; the formation of the latter could be inhibited by adding tunicamycin. Pulse chase experiments with *Aedes albopictus* C6/36 mosquito cells led to the identification of a large oligosaccharide linked to a lipid precursor [6]. After delipidation by mild acid hydrolysis, the oligosaccharide co-migrated with Glc$_3$Man$_9$GlcNAc$_2$ on gel filtration. After its transfer to protein, the tetradecasaccharide underwent a series of modifications to yield a set of high-mannose type chains with Man$_3$GlcNAc$_2$ as the most extensively trimmed variant. All these data were taken as an indication that the precursor processing pathway leading to protein N-glycosylation in insects should consist of a very similar, if not identical, series of reactions as the one elucidated for mammalian organisms.

Other dolichol derivatives, among them dolichyl phosphate, Glc-P-Dol and oligosaccharylpyrophosphoryldolichol, were detected in the course of short term and steady state labeling experiments using the embryonic *Drosophila* K_c cell line [27]. Low levels of the early intermediates of glycoprotein N-glycan biosynthesis pointed at a rapid metabolic flux through these stages, probably to prevent accumulation and transfer of incomplete oligosaccharides to protein. Parker et al. [28] obtained additional evidence for the similarity between insect and vertebrate N-glycan precursor processing. They isolated a lipid-linked oligosaccharide from *Drosophila melanogaster DM* 3 cells and showed, employing mammalian processing α-glucosidases and a plant α-mannosidase, that this insect precursor glycan is undistinguishable from its vertebrate counterpart. It was concluded that this intermediate is ubiquitous to eucaryotes, with the documented exception of some protozoan parasites [29].

2.3.2. The enigma of fucosylated high-mannose glycans

Our knowledge of the glycosyltransfer capacity of insects is particularly poor. From the structural data reviewed so far it seems reasonable to conclude that insects lack the capacity to elongate N-glycan chains beyond the $Man_3GlcNAc_2$ core pentasaccharide to generate complex or hybrid N-glycans.

The recent demonstration of Fuc residues on high-mannose type N-glycans without the concomitant presence of outer-arm GlcNAc poses an interesting question, as it is still held that fucosylation of the asparagine-bound GlcNAc (GlcNAc-1) requires the prior action of GlcNAc-transferase I [30]. However, this phenomenon is not restricted to insect glycoproteins, and such structures were reported in mammalian glycoproteins by Kozutsumi et al. [31] and Mutsaers et al. [32]. Some years earlier, Vitale and Chrispeels [33] had demonstrated in cotyledons of the common bean *Phaseolus vulgaris* L. that an outer-arm GlcNAc is added during the maturation of the glycoprotein lectin phytohemagglutinin in the Golgi complex, but is subsequently removed in the protein bodies. The authors concluded that this GlcNAc residue plays a "transient" role in the glycan processing of phytohemagglutinin. GlcNAc-transferase I was indeed found by Szumilo et al. [34] in mung bean seedlings.

The idea of a transiently present GlcNAc could reconcile the fucosylation of high-mannose type N-glycans in insect glycoproteins with the specificity of fucosyltransferases acting upon GlcNAc-1 as it is known from vertebrates. It was the work on honeybee phospholipase A_2 [35], and on *Locusta migratoria* apolipophorin III [36], which provided first structural evidence that insects do possess the potential for transfer of GlcNAc to the non-reducing terminus. Additional light was recently shed on the issue of insect glycosyltransfer in the context of the work with insect cell lines in combination with the baculovirus vector, aiming at the expression of recombinant (glyco)proteins, and this subject will be addressed below (Section 4.5.).

3. New patterns, new insights — naturally occurring glycoproteins

It is indicative of how scattered our knowledge of protein glycosylation in insects still is that the first complete elucidations of the primary structures of the carbohydrate moieties of natural insect glycoproteins have only recently been published [35,36]. Both studies unraveled unexpected and unusual structural features.

GlcNAc(β1–2)Man(α1–6)   Fuc(α1–6)

            \\   \\

        Man(β1–4)GlcNAc(β1–4)GlcNAc

       /

GlcNAc(β1–2)Man(α1–3)

            GDP-Fuc | honeybee venom gland
               GDP ↵ | homogenate

GlcNAc(β1–2)Man(α1–6)   ↓   Fuc(α1–6)

            \\   \\

        Man(β1–4)GlcNAc(β1–4)GlcNAc

       /   /

GlcNAc(β1–2)Man(α1–3)   **Fuc(α1–3)**

3.1. Honeybee venom phospholipase A2 (PLA)

3.1.1. Difucosylation of the Asn-bound GlcNAc
Staudacher et al. [37] had shown that honeybee venom glands possess the capacity to convert an N-glycan acceptor substrate, which was α-1,6-fucosylated at GlcNAc-1, into a difucosylated structure. The new Fuc was α-1,3-linked to the same residue (see scheme above).

Subsequently, the same authors isolated two PLA oligosaccharides by lentil (*Lens culinaris*) lectin chromatography and treatment with N-glycosidase A [38] of chymotryptic glycopeptides, followed by HPLC fractionation on aminopropyl silica [39]. The oligosaccharides were identified by 500 MHz ^{1}H-NMR to represent α-1,3-mono- and α-1,3-, α-1,6-difucosylated Man3GlcNAc2 (Table I). While α-1,6-fucosylation of GlcNAc-1, which is typical for mammalian glycoproteins, has already been demonstrated for insect glycoproteins, the occurrence of an α-1,3-linked fucose, either alone or in combination with α-1,6-linked fucose, was a remarkable new feature hitherto believed to be confined to plant glycoproteins. The functional significance of this structural element will be discussed later (Section 5.4.).

3.1.2. The GalNAc(β1–4)[Fuc(α1–3)]GlcNAc(β1–2)-moiety
To account for the entire spectrum of carbohydrate variants at the single N-glycosylation site of PLA, the oligosaccharides were liberated from the protein with N-glycosidase A and reductively aminated with 2-aminopyridine to introduce a fluorescent label.

TABLE I

N-linked oligosaccharide structures found in honeybee venom phospholipase A2 (from Ref. [35])

Structure	Man(α1–3)	Relative abundance of variants (%) Fucosylation			
		None	α-1,3-	α-1,6-	Both
Man(α1–6) \ Man(β1–4)GlcNAc(β1–4)GlcNAc / Man(α1–3) [Man(α1–3)]$_{0-1}$	–	5.0			
	+	9.0			
Fuc(α1–6)]$_{0-1}$ \ Man(α1–6) \ Man(β1–4)GlcNAc(β1–4)GlcNAc / [Man(α1–3)]$_{0-1}$	–	13.6	6.1	3.2	2.2
[Fuc(α1–3)]$_{0-1}$	+	31.8	6.7	8.6	4.2
[Fuc(α1–6)]$_{0-1}$ \ Man(α1–6) \ Man(β1–4)GlcNAc(β1–4)GlcNAc / [Fuc(α1–3)]$_{0-1}$ GalNAc(β1–4)GlcNAc(β1–2)Man(α1–3) / Fuc(α1–3)		5.6	1.8	1.3	0.9

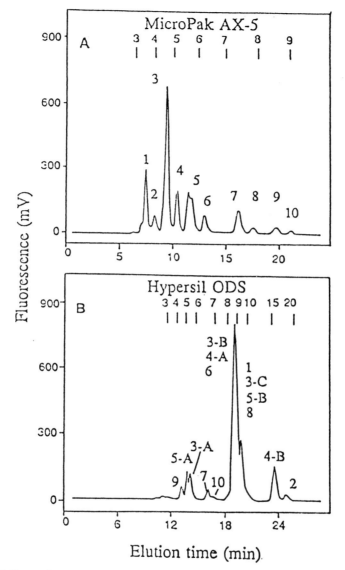

Fig. 1. 2D-HPLC mapping of oligosaccharides from honeybee PLA. 2 nmol of a mixture of the pyridylaminated oligosaccharides from PLA was separated on MicroPak-AX5 (A) and on Hypersil ODS (B). Numbers at the top indicate the elution positions of isomaltose standards (Glc$_n$). Taken from Ref. [35].

Sequential fractionation on HPLC by size and polarity (2D-HPLC) afforded fourteen oligosaccharide fractions (Fig. 1), which underwent monosaccharide and methylation analysis [35]. Ten glycans merely contained Man, Fuc and GlcNAc; their structures were identified by comparing HPLC elution shifts after treatment with α-mannosidase and α-fucosidase under linkage-selective conditions with standard compounds. The other four oligosaccharides, which together accounted for roughly 10% of the total glycan pool, contained GalNAc. 500 MHz ^{1}H-NMR spectroscopy was applied to identify a structural feature, which was until recently not known to occur in glycoproteins, i.e. the GalNAc(β1–4)[Fuc(α1–3)]GlcNAc(β1–2)-moiety [35]. Within a very short time, this trisaccharide has also been discovered in the N-glycans of human urokinase [40], recombinant human Protein C expressed in human kidney cells [41], *Schistosoma mansoni* glycoproteins [42], and also in O-glycans of sea squirt (*Styela plicata*) glycoprotein antigens [43]. Table I presents the structures of all fourteen oligosaccharides found at Asn-13 of PLA, together with their relative abundances [35].

3.2. Apolipophorin III from Locusta migratoria

Apolipophorin III is a glycoprotein from the hemolymph of the migratory locust and consists of two isoforms. Oligosaccharides were enzymatically released, and fast atom bombardment mass spectrometry, ^{1}H- and ^{31}P-NMR spectroscopy, and methylation analysis were applied to identify the following structure(s) [36]:

$$
\begin{array}{l}
[AEP-6]_{0-1} \\
\qquad\backslash \\
[GlcNAc(\beta1-2)]_{0-1}Man(\alpha1-6) \qquad\qquad\qquad Fuc(\alpha1-6) \\
\qquad\qquad\qquad\qquad\backslash \qquad\qquad\qquad\qquad\qquad | \\
[AEP-6]_{0-1} \qquad\qquad\qquad Man(\beta1-4)GlcNAc(\beta1-4)GlcNAc \\
\qquad\backslash \qquad\qquad\qquad / \\
\qquad GlcNAc(\beta1-2)Man(\alpha1-3)
\end{array}
$$

Two unusual features were discovered:
- the presence of 2-aminoethylphosphonate (AEP) at the 6-position of the α-1,6-linked Man and/or α-1,3-branch GlcNAc(β1–2) residue;
- GlcNAc residues, β-1,2-linked to the α-Man residues of the trimannosyl core, clearly the products of the action of GlcNAc transferases I and II.

4. Glycosylation potential of cultured insect cells

The use of insect cell lines in combination with the *Autographa californica* nuclear polyhedrosis virus to express proteins or glycoproteins has over the past ten years been well established; this system seems well suited for high-level protein production. Most widely used are lepidopteran cells [44], in particular *Spodoptera frugiperda Sf*-21, derived from cultured ovaries, or a clonal isolate from this line, *Sf*-9 cells [45]. More recently, cells from *Bombyx mori* and *Mamestra brassicae* and other species [44] have also been employed. A large number of (glyco)proteins, many of them of potential use in human and veterinary medicine, have in the meantime been expressed using this technology [46]. With the growing awareness of the importance to obtain, with respect to post-translational modifications and especially the glycosylation patterns, recombinant products which are as close to the native glycoprotein as possible, and as it became clear that even subtle deviations may result in adverse biological reactions [47,48], the glycosylation potential of insect cells gained increasing attention.

4.1. O-Glycans of insect cell-derived recombinant glycoproteins

Chen et al. [20] analysed the N- and O-glycans of recombinant human chorionic gonadotropin (r-hCGβ) produced in *Sf-9* cells. In contrast to the sialylated O-glycans of the native glycoprotein, the recombinant product contained only neutral disaccharides composed of Gal and GalNAc. In three other recombinant glycoproteins produced in *Sf-9* cells, Gal-GalNAc and GalNAc only were found, i.e. (a) pseudorabies virus glycoprotein gp50 [19]; a chimeric product composed of the extracellular domains of respiratory syncytial virus F and G proteins [21] and (c) human interferon-α2 [23]. These O-glycans were also found on human interleukin 2 expressed in *Sf-21* cells [22] and in this case the structure was determined to be Gal(β1–3)GalNAc. Thus, it seems that lepidopteran cells are in principle capable of O-glycosylating at the same positions as do mammalian host cells [23], but only monosaccharides and non-sialylated disaccharides were detected.

4.2. N-Glycans of recombinant glycoproteins

While the earliest reports on the use of recombinant insect cell lines simply concluded that the (glyco)proteins expressed are "antigenically, immunogenically and functionally similar to their native counterparts" [46], subsequent investigations suggested significant differences between the glycosylation potentials of insect and mammalian cells. First indications were obtained, when influenza virus hemagglutinin was expressed in *Spodoptera frugiperda* cells to afford a product with a molecular mass somewhat lower than that of the native glycoprotein [49,50]. This effect was ascribed to a changed glycosylation pattern. Similar results were reported by Greenfield et al. [51] with human epidermal growth factor receptor and by Wojchowski et al. [52], who expressed recombinant human erythropoietin in *Sf* cells. Although of smaller molecular size, the products retained full biological activity, unless the carbohydrate units were removed. Jarvis and Summers [53] probed the N-glycans of recombinantly expressed human tissue plasminogen activator (t-PA) with endo-β-*N*-acetyl-D-glucosaminidase H (Endo H) and concluded that they were of the high-mannose structural type. A subset, however, was found resistant to Endo H; in view of the specificity of this enzyme and on the basis of the data of Hsieh and Robbins [6], they postulated that this subset represented the $Man_3GlcNAc_2$ core pentasaccharide. Tunicamycin, but not the inhibitors of N-glycan processing, prevented the secretion of t-PA thereby suggesting a role of N-glycosylation in this particular event [53].

More recently, Kuroda et al. [54] undertook to compare the N-glycans of vertebrate-derived influenza virus hemagglutinin with those obtained using *Spodoptera frugiperda* cells. Chromatographic analysis of enzymatically released oligosaccharides in combination with exoglycosidase treatment revealed the presence of high mannose (Man_9- to Man_5-) chains and partially fucosylated $Man_3GlcNAc_2$ on insect cell-derived hemagglutinin. Analysis of the N-glycans of r-hCGβ expressed in *Sf-9* cells revealed partly fucosylated, truncated high-mannose type chains instead of the sialylated complex glycans of the native glycoprotein [20]. As in the case of the aforementioned influenza virus hemagglutinin, the linkage of Fuc was not determined.

Furthermore, r-hCGβ displayed non-sialylated, O-linked disaccharides (see Section 4.1.). In the case of *Sf*-9 cell-expressed FG protein, fucose was shown to be α-1,6-linked to the reducing GlcNAc of $Man_3GlcNAc_2$ [21]. Very recently, *Sf-21* derived human interleukin 2 [22] and *Sf-9* derived human interferon ω1 [55] were shown to contain α-1,6-fucosylated $Man_3GlcNAc_2$ and $Man_2GlcNAc_2$.

In a thorough study on the glycosylation of recombinant HIV-1 gp120 [56] it was found that N-glycosylation in the baculovirus system occurred at all potential sites and was site specific in terms of oligosaccharide structure, i.e. slightly more processed high-mannose glycans ($Man_{5-7}GlcNAc_2$) were detected on glycosylation sites, which in CHO-cell derived recombinant gp120 carried complex type sugar chains.

So far this simple picture correlates well with the earlier reports on insect protein glycosylation, which have been reviewed in the first part of this treatise. They support the view that insect cells lack the capacity to elongate N-glycans beyond mannosidic structures. Clearly, the findings of Davidson, Castellino and coworkers [57–59] (see Section 4.3.) that insect cells do possess the potential to synthesise complex type N-glycans, conflict with this view.

4.3. Insect cell-derived complex type N-glycans? The case of human plasminogen

Human plasminogen (hPg) exists in two glycoforms. While one carries an O-glycan only, the other (type 1) is in addition N-glycosylated at Asn-289. The carbohydrate moiety of hPg was reported to contribute to a number of effects relating to the function of this zymogen, e.g. its binding to lysin-like activator molecules, its activation and deactivation, its enzymatic activity, and its proteolytic degradability [57–60]. The N-glycan of native hPg is a partly bisialylated, biantennary, complex type chain [60]. Recombinant hPg type 1 (r-hPg) was expressed in *Spodoptera frugiperda* IPLB-SF21AE and other insect cells using a recombinant baculovirus/hPg cDNA construct [57–59].

An elaborate methodology, which chiefly consisted of high performance anion exchange chromatography, monosaccharide analysis, exoglycosidase treatments for sequencing and assignment of linkages, was used to compare the glycosylation patterns of r-hPg preparations obtained from insect cells with those of native hPg [57] and CHO cell-derived r-hPg [60].

Contrasting earlier reports and the general belief, the oligosaccharide pool isolated from insect cell-r-hPg contained a significant proportion of sialylated complex type carbohydrate, in addition to the expected high-mannose type chains, which ranged from Man_3- to $Man_9GlcNAc_2$ [57]. Furthermore, the percentage of complex type oligosaccharides was found to increase with the length of infection with the recombinant construct to almost 100% of the total pool [58], with a concomitant increase of branching and outer-arm completion. Similar results were obtained with two other lepidopteran cell lines, *Mamestra brassicae* IZD-MB0503 and *Manduca sexta* CM-1 [59].

For comparison, CHO cell-expressed r-hPg contained a variety of high-mannose structures as well as bi-, tri-, and tetraantennary complex type chains, which were all partly sialylated [60]. Several conclusions were drawn from these data:

(1) As is most pertinent to this particular review, insect cells appear under certain conditions to be able to process nascent N-glycans to sialylated, complex structures with various degrees of branching.

(2) Not only the species or the cell-type governs the glycosylation of a protein. Also a transfection-based, time-dependent activation, either at the gene or the protein level, of normally silent glycosyltransferases can take place to result in the steering of the glycosylation machinery by the recombinant construct or the heterologous protein to yield carbohydrate-structural patterns usually unseen in these cells, but characteristic for the 'authentic' glycoprotein.

Another example for such a protein-induced change of glycosyl transfer capacity was obtained by the same authors from their work on CHO-expressed r-hPg [60]. The 'CHO-typical' α-2,3-linkage of sialic acid to penultimate Gal was found in these studies to be partly replaced by the α-2,6-linkage, which was hitherto not detected in CHO cells [60]. It should be noted, however, that this observation of 'linkage-switching' was solely based on retention times in oligosaccharide maps, lectin binding and resistance to linkage-specific neuraminidase; more direct evidence, e.g. measurement of specific sialyltransferase activities was not presented.

(3) The presence of an "alternate pathway" besides the regular route of N-glycan processing was postulated to explain the occurrence, on insect and CHO cell-derived r-hPg, of $Man_3GlcNAc_2$(+/–Fuc) and $Man_4GlcNAc_2$ [57, 60]. These structures could, however, also be accounted for by a 'transient' GlcNAc residue, transferred by GlcNAc transferase I to $Man_5GlcNAc_2$ and subsequently removed by hexosaminidase action (see Sections 2.3.2. and 4.5.).

To identify the potential key event for the switching of the biosynthetic routes characteristic for insect cells towards the generation of complex type N-glycans as a consequence of infection with wild-type baculovirus or a recombinant construct, Davidson et al. [61] incubated the oligosaccharide $Man_9GlcNAc_2$ with extracts of uninfected and infected *Sf*-21 cells. While uninfected cell extracts displayed a very low capacity to process the substrate oligosaccharide beyond $Man_6GlcNAc_2$, those of infected cells were able to rapidly convert Man_6- to $Man_5GlcNAc_2$, the classical acceptor substrate for GlcNAc transferase I and the key intermediate *en route* to the assembly of complex type N-glycans [61].

4.4. N-Glycans of endogenous insect cell glycoproteins

Kubelka et al. [62] investigated the oligosaccharide spectrum of membrane glycoproteins from *MB*-0503, *Sf*-21 and *Bm*-N cells by 2D-HPLC mapping of pyridylaminated oligosaccharides. An essentially identical pattern of structures was found in all three cell lines. It included a set of high mannose structures ranging from Man_5- to $Man_9GlcNAc_2$ and truncated ("paucimannose") glycans, i.e. $Man_3GlcNAc_2$ and $Man_2GlcNAc_2$. The bulk of these "paucimannose" glycans was fucosylated at GlcNAc-1 (Table II). In addition, small amounts of fucosylated, GlcNAc-terminated glycans were identified (Table II), remarkably, with one β-1,2-linked GlcNAc either at the α-1,3- or at the α-1,6-branch.

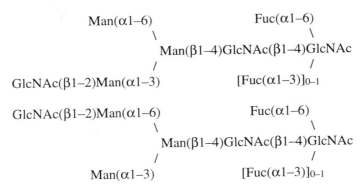

A number of observations and conclusions result from the data shown in Table II:

(1) Following the reasoning of Davidson et al. [61] that the trimming step yielding $Man_5GlcNAc_2$ proceeds in uninfected cells at a very low rate (see above), we should anticipate the accumulation of $Man_6GlcNAc_2$ in the cells investigated. However, approximately equal proportions of Man_9- to $Man_5GlcNAc_2$ oligosaccharides were found in all three glycan pools, each amounting to roughly 10%. Additionally, roughly 40% of the structures were processed beyond the $Man_5GlcNAc_2$ stage and therefore they must have experienced the removal of the last α-1,2-mannosyl residue [62].

(2) The occurrence of GlcNAc-terminated chains demonstrates the presence of GlcNAc transferases I and II in the three lepidopteran cell lines. It may be deduced from the amounts of structures which are likely to require for their synthesis the action of GlcNAc transferase I, i.e. all fucosylated structures (see below), that the actual *in vivo* activity of GlcNAc transferase I is significantly higher than indicated by the low proportion of oligosaccharides with antennal GlcNAc.

(3) Again, the enigma of the fucosylated high mannose glycans is encountered. Actually, even the occurrence of 'paucimannose' structures having only two or three mannose residues remains unexplained by the biosynthetic pathways established for mammalian cells, since α-mannosidase II, like fucosyltransferase, requires the presence of an antennal GlcNAc [29,30].

(4) Finally, while *MB*-0503 cells had, in accordance with an earlier report [63], the highest content of α-1,3- and difucosylated glycans, also *Sf*-21 and *Bm*-N cells displayed these moieties (Fig. 2). Surprisingly, α-1,3-fucosylated oligosaccharides have never been found on recombinant glycoproteins.

4.5. N-glycan processing in insect cells

Fucosyltransfer. In two recent studies [63,64] the fucosyltransferase and GlcNAc-transferase activities of biotechnologically important lepidopteran cell lines were investigated. In a preceding experiment, Staudacher et al. [37] have demonstrated the potential of honeybee venom glands to synthesise two fucosyl linkages, α-1,3 and

TABLE II

N-Glycans found on membrane glycoproteins of cultured insect cells [from 62]

Structure		Relative abundance (%) in cell lines		
		Mb-0503	Sf-21	Bm-N
Man(α1–6) \ Man(β1–4)GlcNAc(β1–4)GlcNAc / [Man(α1–3)]$_{0-1}$ Fuc(α1–6) |		8.8	15.2	26.6
Man(α1–6) \ Man(β1–4)GlcNAc(β1–4)GlcNAc / [Man(α1–3)]$_{0-1}$ Fuc(α1–3) |		4.5	2.8	0.7
Man(α1–6) \ Man(β1–4)GlcNAc(β1–4)GlcNAc / [Man(α1–3)]$_{0-1}$ Fuc(α1–6) | Fuc(α1–3) |		22.0	7.8	2.6
GlcNAc(β1–2) [Man(α1–6) \ Man(β1–4)GlcNAc(β1–4)GlcNAc / Man(α1–3) Fuc(α1–6) | [Fuc(α1–3)]$_{0-1}$ /]		4.0	1.0	2.5
Man$_{2-4}$GlcNAc$_2$		7.7	18.7	6.9
Man$_{5-9}$GlcNAc$_2$		53.0	55.5	61.5

α-1,6 to GlcNAc-1 (see Section 3.1.). In order to assess the fucosylation potential of cultured lepidopteran cells, cell homogenates were incubated with GDP-[^{14}C]Fuc and asialo-agalacto glycopeptides from either human IgG (α-1,6-fucosylated) or from bovine fibrin. The oligosaccharides were then released by N-glycosidase A, pyridylaminated and analysed by 2D-HPLC. All three cell lines tested in this study, i.e. *Mb*-0503, *Sf*-9 and *Bm*-N cells were able to transfer fucose into α-1,6-linkage to the innermost GlcNAc residue of the acceptor glycopeptide, which noteworthy contained terminal GlcNAc residues. Transfer into α-1,3-linkage and formation of a difucosylated glycan

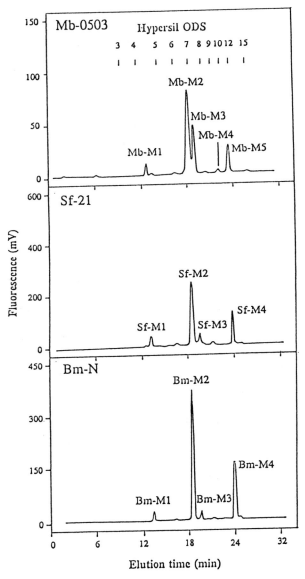

Fig. 2. Fucosylation pattern of the asparagine-linked GlcNAc residue in glycans from Mb-0503, Sf-21, and Bm-N cells. Pyridylaminated oligosaccharides from each of the three cell lines were exhaustively digested with α-mannosidase. The digests were directly applied to reversed-phase HPLC. The peaks designated Mb-, Sf-, and Bm-M2 correspond to the unfucosylated trisaccharide Man(β1–4)GlcNAc(β1–4)GlcNAc; Mb-, Sf-, and Bm-M1 represent the α-1,3-fucosylated, Mb-, Sf-, and Bm-M3 the α-1,3- and α-1,6-difucosylated, and Mb-M5, Sf-M4, and Bm-M4 the α-1,6-fucosylated analog of the core-trisaccharide. Numbers at the top indicate the elution positions of isomaltose standards of the respective degree of polymerization. Taken from Ref. [62].

could only be demonstrated for *Mb*-0503 cells. In the light of the structures found in the insect cell lines (see Section 4.4.) [62], it is reasonable to assume that α-1,3-fucosyltransferase activities in *Sf*-9 and *Bm*-N cells, although present, were below the detection threshold of the applied method.

Considering the structural findings reviewed above (Section 3.1.2.) [35], it is clear that at least honeybees, which belong to the *Hymenoptera*, must express an additional fucosyltransferase responsible for the antennal fucose residue in some glycans of honeybee PLA.

GlcNAc-transfer: As presented above, several N-glycan variants containing antennary GlcNAc were identified in insect glycoproteins, e.g. honeybee PLA, locust apolipophorin III and membrane glycoproteins from lepidopteran cells (see Sections 3.1.2., 3.2., and 4.4.), and similar indications were recently obtained with a mosquito cell-grown virus [65] — altogether good evidence to postulate GlcNAc transferases I and II activities in insect cells.

Indeed, the presence of GlcNAc transferase I was unequivocally established in two recent studies [64,66]. Velardo et al. [66] used Golgi preparations from IPLB-*Sf*-21AE cells as the enzyme source, Man$_5$GlcNAc$_2$ as acceptor and high pH anion exchange chromatography and advanced mass spectrometric techniques for product identification. They found a fourfold increase of activity following infection with baculovirus. Altmann et al. [64] used homogenates of *Sf*-21, *Sf*-9, *Bm*-N, and *Mb*-0503 cells as enzyme sources and pyridylaminated oligosaccharides as acceptors. The products were identified and quantitated by reversed phase HPLC. It turned out that insect cell GlcNAc-transferase I had a much higher preference for Man$_5$GlcNAc$_2$ compared to Man$_3$GlcNAc$_2$ than the enzyme from a mammalian reference cell line. However, when measured with the appropriate substrate Man$_5$GlcNAc$_2$, the specific activities were essentially the same in insect and human HepG2 cells [64]. In contrast to the results of Velardo et al. [66], a significant change of GlcNAc-transferase activity as a result of infection with wild-type baculovirus could not be observed.

The same study [64] demonstrated that a GlcNAc residue linked to the α-1,3-antenna of the core pentasaccharide by GlcNAc-transferase I was essential for the action of insect cell fucosyltransferase.

GlcNAc-transferase II activity, although low, could likewise be observed in all four insect cell lines [64].

In summary, these observations allow the formulation of the following hypotheses.

(1) N-glycan processing in insects follows to a remarkable extent the pathways established for mammals. This means that Man$_5$GlcNAc$_2$ is transformed by GlcNAc-transferase I to permit the action of fucosyltransferases and, although this part of the hypothesis still has to be tested, probably also of mannosidase II, which in turn gives way to GlcNAc-transferase II.

(2) The lack of terminal GlcNAc in most insect cell-derived fucosylated N-glycans must then be attributed to the 'transient' nature of this GlcNAc residue, in other words, rather to the presence of an as yet unidentified 'processing' β-hexosaminidase [64] instead of unusual specificities of insect cell fucosyltransferases as previously speculated. The presence of roughly equal amounts of terminal GlcNAc linked either to the α-1,3- or the α-1,6-linked mannose residue, despite the large difference in GlcNAc-transferase I and II

activities, may indicate a branch specificity of this 'processing' β-hexosaminidase.

(3) Several of the enzymes required for the processing towards complex glycans have been shown by several laboratories to exist in normal, non-infected cells. Thus, the abyss between the usual findings ('paucimannose' structures) and the detection of complex glycans on insect cells may not be as broad as hitherto thought, as it is confined to the expression of β-1,4-galactosyltransferase and α-2,3 and/or α-2,6-sialyltransferases.

(4) Finally, if insect cells are indeed capable of switching from high-mannose towards sialylated, complex type N-glycans, what is the committed step in the biosynthetic route? Does viral infection induce a Man_6-mannosidase and GlcNAc-transferase I as has been postulated [61,66]? Other studies [62] did not reveal an accumulation of $Man_6GlcNAc_2$ in non-infected cells (see Section 4.4.). Considering the above presented results on glycan structures (Section 4.4.) and enzyme activity (Section 4.5.), GlcNAc-transferase I likewise does not appear to be the bottleneck in N-glycan processing in insect cells. Further studies are necessary to elucidate this issue.

5. Biological phenomena

5.1. Introduction

The biological function of protein glycosylation, suggested by high evolutionary conservation of structural patterns and the complexity of the biosynthetic machinery, which appears designed to create enormous diversity, is still not fully understood. As documented in these volumes, commanding evidence supports the view that glycoprotein glycans contribute profoundly to recognitional processes, either functioning as biological masks or as essential determinants [67,68], and there is no reason why insects should pose an exception.

Evidence has so far been accumulated, which demonstrates that (1) protein-bound carbohydrate forms important, developmentally regulated, surface epitopes of insect tissue, and that (2) insect glycoprotein glycans possess structural features, which constitute antigenic, possibly even allergenic, determinants.

5.2. Developmental aspects

First evidence for the developmental importance of insect glycoprotein glycans was obtained, when polyclonal antisera, raised against the plant glycoprotein, horseradish peroxidase (HRP), were used to probe a neural cell surface antigen in *Drosophila* and other insects [10] and to assess the growth of pioneer neuron fibers. Using anti-HRP antibodies in immunoprecipitation studies on *Drosophila* and grasshopper embryos, Snow et al. [69] detected two glycoproteins, fasciclin I and II, on specific subsets of axon pathways. These two glycoproteins are dynamically expressed during growth cone guidance in axon growth and at other embryogenetic stages and may thus serve as pathway labels. Subsequently, evidence was provided for the carbohydrate nature of a neural-specific epitope expressed on all axon pathways [11].

These findings led to a speculation that, as in vertebrate organisms, structural alterations in this insect tissue carbohydrate epitope may profoundly influence its functional contribution as a mediator and/or modulator of cell recognition and adhesion.

More direct evidence established the carbohydrate nature of the neural cell surface antigen, and the data suggested that it contains structural features shared by plant glycoproteins [70].

In another study, Dennis et al. [71] found that monoclonal antibodies directed against the L2/HNK-1 carbohydrate epitope of adhesion molecules of the mammalian nervous system, which was identified to carry a sulfated, glucuronic acid-containing moiety, reacted with brain glycoproteins of the fly, *Calliphora vicina*. Their results indicated that this epitope is however not developmentally regulated and it was concluded from its wide distribution that it is phylogenetically highly conserved.

The results of Roth et al. [16] seem to bear important implications for the overall picture of insect protein glycosylation. They found, during their successful attempt to detect sialic acid in insect tissue, that the appearance of α-2,8-homopolymers is developmentally regulated, as they are only expressed in 14–18 h embryos.

5.3. Common antigenic determinants of insect and plant glycoproteins

Fuc, α-1,3-linked to the GlcNAc-1, and Xyl, β1–2-linked to the β-Man residue of the Man$_3$GlcNAc$_2$ core pentasaccharide, were and are still considered characteristic features of plant glycoprotein N-glycans [72] as e.g. horseradish peroxidase:

Man(α1–6)
$\qquad\qquad$ \
$\qquad\qquad$ Man(β1–4)GlcNAc(β1–4)GlcNAc
$\qquad$ / |$\qquad\qquad\qquad\qquad\qquad$ /
Man(α1–3) |$\qquad\qquad$ **Fuc(α1–3)**
$\quad$ **Xyl(β1–2)**

These two residues were, on the basis of the evidence summarised above, addressed as likely candidates to form a common cross-reactive determinant. When a positive cross-reaction was also observed between a polyclonal antibody preparation raised against carrot cell β-fructosidase and PLA, the occurrence of hitherto undetected Xyl-containing N-glycans in this insect glycoprotein enzyme was suggested [73]. But, as has been reviewed above, Xyl does not occur in PLA [35].

In an important contribution, Kurosaka et al. [74] postulated that the Man(α1–6)-residue of the core pentasaccharide and Fuc, α-1,3-linked to the reducing, terminal, GlcNAc, but not Xyl, represent the predominant cross-reactive features, shared by plant glycoproteins and the insect neural tissue surface antigen oligosaccharide. They based their suggestion on the effects of selective monosaccharide removal from the major HRP N-glycan fraction on its reactivity with anti-HRP antiserum and the recognition of the insect neural epitope by this antiserum. The validity of this hypothesis was established by the discovery of the fucosylation pattern at GlcNAc-1 in PLA (Section 3.1.1.).

5.4. α-1,3-Fucosylation of the Asn-bound GlcNAc: an antigenic/allergenic determinant

5.4.1. Antigenicity

If antisera against plant glycoproteins, e.g. HRP, which contain significant concentrations of anti-carbohydrate antibodies and which react with extracts of insect tissue (see above), do recognise a common determinant, it should be possible to demonstrate the converse reaction of antibodies against an insect glycoprotein with plant glycoproteins. Indeed, a rabbit polyclonal antiserum raised against honeybee venom PLA, which was shown to contain antibodies directed against the carbohydrate part, cross-reacted with fucosylated plant glycoproteins [75]. E.l.i.s.a. binding and inhibition experiments, employing glycoproteins and glycopeptides of plant and animal origin with known N-glycan structures, in combination with chemical and enzymic deglycosylation, identified α-1,3-Fuc at GlcNAc-1 as the antigenic determinant [75].

Faye et al. [76] fractionated antibodies against plant glycoproteins (β-fructosidase and HRP) over immobilised PLA, and showed that the retained antibody fraction specifically recognised N-glycans which were α-1,3-fucosylated at GlcNAc-1, whereas the unretained fraction contained Xyl-specific antibodies.

5.4.2. Allergenicity

PLA is a major allergen of honeybee venom. In an earlier study, Weber et al. [77] had shown that a significant percentage of PLA-directed sera of bee-sting allergic patients contain IgE antibodies, which reacted with the carbohydrate part of PLA. In a recent study [78] neoglycoproteins prepared from bromelain and PLA glycopeptides were used in e.l.i.s.a. experiments to pin down the determinant of the interaction between IgE and carbohydrate. With the applied methodology, 33 of the 122 sera were seen to react with the PLA-carbohydrate and to cross-react with bromelain. After defucosylation, bromelain glycopeptide was no longer reactive. It was concluded that the Fuc(α1–3)GlcNAc-moiety is an essential part of the IgE-reactive determinant [78]. The wide distribution of this substitution in plant glycoprotein and its occurrence in insect glycoproteins suggest that it may profoundly contribute to a "common carbohydrate determinant (CCD)", postulated for food, animal and plant allergens [79]. In this context it should be added that the O-glycans of the sea squirt antigens mentioned above, which, like PLA, contain the GalNAc(β1–4)[Fuc(α1-3)]GlcNAc(β1–2)-moiety as a part structure, have been found to be allergenically active [43].

6. Conclusion and summary

Although many questions remain to be answered, it is clear that insects possess the capacity for O-glycosylation of proteins and that their N-glycosylation potential is far less simple than previously assumed. Firm evidence is available for the capacity to elongate high-mannose type chains via the action of GlcNAc transferase I. Furthermore, while the process of trimming nascent N-glycans from the Man_9 through the Man_5 stage seems to proceed along the same route as established for higher animals, in particular mammals, the subsequent events appear quite different and characteristic. They lead to the generation of structural features like the difucosylation of the Asn-bound GlcNAc and the GalNAc(β1–4)[Fuc(α1–3)] GlcNAc(β1–2)-moiety. It has been shown that some of these unusual part structures are surface antigens in insect tissues. Their potential antigenicity adds a note of caution, but also a new challenge to the use of insect cells for the expression of recombinant glycoprotein for pharmaceutical application. Indications for the possibly developmentally regulated occurrence of sialic acid in insects and for the elicitation of the synthesis of complex type structures by transfective processes render the attempt worthwhile to modulate the biosynthetic potential of insect cells towards the generation of wanted and the elimination of un-wanted structural features.

Acknowledgments

The work from the authors' laboratory, which has been reported here, was supported by the Austrian Ministry of Science and by the Austrian Fonds zur Förderung der wissenschaftlichen Forschung.
We wish to take this opportunity to thank Dr. Harry Schachter for continuous interest and support and Drs. Kamerling and Vliegenthart for their excellent cooperation.

Abbreviations

AEP	2-aminoethylphosphonate
2D-HPLC	two-dimensional high performance liquid chromatography
HRP	horseradish peroxidase
PLA	phospholipase A$_2$
hCGβ	human chorionic gonadotropin β
hPg	human plasminogen

References

1. Law, J.H. and Wells, M.A. (1989) J. Biol. Chem. 264, 16335–16338.
2. Warren, L. (1963) Comp. Biochem. Physiol. 10, 153–171.
3. Butters, T.D. and Hughes, R.C. (1978) Carbohyd. Res. 61, 159–168.
4. Butters, T.D. and Hughes, R.C. (1981) Biochim. Biophys. Acta 640, 655–671.
5. Butters, T.D., Hughes, R.C. and Vischer, P. (1981) Biochim. Biophys. Acta 640, 672–686.
6. Hsieh, P. and Robbins, P.W. (1984) J. Biol. Chem. 259, 2375–2382.
7. Stollar, V., Stollar, B.D., Koo, R., Harrap, K.A., Schlesinger, R.W. (1976) Virology 69, 104–115.
8. Ryan, R.O., Anderson, D.R., Grimes, W.J. and Law, J.H. (1985) Arch. Biochem. Biophys. 243, 115–124.
9. Williams, P.J., Wormald, M.R., Dwek, R.A., Rademacher, T.W., Parker, G.F. and Roberts, D.R. (1991) Biochim. Biophys. Acta 1075, 146–153.

10. Jan, L.Y. and Jan, Y.N. (1982) Proc. Natl. Acad. Sci. USA 79, 2700–2704.

11. Snow, P.M., Patel, N.H., Harrelson, A.L. and Goodman, C.S. (1987) J. Neurosci. 7, 4137–4144.

12. Sturm A. (1995) in: J. Montreuil, H. Schachter and J.F.G. Vliegenthart (Eds.), Glycoproteins. New Comprehensive Biochemistry, Vol. 29a. Elsevier, Amsterdam, pp. 521–541.

13. Kress, H. (1982) Dev. Biol. 93, 231–239.

14. Campbell, A.G., Fessler, L.I., Saalo, T. and Fessler, J.H. (1987) J. Biol. Chem. 262, 17605–17612.

15. Hafer, J. and Ferenz, H.-J. (1991) Comp. Biochem. Physiol. 100B, 579–586.

16. Roth, J., Kempf, A., Reuter, G., Schauer, R. and Gehring, W.J. (1992) Science 256, 673–675.

17. Svoboda, M. and Przybylski, M. (1991) J. Chromatogr. 562, 403–419.

18. Sridhar, P., Panda, A.K., Pal, R., Talwar, G.P. and Hasnain, S.E. (1993) FEBS Lett. 315, 282–286.

19. Thomsen, D.R., Post, L.E. and Elhammer, Å. (1990) J. Cell. Biochem. 43, 67–79.

20. Chen, W., Shen, Q.-X. and Bahl, O.P. (1991) J. Biol. Chem. 266, 4081–4087.

21. Wathen, M.W., Aeed, P.A. and Elhammer, A.P. (1991) Biochemistry 30, 2863–2868.

22. Grabenhorst, E., Hofer, B., Nimtz, M., Jäger, V. and Conradt, H.S. (1993) Eur. J. Biochem. 215, 189–197.

23. Sugyiama, K., Ahorn, H., Maurer-Fogy, I. Voss, T. (1993) Eur. J. Biochem. 217, 921–927.

24. Kelly W.G. and Hart, G.W. (1989) Cell 57, 243–251.

25. Hart, G. (1995) Nuclear and cytoplasmic glycoproteins, in: J. Montreuil, H. Schachter and J.F.G. Vliegenthart (Eds.), Glycoproteins. New Comprehensive Biochemistry, Vol. 29b. Elsevier, Amsterdam.

26. Quesada Allue, L.A. and Belocopitow, E. (1978) Eur. J. Biochem. 88, 529–541.

27. Sagami, H. and Lennarz, W.J. (1987) J. Biol. Chem. 262, 15610–15617.

28. Parker, G.F., Williams, P.J., Butters, T.D. and Roberts, D.B. (1991) FEBS Lett. 290, 58–60.

29. Kornfeld, R. and Kornfeld, S. (1985) Annu. Rev. Biochem. 54, 631–664.

30. Longmore, G.D. and Schachter, H. (1982), Carbohyd. Res. 100, 365–392.

31. Kozutsumi, Y., Nakao, Y., Teramura, T., Kawasaki, T., Yamashina, I., Mutsaers, J.H.G.M., Van Halbeek, H. and Vliegenthart, J.F.G. (1986) J. Biochem. 99, 1253–1265.

32. Mutsaers, J.H.G.M., Van Halbeek, H., Vliegenthart, J.F.G., Tager, J.M., Reuser, A.J.J., Kroos, M. and Galjaard, H. (1987) Biochim. Biophys. Acta 911, 244–251 .

33. Vitale, A. and Chrispeels, M.J. (1984) J. Cell Biol. 99, 133–140.

34. Szumilo, T., Kaushal, G.P. and Elbein, A.D. (1986) Biochem. Biophys. Res. Commun. 134, 1395–1403.

35. Kubelka, V., Altmann, F., Staudacher, E., Tretter, V., März, L., Hård, K., Kamerling, J.P. and Vliegenthart, J.F.G. (1993) Eur. J. Biochem. 213, 1193–1204 .

36. Hård, K., Van Doorn, J.M., Thomas-Oates, J.E., Kamerling, J.P. and Van der Horst, D.J. (1993) Biochemistry 32, 766–775.

37. Staudacher, E., Altmann, F., Glössl, J., März, L., Schachter, H., Kamerling, J.P., Hård, K. and Vliegenthart, J.F.G. (1991) Eur. J. Biochem. 199, 745–751.

38. Tretter, V., Altmann, F. and März L. (1991) Eur. J. Biochem. 199, 647–652.

39. Staudacher, E., Altmann, F., März, L., Hård, K., Kamerling, J.P. and Vliegenthart, J.F.G. (1992) Glycoconj. J. 9, 82–85.

40. Bergwerff, A.A., Thomas-Oates, J.E., Van Oostrum, J., Kamerling, J.P. and Vliegenthart, J.F.G. (1992) FEBS Lett. 314, 389–394.

41. Yan, S.B., Chao, Y.B. and Van Halbeek, H. (1993) Glycobiology 3, 597–608.

42. Srivatsan, J., Smith, D.F. and Cummings, R.D. (1992) Glycobiology 2, 445–452.

43. Ohta, M., Matsuura, F., Kobayashi, Y., Shigeta, S., Ono, K. and Oka, S. (1991) Arch. Biochem. Biophys. 290, 474–483.

44. Hink, W.F., Thomsen, D.R., Davidson, D.J., Meyer, A.L. and Castellino F.J. (1991) Biotechnol. Prog. 7, 9–14.

45. Cameron, I.R., Possee, R.D. and Bishop, D.H.L. (1989) Trends Biotechnol. 7, 66–70.

46. Luckow, V.A. and Summers, M.D. (1988) Bio/Technology 6, 47–55.

47. Knight, P. (1989) Bio/Technology 7, 35–40.

48. Parekh, R.B., Dwek, R.A., Edge, C.J. and Rademacher, T.W. (1989) Trends Biotechnol. 7, 117–121.

49. Possee, R.D. (1986) Virus Res. 5, 43–59.

50. Kuroda, K., Hauser, Ch., Rott, R. Klenk, H.-D. and Doerfler, W. (1986) EMBO J. 5, 1359–1365.

51. Greenfield, C., Patel, G., Clark, S., Jones, N. and Waterfield, M.D. (1988) EMBO J. 7, 139–146.

52. Wojchowski, D.M., Orkin, S.H. and Sytkowski, A.J. (1987) Biochim. Biophys. Acta 910, 224–232.

53. Jarvis, D.L. and Summers, M.D. (1989) Molec. Cell. Biol. 9, 214–223.

54. Kuroda, K., Geyer, H., Geyer, R., Doerfler, W. and Klenk, H.-D. (1990) Virology 174, 418–429.

55. Voss, T., Ergülen, E., Ahorn, H., Kubelka, V., Sugiyama, K., Maurer-Fogy, I. and Glössl, J. (1993) Eur. J. Biochem. 217, 913–919.

56. Yeh, J., Seals, J.R., Murphy C.I., Van Halbeek, H. and Cummings, R.D. (1993) Biochemistry 32, 11087–11099.

57. Davidson, D.J., Fraser, M.J. and Castellino, F.J. (1990) Biochemistry 29, 5584–5590.

58. Davidson, D.J. and Castellino, F.J. (1991) Biochemistry 30, 6167–6174.

59. Davidson, D.J. and Castellino, F.J. (1991) Biochemistry 30, 6689–6696.

60. Davidson, D.J. and Castellino, F.J. (1991) Biochemistry 30, 625–633.

61. Davidson, D.J., Bretthauer, R.K. and Castellino, F.J. (1991) Biochemistry 30, 9811–9815.

62. Kubelka, V., Altmann, F., Kornfeld, G. and März, L. (1994) Arch. Biochem. Biophys. 308, 148–157.

63. Staudacher, E., Kubelka, V. and März, L. (1992) Eur. J. Biochem. 207, 987–993.

64. Altmann, F. Kornfeld, G., Dalik, T., Staudacher, E. and Glössl, J. (1993) Glycobiology 3, 619–625.

65. Naim, H.Y. and Koblet, H. (1992) Arch. Virol. 122, 45–60.

66. Velardo, M.A., Bretthauer, R.K., Boutaud, A., Reinhold, B., Reinhold, V.N. and Castellino, F.J. (1993) J. Biol. Chem. 268, 17902–17907.

67. Feizi, T. and Childs, R.A. (1985) Trends Biochem. Sci. 10, 24–29.

68. Schauer, R. (1985) Trends Biochem. Sci. 357–360.

69. Bastiani, M.J., Harrelson, A.L., Snow, P.M. and Goodman, C.S. (1987) Cell 48, 745–755.

70. Katz, F., Moats W. and Jan, Y.N. (1988) EMBO J. 7, 3471–3477.

71. Dennis, R.D., Antonicek, H., Wiegandt, H. and Schachner, M. (1988) J. Neurochem. 51, 1490–1496.

72. Faye, L., Johnson, K.D., Sturm, A. and Chrispeels, M.J. (1989) Physiol. Plant. 75, 309–314 .

73. Faye, L. and Chrispeels, M.J. (1988) Glycoconj. J. 5, 245–256.

74. Kurosaka, A., Yano, A., Itoh, N., Kuroda, Y., Nakagawa, T. and Kawasaki T. (1991) J. Biol. Chem. 266, 4168–4172.

75. Prenner, Ch., Mach, L., Glössl, J. and März, L. (1992) Biochem. J. 284, 377–380.

76. Faye, L., Gomord, V., Fitchette-Lainé, A.-C. and Chrispeels, M.J. (1993) Anal. Biochem. 209, 104–108.

77. Weber, A., Schröder, H., Thalberg, K. and März, L. (1987) Allergy 42, 464–470.

78. Tretter, V., Altmann, F., Kubelka, V., März, L. and Becker, W.M. (1993) Int. Arch. All. Immunol. 102, 259–266.

79. Aalberse, R.C., Koshte, V. and Clemens, J.G.J. (1981) Int. Archs. Allergy Appl. Immunol. 66, 259–260.

J. Montreuil, H. Schachter and J.F.G. Vliegenthart (Eds.), *Glycoproteins*

The glycoprotein hormone family: structure and function of the carbohydrate chains

MALGORZATA BIELINSKA and IRVING BOIME

Department of Molecular Biology and Pharmacology and Obstetrics and Gynecology, Washington University School of Medicine, 660 S. Euclid Avenue, St. Louis, MO 63110, USA

1. Introduction

The family of glycoprotein hormones consists of lutropin (LH), follitropin (FSH), and thyrotropin (TSH) synthesized in the anterior pituitary and chorionic gonadotropin which is synthesized only in the placenta of human, primate, and equine. The hormones are heterodimers composed of two non-covalently joined α and β subunits. Both subunits are glycosylated and contain N- and, in the case of the CGβ subunit, O-linked oligosaccharides which account for 15–40% of their mass [1]. Only heterodimers and not the subunits are biologically active [2,3]. Assembly of the common α subunit with either of the four β subunits results in unique properties of each hormone that are reflected in binding to hormone-specific receptor and subsequent signal transduction.

Gonadotropins regulate reproductive functions: LH stimulates steroidogenesis in ovarian granulosa and theca cells and testicular Leydig cells [4]. One function of hCG is to maintain high progesterone production in the corpus luteum to maintain pregnancy [5] and hCG and LH bind to the same receptor. FSH also stimulates steroidogenesis in the ovarian follicular cells and testicular Sertoli cells [4,6]. TSH binds to receptors on the thyroid gland and regulates thyroid hormone synthesis [7]. Glycoprotein hormone receptors have been recently cloned and are members of the G protein-coupled receptor family with unusually long extracellular domains presumably to accommodate large ligands [8–12]. A region homologous to the soy bean lectin was found in the human LH/CG receptor [8], but such sequence is not present in the TSH or FSH receptors. Although there is extensive information on the structure of LH/CG and TSH receptors [13–15] and site-directed mutagenesis of the recombinant receptors identified some putative domains involved in ligand binding [10,16,17], the mechanism of the hormone:receptor interaction is unknown. Whereas rat and porcine receptors do not show species specificity, the human LH/CG and FSH receptors recognize only human hormones [18–21].

Structure–function studies of gonadotropins have been hampered by the lack of crystallographic data due to high content and heterogeneity of the negatively charged carbohydrate moieties. Attempts to crystallize glycoprotein hormones involved desialylated or deglycosylated hCG [22,23], and recently recombinant non-sialylated hCG and its selenomethionyl analog have been expressed in insect cells to generate substrates suitable for crystallographic studies [24,25]. Despite the lack of three-dimensional structure, there is evidence for the roles of both polypeptide and carbohydrate component in particular events leading to expression of biologic activity. Here we review the current data on the role of the carbohydrate moieties in glycoprotein hormone function.

2. Structure of the glycoprotein hormone subunits

2.1. The common α subunit

The human α subunit is a 92 amino acid long polypeptide (96 amino acids in other species) with two N-linked oligosaccharides attached to Asn 52 and 78 (56 and 82 in other species) (Fig. 1). The α subunit within a species is identical for all hormones and it is a product of a single gene [26–28]. Between the species the α subunit primary sequences show remarkable similarity [1]. Whereas the amino-terminal sequences among the different species have deletions within first eight amino acids, the carboxy-terminal sequence is similar and is apparently one of the receptor binding domains of the glycoprotein hormones [29–30].

All α subunits contain ten invariant cysteine residues which maintain the structural integrity of the subunit by forming five disulfide bonds [31]. The α subunit combines with each of the four β subunits within a species, and with the β subunits from other species; the resulting dimers express the specificity of the β subunit [1,32].

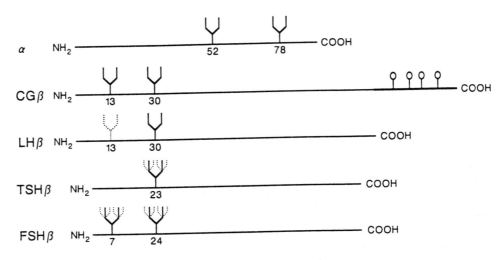

Fig. 1. Schematic representation of the human gonadotropin subunits. Positions of the N-linked and O-linked oligosaccharides (hCGβ) are shown. A dotted tree in the LHβ subunit indicates the glycosylation site in other species. Dotted trees in TSHβ and FSHβ indicate heterogeneity in branching pattern.

2.2. Hormone-specific β subunit

The β subunits are hormone-specific, although they share a high degree of sequence similarity [1]. Each of the glycoprotein hormone β subunits is a product of a different gene. A single gene encodes the LHβ [33–35], FSHβ [36,37] and TSHβ [38,39] subunit, whereas the hCGβ subunit is encoded by a multigene family [33,35]. This contrasts with the equine, where both the LHβ and CGβ subunits are encoded by a single gene [40]. The length of the β subunit polypeptide varies from 110 amino acids for bovine and rat FSHβ subunit to 145 and 149 amino acids for the hCGβ and eLH/CGβ subunits, respectively; the diversity results from deletions and extensions at both amino- and carboxytermini. Similar to the α subunit, the amino acid sequence of the β subunit from all species can be aligned according to the invariant 12 cysteine residues forming 6 disulfide bonds [41].

In most animal species there is one N-linked oligosaccharide at Asn 13 of the LHβ and at Asn 23 of the TSHβ subunit. The human LHβ subunit is glycosylated at Asn 30. The hCGβ and the FSHβ subunits contain two N-linked sugars at Asn 13 and 30, and 7 and 24, respectively (Fig. 1) [1]. The eCGβ/eLHβ subunit is glycosylated at Asn 13 [42–44]. The hCGβ and eCGβ/eLHβ subunits contain a carboxyterminal extension of 29 and 33 amino acids, respectively [45,46], due to a frameshift mutation at codon 114 in the human LH/CG gene [33]. This carboxyterminal extension of hCGβ bears four O-linked oligosaccharides attached to serine 121, 127, 132, and 138 [47] and O-linked oligosaccharides have also been found in the eCGβ subunit [48,49].

3. N-linked oligosaccharides of glycoprotein hormones

3.1. Structure of the N-linked oligosaccharides

3.1.1. HCG
Despite the sequence similarity of the polypeptide chains, the N-linked oligosaccharides of the glycoprotein hormones show extensive structural diversity. Initial studies on the structure of the N-linked oligosaccharides were limited due to insufficient quantity and purity of the pituitary hormone preparations. HCG oligosaccharides were the first to be characterized due to the availability of large quantities of hormone isolated from urine of pregnant women. Endo et al. [50] and Kessler et al. [51] proposed variations of two structures on hCG: disialylated diantennary and sialylated monoantennary, partially fucosylated with NeuAc present in (α2–3)Gal linkage (Fig. 2).

HCG or its subunits are synthesized in choriocarcinoma and several forms of invasive and non-invasive hydatidiform moles providing useful tumor markers and enabling studies on carbohydrate processing in malignant tissues. Whereas oligosaccharides isolated from benign hydatidiform mole were identical to normal urinary hCG [52], the N-linked oligosaccharides of choriocarcinoma hCG are almost devoid of sialic acid and contain triantennary structures which were not observed in hCG derived from normal pregnancy (structures V and VI in Fig. 2) [52–56]. These studies have been summarized by Kobata [57]. Analysis of the carbohydrate structures on dissociated

568

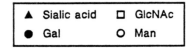

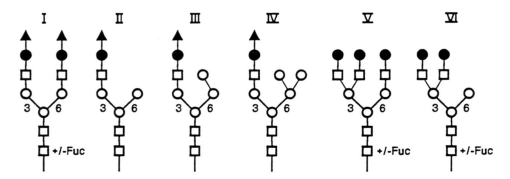

Fig. 2. Structure of hCG N-linked oligosaccharides [50–59]. Structures V and VI are found in hCG derived from choriocarcinoma cells [52–57]. +/– Fuc refers to the presence or absence of fucose. The designations 3 and 6 indicate α-1,3- and α-1,6- antennae, respectively.

subunits from placental hCG have indicated that the oligosaccharide population of each subunit differs: the α subunit bears more heterogeneous complex mono- and diantennary structure than the β subunit, which bears mostly diantennary fucosylated species [58].

Later studies demonstrated that the N-linked oligosaccharides of the hCGβ subunit do not show significant structural variations at either of the two glycosylation sites. They consist predominantly of structure I in Fig. 2 [59]. In contrast, the oligosaccharides of the α subunit consist of a heterogenous population and differ at a particular site. The variations are most prominent at Asn 52. This glycosylation site contains less extensively processed oligosaccharides [59] with structures I–IV in a ratio of 2:5:2:1 (Fig. 2) than the Asn 78 glycosylation site on which structures I and II have been found in a ratio 3:2 [59]. Glycosylation at Asn 52 is critical for maximal expression of hCG biologic activity [60,61] and the heterogeneity of structure at this glycosylation site may modulate the bioactivity of hCG.

3.1.2. Pituitary glycoprotein hormones

The N-linked oligosaccharides of most of the pituitary glycoprotein hormones from a variety of species contain sulfate on their non-reducing termini. This modification was discovered in bovine and ovine LH by Parsons and Pierce [62] and later confirmed by metabolic labeling of bovine and rat pituitary slices with $Na_2[^{35}S]O_4$ [63,64].

3.1.2.1. Sulfation of pituitary glycoprotein hormones

Sulfation of pituitary N-linked oligosaccharides was characterized *in vitro* using a cell-free sulfation system in which the glycoprotein hormone substrates were incubated in the presence of the Golgi membranes as a source of sulfotransferase and 3'-phosphoadenosine 5'-phosphosulfate as a sulfate donor [65,67]. Sulfation requires N-acetylgalactosamine instead of galactose present in sialylated chains at the penultimate position of the oligosaccharide chain. The pituitary 4-O-N-acetylgalactosamine sulfotransferase does not require any specific peptide recognition sequence and can use a synthetic trisaccharide as a substrate [67,68]. The N-acetylgalactosamine transferase is present in pituitary, but not in human placenta which also lacks sulfotransferase [66,69]; thus hCG oligosaccharides in contrast to pituitary hormones lack sulfate. However, the hCG N-linked oligosaccharides could be sulfated since agalactosyl hCG subunits are substrates for N-acetylgalactosamine transferase *in vitro* [70,71]. Studies on the enzyme specificity led to the proposal that the enzyme recognizes a specific motif, -Pro-X-Arg- in the primary sequence of the gonadotropin subunits upstream from the glycosylation site [71]. Such sequence is present in hCG but not in the FSHβ subunit which is poorly sulfated [71]. Similar to human, the horse placenta apparently lacks N-acetylgalactosamine transferase. Smith et al. [72] have shown that whereas the N-linked oligosaccharides in equine LH are sulfated, eCG lacks sulfate and contains di- and triantennary oligosaccharides that terminate with NeuAc(α2–3) or 6Gal(β1,4) [47,72,73].

3.1.2.2. Structure of pituitary hormone oligosaccharides

The comparative studies on sulfated and sialylated oligosaccharides of LH, FSH, and TSH from several species showed an array of structures differing in the extent of processing, branching and sulfation/sialylation pattern as summarized in Fig. 3 [65,74–78]. The majority of LH and TSH N-linked oligosaccharides are sulfated, while FSH contains less sulfate and hFSH oligosaccharides are sialylated. Human pituitary hormones are in general less sulfated than the corresponding hormones in other species [75,76]. The distribution of sulfated and sialylated oligosaccharides appears subunit specific; the hLHβ, hTSHβ and porcine/ovine LHβ subunits are sulfated, while the α subunit contains a population of mixed sulfated/sialylated oligosaccharides [79–82, 83 for review]. The α Asn 78 oligosaccharides are exclusively sialylated [81]. A fraction of hTSH oligosaccharides contains terminal 3-O sulfated galactose (SO$_4$-3Gal), and Fuc(α1–3)GlcNAc in the Man(α1–6) branch [80]. Small amounts of tri- and tetra-negatively charged chains are also seen, but they have been not characterized [80].

Human FSH contains diantennary and tri- tri'- and tetraantennary structures [75,76, 84] with sialic acid in both α-2,3- and α-2,6- linkages. Some of the oligosaccharides on hFSH contain bisecting GlnNAc [84]. Thus, although FSH and LH are synthesized in the same cell, and share the same α subunit, their oligosaccharide chains differ both in branching and terminal modification.

3.1.3. Hormone-specific oligosaccharide processing

Glycoprotein hormone subunit assembly occurs early during synthesis when the N-linked oligosaccharides are still endoglycosidase H sensitive [85,86]. Diversity of the oligosaccharide structure suggests that assembly of the common α subunit with a hormone specific β subunit determines further steps in specific processing of N-linked oligosaccharides. This results in specific processing pattern evens if the hormones are made in the same cell and thus have access to the same set of processing enzymes. This

A

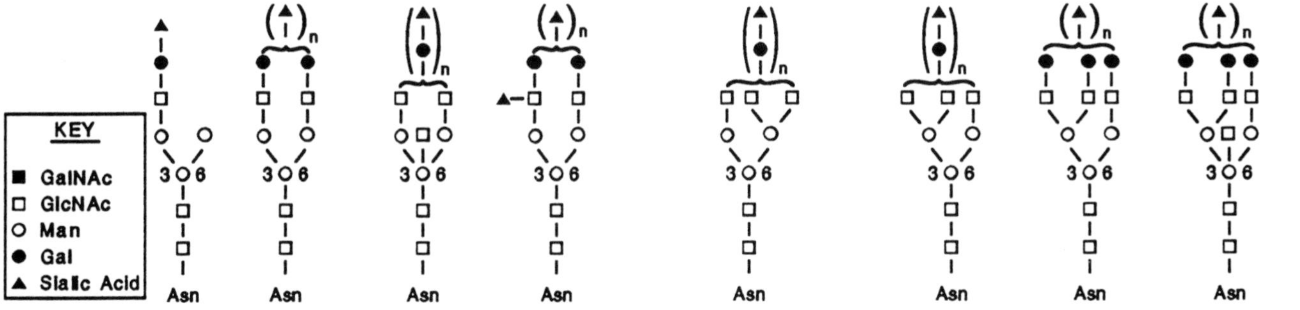

KEY
■ GalNAc
□ GlcNAc
○ Man
● Gal
▲ Sialic Acid

Designation:	N-1(A)	N-1(B) n=1	N-2(B) n=2	N-1(C) n=1	N-2(C) n=2	N-2(D) n=1	N-3(D) n=2	N-1(E) n=1	N-2(E) n=2	N-3(E) n=3	N-1(F) n=1	N-2(F) n=2	N-2(G) n=2	N-3(G) n=3	N-2(H) n=2	N-3(H) n=3
Hormone																
bLH																
bFSH	2	3	33			6	11									
bTSH																
oLH		1														
oFSH	1	2	5					5	4	4	2	2		3		
hLH	10	6	10	3	6											
hFSH	3	3	12	9	7						4	7	7	13	6	12
hTSH	5	12														12

[% OF TOTAL OLIGOSACCHARIDES]

B

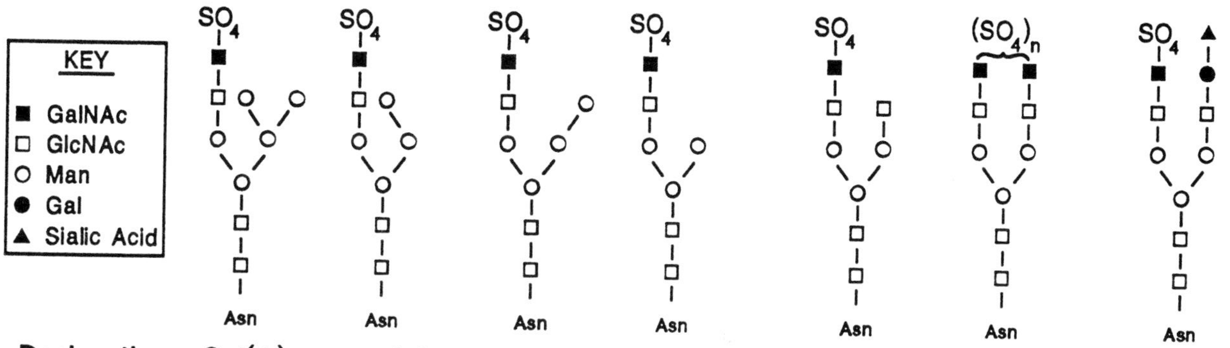

KEY

■ GalNAc
□ GlcNAc
○ Man
● Gal
▲ Sialic Acid

Hormone	S-1(A)	S-1(B)	S-1(C)	S-1(D)	S-1(E)	S-1(F) n=1	S-2 n=2	S-N
bLH	15	8	2	2	2	16	22	
bFSH	3	2	1	1	1	3	1	1
bTSH	12	5	1	2	1	11	48	2
oLH	27	9	9	5		6	13	4
oFSH	1	1	2	2		8	16	10
hLH	3	3	4	8	1		7	23
hFSH	1			1				5
hTSH	5	10		10			18	21

[% OF TOTAL OLIGOSACCHARIDES]

Fig. 3. Heterogeneity of sialylated (A) and sulfated (B) N-linked oligosaccharides of pituitary hormones from different species (reproduced from Baenziger and Green [77]). α-1,3- and α-1,6- antennae are in the left and right, respectively.

conclusion is supported by the results from mammalian cells transfected with gonadot-ropin subunit genes [87]. In such a system the only variable is the β subunit introduced with the common α subunit providing a convenient model to study the effect of subunit assembly on oligosaccharide processing. The N-linked oligosaccharides of both hCG subunits secreted from C127 cells were partially endoglycosidase H sensitive in con-trast to hLH oligosaccharides which were resistant [87]. Hormone-specific processing patterns are conserved in heterologous cells. Similar to pituitary FSH [75,84], the extensive branching of the N-linked oligosaccharides have been found in recombinant hFSH synthesized in CHO cells [88]. However, bisecting GlcNAc is not seen and sialic acid is attached only by α-2,3- linkage due to the prevalence of this type of linkage in CHO cells [89]. Similarly, despite lack of sulfation, the branching pattern of βLH expressed in CHO cells resembles that of pituitary bLH oligosaccharides [89]. Thus, recognition of the determinants which govern hormone-specific oligosaccharide proc-essing is common in different cell-types.

3.1.4. The free α subunit

The influence of subunit assembly on the oligosaccharide processing pattern is also illustrated by differential glycosylation of the free and the dimer forms of the α subunit. The free form of the α subunit is secreted by all tissues synthesizing glycoprotein hormones. It is distinguished from the dimer α by apparent higher molecular weight on SDS-PAGE [90–92]. In the case for the bovine α subunit the modification is due to O-glycosylation of threonine 43 [90,92]. However, little of the human α subunit is modified by O-glycosylation [93,94], rather the higher molecular weight of the free human α subunit is due to different N-linked oligosaccharide processing. The oligosac-charides of the placental [95–97] and the recombinant free α subunit [93,94,98] are more branched and sialylated than those attached to the dimer α subunit. Comparison of the N-linked oligosaccharide at each of the glycosylation sites using mutants in which the Asn 52 or Asn 78 sites were abolished, revealed that processing of the Asn 52 population was more affected by subunit combination than those attached to Asn 78 [98]. This can be attributed to differences in accessibility for the processing enzymes due to changes in the α subunit conformation after α:β assembly [99,100] or by the proximity of the β subunit to the α Asn 52 glycosylation site [101]. In the dimer hCG α subunit threonine 39 (43 in bovine α) is located at the α:β interface [101]; apparently the attachment of an O-linked glycan and altered processing of the Asn 52 oligosac-charide can occur only in the absence of the β subunit. The free α cannot associate with the β subunit unless the O- or N-linked oligosaccharides are removed *in vitro* [90,93].

The physiological significance of the free α subunit has not been established. It has been suggested that it may control the levels of intact hormone *in vivo* [90]; however O-glycosylation and N-linked oligosaccharide processing occurs after subunit assem-bly which is initiated in the ER [85,86]. Intracellularly it may facilitate the exit of the excess α subunit from the cell, since the free bovine pituitary α subunit is secreted faster than LH [92]. Recent studies implicate a paracrine role for the α subunit during pregnancy; both the free and the dimer α subunit stimulate decidual prolactin secretion in isolated decidual cells in culture [102].

3.2. Function of the N-linked oligosaccharides

Oligosaccharides have been implicated in the intracellular events such as folding, subunit assembly, secretion, and in the biological activity of glycoprotein hormones.

3.2.1. Subunit folding, assembly, and secretion

The N-linked oligosaccharides of the α subunit facilitate subunit folding. *In vitro* translation of the bovine α subunit mRNA in the absence of microsomal membranes yielded non-glycosylated product which folded poorly compared to the glycosylated subunit synthesized in the system supplemented with microsomal membranes in which high-mannose oligosaccharides are transferred to the growing nascent chain [103]. Similarly the non-glycosylated recombinant bovine α subunit expressed in *E. coli* was not secreted and only 3% of the reduced and reoxidized subunit derived from bacterial cells was folded correctly and attained an assembly-competent conformation; this was assessed by conformation-specific antibodies and ability to combine with the β subunit [104]. Non-glycosylated TSHα subunit derived from tunicamycin treated mouse thyrotropic tumor cells did not combine with the native TSHβ subunit [105].

To study the intracellular role of oligosaccharides, Matzuk et al. [106,107] prepared mutants of hCG subunit genes in which each of the four N-linked glycosylation sites were eliminated by site-directed mutagenesis of either the asparagine or threonine residues in the Asn-X-Thr glycosylation signal. The α and β mutant subunits were expressed separately or together to obtain hCG dimer in Chinese hamster ovary (CHO) cells. Elimination of the α subunit Asn 78 glycosylation site resulted in increased intracellular degradation and poor secretion of the free subunit compared to the wild-type α [106]. Alteration of the Asn 52 glycosylation site did not affect secretion of the free α subunit but markedly reduced the extent of α:β assembly [106]. Thus, while the Asn 78 oligosaccharide is important for intracellular stability of the α subunit, the Asn 52 oligosaccharide maintains a proper conformation for assembly.

Removal of the N-linked glycosylation sites from the hCGβ subunit revealed that glycosylation of Asn 30 is critical for efficient secretion of the uncombined β subunit [107]. The extent of α:β assembly is apparently not influenced by glycosylation of the β subunit but less of the mutant devoid of both glycosylation sites assembled with the α subunit compared to the wild-type β subunit [107]. Mutation of the single glycosylation site of the bovine LHβ subunit did not affect subunit assembly [108]. However, the mutant LHβ subunit when expressed alone is poorly secreted and slowly degraded with an extended intracellular half-life compared to the wild-type subunit [108].

Thus, the metabolic studies described above suggest that the N-linked oligosaccharides are important for the correct conformation of the glycoprotein hormone subunits. There are site-specific characteristics of the individual glycosylation sites regarding stability of the subunit and dimer assembly (Table I).

In contrast, deglycosylation of the native hCG (or the dissociated subunits) which leaves one proximal *N*-acetylglucosamine at each glycosylation site, apparently does not have a dramatic effect on folding and assembly, since the deglycosylated subunits can refold and recombine [109]. The data suggest that the presence of at least one

TABLE I

Intracellular behavior and bioactivity of hCG dimers containing site–specific oligosaccharide deletions

Mutations	Secretion[a]	Dimer assembly[a]	In vitro signal transduction[b]
α			
ΔAsn52	normal	low	decrease
ΔAsn78	low (decreased stability)	all of secreted mutant assembled	normal
ΔAsn52 & 78	slow	slow	decrease
hCGβ			
ΔAsn13	normal	normal	normal
ΔAsn30	slow	normal	normal
ΔAsn13 & 30	slow	slow	normal

[a]Matzuk and Boime [106] — α subunit; Matzuk and Boime [107] — β subunit.
[b]Matzuk et al. [60].

N-acetylglucosamine attached to the peptide backbone is necessary for folding and
assembly of glycoprotein hormone subunits during synthesis.

3.2.2. Receptor binding and signal transduction

The role of the N-linked oligosaccharides in the bioactivity of the glycoprotein hormones has been studied extensively. Deglycosylation by either chemical (anhydrous
hydrogen fluoride or trifluoromethanesulfonic acid) or enzymatic methods did not
reduce receptor binding; on the contrary, in some cases deglycosylated hormones
showed enhanced affinity compared to the native ligands [110–112]. However, Sairam
et al. [113] reported reduced binding affinity of deglycosylated ovine or bovine lutropin
to the sheep testis receptor. This phenomenon apparently is species specific since these
analogs bound efficiently to the rat or porcine LH/CG receptor [113]. Progressive
enzymatic deglycosylation of hCG revealed that while receptor binding in rat testicular
cells remained unchanged, the biological response measured by stimulation of cAMP
production or steroidogenesis was greatly impaired [109,114]. In addition, the unglycosylated hormone behaved as an antagonist to native hCG. Similar effects on signal
transduction were shown in other studies using chemically deglycosylated hCG [110,
115–120]. The in vivo activity measured by increase of the uterine weight or ovarian
ascorbic acid depletion was also abolished by deglycosylation [110,115,116,121,122]
and the antagonistic activity of the deglycosylated analogs was preserved in vivo even
though they were cleared from circulation more rapidly than the corresponding glycosylated forms. Thus, the absence of the carbohydrate component dissociates receptor
binding from signal transduction and the N-linked oligosaccharides are obligatory for

maximal biological activity of glycoprotein hormones.

Studies on the bioactivity of the hormones in which only one of the subunits has been deglycosylated either by chemical or enzymatic methods, led to the recognition that the α subunit oligosaccharides are more critical for signal transduction than the β subunit oligosaccharides [109,110]. This was demonstrated by Sairam and Bhargavi [123], who used mixed species of chemically deglycosylated oLH and oFSH subunits and showed that dimers containing deglycosylated α subunit did not stimulate cAMP production in rat testicular cells. Deglycosylated analogs showed antagonistic effects on LH- or FSH-stimulated cAMP production [123].

Similar to hCG/LH, the FSH and TSH oligosaccharides are not essential for receptor binding [124–130]. There are discrepancies concerning agonistic and antagonistic properties of deglycosylated ovine and human FSH in signal transduction, and oligosaccharides on both subunits may be necessary for full expression of steroidogenic activity [124,125]. Chemical or enzymatic deglycosylation reduces biological activity of TSH [126–129]. Similar to hCG, deglycosylation of the TSHα subunit has a more profound effect on bioactivity of TSH than deglycosylation of the TSHβ subunit [127, 130].

Deglycosylation by chemical methods or by exo- and endoglycosidase digestion have limitations; chemical treatment can damage the polypeptide, whereas enzymatic deglycosylation may not be complete and proteolytic cleavage can occur during prolonged digestion. Moreover, these methods do not permit study of individual glycosylation sites. These problems are overcome by recombinant DNA technology. Deglycosylation can be controlled by introducing mutations into the gene sequence which after expression in mammalian cells result in homogenous population of the desired mutant.

Matzuk et al. [131] ablated the individual Asn-X-Thr glycosylation signal at each of the four N-linked hCG oligosaccharides using site-directed mutagenesis of recombinant hCG subunit genes [131]. One potential problem encountered with site-directed mutagenesis is that the amino acid substitutions may have phenotypic effects independent of the loss of carbohydrate. To address this point the Asn-X-Thr glycosylation signal was abolished by two independent mutations: in one asparagine was changed to aspartic acid (α subunit) or glutamine (β subunit) and in the other threonine was replaced by alanine [131]. Both substitutions gave similar results in assays measuring secretion, heterodimer formation, or biological activity; although due to its instability in the bioassay, the activity of hCG containing the α mutant wherein the Asn 52 glycosylation site was eliminated by change of Thr 54 to Ala 54 was not determined [131]. Elimination of either of the four N-linked sugar chains of hCG separately or in any combination was without effect on the receptor binding activity [131]. Mutations at either one or both glycosylation sites on the hCGβ subunit or the α Asn 78 glycosylation site had no effect on cAMP accumulation and steroidogenesis, whereas both activities of hCG containing the α Asn 52 glycosylation mutant were markedly diminished. When this α mutant was expressed together with the hCGβ subunit lacking the Asn 13 glycosylation site, steroidogenic activity was further reduced (Table I). Whereas the α Asn 52 oligosaccharide is indispensable for maximal effect, the cooperation of one glycosylation site on each subunit may be necessary to elicit full activity of hCG [131]. All non-glycosylated analogs generated by site-directed mutagenesis behave as competitive antagonists [131].

Removal of the N-linked oligosaccharide of the bovine LHβ by site-directed mutagenesis did not affect signal transduction [108]. This result is consistent with the previous findings that the β subunit oligosaccharides are less important in biologic activity of gonadotropins.

To study if the bioactivity of hCG can be altered by the differences in the underlying oligosaccharide structure, recombinant hCG was expressed in CHO glycosylation mutant cell lines defective in processing of the N-linked oligosaccharides [132]. One of the cell lines, 1021, is deficient in CMP-sialic acid transferase, and therefore asialo hCG is synthesized [133]. HCG expressed in another cell line, 15B, which is deficient in N-acetylglucosaminyl transferase I [134] would contain exclusively high-mannose oligosaccharides. Asialo hCG evoked much less adenylate cyclase activation and steroidogenic response in mouse MA-10 cells than the mutant containing high mannose oligosaccharides [132]. Thus, sialic acid may play a role in modulating biological activity. Earlier studies showed that the bioactivity of sialidase treated hCG or hCG containing asialo β subunit was reduced compared to the native hormone [114, 135–137]. Sialic acid apparently does not influence hFSH bioactivity. When hFSH was expressed in CHO glycosylation mutant cell lines, 1021 and 15B, the *in vitro* bioactivity of the dimers containing either asialo or high-mannose oligosaccharides was unchanged compared to the wild-type hormone. Only the half-life in circulation was affected which in turn reduced *in vivo* bioactivity [138].

In contrast to sialylated hCG, the bioactivity of the sialylated bLH expressed in CHO was threefold less than the sulfated native LH. Sialidase digestion restored it to the level of the sulfated hormone [72] and subsequent removal of galactose had no effect on bioactivity. Sialylated recombinant hTSH showed reduced bioactivity, which could be restored after desialylation [139]. Desialylation of eCG increased steroidogenic response in Leydig tumor cells [140]. Thus, sialic acid may affect bioactivity depending on the hormone.

3.2.2.1. Receptor:hormone interaction

Both hCG subunits bind to the receptor [141–145] and the α subunit appears to be located at the receptor side since anti-α antibodies do not bind to the hCG receptor complex [143,145]. Based on studies using synthetic peptides [146,149] and site-directed mutagenesis of recombinant hCG [150,151], several receptor binding domains have been identified within the amino acid sequence of both α and β subunits of glycoprotein hormones. These data imply that a peptide component is directly involved in receptor binding. However, the domain involved in signal transduction has not been identified. There is evidence that high concentrations of synthetic peptides containing the presumed binding domains of the hCG and FSH subunits can evoke a biologic response after binding to LH/CG and FSH receptor [149,152,153], and the biopotency of synthetic peptides can be significantly improved by combining sequences from both α and the FSHβ subunit [154]. This suggests that signal transduction results from direct interaction of receptor with a peptide component. However, as discussed above, the hCG oligosaccharides, especially those of the α subunit, are critical for maximal bioactivity, and direct interaction of oligosaccharide receptor can not be excluded. Glycopeptides and oligosaccharides derived from non-related glycoproteins can prevent receptor binding [155]. It has been suggested that in addition to the receptor, another oligosaccharide binding membrane component may be involved in hCG signal transduction since it can be blocked by hCG glycopeptides [156].

It is not known how the N-linked oligosaccharides are positioned in the receptor–hormone complex but they are at least partially accessible to glycosidases, since sialic acid can be removed from receptor-bound hCG subunits [157]. Deglycosylation apparently changes the nature of hormone receptor interaction because dghCG binds differently to receptor than the native hormone. Antibodies specific for both deglycosylated and native hormone which still recognize receptor-bound hCG, did not bind to the dghCG receptor complex [158]. Binding studies of deglycosylated hCG to either heterologous cells expressing recombinant human LH/CG receptor or to rat testis receptor show that it behaves as noncompetitive antagonist and remains irreversibly bound to receptor [159]. Moreover, dghCG apparently does not bind to the same receptor domain as the native or asialo hCG [160,161]. Thus reduced activity of dghCG may be due to conformational changes introduced by oligosaccharide removal and results in aberrant receptor binding, preventing signal transduction [161]. Small conformational differences between deglycosylated and native hCG and FSH were found by physico-chemical measurements [117,118,125,162] and differences in immunological properties [112,117,163–165]. These changes are apparently reversible since agonist activity of dghCG is restored by binding of anti-hCGβ subunit antibodies to the dghCG-receptor complex [165–167].

Thus, oligosaccharides may not have a direct link to a signal transducing domain in the receptor but rather they maintain the proper interaction of the peptide component with the receptor transducing domain. This hypothesis is consistent with data showing that the Asn 52 oligosaccharide of the α subunit is required for intracellular subunit assembly [106], and for maximal hCG biologic activity [131]. The processing of this carbohydrate is different in hCG dimer and in the uncombined α subunit [98] and it is more heterogeneous compared with other glycosylation sites on hCG [59]. The data suggest that the Asn 52 oligosaccharide is close to the α:β interface which is a part of or adjacent to the receptor binding domain [101]. This oligosaccharide could influence interaction of both subunits with the receptor, and its heterogeneity may modulate the potency of subsequent signal transduction. The irreversibility of deglycosylated hCG receptor binding implies that oligosaccharides are also required for dissociation of hormone–receptor complex [159].

3.2.3. Half-life of circulating hormones

The extracellular half-life of the glycoprotein hormones depends on the terminal modification of their N-linked oligosaccharides. Sialylation of the N-linked oligosaccharides protects the circulating glycohormones against clearance by hepatic asialoglycoprotein receptors [135,136,168]. However, sulfation leads to more rapid clearance of pituitary hormones; sulfated βLH has 4–5-fold greater clearance rate compared with the sialylated recombinant βLH [169]. Fiete et al. [170] demonstrated that sulfated LH is removed from circulation after binding to the S4GGnM receptor on the surface of liver endothelial cell. This receptor recognizes a structure $SO_4(-4-)GalNAc(\beta1-4)GlcNAc$ $(\beta1-2)Man\alpha$ in both α-1,3- and α-1,6- antennae. Desulfated LH did not show changes in receptor binding or signal transduction in mouse MA-10 cells but similar to desialy-lated hormone, larger doses were required to stimulate ovulation *in vivo* [169]. Thus, the apparent role of sulfate is to regulate the circulatory half-life of LH [72,169,170]. Similarly, sialylated recombinant hTSH has a longer half-life in circulation than pitui-tary-derived TSH [139]. The sulfation/sialylation pattern of TSH changes during de-velopment [171,172] and can be regulated by thyrotropin releasing hormone [173]. In hypothyroid patients TSH is more sialylated than in normal tissue [174] and is cleared with a slower rate [175]. The release of LH and TSH is controlled by surges of hypothalamic releasing hormones, GnRH and TRH, respectively [176,177]. A rapid clearance system may be necessary to maintain pulsatile levels of hormones and to prevent down regulation of the receptor from a sustained level of the glycoprotein hormones. A similar system apparently exists in the equine, since the S4GGnM liver receptor recognizes also sulfated eLH and its circulatory half-life is 6-fold shorter than that of sialylated eCG which may account for the different biopotency of eLH and eCG [72].

The variation in the sialylation and sulfation patterns contribute to the heterogeneity of the pituitary hormones from different species analyzed by isoelectric-focusing or chromato-focusing. These isoforms differ in biopotency, circulatory half-life, and they are apparently under endocrine control [178–187]. The profile changes during devel-opment. More basic biologically potent species appear at puberty [188–190] and after administration of gonadotropin releasing hormone or steroids in castrated animals [191–194]. Some of the TSH [195] and FSH isoforms [196] behave as antagonists. The isoforms received much attention, since regulating the final steps of N-linked oligosac-charide processing may represent a physiological way to control potency and longevity of the pituitary hormones as discussed above.

4. O-linked glycosylation of glycoprotein hormones

4.1. Structure of the O-linked oligosaccharides

The hCGβ and the equine CGβ subunits contain O-linked oligosaccharides. In the hCGβ subunit these are linked to four serine residues 121, 127, 132, and 138 (Fig. 2). The structure of O-linked oligosaccharides was identified by Kessler et al. [47], Cole et al. [197], and Amano et al. [198] on hCG secreted during normal pregnancy and in patients with trophoblastic disease (Fig. 4). While the disaccharide core-containing structures are prevalent on hCG derived from normal pregnancies, the choriocarcinoma hCGβ subunit contains more tetrasaccharides [198]. The hCGβ subunit from the BeWo choriocarcinoma cell line also contains heptasaccharides and fucosylated tetrasaccha-rides [199].

At least four sites are glycosylated on the equine CGβ subunit and the O-linked oligosaccharides of eCGβ subunit comprise a population of heterogeneous species, the majority of which are monosialylated [49]. Among them oligo (N-acetyllactosamine) units have been found [49].

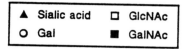

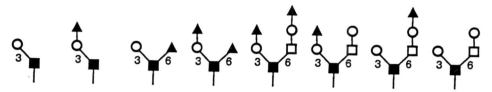

Fig. 4. Structure of the hCGβ subunit serine O-linked oligosaccharides [47,197,198].

4.2. Function of the O-linked oligosaccharides in the hCGβ subunit

Studies on the function of the O-linked oligosaccharides on the hCGβ subunit are possible by recombinant DNA methods, which permit deleting the O-linked glycans and/or the O-linked bearing peptide [200–202].

4.2.1. Intracellular role in subunit assembly and secretion

To assess the effect of O-glycosylation on hCG assembly and secretion, the hCG subunit genes were transfected into CHO mutant cells, ldlD. These cells are deficient in UDP-Gal/UDP-GalNAc 4-epimerase which cannot generate GalNAc and O-glycosylation is prevented [203]. The assembly and secretion of hCG devoid of O-linked oligosaccharides was not impaired indicating that glycosylation of the carboxyterminal extension does not significantly affect these intracellular events [200].

4.2.2. Bioactivity of the O-glycosylated carboxyterminal peptide of placental gonadotropins

The carboxyterminal O-glycosylated peptide was removed from the hCGβ subunit gene by introducing a termination codon following amino acid 114 [201]. Receptor binding and signal transduction of hCG containing the truncated β subunit studied *in vitro* remained unchanged compared to the native hormone [201,202]. However, truncated hCG was threefold less potent *in vivo* than the native hormone in inducing ovulation in primed female rats [201]. Thus, the data indicated that, as suggested earlier [116,204–206], the O-glycosylated carboxyterminal extension of the hCGβ subunit increased the half-life of hCG in circulation.

Based on the above results it became feasible to transfer the O-glycosylated carboxyterminal extension of the hCGβ subunit to other glycoprotein hormones to prolong their circulatory half-life; this could have important clinical applications. Recently, a chimera was constructed in which the sequence coding for the hCGβ carboxyterminal peptide was attached to the –COOH terminus of the hFSHβ subunit gene [207]. This chimera was transfected into CHO cells and the protein named FC was expressed [207]. Its *in vivo* potency and circulatory half-life were dramatically increased compared to

FSH [208]. The presence of sialylated O-linked oligosaccharides results in a significant retardation of the electrophoretic migration of the O-glycosylated glycoprotein [87]. The shift in electrophoretic mobility of FC on SDS-PAGE when compared with FSH indicated that the carboxyterminal extension is glycosylated with O-linked units. Thus the carboxyterminal peptide of the hCGβ subunit contains sufficient information for O-glycosylation to occur. The number and structure of O-linked units on the chimera is not known, but the system in which the O-glycosylated part can be transferred to heterologous proteins and modified by site-directed mutagenesis, creates an opportunity to study the recognition requirements for O-glycosylation.

The above data supported the conclusion that the presence of the sialylated O-linked sugars maintain a high level of hCG in circulation. Placental gonadotropins, hCG and eCG are unique among the glycoprotein hormones, since their synthesis and secretion is limited to a short period. The addition of O-linked glycans on a relatively short peptide extension may represent an adaptative response for maintaining a high level of circulating gonadotropin during pregnancy.

5. Conclusions

The overall picture emerging from the data collected from different experimental systems suggests that the role of the individual components of glycoprotein hormone should not be considered as a separate entity but rather in a context of interactions between subunits and their polypeptides and oligosaccharides. The role of N-linked oligosaccharides in glycoprotein hormones is exerted at several levels. Intracellularly they are involved in subunit folding, heterodimer formation assembly and secretion; extracellularly, the oligosaccharides modulate hormonal clearance and signal transduction. Site-directed mutagenesis studies suggest functional differences in the oligosaccharides at the individual glycosylation sites for intracellular and extracellular events (Table I). Diversity in the N-linked oligosaccharides, especially the different ratio of sialylated sulfated species, can affect the bioavailability and bioactivity of glycohormones. At least one role for the O-linked oligosaccharides in the hCGβ and eCGβ subunits is to increase the circulatory half-life of the hormones where there is a demand for sustained levels during pregnancy. The availability of the three-dimensional structure of glycoprotein hormones together with recombinant DNA methods should permit the construction of agonists or antagonists as already exemplified by longer acting hFSH analog.

Acknowledgments

We are grateful to Dr. J.U. Baenziger for critical reading of the manuscript. M.B. is grateful to Dr. Mesut Muyan for helpful discussions.

Abbreviations

hCG	human chorionic gonadotropin
eCG	equine chorionic gonadotropin
dghCG	deglycosylated hCG
FSH	follicle stimulating hormone
LH	luteinizing hormone
TSH	thyroid stimulating hormone
h	human
b	bovine
Glc	glucose
Gal	galactose
GlcNAc	N-acetylglucosamine
GalNAc	N-acetylgalactosamine
Man	mannose
Sia or NeuAc	sialic acid or N-acetylneuraminic acid
S4GGnM	$SO_4(4)GalNac(\beta1-4)GlcNAc(\beta1-2)Man\alpha$
Asn	asparagine
Thr	threonine

References

1. Pierce, J.G. and Parsons, T.S. (1981) Annu. Rev. Biochem. 50, 465–495.
2. Rayford, P.L., Vaitukaitis, J.L., Ross, G.T., Morgan, F.J. and Canfield, R.E. (1972) Endocrinology 91, 144–146.
3. Catt, K.J., Dufau, M.L. and Tsuruhara, T. (1973) J. Clin. Endocrinol. Metab. 36, 73–80.
4. Catt, K.J. and Dufau, M.L. (1991) In: S.S.C. Yen and R.B. Jaffe (Eds.), Reproductive Endocrinology, W.S. Saunders Company, Philadelphia, PA, Ch. 4 pp. 105–155.
5. Hodgen, G.D., Itskowitz, J. (1988) In: E. Knobil, J.D. Neill, L.L. Ewing, G.S. Greenwald, C.L. Markert and D.W. Pfaff (Eds.), The Physiology of Reproduction. Raven Press, New York, Vol. II, pp. 1995–2021.
6. Hsueh, A.J.W., Bicsak, T.A., Jai, X-C., Dalh, K.D., Fauser, B.C.J.M., Galway, A.B., Czekala, N., Pavlou, S.N., Papkoff, H., Keene, J. and Boime, I. (1989) Rec. Prog. Horm. Res. 45, 209–277.
7. Burrow, G.N. (1991) In: S.S.C. Yen and R.B. Jaffe (Eds.), Reproductive Endocrinology, 3rd Edn., W.S. Saunders Company, Philadelphia, PA, Ch. 16, pp.555–575.
8. McFarland, K.C., Sprengel, R., Phillips, H.S., Kohler, M., Rosenblit, N., Nikolics, K., Segaloff, D.L. and Seeburg, P.H. (1989) Science 245, 494–499.
9. Loosfeld, H., Misrahi, M., Atger, M., Salesse, R., Vu Hai-Luu Thi, M.T., Jolivet, A., Guiochon-Mantel, A., Sar, S., Jallal, B., Garnier, J. and Milgrom, E. (1989) Science, 245, 525–528.
10. Nagayama, Y. and Rapoport, B. (1992) Molec. Endocrinol. 6, 145–156.
11. Sprengel, R., Braun, T., Nikolics, K., Segaloff, D.L. and Seeburg, P.H. (1990) Molec. Endocrinol. 4, 525–530.
12. Minegish, T., Nakamura., K., Takakura, Y., Ibuki, Y., Igarashi, M. (1991) Biochem. Biophys. Res. Commun. 175, 1125–1130.
13. Ascoli, M. and Segaloff, D.L. (1989) Endocrine Rev. 10, 27–44.

14. Rajanieni, H.J., Keinanen, K.P., Kellokumpu, S., Petaja-Repo, U.E. and Metsikko, M.K. (1989) Biol. Reprod. 40, 1–12.

15. Segaloff, D.L., Sprengel, R., Nikolics, K., Ascoli, M. (1990) Recent Prog. Horm. Res. 46, 261–303.

16. Xie, Y-B., Wang, H. and Segaloff, D.L. (1990) J. Biol. Chem. 265, 21411–21414.

17. Moyle, W.R., Bernard, M.P., Myears, R.V., Marko, O.M. and Strader C.D. (1991) J. Biol. Chem. 266, 10807–10812.

18. Cole, F.E., Weed, J.C., Schneider, G.T., Holland, J.B., Geary, W.L., Levy, D.L., Huseby, R.A. and Rice, B.F. (1976) Fertil. Steril. 27, 921–928.

19. Cameron, J.L. and Stouffer, R.L. (1981) Biol. Reprod. 25, 568–572.

20. Jia, X.C., Oikawa, M., Bo, M., Tanaka, T., Ny, T., Boime, I. and Hsueh, A.J.W. (1991) Mol. Endocrinol. 5, 759–768.

21. Till, N.L., Aihara, T., Nishimori, K., Jia, X-C., Billig, H., Kowalski, K.I., Perlas, E.A. and Hsueh, A.J.W. (1992) Endocrinology, 131, 799–806.

22. Harris, D.C., Machin, K.J., Evin, G.M., Morgan, F.J. and Isaacs, N.W. (1989) J. Biol. Chem. 264, 6705–6706.

23. Lustbader, J.W., Birken, S., Pillegi, N.F., Gawinowicz-Kolks, M.A., Pollak, S., Cuff, M.E., Yang, W., Hendrickson, W.A. and Canfield, R.E. (1989) Biochemistry 28, 9239–9243.

24. Chen, W. and Bahl, O.P. (1991) J. Biol. Chem. 266, 9355–9358.

25. Chen, W., Shen, Q-X. and Bahl, O.P. (1991) J. Biol. Chem. 266, 4081–4087.

26. Fiddes, J.C. and Goodman, H.M. (1981) J. Mol. Appl. Genet. 1, 3–18.

27. Boothby, M., Ruddon, R.W., Anderson, C., McWiliams, D. and Boime, I. (1981) J. Biol. Chem. 256, 5121–5127.

28. Stewart, F., Thompson, J.A., Leigh, S.E.A. and Warwick, J.M. (1987) J. Endocrinol. 115, 341–346.

29. Ryan, R.J., Charlesworth, M.C., McCormick, D.J., Milius, R.P. and Keutman, H. (1988) FASEB J. 2, 2661–2669.

30. Gordon, W.L. and Ward, D.N. (1985) In: M. Ascoli (Ed.) Luteinizing Hormone Action and Receptors. CRC Press, Inc., Boca Raton, FL, pp. 173–197.

31. Mise, T. and Bahl, O.P. (1980) J. Biol. Chem. 255, 8516–8522.

32. Strickland, T.W. and Puett, D. (1981) Endocrinology 109, 1933–1942.

33. Boorstein, W.R., Vamvakapoulos, N.C. and Fiddes, J.C. (1982) Nature, 300, 419–422.

34. Talmadge, K., Boorstein, W.R. and Fiddes, J.C. (1983) DNA 2, 281–289.

35. Policastro, P.F., Daniels-McQueen, S., Carle, G. and Boime, I. (1986) J. Biol. Chem. 261, 5907–5916.

36. Watkins, P.C., Eddy, R., Beck, A.K., Vellucci, V., Leverone, B., Tanzi, R.E., Gusalla, J.F. and Shon, T.B. (1987) DNA 6, 205–212.

37. Keene, J.L., Matzuk, M.M., Otani, T., Fauser, B.C.J.M., Galway, A.B., Hsueh, A.J.W., Boime, I. (1989) J. Biol. Chem. 264, 4769–4775.

38. Whitfield, G.K., Powers, R.E., Gurr, J.A., Wolf, O. and Kourides, I.A. (1986) In: G. Medeiros-Neto and E. Gaitan (Eds.), Frontiers in Thyroidology, Vol. I. Plenum Publishing Co., New York, pp. 173–176.

39. Hayashizaki, Y., Miyai, K., Kata, K., Matsubara, K. (1985) FEBS Lett. 188, 394–400.

40. Sherman, G.B., Wolfe, M.W., Farmerie, T.A., Clay, C.M., Threadgill, D.S., Sharp, D.C. and Nilson, J.H. (1992) Mol. Endocrinol. 6, 951–959.

41. Mise, T. and Bahl, O.P. (1981) J. Biol. Chem. 256, 6587–6592.

42. Anumula, K.R. and Bahl, O.P. (1986) Fed. Proc. 45, 1843, Abstract 2117.

43. Bousfield, G.R., Liu, W.-K., Sugino, H. and Ward, D.N. (1987) J. Biol. Chem. 262, 8610–8620.

44. Murphy, B.D. and Martinuk, S.D. (1991) Endocrine Rev. 12, 27–44.

45. Birken, S., Canfield, R.E. (1977) J. Biol. Chem. 252, 5386–5392.
46. Bousfield, G.R., Sugino, H. and Ward, D.N. (1985) J. Biol. Chem. 260, 9531–9533.
47. Kessler, M.J., Mise, T., Ghai, R.D. and Bahl, O.P. (1979) J. Biol.Chem. 254, 7909–7914.
48. Bahl, O.P. and Anumula, K.R. (1986) Fed. Proc. 45, 1818, Abstract 1973.
49. Damm, J.B., Hard, K., Kammerling, J.P., Van Dedem, G.W.K., Vliegenthart, J.F.G. (1990) Eur. J. Biochem. 189, 175–183.
50. Endo, Y., Yamashita, K., Tachibana, Y., Tojo, S. and Kobata, A. (1979) J. Biochem. (Tokyo) 85, 669–679.
51. Kessler, M.J., Reddy, M.S., Shah, R.H. and Bahl, O.P. (1979) J. Biol. Chem. 254, 7901–7908.
52. Mizouchi, T., Nishimura, R., Taniguchi, T., Utsunomiya, T., Mochizuki, M., Derappe, C. and Kobata, A. (1985) Jpn. J. Cancer Res. (Gann) 76, 752–759.
53. Nishimura, R., Endo, K., Tanabe, Y. and Tojo, S. (1981) J. Endocrine Invest. 4, 349–358.
54. Mizuochi, T., Nishimura, R., Derappe, C., Taniguchi, T., Hamamoto, T., Mochizuki, M. and Kobata, A. (1983) J. Biol. Chem. 258, 14126–14129.
55. Endo, T., Nishimura, R., Kawano, T., Mochizuki, M. and Kobata, A. (1987) Cancer Res. 47, 5242–5245.
56. Endo, T., Nishimura, R., Mochizuki, M., Kochibe, N. and Kobata, A. (1988) J. Biochem. (Tokyo) 103, 1035–1038.
57. Kobata, A. (1988) J. Cell. Biochem. 37, 79–90.
58. Mizuochi, T. and Kobata, A. (1980) Biochem. Biophys. Res. Commun. 97, 772–778.
59. Weisshaar, G., Hiyama, J. and Renwick, A.C.G. (1991) Glycobiology 1, 393–404.
60. Matzuk, M.M., Keene, J.L. and Boime, I. (1989) J. Biol. Chem. 263, 2409–2414.
61. Matzuk, M.M. and Boime, I. (1989) Biol. Reprod. 40, 48–53.
62. Parsons, T.F. and Pierce, J.G. (1980) Proc. Natl. Acad. Sci. U.S. A. 77, 7089–7093.
63. Hortin, G., Natowicz, M., Pierce, J., Baenziger, J., Parsons, T. and Boime, I. (1981) Proc. Natl. Acad. Sci. USA 78, 7468–7472.
64. Anumula, K.R. and Bahl, O.P. (1983) Arch. Biochem. Biophys. 220, 645–651.
65. Green, E.D., Boime, I. and Baenziger, J. (1986) J. Biol. Chem. 261, 16309–16316.
66. Green, E.D., Gruenebaum, J., Bielinska, M., Baenziger, J.U. and Boime, I. (1984) Proc. Natl. Acad. Sci. USA 81, 5320–5324.
67. Green, E.D., Morishima, C., Boime, I. and Baenziger, J.U. (1985) Proc. Natl. Acad. Sci. USA 82, 7850–7854.
68. Skelton, T.P., Hooper, L.U., Srivastava, V.V., Hindsgaul, O. and Baenziger, J.U. (1991) J. Biol. Chem. 266, 17142–17150.
69. Smith, P.L. and Baenziger, J.U. (1988) Science 242, 930–933.
70. Smith, P.L. and Baenziger, J.U. (1990) Proc. Natl. Acad. Sci. USA 87, 7275–7279.
71. Smith, P.J. and Baenziger, J.U. (1992) Proc. Natl. Acad. Sci. USA 89, 329–333.
72. Smith, P., Bousfield, G.R., Kumar, S., Fiete, D. and Baenziger, J.U. (1993) J. Biol. Chem. 268, 795–802.
73. Matsui, T., Sugino, H., Miura, M., Bousfield, G.R., Ward, D.N., Titani, K. and Mizouchi, T. (1991) Biochem. Biophys. Res. Commun. 174, 940–945.
74. Green, E.D., Van Halbeek, H., Boime, I. and Baenziger, J.U. (1985) J. Biol. Chem. 260, 15623–15630.
75. Green, E.D. and Baenziger, J.U. (1988) J. Biol. Chem. 263, 25–35.
76. Green, E.D. and Baenziger, J.U. (1988) J. Biol. Chem. 263, 36–44.
77. Baenziger, J.U. and Green, E.D. (1988) Biochim. Biophys. Acta 947, 287–306.
78. Green, E.D., Boime, I. and Baenziger, J.U. (1986) Molec. Cell. Biochem. 72, 81–100.
79. Weisshaar, G., Hiyama, J. and Renwick, A.G.C. (1990) Eur. J. Biochem. 192, 741–751.
80. Hiyama, J., Weisshaar, G. and Renwick, A.G.C. (1992) 2, 401–409.

81. Weisshaar, G., Hiyama, J., Renwick, A.G.C. and Nimtz, M. (1991) Eur. J. Biochem. 195, 257–268.
82. Ujihara, M., Yamamoto, K., Nomura, K., Toyoshima, S., Demura, H., Nakamura, Y., Ohmura, K. and Osawa, T. (1992) Glycobiology 2, 225–231.
83. Stokwell Hartree, A. and Renwick, A.G.C. (1992) Biochem. J. 287, 665–679.
84. Renwick, A.C.G., Mizuochi, T., Kochibe, N. and Kobata, A. (1987) J. Biochem. 101, 1209–1221.
85. Hoshina, H. and Boime, I. (1982) Proc. Natl. Acad. Sci. USA 79, 7649–7653.
86. Peters, B.P., Krzesicki, R.F., Hartle, R.J., Perini, F. and Ruddon, R.W. (1984) J. Biol. Chem. 259, 15123–15130.
87. Corless, C.L., Matzuk, M.M., Ramabhadran, T.V., Krichevsky, A. and Boime, I. (1987) J. Cell Biol. 104, 1173–1181.
88. Hard, K., Mekking, A., Damm, J.B.L., Kammerling, J.P., De Boer, W., Wijnards, R.A. and Vliegenhart, J.F.G. (1990) Eur. J. Biochem. 193, 263–271.
89. Smith, P.L., Kaetzel, D., Nilson, J. and Baenziger, J.U. (1990) J. Biol. Chem. 265, 874–881.
90. Parsons T.F. and Pierce, J.G. (1984) J. Biol. Chem. 259, 2662–2666.
91. Cole, L.A., Perini, F., Birken, S. and Ruddon, R.W. (1984) Biochem. Biophys. Res. Commun. 122, 1260–1267.
92. Corless, C.L. and Boime, I. (1985) Endocrinology, 117, 1699–1706.
93. Corless, C.L., Bielinska, M., Ramabhadran, T.V., Daniels-McQueen, S., Otani, T., Reitz, B.A., Tiemeier, D.C. and Boime, I. (1987) J. Biol. Chem. 262, 14197–14203.
94. Lustbader, J.W., Birken, S., Pollak, S., Levinson, L., Bernsine, E., Hsiung, N. and Canfield, R.E. (1987) J. Biol. Chem. 262, 14204–14212.
95. Blithe, D.L. and Nisula, B.C. (1985) Endocrinology 117, 2218–2228.
96. Blithe, D.L. (1990) J. Biol. Chem. 265, 1951–1956.
97. Blithe, D.L. (1990) Endocrinology 126, 2788–2799.
98. Bielinska, M., Matzuk, M. and Boime, I. (1989) J. Biol. Chem. 264,17113–17118.
99. Strickland, T.W. and Puett, D. (1982) Endocrinology 11, 95–100.
100. Strickland, T.W. and Puett, D. (1983) Int. J. Peptide Protein Res. 21, 374–380.
101. Bielinska M. and Boime, I. (1992) Molec. Endorinol. 6, 267–271.
102. Blithe, D., Richards, R.G. and Skarulis, M.C. (1991) Endocrinology 129, 2257–2259.
103. Strickland, T.W. and Pierce, J.G. (1983) J. Biol. Chem. 258, 5927–5932.
104. Strickland, T.W., Thomason, A.R., Nilson, J.H. and Pierce, J.G. (1985) J. Cell. Biochem. 29, 225–237.
105. Weintraub, B.D., Stannard, B.S., Linnekin, D. and Marshall, M. (1980) J. Biol. Chem. 255, 5715–5723.
106. Matzuk, M.M. and Boime, I. (1988a) J. Cell Biol. 106, 1049–1059.
107. Matzuk, M.M. and Boime, I. (1988b) J. Biol. Chem. 263, 17106–17111.
108. Kaetzel, D.M., Virgin, J.B., Clay, C.M. and Nilson, J.H. (1989) Mol. Endocrinol. 3, 1765–1774.
109. Goverman, J.M., Parson, T.F. and Pierce, J.G. (1982) J. Biol. Chem. 257, 15059–15064.
110. Chen, H.-C., Shimohigashi, Y., Dufau, M.L. and Catt, K.J. (1982) J. Biol. Chem. 257, 14446–14452.
111. Sairam, M.R. (1985) In: P.M. Conn (Ed.), The Receptors, Vol. II. Academic Press, New York, pp. 307–340.
112. Sairam, M.R. (1989) FASEB J. 3, 1915–1926.
113. Sairam, M.R., Bhargavi, G.N. and Yarney, T.A. (1990) FEBS Lett. 276, 143–146.
114. Moyle, W.R., Bahl, O.P. and Marz, L. (1975) J. Biol. Chem. 250, 9163–9169.
115. Kalyan, N.K., Lippes, H.A. and Bahl, O.P. (1982) J. Biol. Chem. 257, 12624–12631.
116. Kalyan, N.K. and Bahl, O.P. (1983) J. Biol. Chem. 258, 67–74.
117. Keutmann, H.T., McIlroy, P.J., Bergert, E.R. and Ryan, R.J. (1983) Biochemistry 22, 3067–3072.

118. Sairam, M.R. and Manjunath, P. (1983) J. Biol. Chem. 258, 445–449.
119. Manjunath, P. and Sairam, M.R. (1982) J. Biol. Chem. 257, 7109–7115.
120. Manjunath, P., Sairam, M. R. and Sairam, J. (1982) Mol. Cell. Endocrinol. 28, 125–138.
121. Kato, K., Sairam, M.R. (1983) Contraception, 27, 515–520.
122. Kato, K., Sairam, M.R. and Manjunath, P. (1983) Endocrinology 113, 195–199.
123. Sairam, M.R. and Bhargavi, G.N. (1985) Science 229, 65–67.
124. Pabmanabhan, V., Sairam, M.R., Hassing, J.M., Brown, M.B., Ridings, J.W. and Beitins, I.Z. (1991) Mol. Cell. Endocrinol. 79, 119–128.
125. Calvo, F.O., Keutmann, H.T., Bergert, E.R. and Ryan, R.J. (1986) Biochemistry 25, 3938–3943.
126. Amir, S.M., Kubota, K., Tramontano, D., Ingbar, S.H. and Keutmann, H.T. (1987) Endocrinology 120, 345–352.
127. Amr, S., Menezes-Ferreira, M., Shimohigashi, Y., Chen, H.C., Nisula, B. and Weitraub, B.D. (1986) J. Endocrinol. Invest. 8, 537–541.
128. Thotakura, N.R., LiCalzi, L. and Weintraub, B. (1990) J. Biol. Chem. 265, 11527–11534.
129. Thotakura, N.R., Desai, R.K., Szkudlinski, M. and Weintraub, B.D. (1992) Endocrinology, 131, 82–88.
130. Papandreou, M.-J., Sergi, I., Medri, G., Labbe-Jullie, C., Braun, J.M., Cannone, C. and Ronin, C. (1991) Mol. Cell. Endocrinol. 78, 137–150.
131. Matzuk, M.M., Keene, J.L. and Boime, I. (1989) J. Biol. Chem. 264, 2409–2414.
132. Keene, J.L., Matzuk, M.M. and Boime, I (1989) Molec. Endocrinol. 3, 2011–2017.
133. Deutscher, S.L., Nuwaybid, N., Stanley, P., Briles, E.I.B. and Hirschberg, C.J. (1984) Cell 39, 295–299.
134. Gottlieb, C., Baenziger, J. and Kornfeld, S. (1975) J. Biol. Chem. 250, 3303–3309.
135. Van Hall, E.V., Vaitukaitis, J.L., Ross, G.T., Hickman, J.W. and Ashwell, G. (1971) Endocrinology 88, 456–464.
136. Van Hall, E.V., Vaitukaitis, J.L., Ross, G.T., Hickman, J.W. and Ashwell, G. (1971) Endocrinology 89, 11–15.
137. Amir, S.M., Kasagi, K., Ingbar, S.H. (1987) Endocrinology 121, 160–166.
138. Galway, A.B., Hsueh, A.J.W., Keene, J.L., Yamato, M., Fauser, B.C.J.M. and Boime, I. (1990) Endocrinology 127, 93–100.
139. Thotakura, N.R., Desai, R.K., Bates, L.G., Cole, E.S., Pratt, B.M. and Weintraub, B.D. (1991) Endocrinology 128, 341–348.
140. Moore Jr., T.W. and Ward, D.N. (1980) J. Biol. Chem. 255, 6930–6936.
141. Pierce, J.G., Bloomfield, G.A. and Parsons, T.F. (1979) Int. J. Peptide Protein Res. 13, 54–61.
142. Ji, I. and Ji, T.H. (1981) Proc. Natl. Acad. Sci. USA 78, 5465–5469.
143. Milius, R.P., Midgley, Jr., A.R. and Birken S. (1983) Proc. Natl. Acad. Sci. USA 80, 7359–7379.
144. Ascoli, M. and Segaloff, D.L. (1986) J. Biol. Chem. 261, 3807–3815.
145. Moyle, W.R., Erlich, P.H. and Canfield, R.E. (1982) Proc. Natl. Acad. Sci. USA 79, 2245–2249.
146. Morris III, J.C., Jiang, N.-S., Charlesworth, M.C., McCormick, D.J. and Ryan, R.J. (1988) Endocrinology 123, 456–462.
147. Charlesworth, M.C., McCormick, D.J., Madden, B.J. and Ryan, R.J. (1987) J. Biol. Chem. 262, 13409–13415.
148. Reed, D.K., Ryan, R.J., McCormick, D.J. (1991) J. Biol. Chem. 266, 14251–14255.
149. Keutman, H.T., Charlesworth, M.C., Mason, K.A., Ostrea, T., Johnson, L. and Ryan, R.J. (1987) Proc. Natl. Acad. Sci. USA 84, 2038–2042.
150. Bielinska, M., Pixley, M.R. and Boime, I. (1990) J. Cell Biol. 111, 330a (Abstract 1844).

586

151. Yoo, J., Ji, I., Ji, T.H. (1991) J. Biol. Chem. 266, 17741–17743.
152. Santa Coloma, T.A., Dattatreyamurty, B. and Reichert, Jr., L.E. (1990) Biochemistry 29, 1194–1200.
153. Erickson, L.D., Rizza, S.A., Bergert, E.R., Charlesworth, M.C., McCormick, D.J. and Ryan, R.J. (1990) Endocrinology 126, 2555–2560.
154. Hage-van Noort, M., Puijk, W.C., Plasman, H.H., Kuperus, D., Schaaper, W.M.M., Beekman, N.J.C.M., Grootegoed, J.A. and Meloen, R.H. (1992) Proc. Natl. Acad. Sci. USA 89, 3922–3926.
155. Thotakura N.R., Weintraub, B.D. and Bahl, O.P. (1990) Mol. Cell. Endocrinol. 70, 263–272.
156. Calvo, F.O. and Ryan, R.J. (1985) Biochemistry 24, 1953–1959.
157. Petaja-Repo, U.E., Merz, W.E., Rajaniemi, H. (1991) Endocrinology 128, 1209–1217.
158. Schwarz, S., Krude, H., Merz, W.E., Lottersberger, C., Wick, G. and Berger, P. (1991) Biochem. Biophys. Res. Commun. 178, 699–706.
159. Dunkel, L., Jia, X.-C., Nishimori, K., Boime, I. and Hsueh A.J.W. (1993) Endocrinology 132, 765–769.
160. Nishimura, R., Raymond, M.J., Ji, I., Rebois, V.R. and Ji, T.H. (1986) Proc. Natl. Acad. Sci. USA 83, 6327–6331.
161. Ji, I. and Ji, T.H. (1990) Proc. Natl. Acad. Sci. USA 87, 4396–4400.
162. Merz, W.E. (1988) Biochem. Biophys. Res. Commun. 156, 1271–1278.
163. Keutman, H.T., Johnson, L. and Ryan, R.J. (1985) FEBS Lett. 185, 333–338.
164. Papandreou, M.-J., Sergi, I., Benkirane, M. and Ronin, C. (1990) Mol. Cell. Endocrinol. 73, 15–26.
165. Rebois, R.V. and Fishman, P.H. (1984) J. Biol. Chem. 259, 8087–8090.
166. Rebois, R.V. and Liss, M.T. (1987) J. Biol. Chem. 262, 3891–3896.
167. Hattori, M., Hachisu, T., Shimohigashi, Y. and Wakabayashi, K. (1988) Mol. Cell. Endocrinol. 57, 17–23.
168. Morrell, A.G., Gregoriadis, G. and Scheinberg, I.H. (1971) J. Biol. Chem. 246, 1461–1467.
169. Baenziger, J.U., Kumar, S., Brodbeck, R.M., Smith, P.L. and Beranek, M.C. (1992) Proc. Natl. Acad. Sci. USA 89, 334–338.
170. Fiete, D., Srivastava, V., Hindsgaul, O. and Baenziger, J.U. (1991) Cell 67, 1103–1110.
171. Gyves, P.W., Gesundheit, N., Stannard, B.S., De Cherney, G.S. and Weintraub, B.D. (1989) J. Biol. Chem. 264, 6104–6110.
172. Gyves, P.W., Gesundheit, N., Taylor, T., Butler, J.B. and Weintraub, B.D. (1987) Endocrinology 121, 133–140.
173. Gesundheit, N., Magner, J.A., Chen, T. and Weintraub, B.D. (1986) Endocrinology 119, 455–463.
174. Gyves, P.W., Gesundheit, N., Thotakura, N.R., Stannard, B.S., De Cherney, G.S. and Weintraub, B.D. (1990) Proc. Natl. Acad. Sci. USA 87, 3792–3796.
175. Constant, R.B. and Weintraub, B.D. (1986) Endocrinology 119, 2720–2727.
176. Clayton, R.N. (1984) Clinic. Endocrinol. 26, 361–384.
177. Brabant, G., Prank, K., Huang-Vu, C., Hesch, R. D. and von zur Muhlen, A. (1991) J. Clin. Endocrinol. Metab. 72, 145–150.
178. Wakabayashi, K. (1977) Endocrinol. Jpn. 24, 473–485.
179. Chapel, S.C., Ulloa-Aguirre, A. and Ramalay, J.A. (1983) Biol. Reprod. 28, 196–205.
180. Chapel, S.C. and Ramalay, J.A. (1985) Biol. Reprod. 32, 567–573.
181. Wide, L. (1985) Acta Endocrinol. (Copenh.) 109, 181–189.
182. Wide, L. (1985) Acta Endocrinol. (Copenh.) 109, 190–197.
183. Ulloa-Aquire, A., Espinoza, R., Damian-Matsumura, P. and Chapel, S.C. (1988) Human Reprod. 3, 491–501.

184. Matteri, R.L. and Papkoff, H. (1988) Biol. Reprod. 28, 13–22.
185. Keel, B.A. and Grotjan, H.E., Jr. (1989) In: B.A. Keel and H.E. Grotjan (Eds.), Microhetero-geneity of Glycoprotein Hormones. CRC Press, Boca Raton, FL, pp. 149–184.
186. Wilson, C.A., Leigh, A.J., Chapman, A.J. (1990) J. Endocrinol. 125, 3–14.
187. Sergi, I., Papandreou, M.-J., Medri, G., Canonne, C., Vervier, B. and Ronin, C. (1991) Endocrinology 128, 3259–3268.
188. Chapel, S.C., Cautifaris, C. and Jacobs, S.J. (1982) Endocrinology 110, 847–854.
189. Chapel, S.C. and Ramalay, J.A. (1985) Biol. Reprod. 32, 567–573.
190. Ulloa-Aquirre, A., Mejia, J.J., Dominguez, R., Guevara-Aquirre, J., Diaz-Sanchez, V. and Larrea, F. (1986) J. Endocrinol. 110, 539–549.
191. eckham, W.D. and Knobil, E. (1976) Endocrinology 98, 1054–1060.
192. alle, P.C., Ulloa-Aguirer, A. and Chapel, S.C. (1983) J. Endocrinol. 99, 31–39.
193. Veldhuis, J.D., Beitins, I.Z., Johnson, M.L., Serabian, M.A. and Dufau, M.L. (1984) J. Clinic. Endocrinol. Metab. 58, 1050–1058.
194. Baldwin, D.M., Highsmith, R.F., Ramey, J.W., Krummenn, L.A., (1986) Biol. Reprod. 34, 304–315.
195. Joshi, L.R. and Weintraub, B.D. (1983) Endocrinology 113, 2145–2151.
196. Dahl, K.D., Bicsak, T.A. and Hsueh, A.J.W. (1988) Science 239, 72–74.
197. Cole, L.A., Birkin, S. and Perini, F. (1985) Biochem. Biophys. Res. Commun. 126, 333–339.
198. Amano, J., Nishimura, R., Mochizuki and Kobata, A. (1988) J. Biol. Chem. 263, 1157–1163.
199. Hard, K., Damm, J.B.C., Spruijt, M.P.N., Bergwerff, A.A., Kamerling, J.P., Van Dedem, G.W.K. and Vliegenthart, J.F.G. (1992) Eur. J. Biochem. 205, 785–798.
200. Matzuk, M.M., Krieger, M., Corless, C.L. and Boime, I. (1987) Proc. Natl. Acad. Sci. USA 84, 6354–6358.
201. Matzuk, M.M., Hsueh, A.J.W., LaPolt, P., Tsafriri, A., Keene, J.L. and Boime, I. (1990) Endocrinology 126, 376–383.
202. El-Deiry, S., Kaetzel, D., Kennedy, G., Nilson, J. and Puett, D. (1989) Mol. Endocrinol. 3, 1523–1528.
203. Kingsley, D.M., Kozarsky, K.F., Hobbie, L. and Krieger, M. (1986) Cell, 44, 749–759.
204. Braunstein, G.D., Vaitukaitis, J.L. and Ross, G.T. (1972) Endocrinology 91, 1030–1036.
205. de Kretser, D.M., Atkins, R.C. and Paulsen, C.A. (1973) J. Endocrinol. 58, 425–434.
206. Sowers, J.R., Pekary, A.E., Hershman, J.M., Kanter, M. and DiStefano III, J.J. (1979) J. Endocrinol. 80, 83–89.
207. Fares, F.A., Suganuma, N., Nishimori, K., LaPolt, P.S., Hsueh, A.J.W. and Boime, I. (1992) Proc. Natl. Acad. Sci. USA 89, 4304–4308.
208. LaPolt, P.S., Nishimori, K., Fares, F.A., Perlas, E., Boime, I. and Hsueh, A.J.W. (1992) Endocrinology 131, 2514–2520.
209. Ashwell, G. and Morrell, A.G. (1974) Adv. Enzymol. 41, 99–128.

J. Montreuil, H. Schachter and J.F.G. Vliegenthart (Eds.), *Glycoproteins*

589

Carbohydrate moiety of vertebrate collagens

MILTON E. NOELKEN and BILLY G. HUDSON

Department of Biochemistry and Molecular Biology, University of Kansas Medical Center,
Kansas City, KS 66160-7421, USA

1. Introduction

The collagen superfamily of glycoproteins is comprised of eighteen known types (I–XVIII) consisting of 33 genetically distinct polypeptide chains. These types have common structural features along with considerable structural diversity [1–8]. The hallmark feature common to all collagen types is one or more regions consisting of various Gly-Xaa-Yaa sequences in tandem, referred to as collagenous sequences. These regions associate intermolecularly to generate a three-chained triple-helical protomer. All collagens also contain noncollagenous sequences (i.e., non-Gly-Xaa-Yaa) that border on or are interspersed among the collagenous sequences of the polypeptide chain. They form noncollagenous domains (non-triple-helical) in the three-chained protomers. Among all collagen types, the percentage of collagenous sequences varies from less than 10% to greater than 95%. The three-chained protomers of each collagen type associate and form a supramolecular structure that is characterized as either a fibrillar or nonfibrillar structure.

Recent sequence comparisons [9] of collagen I–XIV to other classes of proteins indicate that a number of the noncollagenous domains contain sequences, called modules, that are homologous to sequences found in other proteins. The noncollagenous domains contain modules found in thrombospondin, complement component C1q, fibronectin, von Willebrand factor, Kunitz type inhibitor and the S domain of salivary proteins [9]. These modules may confer particular subfunctions to proteins in which they are located.

Among the eighteen types of collagens, there are four known kinds of carbohydrate moieties that are covalently linked to the polypeptide chains. These are: a glucosyl-galactose disaccharide β-linked to the hydroxyl group of hydroxylysine, $Glc(\alpha1\rightarrow2)$ $Gal(\beta1\rightarrow O)Hyl$, and the related $Gal(\beta1\rightarrow O)Hyl$ [10,11]; a complex oligosaccharide N-linked to asparagine [12]; and chrondroitin sulfate chains of varying sizes O-linked to serine [13,14]. The best characterized moieties are hydroxylysine-linked disaccharides, which appear to occur in all collagen types, and the asparagine-linked oligosaccharide of type IV collagen. The disaccharide unit is postulated to function as an inhibitor of fibril formation [12] and the Asn-linked unit is postulated to play a key role in determining the geometry and stability of the supramolecular structure of collagen IV [15].

In this chapter, we will briefly review the eighteen types of collagen with respect to structure, chain organization, and tissue distribution, and then focus on the nine types that are characterized, to a varying extent, with respect to amount, location, structure and function of the carbohydrate moieties.

2. Classification of collagens on the basis of supramolecular structure

Collagens can be divided into two major classes on the basis of their primary structure and supramolecular structure: fibril-forming and non-fibril-forming collagens (Tables I and II) [1]. The fibril-forming collagens contain a long central triple-helical domain consisting of about 1,000 residues per chain. This class includes types I, II, III, V, and XI. The remaining collagens (IV, VI–X, and XII–XVI and XVIII) are in the non-fibril-forming class. Within this class, types IX, XII, XIV, and perhaps XV, XVI and XVIII, form a sub-group called the *fibril-associated collagens with interrupted triple helices* (*FACIT*) in which there are interruptions of the *Gly-Xaa-Yaa* repeat [1]. The FACIT collagens are associated with type I or II collagen fibrils, are involved in interactions with other matrix components, and play a role in determining fibril size. The classification of collagen XVII is not clear.

TABLE I

Fibrillar collagens: chain composition and tissue distribution

Type	Chains	Molecules	Representative tissues
Associated with type I collagen			
I	α1(I), α2(I)	[α1(I)]₂α2(I)	Skin, bone, tendon, dentin, arterial wall; widespread
		[α1(I)]₃	Dentin, skin (minor form)
III	α1(III)	[α1(III)]₃	Skin, arterial wall, absent in bone
V	α1(V), α2(V) α3(V), α4(V)	[α1(V)]₃	Hamster lung cell cultures
		[α1(V)]₂α2(V)	Fetal membranes, skin, bone, blood vessels
		α1(V)α2(V)α3(V)	Placenta
Associated with type II collagen			
II	α1(II)	[α1(II)]₃	Hyaline cartilage, vitreous body
XI	α1(XI), α2(XI) α3(XI)*	α1(XI)α2(XI)α3(XI)	Hyaline cartilage

*α3(XI) is probably identical to α1(II), except for post-translational modifications.

TABLE II

Nonfibrillar collagens: chain composition and tissue distribution

Type	Chains	Molecules	Representative tissues
IV	α1(IV), α2(IV) α3(IV), α4(IV) α5(IV), α6(IV)	[α1(IV)]₂α2(IV) ?	Basement membranes
VI	α1(VI), α2(VI) α3(VI)	α1(VI)α2(VI)α3(VI)	Blood vessels, skin, intervertebral disc
VII	α1(VII)	[α1(VII)]₃	Dermoepidermal junction
VIII	α1(VIII), α2(VIII)	?	Descemet's membrane, endothelial cells
IX[a,b]	α1(IX), α2(IX) α3(IX)	α1(IX)α2(IX)α3(IX)	Hyaline cartilage, vitreous humor
X	α1(X)	α1(X)₃	Growth plate (hypertrophic chondrocytes)
XII[b]	α1(XII)	[α1(XII)]₃	Embryonic tendon and skin, periodontal ligament
XIII	α1(XIII)	?	Endothelial cells
XIV[b]	α1(XIV)	[α1(XIV)]₃	Fetal skin and tendon
XV[b,c]	α1(XV)	?	Placenta
XVI[b,d]	α1(XVI)	?	Human skin fibroblasts
XVII[e,f]	α1(XVII)	?	Skin, eyes, mucous membranes
XVIII[g]	α1(XVIII)	?	Liver, kidney, placenta

[a]Type IX is associated with type II collagen.
[b]Types IX, XII, XIV, and perhaps XV, XVI and XVIII, form a subgroup called the fibril-associated collagens with interrupted triple helices (FACIT). These collagens are associated with type I or type II collagen fibrils.
[c]Meyers et al. [2].
[d]Pan et al. [3].
[e]Li et al. [4].
[f]The precise classification has not been established.
[g]Oh et al. [5].

3. Chain composition and tissue distribution of collagens

There are 33 genetically distinct α chains comprising the 18 types of collagen. These are listed in Tables I and II along with the various collagen molecules and the representative, but not complete, tissue distributions [1,8,9,16]. In cases where there are several kinds of chain for a given collagen type, there is a variety of collagen molecules possible. For example, three different α chains can potentially associate in a maximum of 20 different combinations of three chains to form 20 subtypes of a given type of collagen molecule. The exact number found, however, is expected to be less than the maximum, and will depend on the kinds of chains available during biosynthesis in a given tissue and on possible specificity for association. At present, the number of known collagen molecule subtypes ranges from one for several types of collagen to three for collagen V (Tables I and II).

4. Carbohydrate moieties of collagens

4.1. Glucosylgalactose and galactose

Glc-Gal-Hyl and Gal-Hyl are found in all collagens for which the carbohydrate composition has been determined. The structures of Glc-Gal-Hyl and Gal-Hyl are 2-O-α-D-glucosyl-O-β-D-galactosylhydroxylysine (Fig. 1) and O-β-D-galactosylhydroxylysine, respectively [10–12]. In the biosynthesis of Glc-Gal-Hyl and Gal-Hyl, the Hyl residue is produced by hydroxylation of Lys residues in Gly-Xaa-Lys triplets of the collagenous sequences. The percentage of the Hyl residues that become glycosylated varies among the collagens.

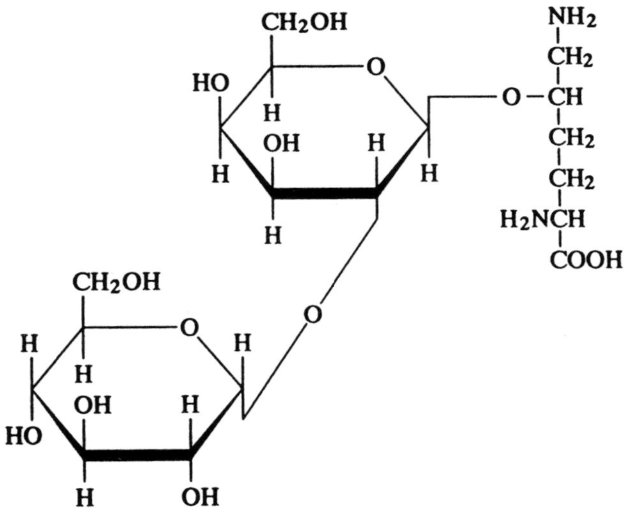

Fig. 1. Structural formula for 2-O-α-D-glucosyl-O-β-D-galactosylhydroxylysine. Glc-Gal and Gal are attached to many of the Hyl residues of collagens [10,11].

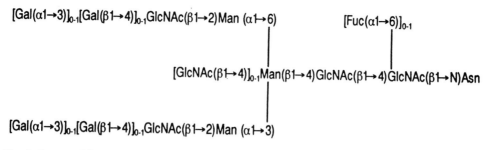

Fig. 2. Structural formula for the Asn-linked oligosaccharide unit of the 7 S cross-linking domain of collagen IV isolated from bovine glomerular basement membrane [15]. The oligosaccharide is of a diantennary type with heterogeneity in the presence of the five sugar residues at the non-reducing termini, as is indicated by 0–1.

4.2. Asn-linked oligosaccharides

Asn-linked oligosaccharides are present in collagens IV [12,15,17–19], V [20], and VI [19] and in procollagens I and III [21,22]. The oligosaccharide found in the 7 S cross-linking domain of collagen IV has been shown to be a heterogeneous diantennary N-acetyllactosamine type of oligosaccharide (Fig. 2) [15,18,19].

4.3. Chondroitin sulfate

Chondroitin sulfate has been found in collagen IX [1,13,14], and there is evidence for its presence in collagens XII and XIV from bovine fetal cartilage [23]. Vaughan et al. [13] found that the two repeating disaccharide units, D-glucopyranosyluronic acid-$(1{\rightarrow}3)$-2-acetoamido-2-deoxy-4-O-sulfo-β-D-galactopyranose and D-glucopyranosyl-uronic acid-$(1{\rightarrow}3)$-2-acetoamido-2-deoxy-6-O-sulfo-β-D-galactopyranose, of chond-roitin sulfate of type IX collagen isolated from chick embryo sternum, occurred in the ratio of 3:1 (4-O-sulfo: 6-O-sulfo). The structure of the two disaccharides is presented in Fig. 3.

5. Collagens with characterized carbohydrate moieties

5.1. Collagens I, II, III and V

Collagens I, II, III, and V (and collagen XI which is closely related to collagen V) are fibril-forming collagens (Table I). The constituent chains of collagens I, II, III, and V are similar in domain structure and length of the triple-helical domain [8 (review), 25,26]. Their amino acid sequences begin with a signal peptide 22–26 residues long, followed by a 57–522 residue amino-terminal propeptide, an 11–19 residue N-telopep-tide, a large central triple-helical domain 1014–1029 residues long, an 11–27 residue C-telopeptide, and a carboxyl-terminal propeptide (Fig. 4). The signal peptides and

594

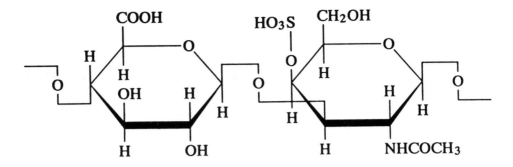

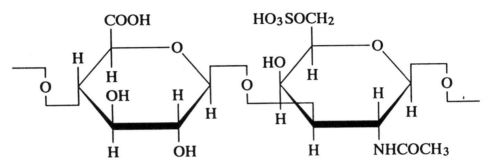

Fig. 3. Structural formulas for the two repeating disaccharide units of chondroitin sulfate of type IX collagen isolated from chick embryo sternum [13]. Top: D-glucopyranosyluronic acid-(1→3)-2-aceto-amido-2-deoxy-4-*O*-sulfo-β-D-galactopyranose. Bottom: D-glucopyranosyluronic acid-(1→3)-2-aceto-amido-2-deoxy-6-*O*-sulfo-β-D-galactopyranose. The disaccharides occurred in the ratio of 3:1 (4-*O*-sulfo: 6-*O*-sulfo).

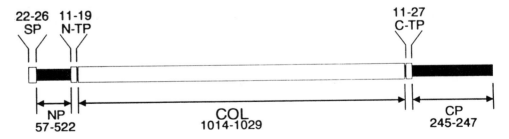

Fig. 4. Schematic illustrating the common elements of structure of pre-procollagens I, II, III and V [8,25,26]. Each of the α chains has a putative signal peptide, SP, in the 22-26 residue size range. The other domains are: N-P, amino-terminal propeptide; N-TP, amino-terminal telopeptide; COL, col-lagenous domain; C-TP, carboxyl-terminal telopeptide; and C-P, carboxyl-terminal propeptide. The size range is given for each kind of domain. The SP, N-P and C-P domains are cleaved away by specific peptidases in the biosynthesis of these collagens, allowing the self-assembly of the resulting collagen molecule into fibrils.

propeptides are cleaved during processing of the collagens. However, some collagen III molecules retain propeptides [27 and references therein] and the *N*-propeptides of collagen V are not completely processed [20].

The carbohydrate composition of collagens I–III and V was reviewed by Miller in 1984 [27]. There is one Glc-Gal-Hyl residue per thousand for the α1(I), α1(III) and α2(I) chain and the latter chain has one Gal-Hyl residue per thousand residues. On the other hand, collagen V is heavily glycosylated. The total of Glc-Gal-Hyl and Gal-Hyl residues for the α(V) chains is as follows: α1(V), 34; α2(V), 8; and α3(V), 24; with the majority of the sugars being disaccharides. The role of glycosylation is not fully understood, but Spiro, in a comparison of collagen I with collagen IV, which is heavily glycosylated, suggested that a high level of glycosylation could interfere with fibril formation [12]. The results of Silver and Birk support this idea [28]. Light-scattering studies on collagens I, II, III and V in acid solution at 10°C indicated that collagens I, II and III participate in a monomer-aggregate equilibrium, whereas collagen V molecules apparently attract each other but do not form aggregates.

In addition to Glc-Gal and Gal, procollagens I and III contain Asn-linked oligosaccharides that are linked to propeptides. The α1(I)- and α2(I) procollagen chains contain 'high mannose' oligosaccharides in the carboxyl-terminal propeptide domain [21]. Dermatosporactic pN-collagen III contains an Asn-linked oligosaccharide in its amino-terminal propeptide [22]. In addition, collagen V contains Asn-linked oligosaccharides.

The structure of collagen V is more complicated than that of collagens I, II and III, and discussion of its carbohydrate composition requires a review of the protein structure. Fessler and Fessler have written a comprehensive review that covers the biosynthesis of type V collagen, the structure of the procollagen V chains and their processing, the fiber forms, occurrence in various tissues, and susceptibility to degradative enzymes [20].

There are four known chains: α1(V), α2(V), α3(V) and α4(V) (Table I). The α4(V) chain is also called α1(V)′ because it is very similar to the α1(V) chain; however, it is possible that it is an α(XI) chain [1]. Three kinds of type V collagen molecules have been found: the major species [α1(V)]$_2$α2(V), as well as the species [α1(V)]$_3$ and α1(V)α2(V)α3(V). A hybrid molecule comprised of mixed type V and XI chains has been proposed [1].

The sequence of human and hamster prepro-α1(V) collagen and the sequence of human prepro-α2(V) have been determined from the nucleotide sequence of cDNA clones [25,26,29-31]. The hamster and human prepro-α1(V) collagen sequences are very similar and we will consider only the human sequence in discussing the carbohydrate composition and location. Certain structural aspects of the human pro-α1(V) and pro-α2(V) chains are portrayed in Fig. 5 (which does not include the 36-residue α1(V)- or the 26-residue α2(V) signal peptide). The pro-α1(V) chain has a 1014-residue major collagenous domain (COL) composed of 338 uninterrupted Gly-Xaa-Yaa triplets. Preceding the COL domain is a 522-residue amino-terminal domain that includes a 96-residue interrupted collagenous sub-domain (not shown) that begins at residue 408 and contains 25 interrupted Gly-Xaa-Yaa repeats. Following the COL domain is the 266-residue carboxyl-terminal domain. The amino acid sequence contains two putative sites (Ala-Gln) for cleavage by N-proteinase, located at peptide bond positions 505–

PROALPHA 1(V)

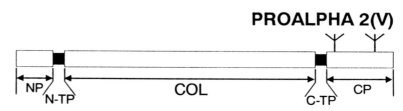

PROALPHA 2(V)

Fig. 5. Schematic representation of the structure of pro-α(V) collagen chains based on amino acid sequences. COL represents a major collagenous domain of about 1000 residues that is flanked by an amino-terminal- and a carboxyl-terminal domain. The pro-α1(V) chain and pro-α2(V) chains are 1802 and 1463 residues, respectively. The Y-shaped structures represent putative sites for attachment of Asn-linked oligosaccharides. There is at least one oligosaccharide shown to be present in the amino-terminal domain of the pro-α1(V) chain and the carboxyl-terminal domain of the pro-α1(V) and pro-α2(V) chains [20]. The collagenous domain of the pro-α1(V)- and pro-α2(V) chains contain 30 and 14 Gal or Glc-Gal units, respectively [33].

506 and 510–511 and a putative site (Ala-Asp) for cleavage by C-proteinase at position 1569–1570. The pro-α2(V) chain has a 1017-residue major collagenous domain flanked by a 270 carboxyl-terminal domain and a 186-residue amino-terminal domain, which is considerably shorter than the amino-terminal domain of the pro-α1(V) chain. The pro-α2(V) amino-terminal domain includes a 79-residue interrupted collagenous sub-domain that contains 24 Gly-Xaa-Yaa triplets. The pro-α2(V) interrupted collagenous sub-domain has essentially the same location relative to the COL domain as has its pro-α1(V) IC counterpart with respect to the pro-α1(V) COL domain. The pro-α2(V) chain has a putative site for cleavage by N-proteinase at position 167–168 and a putative site for cleavage by C-proteinase at position 1227–1228.

When type V procollagen is processed, the putative N-proteinase cleavage sites of the pro-α1(V) and pro-α2(V) domains are scarcely hydrolyzed [20 and references therein], perhaps because of partial triple helix formation by the IC sub-domains. It is not clear whether the C-proteinase cleavage site of the pro-α1(V) and pro-α2(V) collagen chains is used in processing. Peptides with sizes similar to the predicted sizes are released by processing [20], but their amino-terminal sequences have not been reported.

The presence of Asn-linked complex oligosaccharide in the pro-α1(V) amino-terminal domain was indicated by a reduction in size after any addition of Asn-linked oligosaccharide had been inhibited by tunicamycin [20]. The carboxyl-terminal pro-peptides of the pro-α1(V)- and pro-α2(V) chains both contain Asn-linked oligosaccharides [20]. The amino-acid sequence results (Fig. 5) show putative attachment sites at five locations in the pro-α1(V) chain, three in the amino-terminal domain (Asn-122, Asn-140 and Asn-349), and two in the carboxyl-terminal domain (Asn-1636 and 1705).

There are no putative attachment sites in the amino-terminal domain of the pro-α2(V) chain, but there are two putative sites in the carboxyl-terminal domain (Asn-1239 and Asn-1374). Each of the putative sites is the beginning of a three-residue sequence, Asn-Xaa-Ser/Thr, that favors linkage of a complex oligosaccharide to the Asn residue [32]. Thus, the biochemical studies [20] show that the amino-terminal domain of the pro-α1(V) chain and the carboxyl-terminal domain of the pro-α1(V) and pro-α2(V) chains all have at least one attached oligosaccharide.

The COL domains of the α1(V)-, α2(V)- and α3(V) chains have a high content of 5-hydroxylysine residues, the majority of which are glycosylated. Burgeson et al. [33] found that the collagenous fragment of the α1(V) chain, which is essentially equivalent to the COL domain, contains 35 residues/1000 of 5-hydroxylysine, 30 of which have Gal or Glc-Gal bound. Similarly, the collagenous fragment of the α2(V) chain contains 24 Hyl residues/1000, 14 of which have Gal or Glc-Gal bound. Sage and Bornstein [34] have found that the major collagenous domain of the α3(V) chain has 30 Gal or Glc-Gal units/1000 residues.

The distribution of the various collagens in fibrillar structures is complex, as illustrated by the study of Adachi et al. [35] of the localization of collagens I, III and V in monkey liver. Immunofluorescent microscopy and immuno-electron microscopy results on liver sections showed that collagens I, III (and/or its biosynthetic precursor pN-collagen III) and V are found in coarse reticular fibers (diameter ≈1 μm) of lobules and in cross-striated collagen fibers. However, fine reticular fibers of lobules (diameter ≈ 0.5 μm) contain collagens III (and/or its biosynthetic precursor pN-collagen III) and V, but apparently lack collagen I. In the cross-striated collagen fibers, collagens III and/or pN-collagen III and V surround the constituent collagen I fibrils (diameter ≈ 45 nm). The co-distribution of collagens III and V with collagen I and their surface location shown in the study of Adachi et al. [35] and other studies [1] suggests that they are involved in regulating the diameter of collagen I fibrils. The mechanism by which collagen III is involved in this process is being investigated [36 and references therein] and may involve copolymerization of collagens I and III [36].

Perhaps the mode of interaction of collagen V with collagen I and/or collagen III is determined in part by the Hyl-linked di- and monosaccharides and the Asn-linked oligosaccharides. An [α1(V)]₂α2(V) molecule has 74 Glc-Gal-Hyl and Gal-Hyl units in the major collagenous domain and an Asn-linked oligosaccharide unit in the amino-terminal domain of each pro-α1(V) chain (assuming only partial proteolytic cleavage during processing [20]). If it is assumed that these rather bulky and hydrophilic carbohydrates are constrained to maintain a surface location after binding to collagen I and/or collagen III, then it would follow that the part of the collagen V molecule involved in binding would contain few of the Hyl-linked carbohydrate units and neither of the Asn-linked oligosaccharides. In connection with this, it would be of interest to develop models of the major collagenous domain in order to determine whether an asymmetric spatial distribution of the Hyl-linked carbohydrate units around the surface of the triple helix is possible. Such an asymmetric distribution has been shown for the 7 S amino-terminal tetramerization domain of collagen IV (Ref. [15] and next section of this chapter). In that case, the distribution of saccharides places most of them outside of a hydrophobic reaction zone encompassing one-fourth of the surface.

598

5.2. Collagen IV

Collagen IV, the major constituent of basement membranes, is a family of six chains, the α1(IV) and α2(IV) chains, and the more recently discovered α3(IV), α4(IV), α5(IV) and α6(IV) chains [37–42]. The chains form an assortment of trimers, with the major one being [α1(IV)]₂α2(IV). The complete amino acid sequence of the α1(IV)-[43,44], α2(IV)- [45,46] and α5(IV) [40,41,47] chains is known and their sequences are homologous. The sequence of each chain (Fig. 6) is characterized by: (1) a noncollagenous domain (NC1) of ~230 residues at the carboxyl terminus; (2) a long collagenous region of ~1400 residues that forms a triple helix with two other chains; and (3) several non-triple-helical interruptions within the triple-helical region. The first ~130 residues of the collagenous domain constitute the 7 S domain. The NC1 and 7 S domains are involved in the association of protomers in the formation of a network-type supramolecular structure, with the NC1 domain involved in head-to-head dimerization and the 7 S domain involved in tail-to-tail tetramerization of protomers, respectively [48,49]. In the process, the NC1 domains of two protomers form an NC1 hexamer, and their 7 S domains form a dodecamer.

The α3(IV)-, α4(IV)-, α5(IV)- and α6(IV)-containing protomers are not as well characterized as the [α1(IV)]₂α2(IV) protomer because: (1) only the latter protomer has been isolated and purified; and (2) the complete amino acid sequence of its constituent chains has been determined. To date, only the sequence of the ~500 carboxyl-terminal residues of the α3(IV)- [37,50,51]) and α4(IV) [39,52,53] chains has been reported. The formation of NC1 hexamer domains as a step in the formation of the collagen IV supramolecular structure involves specificity in the association of

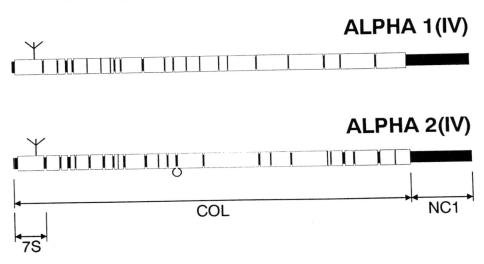

Fig. 6. Schematic diagram of the human α1(IV)- and α2(IV) chains. COL is the collagenous domain, NC1 is the noncollagenous domain, 7 S is part of the COL domain, the vertical bars represent noncollagenous sequences that interrupt the collagenous sequence, and the Y-shaped structures represent the Asn-linked oligosaccharide whose formula [15] is presented in Fig. 2.

NC1 domains. Gunwar et al. [54] found that the dimeric components of bovine GBM NC1 domain hexamer were homodimers of the α1(IV)NC1-, α2(IV)NC1-, α3(IV)NC1-, α4(IV)NC1- and, α5(IV)NC1 domains and an α1(IV)NC1-α3(IV)NC1 heterodimer. Dimers are formed between NC1 subunits of different collagen IV protomers and not within protomers.

Kleppel et al. [55] have found that collagen IV NC1 hexamers from bovine glomerular basement, lens capsule basement membrane and Descemet's membrane can be fractionated into two populations, one containing only NC1 domains of the α1(IV)- and α2(IV) chains and the other containing only NC1 domains of the α3(IV)-, α4(IV)- and Alport-antigen-containing (perhaps α5(IV)) chains. This, and other evidence that they cited, such as specific co-localization, within several tissues, of the α1(IV) chain with the α2(IV) chain and the α3(IV) chain with the α4(IV) chain, led to the suggestion that the α1(IV)- and α2(IV) chains can form a network separate from that of the other chains. However, definitive proof for separate networks would require obtaining equivalent results for specific co-localization of chains if the 7 S domains or major collagenous domains were analyzed instead of NC1 hexamers.

The only 7 S domain characterized to date is that obtained from basement membranes containing the $[\alpha1(IV)]_2\alpha2(IV)$ protomer. The complete amino acid sequence and carbohydrate composition have been determined, and these have been incorporated into a model for tetramerization of the protomers [15,49] that takes into account the location of hydrophobic amino acid sidechains and carbohydrate moieties on the surface of the triple-helical protomers.

The 7 S domain consists of residues 28–162 of the α1(IV) chain and residues 29–171 of the α2(IV) chain [43,45]. The 7 S domain begins with an amino-terminal noncollagenous region that is 15 residues long in the α1(IV) chain and 18 residues long in the α2(IV) chain. The following 117-residue triple-helical region in each chain is responsible for the overlapping tetramerization of the $[\alpha1(IV)]_2\alpha2(IV)$ protomer as well as cross-linking by means of intermolecular disulfide bond formation. The boundary between the 7 S domain and the major triple-helical domain is included in a second noncollagenous, flexible region, 13 residues long. This region creates an ~40° bend and allows the following region of the protomers of a tetramer to be removed from each other. The 7 S domain of the tetramer is referred to herein as the 7 S tetramer. The tetramer is resistant to collagenase digestion and can be excised from basement membrane for study.

The carbohydrate units of the 7 S tetramer are of two types: Asn-linked oligosaccharides and Hyl-linked disaccharide units. Each of the α1(IV)- and α2(IV) chains contains in its 7 S domain one Asn-Xaa-Thr sequence that is N-glycosylated at Asn, corresponding to position 126 of the α1(IV) chain and 138 of the α2(IV) chain [15]. Each oligosaccharide is a diantennary N-acetyllactosamine type of Asn-linked oligosaccharide with a broad heterogeneity in the presence of the sugar residues at their nonreducing termini [15,18,19] as indicated in Fig. 2. ^{1}H-NMR spectroscopic analysis of the oligosaccharide moiety from each chain indicated essentially identical structures [15]. The α1(IV)- and α2(IV) chains contain an average of ~50 Glc-Gal-Hyl units each [18]. The 7 S domain of the α1(IV) chain contains 6 Glc-Gal-Hyl units (at positions 45, 48, 78, 90, 129, and 156) as does the α2(IV) chain (positions 57, 86, 90, 102, 165 and 168 [49,56].

Langeveld et al. [15] used the carbohydrate structures and amino acid sequences to extend a model for tetramerization developed by Siebold et al. [49]. Their results indicate that the oligosaccharide and disaccharide units of the 7 S domain have a role in determining the geometry and stability of the 7 S tetramer because of their abundance, structure, hydrophilic properties and their strategic distribution about the surface of the triple helix [15].

The study of Langeveld et al. [15] entailed calculating the three-dimensional positions of the beta carbons of Xaa and Yaa residues of the (Gly-Xaa-Yaa)$_n$ repeating sequence by use of a model developed for triple-helical collagen I by Hofmann et al. [57] and applied to the collagen IV 7 S domain by Siebold et al. [49]. In this model, the $\alpha 2$(IV) chain is the reference chain for relative chain location. Pro-55 of the $\alpha 2$(IV) chain is the first residue of the triple-helical domain and its Cα atom is assigned the reference cylindrical coordinates, $z = 0$ nm, and $\phi = 0°$, where z is the axial distance and ϕ is the azimuthal angle, measured anti-clockwise when viewing the triple helix from its carboxyl terminus. The relative positions of the Cβ atoms of the Xaa and Yaa residues were calculated from the atomic coordinates of (Gly-Pro-Pro)$_n$. Equivalent atoms in the other two chains are located at $z + 0.286$, $\phi - 108°$; and $z + 0.574$, $\phi - 216°$, respectively. To determine potential interaction between hydrophobic residues of different protomers, the distances between β-carbons of hydrophobic residues of different associated protomers were calculated at various relative orientations and amount of longitudinal overlap of the protomers. Hydrophobic interaction between a pair of sidechains is predicted if their β-carbons are within a particular minimum distance from each other, the distance depending on the size of the two sidechains [57]. In this way, Langeveld et al. [15] verified a putative hydrophobic reaction zone [49] and located the saccharide-bearing residues and cysteine residues with respect to it. The results of this analysis are shown in Fig. 7.

In Fig. 7 are plotted values of z, the axial translation in nm, and ϕ, the azimuthal angle in degrees for the beta carbons of hydrophobic residues, residues containing saccharides, and Cys residues. The figure shows a hydrophobic reaction zone, bounded by the longitudinal lines A and B, containing the hydrophobic sidechains are that are the most likely to be involved in tetramerization of the protomers [15,49]. The width of the hydrophobic interaction zone ($\geq 90°$) is large enough to be consistent with the association of the protomers being limited to tetramer formation.

The location (Fig. 7) of the oligosaccharides indicates that they play a key role in determining the geometry of the 7 S tetramer, especially because the oligosaccharides have a size considerably larger than the diameter of a triple helix. Indeed, it can be seen in Fig. 7 that one of the oligosaccharides, located at $z = 24.5$ nm, $\phi = 23.6°$, is located in the ϕ range of the reaction zone near the B boundary. This oligosaccharide would sterically restrict the longitudinal overlap of the triple-helical domains of monomers in the tetramer. The maximum value for the longitudinal overlap of 24.5 nm (corresponding to 84 amino acid residues for a chain) is consistent with the dimensions of the 7 S tetramer determined by electron microscopy [15]. The location of the disaccharide and oligosaccharide units on the surface of the triple helix is presented in a pseudo three-dimensional mode in Fig. 8.

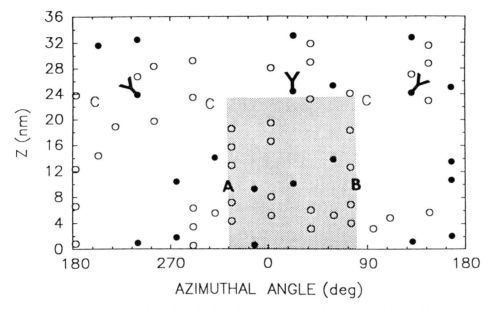

Fig. 7. Location and orientation of saccharides, hydrophobic- and cysteine residues with respect to the surface of the 7 S triple-helical domain. Two-dimensional representation of the surface that shows the calculated axial distance in nm, z, and the azimuthal angle in degrees, ϕ, for the beta carbons of four classes of residues. The beta carbons of residues with covalently bound disaccharides are indicated by filled circles; Asn-linked oligosaccharides by Y's contacting filled circles; hydrophobic residues by open circles, and cysteine residues by C's alongside open circles. The shaded area bounded on two sides by A and B represents a hydrophobic zone involved in tetramer formation [15,49]. The upper boundary of the reaction zone has a maximum value of about 24 nm and is determined by the large hydrophilic oligosaccharide unit attached to an Asn residue whose beta carbon is at $z = 24.5$ nm, $\phi = 23.6°$.

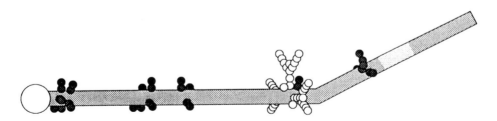

Fig. 8. Schematic three-dimensional representation of the 7 S tetramerization domain of an $[\alpha 1(IV)]_2\alpha 2(IV)$ protomer. Small spheres (black or white) represent hexose residues; the Y-shaped structures represent Asn-linked oligosaccharides; the two-beaded structures represent Glc-Gal units O-linked to Hyl residues; the large white sphere represents the 15 amino-terminal noncollagenous residues of each $\alpha 1(IV)$ chain and plus those (20) of the $\alpha 2(IV)$ chain. An $\sim 90°$ hydrophobic zone (not shown) located between the large white sphere and the Y-shaped structures is proposed to be involved in tetramerization [15,49]. The 7 S domain ends in the lightly shaded region. The position of the bend is approximate.

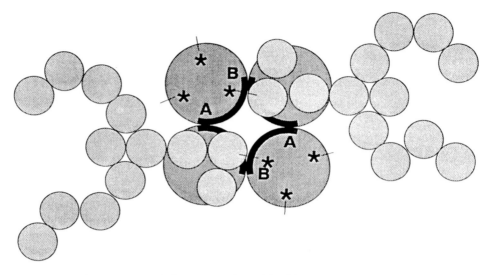

Fig. 9. Schematic (to scale) depicting an end view of the 7 S tetramer and the steric restrictions on assembly caused by Asn-linked oligosaccharides. The four 7 S triple-helical aggregation domains (large circles) interact by means of the hydrophobic zones located between A and B [15,49]. The diameter of the triple helix is 1.3 nm. Adjacent triple helices are antiparallel to each other, whereas diagonally related pairs are parallel to each other. The relative azimuthal positions of Asn-linked oligosaccharides are indicated by asterisks. Schematics of two of the oligosaccharides are shown, with their monomer residues indicated by small circles of diameter 0.67 nm [15]. The conformation of the oligosaccharides is unknown. The oligosaccharides restrict the longitudinal overlap of interacting triple helices, requiring the ends of the shaded triple helices to lie below the oligosaccharides.

The location of the α1(IV)-chain oligosaccharide on the surface of the triple helix, relative to the hydrophobic reaction zone, its size relative to the diameter of the triple helix, and its orientation and location on the tetramer are depicted in Fig. 9. The schematic diagram is based on the tetramer model described by Siebold et al. [49]. in which four protomers are arranged in an antiparallel fashion with their amino-terminal triple-helical domains overlapping longitudinally. In the tetramer structure, the oligosaccharide of one protomer is positioned next to the amino-terminal end of a second protomer. Steric hindrance due to the oligosaccharide is predicted to determine the maximum longitudinal overlap of two protomers as shown in Fig. 9 and in the three-dimensional models shown in Fig. 10.

In the 7 S tetramerization model, hydrophobic residues near the A boundary in one protomer interact with their "A" counterparts in the other three protomers, and analogous interactions occur between residues near the B boundaries; only a small number of the interactions are of an A:B type. Thus, as an approximation, the protomer can be thought of as having A and B hydrophobic faces. The contacts between A faces and between B faces are shown in Fig. 9. These A:A and B:B contacts are also illustrated in Figs. 10A and 10B. In a hypothetical reaction sequence, two protomers interact through

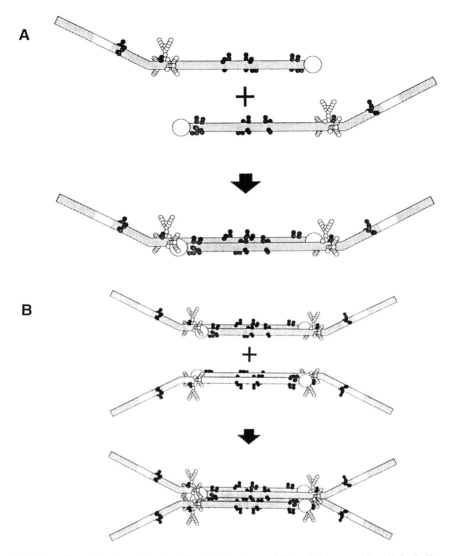

Fig. 10. Three-dimensional models depicting the locations of disaccharides and *A*sn-linked oligosaccharides of the 7 S domain in its dimeric and tetrameric forms and illustrating the mode of association. Panel A: Hypothetical formation of dimers through the interaction of the "B face" (a part of the hydrophobic zone of the *B* boundary) of one protomer with the B face of a second protomer (Figs. 7 and 9, and Ref. [15]). An oligosaccharide on each triple helix determines the maximum extent of longitudinal overlap of the helices when they form an antiparallel dimer, as shown by a white sphere in contact with an oligosaccharide. The "A face" (part of the hydrophobic region near the *A* boundary of the hydrophobic interaction zone) of each protomer is at the bottom side of the dimer. Panel B: Hypothetical formation of a tetramer from two dimers. The tetramer is formed from interaction of "A faces" (Figs. 7 and 9). Note that the oligosaccharides and amino-terminal noncollagenous also have the same steric functions in tetramer formation that they do in dimer formation. Panels A and B are used to illustrate the geometry of tetramer formation and it is not known what dimers, if any, are intermediates.

their *B* faces to form a dimer (Fig. 10A), and the *A* faces of the dimer are at the bottom of the dimer. Then, interaction of two dimers (Fig. 10B) through their *A* faces produces a tetramer.

In addition to effects on the geometry of the 7 S tetramer, it is likely that the oligosaccharide and disaccharide units contribute to its stability because of: (a) their occurrence within the longitudinal overlap region (i.e., between $z = 0$ and $z = 24.5$ nm) and at its boundary, (b) their confined distribution within the overlap region, and (c) their hydrophilic nature. The three oligosaccharides of each protomer are located in a narrow region at the boundary of the overlap region (Figs. 7 and 10B). Twelve disaccharides of the protomer are located within the overlap region (Fig. 7). In the tetramer, eight of these are distributed about the exterior and four are buried inside (Figs. 7 and 10B). Thus, the exterior surface of the tetramer, in the overlap region, contains 32 disaccharide units and 12 oligosaccharide units, which corresponds to 220 monosaccharide residues, assuming no heterogeneity, whereas the interior contains 16 disaccharides, which corresponds to 32 residues. This asymmetrical distribution of carbohydrate units about the surface of the triple helix and the hydrophobic faces of the interior surface confers to the protomer an amphipathic character which contributes to the stability of the tetramer.

Furthermore, the bulky oligo- and disaccharides, by steric hindrance, may restrict the number of ways in which protomers can associate, thus promoting their antiparallel arrangement in the tetramer, as postulated [49]. This feature may also be important in limiting the association of protomers to tetramerization. Likewise, the interior disaccharide units, because of their bulk, may limit the number of ways in which the dimers can associate and thus be a determinant of the geometry of the tetramer.

In addition to tetramerization through 7 S domains and dimerization through NC1 domains, collagen IV undergoes lateral association through interaction of NC1 domains with the major triple-helical domains and interactions between the major triple domains [58–62]. The $\alpha 1(IV)$- and $\alpha 2(IV)$ chains have an average of ~45 Glc-Gal disaccharides in the major triple domain and it is possible that the ~135 disaccharides restrict the interactions to particular locations and are thus involved in determining the geometry of the irregular polyhedra formed by the triple-helical domains. In this context, the disaccharide units would have the function suggested for them by Spiro [12], in rationalizing the lack of fibril formation of collagen IV, at a time before the 7 S and NC1 domains were discovered.

5.3. Collagen VI

Comprehensive reviews of the structure and function of type VI collagen have been published by Timpl and Engel [63] and van der Rest and Garrone [1]. The protomer of collagen VI collagen is a heterotrimer composed of three chains, $\alpha 1(VI)$, $\alpha 2(VI)$ and $\alpha 3(VI)$ (Table II). The amino-acid sequence of each of the chains of chicken- [64–66] and human collagen VI [67–69] has been determined. The sequence of each chicken chain is very similar to that of its human counterpart, and in either chicken- or human collagen VI, the $\alpha 1(VI)$-, $\alpha 2(VI)$- and $\alpha 3(VI)$ chains are similar to each other. In each chain, the collagenous domain is 335 or 336 residues long and encompasses a relatively small part of the whole chain, only about one-third of the α1(VI)- and α2(VI) chains and about one-eighth of the α3(VI) chain (Fig. 11). The collagenous domain of each chain is flanked by two noncollagenous domains. The NC domains have modular structures [9,64–66,68,69]. The ~230-residue amino-terminal NC domain of the α1(VI)- and α2(VI) chains contain an ~180-residue VWA module, and the ~1800-residue amino-terminal NC domain of the α3(VI) chain contains nine VWA modules. The carboxyl-terminal NC domain of the α1(VI)- and α2(VI) chain and the amino-terminal half of the carboxyl-terminal NC domain of the α3(VI) chain each contain two VWA modules. In addition, the carboxyl-terminal half of the α3(VI) NC domain contains an S module, an Fn3 module and a K module. The VWA modules share sequence identity with collagen-binding domains of the von Willebrand factor and may be involved in the binding of microfilaments of collagen VI to fibrils of collagen I [66].

TYPE VI COLLAGEN

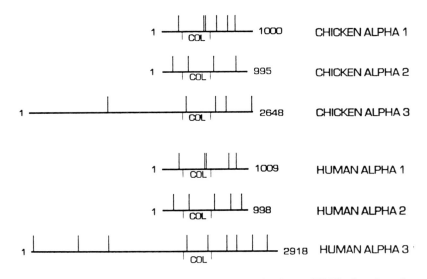

Fig. 11. Schematic diagram of the structure of the chains of collagen VI. The location of potential Asn-linked oligosaccharide binding sites in the polypeptide chains of chicken and human type VI collagen is indicated by vertical lines above the horizontal line. Signal peptides are not included in the figure; the number of residues is indicated on the right-hand side of each chain. Each chain has a 335-residue collagenous domain (COL), except human α3(VI), which has a 336-residue domain. The collagenous domain of each chain is flanked by an amino-terminal and a carboxyl-terminal noncollagenous domain (NC). The length of the amino-terminal NC domain is 235, 228, 1819, 237, 234 and 1804 residues for the chicken α1(VI)-, α2(VI)- and α3(VI)-, human α1(VI)-, α2(VI)- and α3(VI) chains, respectively. The locations of the Asn residues that potentially bind Asn-linked oligosaccharides are: chicken α1(VI) — 193, 495, 516, 647, 780 and 868; chicken α2(VI) — 114, 178, 300, 603 and 870; chicken α3(VI) — 1861, 2213, 2340, 2358, 2460 and 2644; human α1(VI) — 193, 497, 518, 785 and 877; human α2(VI) — 120, 307, 609, 804 and 933; and human α3(VI) — 20, 559, 919, 1846, 2098, 2325, 2444, 2629 and 2803.

The carbohydrate composition and structure of mammalian collagen VI has been investigated extensively. Amino-acid sequence results have revealed 19 potential sites for attachment of Asn-linked oligosaccharides to the three chains of human [67–69] type VI collagen (Fig. 11). The structure of the Asn-linked oligosaccharide from the pepsin-solubilized collagen VI fragment from human placenta has been determined by Fujiwara et al. [19]. The oligosaccharide is a heterogeneous diantennary N-acetyllac-tosamine type of oligosaccharide like that found in the 7 S domain of collagen IV (Fig. 2). The number of Asn-linked oligosaccharides and Glc-Gal-Hyl and Gal-Hyl units in the collagenous domain of human collagen VI can be calculated from the carbohydrate composition of pepsin-solubilized fragments of type VI collagen from aortic intima [70]. If it is assumed that the heterogeneous oligosaccharide units contain either 4 or 5 Glc-NAc residues [15], which yield $GlcNH_2$ residues on hydrolysis of the peptic fragments, then collectively, the $\alpha1(VI)$-, $\alpha2(VI)$- and $\alpha3(VI)$ collagenous domains contain a total of 4.4 to 5.5 oligosaccharide units, in good agreement with the theoreti-cal maximum of 5 units (Fig. 11). Thus, it appears that most or all of the potential N-glycosylation sites are occupied. SDS-polyacrylamide gel electrophoresis studies of rat myocardium collagen VI $\alpha1$ and $\alpha2$ chains indicate that each chain is extensively glycosylated in the noncollagenous domains as well as in the collagenous domain [71]. The amount by which the apparent molecular weight decreased after removal of the Asn-linked oligosaccharide units by use of N-glycanase as well as the carbohydrate content is consistent with a content of 5 to 6 Asn-linked oligosaccharide units per $\alpha1(VI)$- or $\alpha2(VI)$ chain. These values are close to the theoretical maximum of 5 per chain (Fig. 11) predicted for the human chains. The rat $\alpha3(VI)$ chain has a weight percent of $GlcNH_2$ and Man comparable to that of the $\alpha1(VI)$- and $\alpha2(VI)$ chains, indicating that it too is heavily N-glycosylated. The rat $\alpha3(VI)$ chain has an apparent molecular weight of 205,000, which indicates that it is considerably shorter than the human procollagen $\alpha3(VI)$ chain and thus would not have all of the 9 putative N-glycosylation sites (Fig. 11).

The location of potential sites for attachment of Asn-linked oligosaccharides in chicken and human type VI collagen is presented in Fig. 11. Chicken collagen VI has not been well characterized with respect to carbohydrate composition. The amino acid sequences deduced for the chicken $\alpha(VI)$ chains indicate that the chicken $\alpha1(VI)\alpha2(VI)\alpha3(VI)$ heterotrimer should be similar to the human heterotrimer. For example, the chicken heterotrimer is predicted to contain a maximum of 17 N-glyco-sylation sites compared to 19 for the human heterotrimer.

Considering the number and location of Glc-Gal-Hyl and Gal-Hyl units in collagen VI, the amino acid composition of the collagenous domain of human type VI collagen, excised from placenta by pepsin and re-treated with pepsin after reduction of disulfides, indicates about 16, 25 and 15 Hyl residues for the $\alpha1(VI)$-, $\alpha2(VI)$- and $\alpha3(VI)$ collagenous domains, respectively [64]. Thus the $\alpha1(VI)\alpha2(VI)\alpha3(VI)$ heterotrimer should have a total of 56 Hyl residues. In comparison, the collagenous domain of human aortic intima collagen VI contains 41 Glc-Gal-Hyl and 4 Gal-Hyl units [70], indicating an overall degree of glycosylation of Hyl residues of 80%.

The human collagen VI molecule differs from other well-characterized collagens in that it has a relatively small triple-helical domain and is heavily glycosylated, both in the triple-helical domain (by Glc-Gal, Gal and Asn-linked oligosaccharides) and in the noncollagenous domains (by Asn-linked oligosaccharides). These features prevent it from forming fibrils. Instead, in its self-assembly process, the collagen VI molecule dimerizes in a side-to-side, antiparallel mode with a stagger of about one-third of the molecular length [73]. The dimer is stabilized by disulfide bonds between globular noncollagenous domains and triple-helical domains of different molecules and prob-ably by noncovalent intermolecular interactions between triple-helical domains. The dimer then dimerizes in a side-to-side, parallel mode to form a tetramer. The tetramer is stabilized by disulfide bonds between the end regions of the dimers and perhaps by noncovalent interactions between the triple helices. The tetramer then associates in an end-to-end fashion to form a microfilament that has a beaded appearance because of the size, location and globular shape of the noncollagenous domains. The geometrical restrictions for tetramerization suggest that the triple-helical domain has one or two sectors that provide the amino-acid side chains that are involved in intermolecular interactions. In connection with this, the bulky, hydrophilic carbohydrate moieties of the triple-helical domain must be situated outside of the interacting sectors to avoid unfavorable energetic and steric conditions for association.

5.4. Collagens IX, XII and XIV

5.4.1. Collagen IX
Comprehensive reviews of the structure and function of collagen IX have been pub-lished by van der Rest and Mayne [74], and van der Rest and Garrone [1]. Collagen IX is a heterotrimer with the formula: $\alpha 1(IX)\alpha 2(IX)\alpha 3(IX)$. A schematic representation of the three chains from chicken cartilage, based on the reported amino acid sequences [75-77], is presented in Fig. 12. Each chain has three collagenous (COL) and four noncollagenous (NC) domains. Collagen IX is found in extracellular matrices, such as vitreous humor and corneal stroma of the eye and hyaline cartilage, that contain collagen II as the major fibril-forming collagen. Collagen IX has been found on the exterior of type II collagen fibrils, and, in cartilage, it is bound to collagen II by a trivalent hydroxypyridinium link involving Hyl residue 169(X) in the $\alpha 2(IX)$ COL2 domain (Fig. 12) and a Hyl residue from the amino-terminal telopeptide of each of two $\alpha(II)$ chains [78]. A second site in type IX collagen for formation of a hydroxypyrid-inium cross-link is located in the COL2 domain, although the exact site has not been determined [79]. Two potential $\alpha 3(IX)$ sites are marked X in Fig. 12.

The size of the $\alpha 1(VI)NC4$ domain is tissue-dependent, with a length of 243 residues for cartilage $\alpha 1(VI)NC4$, whereas in corneal stroma it is very short or perhaps absent [80–82]. The size difference is due to use of an downstream promoter in corneal stroma, instead of an upstream promoter, in the transcription of the $\alpha 1(IX)$ collagen gene [82]. The 243-residue cartilage $\alpha 1(VI)NC4$ domain is predicted to be quite cationic, with a calculated pI of 9.7, and potentially participates in electrostatic interactions with polyanionic glycosaminoglycans in the cartilage matrix [80]. It is of interest that the $\alpha 1(VI)NC4$ domain contains a 217-residue sequence that is homologous to the hepa-rin-binding domain of thrombospondin [9]. The cationic nature of the $\alpha 1(VI)NC4$ domain has been incorporated into a model for stabilization of cartilage matrix in which

TYPE IX COLLAGEN

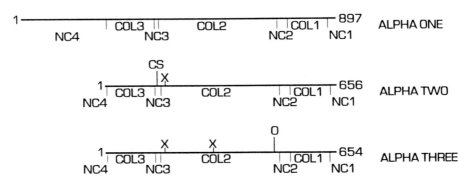

Fig. 12. Schematic diagram of the structure of the chains of chicken collagen IX. The numbering of residues does not include signal peptides. Each chain has seven domains, three of which are collagenous (COL1–COL3) and four of which are noncollagenous (NC1–NC4), with the numbering beginning at the carboxyl-terminal end of the chains. The α1(IX) chain is considerably longer than the α2(IX)- and α3(IX) chains because of the greater size of the α1(IX)NC4 domain (243 residues compared with 3 residues). In the intact protomer, like-numbered COL domains associate to form triple helices.

the α1(VI)NC4 domains on different fibers bind to anionic proteoglycans located between them [83].

Collagen IX, depending on the tissue source, may contain a covalently bound chondroitin sulfate (CS) molecule [13,84–88]. The percentage of collagen IX molecules with bound chondroitin sulfate varies among tissues, with a value of 100% for chick vitreous humor [86], ~80% for chick embryo cartilage [14], ~16% for rat chondrosarcoma [88] and < 5% for bovine cartilage [85]. In chick embryo cartilage collagen IX, the bound chondroitin sulfate molecule has an M_r ~ 40,000 whereas in chick embryo vitreous humor collagen IX, the M_r ~ 350,000 [86]. It has been proposed that the long chondroitin sulfate molecule of collagen IX contributes to the gel structure of chicken vitreous humor and thus compensates for the low amount of hyaluronic acid [86,87].

The chondroitin sulfate molecule of cartilage collagen IX is attached to a serine residue, Ser 146, in the NC3 domain of the α2(IX) chain [14,89,90]. The α2(IX)NC3 domain is longer than the α1(IX)NC3 and α3(IX)NC3 domains in the glycosaminoglycan-binding region because of an extra five-residue sequence, Val-Glu-Gly-Ser-Ala, and the Ser in the sequence is bound to the glycosaminoglycan [14]. This region of the α2 chain(IX) and corresponding regions of the α1(IX) and α3(IX) chains are depicted in Fig. 13.

Little is known about the content of carbohydrates other than chondroitin sulfate in type IX collagen. Both the 84 and 68 kDa chains of type IX collagen (corresponding to the α1(IX)- and α3(IX) chains, respectively [74]) from chick embryo sternum incorporated radiolabelled mannose during biosynthesis, with the latter incorporating most of the label [13]. This suggested the presence of Asn-linked oligosaccharides in

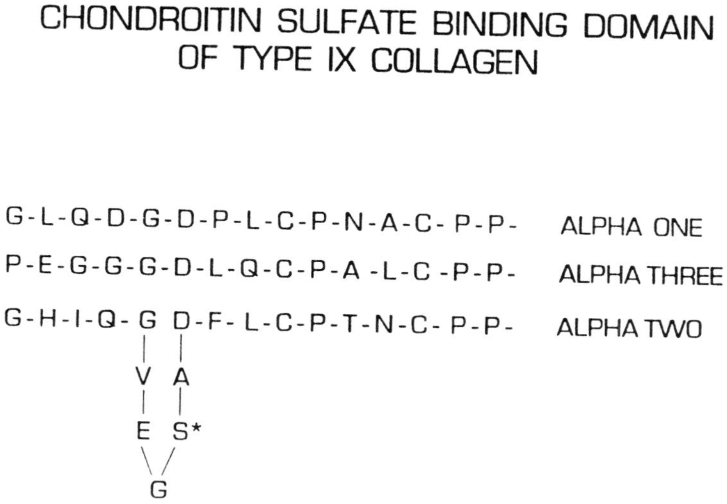

CHONDROITIN SULFATE BINDING DOMAIN OF TYPE IX COLLAGEN

```
G-L-Q-D-G-D-P-L-C-P-N-A-C- P-P-    ALPHA ONE

P-E-G-G-G-D-L-Q-C-P-A -L-C -P-P-   ALPHA THREE

G-H-I-Q- G D-F- L-C-P-T-N-C- P-P-  ALPHA TWO
          | |
          V A
          | |
          E S*
           \ /
            G
```

Fig. 13. Amino acid sequence of the glycosaminoglycan-binding region of collagen IX. The chondroitin sulfate glycosaminoglycan is bound to Ser 146 (*S**), which is in the α2(IX)NC3 domain.

these chains. The α3(IX) chain has a putative binding site for an Asn-linked oligosaccharide (Asn-Gly-Thr, starting with Asn 479). However, the α1(IX) chain lacks an Asn-Xaa-Ser/Thr sequence. There is apparently no published direct evidence for the presence of Glc-Gal-Hyl or Gal-Hyl units in collagen IX, although the presence of Hyl residues [91,92] suggests that these units could be present.

5.4.2. Collagens XII and XIV

Collagens XII and XIV are closely related collagens that have been found in low amount in embryonic skin and tendon, and fetal skin and tendon, respectively [93]. Immunolocalization studies on type XII collagen in chick embryo show that it is present in non-mineralized matrices containing collagen I as the major fibrillar constituent. Collagens XII and XIV are homotrimers, i.e., [α1(XII)]$_3$ or [α1(XIV)]$_3$.

The cDNA-based amino acid sequence of collagen XII from chick embryonic fibroblasts has been determined by Yamagata et al. [94], and is represented in Fig. 14. The polypeptide chain is 3124 residues long and contains distinct, repeated modules. The chain can be divided into 14 putative domains. Starting with the amino terminus, they are designated S, IIIA, VA, IIIB, VB, IIIC, VC, IIID, IXP, COL2, NC2, COL1, and NC1, respectively. S is a signal peptide (residues 1–24), the four III domains contain one or more ~90-residue sequences that are homologous to the sequence of the type III motif of fibronectin, the four V domains contain one or more ~190-residue sequences that are homologous to the sequence of the von Willebrand factor A domain and the IXP domain (residues 2509–2750) has a sequence homologous to that of the NC4 domain of collagen IX. The COL2, NC2, COL1 and NC1 sequences are unique to collagen XII but are homologous to their counterparts in collagen XIV, which has

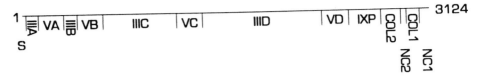

Fig. 14. Schematic diagram of the polypeptide chain of embryonic chicken fibroblast collagen XII. The chain can be divided into 14 domains; starting with the amino terminus, the domains are designated: S, IIIA, VA, IIIB, VB, IIIC, VC, IIID, IXP, COL2, NC2, COL1, and NC1, respectively. S is a signal peptide (residues 1–24), the four III domains contain one or more ~90-residue sequences that are homologous to the sequence of the type III motif of fibronectin, the four V domains contain one or more ~190-residue sequences that are homologous to the sequence of the von Willebrand factor A domain, and the IXP domain (residues 2509–2750) has a sequence homologous to that of the NC4 domain of collagen IX. The COL2 (residues 2751–2902), NC2 (2903–2945), COL1 (residues 2946–3048) and NC1 (residues 3049–3124) domains have homologous counterparts in collagen XIV. Domains IIIA–IIIB (residues 24–114, 332–425, 629–1178 and 1384–2295, respectively) contain 1, 1, 6 and 10 fibronectin type III motifs, respectively. Domains VA–VD (residues 128–320, 426–619, 1187–2295 and 2317–2508, respectively) each contain one von Willebrand factor A domain.

had part of its sequence determined [95]. The COL1 domains may contain binding sites for collagen I fibrils [93].

Neither collagen XII nor collagen XIV has been extensively characterized with respect to carbohydrate composition. However, collagens type XII and XIV from bovine cartilage are both sensitive to chondroitinase [23] and are thus chondroitin sulfate proteoglycans.

Acknowledgements

Contributions from the authors' laboratory were supported by the National Institutes of Health, Grant DK 18381. The authors thank Bill Paige for the artwork in Figs. 8 and 10.

Abbreviations

COL	collagenous
IC	interrupted collagenous
FACIT	fibril-associated collagens with interrupted triple helices
NC	noncollagenous
pN collagen	A collagen that retains the amino-terminal propeptide
Xaa	Any amino acid residue
Yaa	Any amino acid residue
7 S	The amino-terminal tetramerization domain of collagen IV

References

1. Van der Rest, M. and Garrone, R. (1991) Collagen family of proteins. FASEB. J. 5, 2814–2823.
2. Myers, J.C., Kivirikko, S., Gordon, M.K. and Dion, A.S. (1992) Identification of a previously unknown human collagen chain, α1(XV), characterized by extensive interruptions in the triple-helical region. Proc. Natl. Acad. Sci. USA 89, 10144–10148.
3. Pan, T.-C., Zhang, R.-Z., Mattei, M.-G., Timpl, R., and Chu, M-L. (1992) Cloning and chromosomal location of human α1(XVI) collagen. Proc. Natl. Acad. Sci. USA 89, 6565–6569.
4. Li, K., Tamai, K., Tan, E.M.L. and Uitto, J. (1993) Cloning of type XVII collagen. J. Biol. Chem. 268, 8825–8834.
5. Oh, S.P., Warman, M.L., Seldin, M.F., Cheng, S.-D., Knoll, J.H.M., Timmons, S. and Olsen, B. R. (1994) Cloning of cDNA and genomic DNA encoding human type XVIII collagen and localization of the α1(XVIII) collagen gene to mouse chromosome 10 and human chromosome 21. Genomics 19, 494–499.
6. Fleischmajer, R., Olsen, B.R. and Kühn, K. (Eds.) (1990) The structure, molecular biology, and pathology of collagen. Ann. N.Y. Acad. Sci. 580, 1–587.
7. Mayne, R. and Burgeson, R.E. (Eds.) (1987) Structure and Function of Collagen Types. Academic Press, Orlando, pp. 81–103.
8. Vuorio, E. and de Crombrugghe, B. (1990) The family of collagen genes. Annu. Rev. Biochem. 59, 837–872.
9. Bork, P. (1992) The modular architecture of vertebrate collagens. FEBS Lett. 307, 49–57.
10. Butler, W.T. and Cunningham, L.W. (1966) Evidence for the linkage of a disaccharide to hydroxylysine in tropocollagen. J. Biol. Chem. 241, 3872–3878.
11. Spiro, R.G. (1967) The structure of the disaccharide unit of the renal glomerular basement membrane. **JOURNAL ???? 242,** 4813–4818.
12. Spiro, R.G. (1972) Basement membranes and collagens. In: A. Gottschalk (Ed.), Glycoproteins. Elsevier, Amsterdam, pp. 964–999.
13. Vaughan, L., Winterhalter K.H. and Bruckner, P. (1985) Proteoglycan Lt from chicken embryo sternum identified as type IX collagen. J. Biol. Chem. 260, 4758–4763.
14. Huber, S., Winterhalter K.H. and Vaughan, L. (1987) Isolation and sequence analysis of the glycosaminoglycan attachment site of type IX collagen. J. Biol. Chem. 263, 752–756.
15. Langeveld, J.P.M., Noelken, M.E., Hård, K., Todd, P., Vliegenthart, J.F.G., Rouse, J. and Hudson, B.G. (1991) Bovine glomerular basement membrane: location and structure of the Asn-linked oligosaccharide units and their potential role in the assembly of the 7 S collagen IV tetramer. J. Biol. Chem. 266, 2622–2631.
16. Miller, E.J. (1976) Biochemical characteristics and biological significance of the genetically-distinct collagens. Mol. Cell. Biochem. 13, 165–192.
17. Leushner, J.R.A. (1987) Partial characterization of the heteropolysaccharide associated with the 7S domain of type IV collagen from placenta. Biochem. Cell Biol. 65, 501–506.
18. Nayak, B.R. and Spiro, R.G. (1991) Localization and structure of the asparagine-linked oligosaccharides of type IV collagen from glomerular basement membrane and lens capsule. J. Biol. Chem. 266, 13978–13987.
19. Fujiwara, S., Shinkai, H. and Timpl, R. (1991) Structure of N-linked oligosaccharide chains in the triple-helical domains of human type VI and mouse type IV collagen. Matrix 11, 307–312.
20. Fessler, J.H. and Fessler, L.I. (1987) Type V collagen. In: R. Mayne and R.E. Burgeson (Eds.) Structure and Function of Collagen Types. Academic Press, Orlando, pp. 81–103.
21. Clark, C.C. (1979) The distribution and initial characterization of oligosaccharide units on the COOH-terminal propeptide extensions of the pro-α1 and pro-α2 chains of type I procollagen. J. Biol. Chem. 254, 10798–10802.

22. Shinkai, H. and LaPiere, C.M. (1983) Characterization of oligosaccharide units of p-N-collagen type III from dermatosparactic bovine skin. Biochim. Biophys. Acta 758, 30–36.

23. Watt, S.L., Lunstrum, G.P., McDonough, A.M., Keene, D.R., Burgeson, R.E. and Morris, N.P. (1992) Characterization of collagen types XII and XIV from bovine fetal cartilage. J. Biol. Chem. 267, 20093–20099.

24. Vaughan, L., Winterhalter, K.H. and Bruckner, P. (1985) Proteoglycan Lt from chicken embryo sternum identified as type IX collagen. J. Biol. Chem. 260, 4758–4763.

25. Takahara, K., Sato, Y., Okazawa, K., Okamoto, N., Noda, A., Yaoi, Y. and Kato, I. (1991) Complete primary structure of human collagen α1(V) chain. J. Biol. Chem. 266, 13124–13129.

26. Greenspan, D.S., Cheng, W. and Hoffman, G.G. (1991) The pro-α1(V) collagen chain. J. Biol. Chem. 266, 24727–24763.

27. Miller, E.J. (1984) Chemistry of the collagens and their distributions. In: K.A. Piez and A.H. Reddi (Eds.), Extracellular Matrix Biochemistry. Elsevier, New York, pp. 41–81.

28. Silver, F.H. and Birk, D.E. (1984) Molecular structure of collagen in solution: comparison of types I,II,III and V. Int. J. Biol. Macromolecules 6, 125–132.

29. Myers, J.C., Loidl, H.R., Seyer, J.M. and Dion, A.S. (1985) Complete primary structure of the human α2 type V procollagen COOH-terminal propeptide. J. Biol. Chem. 260, 11216–11222.

30. Weil, D., Bernard, M., Gargano, S. and Ramirez, F. (1987) The pro α2(V) collagen gene is evolutionarily related to the major fibrillar-forming collagens. Nucleic Acids Res. 15, 181–198.

31. Woodbury, D., Benson-Chanda, V. and Ramirez, F. (1989) Amino-terminal propeptide of human pro-α2(V) collagen conforms to the structural criteria of a fibrillar procollagen molecule. J. Biol. Chem. 264, 2735–2738.

32. Kornfeld, R., and Kornfeld, S. (1985) Assembly of asparagine-linked oligosaccharides. Annu. Rev. Biochem. 57, 631–664.

33. Burgeson, R.E., Hebda, P.A., Morris, N.P. and Hollister, D.W. (1982) Human cartilage collagens. J. Biol. Chem. 257, 7852–7856.

34. Sage, H. and Bornstein, P. (1979) Characterization of a novel collagen chain in human placenta and its relation to AB collagen. Biochemistry 18, 3815–3822.

35. Adachi, E., Hayashi, T. and Hashimoto, P.H. (1991) A comparison of the immunofluorescent localization of collagen types I, III, and V with the distribution of reticular fibers on the same liver sections of the snow monkey (*Macaca fuscata*) Cell Tissue Res. 264, 1–8.

36. Romanic, A.M., Adachi, E., Kadler, K.E., Hojima, Y. and Prockop, D.J. (1991) Copolymerization of pNcollagen III and collagen I. J. Biol. Chem. 266, 12703–12709.

37. Butkowski, R.J., Langeveld, J.P.M., Wieslander, J., Hamilton, J. and Hudson, B.G. (1987) Localization of the Goodpasture epitope to a novel chain of basement membrane collagen. J. Biol. Chem. 262, 7874–7877.

38. Saus, J., Wieslander, J., Langeveld, J.P.M., Quinones, S. and Hudson, B.G. (1987) Identification of the Goodpasture antigen as the α3(IV) chain of collagen IV. J. Biol. Chem. 263, 13374–13380.

39. Gunwar, S., Saus, J., Noelken, M.E. and Hudson, B.G. (1990) Glomerular basement membrane. Identification of a fourth chain, α4, of type IV collagen. J. Biol. Chem. 265, 5766–5769.

40. Hostikka, S.L., Eddy, R.L., Hoyhta, M., Shows, T.B. and Tryggvason, K. (1990) Identification of a distinct type IV chain with restricted kidney distribution and assignment of its gene to the locus of X chromosome-linked Alport syndrome. Proc. Natl. Acad. Sci. USA 87, 1606–1610.

41. Pihlajaniemi, T., Pohjolainen, E.-R., and Myers, J.C. (1990) Complete primary structure of the triple-helical region and the carboxyl-terminal domain of a new type IV collagen chain, α5(IV). J. Biol. Chem. 265, 13758–13766.

42. Zhou, J., Mochizuki, T., Smeets, H., Antignac, C., Laurila, P., de Paepe, A., Tryggvason, K. and Reeders, S.T. (1993) Deletion of the paired α5(IV) and α6(IV) collagen genes in inherited

smooth muscle tumors. Science 261, 1167–1169.

43. Soininen, R., Haka-Risku, T., Prockop, D.J. and Tryggvason, K. (1987) Complete primary structure of the α1-chain of human basement membrane (type IV) collagen. FEBS Lett. 225, 187–194.

44. Muthukumaran, G., Blumberg, B. and Kurkinen, M. (1989) The complete primary structure for the α1-chain of mouse collagen IV. J. Biol. Chem., 264, 6310–6317.

45. Hostikka, S.L. and Tryggvason, K. (1987) The complete primary structure of the α2 chain of human type IV collagen and comparison with the α1(IV) chain. J. Biol. Chem. 263, 19487–19493.

46. Saus, J., Quinones, S., MacKrell, A., Blumberg, B., Muthukumaran, G., Pihlajaniemi, T. and Kurkinen, M. (1989) The complete primary structure of mouse α2(IV) collagen. J. Biol. Chem. 264, 6318–6324.

47. Zhou, J., Hertz, J. M., Leinonen, A. and Tryggvason, K. (1992) Complete amino acid sequence of the human α5(IV) collagen chain and identification of a single-base mutation in exon 23 converting glycine 551 in the collagenous domain to cysteine in an Alport syndrome patient. J. Biol. Chem. 267, 12475–12481.

48. Weber, S., Engel, J., Wiedemann, H., Glanville, R.W. and Timpl, R. (1984) Subunit structure and assembly of the globular domain of basement membrane collagen type IV. Eur. J. Biochem. 168, 401–410.

49. Siebold, B., Qian, R.-Q., Glanville, R.W., Hofmann, H., Deutzmann, R. and Kühn, K. (1987) Construction of a model for the aggregation and cross-linking region (7S domain) of type IV collagen based upon an evaluation of the primary structure of the α1 and α2 chains in this region. Eur. J. Biochem.. 168, 569–575.

50. Morrison, K.E., Germino, G.G. and Reeders, S.T. (1991) Use of the polymerase chain reaction to clone and sequence a cDNA encoding the bovine α3 chain of type IV collagen. J. Biol. Chem. 266, 34–39.

51. Morrison, K.E., Mariyama, M., Yang-Feng, T.L. and Reeders, S.T. (1991) Sequence and localization of a partial cDNA encoding the human α3 chain of type IV collagen. Am. J. Hum. Genet. 49, 545–554.

52. Matsukura, H., Michael, A.F., Fish, A.J. and Butkowski, R.J. (1992) Partial protein sequence of the globular domain of α4(IV) collagen: sites of sequence variability and homology with α2(IV). Conn. Tissue Res. 28, 231–244.

53. Mariyama, M., Kalluri, R., Hudson, B.G. and Reeders, S. (1992) The α4(IV) chain of basement membrane collagen: Isolation of cDNAs encoding bovine α4(IV) and comparison with other type IV collagens. J. Biol. Chem. 267, 1253–1258.

54. Gunwar, S., Ballester, F., Kalluri, R., Timoneda, J., Chonko, A.M., Edwards, S.J., Noelken, M.E. and Hudson, B.G. (1991) Glomerular basement membrane: Identification of dimeric subunits of the noncollagenous domain (hexamer) of collagen IV and the Goodpasture antigen. J. Biol. Chem. 266, 15318–15324.

55. Kleppel, M.M., Fan, W.W., Cheong, H.I. and Michael, A.F. (1992) Evidence for separate networks of classical and novel basement membrane collagen. J. Biol. Chem. 267, 4137–4142.

56. Glanville, R.W., Qian, R.-Q., Siebold, B., Risteli, J. and Kühn, K. (1985) Amino acid sequence of the N-terminal aggregation and cross-linking region (7 S domain) of the α1(IV) chain of human basement membrane collagen. Eur. J. Biochem. 152, 213–219.

57. Hofmann, H., Fietzek, P.P. and Kühn, K. (1978) The role of polar and hydrophobic interactions for the molecular packing of type I collagen: A three-dimensional evaluation of the amino acid sequence. J. Mol. Biol. 125, 137–165.

58. Timpl, R., Wiedemann, H., Van Deeden, V., Furthmayr, H. and Kühn, K. (1984) A network model for the organization of type IV collagen molecules in basement membranes. Eur. J. Biochem. 120, 203–211.

59. Yurchenco, P.D., Tsilibary, E.C., Charonis, A.S. and Furthmayr, H. (1986) Models for the self-assembly of basement membrane. J. Histochem. Cytochem. 34, 93–102.

60. Tsilibary, E.C. and Charonis, A.S. (1986) The role of the main noncollagenous domain (NC1) in type IV collagen self-assembly. J. Cell Biol. 103, 2467–2473.

61. Yurchenco, P.D. and Ruben, G.C. (1987) Basement membrane structure in situ: Evidence for lateral associations in the type IV collagen network. J. Cell Biol. 105, 2559–2568.

62. Yurchenco, P.D. and Ruben, G.C. (1987) Type IV collagen lateral associations in the EHS tumor matrix. Am. J. Pathol. 132, 278–291.

63. Timpl, R., and Engel, J. (1987) Type VI collagen. In: R. Mayne and R.E. Burgeson (Eds.), Structure and Function of Collagen Types. Academic Press, Orlando, FL, pp. 105–143.

64. Bonaldo, P., Russo, V., Bucciotti, F., Bressan, G.M. and Colombatti, A. (1989) α_1 chain of chick type VI collagen: The complete cDNA sequence reveals a hybrid molecule made of one short collagen and three von Willebrand factor type A-like domains. J. Biol. Chem. 264, 5575–5580.

65. Koller, E., Winterhalter, K.H., and Trueb, B. (1989) The globular domains of type VI collagen are related to the collagen-binding domains of cartilage matrix protein and von Willebrand factor. EMBO J. 8, 1073–1077.

66. Bonaldo, P., Russo, V., Bucciotti, F., Doliana, R. and Colombatti, A. (1990) Structural and functional features of the $\alpha 3$ chain indicate a bridging role for chicken collagen VI in connective tissues. Biochemistry 29, 1245–1254.

67. Chu, M.-L., Conway, D., Pan, T.-C., Baldwin, C., Mann, K., Deutzmann, R. and Timpl, R. (1987) Amino acid sequence of the triple-helical domain of human collagen type VI. J. Biol. Chem. 263, 18601–18606.

68. Chu, M.-L., Pan, T.-C., Conway, D., Kuo, H.-J., Glanville, R.W., Timpl, R., Mann, K. and Deutzmann, R. (1989) Sequence analysis of $\alpha 1$(VI) and $\alpha 2$(VI) chains of human type VI collagen reveals internal triplication of globular domains similar to the A domains of von Willebrand factor and two $\alpha 2$(VI) chains that differ in the carboxy terminus. EMBO J. 8, 1939–1946.

69. Chu, M.-L., Zhang, R.-Z., Pan, T.-C., Stokes, D., Conway, D. Kuo, H.-J., Glanville, R., Mayer, U., Mann, K., Deutzmann, R. and Timpl R. (1990) Mosaic structure of globular domains in the human type VI collagen $\alpha 3$ chain: similarity to von Willebrand factor, fibronectin, actin, salivary proteins and aprotinin type protease inhibitors. EMBO J. 9, 385–393.

70. Chung, E., Rhodes, R.K. and Miller, E.J. (1976) Isolation of three collagenous components of probable basement membrane origin from several tissues. Biochem. Biophys. Res. Commun. 71, 1167–1174.

71. Spiro, M.J., Kumar, B.R.R. and Crowley, T.J. (1992) Myocardial glycoproteins in diabetes: Type VI collagen is a major PAS-reactive extracellular matrix protein. J. Mol. Cell Cardiol. 24, 397–410.

72. Jander, R., Rauterberg, J. and Glanville, R.W. (1983) Further characterization of the three polypeptide chains of bovine and human short-chain collagen (intima collagen). Eur. J. Biochem. 133, 39–46.

73. Furthmayr, H., Wiedemann, H., Timpl, R., Odermatt, E. and Engel, J. (1983) Electron-microscopical approach to a structural model of intima collagen. Biochem. J. 211, 303–311.

74. van der Rest, M. and Mayne, R. (1987) Type IX collagen. In: R. Mayne and R.E. Burgeson (Eds.), Structure and Function of Collagen Types. Academic Press, Orlando, FL, pp. 195–221.

75. Har-El, R., Sharma, Y.D., Aguilera, A., Ueyama, N., Wu, J.-J., Eyre, D.R., Juricic, L., Chandrasekaran, S., Li, M., Nah, H.-D., Upholt, W.B. and Tanzer, M.L. (1992) Cloning and developmental expression of the $\alpha 3$ chain of chicken type IX collagen. J. Biol. Chem. 267, 10070–10076.

76. Brewton, R.G., Ouspenskaia, M.V., van der Rest, M. and Mayne, R. (1992) Cloning of the chicken α3(IX) collagen chain completes the primary structure of type IX collagen. Eur. J. Biochem. 205, 443–449.

77. Ninomiya, Y., Castagnola, P., Gerecke, D., Gordon, M.K., Jacenko, O., LuValle, P., McCarthy, M., Murigaki, Y., Nishimura, I., Oh, S., Rosemblum, N., Sato, N., Sugrue, S., Taylor, R., Vasios, G., Yamaguchi, N. and Olsen, B. R. (1990) Molecular biology of collagens with short triple helical domains. In: L.J. Sandell and C.D. Boyd (Eds.), Extracellular Matrix Genes. Academic Press, San Diego, CA, pp. 79–114.

78. Van der Rest, M. and Mayne, R. (1987) Type IX collagen proteoglycan from cartilage is covalently cross-linked to type II collagen. J. Biol.Chem. 263, 1615–1618.

79. Shimokomaki, M., Wright, D.W., Irwin, M.H., van der Rest, M. and Mayne, R. The structure and macromolecular organization of type IX collagen in cartilage. (1990) Ann. N.Y. Acad. Sci. 580, 1–7.

80. Vasios, G., Nishimura, I., Konomi, H., van der Rest, M., Ninomiya, Y. and Olsen, B.R. (1987) Cartilage type IX collagen-proteoglycan contains a large amino-terminal globular domain encoded by multiple exons. J. Biol. Chem. 263, 2324–2329.

81. Svoboda, K.K., Nishimura, I., Sugrue, S.P., Ninomiya, Y. and Olsen, B.R. (1987) Embryonic chicken cornea and cartilage synthesize type IX collagen molecules with different amino-terminal domains. Proc. Natl. Acad. Sci. USA 85, 7496–7500.

82. Nishimura, I., Muragaki, Y. and Olsen, B.R. (1989) Tissue-specific forms of type IX collagen-proteoglycan arise from the use of two widely separated promoters. J. Biol. Chem. 264, 20033–20041.

83. Smith, G.N., Jr. and Brandt, K.D. (1992) Hypothesis: Can type IX collagen "glue" together intersecting type II fibers in articular cartilage matrix? A proposed mechanism. J. Rheumatol. 19, 14–17.

84. Wright, D.W. and Mayne, R. (1987) Vitreous humor of chicken contains two fibrillar systems: An analysis of their structure. J. Ultrastruct. Mol. Struct. Res. 100, 224–234.

85. Bruckner, P., Mendlar, M., Steinmann, B., Huber, S. and Winterhalter, K.H. (1987) The structure of human collagen type IX and its organization in fetal and infant cartilage fibrils. J. Biol. Chem. 263, 16911–16917.

86. Yada, T., Suzuki, S., Kobayashi, K., Kobayashi, M., Hoshino, T., Horie, K. and Kimata, K. (1990) Occurrence in chick embryo vitreous humor of a type IX collagen proteoglycan with an extraordinarily large chondroitin sulfate chain and short α1 polypeptide. J. Biol. Chem. 265, 6992–6999.

87. Brewton, R.G., Wright, D.W. and Mayne, R. (1991) Structural and functional comparison of type IX collagen-proteoglycan from chicken cartilage and vitreous humor J. Biol. Chem. 266, 4752–4757.

88. Arai, M., Yada, T., Suzuki, S. and Kimata, K. (1992) Isolation and characterization of type IX collagen-proteoglycan from the Swarm rat chondrosarcoma. Biochim. Biophys. Acta 1117, 60–70.

89. Irwin, M.H. and Mayne, R. (1986) Use of monoclonal antibodies to locate the chondroitin sulfate chain(s) in type IX collagen. J. Biol. Chem. 261, 16281–16283.

90. McCormick, D., van der Rest, M., Goodship, J., Lozano, G., Ninomiya, Y. and Olsen, B.R. (1987) Structure of the glycosaminoglycan domain in the type IX collagen-proteoglycan. Proc. Natl. Acad. Sci. USA 84, 4044–4048.

91. Reese, C.A., and Mayne, R. (1981) Minor collagens of chicken hyaline cartilage. Biochemistry 20, 5443–5448.

92. Reese, C.A., Wiedemann, H., Kühn, K. and Mayne, R. (1982) Characterization of a highly soluble collagenous molecule isolated from chicken hyaline cartilage. Biochemistry 21, 826–830.

93. Van der Rest, M., Aubert-Foucher, E., Dublet, B., Eichenberger, D., Font, B. and Goldschmidt, D. (1991) Structure and function of the fibril-associated collagens. Biochem. Soc. Trans. 19, 820–824.

94. Yamagata, M., Yamada, K.M., Yamada, S.S., Shinomura, T., Tanaka, H., Nishida, Y., Obara, M. and Kimata, K. (1991) The complete primary structure of type XII collagen shows a chimeric molecule with reiterated fibronectin type III motifs, von Willebrand factor A motifs, a domain homologous to a noncollagenous region of type IX collagen, and short collagenous domains with an Arg-Gly-Asp site. J. Cell Biol. 115, 209–221.

95. Gordon, M.K., Castagnola, P., Dublet, B., Linsenmayer, T.F., van der Rest, M., Mayne, R. and Olsen, B.R. (1991) Cloning of a cDNA for a new member of the class of fibril-associated collagens with interrupted triple helices. Eur. J. Biochem. 201, 333–338.

Subject Index

Recent volumes in the *New Comprehensive Biochemistry* series